Intelligent Coordination of UAV Swarm Systems

Intelligent Coordination of UAV Swarm Systems

Editors

Xiwang Dong
Mou Chen
Xiangke Wang
Fei Gao

Basel • Beijing • Wuhan • Barcelona • Belgrade • Novi Sad • Cluj • Manchester

Editors

Xiwang Dong
Institute of Artificial
Intelligence
Beihang University
Beijing
China

Mou Chen
College of Automation
Engineering
Nanjing University of
Aeronautics and Astronautics
Nanjing
China

Xiangke Wang
College of Intelligence
Science and Technology
National University of
Defense Technology
Changsha
China

Fei Gao
College of Control Science
and Engineering
Zhejiang University
Hangzhou
China

Editorial Office
MDPI
St. Alban-Anlage 66
4052 Basel, Switzerland

This is a reprint of articles from the Special Issue published online in the open access journal *Drones* (ISSN 2504-446X) (available at: www.mdpi.com/journal/drones/special_issues/uav_coordination).

For citation purposes, cite each article independently as indicated on the article page online and as indicated below:

Lastname, A.A.; Lastname, B.B. Article Title. *Journal Name* **Year**, *Volume Number*, Page Range.

ISBN 978-3-0365-8659-5 (Hbk)
ISBN 978-3-0365-8658-8 (PDF)
doi.org/10.3390/books978-3-0365-8658-8

Contents

Preface

In recent years, the field of unmanned aerial vehicles (UAVs) has witnessed remarkable advancements, particularly in the area of UAV swarm systems. These systems, consisting of multiple UAVs working together in a coordinated manner, have attracted significant attention due to their potential to revolutionize various applications, ranging from disaster response and surveillance to precision agriculture and logistics. The intelligent coordination of UAV swarm systems holds the key to unlocking their full potential and achieving efficient and effective mission execution.

This special issue reprint, titled "Intelligent Coordination of UAV Swarm Systems," aims to explore the latest research and developments in this exciting field. It brings together researchers and practitioners from diverse disciplines to share their insights, methodologies, and innovative solutions related to UAV swarm systems. The issue covers a wide range of topics, including intelligent perception and cognition, swarm navigation and localization, autonomous decision and planning, cooperative guidance and control, UAV simulation and experiment, and swarm intelligence.

This special issue reprint provides a platform for researchers and practitioners to present and discuss the latest advancements, challenges, and opportunities in the field of intelligent coordination of UAV swarm systems. We hope that the contributions in this issue will inspire further research and innovation, leading to the development of more robust, efficient, and intelligent UAV swarm systems. We extend our gratitude to all the authors who have contributed to this special issue and to the reviewers for their valuable insights and feedback. Together, let us explore the frontiers of intelligent coordination in UAV swarm systems and shape the future of aerial robotics.

Xiwang Dong, Mou Chen, Xiangke Wang, and Fei Gao
Editors

Article

A GPS-Adaptive Spoofing Detection Method for the Small UAV Cluster

Lianxiao Meng [1,2,3], Long Zhang [3], Lin Yang [3] and Wu Yang [2,*]

1   Cyberspace Security Academy, Information Engineering University, Zhengzhou 450001, China
2   Information Security Research Center, Harbin Engineering University, Harbin 150000, China
3   National Key Laboratory of Science and Technology on Information System Security, Systems Engineering Institute, Beijing 100000, China
*   Correspondence: yangwu@hrbeu.edu.cn

Abstract: The small UAV (unmanned aerial vehicle) cluster has become an important trend in the development of UAVs because it has the advantages of being unmanned, having a small size and low cost, and ability to complete many collaborative tasks. Meanwhile, the problem of GPS spoofing attacks faced by submachines has become an urgent security problem for the UAV cluster. In this paper, a GPS-adaptive spoofing detection (ASD) method based on UAV cluster cooperative positioning is proposed to solve the above problem. The specific technical scheme mainly includes two detection mechanisms: the GPS spoofing signal detection (SSD) mechanism based on cluster cooperative positioning and the relative security machine optimal marking (RSOM) mechanism. The SSD mechanism starts when the cluster enters the task state, and it can detect all threats to the cluster caused by one GPS signal spoofing source in the task environment; when the function range of the mechanism is exceeded, that is, there is more than one spoofing source and more than one UAV is attacked by different spoofing sources, the RSOM mechanism is triggered. The ASD algorithm proposed in this work can detect spoofing in a variety of complex GPS spoofing threat environments and is able to ensure the cluster formation and task completion. Moreover, it has the advantages of a lightweight calculation level, strong applicability, and high real-time performance.

Keywords: GPS spoofing; collaborative positioning; rigid structure; complex scene; yaw

Citation: Meng, L.; Zhang, L.; Yang, L.; Yang, W. A GPS-Adaptive Spoofing Detection Method for the Small UAV Cluster. *Drones* **2023**, *7*, 461. https://doi.org/10.3390/drones7070461

Academic Editors: Xiwang Dong, Mou Chen, Xiangke Wang and Fei Gao

Received: 28 April 2023
Revised: 16 May 2023
Accepted: 29 May 2023
Published: 11 July 2023

1. Introduction

The term unmanned aerial vehicle (UAV for short) refers to an unmanned aircraft operated by radio remote control equipment and self-contained program control device, which can provide services in places that are difficult for humans to reach. In the early stage, the application of UAVs was limited to the military field. In recent years, with the rapid improvement of sensing, remote sensing, flight control, computational vision, image transmission, and other related technologies, the development of UAV has entered the fast lane [1]. Especially since 2015, with the continuous improvement of civil UAV technology, its application in agriculture, forestry and plant protection, power inspection, geographic mapping, aerial photography, and other aspects has become more and more normal. After 2019, UAV autonomous control and application technology has made great progress, showing some new development trends. Because a single UAV can only carry a single mission load and has limited mission execution capacity, the efficiency of the whole system can be improved through the complementary ability and action coordination of multiple UAVs. Therefore, the application of UAVs is gradually developing from a single platform to multiple platforms [2].

By learning from the self-organization mechanism of nature, UAV cluster consisting of multiple UAVs with limited autonomous ability is able to achieve an overall performance gain through mutual information communication without relying on centralized command and control. As a result, UAV cluster often possesses a higher degree of autonomous cooperation and requires little human intervention to complete the expected task objectives [3].

The wide application of UAV in different fields exacerbates its security issues, e.g., network attack [4], channel attack [5], and signal attack [6]. Among these attacks, GPS spoofing as a kind of signal attack has been the most urgent threat [7,8], given the fact that modern UAV positioning and navigation systems have become highly dependent on GPS signals. If the positioning and navigation information has been deceived, UAVs may deviate from the normal flight route and in more serious cases end up in a catastrophic crash.

1.1. Problem Statement

Detecting a GPS spoofing attack is a challenging problem. In a confrontation environment, the adversary usually causes a more complex and bad impact on the cluster by deploying more spoofing sources. According to how many spoofing sources there are, the possible GPS spoofing attach faced by an UAV cluster can be divided into two categories: the single GPS spoofing source attack and the multiple GPS spoofing source attack, different deployment strategies have different effects on clusters. When there is only one spoofing source, it may attack only one UAV or multiple UAVs. Considering the case where only one target UAV has been attacked, although the target UAV may have the capability to detect the existence of spoofing by itself, the detection can hardly be reliable given the uncertainty of the environment. If multiple UAVs have been attacked by one spoofing sources, although the attack is obvious, it is still difficult to determine whether there is only one spoofing source or not. When there are multiple spoofing sources, the problem becomes even more challenging due to the complicated interactions of the spoofing sources.

1.2. Contribution

In this work, we aim at solving the GPS spoofing detection problem for the case of multiple spoofing sources and we propose a cluster cooperative positioning-based algorithm that can successfully detect the existence of spoofing for UAV clusters, no matter how complex the threat environment is. The algorithm includes two mechanisms. Under the guidance of distributed computing, we design the GPS spoofing signal detection (SSD) mechanism. Furthermore, "no longer considering who is cheated, we should pay attention to who is safe", which is the core of the relative security machine optimal marking (RSOM) mechanism. Thus, it is worth mentioning that in order to ensure the autonomous recovery of the formation in an unsafe environment, we assume that not all members of the UAV cluster are deceived and at least one UAV in the cluster is safe. The main contributions of this paper can be summarized as follows:

- The GPS spoofing attacks for the UAV cluster are analyzed and classified, and the various complex attack scenarios under a cluster environment are simulated. To the best of our knowledge, research into the problem of spoofing attacks on the UAV cluster from multiple spoofing sources, as considered in this paper, is novel.
- A novel GPS-adaptive spoofing detection (ASD) algorithm which includes two detection mechanisms, GPS Spoofing Signal detection (SSD) mechanism and Relative Security UAV Optimal Marking (RSOM) mechanism, is proposed. The algorithm can switch between different detection mechanisms to effectively detect GPS spoofing signals according to the characteristics of GPS spoofing attack initiated by the attacker in different attack scenarios.
- A modeling and hardware simulation based technique has been studied to ensure the mission safety of UAV cluster. In fact, how to ensure the mission safety of UAV cluster in GPS spoofing environment is still in its infant stage. This work provides theoretical support and an application guidance for the development and application of this new task model.

The rest of this paper is organized as follows: Section 2 mainly summarizes the relevant research work. Section 3 discusses the establishment of the small smart UAV cluster model, the principle of the GPS spoofing attack, and its impact on the cluster task state. Section 4 introduces the detailed design of ASD. Section 5 presents the simulation experiments and compares with the latest results in the same domain. Section 6 summarizes this work and concludes with the potential impacts and prospects.

2. Related Work

As reported in [reference to the Volpe report], the U.S. Department of Transportation has performed a thorough security evaluation of civil GPS signal applications and concluded that "GPS has further penetrated into civil infrastructure. It has become an attractive target and can be used by individuals, groups or countries hostile to the United States". Malicious attacks on GPS signals mainly include intentional interference and deception, where the consequence of deception is often considered more severe than that of intentional interference. As a result, the detection of GPS deception has become a hot topic and been investigated intensively [9,10].

Some recent research has shown that civil UAVs can be easily deceived [11–13]. A simple GPS spoofing attack has been successfully implemented by researchers from Los Alamos National Laboratory [10]. Later, the Iranian army has claimed that they successfully controlled an American rq-170 sentinel UAV, when it was flying about 140 miles from the border between Iran and Afghanistan [14]. In [15], the authors showed that they can deceive the UAV by sending false position data to their GPS receiver, thus misleading the UAV to crash on the sand.

Regarding the detection and response schemes for GPS deception, the work in [16] has made a complete overview of the effort on combating GPS deception and jamming. A method to further improve the detectability of false GPS spoofing signal by encrypting the signature of navigation message was proposed in [17]. An algorithm for monitoring GPS deception based on power measurement and automatic gain control behavior observation has been proposed in [18]. The effectiveness of this algorithm has been verified by using commercial GPS receivers. In [19], the authors have proposed a GPS deception detection and protection scheme, leveraging the calculation of moving variance based on Doppler offset and consistency test of PVT calculation. In [20], the authors claimed that the forged GPS deception signal could not completely cover the real GPS signal, and proposed a method to detect GPS deception in the signal tracking stage through the detection technology of its residual signal. In [8], automatic gain control is used within the GPS receiver to detect and flag potential spoofing attacks within a low computational complexity framework. Moreover, [21] proposed a technique that allows UAVs to detect GPS spoofing by using an independent ground infrastructure that continuously analyzes the contents and times of arrival of the estimated UAV positions. The proposed technique is able to detect the spoofing attacks in less than two seconds and further determine the spoofing location after 15 min of monitoring time with an accuracy of up 150 m.

Notably, some other work have studied the use of multiple receivers to detect GPS spoofing attacks [11,22,23]. In [22], the authors demonstrated the ability of detecting GPS spoofing using a dual antenna receiver. Their technique relies on observing the carrier difference between different antennas under the same oscillator. In this configuration, the attacker needs to add a transmitting antenna every time when a receiving antenna is added, which makes the attacking task more complex. In [11], multiple receivers are used to authenticate GPS signal by using the correlation between GPS signal and military GPS signal. Among these receivers, a cross check receiver is used to determine whether its GPS signal is true. The technique has been tested on stationary and mobile GPS receivers and it can effectively detect spoofing attacks. In [23], multiple independent GPS receivers are used to detect GPS spoofing attacks. This technique relies on fixing the distance between receivers and then measuring the distance between the positions reported by the receivers. Under the real GPS signal, the measured distance is similar to the previous fixed distance.

However, under the GPS spoofing attack, the measured distance can be close to zero. This is because all receivers are cheated of the same false position. Currently, there are still some scholars who use machine learning methods to solve this problem. Their research focus is mainly on extracting ground features, and they are committed to how to extract the accuracy of features. Although it is equally effective in application, it does not have strong interpretability [24–26].

There are not many existing research results based on cluster deployment to detect deception signals and ensure the safety of drone missions. Among them, a game-based detection method for drone clusters was proposed in reference [27], which utilizes the relative position relationships of members in the cluster to effectively detect spoofing attacks. However, there are strong limitations on the size and threat scenarios of the cluster. Furthermore, a method based on task prior knowledge and formation rigid structure proposed by Liang Chen takes 8 s to achieve spoofing detection [28]. The Euclidean distance between members in a cluster calculated from different data sources in reference [29] is used to determine deception. This paper enriches threat scenarios and adversary capabilities, but has a strong dependence on security thresholds. Moreover, existing achievements all share a common problem, as they do not provide a method to determine the true position of drones or ensure the continuation of missions after detecting attacks [30]. In fact, these previous works mainly focused on detection technology and did not provide mature and effective autonomous attack mitigation or defense mechanisms.

3. System Models

3.1. The Small UAV Cluster Model

Given a set of UAVs, M, performing a common mission, each of which is equipped with a GPS receiver, a wireless communication module, and some sensors for specific applications. According to the GPS signal characteristics, we use three-dimensional (3D) data to specify their locations. Let the location of UAV m at time t be $u_m(t) = [x_m(t), y_m(t), z_m(t)]^T$, where $m \in N^+$. The UAV cluster model uses the flooding broadcast mode, which is commonly used in an ad hoc network to realize the communication between UAVs. That is, each UAV in the cluster shares the location information of all the others within the effective distance of broadcast. As shown in Figure 1, d_{max} is the largest distance between UAVs in the cluster, and e_{max} is the maximum effective range of UAV broadcasting. When designing the cluster formation, the condition $d_{max} < e_{max}$ ensures that each UAV in the cluster can receive the location information from the other UAVs.

The relative position between UAVs is one of the key bases for the formation design. When the navigation information of an UAV is detected to be dishonest, its position can be obtained through the relative positions between the other UAVs. Therefore, when designing the model, the relative position to the other UAVs is known to each UAV in the cluster.

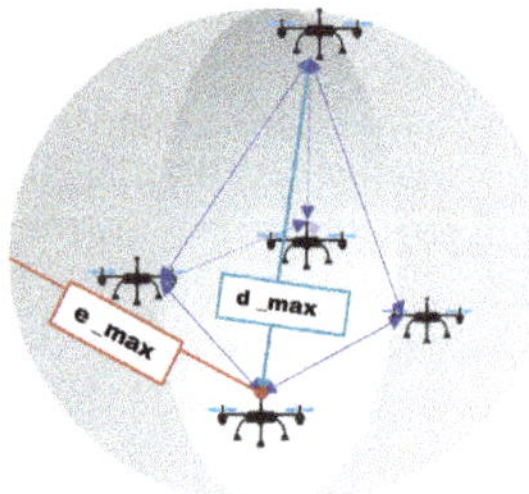

Figure 1. Relationship between the largest relative distance in an UAV cluster and the maximum effective distance of flooding communication.

1. UAV Position representation in the Cartesian coordinate system

    It is known that in the position calculation, the original output data of the GPS receiver cannot directly be used in the calculation. Instead, it needs to be transformed from the spherical coordinate system to the Cartesian coordinate system.

    Suppose that D is a point on the Earth's surface and the spherical coordinate of D is (lat, lon, r), where r is the radius of the earth. It is shown in Figure 2 that $\angle AOB = lat$, $\angle DOB = lon$, and the point D is expressed as follows:

$$D = \begin{bmatrix} x_D \\ y_D \\ z_D \end{bmatrix} = \begin{bmatrix} r \cdot cos(lon) \cdot sin(lat) \\ r \cdot sin(lon) \\ r \cdot cos(lon) \cdot cos(lat) \end{bmatrix} \tag{1}$$

If an UAV in the cluster reaches the specified position at H, which is vertically above point D, then it broadcasts the position D':

$$D' = \begin{bmatrix} x_{D'} \\ y_{D'} \\ z_{D'} \end{bmatrix} = \begin{bmatrix} (r + H) \cdot cos(lon) \cdot sin(lat) \\ (r + H) \cdot sin(lon) \\ (r + H) \cdot cos(lon) \cdot cos(lat) \end{bmatrix} \tag{2}$$

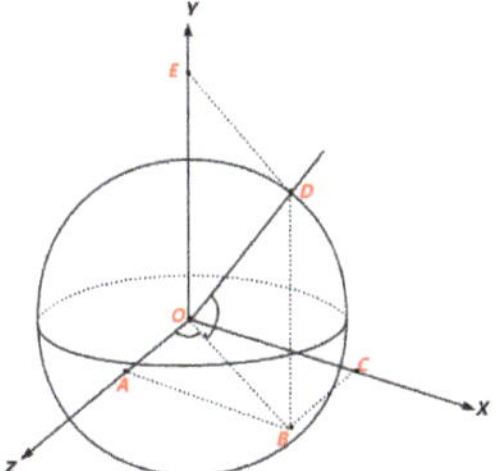

Figure 2. Schematic diagram of the conversion between the spherical coordinate system and ground coordinate system.

2. Indication of the relative position between UAVs

    The object of formation design is mainly to achieve a small cluster of UAVs. Thus, the full connection mode is adopted for the information interaction between UAVs. For model $M = \{m \in N + |u_1, u_2, u_3, \ldots, u_m\}$, as shown in Figure 3, the position relationship between any two UAVs can be expressed as a four-dimensional vector:

$$u_1 u_2 = \begin{bmatrix} \alpha \\ \beta \\ \theta \\ l \end{bmatrix} \tag{3}$$

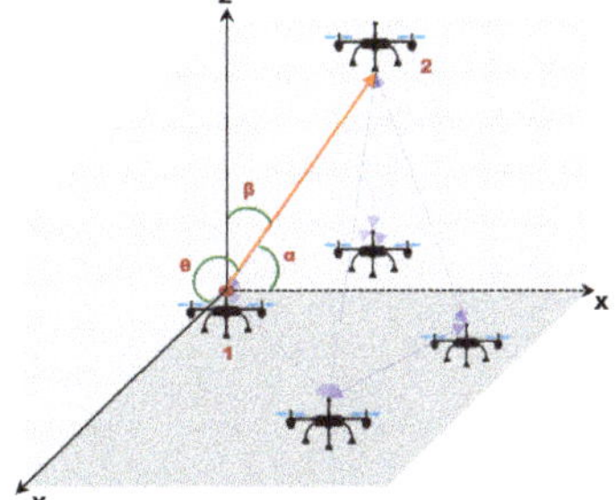

Figure 3. Schematic diagram of the relative position between UAVs in a cluster.

Thus, if $u_1 = \{x_1, y_1, z_1\}$, the following equation holds:

$$u_2 = \begin{bmatrix} x_2 \\ y_2 \\ z_2 \end{bmatrix} = \begin{bmatrix} x_1 + l\cos\alpha \\ y_1 + l\cos\beta \\ z_1 + l\cos\theta \end{bmatrix} \tag{4}$$

3.2. Adversary Model: GPS Spoofing Principle

The principle of GPS spoofing on the target UAV is as follows: the position spoofing attack will not change the UAV's position, but change the UAV's belief in its position. Thus, while the UAV is still in its real position when attacked, the perception of its location by its navigation system will be given by the attacker. Then, the UAV plans its route to the final destination according to the instructions transmitted to the controller by the navigation cognition.

The purpose of a GPS spoofer is to control the GPS antenna, in order to send the customized GPS positioning information to make the UAV navigation system believe that it is deceiving the expected position. According to the concealment and strategy of the attack, GPS spoofing attacks can be divided into the following two categories.

- Public: the spoofer does not try to cover up the attack, no matter whether the change between the customized deceptive GPS positioning information and the real GPS positioning information is within a reasonable range. It only tries to capture the target faster.
- Covert: the spoofer tries to avoid detection by sending cleverly crafted deceptive signals that match the actual signal in terms of output power and other parameters. Thus, the spoofer can prevent the target from triggering a fault detection alarm.

Since the spoofer can attack the target publicly or covertly, we consider that UAV is equipped with a fault detector, which can filter out the navigation signal with large mutations. Therefore, for the spoofer design, we would like to keep its attack covert by adjusting the parameters of the forged GPS signal, in order to avoid being found. The specific setting rules for parameter requirements can be found in [31]. The main idea is that the change between the spoofing signal sent by the spoofer and the signal received by the UAV GPS receiver at the previous time will be limited to a threshold, so that these applied positions will not trigger the fault detector in the UAV. Such a threshold between the current position and the position where the spoofing is applied is called the instance drift distance [32,33].

Let E_{max} be the instance drifted distance that limits the attack, $\hat{x}_m(t) = [\hat{x}_m(t), \hat{y}_m(t), \hat{z}_m(t)]^T$ be the attacker's imposed location on UAV m, and $E_m(t) = [E_{x_m}(t), E_{y_m}(t), E_{z_m}(t)]^T$ be a vector whose individual elements represent the distance difference between the UAV's actual location and the attacker's imposed location. Then, we have the following equation:

$$\|E_m(t)\|_2 = \|x_m(t) - \hat{x}_m(t)\|_2 \leq E_{max} \tag{5}$$

Explanations of all variables mentioned in this section are summarized in Table 1.

Table 1. Explanations of all variables mentioned in Section 3.

Variables	Explanation
M	A set of UAVs
u_m	One of the members in M
d_{max}	The largest distance between each two UAVs in M
e_{max}	The maximum effective range of UAV broadcasting
D	The parking position of UAV on the ground
D'	The hovering position of UAV in the air
$u_1 u_2$	The position relationship between any two UAVs in M
E_{max}	The instance maximum drifted distance

4. Proposed Method

In this section, the ASD method based on UAV cluster cooperative positioning is proposed, The workflow description is detailed in the Appendix A, which includes two detection mechanisms: the SSD mechanism based on the cluster cooperative positioning and the RSOM mechanism.

4.1. SSD Mechanism Based on Cluster Cooperative Positioning

During the execution of public tasks by the UAV cluster N, all members of the cluster broadcast the real-time position obtained by GPS receiver to the team through their respective wireless communication module at each time. The design principle of SSD is: at each broadcast time, when the signals broadcast by the cluster have the same location information, one can determine that there is at least one spoofing source in the mission airspace, and the RSOM mechanism of the ASD algorithm is triggered at this time; when the broadcast signals are different, we randomly select a submachine in the cluster, U_n, and extract its location information, P_{U_n}. Then, we use the real-time location information broadcasted by other members and the relative location information between other members and U_n in the formation to calculate where the other members think U_n should be. For example, based on the location information broadcasted by U_1, the position where U_1 thinks U_n should be located can be obtained by Formulas (2)–(4). If there is only one spoofing source in the mission airspace, the SSD mechanism can accurately locate the spoofing attack submachine in the cluster; otherwise, the RSOM mechanism will be triggered. Figure 4 shows the workflow of the SSD mechanism.

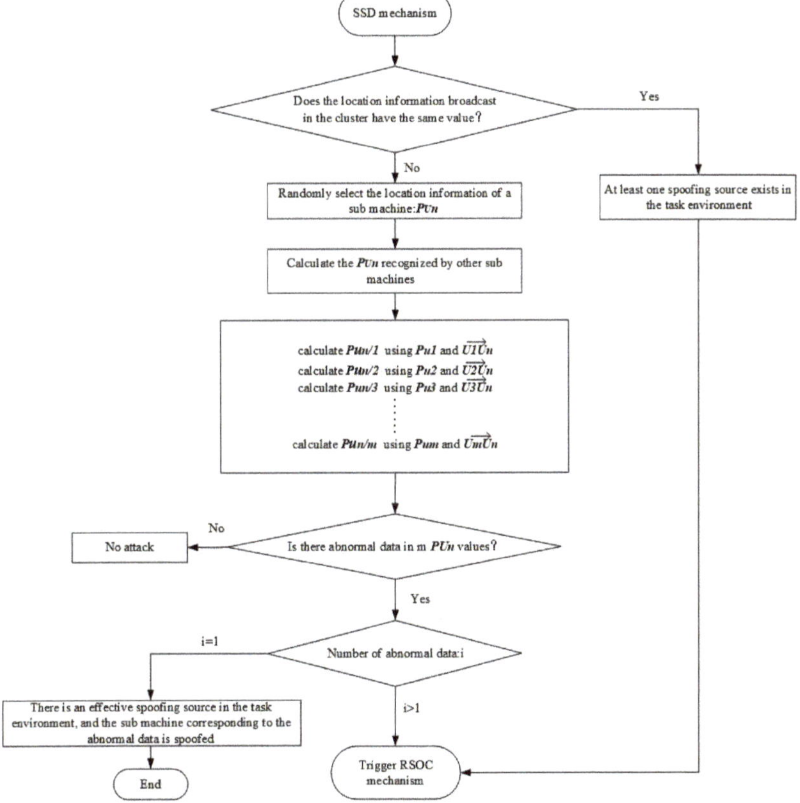

Figure 4. The workflow of the SSD mechanism.

4.2. RSOM Mechanism

The RSOM mechanism is triggered when there are multiple spoofing sources in the mission airspace and the GPS signal security status of each submachine in the cluster cannot be accurately determined. Compared to the assumption of [27], i.e., "at least one UAV in the cluster is safe", our RSOM can detect the case of a full cluster spoof. However, this assumption is still followed in our designed algorithm. Our purpose in doing so differs from that of [27] in that their spoofing detection has to be implemented under this assumption, whereas we do so to guarantee that the UAV cluster has the ability to recover autonomously in case of a spoofing attack. By letting go of this restriction, RSOM can call the ground station to achieve an artificial takeover of the cluster mission in the event of a full overrun being detected. In this attack scenario, in order to ensure the self-recovery capability of the cluster, the premise of RSOM is that at least one aircraft in the cluster is safe. Therefore, the threats faced by the UAV cluster can be summarized as follows: if two or more UAVs are attacked by different GPS spoofing signals, how can they be detected?

RSOM is designed with the idea that there is no need to face this problem directly. Specifically, at least one aircraft in the cluster is safe, so in such a complex threat scenario, we should accurately find the safe one. The details of the design idea are as follows: RSOM selects a virtual central machine for the UAV cluster to provide us with reference information representing the motion state of the whole cluster. Considering the loose coupling between the GPS measurement and the strapdown inertial navigation system (INS), the altitude dynamics of UAVs will not be affected by GPS spoofing attacks at the fist moment of spoofing, which has been confirmed by Kerns et al. [34] through a field test. At the same time, a large number of studies have shown that the relative controllability of altitude dynamics can maximize the asymptotic stability of closed-loop systems when applying optimal control signals in the event of GPS failure. Therefore, in the RSOM mechanism, the optimal marking of relative security machine is realized by using the deviation of the altitude information obtained by each member of UAV cluster from the GPS relative to the flight altitude obtained by altitude dynamics of the virtual central machine. In the RSOM mechanism, the yaw information is the core factor in determining the altitude of the UAV, so we simplify and divide the altitude model of the UAV, and finally, obtain the independent yaw model.

The RSOM mechanism includes three altitude models: the independent yaw model of the submachine, the independent yaw model of the virtual central machine, and the marking model. The workflow of the RSOM mechanism is shown in Figure 5.

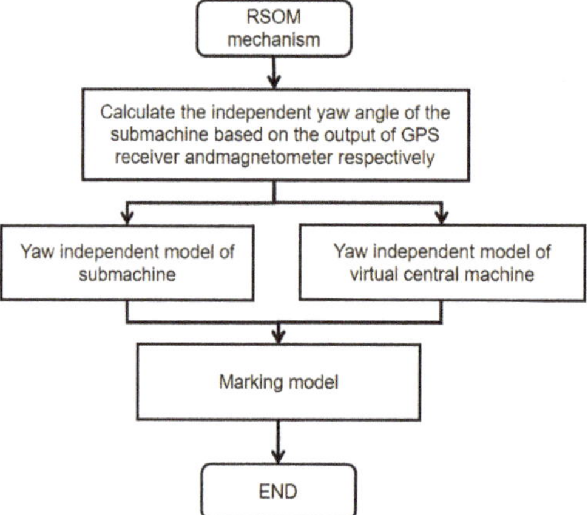

Figure 5. The design and work principle of the RSOM mechanism.

For each UAV, the yaw angle is given by the GPS receiver and the magnetometer, which are expressed as ψ_{GPS} and ψ_{mag}, the obtain algorithm are shown as Algorithm 1 and Algorithm 2, respectively [35].

Algorithm 1 Algorithm for obtaining obtainψ_{GPS} based on GPS receiver input data

---

1: Input the position information of the current time and the previous time:(lat_1, lon_1, alt_1) and(lat_2, lon_2, alt_2);
2: Take (lat_2, lon_2, alt_2) as the representation of a Cartesian coordinate system:(x_2, y_2, z_2);
3: Based on (lat_1, lon_1, alt_1), take (x_2, y_2, z_2) as the representation of a ENU system:(d_{e2}, d_{n2}, d_{u2});
4: Constraint $\psi_{GPS} \in [-\pi, \pi]$;
5: $\psi_{GPS} = arctan2(d_{e2}, d_{n2})$;

Algorithm 2 Algorithm for obtaining the ψ_{mag} based on magnetometer attitude measurement data

---

1: Suppose that the measured value of the magnetometer in the body coordinate system, (x_b, y_b, z_b), is ${}^b m_m = [m_{x_b}\ m_{y_b}\ m_{z_b}]^T$
2: Considering that the magnetometer may not be placed horizontally during the UAV mission, it is necessary to use the two axis inclination sensors to measure the pitch angle, θ, and the roll angle, ϕ, and then project the measured values on the horizontal plane. Therefore,
$$\begin{cases} \overline{m}_{x_e} = m_{x_b}cos\theta_m + m_{y_b}sin\phi_m sin\theta_m \\ \qquad + m_{z_b}cos\phi_m sin\theta_m \\ \overline{m}_{y_e} = m_{y_b}cos\phi_m - m_{z_b}sin\phi_m \end{cases}$$
where $\overline{m}_{x_e}, \overline{m}_{y_e} \in \mathbb{R}$ indicates the projection of the magnetometer reading on the horizontal plane.
3: Constraint $\psi_{mag} \in [-\pi, \pi]$
4: $\psi_{mag} = arctan2(\overline{m}_{y_e}, \overline{m}_{x_e})$

---

1. Independent yaw model of the submachine
   In the independent yaw model of submachine, the yaw angle of the submachine in the cluster is defined as:

$$\psi = (1 - \mu_\psi)\psi_{GPS} + \mu_\psi \psi_{mag} \tag{6}$$

where ψ_{GPS} and ψ_{mag} can be obtained by algorithms 2 and 3. $\mu_\psi \in [0,1]$ is a weighting factor.

The basic idea of the linear complementary filter is to use their complementary features to obtain more accurate altitude angle. In this model, the linear complementary filter [36–38] is only used as a known tool, so it is only briefly explained without showing the detailed reasoning process. At time k, after obtaining $\psi(k)$, the yaw angle is estimated as:

$$\hat{\psi}(k) = \frac{\tau}{\tau + T_s}(\hat{\psi}(k-1) + T_s \omega_{z_b}(k)) + \frac{T_s}{\tau + T_s}\psi(k) \tag{7}$$

where $\tau \in \mathbb{R}^+$ represents the time constant, $T_s \in \mathbb{R}^+$ represents the sampling period used by the filter, and ω_{z_b} represents the component of the angular velocity in the z direction in the earth fixed coordinate system [39]. Take $\frac{\tau}{\tau + T_s} = 0.95$, then $\frac{T_s}{\tau + T_s} = 0.05$. The complementary filter of the yaw angle is expressed as follows:

$$\hat{\psi}(k) = 0.95(\hat{\psi}(k-1) + T_s \omega_{z_b}(k)) + 0.05\psi(k) \tag{8}$$

2. Independent yaw model of the virtual central machine
   GPS provides external information to the UAV. It belongs to the experimental group of this subject and needs to be verified. Therefore, we need a control group in the

model. For the flight altitude estimation of the whole cluster, we only use the internal information of the UAV, namely the magnetometer. The yaw representation of the flight altitude of the whole cluster is realized by fusing the yaw altitude of each member machine with a weighted average method to form a new yaw altitude model. It can be considered that we have selected a virtual central machine for the cluster, and the new yaw altitude model is the yaw representation of the virtual central machine; its physical meaning is to represent the flight altitude of the cluster to the greatest extent.

Here:

$$\psi' = \psi_{mag}. \tag{9}$$

Input $\psi'(k)$ to Equations (7) and (8) to obtain $\hat{\psi}'(k)$. Then, the independent yaw model of the virtual central machine, $\Psi(k)$, can be expressed as follows:

$$\begin{cases} \Psi(k) = \sum_{n=1}^{m} \psi'_n(k)\rho_n(k) \\ \rho(k) = \dfrac{1}{2}log\dfrac{1-\varepsilon(k)}{\varepsilon(k)} \end{cases} \tag{10}$$

where $\varepsilon(k)$ is the error confidence obtained by the exponential standardization of the *softmax* function to the current error of each submachine magnetometer, and $\varepsilon_1(k) + \varepsilon_2(k) + \varepsilon_3(k) + \ldots + \varepsilon_m(k) = 1$. $\rho(k)$ is the final weight coefficient of each submachine.

3. Marking model

The difference between the results of the independent yaw model of the virtual central machine and that of the submachine is used as the basis for the results of the calibration model:

$$d_n = |\Psi(k) - \hat{\psi}_n(k)| \tag{11}$$

Note that $d_{min} = (d_1, d_2, \ldots, d_m)$, d_{min} corresponding to the submachine is the optimal marking of the making model to the relative security of the UAV.

4.3. Time Complexity Analysis

According to the big O representation, $O(n)$, the algorithm grows as the data size n increases. The ASD algorithm designed in this paper does not contain loops and recursive statements, so the time complexity is $O(1)$. It should be noted that it does not fully represent the actual execution time. The actual execution time of the algorithm is also closely related to the performance of the hardware device.

To sum up, it can be concluded that the two mechanisms of the ASD algorithm have a serial relationship in the working process. Last, but not least, at the end of the algorithm design, we added a straightforward defense, the "Leader-follower mode". This mode is triggered when the ASD algorithm detects GPS spoofing. That is, the relatively safe submachine selected by RSOM will enter the leader mode and the other submachines will enter the follower mode. Generally speaking, under the premise that "at least one submachine in the cluster is safe", the ASD algorithm can solve various threats faced by UAV cluster in the mission environment, and has the ability to guarantee the formation and flight mission at the same time. This study proposes the constraint that "at least one submachine in the cluster is safe", and its application background is the fully autonomous task of the UAV cluster. With manual monitoring and intervention during the task, this restriction can be released and the cluster submachines can be switched to manual takeover when all of them are under attack.

5. Simulation and Evaluation

To verify the effectiveness of the ASD algorithm proposed in this paper, simulation experiments are carried out in this section. The experiments are performed on Gazebo and MATLAB platforms. We built the UAV cluster system model on the Gazebo platform, and connected the MATLAB-based ASD algorithm to the Gazebo flight control through cross-platform combination. The verification process and result analysis are as follows.

5.1. Experimental Configuration

In this experiment, during the task of the UAV cluster system model, the motion heights of all submachines are always the same, and the subsequent spoofing signal generation is only also based on longitude and latitude. Therefore, when designing the formation, $\beta = 0$ and $\theta = 0$ are in the relative position relationship between the cluster submachines, and the overall structure is a pentagon. The specific motion parameters of the small smart UAV cluster after entering the stable flight are as follows:

- Cluster size: 5;
- Relative position relationship between machines: $[\alpha \ \beta \ \theta \ 1]$;
- Cluster velocity: 5 m/s;
- Cluster motion height: 50 m;
- Cluster motion direction: all submachines are consistent;
- Maximum distance between machines: 20 m;
- Maximum effective range of communication: 500 Hz.

Correspondingly, to verify the detection efficiency of the ASD algorithm proposed in this study, we modeled the enemy according to the GPS spoofing principle on the Gazebo simulation platform. Five spoofing sources (S1, S2, S3, S4, S5) are set up; following the movement of the cluster, they are randomly distributed around the cluster and the distance from the cluster is always within the effective range of the spoofing signal. Section 3.2 mentions both public and covert spoofing, but the detection principle of the ASD proposed does not specifically target a certain type of spoofing. However, in the experimental deployment, the enemy models all used covert deception, as it is a more advanced spoofing ability.

5.2. Experimental Deployment

The initial state of the UAV cluster system model at the beginning of each scenario: the submachines are lined up on the ground. The UAV cluster is manually controlled to take off vertically one by one, reaching a specified altitude of 50 m. The cluster then enters the fully autonomous mode. Each submachine adjusts its position according to the preset positional relationship between the aircraft, forms a formation, and enters the flight mission.

- Scenario 1:Baseline model test: This case is to obtain the normal movement log of the UAV cluster in the mission scenario without any attack or threat, which can be used as a baseline to detect the threat later.
- Scenario 2: Adversary model test: Note that the five spoofing sources work exactly the same, so only one of them is randomly selected for validity testing. In this scenario, the deployment location of S4 is shown in Figure 6. In Figure 6a, there is only one submachine in the signal radiation range of S4, while in Figure 6b, there are more submachines in its signal radiation range. Such a setup can test not only the effectiveness of the spoofing source, but also whether the spoofing source can spoof all submachines within its signal radiation range. In the experiment, after the cluster enters a stable mission state, we do not start the ASD algorithm, but we start S4, after which we observe the movement state of the cluster and save the flight logs.
- Scenario 3: Contrast experiment of scenario 2: In this case, the deployment location of S4 is shown in Figure 6a; that is, there is only one submachine in the signal radiation range of S4. Different from the setting of scenario 2, after the cluster enters a stable mission state, we first start the ASD algorithm and then start S4. After that, we record the movement state of the cluster and save the flight logs.
- Scenario 4: Testing of two spoofing sources: In this case, the deployment locations of the two spoofing sources (i.e., S1 and S2) are shown in Figure 6c. It can be observed from Figure 6c that the signal radiation range of these two spoofing sources contains three submachines. Here, we will use S1 and S2 to attack them. In the experiment,

after the cluster enters a stable mission state, we start the ASD algorithm and turn on the two spoofing sources. Then, the movement state of the cluster and the flight logs will be recorded.

- Scenario 5: Testing of three spoofing sources: In this case, the deployment locations of the three spoofing sources (i.e., S1, S2, and S3) are shown in Figure 6d. The rest of the operation is the same as in Scenario 4.
- Scenario 6: Testing of four spoofing sources: In this case, the deployment locations of the three spoofing sources (i.e., S1, S2, S3, and S4) are shown in Figure 6e. The rest of the operation is the same as in Scenario 4.
- Scenario 7: Testing of the full cluster spoofed: In this case, we directly considered and deployed the most complex attack scenario with five spoofing sources (i.e., S1, S2, S3, S4, and S5); as shown in Figure 6f, after the cluster enters a stable mission state, we start the ASD algorithm and the five spoofing sources. It can be observed that the cluster suddenly oscillates in formation after a period of time, but soon returns to its original form; however, the overall motion direction is off the expected trajectory. At that point, the UAV cluster sent a distress signal to the ground station. Again, we keep the flight logs.

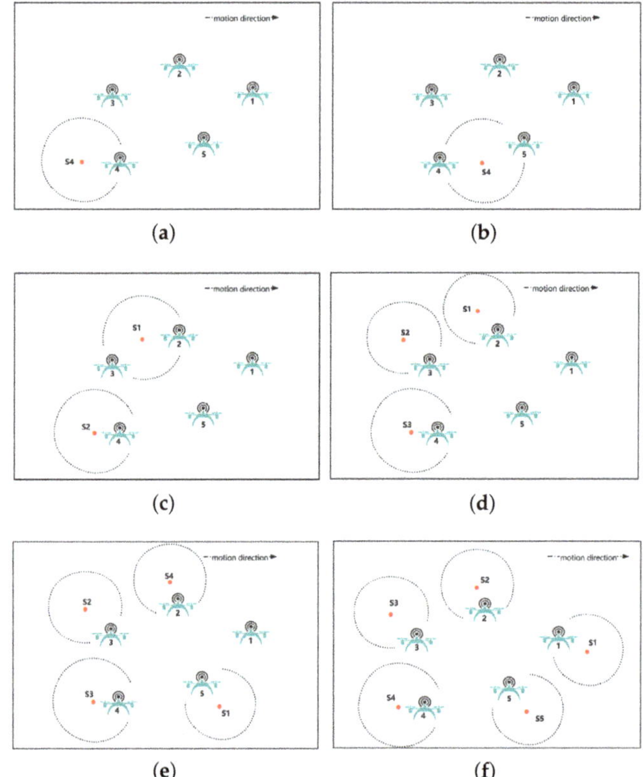

Figure 6. Tactics of an adversary: deploying the deception source. (**a**) Deploy S4 to attack one of the submachines to verify the effectiveness of the spoofing source. (**b**) Verify whether the S4 spoofing source has the ability to spoof all submachines within its signal radiation range. (**c**) Deploy two different spoofing sources to launch a GPS spoofing signal attack on three submachines in the cluster. (**d**) Deploy three different spoofing sources to launch a GPS spoofing signal attack on three submachines in the cluster. (**e**) Deploy four different spoofing sources to attack four submachines in the cluster. (**f**) Deploy five different spoofing sources to attack five submachines in the cluster.

5.3. Experimental Results and Analysis

Scenarios 1 and 2 of the experimental deployment belong to equipment testing and the others belong to algorithm verification.

In each scenario, we conduct multiple sets of experiments. Throughout the experiment, we observed that, in the state of cluster motion, at some point after the spoofing source was turned on, individual submachines did shake abnormally or leave the team, but the final observation result was that the cluster corrected the formation and finished the flight mission. The specific result analysis can be obtained through the retained flight logs, as shown below.

Model testing results

Figure 7a shows the state diagram of the UAV cluster system model completing a flight mission in a safe environment, i.e., scenario 1. Figure 7b,c are the results of the verification of the enemy model, i.e., scenario 2. Among them, Figure 7b shows the output of the GPS receiver deploying a spoofing source and the S4 spoofing a submachine, No. 4. Figure 7c shows the output of the GPS receiver deploying S4 to deceive two submachines, i.e., 4 and 5, simultaneously.

It can be seen that the formation of the UAV cluster has been disrupted and the output conforms to the spoofing principle. This phenomenon implies that S4 does effectively attack the submachines within its signal radiation range. Furthermore, it demonstrates that the enemy model we designed is effective, which can support the construction of the GPS spoofing countermeasure environment required for the experiment.

Algorithm verification results

Since no abnormality was observed in the overall motion state of the UAV cluster, we chose to use the data for a more intuitive interpretation. In the table recording data information, we use the same color to indicate the corresponding relationship between the spoofing source and the target. Moreover, ▶ marks the reference machine selected by ASD, while ★ marks the target selected.

Table 2 shows the record of current spoofing sources, and the flight logs of each submachine in scenario 3. According to the ASD algorithm design, the RSOM mechanism will not be triggered when only one aircraft suffers a spoofing attack. In fact, the final output of the ASD algorithm is the detection result of the SSD mechanism. According to the log information of the submachine GPS receiver, it can be seen that No. 4 was attacked; the SSD randomly selected No. 1 at this time, and only No. 4 had abnormal cognition of the position of No. 1 of the other four racks. Furthermore, we can see from the logs that after detecting a spoofing attack on No. 4, the system tells No. 4 to disable the GPS receiver and go into the leader mode in the cluster. Similarly, we can also see in the logs that the algorithm detected the threat at the second moment after being spoofed.

Table 3 shows the record of spoofing sources currently, and the flight logs for each submachine in scenario 4. Unlike Table 2, the final output of the ASD algorithm is no longer the result of the SSD mechanism, but rather RSOM. Based on the analysis of the spoofing sources data and GPS receiver information, it is not difficult to see that No. 2 and 3 were attacked by the same spoofing source, S2, and No. 4 was attacked by a different spoofing source, S1, from the previous signal. This situation cannot be solved by SSD, which triggers RSOM. In No. 1 and No. 5, which are safe in the cluster, RSOM finally chooses No. 5 as the leader of the safety machine according to the idea of algorithm design. Similarly, in scenario 5, these three submachines are also subject to a spoofing attack, the difference being that these three submachines receive spoofing signals from three different spoofing sources, respectively, which can be obtained from Table 4. The RSOM mechanism also works perfectly; it selected No. 1.

Scenario 6 is the most complicated of all. To ensure that each spoofing source deployed achieves the expected efficiency, we iteratively adjust their location and signal strength, and finally, achieve one-to-one spoofing, as shown in Table 5. Of course, scenario 6 is also the strongest proof of the effectiveness of the ASD algorithm. In our deployment, No. 1 is outside the effective range of all spoofing signals. From the table we can see that the RSOM does calibrate it accurately, making it the leader of the cluster. During the experimental observation, we saw that the formation of the UAV cluster vibrated obviously when attacking, but it quickly recovered and adjusted as before, and finally completed the task.

Table 6 shows the record of spoofing sources and the flight logs for each submachine in scenario 7. In the validation work of this scenario, we liberalized the "at least one drone safe" restriction and deployed five different spoofing sources to spoof each of the five submachines separately. As you can see from the information in Table 6, the RSOM still selected the submachine it thought could be the leader out of the five submachines: No. 2. However, the fact is that No. 2 has also been attacked by the spoofer S2. Its yaw information relative to that of the virtual central machine was already far greater than the normal drift range of the magnetometer. At this point, the UAV cluster no longer had completely reliable navigation information and the ASD eventually sent a distress command to the ground station.

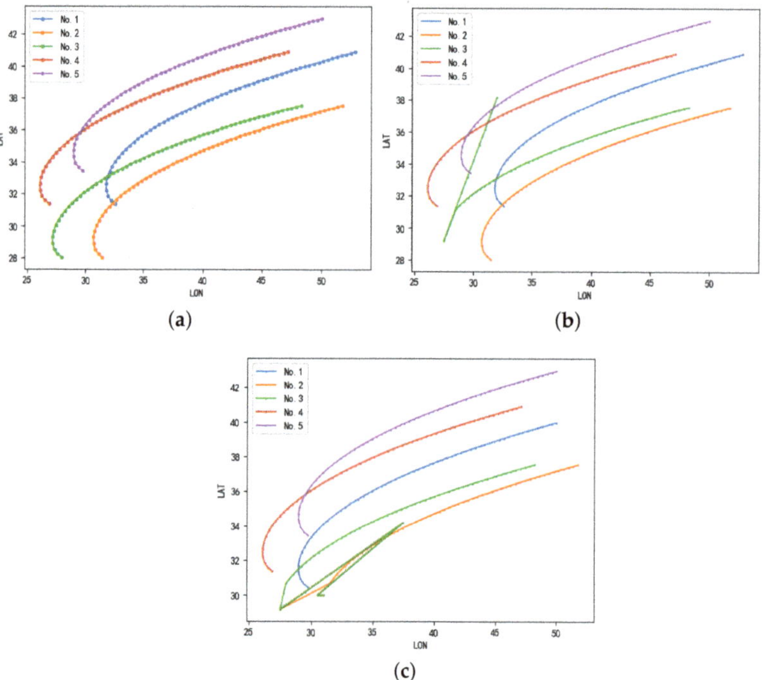

Figure 7. Illustration of effectiveness verification of the UAV cluster system model and the enemy model. (**a**) The trajectory information output by GPS receivers of the UAV cluster system model in the safe mission environment. (**b**) Deploy spoofing source 4 to attack No. 4 in the cluster without any detection and defense measures. The motion trajectory output by UAV cluster GPS receivers. (**c**) Deploy spoofing source 4 to attack No. 4 and 5 in the cluster without any detection and defense measures. The motion trajectory output by UAV cluster GPS receivers.

Table 2. The spoofing data record and flight log information obtained by scenario 3.

Enemy deployment strategy and spoofing data record				
S1	S2	S3	S4	S5
-	-	-	25.5 28.2	-
-	-	-	27.5 29.2	-
-	-	-	29.5 30.2	-
-	-	-	31.5 31.2	-
Position information received by the GPS receiver of each submachine in the cluster				
No. 1	No. 2	No. 3	No. 4	No. 5
32.87953077 34.30943511	30.02636122 36.38238413	27.17319167 34.30943511	28.26300546 30.95533315	31.78971698 30.95533315
32.59085634 34.03254691	29.73768679 36.10549593	26.88451724 34.03254691	27.97433104 30.67844495	31.50104255 30.67844495
32.32634117 33.73249404	29.47317162 35.80544306	26.62000208 33.73249404	27.5 29.2	31.23652738 30.37839208
32.09622408 33.405315	29.24305453 35.47826402	26.38988498 33.405315	29.5 30.2	31.00641029 30.05121304
31.91688606 33.04777078	29.06371651 35.1207198	26.21054696 33.04777078	-	30.82707227 29.69366881
31.81412885 32.66119484	28.9609593 34.73414386	26.10778976 32.66119484	-	30.72431506 29.30709287
Output of the SSD mechanism in the ASD algorithm under the same timestamp				
▶No. 1	No. 2	No. 3	★No. 4	No. 5
32.87953077 34.30943511	32.87956122 34.30948413	32.87949167 34.30943511	32.87950546 34.30943315	32.87951698 34.30943315
32.59085634 34.03254691	32.59088679 34.03259593	32.59081724 34.03254691	32.59083104 34.03254495	32.59084255 34.03254495
32.32634117 33.73249404	32.32637162 33.73254306	32.32630208 33.73249404	32.1165 32.5541	32.32632738 33.73249208
32.09622408 33.405315	32.09625453 33.40536402	32.09618498 33.405315	★34.1165 33.5541	32.09621029 33.40531304
31.91688606 33.04777078	31.91691651 33.0478198	31.91684696 33.04777078	-	31.91687227 33.04776881
31.81412885 32.66119484	31.8141593 32.66124386	31.81408976 32.66119484	-	31.81411506 32.66119287

The colored section emphasizes the successful entry of the deception source into the GPS receiver's data.

Table 3. The spoofing data record and flight log information obtained by scenario 4.

Enemy deployment strategy and spoofing data record				
S1	S2	S3	S4	S5
25.5 28.2	26 30.2	-	-	-
27.5 29.2	28 31.2	-	-	-
29.5 30.2	30 32.2	-	-	-
31.5 31.2	32 33.2	-	-	-

Position information received by the GPS receiver of each submachine in the cluster				
No. 1	No. 2	No. 3	No. 4	No. 5
32.87953077 34.30943511	30.02636122 36.38238413	27.17319167 34.30943511	28.26300546 30.95533315	31.78971698 30.95533315
32.59085634 34.03254691	29.73768679 36.10549593	26.88451724 34.03254691	27.97433104 30.67844495	31.50104255 30.67844495
32.32634117 33.73249404	28 31.2	28 31.2	27.5 29.2	31.23652738 30.37839208
32.09622408 33.405315	30 32.2	30 32.2	29.5 30.2	31.00641029 30.05121304
-	-	-	-	30.82707227 29.69366881

Output of the RSOM mechanism in the ASD algorithm under the same timestamp					
	▶virtual central machine	No. 1	No. 2	No. 3	No. 4	★No. 5
$\Psi(k)\&\hat{\psi}_n(k)$	−2.083049636	−2.103285455	0.400434363	0.400434363	0.40238313	−2.098250379
d_n	-	0.020235819	2.483483999	2.483483999	2.485432766	0.015470743

The colored section emphasizes the successful entry of the deception source into the GPS receiver data, representing the different colors of each deception source. The corresponding parts of the color blocks in the table can reflect which drone the deception source attacked.

Table 4. The spoofing data record and flight log information obtained by scenario 5.

Enemydeployment strategy and spoofing data record				
S1	S2	S3	S4	S5
26 31.3	24.5 30.6	25.5 28.4	-	-
28 31.3	26.5 30.6	27.5 29.4	-	-
30 32.3	28.5 31.6	29.5 30.4	-	-
32 33.3	30.5 32.6	31.5 31.4	-	-

Position information received by the GPS receiver of each submachine in the cluster				
No. 1	No. 2	No. 3	No. 4	No. 5
31.91688606 33.04777078	29.06371651 35.1207198	26.21054696 33.04777078	27.30036075 29.69366881	30.82707227 29.69366881
31.81412885 32.66119484	28.9609593 34.73414386	26.10778976 32.66119484	27.19760355 29.30709287	30.72431506 29.30709287
31.82313175 32.26129617	28 31.3	26.5 30.6	27.5 29.4	30.73331796 28.9071942
31.97080846 31.88955487	30 32.3	28.5 31.6	29.5 30.4	30.88099467 28.53545291
32.24110927 31.59470327	-	-	-	-
32.58404982 31.38881062	-	-	-	-

Output of the RSOM mechanism in the ASD algorithm under the same timestamp					
	▶virtual central machine	★No. 1	No. 2	No. 3	No. 4	No. 5
$\Psi(k)\&\hat{\psi}_n(k)$	−1.128328325	−1.131434769	0.408047301	0.413437933	0.409874591	−1.135919561
$\dot{d}_n$	-	0.003103444	1.536375626	1.541766258	1.538202916	0.007591236

The colored section emphasizes the successful entry of the deception source into the GPS receiver data, representing the different colors of each deception source. The corresponding parts of the color blocks in the table can reflect which drone the deception source attacked.

Table 5. The spoofing data record and flight log information obtained by scenario 6.

Enemy deployment strategy and spoofing data record				
S1	S2	S3	S4	S5
25.5 28.2	26.5 30.3	24.5 28.5	25 29.6	-
27.5 29.2	28.5 31.3	26.5 29.5	27 30.6	-
29.5 30.2	30.5 32.3	28.5 30.5	29 31.6	-
31.5 31.2	32.5 33.3	30.5 31.5	31 32.6	-
Position information received by the GPS receiver of each submachine in the cluster				
No. 1	No. 2	No. 3	No. 4	No. 5
32.32634117 33.73249404	29.47317162 35.80544306	26.62000208 33.73249404	27.70981587 30.37839208	31.23652738 30.37839208
32.09622408 33.405315	29.24305453 35.47826402	26.38988498 33.405315	27.47969878 30.05121304	31.00641029 30.05121304
31.91688606 33.04777078	27 30.6	28.5 31.3	26.5 29.5	27.5 29.2
31.81412885 32.66119484	29 31.6	30.5 32.3	28.5 30.5	29.5 30.2
31.82313175 32.26129617	-	-	-	-
Output of the RSOM mechanism in the ASD algorithm under the same timestamp					

	▶virtual central machine	★No. 1	No. 2	No. 3	No. 4	No. 5
$\Psi(k)\&\hat{\psi}_n(k)$	-1.868540485	-1.872339647	0.411671473	0.406189617	0.413437933	0.409874591
d_n	-	0.006799162	2.280211958	2.274730102	2.281978418	2.278415.76

The colored section emphasizes the successful entry of the deception source into the GPS receiver data, representing the different colors of each deception source. The corresponding parts of the color blocks in the table can reflect which drone the deception source attacked.

Table 6. The spoofing data record and flight log information obtained by scenario 7.

Enemy deployment strategy and spoofing data record				
S1	S2	S3	S4	S5
24.5 28.5	25 29.6	26.5 30.3	25.5 28.2	31.5 32.2
26.5 29.5	27 30.6	28.5 31.3	27.5 29.2	33.5 33.2
28.5 30.5	29 31.6	30.5 32.3	29.5 30.2	35.5 34.2
30.5 31.5	31 32.6	32.5 33.3	31.5 31.2	37.5 35.2
Position information received by the GPS receiver of each submachine in the cluster				
No. 1	No. 2	No. 3	No. 4	No. 5
35.57992834 36.00852281	32.7267588 38.08147183	29.87358925 36.00852281	30.96340304 32.65442085	34.49011455 32.65442085
35.22171153 35.83053205	32.3685419 37.90348106	29.51537243 35.83053205	30.60518622 32.47643008	34.13189774 32.47643008
26.5 29.5	27 30.6	28.5 31.3	27.5 29.2	33.5 33.2
8.5 30.5	29 31.6	30.5 32.3	29.5 30.2	35.5 34.2
-	31 32.6	-	-	-
Output of the RSOM mechanism in the ASD algorithm under the same timestamp					
	▶virtual central machine	No. 1	★No. 2	No. 3	No. 4	No. 5

	▶virtual central machine	No. 1	★No. 2	No. 3	No. 4	No. 5
$\Psi(k)\&\hat{\psi}_n(k)$	−2.24089981	0.41928125	0.41256874	0.418652482	0.4198514278	0.500265478
d_n	-	2.66018106	**2.65346855**	2.659552292	2.6607512378	2.741165288

The colored section emphasizes the successful entry of the deception source into the GPS receiver data, representing the different colors of each deception source. The corresponding parts of the color blocks in the table can reflect which drone the deception source attacked.

From the above series of experimental results, the ASD algorithm can detect the attack behavior at the second moment of spoofing. The acquisition frequency of UAV flight logs is 5 Hz, which means that the time required to detect deception is 0.4 s. This is because the ASD algorithm contains two mechanisms, which take time to judge, trigger, and switch. From the information output frequency of the flight log, ASD is known as a very efficient real-time detection algorithm, which is not affected by the time delay of one recording. On the other hand, the RSOM mechanism does not seem to focus on detecting spoofing intuitively, but this is not the case. When a secure submachine is selected, all information it provides is trusted by default. Then, based on the geometric relationship between the submachines, it is easy to obtain the location where other submachines should be. At this time, if there is a non-negligible error between the information output by the GPS receiver of which submachine and the information provided by the secure submachine, it can be determined that the information has been spoofed. Because this problem is obvious, it is not emphasized. The "Lead-follower" mode is a small defense set up for the cluster to ensure that at least one submachine is safe to complete the task.

5.4. Comparative Analysis of the Method's Performance

Regardless of whether used in a simulation environment or a real physical environment, it is difficult to fully reproduce the theoretical results of existing research in UAV flight experiments due to the uncertainty brought by atmospheric disturbances and motion time drift in the environment on the output of UAV sensors. Therefore, in this section, the original authors' analysis of the original performance data of the methods proposed by them is directly referenced and compared with the methods proposed in this chapter in different performance dimensions.

Comparing the ASD method proposed in this article with the detection method proposed by Liang, Chen et al. [28] in Table 7, our method only took 0.4 s in a task, which can be called a very effective real-time detection method that is not affected by the time delay of a single record; concurrently, ASD is, without requiring prior knowledge, suitable for random flight missions and also better at detecting accuracy.

AR Eldosouky, A Ferdowsi, et al. [27], when analyzing their proposed method, did not analyze the performance of the method such as timeliness and detection accuracy. They paid more attention to the effectiveness of a simulation experiment, and their method can solve a narrow problem domain, which not only has strong limitations on the threat scenarios where deception occurs, but also specifies the applicable cluster size. By relaxing these limitations, the proposed ASD method can face complex threat scenarios with the same detection capabilities.

The method Pavlo Mykytyn (2023) [29] proposed does not limit the types of threats that occur, and also designs complex adversarial scenarios. However, the design of the method to determine whether the spoofing attack occurs based on the distance difference has a strong dependence on the security threshold, but there is currently no authoritative setting rule for the security threshold. In addition, the infrared ranging method introduces additional hardware equipment. In ASD method proposed, there is no such issue, as there is no need for auxiliary values or equipment.

Finally, the confrontation environment that ASD proposed in this chapter can face is complex, and it is worth mentioning that the ASD method does not require any prior knowledge, and the assistance of any other additional equipment and does not increase the load burden on unmanned aerial vehicles. Moreover, the ASD method has small computational complexity, has high efficiency, and is timely and accurate.

Table 7. Comparison of similar methods.

Methods	Detection Accuracy	Detection Time	Method Characteristics
Liang, Chen (2019) [28]	98.6%	8 s	Requires prior knowledge of a given task, and other members within the communication range must be greater than 3
AR Eld. etc. (2020) [27]	Undefined and not analyzed	Undefined and not analyzed	There is only one deception source, only one aircraft is deceived at a time, and one aircraft is absolutely safe; the method is applicable to clusters with a scale of 5 or more
Pavlo Mykytyn (2023) [29]	Undefined and not analyzed	Undefined and not analyzed	One distance ranging technology; the execution of this method strongly relies on security thresholds
method proposed	100%	0.4 s	1. The cluster size is greater than or equal to 3 and is suitable for random flight missions; 2. There can be multiple deception sources in the flight environment that launch indiscriminate attacks against the cluster; 3. There are no constraints required for the execution of deception detection in the method, and during the task, after implementation of detection, it follows the safe machine concept, but not a strong constraint.

6. Conclusions

At present, in view of the impact of GPS spoofing on UAVs, the existing detection methods mainly focus on the single-machine problem. Machine learning methods are the most popular of these methods. In the practical application of UAVs, timeliness is an issue that cannot be ignored. The detection mechanism in the ASD algorithm has good detection efficiency in the simulation environment; accurate detection can be achieved almost immediately when a spoofing attack occurs. On the other hand, at present, how to solve the UAV cluster in the face of GPS spoofing attack is still a new problem. Among the few research results that address the same problem [27,29], the execution of methods requires the execution under various constraints.

Obviously, the confrontation environment faced by the ASD method proposed in this study is more complex. It is worth mentioning that the ASD method does not use any other equipment except the most basic airborne equipment, and the computation sequence is simple. In the experimental design of this article, in order to accurately grasp and analyze the objective performance of the method, atmospheric disturbance factors were not added to the simulation environment. Furthermore, the autonomous performance of ROSM mechanism is established under a constraint condition of "at least one secure drone exists in the cluster". Thus, in the next research step, we will find problems based on practical applications, hoping to improve the robustness of ASD. In addition, in future research, we will consider using visual ranging among UAV cluster members to determine the true location of the submachines attacked by spoofing. In this way, the algorithm will become more complete and intelligent, enabling better cluster control.

Author Contributions: Conceptualization, L.M.; methodology, L.M.; validation, L.M. and L.Z.; writing—original draft preparation, L.M.; writing—review and editing, W.Y. and L.Y. All authors have read and agreed to the published version of the manuscript.

Funding: This work was supported in part by the National Natural Science Foundation of China under Grant 61831007 and 61971154 and Basic Scientific Research Projects under Grant JCKY2020604B004.

Data Availability Statement: Due to some data being classified, it cannot be made public.

Conflicts of Interest: We hereby declare that there is no conflict of interest in the research content and process of this article.

Appendix A

This is the flowchart of our proposed method, ASD, and the mechanisms description included is in the main text.

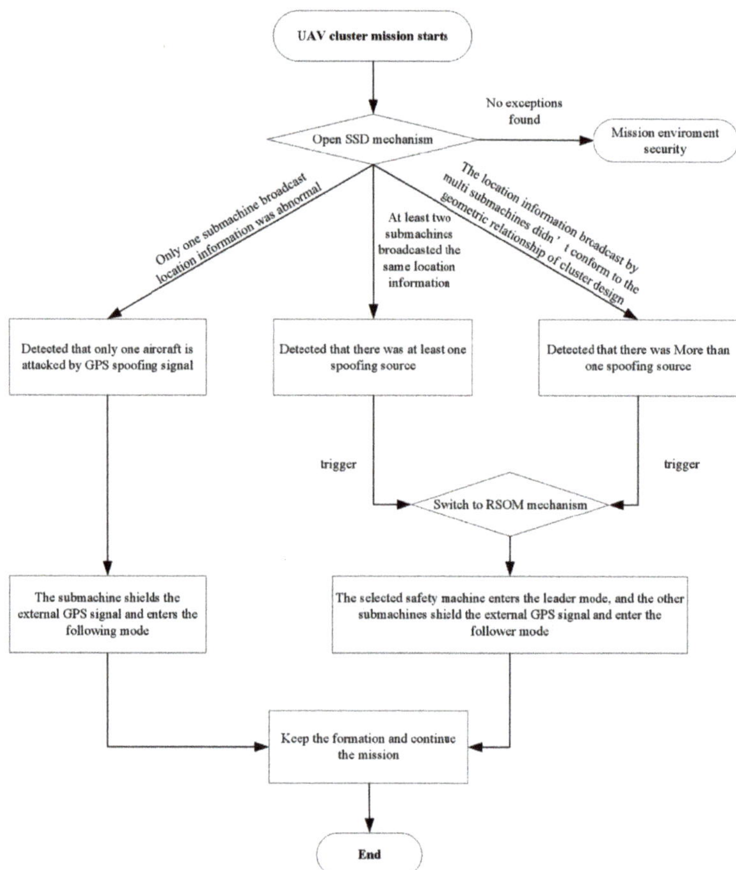

Figure A1. Flowchart: outline of the ASD's process.

References

1. Gaspar, J.; Ferreira, R.; Sebastião, P.; Souto, N. Capture of UAVs Through GPS Spoofing Using Low-Cost SDR Platforms. *Wirel. Pers. Commun.* **2020**, *115*, 2729–2754. [CrossRef]
2. Na, Z.; Liu, Y.; Shi, J.; Liu, C.; Gao, Z. UAV-Supported Clustered NOMA for 6G-Enabled Internet of Things: Trajectory Planning and Resource Allocation. *IEEE Internet Things J.* **2021**, *8*, 15041–15048. [CrossRef]
3. Yi, W.; Liu, Y.; Deng, Y.; Nallanathan, A. Clustered UAV Networks With Millimeter Wave Communications: A Stochastic Geometry View. *IEEE Trans. Commun.* **2020**, *68*, 4342–4357. [CrossRef]
4. Basan, E.; Basan, A.; Nekrasov, A.; Fidge, C.; Sushkin, N.; Peskova, O. GPS-Spoofing Attack Detection Technology for UAVs Based on Kullback–Leibler Divergence. *Drones* **2022**, *6*, 8. [CrossRef]
5. Jetto, J.; Gandhiraj, R.; Sundaram, G.; Soman, K.P. Software Defined Radio-Based GPS Spoofing Attack Model on Road Navigation System. In *Soft Computing and Signal Processing*; Spinger: Singapore, 2022.
6. Huang, K.W.; Wang, H.M. Combating the Control Signal Spoofing Attack in UAV Systems. *IEEE Trans. Veh. Technol.* **2018**, *67*, 7769–7773. [CrossRef]
7. Mekdad, Y.; Aris, A.; Babun, L.; Fergougui, A.E.; Conti, M.; Lazzeretti, R.; Uluagac, A.S. A Survey on Security and Privacy Issues of UAVs. *arXiv* **2021**, arXiv:2109.14442.
8. Akos, D.M. Who's Afraid of the Spoofer? GPS/GNSS Spoofing Detection via Automatic Gain Control (AGC). *Navigation* **2012**, *59*, 281–290. [CrossRef]

9. Pardhasaradhi, B.; Srihari, P.; Aparna, P. Spoofer-to-Target Association in Multi-Spoofer Multi-Target Scenario for Stealthy GPS Spoofing. *IEEE Access* **2021**, *9*, 108675–108688. [CrossRef]

10. Manesh, M.R.; Kenney, J.; Hu, W.; Devabhaktuni, V.K.; Kaabouch, N. Detection of GPS Spoofing Attacks on Unmanned Aerial Systems. In Proceedings of the 16th IEEE Annual Consumer Communications & Networking Conference, CCNC 2019, Las Vegas, NV, USA, 11–14 January 2019; IEEE: Piscataway, NJ, USA, 2019; pp. 1–6. [CrossRef]

11. Williamson, M. GPS Spoofing. Ph.D. Thesis, Utica College, Utica, NY, USA, 2014.

12. She, F.; Zhang, Y.; Shi, D.; Zhou, H.; Xu, T. Enhanced Relative Localization Based on Persistent Excitation for Multi-UAVs in GPS-Denied Environments. *IEEE Access* **2020**, *8*, 148136–148148. [CrossRef]

13. Shafique, A.; Mehmood, A.; Elhadef, M. Detecting Signal Spoofing Attack in UAVs Using Machine Learning Models. *IEEE Access* **2021**, *9*, 93803–93815. [CrossRef]

14. Meng, L.; Yang, L.; Ren, S.; Tang, G.; Zhang, L.; Yang, F.; Yang, W. An Approach of Linear Regression-Based UAV GPS Spoofing Detection. *Wirel. Commun. Mob. Comput.* **2021**, *2021*, 5517500. [CrossRef]

15. Bada, M.; Boubiche, D.E.; Lagraa, N.; Kerrache, C.A.; Imran, M.; Shoaib, M. A policy-based solution for the detection of colluding GPS-Spoofing attacks in FANETs. *Transp. Res. Part A Policy Pract.* **2021**, *149*, 300–318. [CrossRef]

16. Humphreys, T.E.; Ledvina, B.M.; Psiaki, M.L.; O'Hanlon, B.W.; Kintner, P.M. Assessing the Spoofing Threat: Development of a Portable GPS Civilian Spoofer. In Proceedings of the International Technical Meeting of the Satellite Division of the Institute of Navigation, Savannah, GA, USA, 16–19 September 2008.

17. Guenther, C. A Survey of Spoofing and Counter-Measures. *Navigation* **2015**, *61*, 159–177. [CrossRef]

18. Manfredini, E.G.; Akos, D.M.; Chen, Y.H.; Lo, S.; Enge, P. Effective GPS Spoofing Detection Utilizing Metrics from Commercial Receivers. In Proceedings of the 2018 International Technical Meeting of the Institute of Navigation, Reston, VA, USA, 29 January–1 February 2018.

19. Jovanovic, A.; Botteron, C.; Fariné, P.A. Multi-test Detection and Protection Algorithm Against Spoofing Attacks on GNSS Receivers. In Proceedings of the Position, Location & Navigation Symposium-Plans, IEEE/ION, Monterey, CA, USA, 5–8 May 2014.

20. Wesson, K.; Rothlisberger, M.; Humphreys, T. Practical Cryptographic Civil GPS Signal Authentication. *Navigation* **2012**, *59*, 177–193. [CrossRef]

21. Kai, J.; Schafer, M.; Moser, D.; Lenders, V.; Schmitt, J. Crowd-GPS-Sec: Leveraging Crowdsourcing to Detect and Localize GPS Spoofing Attacks. In Proceedings of the 2018 IEEE Symposium on Security and Privacy (SP), San Francisco, CA, USA, 20–24 May 2018.

22. Montgomery, P.Y.; Humphreys, T.E.; Ledvina, B.M. Receiver-Autonomous Spoofing Detection: Experimental Results of a Multi-Antenna Receiver Defense Against a Portable Civil GPS Spoofer. In Proceedings of the 2009 International Technical Meeting of The Institute of Navigation, Anaheim, CA, USA, 26–28 January 2009.

23. Jansen, K.; Tippenhauer, N.O.; Ppper, C. Multi-receiver GPS spoofing detection: Error models and realization. In Proceedings of the the 32nd Annual Conference, Los Angeles, CA, USA, 5–8 December 2016.

24. Rao, H.; Wang, S.; Hu, X.; Tan, M.; Da, H.; Cheng, J.; Hu, B. Self-Supervised Gait Encoding with Locality-Aware Attention for Person Re-Identification. In Proceedings of the International Joint Conference on Artificial Intelligence, Yokohama, Japan, 11–17 July 2020.

25. Xu, S.; Rao, H.; Peng, H.; Jiang, X.; Guo, Y.; Hu, X.; Hu, B. Attention-Based Multilevel Co-Occurrence Graph Convolutional LSTM for 3-D Action Recognition. *IEEE Internet Things J.* **2021**, *8*, 15990–16001. [CrossRef]

26. Rao, H.; Xu, S.; Hu, X.; Cheng, J.; Hu, B. Augmented Skeleton Based Contrastive Action Learning with Momentum LSTM for Unsupervised Action Recognition. *Inf. Sci.* **2021**, *569*, 90–109. [CrossRef]

27. Eldosouky, A.R.; Ferdowsi, A.; Saad, W. Drones in Distress: A Game-Theoretic Countermeasure for Protecting UAVs Against GPS Spoofing. *IEEE Internet Things J.* **2020**, *7*, 2840–2854. [CrossRef]

28. Liang, C.; Miao, M.; Ma, J.; Yan, H.; Li, T. Detection of GPS Spoofing Attack on Unmanned Aerial Vehicle System. In Proceedings of the Machine Learning for Cyber Security: Second International Conference, ML4CS 2019, Xi'an, China, 19–21 September 2019.

29. Mykytyn, P.; Brzozowski, M.; Dyka, Z.; Langendoerfer, P. GPS-Spoofing Attack Detection Mechanism for UAV Swarms. *arXiv* **2023**, arXiv:2301.12766.

30. Chen, M.; Mozaffari, M.; Saad, W.; Yin, C.; Debbah, M.; Hong, C.S. Caching in the Sky: Proactive Deployment of Cache-Enabled Unmanned Aerial Vehicles for Optimized Quality-of-Experience. *IEEE J. Sel. Areas Commun.* **2017**, *35*, 1046–1061. [CrossRef]

31. Su, J.; He, J.; Cheng, P.; Chen, J. A Stealthy GPS Spoofing Strategy for Manipulating the Trajectory of an Unmanned Aerial Vehicle. *IFAC-PapersOnLine* **2016**, *49*, 291–296. [CrossRef]

32. Zeng, K.C.; Shu, Y.; Liu, S.; Dou, Y.; Yang, Y. A Practical GPS Location Spoofing Attack in Road Navigation Scenario. In Proceedings of the the 18th International Workshop, Sonoma, CA, USA, 21–22 February 2017.

33. Liu, Q.; Chen, S.; Wang, G.; Lan, Y. Drift Evaluation of a Quadrotor Unmanned Aerial Vehicle (UAV) Sprayer: Effect of Liquid Pressure and Wind Speed on Drift Potential Based on Wind Tunnel Test. *Appl. Sci.* **2021**, *11*, 7258. [CrossRef]

34. Kerns, A.J.; Shepard, D.P.; Bhatti, J.A.; Humphreys, T.E. Unmanned Aircraft Capture and Control Via GPS Spoofing. *J. Field Robot.* **2014**, *31*, 617–636. [CrossRef]

35. Kj, M.; Wittenmark, B. *Computer-Controlled Systems: Theory and Design*, 2nd ed.; Elsevier: New York, NY, USA, 1990.

36. Wu, J.; Zhou, Z.; Fourati, H.; Li, B.; Liu, M. Generalized Linear Quaternion Complementary Filter for Attitude Estimation From Multisensor Observations: An Optimization Approach. *IEEE Trans. Autom. Sci. Eng.* **2019**, *16*, 1330–1343. [CrossRef]

37. Kottath, R.; Narkhede, P.; Kumar, V.; Karar, V.; Poddar, S. Multiple Model Adaptive Complementary Filter for Attitude Estimation. *Aerosp. Sci. Technol.* **2017**, *69*, 574–581. [CrossRef]
38. Yoo, T.S.; Hong, S.K.; Yoon, H.M.; Park, S. Gain-Scheduled Complementary Filter Design for a MEMS Based Attitude and Heading Reference System. *Sensors* **2011**, *11*, 3816–3830. [CrossRef] [PubMed]
39. Kang, C.W.; Chan, G.P.; Filter, K. Attitude estimation with accelerometers and gyros using fuzzy tuned Kalman filter. In Proceedings of the 2009 European Control Conference (ECC), Budapest, Hungary, 23–26 August 2009.

Article

Minimum-Effort Waypoint-Following Differential Geometric Guidance Law Design for Endo-Atmospheric Flight Vehicles

Xuesheng Qin [1,2], Kebo Li [1,2,*], Yangang Liang [1,2] and Yuanhe Liu [1,2]

[1] College of Aerospace Science and Engineering, National University of Defense Technology, Changsha 410073, China; qinxuesheng@alumni.nudt.edu.cn (X.Q.)

[2] Hunan Key Laboratory of Intelligent Planning and Simulation for Aerospace Missions, Changsha 410073, China

* Correspondence: likeboreal@nudt.edu.cn; Tel.: +86-18684658210

Abstract: To improve the autonomous flight capability of endo-atmospheric flight vehicles, such as cruise missiles, drones, and other small, low-cost unmanned aerial vehicles (UAVs), a novel minimum-effort waypoint-following differential geometric guidance law (MEWFDGGL) is proposed in this paper. Using the classical differential geometry curve theory, the optimal guidance problem of endo-atmospheric flight vehicles is transformed into an optimal space curve design problem, where the guidance command is the curvature. On the one hand, the change in speed of the flight vehicle is decoupled from the guidance problem. In this way, the widely adopted constant speed hypothesis in the process of designing the guidance law is eliminated, and, hence, the performance of the proposed MEWFDGGL is not influenced by the varying speed of the flight vehicle. On the other hand, considering the onboard computational burden, a suboptimal form of the MEWFDGGL is proposed to solve the problem, where both the complexity and the computational burden of the guidance law dramatically increase as the number of waypoints increases. The theoretical analysis demonstrates that both the original MEWFDGGL and its suboptimal form can be applied to general waypoint-following tasks with an arbitrary number of waypoints. Finally, the superiority and effectiveness of the proposed MEWFDGGL are verified by a numerical simulation and flight experiments.

Keywords: waypoint-following guidance; varying speed; differential geometric curve theory; global energy optimization; suboptimal form; flight experiments

Citation: Qin, X.; Li, K.; Liang, Y.; Liu, Y. Minimum-Effort Waypoint-Following Differential Geometric Guidance Law Design for Endo-Atmospheric Flight Vehicles. *Drones* **2023**, *7*, 369. https://doi.org/10.3390/drones7060369

Academic Editor: Abdessattar Abdelkefi

Received: 29 March 2023
Revised: 16 May 2023
Accepted: 18 May 2023
Published: 1 June 2023

1. Introduction

A key technology of endo-atmospheric flight vehicles, such as cruise missiles, drones, and other small, low-cost unmanned aerial vehicles (UAVs), is the ability to autonomously reach a destination via an expected path [1]. In terms of general multitarget missions, there are two mainstream methods for endo-atmospheric flight vehicles to realize waypoint-following guidance, by visiting multiple target points at a time. The first is to decompose the waypoint-following task into two parts: path planning and path tracking [2–6]. The second is to individually separate each waypoint on the desired path and transform the path-tracking problem into countless point-to-point guidance problems, using closed-circuit guidance. Missile guidance laws, which have been well-developed over the last few decades, can also be adopted to solve waypoint-following problems. Recent developments in advanced missile guidance laws resulted in many elegant solutions to problems related to waypoint following, such as pure pursuit guidance (PPG) [7], proportional navigation guidance (PNG) [8–11], and their variations [12–15].

Considering certain performance indexes, the first approach typically discovers the energy- or time-optimal path using complex numerical trajectory optimization methods [16–20]. However, numerical simulations require large onboard computing power, so they may not be suitable for the general growth of small endo-atmospheric flight vehicles, such as cruise missiles, drones, and other small, low-cost unmanned aerial vehicles (UAVs). The guidance instructions produced using the second method can be simple and concise, thus eliminating concerns about the capability of an onboard computer. However, under the boundary conditions of multiple waypoints, even the use of the minimum-effort point-to-point guidance law [21–24] between every two adjacent waypoints cannot guarantee the optimal total energy consumption for the duration of a task.

For such problems, inspired by the above two methods, an optimal error dynamics (OED) method for the design of guidance laws for UAVs that need to visit multiple waypoints was proposed in [25]. As mentioned in [25], the challenge of designing guidance laws is defined, firstly, as a finite-time tracking problem in cybernetics. Later, a global minimum-effort waypoint-following guidance law (MEWFGL) that can be applied to situations of arbitrary waypoint numbers was analytically derived in [26]. For the first time, the MEWFGL allowed for the development of a path-planning and tracking set in one single step, and it effectively reduced the complexity of the initial task's analysis and design. However, the MEWFGL still assumed a constant speed and did not consider the influence of a change in a UAV's speed on the performance of the guidance. Although a large number of numerical simulations and flight experiments showed that, under the assumption of a constant speed, the designed guidance laws still work in most practical varying speed guidance scenarios, their performances are degraded to some degree, since the speed of a UAV changes during the guidance process. In practical applications, real-time speeds can be obtained and utilized in guidance and control systems, as almost all endo-atmospheric flight vehicles are equipped with Doppler radars or accelerometers. Therefore, if guidance laws can be directly designed without the assumption of a constant speed, their performances must, hence, be assumed to be better.

In essence, the flight trajectory of an endo-atmospheric flight vehicle can be taken as a space curve, which is more suitably depicted with a Frenet–Serret frame, curvature, and torsion, using the classical differential geometry curve theory [27]. In addition to designing and analyzing guidance laws in the time domain, differential geometry theory can be used to design and derive a variety of guidance laws in the arc-length domain. In [28], the performance of a PNG and the capture region of a missile with time-varying speed that was directed at a stationary target were preliminarily explored in the arc-length domain. In [28], by introducing the differential of the arc length, the influence of the time-varying missile's speed on the performance of the guidance law was eliminated for the first time. In addition, in [29], the PNG's performance in the arc-length domain was further analyzed in detail, which revealed the essence of the guidance law design in theory, and this also showed that the differential geometry curve theory is beneficial for reducing the influence of time-varying speed on the performance of guidance law.

On the one hand, the computational capacities of small, low-cost endo-atmospheric flight vehicles are not powerful enough for the computational burden of the complex numerical calculations required to determine an optimal path. On the other hand, it is difficult to design guidance laws for time-varying speed endo-atmospheric flight vehicles because the remaining flying time cannot be estimated accurately, especially when the time-varying speed of an endo-atmospheric flight vehicle is considered. Motivated by the aforementioned observations, in this paper, a novel minimum-effort waypoint-following differential geometric guidance law (MEWFDGGL) is proposed for endo-atmospheric flight vehicles by combining the MEWFGL concept and the differential geometry curve theory, which can be used to improve the autonomous flight capabilities of small, low-cost flight vehicles. The nonlinear guidance model is presented in the arc-length domain. The MEWFDGGL is derived using the linearized dynamics of a zero-effort miss (ZEM) and adopting the optimal control theory. Next, numerical simulations of the MEWFDGGL with varying-speed endo-atmospheric flight vehicles are conducted, and the results are compared with the original MEWFGL to demonstrate the effectiveness and superiority of the guidance law. Finally, flight experiments based on a small quadrotor UAV are conducted to further verify the effectiveness of the proposed guidance law. For ease of presentation, we focus on small, low-cost UAVs in this paper, but it is worth emphasizing that the guidance law proposed in this paper is designed for numerous endo-atmospheric flight vehicles, including cruise missiles, drones, etc. The key contributions of the MEWFDGGL are threefold:

1. The MEWFDGGL decouples the speed change in a UAV from the guidance problem in theory, rather than directly adopting the constant speed hypothesis. With the help of the classical differential geometry curve theory, the optimal guidance problem is transformed into an optimal space curve design problem, which makes the speed change in the UAV no longer a concern during the guidance law design process, and the optimality of the space curve is independent of the UAV's speed in the process of the guidance law design.
2. The MEWFDGGL is globally energy-optimal. By linearizing the ZEM dynamics and adopting the optimal control theory, the guidance curvature command of the MEWFDGGL can be obtained by solving the linear-quadratic optimal control problem, and then the energy consumption of a UAV throughout the whole guidance process can be minimized.
3. The suboptimal MEWFDGGL can be applied to general waypoint-following tasks with arbitrary waypoint numbers. By adopting just two waypoints at one time to generate the guidance command, the formation of the original MEWFG becomes much simpler, and the computation burden is greatly reduced.

The remainder of this paper is organized as follows: The backgrounds and preliminaries of this paper are stated in Section 1. In Section 2, the derivation of the MEWFDGGL and the suboptimal MEWFDGGL are given in detail. Finally, numeric simulations and experimental verification results are offered.

2. Materials and Methods

2.1. Preliminaries

2.1.1. Nonlinear Kinematics

We begin by taking into account that the UAV needs to visit N-many waypoints. The planar engagement geometry is shown in Figure 1, where XOY is the inertial coordinate system. The UAV is expressed as the symbol U, and the i-th waypoint is represented as W_i. The variables γ and σ_i signify the UAV's flight path angle and LOS (Line-of-Sight) angle, respectively. For the sake of simplicity, the UAV is assumed to be a particle model, that is, the time delay of the autopilot is not considered. The variable r_i, which cannot be zero, denotes the relative range between the UAV and the i-th waypoint. The symbol z represents the zero-effort miss (ZEM), which refers to the nearest distance during the process that the UAV moves to the i-th target waypoint at the current speed without any control input, namely, $\overline{CW_i}$, as shown in Figure 1. Based on the principles of dynamics, as shown in Figure 1, the differential equations to describe the planar engagement geometry in the arc-length domain can be formulated as

$$
\begin{aligned}
r_i' &= -\cos(\gamma - \sigma_i), \\
\sigma_i' &= -\frac{\sin(\gamma - \sigma_i)}{r_i}, \\
\gamma' &= \frac{a}{V^2} = \kappa, i \in \{1, 2, 3, \dots, N\}.
\end{aligned}
\tag{1}
$$

where a and V represent the speed and acceleration of the UAV, respectively; κ is the guidance curvature; and the prime symbol ($'$) indicates the derivative of the variables with respect to s, the arc length of the UAV's flight trajectory, such as how γ' denotes the derivative of γ with respect to s. In this paper, κ is used as the guidance input command to design the guidance law, which can improve the performance of the guidance, to some extent, by avoiding using the speed value of the UAV during the process of the guidance law design, especially when the speed is time-varying.

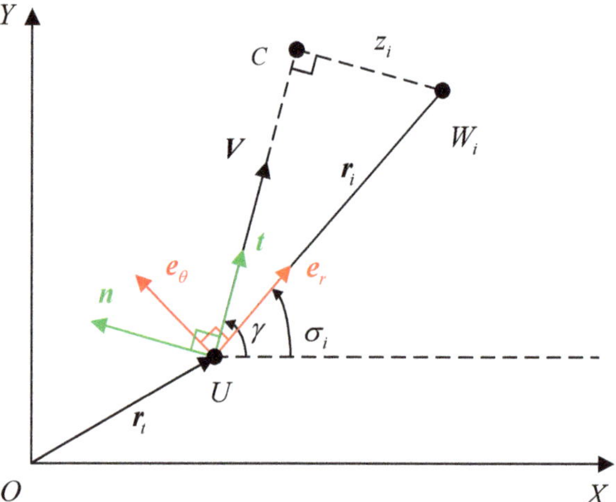

Figure 1. Planar engagement geometry.

Omitting the derivation process, the dynamic equation of the varying-speed missile is directly given as follows [30]:

$$\mathbf{r}_i'' = \left(r_i'' - r_i\sigma_i'^2\right)\mathbf{e}_r + \left(r_i\sigma_i'' + 2r_i'\sigma_i'\right)\mathbf{e}_\theta = -\kappa\mathbf{n}. \tag{2}$$

Decomposing the above formula in the LOS frame $(\mathbf{e}_r, \mathbf{e}_\theta)$, we have

$$\begin{cases} r_i'' - r_i\sigma_i'^2 = \kappa\sin(\gamma - \sigma_i) \\ r_i\sigma_i'' + 2r_i'\sigma_i' = -\kappa\cos(\gamma - \sigma_i) \end{cases}. \tag{3}$$

As mentioned before, the guidance command in the arc-length domain is κ. The transformational relation between the curvature command and the lateral acceleration command can be given directly as

$$\kappa = a/V^2. \tag{4}$$

2.1.2. Problem Formulation

Without losing generality, we can suppose that the order of the waypoints is numbered according to the corresponding path length $s_{f,i}$ as $s_{f,i} < s_{f,i+1}$. The corresponding path length of the ith waypoint can be written directly as

$$s_{f,i} = s + s_{go,i}, \tag{5}$$

where $s_{go,i}$ stands for the corresponding remaining path length to visit the ith waypoint.

Since it is difficult to exactly calculate the value of the remaining path length, it is generally replaced by its estimated value, $\hat{s}_{go}$. As shown in Figure 2, since the purpose of guidance design is to minimize the ZEM, that is, to make the leading angle tend to zero, the next flight trajectory of the UAV is generally within ΔUCW_i, which means that the path to travel is larger than $\overline{UW_i}$. If $\overline{UD}$ is the estimated value of $\hat{s}_{go}$, the remaining path length is expressed by

$$\hat{s}_{go,i} = \frac{r_i}{\cos(\gamma - \sigma_i)}. \tag{6}$$

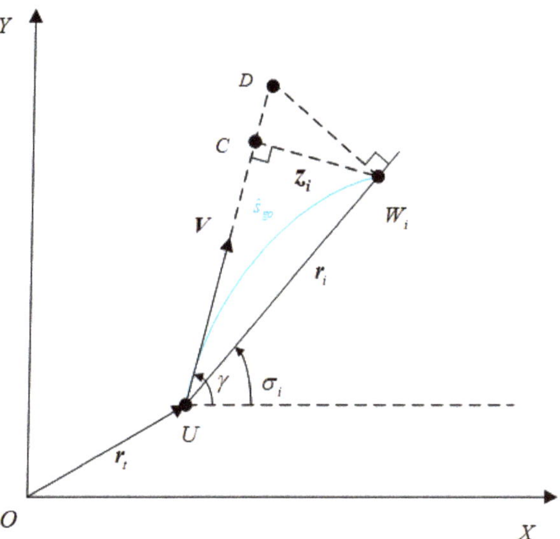

Figure 2. ZEM geometric model.

As a matter of fact, energy consumption is extremely important for UAVs because it determines the endurance of a UAV, that is, it determines how long a vehicle can fly. For this reason, the performance index is taken into consideration as follows:

$$
J = \int_s^{s_{f,N}} \kappa^2(\delta)d\delta = \begin{cases} \int_s^{s_{f,1}} \kappa^2(\delta)d\delta + \sum_{i=1}^{N-1} \int_{s_{f,i}}^{s_{f,i+1}} \kappa^2(\delta)d\delta, \; s \leq s_{f,1} \\ \int_s^{s_{f,2}} \kappa^2(\delta)d\delta + \sum_{i=1}^{N-1} \int_{s_{f,i}}^{s_{f,i+1}} \kappa^2(\delta)d\delta, \; s_{f,1} < s \leq s_{f,2} \\ \vdots \\ \int_s^{s_{f,N}} \kappa^2(\delta)d\delta, \; s_{f,N-1} < s \leq s_{f,N} \end{cases} \tag{7}
$$

Similarly, minimizing the performance index is a valuable goal for the multi-objective optimization problem. It is worth noting that the performance index is a sum of the energy functions computed in each path length. Hence, finding an analytical solution for this problem is the main purpose of this paper, that is, we are adopting the given nonlinear dynamics model and finding the guidance curvature k, which ensures an optimal performance index and perfect ZEM constraints, as follows:

$$
z_i\left(s_{f,i}\right) = 0, \quad i \in [1,2,\ldots,N]. \tag{8}
$$

2.2. Guidance Law Design

2.2.1. Derivation

This section derives the energy-optimal differential geometric guidance law to solve the generalized optimal waypoint-following problem. As can be seen from Figure 2, the ZEM can be expressed as

$$
z_i = \begin{cases} r_i \sin(\gamma - \sigma_i), \; s \leq s_{f,i} \\ z_i\left(s_{f,i}\right), \; s > s_{f,i} \end{cases} \tag{9}
$$

Substituting the second Equation of (1) into (9), the ZEM is obtained as

$$z_i = \begin{cases} -r_i^2 \sigma_i', & s \le s_{f,i} \\ z_i\left(s_{f,i}\right), & s > s_{f,i} \end{cases},\tag{10}$$

Taking the arc length derivative of (10) results in

$$z_i' = \begin{cases} -r_i\left(2r_i'\sigma_i' + r_i\sigma_i''\right), & s \le s_{f,i} \\ 0, & s > s_{f,i} \end{cases},\tag{11}$$

Substituting the second Equation of (3) into (11) provides

$$z_i' = \begin{cases} r_i\cos(\gamma - \sigma_i)\kappa, & s \le s_{f,i} \\ 0, & s > s_{f,i} \end{cases},\tag{12}$$

Then, by combining (12) and (6), the nonlinear ZEM dynamics can be acquired as follows:

$$z_i' = \begin{cases} \hat{s}_{go,i}\cos^2(\gamma - \sigma_i)\kappa, & s \le s_{f,i} \\ 0, & s > s_{f,i} \end{cases}.\tag{13}$$

Taking the assumption of a small leading angle into consideration, i.e., $\gamma - \sigma_i$ is small, we have the approximation as

$$\sin(\gamma - \sigma_i) \approx \gamma - \sigma_i, \ \cos(\gamma - \sigma_i) \approx 1,\tag{14}$$

The linear ZEM dynamics can be obtained by substituting (14) into (13), as follows:

$$z_i' = \begin{cases} \hat{s}_{go,i}\kappa, & s \le s_{f,i} \\ 0, & s > s_{f,i} \end{cases},\tag{15}$$

Further, the system in (15) can be written as follows:

$$z_i\left(s_{f,i}\right) - z_i(s) = \int_s^{s_{f,i}} -\left(s_{f,i} - \delta\right)\kappa(\delta)d\delta, \ s \le s_{f,i}.\tag{16}$$

Introducing the terminal constraint terminal constraints in (8) and substituting them into (16) provides

$$z_i(s) = \int_s^{s_{f,i}} \left(s_{f,i} - \delta\right)\kappa(\delta)d\delta, \ s \le s_{f,i}.\tag{17}$$

According to [27], if the guidance curvature command k is optimal in the energy consumption, then there are N Lagrange multipliers (λ_i, $i \in \{1,2,\ldots,N\}$), so the guidance curvature k can be expressed as follows:

$$\kappa = \begin{cases} \sum\limits_{i=1}^{N} \lambda_i\left(s_{f,i} - s\right), & s \le s_{f,1} \\ \sum\limits_{i=2}^{N} \lambda_i\left(s_{f,i} - s\right), & s_{f,1} < s \le s_{f,2} \\ \vdots \\ \lambda_N\left(s_{f,N} - s\right), & s_{f,N-1} < t \le s_{f,N} \end{cases}.\tag{18}$$

By substituting (18) into (17) and solving the corresponding equation, the Lagrange multiplier can be obtained. It is worth mentioning that this method can be regarded as an extension of the Schwartz inequality method for any number of terminal constraints.

Hence, we consider only the case of $s \leq s_{f,1}$ in the following derivation, without the loss of generality. Therefore, substituting (17) into (18) provides

$$
\begin{aligned}
z_i(s) &= \int_s^{s_{f,i}} \left(s_{f,i} - \delta\right) \kappa(\delta) d\delta \\
&= \int_s^{s_{f,1}} \left(s_{f,i} - \delta\right) \sum_{j=1}^{N} \lambda_j \left(s_{f,i} - \delta\right) d\delta + \int_{s_{f,1}}^{s_{f,2}} \left(s_{f,i} - \delta\right) \sum_{j=2}^{N} \lambda_j \left(s_{f,i} - \delta\right) d\delta \\
&\quad + \ldots + \int_{s_{f,i-1}}^{s_{f,i}} \left(s_{f,i} - \delta\right) \sum_{j=i}^{N} \lambda_j \left(s_{f,i} - \delta\right) d\delta \\
&= \sum_{j=1}^{N} \lambda_j \left[\frac{s_{go,\max(i,j)} s_{go,\min(i,j)}^2}{2} - \frac{s_{go,\min(i,j)}^3}{6} \right].
\end{aligned}
\tag{19}
$$

Defining the Lagrange multiplier vector as $\lambda = [\lambda_1, \lambda_2, \ldots, \lambda_N]^T$ and the ZEM vector as $Z = [z_1, z_2, \ldots, z_N]^T$, (19) can be given as a compact matrix format, as follows:

$$
G\lambda = Z,
\tag{20}
$$

where the symmetric matrix $G \in \mathbb{R}^{N \times N}$ can be obtained as

$$
G = \begin{bmatrix}
\frac{s_{go,1}^3}{3} & \frac{s_{go,2} s_{go,1}^2}{2} - \frac{s_{go,1}^3}{6} & \cdots & \cdots & \frac{s_{go,N} s_{go,1}^2}{2} - \frac{s_{go,1}^3}{6} \\
 & \frac{s_{go,2}^3}{3} & \frac{s_{go,3} s_{go,2}^2}{2} - \frac{s_{go,2}^3}{6} & \cdots & \frac{s_{go,N} s_{go,2}^2}{2} - \frac{s_{go,2}^3}{6} \\
\ddots & & \ddots & & \\
 & & & \frac{s_{go,N-1}^3}{3} & \frac{s_{go,N} s_{go,N-1}^2}{2} - \frac{s_{go,N-1}^3}{6} \\
\ddots & & & & \frac{s_{go,N}^3}{3}
\end{bmatrix}.
\tag{21}
$$

Then, the Lagrange multiplier vector is given by inverting both sides of (20), as follows:

$$
\lambda = G^{-1} Z.
\tag{22}
$$

Substituting (22) into (18), the guidance curvature command k in the case of $s \leq s_{f,1}$ can be written as

$$
\begin{aligned}
\kappa &= \lambda^T \left[s_{go,1}, s_{go,2}, \ldots, s_{go,N} \right]^T \\
&= \left(G^{-1} Z \right)^T \left[s_{go,1}, s_{go,2}, \ldots, s_{go,N} \right]^T.
\end{aligned}
\tag{23}
$$

Remark 1. Following a similar procedure, it is easy to obtain the solution in the case of $s > s_{f,1}$. For instance, when $s_{f,1} < s \leq s_{f,2}$, the ZEM vector will be reduced by one dimension and become $Z = [z_2, z_3, \ldots, z_N]^T$, and then the matrix $G \in \mathbb{R}^{(N-1) \times (N-1)}$ will reduce to

$$
G = \begin{bmatrix}
\frac{s_{go,2}^3}{3} & \frac{s_{go,3} s_{go,2}^2}{2} - \frac{s_{go,2}^3}{6} & \cdots & \cdots & \frac{s_{go,N} s_{go,2}^2}{2} - \frac{s_{go,2}^3}{6} \\
 & \frac{s_{go,3}^3}{3} & \frac{s_{go,4} s_{go,3}^2}{2} - \frac{s_{go,3}^3}{6} & \cdots & \frac{s_{go,N} s_{go,3}^2}{2} - \frac{s_{go,3}^3}{6} \\
\ddots & & \ddots & & \\
 & & & \frac{s_{go,N-1}^3}{3} & \frac{s_{go,N} s_{go,N-1}^2}{2} - \frac{s_{go,N-1}^3}{6} \\
\ddots & & & & \frac{s_{go,N}^3}{3}
\end{bmatrix},
\tag{24}
$$

Hence, the guidance curvature k in the case of $s_{f,1} < s \leq s_{f,2}$ can be written as

$$\kappa = \left(G^{-1} Z \right)^{T} \left[s_{go,2}, s_{go,3}, \dots, s_{go,N} \right]^{\mathrm{T}}. \tag{25}$$

Remark 2. In order to effectively eliminate the deviation caused by the linearization process, the ZEM is written in a nonlinear form in practice, as follows:

$$z_i = r_i \sin(\gamma - \sigma_i), \tag{26}$$

By substituting (26) into (23), the guidance curvature command k, composed of the measured signals r_i, γ, and σ_i, is obtained. It is worth noticing that (23) converts the linear terms of the substitution of (26) into the nonlinear expressions, which provides support for the engineering application of the guidance mentioned above.

2.2.2. Particular Cases

$N = 1$

When the UAV is only required to visit one waypoint, the problem is transformed into an energy-optimal interception problem. For such special cases, the guidance curvature (23) can be expressed as

$$\kappa = \lambda_1 \left(s_{f,1} - s \right). \tag{27}$$

The matrix G can reduce to a scalar form when $N = 1$, as follows:

$$G = \frac{s_{go,1}^3}{3}, \tag{28}$$

and the Lagrange multiplier can easily be expressed as

$$\lambda_1 = \frac{3z_1}{s_{go,1}^3}, \tag{29}$$

Then, the clear guidance curvature command can be obtained by substituting (29) into (27), as follows:

$$\kappa = \frac{3z_1}{s_{go,1}^2}. \tag{30}$$

In order to more conveniently analyze the MEWFDGGL, the guidance acceleration command can be given by substituting (4) into (30), as follows:

$$a = \frac{3z_1}{t_{go,1}^2}. \tag{31}$$

In the time domain, the ZEM could be signified as the following form:

$$z_i = V \dot{\sigma}_i t_{go,i}^2. \tag{32}$$

Combining (32) and (31) results in

$$a = 3V \dot{\sigma}_1. \tag{33}$$

This is consistent with the classical energy optimal PNG. According to the statement in [31,32], the energy consumption of the PNG with a navigation gain of three is minimal when there is only a single waypoint to be visited. However, it is also clear from the previous derivation that when the number of waypoints is greater than two, merely adopting the PNG to visit each waypoint in turn cannot ensure that the energy consumption is optimal in the whole process of the guidance.

$N = 2$

Similarly, when a UAV is required to visit two waypoints, the guidance curvature command k, in the case of $s \leq s_{f,1}$, can be expressed as follows:

$$\kappa = \lambda_1 \left(s_{f,1} - s\right) + \lambda_2 \left(s_{f,2} - s\right). \tag{34}$$

The matrix G can be written as

$$G = \begin{bmatrix} \frac{s_{go,1}^3}{3} & \frac{s_{go,2}s_{go,1}^2}{2} - \frac{s_{go,1}^3}{6} \\ \frac{s_{go,2}s_{go,1}^2}{2} - \frac{s_{go,1}^3}{6} & \frac{s_{go,2}^3}{3} \end{bmatrix}, \tag{35}$$

and the corresponding Lagrange multiplier can be easily given from (22), as follows:

$$\begin{aligned} \lambda_1 &= \frac{6\left(2s_{go,2}^3 z_1 + s_{go,1}^3 z_2 - 3s_{go,1}^2 s_{go,2} z_2\right)}{s_{go,1}^3 \left(s_{go,2} - s_{go,1}\right)^2 \left(4s_{go,2} - s_{go,1}\right)} \\ \lambda_2 &= \frac{6\left(s_{go,1} z_1 - 3s_{go,2} z_1 + 2s_{go,1} z_2\right)}{s_{go,1} \left(s_{go,2} - s_{go,1}\right)^2 \left(4s_{go,2} - s_{go,1}\right)}, \end{aligned} \tag{36}$$

Then, substituting (36) into (34), the clear guidance curvature command can be obtained as follows:

$$\begin{aligned} \kappa &= \frac{6\left(2s_{go,2}^2 z_1 - s_{go,1}s_{go,2} z_1 - s_{go,1}^2 z_2\right)}{s_{go,1}^2 \left(s_{go,2} - s_{go,1}\right)\left(4s_{go,2} - s_{go,1}\right)} \\ &= \frac{6\left(2s_{go,2} - s_{go,1}\right)s_{go,2}}{s_{go,1}^2 \left(s_{go,2} - s_{go,1}\right)\left(4s_{go,2} - s_{go,1}\right)} z_1 \\ &\quad - \frac{6}{\left(s_{go,2} - s_{go,1}\right)\left(4s_{go,2} - s_{go,1}\right)} z_2. \end{aligned} \tag{37}$$

From (32), the ZEM can be denoted in the arc-length domain as

$$z_i = \sigma_i' s_{go,i}^2, \tag{38}$$

Then, substituting (38) into (30) and (37), the clear guidance curvature command, in the case of $N = 2$, can be obtained as follows:

$$\kappa = \begin{cases} K_1 \sigma_1' + K_2 \sigma_2', & s \leq s_{f,1} \\ 3\sigma_2', & s_{f,1} < s \leq s_{f,2} \end{cases}, \tag{39}$$

where

$$\begin{aligned} K_1 &= \frac{6\left(2s_{go,2} - s_{go,1}\right)s_{go,2}}{\left(s_{go,2} - s_{go,1}\right)\left(4s_{go,2} - s_{go,1}\right)} \\ K_2 &= \frac{6s_{go,2}^2}{\left(s_{go,2} - s_{go,1}\right)\left(4s_{go,2} - s_{go,1}\right)}. \end{aligned} \tag{40}$$

In other words, when $s \leq s_{f,1}$, the MEWFDGGL can be regarded as a biased proportional guidance (BPN) that has a time-varying navigation gain. As mentioned earlier in (39), the first term represents the proportional guidance term for the first waypoint, and the second term represents the influence term of the second waypoint on the first one.

Remark 3. Following a similar procedure, it is easy to obtain the solution, in the case of $N > 2$. For instance, when $N = 3$, the clear guidance curvature command is given as

$$
\kappa = \begin{cases} K_1\sigma_1' + K_2\sigma_2' + K_3\sigma_3', & s \leq s_{f,1} \\ K_{2'}\sigma_2' + K_{3'}\sigma_3', & s_{f,1} < s \leq s_{f,2} \\ 3\sigma_3', & s_{f,2} < s \leq s_{f,3} \end{cases}.
\tag{41}
$$

According to Equation (41), the final form of the MEWFDGGL is a piecewise BNG law based on the waypoints segment, and its proportional navigation gain changes with time. This also shows that the MEWFDGGL is generic because it reduces to the classical PNG when there is only one waypoint to be visited.

2.2.3. Improvement

For a waypoint-following task with N waypoints, it follows from Equation (21) that implementing the MEWFDGG requires the calculation of the inverse of the matrix G. We notice that the size of matrix G is proportional to the number of waypoints to be traveled. Therefore, a large number of waypoints poses a great challenge to the computing power of an airborne computer. Hence, an algorithm based on finite waypoint information is proposed to improve the applicable scope of the guidance law and make it more suitable for the general situation of an arbitrary number of waypoints.

Without losing generality, we can assume that the UAV has visited the previous *i-1*th waypoint at the current moment and is required to visit the *i*-th waypoint to the nth waypoint sequentially. After several simulation analyses, it is found that the energy consumption is not significantly different from that of the MEWFDGG when adopting only two waypoints at one time to generate the guidance command. Therefore, the method proposed in this paper can be improved to design a suboptimal minimum-effort waypoint-following differential geometric guidance law (SMEWFDGGL), by considering only two waypoints.

Therefore, when $i < N$, the guidance command of SMEWFG considers only two waypoints, and it can be expressed as follows:

$$
k = \left(G^{-1}Z \right)^{\mathrm{T}} \left[s_{go,i}, s_{go,i+1} \right]^{\mathrm{T}},
\tag{42}
$$

where

$$
G = \begin{bmatrix} \dfrac{s_{go,i}^3}{3} & \dfrac{s_{go,i+1}s_{go,i}^2}{2} - \dfrac{s_{go,i}^3}{6} \\ \dfrac{s_{go,i+1}s_{go,i}^2}{2} - \dfrac{s_{go,i}^3}{6} & \dfrac{s_{go,i+1}^3}{3} \end{bmatrix},
\tag{43}
$$

$$
Z = [z_i, z_{i+1}]^{\mathrm{T}}.
\tag{44}
$$

In order to effectively eliminate the deviation caused by the linearization process, the ZEM is written in a nonlinear form in practice, as follows:

$$
Z = [r_i \sin(\gamma - \sigma_i), r_{i+1} \sin(\gamma - \sigma_{i+1})]^{\mathrm{T}},
\tag{45}
$$

or

$$
Z = \left[\sigma_i' s_{go,i}^2, \sigma_j'^{i+1} s_{go,i+1}^2 \right]^{\mathrm{T}}.
\tag{46}
$$

The algorithm can be expressed as follows:

Algorithm 1:
The suboptimal minimum-effort waypoint-following differential geometric guidance

---

Input: The relative range and LOS angle between the UAV and all waypoints,
$r = [r_1, r_2, \ldots, r_i, \ldots, r_N]$ and $\sigma = [\sigma_1, \sigma_2, \ldots, \sigma_i, \ldots, \sigma_N]$, and the UAV speed V.
Require: $r_i < r_{i+1}$.
Denote: $k = 1$.
Step 1: Compute the remaining path length s_{go}. If $k < i$, proceed to Step 2; otherwise, proceed to Step 3.
Step 2: $s_{go,k} = 0, k = k + 1$. Return to Step 1.
Step 3: If $k \leq N$, proceed to Step 4; otherwise, proceed to Step 5.
Step 4: Compute the remaining path length $s_{go,k}$ using Equation (6) ($k = k + 1$). Return to Step 1.
Step 5: $s_{go} = \left[s_{go,1}, s_{go,2}, \ldots, s_{go,N}\right], [N] \triangleq \{1, 2, 3, \ldots, N\}$
Step 6: Determine the UAV's current position. If $i < N$, proceed to Step 7; otherwise, proceed to Step 8.
Step 7: Compute the acceleration command k using Equation (42) and proceed to Step 9.
Step 8: $\kappa = 3z_1/s_{go,1}^2 = \sigma_1' s_{go,1}^2$
Step 9: Return k

According to Step 7 in Algorithm 1, when the UAV passes a waypoint, the guidance command is updated, the information of this waypoint is discarded, and, hence, the guidance command may experience an abrupt change at the very moment of passing. However, this does not cause any unacceptably bad influence on the guidance law, because the classical PN could also be used as a waypoint-following guidance law in contrast, and it would only use the information of one waypoint each time (as its target). The guidance command of PN definitely greatly changes when the UAV passes this waypoint. This is seen in the simulation cases in Section 3.

3. Numerical Simulation Results

The MEWFDGGL designed in Section 2 is analyzed and demonstrated using numerical simulations in this section. The numerical simulations are conducted for two different scenarios: when the speed of the UAV changes periodically and when it is influenced by randomly varying wind. In all the following simulations, the UAV's initial flight path angle is 30°, and there are eight waypoints to be visited. The UAV's initial location is at the origin of the reference frame. The inertial positions of the eight waypoints to be visited are listed in Table 1.

Table 1. Inertial positions of the eight waypoints.

Waypoint Number	Inertial Position (m)
1	(1000, 500)
2	(2000, 750)
3	(2500, 1000)
4	(4000, 1500)
5	(6000, 2000)
6	(7500, 1500)
7	(9000, 1000)
8	(11,000, 0)

3.1. Performance of the UAV under Varying Speeds

This subsection primarily verifies the performance of the MEWFDGGL and the SMEWFDGGL when the speed of the UAV periodically changes. As is known, due to the effect of actuator delay or resistance, the speed of a UAV does not always remain constant during flight. In practice, the actual speed of a UAV typically fluctuates around the expected speed during the implementation of speed control. Hence, in the considered scenario in this subsection, the UAV's speed $V = 30 - 10\cos 0.8t$ periodically changes, as shown in Figure 3.

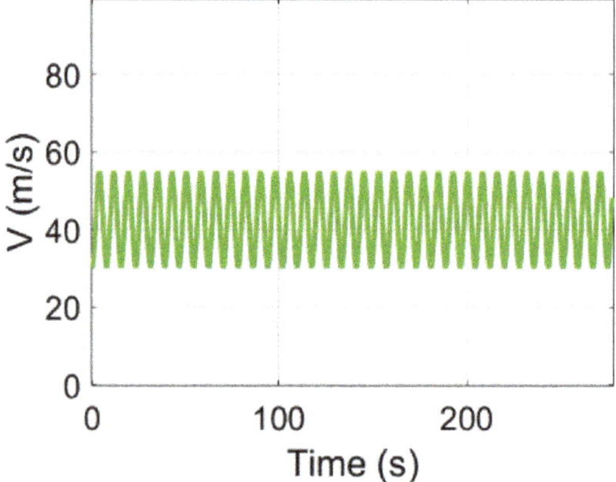

Figure 3. The real-time speed of the UAV.

3.1.1. MEWFDGGL

Figure 4a compares the flight trajectories of the UAV guided by both the MEWFDGGL and the MEWFG. It can be clearly seen in Figure 4a that the UAV can follow the desired waypoints when guided by these two guidance laws. Although the flight trajectories largely coincided, it can still be seen that the MEWFDGGL is better than the MEWFG law, especially near the fifth waypoint. The UAV's flight path angle is presented in Figure 4b, which clearly shows that the UAV guided by the MEWFG takes small, sharp turns, while the flight guided by the MEWFDGGL has a smooth path. For comparison purposes, the guidance curvature command in the arc-length domain is converted into the guidance acceleration command in the time domain. A comparison of the acceleration commands between the MEWFG and the MEWFDGGL is presented in Figure 4c. As exhibited in the diagram, when the UAV guided by the MEWFG passes the waypoint, an acceleration command occurs, which suddenly grows larger, whereas the variation in the command obtained from the MEWFDGGL is quite gentle. The reason for this phenomenon can be found in the comparison shown in Figure 4g,h. The periodic variation in the UAV's speed leads to the oscillation in the remaining flight time, while the linear variation in the remaining path length ensures the stability of the calculation.

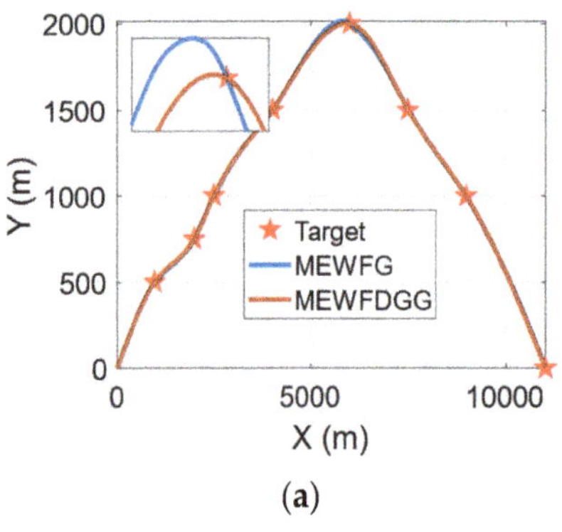

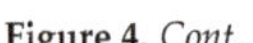

(**a**)

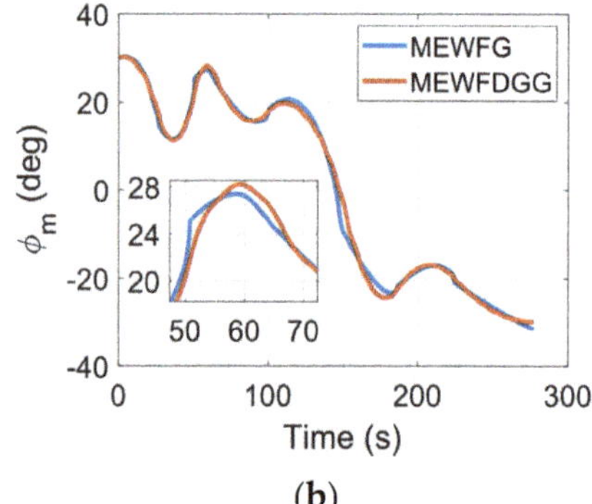

(**b**)

Figure 4. *Cont.*

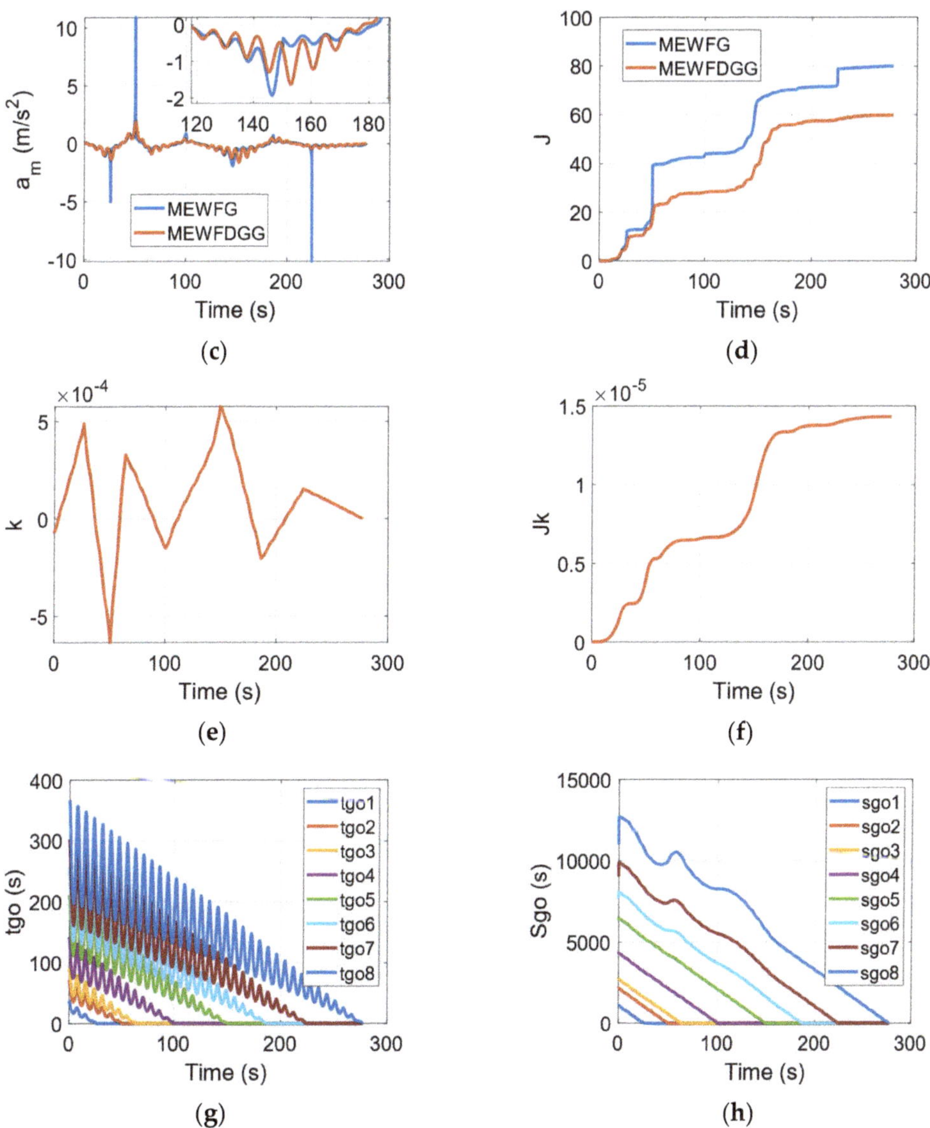

Figure 4. Compared results of the MEWFDGGL and the MEWFG when the UAV's speed changes periodically. (**a**) Flight trajectory. (**b**) Flight path angle. (**c**) Guidance acceleration command. (**d**) Performance index of *a*. (**e**) Guidance curvature command. (**f**) Performance index of *k*. (**g**) Remaining flight time. (**h**) Remaining path length.

The guidance curvature command and its quantitative energy consumption obtained from the MEWFDGGL are exhibited in Figure 4e,f. By comparing it to Figure 4c,e, we can observe that the derivation that is completely independent of the UAV's speed ensures the smoothness of the guidance curvature command. The corresponding quantitative energy consumption of the guidance acceleration commands obtained from the MEWFDGGL and the MEWFG are compared in Figure 4d. As shown in Figure 4d, we can clearly observe that the UAV guided by the MEWFDGGL requires less energy consumption than the flight guided by the MEWFG. Therefore, the UAV guided by the MEWFDGGL can be reasonably considered to have better endurance in scenarios where its speed changes. The comparison results of the performance of the proposed guidance laws are shown in Table 2.

Table 2. Performance comparison of the MEWFDGGL and the MEWFG under the varying speeds of UAVs.

Guidance	MEWFG	MEWFDGGL
Maximum ZEM (m)	0.02	0.0018
Maximum acceleration command (m/s^2)	10.17	1.891
Energy consumption	80.09	59.71

3.1.2. SMEWFDGGL

In order to better prove the effectiveness of the improved guidance law, the MEWFDGGL and the optimal guidance law (OGL) of a single point are used as the control group for comparative analysis. The OGL can be expressed as follows:

$$\kappa_i = 3z_i / s_{go,i}^2 \tag{47}$$

which is essentially the optimal proportional guidance law in the arc-length domain.

In this scenario, the flight trajectories of UAVs under the three guidance laws are shown in Figure 5a. The UAVs guided by the different guidance laws can successively visit the target waypoints, with small error, and they have largely coincident trajectories, which indicates that the velocity variations along the velocity directions have little effect on the performances of all the guidance laws in the arc-length domain. A comparison of the flight path angles is shown in Figure 5b. As shown in Figure 5b, the UAV guided by the OGL makes a sharp turn when passing through the current waypoint, while the UAVs guided by the SMEWFDGGL and the MEWFDGGL show good performances, and all of them can ensure smooth flight trajectories.

A comparison of the acceleration command and the curvature command between the three guidance laws is presented in Figure 5c,e. As can be seen in Figure 5c,e, the curvature command under the OGL shows discontinuity when passing through the way-point. Although the UAV guided by the SMEWFDGGL cannot produce a continuous curvature command compared with the UAV guided by the MEWFDGGL, the amplitude of discontinuity is obviously smaller than that of the UAV guided by the OGL, which has little influence on the implementation of the guidance. The quantitative energy consumption levels of the three guidance laws are compared in Figure 5d,f. In Figure 5d,f, the UAV guided by the OGL consumes the most energy—approximately three times as much as the UAVs guided by the MEWFDGGL and the SMEWFDGGL. The energy consumption levels under the MEWFDGGL and the SMEWFDGGL are not much different, with the consumption under the SMEWFDGGL being slightly more (approximately 10%) than that under the MEWFDGGL. The comparison results of the performances of the proposed guidance laws are shown in Table 3.

Table 3. Performance comparison of the SMEWFDGGL under the varying speeds of UAVs.

Guidance	OGL	MEWFDGGL	SMEWFDGGL
Maximum ZEM (m)	0.00001	0.0018	0.0015
Maximum acceleration command (m/s^2)	4.552	1.891	2.271
Energy consumption	199.5	59.71	66.21

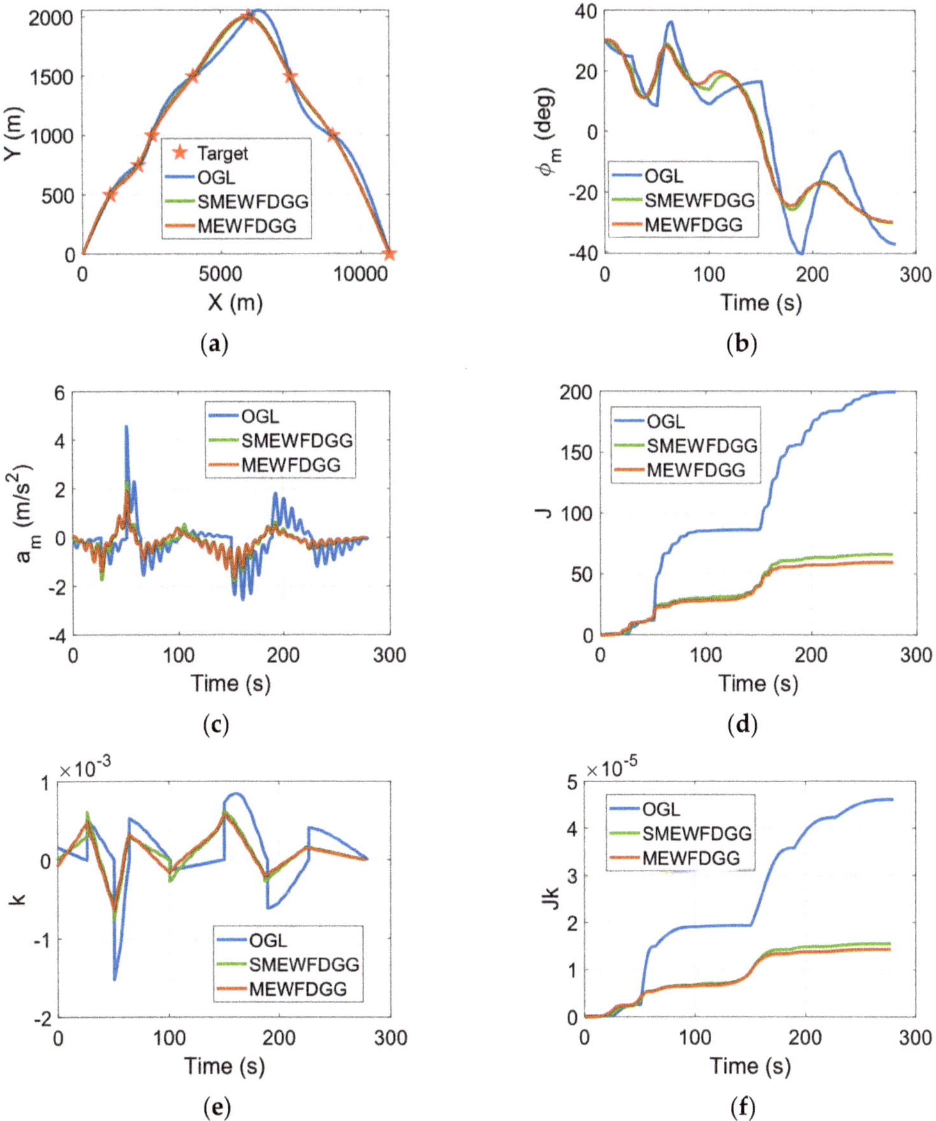

Figure 5. The performance of the SMEWFDGGL under the varying speeds of the UAV. (**a**) Flight trajectory. (**b**) Flight path angle. (**c**) Guidance acceleration command. (**d**) Performance index of *a*. (**e**) Guidance curvature command. (**f**) Performance index of *k*.

3.2. Performance under the Influence of the Wind

This subsection primarily verifies the performance of the MEWFDGGL when the speed of the UAV is influenced by randomly varying wind. As is known, for aerial vehicles, a major factor that can affect their speed is wind. In this subsection, we simulate the direction of the wind with a uniformly generated random number, i.e., east, south, west, and north, and the size of the wind is simulated by a normally distributed random number with a standard deviation of 5 and a variance of 1. The UAV's speed, as influenced by randomly varying wind, is shown in Figure 6.

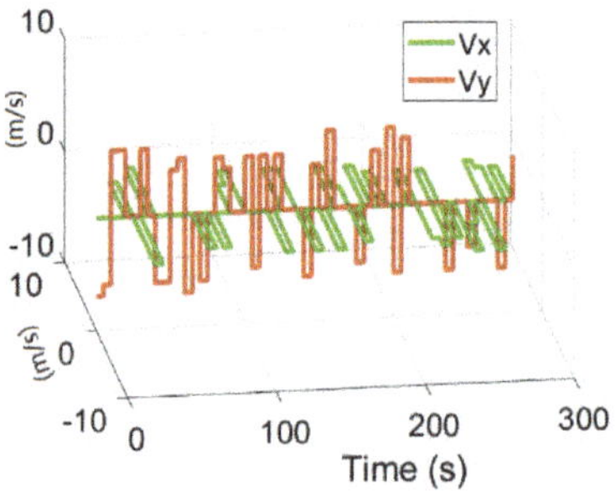

Figure 6. The real-time speed of the UAV, as influenced by randomly varying wind.

3.2.1. MEWFDGGL

Figure 7a compares the flight trajectories of the UAVs guided by the MEWFDGGL and the MEWFG when the UAVs' speeds are influenced by randomly varying wind. It can be clearly seen in Figure 7a that, although both of the UAVs can follow the desired waypoints guided by the two guidance laws, the flight trajectory obtained from the MEWFDGGL is closer to an ideal approaching course, especially near the fifth and sixth waypoints. This UAV's flight path angle is presented in Figure 7b. Due to the influence of randomly varying wind, both of the curves obtained from the MEWFDGGL and the MEWFG are not smooth, which is inevitable. For comparison purposes, the guidance curvature command in the arc-length domain is converted into the guidance acceleration command in the time domain. A comparison of the acceleration commands between the MEWFG and the MEWFDGGL is presented in Figure 7c. As can be seen in the chart, when the UAV guided by the MEWFG passes the waypoint, an acceleration command occurs, which suddenly grows larger, whereas the variation in the command obtained from the MEWFDGGL is quite gentle. The reason for this phenomenon is similar to that in the previous section.

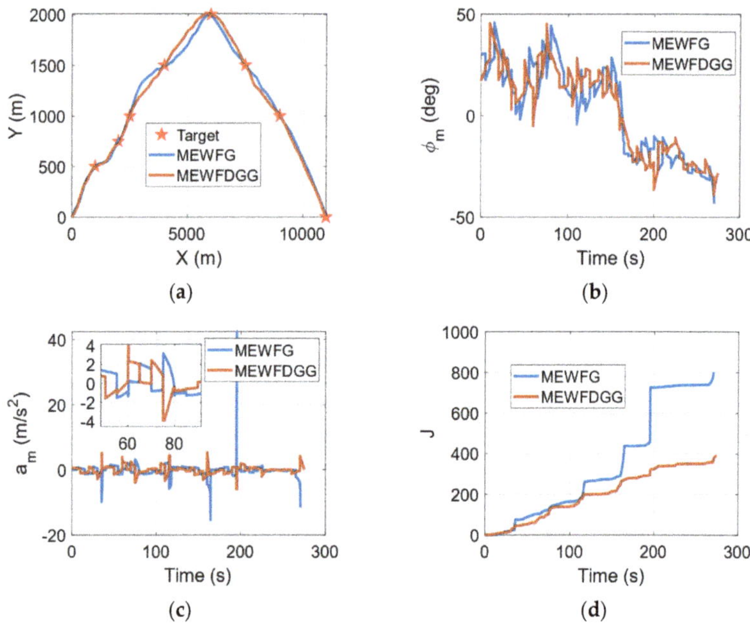

Figure 7. *Cont.*

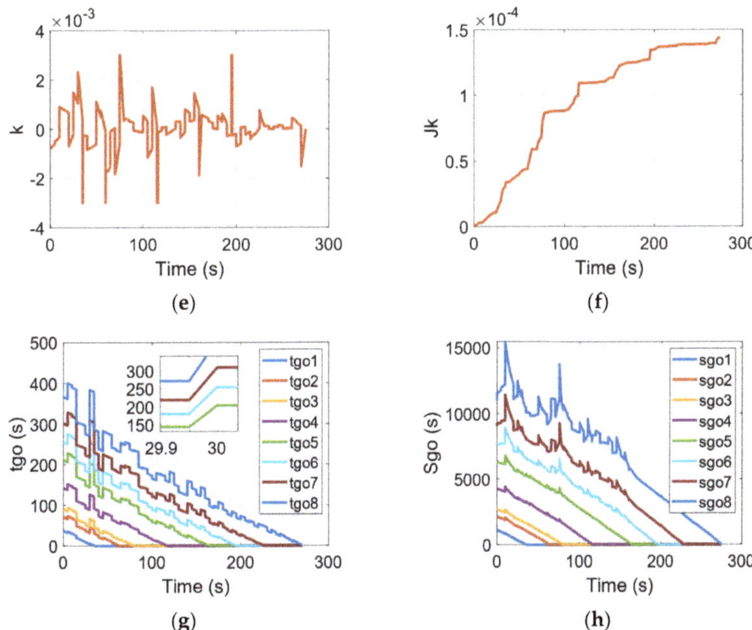

Figure 7. Comparison of the results of the MEWFDGGL and the MEWFG when the UAVs' speeds are influenced by randomly varying wind. (**a**) Flight trajectory. (**b**) Flight path angle. (**c**) Guidance acceleration command. (**d**) Performance index of a. (**e**) Guidance curvature command. (**f**) Performance index of k. (**g**) Remaining flight time. (**h**) Remaining path length.

The guidance curvature command and its quantitative energy consumption obtained from the MEWFDGGL when the UAV's speed is influenced by randomly varying wind are shown in Figure 7e,f, respectively. By comparing Figures 4e and 7e, it can be observed that the guidance curvature commands show random sharp turns, because the directions of the UAVs' speeds are affected by randomly varying wind. The guidance curvature command may require adjustments at any time to ensure that an optimal trajectory is generated. The quantitative energy consumption levels of the guidance acceleration commands obtained from the MEWFDGGL and the MEWFG are compared in Figure 7d. As exhibited in Figure 7d, we can clearly observe that the UAV guided by the MEWFDGGL consumes approximately 50% less energy than the UAV guided by the MEWFG in the considered scenarios. Hence, it is reasonable to consider that the UAV guided by the MEWFDGGL is good at overcoming the effects of randomly varying wind and has a better endurance in the considered scenario. The comparison results of the performances of the proposed guidance laws are shown in Table 4.

Table 4. Performance comparison of the MEWFDGGL and the MEWFG when the UAVs' speeds are influenced by randomly varying wind.

Guidance	MEWFG	MEWFDGGL
Maximum ZEM (m)	0.1507	0.0488
Maximum acceleration command (m/s^2)	42.4	5.247
Energy consumption	802.2	388.1

3.2.2. SMEWFDGGL

When a UAV's speed is influenced by randomly varying wind, the variation in the UAV's velocity is no longer limited to the direction of that UAV's velocity. As shown in

Figure 8b, under the influence of randomly varying wind, the velocity direction of the UAV constantly changes, and the flight path angles of the UAVs under the three guidance laws are in discontinuous fluctuation. As can be seen in Figure 8a, compared with the ideal situation, the flight trajectory when a UAV's speed is influenced by randomly varying wind has obvious jitter and a small turning point, while the flight trajectories under the guidance of the MEWFDGGL and the SMEWFDGGL are relatively smoother.

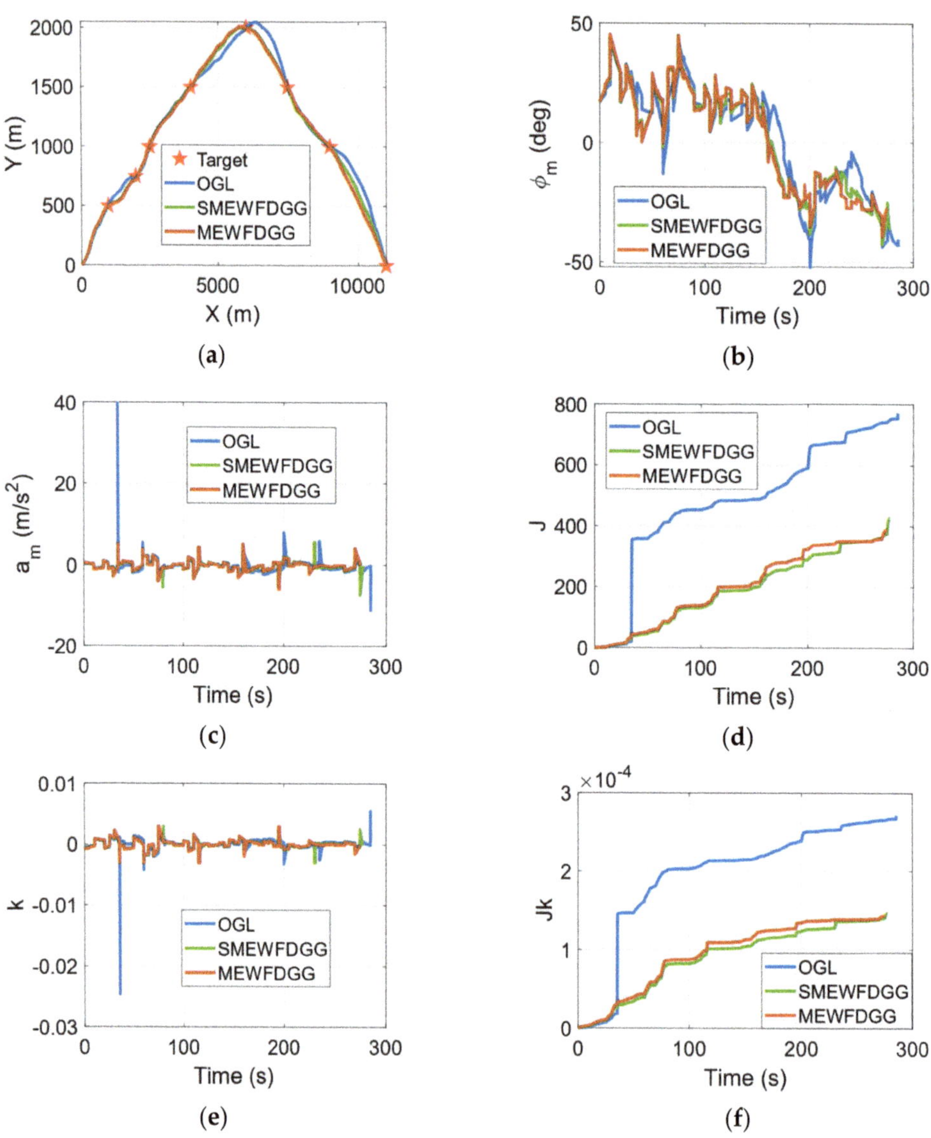

Figure 8. The performance of the UAV guided by the SMEWFDGGL under the influence of randomly varying wind. (**a**) Flight trajectory. (**b**) Flight path angle. (**c**) Guidance acceleration command. (**d**) Performance index of *a*. (**e**) Guidance curvature command. (**f**) Performance index of *k*.

The guidance curvature command and its quantitative energy consumption obtained from the MEWFDGGL are shown in Figure 8e,f, respectively. When a UAV's speed is influenced by randomly varying wind, the curvature instruction of that UAV's guidance also shows discontinuity. If the sudden change value generated during the ill-conditioned solution of the guidance command is ignored, then the guidance energy consumption levels under the three guidance laws have little difference, which is consistent with the theoretical conclusion that the MEWFDGGL's guidance energy consumption level is the lowest and that the SMEWFDGGL's guidance energy consumption level is the second-lowest, although both of them are far lower than the guidance energy consumption level of the OGL. The comparison results of the performances of the proposed guidance laws are shown in Table 5.

Table 5. Performance comparison of the SMEWFDGGL under the influence of randomly varying wind.

Guidance	OGL	MEWFDGGL	SMEWFDGGL
Maximum ZEM (m)	0.001	0.04883	0.0532
Maximum acceleration command (m/s^2)	9.79	5.247	7.336
Energy consumption	754.8	388.1	427.7

4. Experiment Verification

Using a small quadrotor UAV, the experimental verification of the proposed guidance laws is presented in this section. The outdoor experimental field is shown in Figure 9.

Figure 9. The small quadrotor UAV and outdoor experimental field.

Considering the UAV's performance and the limitations of the experimental site, the UAV starts from the take-off point and successively visits three target waypoints. The coordinates of the take-off point are (0,0), the initial flight path angle of the UAV is 60°, and the UAV's initial velocity is 2 m/s. The specific position coordinates of all waypoints are shown in Table 6.

Table 6. Inertial positions of the three waypoints in the experimental scenario.

Waypoint Number	Inertial Position (m)
1	(30, 30)
2	(70, 25)
3	(90, 10)

4.1. MEWFDGGL

As can be seen in Figure 10a, the small quadrotor UAV guided by the two guidance laws can successfully visit the waypoints in the actual flight scenario, but the trajectory of the UAV guided by the MEWFDGG is significantly smoother, and its ZEM is smaller. A comparison of the acceleration commands between the MEWFG and the MEWFDGGL is presented in Figure 10b. Theoretically, the guidance acceleration instructions obtained by the two guidance laws are continuous. However, in the actual flight process, the UAV cannot visit the target waypoint along the completely ideal trajectory, which makes the UAV's acceleration command discontinuous. It is worth emphasizing that both of the guidance laws are applicable to small quadrotor UAvs, but the latter has a more stable performance.

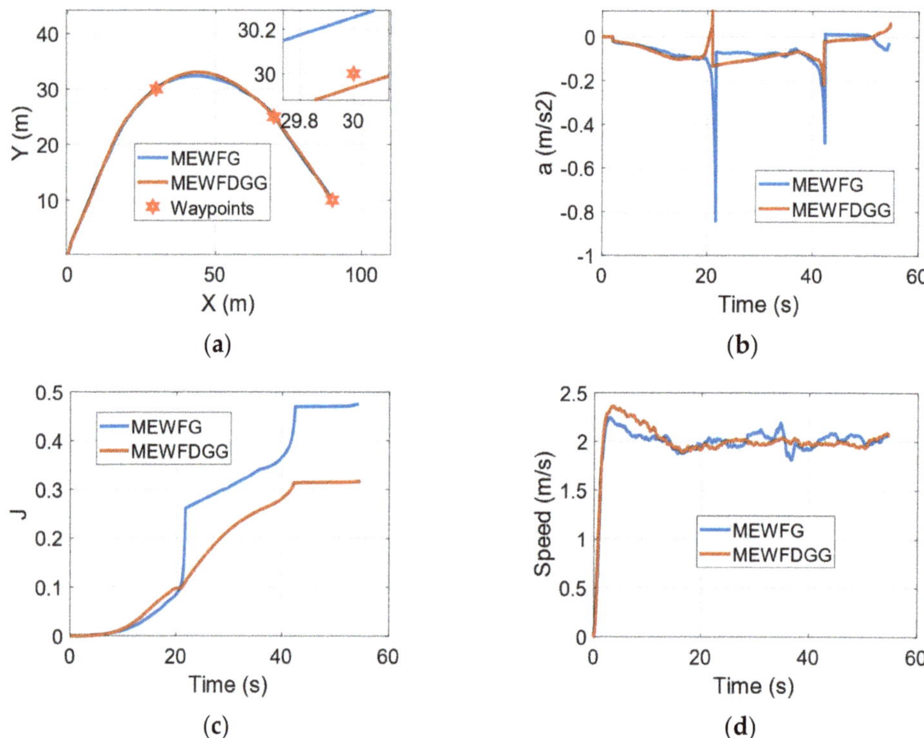

Figure 10. The performance of the MEWFDGGL in the experimental scenario. (**a**) Flight trajectory. (**b**) Guidance acceleration command. (**c**) Performance index of *a*. (**d**) The real-time speed of the UAV.

The quantitative energy consumption levels of the two guidance laws are compared in Figure 10c. As shown in Figure 10c, the energy consumption level of the UAV guided by the MEWFDGGL is reduced by approximately 40% compared to that of the UAV guided by the MEWFGL. The comparison results of the performances of the proposed guidance laws are shown in Table 7.

Table 7. Performance comparison of the MEWFDGGL and the MEWFG in the experimental scenario.

Guidance	MEWFG	MEWFDGGL
Maximum ZEM (m)	0.25	0.06
Maximum acceleration command (m/s^2)	0.843	0.1173
Energy consumption	0.4753	0.3176

4.2. SMEWFDGGL

In this scenario, the flight trajectories of the UAVs under the three guidance laws are shown in Figure 11a. From the flight trajectories, we can see that the UAV guided by the MEWFDGG has a smoother flight trajectory than that of the UAV guided by the SMEWFDGGL, which obviously consumes less energy than the UAV guided by the OGL. The experimental results show that the energy consumption levels of the UAVs guided by the MEWFDGG and the SMEWFDGG are similar, and they are much lower than that of the UAV guided by the OGL. The experimental results are consistent with both the simulation results and the theoretical conclusions.

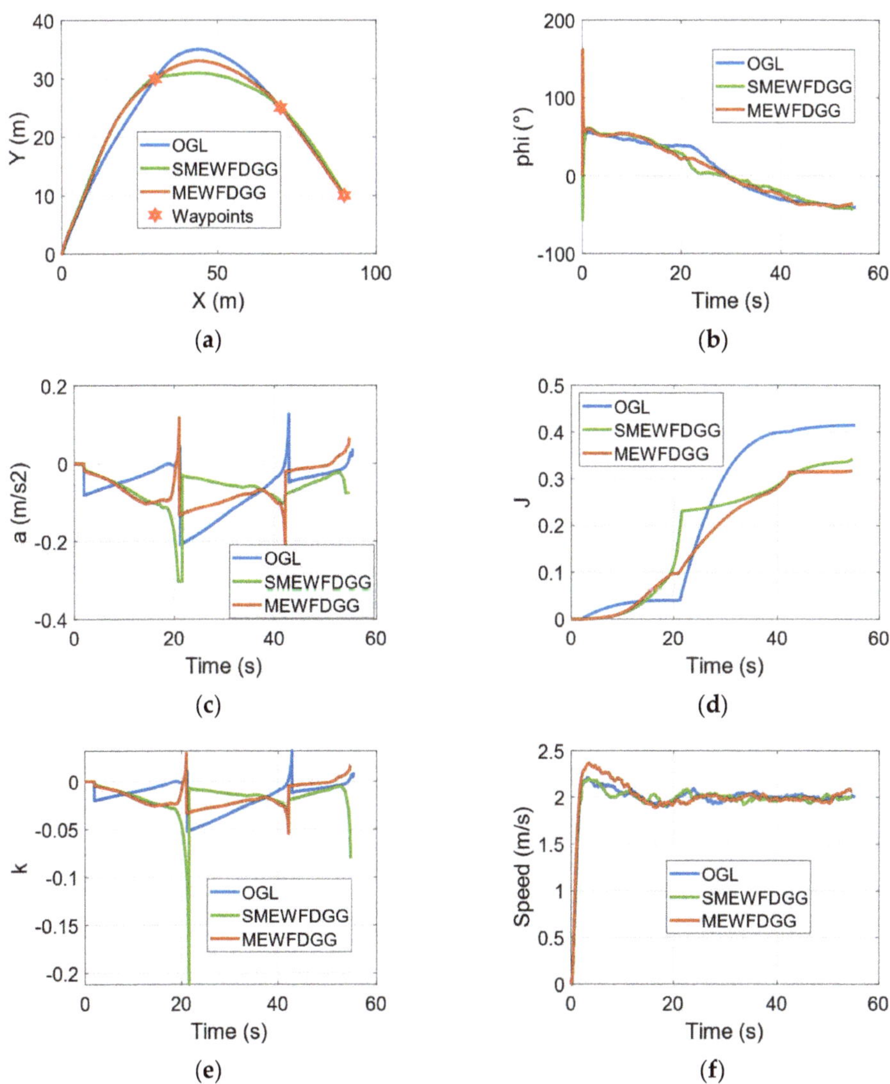

Figure 11. The performance of the SMEWFDGGL in the experimental scenario. (**a**) Flight trajectory. (**b**) Flight path angle. (**c**) Guidance acceleration command. (**d**) Performance index of *a*. (**e**) Guidance curvature command. (**f**) The real-time speed of the UAV.

When carefully observing Figure 11d, it is not difficult to see that the guidance energy consumption level and the track angle of the UAV under the guidance of the MEWFDGGL remain in an approximately straight line for the last 20 s, while the guidance energy consumption levels nearly stay the same and do not increase. This is because the MEWFDGGL takes into account the influence of all waypoint constraints during the guidance process. When the UAV reaches the second waypoint, the speed direction of the UAV is adjusted to the direction of the last waypoint. However, the OGL cannot be used to adjust the flight angle of the UAV between every two waypoints, which is the reason why the MEWFDGGL can achieve global energy optimization. The performance comparison results of three guidance laws are shown in Table 8.

Table 8. Performance comparison of the SMEWFDGGL in the experimental scenario.

Guidance	OGL	MEWFDGGL	SMEWFDGGL
Maximum ZEM (m)	0.02	0.06	0.17
Maximum acceleration command (m/s^2)	0.2086	0.2177	0.3029
Energy consumption	0.4146	0.3174	0.3423

5. Conclusions

A minimum-effort waypoint-following differential geometric guidance law (MEWFDGGL) and its suboptimal form for varying-speed endo-atmospheric flight vehicles were proposed in this paper. The optimal guidance problem was transformed into an optimal space curve design problem using the differential geometric guidance model. The speed changes in endo-atmospheric flight vehicles were theoretically decoupled from the guidance problem, rather than the constant speed hypothesis being directly adopted. It was theoretically proven that the proposed MEWFDGGL is a globally energy-optimal guidance law; the suboptimal MEWFDGGL was proposed in order to solve the problems of complexity and high computation burden, and it is advantageous for improving the autonomous flight capability of small, low-cost endo-atmospheric flight vehicles. Finally, in comparison with the original MEWFG law, the nonlinear numerical simulations and experimental verifications show that the MEWFDDGGL is more efficient for eliminating the adverse influences on the guidance performance caused by a UAV's speed changes. It is worth noting that the MEWFDDGGL proposed in this paper does not break through some of the limitations faced by the MEWFG. For example, the optimality of the MEWFDDGGL is affected by the estimation accuracy of the remaining path length. However, for theoretical research, the proposed guidance law may also be extended to maneuvering target interception scenarios and salvo attack scenarios.

Author Contributions: Conceptualization, K.L.; Methodology, X.Q., K.L., Y.L. (Yangang Liang) and Y.L. (Yuanhe Liu); Software, X.Q.; Formal analysis, X.Q.; Investigation, X.Q.; Resources, X.Q.; Data curation, X.Q.; Writing—original draft, X.Q.; Writing—review & editing, X.Q., K.L., Y.L. (Yangang Liang) and Y.L. (Yuanhe Liu); Visualization, X.Q.; Supervision, K.L., Y.L. (Yangang Liang) and Y.L. (Yuanhe Liu); Project administration, X.Q., K.L. and Y.L. (Yangang Liang). All authors have read and agreed to the published version of the manuscript.

Funding: This research was funded by [The National Natural Science Foundation of China] grant number [No. 12002370] and the APC was funded by [The National Natural Science Foundation of China] grant number [No. 12002370].

Data Availability Statement: The authors don't create a link to the research data for this study.

Conflicts of Interest: The authors declare no conflict of interest.

References

1. He, S.; Lee, C.-H.; Shin, H.-S.; Tsourdos, A. *Optimal Guidance and Its Applications in Missiles and UAVs*, 1st ed.; Springer: Cham, Switzerland, 2020; pp. 151–173.
2. Beard, R.W.; Ferrin, J.; Humpherys, J. Fixed Wing UAV Path Following in Wind With Input Constraints. *IEEE Trans. Control. Syst. Technol.* **2014**, *22*, 2103–2117. [CrossRef]
3. Ullah, N.; Mehmood, Y.; Aslam, J.; Shaoping, W.A.N.G.; Phoungthong, K. Fractional order adaptive robust formation control of multiple quad-rotor UAVs with parametric un-certainties and wind disturbances. *Chin. J. Aeronaut.* **2022**, *35*, 204–220. [CrossRef]
4. Piprek, P.; Hong, H.; Holzapfel, F. Optimal trajectory design accounting for the stabilization of linear time-varying error dynamics. *Chin. J. Aeronaut.* **2022**, *35*, 55–66. [CrossRef]
5. Pang, B.; Dai, W.; Hu, X.; Dai, F.; Low, K.H. Multiple air route crossing waypoints optimization via artificial potential field method. *Chin. J. Aeronaut.* **2020**, *34*, 279–292. [CrossRef]
6. Medagoda, E.D.B.; Gibbens, P.W. Synthetic-Waypoint Guidance Algorithm for Following a Desired Flight Trajectory. *J. Guid. Control. Dyn.* **2015**, *33*, 601–606. [CrossRef]
7. Wang, X.; Tan, G.; Dai, Y.; Lu, F.; Zhao, J. An Optimal Guidance Strategy for Moving-Target Interception by a Multirotor Unmanned Aerial Vehicle Swarm. *IEEE Access* **2020**, *8*, 121650–121664. [CrossRef]
8. Sun, G.; Wen, Q.; Xu, Z.; Xia, Q. Impact time control using biased proportional navigation for missiles with varying velocity. *Chin. J. Aeronaut.* **2020**, *33*, 956–964. [CrossRef]
9. Li, K.-B.; Shin, H.-S.; Tsourdos, A.; Tahk, M.-J. Performance of 3-D PPN Against Arbitrarily Maneuvering Target for Homing Phase. *IEEE Trans. Aerosp. Electron. Syst.* **2020**, *56*, 3878–3891. [CrossRef]
10. Li, K.-B.; Shin, H.-S.; Tsourdos, A.; Tahk, M.-J. Capturability of 3D PPN Against Lower-Speed Maneuvering Target for Homing Phase. *IEEE Trans. Aerosp. Electron. Syst.* **2019**, *56*, 711–722. [CrossRef]
11. Kebo, L.I.; Zhihui, B.A.I.; Hyo-Sang, S.H.I.N.; Tsourdos, A.; Min-Jea, T.A.H.K. Capturability of 3D RTPN guidance law against true-arbitrarily ma-neuvering target with maneuverability limitation. *Chin. J. Aeronaut.* **2022**, *35*, 75–90.
12. Zhao, Y.; Sheng, Y.; Liu, X. Trajectory reshaping based guidance with impact time and angle constraints. *Chin. J. Aeronaut.* **2016**, *29*, 984–994. [CrossRef]
13. Kim, Y.-W.; Kim, B.; Lee, C.-H.; He, S. A unified formulation of optimal guidance-to-collision law for accelerating and decelerating targets. *Chin. J. Aeronaut.* **2022**, *35*, 40–54. [CrossRef]
14. Qi, N.; Sun, Q.; Zhao, J. Evasion and pursuit guidance law against defended target. *Chin. J. Aeronaut.* **2017**, *30*, 1958–1973. [CrossRef]
15. He, S.; Lee, C.-H.; Shin, H.-S.; Tsourdos, A. Optimal three-dimensional impact time guidance with seeker's field-of-view constraint. *Chin. J. Aeronaut.* **2021**, *34*, 240–251. [CrossRef]
16. Kyaw, P.T.; Le, A.V.; Veerajagadheswar, P.; Elara, M.R.; Thu, T.T.; Nhan, N.H.K.; Van Duc, P.; Vu, M.B. Energy-Efficient Path Planning of Reconfigurable Robots in Complex Environments. *IEEE Trans. Robot.* **2022**, *38*, 2481–2494. [CrossRef]
17. Yin, Q.; Chen, Q.; Wang, Z. Energy-Optimal Waypoint-Following Guidance for Gliding-Guided Projectiles. In Proceedings of the 21st International Conference on Control, Automation and Systems (ICCAS), Jeju, Republic of Korea, 12–15 October 2021; IEEE: New York, NY, USA, 2021; Volume 34, pp. 1477–1482.
18. Chen, Y.; Yu, J.; Mei, Y.; Zhang, S.; Ai, X.; Jia, Z. Trajectory optimization of multiple quad-rotor UAVs in collaborative assembling task. *Chin. J. Aeronaut.* **2016**, *29*, 184–201. [CrossRef]
19. Wang, X.; Yang, Y.; Wang, D.; Zhang, Z. Mission-oriented cooperative 3D path planning for modular solar-powered aircraft with energy optimization. *Chin. J. Aeronaut.* **2022**, *35*, 98–109. [CrossRef]
20. Zhou, Y.; Su, Y.; Xie, A.; Kong, L. A newly bio-inspired path planning algorithm for autonomous obstacle avoidance of UAV. *Chin. J. Aeronaut.* **2021**, *34*, 199–209. [CrossRef]
21. Jeon, I.-S.; Lee, J.-I. Optimality of Proportional Navigation Based on Nonlinear Formulation. *IEEE Trans. Aerosp. Electron. Syst.* **2010**, *46*, 2051–2055. [CrossRef]
22. Kim, T.-H.; Park, B.-G. Rapid Homing Guidance Using Jerk Command and Time-Delay Estimation Method. *IEEE Trans. Aerosp. Electron. Syst.* **2022**, *58*, 729–742. [CrossRef]
23. Shaferman, V.; Shima, T. Linear Quadratic Guidance Laws for Imposing a Terminal Intercept Angle. *J. Guid. Control. Dyn.* **2008**, *31*, 1400–1412. [CrossRef]
24. Palumbo, N.F.; Blauwkamp, R.A.; Lloyd, J.M. Modern homing missile guidance theory and techniques. *Johns Hopkins APL Tech. Dig.* **2010**, *29*, 42–59.
25. He, S.; Lee, C.-H. Optimality of Error Dynamics in Missile Guidance Problems. *J. Guid. Control. Dyn.* **2018**, *41*, 1620–1629. [CrossRef]
26. He, S.; Lee, C.-H.; Shin, H.-S.; Tsourdos, A. Minimum-Effort Waypoint-Following Guidance Law. *J. Guid. Control. Dyn.* **2019**, *32*, 151–173. [CrossRef]
27. Kobayashi, S. *Differential Geometry of Curves and Surfaces*, 1st ed.; Springer: Cham, Switzerland, 2019.
28. Lu, P. Intercept of Nonmoving Targets at Arbitrary Time-Varying Velocity. *J. Guid. Control. Dyn.* **1998**, *21*, 176–178. [CrossRef]
29. Li, K.; Liang, Y.; Su, W.; Chen, L. Performance of 3D TPN against true-arbitrarily maneuvering target for exoat-mospheric interception. *Sci. China Technol. Sci.* **2018**, *61*, 1161–1174. [CrossRef]

30. Li, K.B.; Su, W.S.; Chen, L. Performance analysis of differential geometric guidance law against high-speed target with arbitrarily maneuvering acceleration. *Proc. Inst. Mech. Eng. Part G-J. Aerosp. Eng.* **2019**, *233*, 3547–3563. [CrossRef]
31. Li, K.B.; Su, W.S.; Chen, L. Performance analysis of realistic true proportional navigation against maneuvering targets using Lyapunov-like approach. *Aerosp. Sci. Technol.* **2017**, *69*, 333–341. [CrossRef]
32. Shin, H.S.; Li, K.B. An Improvement in Three-Dimensional Pure Proportional Navigation Guidance. *IEEE Trans. Aerosp. Electron. Syst.* **2021**, *57*, 3004–3014. [CrossRef]

Article

A Dual Aircraft Maneuver Formation Controller for MAV/UAV Based on the Hybrid Intelligent Agent

Luodi Zhao [1,2,3], Yemo Liu [4], Qiangqiang Peng [4] and Long Zhao [1,2,3,*]

1   School of Automation Science and Electrical Engineering, Beihang University, Beijing 100191, China; luodizhao@buaa.edu.cn
2   Digital Navigation Center, Beihang University, Beijing 100191, China
3   Science and Technology on Aircraft Control Laboratory, Beihang University, Beijing 100191, China
4   Beijing Aerospace Automatic Control Institute, Beijing 100854, China
*   Correspondence: buaa_dnc@buaa.edu.cn

Abstract: This paper proposes a hybrid intelligent agent controller (HIAC) for manned aerial vehicles (MAV)/unmanned aerial vehicles (UAV) formation under the leader–follower control strategy. Based on the high-fidelity three-degrees-of-freedom (DOF) dynamic model of UAV, this method decoupled multiple-input-multiple-output (MIMO) systems into multiple single-input-single-output (SISO) systems. Then, it innovatively combined the deep deterministic policy gradient (DDPG) and the double deep Q network (DDQN) to construct a hybrid reinforcement learning-agent model, which was used to generate onboard desired state commands. Finally, we adopted the dynamic inversion control law and the first-order lag filter to improve the actual flight-control process. Under the working conditions of a continuous S-shaped large overload maneuver for the MAV, the simulations verified that the UAV can achieve accurate tracking for the complex trajectory of the MAV. Compared with the traditional linear quadratic regulator (LQR) and DDPG, the HIAC has better control efficiency and precision.

Keywords: MAV/UAV; formation control; hybrid reinforcement learning; hybrid intelligent agent

Citation: Zhao, L.; Liu, Y.; Peng, Q.; Zhao, L. A Dual Aircraft Maneuver Formation Controller for MAV/UAV Based on the Hybrid Intelligent Agent. *Drones* **2023**, 7, 282. https://doi.org/10.3390/drones7050282

Academic Editor: Shiva Raj Pokhrel

Received: 28 February 2023
Revised: 12 April 2023
Accepted: 18 April 2023
Published: 22 April 2023

1. Introduction

Aiming at increasingly fast-paced and high-intensity air combat, the use of MAVs as combat operations leaders with a certain number of UAVs as wingers to form a hybrid formation of UAV/MAV has become the development trend for future air confrontations. Among them, the two-aircraft formation consisting of an MAV and a UAV is one of the most typical combat styles. In MAV/UAV formations, the unmanned system must be able to share information and carry out cooperative operations with the manned systems across systematic boundaries [1]. The Fast Lightweight Autonomy (FLA) Program by the Defense Advanced Research Projects Agency (DARPA) has developed an advanced algorithm that enables an MAV or a UAV to operate autonomously without a human operator, the Global Positioning System (GPS), or any data resources. DARPA's Lifelong Learning Machines (L2M) Project also aims to develop new machine learning methods that enable unmanned systems to continuously adapt to new environments and remember what they have learned [2]. Meanwhile, the U.S. Air Force's Loyal Wingman Program aims to enhance the autonomy of UAVs and improve their combat capabilities in complex war environments [3]. Moreover, the recently proposed Skyborg program is working on the combination of manned and unmanned combat aerial vehicles. Therefore, improving the capability of autonomous flight control has become an important direction for the development of future UAV technology.

One of the research hotspots of UAV autonomous control capability is the formation flight-control problem [4]. In terms of the traditional design of the formation controller, Ref. [5] proposed a sliding mode controller for MAV/UAV formation flights based on a layered architecture. However, it makes extensive simplifications on the strong nonlinear dynamic model of MAV/UAVs, and was only validated by simulations for flat trajectories. Ref. [6] considered sensor noise and developed a leader–follower formation PID controller for multi-robots, which can achieve better performance in limiting position deviations. Furthermore, Ref. [7] proposed a parallel approach control law for fixed-wing UAV formations under the leader–follower strategy. Ref. [8] referred to an idea of multi-channel decoupling that split the MIMO system into multiple SISO systems and used sliding mode control to track the reference trajectory, which can be further applied to formation-control problems. Refs. [9,10] proposed a consensus-based multiple aircraft cooperative formation control method, but the consensus theory analysis was highly dependent on the linearized dynamic model, which limited its further application in a complex nonlinear dynamic system. Refs. [11,12] developed a formation controller where the commands were generated independently of the dynamic model, decreasing the control precision in extreme working conditions. Refs. [13,14] considered the confrontation situation and adopted predefined maneuver strategy collections, taking typical maneuvers as the basic units and building a collection of maneuver strategies with free combinations of various basic units. However, due to the model uncertainty and non-cooperative environment, this method hardly dealt with complex working conditions. Therefore, the intelligent agent method has become a novel research trend because of its weak model dependence and strong ability in terms of strategy exploration. Refs. [15,16] adopted deep neural networks to learn aircraft-maneuvering strategies and made progress in enhancing the autonomous maneuvering capability of UAVs. However, UAV formation control is a high-dimensional dynamical control problem with tightly coupled variables. When traditional neural networks learn such complex behaviors, they cause problems such as low training efficiency and difficulty in stable convergence [17]. Among the novel neural networks, the double deep Q network (DDQN) algorithm has shown good performance in control problems with discrete action sets by fitting the value functions of state actions through neural networks [18–20], but it cannot be applied to control problems with continuous variables. Based on the deterministic policy gradient (DPG) algorithm, DeepMind proposed the deep deterministic policy gradient (DDPG) algorithm which is proven to perform well on many kinds of continuous control problems [21–23]. However, in the field of aircraft control, the large variation in the angle of attack commands will increase the load on the attitude control loop [24]. Meanwhile, when it comes to complex tasks with multiple continuous control variables problems, DDPG has problems with unstable networks and low exploration efficiency [25–28]. For the above dilemma, some scholars have turned to hybrid reinforcement learning methods in recent years. By adding discrete "meta-actions" to continuous control problems, Ref. [29] partially solved the reinforcement learning traps and improved exploration efficiency. The experiments verified its superiority to the traditional continuous strategy algorithm in some cases. [30] proposed the parametrized deep Q-network for the hybrid action space without approximation or relaxation, which provides a reference for solving the hybrid control problem.

Based on the above analysis, it is obvious that the formation controller must be able to better adapt to complex flight conditions in future confrontation situations, e.g., continuous large overload maneuvers for the MAV, etc. Therefore, inspired by [29], we propose a hybrid reinforcement intelligent agent controller based on the decoupling of multi-channels, which can effectively solve the problem of formation-tracking under continuous maneuvering conditions. It should be emphasized that when designing controllers based on artificial intelligent methods, especially when the reinforcement learning controller is directly applied to the generation of underlying flight-control commands, the lack of flight dynamic constraints can easily bring about problems. Due to the lack of dynamic constraints, the attitude control system cannot quickly track the commands, leading to flight instability. Therefore, this paper introduced the dynamic inversion controller and the first-order lag filter to the hybrid reinforcement learning agent to enhance the smoothness and executability of control commands.

In summary, the main contributions of this paper are as follows:

(1) A hybrid intelligent agent was designed based on the novel concept of "meta-action" to further enhance formation control performance. The hybrid intelligent agent combined DDPG and DDQN according to the specific formation control targets;

(2) The framework of the HIAC was developed that combined the dynamic inversion controller and the first-order lag filter with the hybrid intelligent agent to effectively overcome the common drawbacks of reinforcement learning;

(3) The superiority of the HIAC method was validated with experiments of nominal conditions. Monte Carlo simulations with different initial conditions were then conducted to verify the adaptability of the HIAC.

The organization of this paper is as follows: Section 2 establishes the UAV dynamic model and formation-control targets. Section 3 designs the novel formation controller HIAC based on the DDPG/DDQN hybrid intelligent agent. The dynamic inversion controller and first-order lag filter are introduced to the framework of the HIAC as well. Section 4 conducts the experiments of nominal conditions and 100 Monte Carlo simulations with varying initial conditions. Finally, we summarize the research conclusion of this paper in Section 5.

2. Mathematical Modeling

2.1. UAV Dynamic Model

The main concern in dual aircraft formation flights is the real-time position, velocity, and attitude of the two aircraft, so it is necessary to establish a dynamic model of the UAV according to the forces on the mass as shown in Figure 1. To simplify the problem, the constraints flight envelope is ignored.

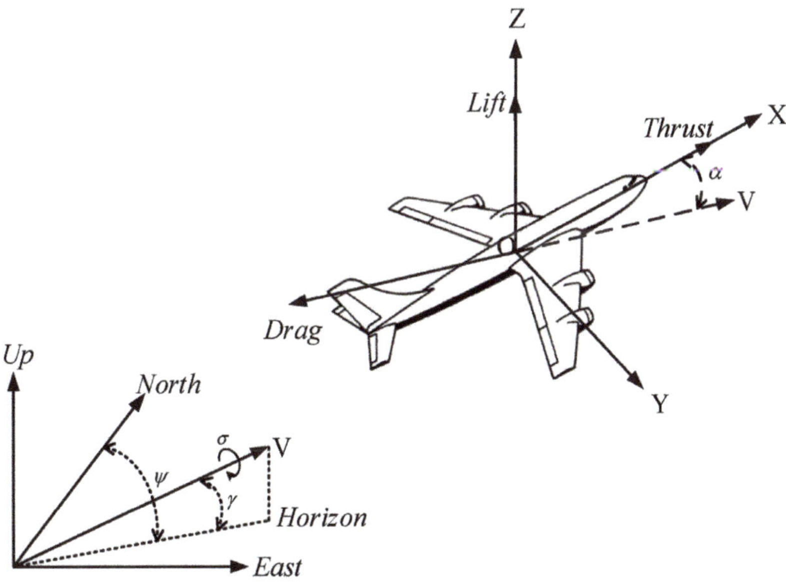

Figure 1. The forces on the center of gravity of the aircraft.

In the ground inertial coordinate system $o - xyz$, V is the UAV flight velocity. γ and ψ are the flight path angle and flight azimuth angle, respectively. The flight adopts the Bank-To-Turn (BTT), which is considered to have no sideslip. α is the attack angle, and σ is the bank angle. The engine thrust and drag of the aircraft are denoted by T and D, respectively. n is the normal overload of the UAV in the velocity coordinate system

$o - x_V y_V z_V$. Ignoring the wind disturbance in the flight, the three-degrees-of-freedom of the dynamic model for the UAV is established as follows [31–33]:

$$H = \begin{cases} \dot{x} = V \cos\gamma \sin\psi \\ \dot{y} = V \cos\gamma \cos\psi \\ \dot{z} = V \sin\gamma \\ \dot{V} = (T - D)/m - g\sin\gamma \\ \dot{\gamma} = g(n\cos\sigma - \cos\gamma)/V \\ \dot{\psi} = -gn\sin\sigma/(V\cos\gamma) \end{cases} \tag{1}$$

where m is the weight of the aircraft, which is considered constant in this paper, and g is the local gravity.

The engine thrust T can be denoted by

$$T = \eta T_{\max}, \tag{2}$$

where η is the throttle manipulator, and its range is defined as $[0, 1]$. $T_{\max}$ is the maximum thrust that the engine can achieve.

The air drag D consists of the parasite drag and the induced drag, which can be expressed as follows [31]:

$$D = C_{D_P}\rho V^2 S/2 + 2C_{D_I} n^2 m^2 g^2/\left(\rho V^2 S\right), \tag{3}$$

where S is the reference area of the UAV. C_{D_P} is the parasite drag coefficient. C_{D_I} is the induced drag coefficient. ρ is the atmospheric density, which varies with the altitude of the aircraft in the stratosphere. It is calculated by [34]

$$\rho = \rho_0 \cdot e^{-z/z_0}, \tag{4}$$

where $\rho_0 = 1.225 \ \text{kg/m}^3$ and $z_0 = 6700 \ \text{m}$.

2.2. Formation Control Targets

In this paper, the formation control target of the UAV was determined based on the leader–follower formation strategy. Taking a typical dual aircraft formation flight as an example, the formation configuration of the MAV/UAV was designed as shown in Figure 2. Since the reference trajectory of the MAV as the leader aircraft is known, the flight velocity, attitude, and position can be obtained from the sensors mounted within the MAV. The winger aircraft can receive real-time flight data from the MAV through the onboard data chain and complete the trajectory tracking and formation control autonomously. During the flight, it is required that the UAV and MAV keep a specific formation throughout the whole flight, as shown in Figure 2.

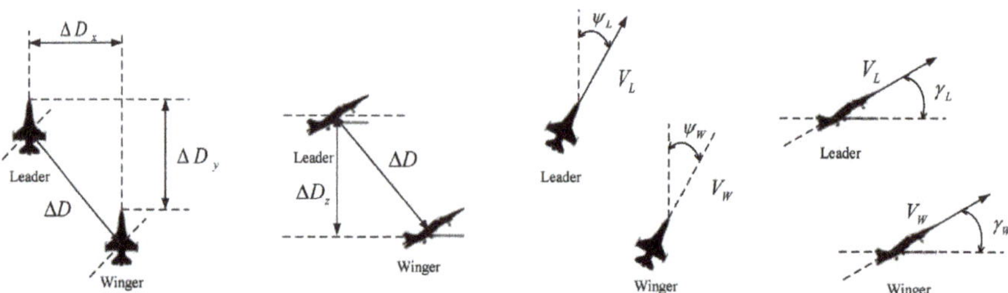

Figure 2. Dual aircraft formation for MAV/UAV.

2.2.1. Flight Velocity Control Targets

The MAV and UAV keep the same formation flight velocity. The reference velocity of the MAV is V_L, and the UAV velocity is V_W, then the velocity deviation ΔV is

$$\Delta V = |V_L - V_W|. \tag{5}$$

The MAV and UAV keep the same flight path angle in formation flight. The MAV flight path angle is γ_L, and the UAV flight path angle is γ_W, then the flight path angle deviation $\Delta \gamma$ is

$$\Delta \gamma = |\gamma_L - \gamma_W|. \tag{6}$$

The MAV and UAV keep the same flight azimuth angle in formation flight. The flight azimuth angle of the MAV is ψ_L, and the flight azimuth angle of the UAV is ψ_W, then the deviation of the flight azimuth angle $\Delta \psi$ is

$$\Delta \psi = |\psi_L - \psi_W|. \tag{7}$$

The flight velocity and attitudes of the UAV should be consistent with the MAV within an allowable error

$$\Delta V \leq V_{\Delta\max}, \ \Delta \gamma \leq \gamma_{\Delta\max}, \ \Delta \psi \leq \psi_{\Delta\max}, \tag{8}$$

where $V_{\Delta\max}$, $\gamma_{\Delta\max}$, $\psi_{\Delta\max}$ represent the error thresholds of the velocity, flight path angle, and flight azimuth angle of the UAV, respectively.

2.2.2. Flight Distance Control Targets

The UAV is located around the MAV and maintains the specified formation distance. ΔD denote the distance between the MAV and the UAV in the ground inertial coordinate system. ΔD_x, ΔD_y and ΔD_z denote the spatial distance of ΔD as follows:

$$\Delta D = \sqrt{\Delta D_x{}^2 + \Delta D_y{}^2 + \Delta D_z{}^2}. \tag{9}$$

Summarily, the UAV should keep a distance larger than the safe flight distance from the MAV, which is as follows:

$$D_{\Delta\min} \leq \Delta D \leq D_{\Delta\max}, \tag{10}$$

where $D_{\Delta\min}$ and $D_{\Delta\max}$ represent the thresholds of the safe distance.

3. Design of the HIAC

The HIAC first adopted a DDPG/DDQN hybrid reinforcement learning method to train the agent model to generate the tracking commands. Then, we further designed a dynamic inversion controller and a first-order lag filter to construct an improved formation flight controller. Overall, the HIAC consists of three parts, i.e., desired state command solver, dynamic inversion controller, and first-order lag filter. The framework of the HIAC is shown in Figure 3.

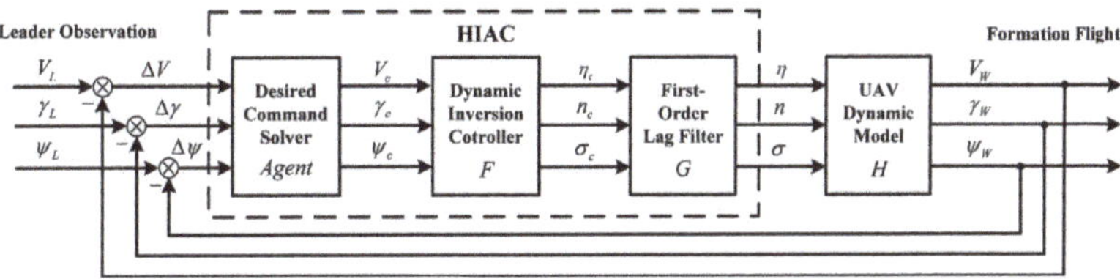

Figure 3. The framework of the HIAC.

In order to track the MAV, the HIAC adopted the current deviation between the states of the UAV and the states of the MAV, i.e., ΔV, $\Delta \gamma$ and $\Delta \psi$ as inputs, and outputs the control commands of the thrust, normal overload, and bank angle, i.e., η, n and σ. The main difference of the HIAC from other traditional controllers is that to further enhance the control accuracy, the HIAC adopted a hybrid intelligent agent as the desired command solver to generate the desired commands, V_c, γ_c and ψ_c. Then, these commands were sent to the dynamic inversion controller to generate the control commands, η_c, n_c and σ_c. Finally, the first-order lag filter further smoothed η_c, n_c and σ_c to improve the executability of these commands. The three parts will be introduced in detail as the order of the information flow.

3.1. Desired Command Solver

Learning from the idea of "meta-action", we partially discretized the control variables in the continuous control problems and developed a continuous–discrete mixed action space according to the characteristics of these control variables. Based on this process, we constructed a hybrid intelligent agent based on DDPG and DDQN to control V, γ, ψ and D of the UAV.

3.1.1. Framework of Hybrid Intelligent Agent Based on DDPG/DDQN

Based on the traditional Q-Learning algorithm, DDQN uses the neural network to fit the value function. It adopts discrete action sets to define the strategy and evaluates the Q value of the generated strategy through the Critic network. Compared with the traditional DQN algorithm [18–20], DDQN decouples the action selection strategy of the Q value and the calculation of the Q value and solves the problem of overestimation of the Q value compared with the traditional methods.

DDPG adopts the Actor–Critic network based on DQN and uses continuous action sets to define the control strategy. The model consists of the Actor–Critic network, where the Critic evaluates the actions generated by the Actor, and the Actor feeds back the evaluation results to the Critic for policy optimization [23]. More proofs and conclusions of the DDQN and DDPG can be found in [18,23], respectively.

However, the DDQN and DDPG suffer from different drawbacks when applied in practical engineering. Although the DDQN is easier to converge when compared with DDPG, it can only deal with discrete and low-dimensional action spaces. However, most of the practical targets, especially physical control targets, have continuous and high-dimensional action spaces. Moreover, even though the continuous space can be transferred into the discrete space, DDQN will generate high high-dimensional action space in this process and finally cause quite low computational efficiency. Meanwhile, although DDPG can solve the problem of continuous and high-dimensional action spaces, it is more likely to diverge than DDQN. Therefore, learning from "meta-action", we proposed a hybrid intelligent agent combining the DDQN and DDPG according to their complementary characteristics. Considering the value range and the control precision of V, γ and ψ, we adopted the idea of multi-channel decoupling to perform partial discretization of the action space. For the velocity control agent V_c, the DDPG was used to generate the set of continuous state commands. Because the value range of V_c is larger than γ_c and ψ_c, discretizing the continuous action space with high precision will lead to dimension explosion. Meanwhile, for the angle control agents γ_c and ψ_c, the DDQN is used to generate the set of discretized state commands. Combining the DDQN and DDPG can improve the capability of convergence when these two agents are trained together.

The framework of the desired commands solver was designed as shown in Figure 4. It includes three agents which process the variation of the state commands V_c, γ_c and ψ_c, respectively. Based on the decoupling principles between different agents, each agent calculates the action A_V, A_γ and A_ψ, and updates the desired state commands respectively.

The outputs are executed by the flight-control system of the UAV and fed back the rewards of each agent. The total reward function R_Σ is expressed by

$$R_\Sigma = R_\Sigma^{(D,V)} + R_\Sigma^{(\gamma)} + R_\Sigma^{(\psi)}, \tag{11}$$

where $R_\Sigma^{(D,V)}$, $R_\Sigma^{(\gamma)}$ and $R_\Sigma^{(\psi)}$ are components of R_Σ in each agent.

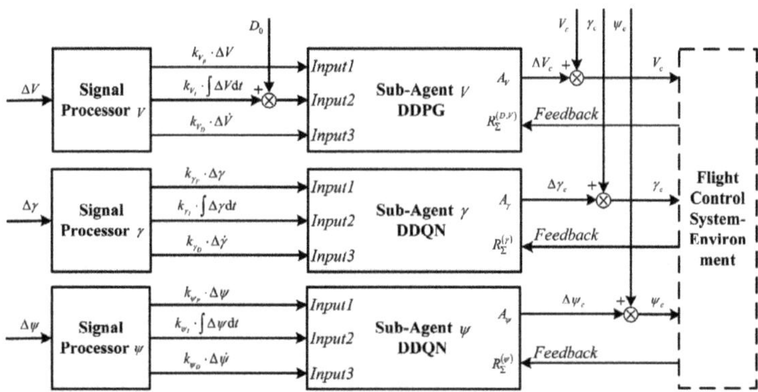

Figure 4. The framework of the desired commands solver.

To construct an intelligent agent based on the DDPG/DDQN hybrid reinforcement learning network, it was necessary to transform the trajectory tracking problem into a Markov decision process, which mainly includes three parts, i.e., the state space, the action space, and the reward function.

3.1.2. State Space S

According to the targets of formation flight control, the state space S is designed as follows:

$$S = [\Delta V, \Delta D, \Delta \dot{V}, \Delta \gamma, \int \Delta \gamma dt, \Delta \dot{\gamma}, \Delta \psi, \int \Delta \psi dt, \Delta \dot{\psi}], \tag{12}$$

where ΔD equals

$$\Delta D = \Delta D_0 + \int \Delta V dt, \tag{13}$$

where ΔD_0 is the flight distance deviation between the MAV and the UAV at the initial epoch. The integral items ΔV, $\Delta \gamma$ and $\Delta \psi$ are the cumulative deviation from the initial epoch till the current epoch. $\Delta \dot{V}$, $\Delta \dot{\gamma}$ and $\Delta \dot{\psi}$ is the deviation rate of the velocity, flight azimuth angle, and flight path angle.

3.1.3. Action Space A

The action space A is defined as follows:

$$A = \left[A_V, A_\gamma, A_\psi\right], \tag{14}$$

where the action A_V denotes the correction value of the UAV velocity commands ΔV_c, the action A_γ denotes the correction value of the UAV flight path angle commands $\Delta \gamma_c$, and the action A_ψ denotes the correction value of the UAV flight azimuth angle commands $\Delta \psi_c$, i.e.,

$$\begin{cases} \Delta V_c = A_V, \Delta \gamma_c = A_\gamma, \Delta \psi_c = A_\psi \\ |\Delta V_c| \le \lambda_{V_c\max c}, |\Delta \gamma_c| \le \lambda_{\gamma_c\max}, |\Delta \psi_c| \le \lambda_{\psi_c\max}, \end{cases} \tag{15}$$

where $\lambda_{V_c\max}$, $\lambda_{\gamma_c\max}$ and $\lambda_{\psi_c\max}$ are the maximum of corrections, respectively. A_V is used to generate the set of continuous velocity commands. A_γ, A_ψ is used to generate the set of discretized angle commands. Specifically, the discretization can be further expressed as follows:

$$
\begin{aligned}
|\Omega_{\gamma_c}| &= 2\lceil \lambda_{\gamma_c\max}/\partial\gamma_c\rceil + 1 \\
|\Omega_{\psi_c}| &= 2\lceil \lambda_{\psi_c\max}/\partial\psi_c\rceil + 1
\end{aligned}
\tag{16}
$$

$$
\begin{aligned}
\Omega_{\gamma_c} &= \{\, A_\gamma | 0, \pm\partial\gamma_c, \pm2\partial\gamma_c, \cdots, \pm(|\Omega_{\gamma_c}|-1)\partial\gamma_c/2, \pm\lambda_{\gamma_c\max}\} \\
\Omega_{\psi_c} &= \{\, A_\psi | 0, \pm\partial\psi_c, \pm2\partial\psi_c, \cdots, \pm(|\Omega_{\psi_c}|-1)\partial\psi_c/2, \pm\lambda_{\psi_c\max}\}.
\end{aligned}
\tag{17}
$$

Then, the update of desired state commands is

$$
\left\{
\begin{aligned}
V_c &\leftarrow V_c + \Delta V_c \\
\gamma_c &\leftarrow \gamma_c + \Delta\gamma_c \\
\psi_c &\leftarrow \psi_c + \Delta\psi_c.
\end{aligned}
\right.
\tag{18}
$$

3.1.4. Reward Function R_Σ

According to the formation control targets of UAVs, the reward function R_Σ was designed as follows:

$$
R_\Sigma = R_P + R_N + R_C
\tag{19}
$$

where R_P is the reward sub-function, which gives a positive response when the flight state of the UAV meets the control targets. R_N is the penalty sub-function, which gives a negative response when the flight states exceed the allowable error of the control target. R_C is the command limiting function, which can limit the values of the control commands η_c, n_c, σ_c. More specifically, R_C can smooth the variation of the control commands to finally reduce energy consumption.

R_P is calculated by

$$
R_P = 10 \times \left(\varepsilon_D^2 + \varepsilon_V^2 + \varepsilon_\gamma^2 + \varepsilon_\psi^2 \right)
\tag{20}
$$

where ε_D, ε_V, ε_γ, and ε_ψ are reward coefficients, which are defined as follows

$$
\begin{aligned}
\varepsilon_D &= \left\{
\begin{aligned}
&1, D_{\Delta\min} \leq \Delta D \leq D_{\Delta\max} \\
&0, \Delta D < D_{\Delta\min} \text{ or } \Delta D > D_{\Delta\max}
\end{aligned}
\right. , \\[4pt]
\varepsilon_V &= \left\{
\begin{aligned}
&1 - \Delta V / V_{\Delta\max}, \Delta V \leq V_{\Delta\max} \\
&0, \Delta V > V_{\Delta\max}
\end{aligned}
\right. , \\[4pt]
\varepsilon_\gamma &= \left\{
\begin{aligned}
&1 - \Delta\gamma / \gamma_{\Delta\max}, \Delta\gamma < \gamma_{\Delta\max} \\
&0, \Delta\gamma \geq \gamma_{\Delta\max}
\end{aligned}
\right. , \\[4pt]
\varepsilon_\psi &= \left\{
\begin{aligned}
&1 - \Delta\psi / \psi_{\Delta\max}, \Delta\psi < \psi_{\Delta\max} \\
&0, \Delta\psi \geq \psi_{\Delta\max}
\end{aligned}
\right. .
\end{aligned}
\tag{21}
$$

R_N is calculated by

$$
R_N = -100 \times \left(e_D^2 + e_V^2 + e_\gamma^2 + e_\psi^2 \right)
\tag{22}
$$

where e_D, e_V, e_γ, and e_ψ are penalty coefficients, which are defined as follows:

$$e_D = \begin{cases} 1, \Delta D < D_{\Delta\min} \text{ or } \Delta D > D_{\Delta\max} \\ 0, D_{\Delta\min} \le \Delta D \le D_{\Delta\max} \end{cases},$$

$$e_V = \begin{cases} 1, \Delta V > 2V_{\Delta\max} \\ \Delta V/V_{\Delta\max} - 1, V_{\Delta\max} \le \Delta V \le 2V_{\Delta\max} \\ 0, \Delta V < V_{\Delta\max} \end{cases},$$

$$e_\gamma = \begin{cases} 1, \Delta\gamma > 2\gamma_{\Delta\max} \\ \Delta\gamma/\gamma_{\Delta\max} - 1, \gamma_{\Delta\max} \le \Delta\gamma \le 2\gamma_{\Delta\max} \\ 0, \Delta\gamma < \gamma_{\Delta\max} \end{cases},$$

$$e_\psi = \begin{cases} 1, \Delta\psi > 2\psi_{\Delta\max} \\ \Delta\psi/\psi_{\Delta\max} - 1, \psi_{\Delta\max} \le \Delta\psi \le 2\psi_{\Delta\max} \\ 0, \Delta\psi < \psi_{\Delta\max} \end{cases}. \tag{23}$$

R_C is calculated by

$$R_C = -0.2(|\eta_c|/\eta_{\max} + |n_c|/n_{\max} + |\sigma_c|/\sigma_{\max}). \tag{24}$$

3.2. Dynamic Inversion Controller

To realize tracking of the commands of a given flight trajectory, the dynamic inversion control law was designed as follows [35]

$$\dot{V}_c = \omega_V(V_c - V)$$
$$\dot{\gamma}_c = \omega_\gamma(\gamma_c - \gamma) \tag{25}$$
$$\dot{\psi}_c = \omega_\psi(\psi_c - \psi)$$

where ω_V, ω_γ, and ω_ψ denote the bandwidth of the controller, respectively. V_c, γ_c, ψ_c denote the desired state commands of the flight velocity, the flight path angle, and the flight azimuth angle, respectively.

Since the UAV commands follow the dynamic constraints by Equation (1), considering Equations (1), (2), and (25) yields

$$T_c = \eta_c T_{\max} = [D + m\omega_V(V_c - V) + mg\sin\gamma],$$
$$N_\gamma = \omega_\gamma V(\gamma_c - \gamma)/g + \cos\gamma, \tag{26}$$
$$N_\psi = \omega_\psi V(\psi_c - \psi)\cos\gamma/g,$$

where N_γ and N_ψ denote the normal overload and lateral overload, respectively. The throttle δ_c, normal overload n_c and bank angle σ_c are selected as the control commands. Then, the UAV control command was designed as follows:

$$F = \begin{cases} \eta_c = [D + m\omega_V(V_c - V) + mg\sin\gamma]/T_{\max} \\ n_c = \sqrt{N_\gamma^2 + N_\psi^2} \\ \sigma_c = \arctan(N_\psi/N_\gamma) \end{cases}. \tag{27}$$

Moreover, the control command must satisfy the constraints:

$$\eta_{\min} \le \eta_c \le \eta_{\max}, 0 \le n_c \le n_{\max}, |\sigma_c| \le \sigma_{\max} \tag{28}$$

where $\eta_{\min}$ and $\eta_{\max}$ is the minimum and maximum values of the throttle commands, respectively. $n_{\max}$ is the maximum value of the normal overload, and $\sigma_{\max}$ is the maximum value of the bank angle.

3.3. First-Order Lag Filter

Considering the fact that the UAV cannot instantly complete the change of the engine thrust, normal overload, and bank angle, a first-order lag filter model was constructed to simulate the delayed variation processes of these three variables:

$$
G = \begin{cases}
\dot{\eta} = (\eta_c - \eta)/\tau_\delta \\
\dot{n} = (n_c - n)/\tau_n \ , \\
\dot{\sigma} = (\sigma_c - \sigma)/\tau_\sigma
\end{cases}
\tag{29}
$$

where η_c, n_c, σ_c represent the control commands of the throttle, normal overload, and bank angle, respectively. τ_δ, τ_n, and τ_σ represent the response time of the UAV control system accordingly.

Summarily, considering Equations (1), (27), and (29), the UAV flight process can be presented by the control equations as follow:

$$
\begin{cases}
F(V_c, \gamma_c, \psi_c)^{\mathrm{T}} = [\eta_c, n_c, \sigma_c]^{\mathrm{T}} \\
G(\eta_c, n_c, \sigma_c)^{\mathrm{T}} = [\dot{\eta}, \dot{n}, \dot{\sigma}]^{\mathrm{T}} \\
H(\eta, n, \sigma)^{\mathrm{T}} = [V, \gamma, \psi]^{\mathrm{T}}
\end{cases}
.
\tag{30}
$$

Equation (30) reveals the calculation process from the desired control commands to the actual control commands. It is clear that the premise to realize the formation flight is to acquire the desired control commands of the UAV V_c, γ_c, ψ_c under the specific formation strategy. Then, the ultimate flight trajectory can be obtained by the Runge–Kutta method.

4. Simulation Validation

4.1. Simulation Design

Based on the 3-DOF dynamic model in this paper, the MAV was designed to make a complex maneuver and provide the reference trajectory and control commands, accordingly. Under the leader–follower formation strategy, the UAV adopts the HIAC, DDPG, and LQR to track the MAV and keep the dual aircraft formation, respectively. LQR is a commonly used guidance method for tracking multi-state trajectories in aerospace engineering and it has been validated by extensive flight tests [36,37]. Therefore, we compared the proposed method with LQR and DDPG to verify its superiority in the following Sections 4.2 and 4.3. The design of DDPG is described in Section 3.1.

First, the experiment of nominal conditions was conducted to analyze the superiority of the proposed method in detail. Meanwhile, the initial values greatly affect the performance of the reinforcement learning models. Therefore, the generalization ability of the model was required to be fully verified. Then, 100 Monte Carlo experiments were conducted to verify the adaptability of this method to different initial conditions.

The simulations were conducted by Matlab2021a and the 3-DOF dynamic model was built by Simulink. The total simulation time was T, the simulation interval was ΔT, and the specific experimental parameters are shown in Table 1.

The training methods of DDPG and DDQN refer to [18,23], respectively. Learning rate, max episode, discount factor, and experience buffer length were set as the same for both DDPG and DDQN. In addition, the batch size of DDPG was set to 256, and the batch size of DDQN was set to 64. The specific parameters are shown in Table 2.

Table 1. The experimental parameter settings.

Parameters	Settings	Parameters	Settings
T (s)	50	η_{min}	0
ΔT (s)	0.1	η_{max}	1
T_{max} (lb)	25,600	n_{max}	6
m (kg)	14,470	σ_{max} (rad)	$\pi/2$
g (m/s^2)	9.81	$D_{\Delta max}$ (m)	600
S (ft^2)	400	$D_{\Delta min}$ (m)	100
C_{D_P}	0.02	$V_{\Delta max}$ (m/s)	50
C_{D_I}	0.1	$\psi_{\Delta max}$ (rad)	0.2
τ_δ (s)	0.6	$\gamma_{\Delta max}$ (rad)	0.2
τ_n (s)	0.5	$\lambda_{V_c max}$ (m/s)	50
τ_σ (s)	0.5	$\lambda_{\gamma_c max}$ (rad)	$\pi/2$
ω_V (s)	0.3	$\lambda_{\psi_c max}$ (rad)	$\pi/2$
ω_γ (s)	0.2	$\partial\gamma_c$ (rad)	$\pi/180$
ω_ψ (s)	0.2	$\partial\psi_c$ (rad)	$\pi/180$

Table 2. The training parameters of DDPG/DDQN.

Parameters	Settings
Learning Rate	0.0001
Max Episode	25,000
Batch Size (DDPG)	256
Batch Size (DDQN)	64
Discount Factor	0.99
Experience Buffer Length	1×10^6

4.2. Basic Principles of LQR

The implementation of LQR mainly includes three parts: linearization of the motion model, design of the tracking controller for the reference trajectory, and solution of the feedback gain matrix.

By linearizing the dynamic model of the UAV in Equation (1) with small deviations, the linear system can be obtained as follows:

$$\dot{X} = AX + Bu. \tag{31}$$

Equation (31) can be expressed by

$$\begin{bmatrix} \delta\dot{x} \\ r\delta\dot{y} \\ \delta\dot{z} \\ \delta\dot{V} \\ \delta\dot{\gamma} \\ \delta\dot{\psi} \end{bmatrix} = \begin{bmatrix} A_{11} & A_{12} & A_{13} & A_{14} & A_{15} & A_{16} \\ A_{21} & A_{22} & A_{23} & A_{24} & A_{25} & A_{26} \\ A_{31} & A_{32} & A_{33} & A_{34} & A_{35} & A_{36} \\ A_{41} & A_{42} & A_{43} & A_{44} & A_{45} & A_{46} \\ A_{51} & A_{52} & A_{53} & A_{54} & A_{55} & A_{56} \\ A_{61} & A_{62} & A_{63} & A_{64} & A_{65} & A_{66} \end{bmatrix} \begin{bmatrix} \delta x \\ \delta y \\ \delta z \\ \delta V \\ \delta\gamma \\ \delta\psi \end{bmatrix} + \begin{bmatrix} B_{11} & B_{12} & B_{13} \\ B_{21} & B_{22} & B_{23} \\ B_{31} & B_{32} & B_{33} \\ B_{41} & B_{42} & B_{43} \\ B_{51} & B_{52} & B_{53} \\ B_{61} & B_{62} & B_{63} \end{bmatrix} \begin{bmatrix} \delta\eta \\ \delta n \\ \delta\sigma \end{bmatrix}. \tag{32}$$

Set the given MAV trajectory as the reference, the state space is defined as follows:

$$\delta x = x_W - x_L, \delta y = y_W - y_L, \delta z = z_W - z_L,$$
$$\delta V = V_W - V_L, \delta\gamma = \gamma_W - \gamma_L, \delta\psi = \psi_W - \psi_L. \tag{33}$$

The control commands are defined as follows:

$$\delta\eta = \eta - \eta_L, \delta n = n - n_L, \delta\sigma = \sigma - \sigma_L, \tag{34}$$

where A and B are the partial derivative coefficient matrix calculated according to the motion differential equation and the feature points of the reference trajectory. The calculation results are as follows:

$$
\begin{aligned}
&A_{11} = A_{12} = A_{13} = 0,\\
&A_{14} = \cos\gamma\sin\psi,\, A_{15} = -V\sin\gamma\sin\psi,\, A_{16} = V\cos\gamma\cos\psi,\\
&A_{21} = A_{22} = A_{23} = 0,\\
&A_{24} = \cos\gamma\cos\psi,\, A_{25} = -V\sin\gamma\cos\psi,\, A_{26} = -V\cos\gamma\sin\psi,\\
&A_{31} = A_{32} = A_{33} = A_{36} = 0,\\
&A_{34} = \sin\gamma,\, A_{35} = V\cos\gamma,\\
&A_{41} = A_{42} = A_{46} = 0,\, A_{43} = D_z/m,\\
&A_{44} = D_V/m,\, A_{45} = -g\cos\gamma,\\
&A_{51} = A_{52} = A_{53} = A_{56} = 0,\\
&A_{54} = -g(n\cos\sigma - \cos\gamma)/V^2,\, A_{55} = g\sin\gamma/V,\\
&A_{61} = A_{62} = A_{63} = A_{66} = 0,\\
&A_{64} = g\sin\sigma n/\left(V^2\cos\gamma\right),\, A_{65} = -gn\sin\sigma\sin\gamma/\left(V\cos^2\gamma\right),\\
&B_{11} = B_{12} = B_{13} = B_{21} = B_{22} = B_{23} = B_{31} = B_{32} = B_{33} = 0,\\
&B_{41} = T_{\max}/m,\, B_{42} = B_{43} = 0,\\
&B_{51} = -D_n/m,\, B_{52} = g\cos\sigma/V,\, B_{53} = -gn\sin\sigma,\\
&B_{61} = 0,\, B_{62} = -g\sin\sigma/(V\cos\gamma),\, B_{63} = -gn\cos\sigma/(V\cos\gamma).
\end{aligned}
\tag{35}
$$

where D_z, D_V and D_n are the partial derivatives of the drag D on the feature point of the reference trajectory to the flight height z, velocity V and normal overload n respectively. Define the optimal control performance index from t_0 to t_f as follows:

$$
J = 0.5\int_{t_0}^{t_f}\left[X^{\mathrm{T}}(t)QX(t) + u^{\mathrm{T}}(t)Ru(t)\right]\,\mathrm{d}t,
\tag{36}
$$

where Q and R are the weight matrices of state and control respectively. Q is positive semi-definite and R is positive-definite. Then, there exists an optimal control law $u^* = -K^*X$ to minimize the above performance index, and the feedback gain matrix K^* is

$$
K^* = \begin{bmatrix} K_{\eta1} & K_{\eta2} & K_{\eta3} & K_{\eta4} & K_{\eta5} & K_{\eta6}\\ K_{n1} & K_{n2} & K_{n3} & K_{n4} & K_{n5} & K_{n6}\\ K_{\sigma1} & K_{\sigma2} & K_{\sigma3} & K_{\sigma4} & K_{\sigma5} & K_{\sigma6} \end{bmatrix},
\tag{37}
$$

$$
K^* = -R^{-1}B^{\mathrm{T}}P
\tag{38}
$$

where P is the solution of the Riccati equation. It is calculated by

$$
-PA - A^{\mathrm{T}}P + PBR^{-1}B^{\mathrm{T}}P - Q = 0.
\tag{39}
$$

Define Q and R as follows:

$$
\begin{aligned}
Q &= \mathrm{diag}[Q_1, Q_2, Q_3, Q_4, Q_5, Q_6],\\
R &= \mathrm{diag}[R_1, R_2, R_3].
\end{aligned}
\tag{40}
$$

To reflect the impact of the flight relative distance in the dual aircraft formation flight. Set $Q_1 = Q_2 = Q_3$ and define $\delta D^2 = \Delta D^2 = \delta x^2 + \delta y^2 + \delta z^2$, then

$$
\begin{aligned}
J = \ & 0.5 \int_{t_0}^{t_f} \big[(Q_1 \delta D^2 + Q_4 \delta V^2 + Q_5 \delta \gamma^2 + Q_6 \delta \psi^2) \\
& + (R_1 \delta \eta^2 + R_2 \delta n^2 + R_3 \delta \sigma^2) \big] \mathrm{d}t.
\end{aligned}
\tag{41}
$$

According to Bryson Law [38], Q and R are set as follows:

$$
\begin{aligned}
Q_1 D^2{}_{\Delta\max} &= Q_4 V^2{}_{\Delta\max} = Q_5 \gamma^2{}_{\Delta\max} = Q_6 \psi^2{}_{\Delta\max} \\
&= R_1 \eta^2{}_{\max} = R_2 n^2{}_{\max} = R_3 \sigma^2{}_{\max}.
\end{aligned}
\tag{42}
$$

Set $Q_1 = 1$, then other parameters can be obtained. According to u^*, the control commands can be obtained:

$$
\begin{aligned}
\eta &= \eta_L - \left(K_{\eta 1} \delta x + K_{\eta 2} \delta y + K_{\eta 3} \delta z + K_{\eta 4} \delta V + K_{\eta 5} \delta \gamma + K_{\eta 6} \delta \psi \right), \\
n &= n_L - \left(K_{n1} \delta x + K_{n2} \delta y + K_{n3} \delta z + K_{n4} \delta V + K_{n5} \delta \gamma + K_{n6} \delta \psi \right), \\
\sigma &= \sigma_L - \left(K_{\sigma 1} \delta x + K_{\sigma 2} \delta y + K_{\sigma 3} \delta z + K_{\sigma 4} \delta V + K_{\sigma 5} \delta \gamma + K_{\sigma 6} \delta \psi \right).
\end{aligned}
\tag{43}
$$

Since the feedback gains obtained at different feature points of the reference trajectory are different, the monotonic flights can be selected as an independent variable, and the feedback gain coefficient of the offline design can be interpolated to obtain the corresponding control commands.

4.3. Experiment of Nominal Conditions

In the experiment of nominal conditions, the initial position of the MAV was $x_{L0} = 0$ m, $y_{L0} = 0$ m, $z_{L0} = 10,000$ m, $V_{L0} = 400$ m/s, $\gamma_{L0} = \pi/6$, $\psi_{L0} = 0$. The initial position of the UAV was $x_{W0} = 100$ m, $y_{W0} = 100$ m, $z_{W0} = 10,000$ m, $V_{W0} = 400$ m/s, $\gamma_{W0} = \pi/6$, $\psi_{W0} = 0$.

The formation flight trajectories of MAV and UAV of the three methods are shown in Figure 5. The MAV is designed to make continuous S-shaped large maneuver with a maximum overload of about 4 g at 1 s, 11 s, 29 s and 41 s, respectively. Figure 5 indicates that the LQR, DDPG, and HIAC can realize the stable tracking of the given trajectory of the MAV under large, overloaded maneuvers and reach the target of the designed formation.

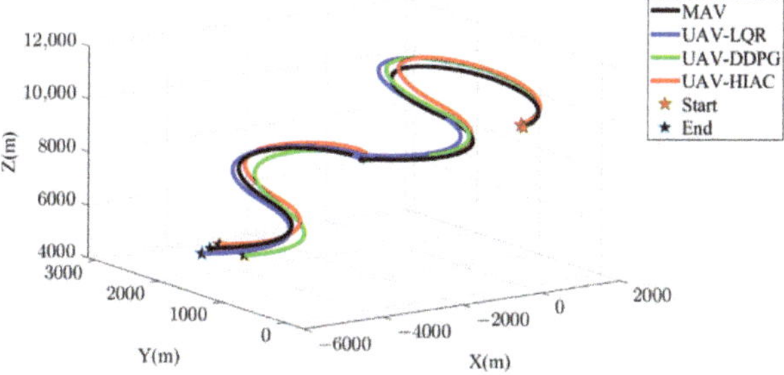

Figure 5. The formation flight trajectory of MAV and UAV of three methods.

Figure 6a–c shows the control commands of the UAV, i.e., the thrust, the normal overload, and the bank angle, generated by the LQR, DDPG and HIAC, respectively, with reference commands of the MAV. Figure 7a–c shows the errors between the control commands of the LQR, DDPG, and HIAC and the reference commands of the MAV. Figure 6 illustrates that there are four peaks in the curves of the control commands due to the four large, overloaded maneuvers. Moreover, compared with the LQR and DDPG, the trend of the control commands of the HIAC can be better consistent with the MAV in thrust, normal overload, and bank angle. Especially in the control of the normal overload, the HIAC has mitigated the sharp change of the commands generated by the reinforcement learning controller to a certain extent. It can provide more smooth and executable control commands under large maneuvers. However, during the large maneuver of the MAV, in order to track the reference commands, it inevitably generates a certain amount of extra adjustment for the thrust, overload, and bank angle for the three methods.

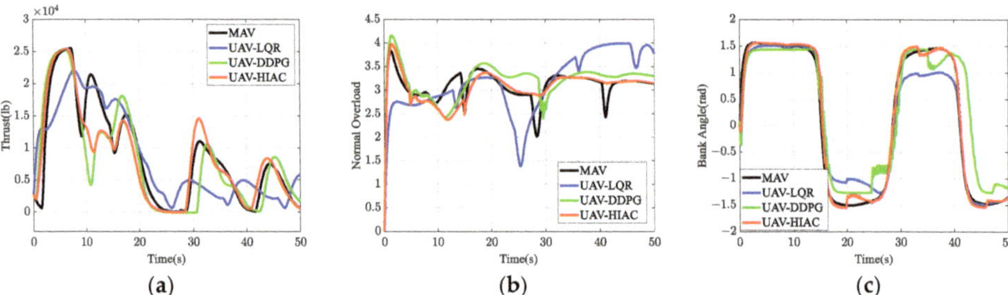

Figure 6. The control commands of the MAV generated by the LQR, DDPG, and HIAC, respectively, with reference commands of the MAV. The results of the thrust, the thrust, the normal overload, and the bank angle are presented in (**a**–**c**), respectively.

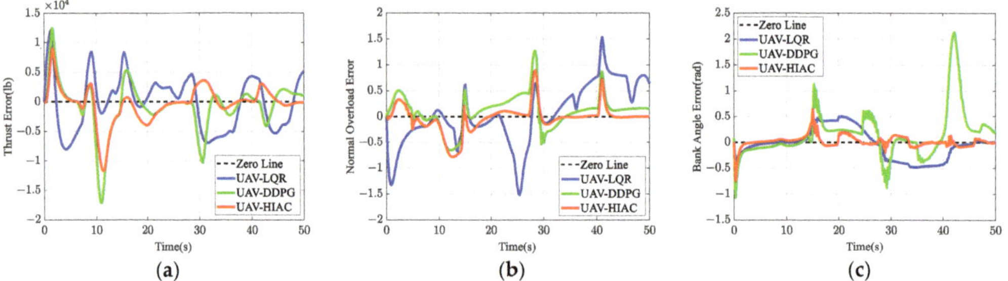

Figure 7. The errors between the control commands of the LQR, DDPG, HIAC and the reference commands of the MAV. The errors of the thrust, the thrust, the normal overload, and the bank angle are presented in (**a**–**c**), respectively.

Figure 8 shows the change of the three controlled states of the UAV, i.e., the velocity, the flight path angle, and the flight azimuth angle, generated by LQR, DDPG and HIAC. Figure 9 shows the deviation of the three controlled states and the relative distance. It can be seen from Figure 8 that the change trend of the controlled state of the HIAC is basically the same as that of the MAV, and the formation maintenance performance is obviously better than that of the LQR and DDPG. Especially, the HIAC can keep up with most of the fluctuations of the MAV in the flight velocity and the flight azimuth angle. Moreover, Figure 9 shows that compared with the LQR and DDPG, the control precision of the HIAC has been significantly improved, and the control deviation can rapidly decrease to nearly 0 under the large maneuver. Figure 9d indicates that the HIAC successfully limits the formation distance within the safe distance between 100 m and 600 m while LQR and DDPG fail. The LQR continuously accumulates distance deviation due to the velocity deviation during the flight, and ultimately, the formation distance reveals a divergent trend. Meanwhile, although the relative distance of the DDPG gradually converges, it still extends beyond the safe distance at the end of the flight.

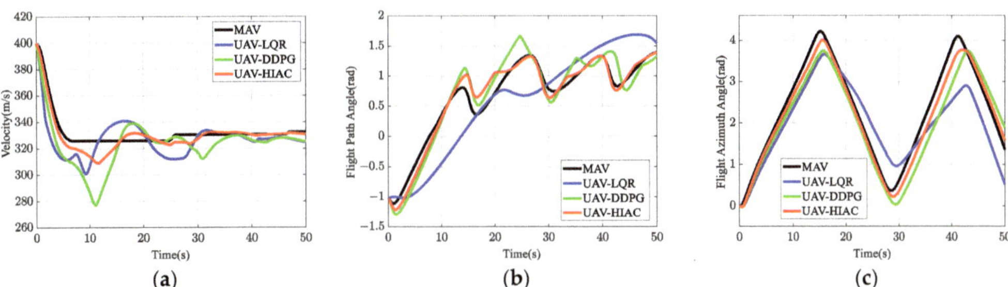

Figure 8. The change of the three controlled state of the UAV generated by the LQR, DDPG and HIAC. The results of the velocity, the flight path angle, and the flight azimuth angle are presented in (**a**–**c**), respectively.

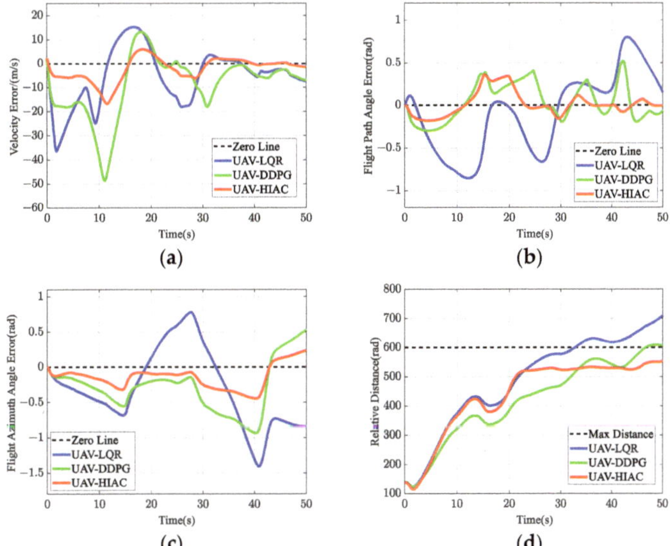

Figure 9. The deviation of the velocity, the flight path angle, the flight azimuth angle and the relative distance are presented in (**a**–**d**), respectively.

Table 3 presents the root mean square (RMS) errors and maximum errors of the four controlled states of the LQR, DDPG and HIAC. It is clear that both the RMS error and maximum error of the HIAC are smaller than those of the LQR and DDPG. Moreover, the HIAC has a reduction of 5.81%, 70.44%, and 64.95%, respectively, in the RMS error of the velocity, flight path angle and flight azimuth angle compared with the LQR, and has a reduction of 60.35%, 55.32% and 69.47% in the maximum error of velocity, flight path angle and flight azimuth angle, respectively, compared with the LQR. The HIAC has a reduction of 36.10%, 35.85% and 51.61%, respectively, in the RMS error of velocity, flight path angle and flight azimuth angle compared with the DDPG, and has a reduction of 54.43%, 31.57% and 55.01% in the maximum error of velocity, flight path angle and flight azimuth angle, respectively, compared with the DDPG.

Table 3. RMS and maximum errors of the four states of the LQR, DDPG, and HIAC in nominal conditions.

Controller		Velocity (m/s)	Flight Path Angle (rad)	Flight Azimuth Angle (rad)	Relative Distance (m) Safe Distance [100, 600]
LQR	RMS	5.6957	0.4737	0.6202	516.7072
	Max.	15.3307	0.8027	0.7833	710.2799
DDPG	RMS	8.3953	0.2183	0.4493	444.1190
	Max.	13.3379	0.5241	0.5315	610.6078
HIAC	RMS	5.3647	0.1401	0.2174	460.0709
	Max.	6.0780	0.3586	0.2391	552.1845

In summary, the proposed HIAC significantly improves the state control performance and guarantees that the flight distance stays within a safe distance as well.

4.4. Monte Carlo Experiments

In order to further test how the HIAC adapts to various initial conditions, 100 Monte Carlo simulations were carried out by adding random deviations to the nominal conditions. The initial position of the MAV is $x_{L0} = 0\,\text{m}$, $y_{L0} = 0\,\text{m}$, $z_{L0} = 10,000\,\text{m}$, $V_{L0} = 400\,\text{m/s}$, $\gamma_{L0} = \pi/6$, $\psi_{L0} = 0$. The baseline of initial values of the UAV is $x_{W0} = 100\,\text{m}$, $y_{W0} = 100\,\text{m}$, $z_{W0} = 10,000\,\text{m}$, $V_{W0} = 400\,\text{m/s}$, $\gamma_{W0} = \pi/6$, $\psi_{W0} = 0$. Then, random deviations which follow the uniform distributions were added to these six baselines, respectively. The specific values of the deviations are presented in Table 4.

Table 4. Uniform distribution of deviations for the six initial values.

Numbers of Monte Carlo Simulations	X (m)	Y (m)	Z (m)	Velocity (m/s)	Flight Path Angle (rad)	Flight Azimuth Angle (rad)
100	[−50, 550]	[−50, 550]	[−1000, 1000]	[−100, 100]	[−π/18, π/18]	[−π/18, π/18]

Figure 10 is the scatterplot of the Monte Carlo simulation results of the velocity errors, flight path angle error, flight azimuth angle, and relative distance for the LQR, DDPG and HIAC. For each evaluation index, the horizontal axis is the RMS error, and the vertical axis is the maximum error. It can be seen that the HIAC can fulfill the control target in the magnitude of velocity. Meanwhile, because the training threshold is set quite strictly in order to achieve better control performance, the maximum error and RMS error of the flight path angle and flight azimuth angle may extend out of the threshold when the extreme deviations are added to the initial values. However, the HIAC can still present a satisfactory control accuracy of the angle compared with the DDPG and LQR. Moreover, in terms of the safety distance, the HIAC can stay within a safe distance of 100 m to 600 m from the MAV, which reaches the distance control target. However, the DDPG and LQR gradually extend out of the safe distance as the initial values vary. Statistically, compared with LQR and DDPG, the HIAC has smaller values in both the RMS error and maximum error of these four evaluation indices. In summary, the performance of the HIAC in formation control is better than that of the other two methods, which is consistent with the simulation results under nominal conditions. It is believed that the HIAC has significant adaptability to the varying initial conditions.

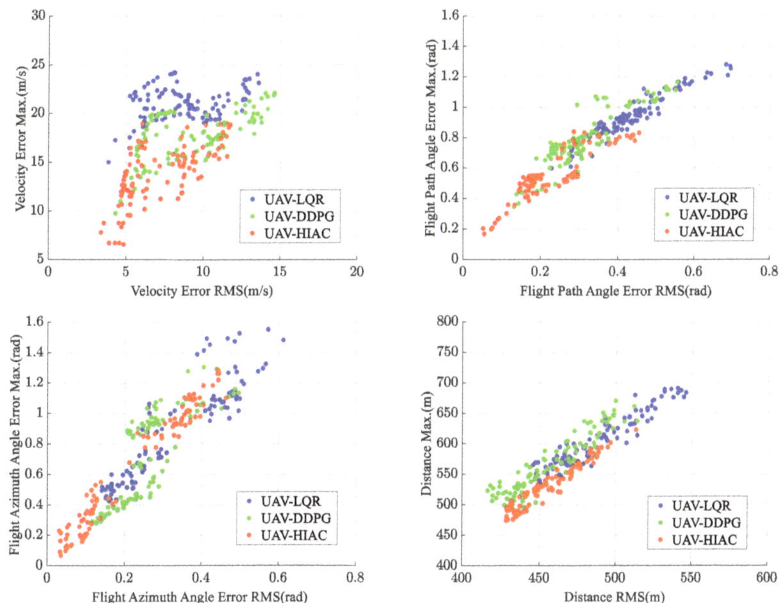

Figure 10. Monte Carlo simulation results of LQR, DDPG and HIAC.

5. Conclusions

In this study, a novel HIAC method was proposed, which is able to enhance the smoothness and executability of control commands and improve the control performance of the MAV/UAV flight formation. First, based on the idea of "meta-action" in hybrid reinforcement learning, the formation control was modeled as a continuous–discrete space control problem. Then, we proposed the framework of the HIAC, and the hybrid intelligent agent model based on the DDPG/DDQN was designed through multi-channel decoupling. Finally, we carried out simulations of nominal conditions and 100 Monte Carlo simulations in varying initial conditions. The simulation results showed that, compared with the traditional LQR and DDPG, the HIAC has better performance of high control precision and rapid convergence. Meanwhile, the adaptability of HIAC to the varying initial conditions was verified as well.

For further practical applications, HIAC can gradually support practical scenarios such as formation military operations and terrain surveys. In particular, two aspects should be considered when applying HIAC. The first is the reliability of the method. HIAC should be preliminarily trained with a large number of ground tests before the real flights, to ensure that intelligent control gradually takes authority over traditional flight-control methods. The second is the portability of the method. At present, the method supports the deployment of reinforcement learning on hardware such as DSP, and FPGA, and can realize airborne portability and the online training of agent models.

Author Contributions: Conceptualization, methodology, software, validation, formal analysis, resources, data curation, writing—original draft preparation, L.Z. (Luodi Zhao); writing—review and editing, Y.L. and Q.P.; visualization, investigation, Y.L.; supervision, project administration, funding acquisition, L.Z (Long Zhao). All authors have read and agreed to the published version of the manuscript.

Funding: This research was funded by the National Science Foundation of China, grant number 42274037 and 41874034, the National key research and development program of China, grant number 2020YFB0505804 and the Beijing Natural Science Foundation, grant number 4202041.

Data Availability Statement: Not applicable.

Conflicts of Interest: The authors declare no conflict of interest.

References

1. Lei, L.; Wang, T.; Jiang, Q. Key Technology Develop Trends of Unmanned Systems Viewed from Unmanned Systems Integrated Roadmap 2017–2042. *Unmanned Syst. Technol.* **2018**, *1*, 79–84.
2. Mishory, J. DARPA Solicits Information for New Lifelong Machine Learning Program. *Inside Pentagon* **2017**, *33*, 10.
3. Pittaway, N. Loyal Wingman. *Air Int.* **2019**, *96*, 12–13.
4. Oh, K.; Park, M.; Ahn, H. A survey of multi-agent formation control. *Automatica* **2015**, *53*, 424–440. [CrossRef]
5. Wang, H.; Liu, S.; Lv, M.; Zhang, B. Two-Level Hierarchical-Interaction-Based Group Formation Control for MAV/UAVs. *Aerospace* **2022**, *9*, 510. [CrossRef]
6. Choi, I.S.; Choi, J.S. Leader-Follower formation control using PID controller. In Proceedings of the International Conference on Intelligent Robotics & Applications, Montreal, QC, Canada, 3–5 October 2012.
7. Gong, Z.; Zhou, Z.; Wang, Z.; Lv, Q.; Xu, Q.; Jiang, Y. Coordinated Formation Guidance Law for Fixed-Wing UAVs Based on Missile Parallel Approach Metho. *Aerospace* **2022**, *9*, 272. [CrossRef]
8. Liang, Z.; Ren, Z.; Shao, X. Decoupling trajectory tracking for gliding reentry vehicles. *IEEE/CAA J. Autom. Sin.* **2015**, *2*, 115–120.
9. Kuriki, Y.; Namerikawa, T. Formation Control of UAVs with a Fourth-Order Flight Dynamics. *J. Control. Meas. Syst. Integr.* **2014**, *7*, 74–81. [CrossRef]
10. Kuriki, Y.; Namerikawa, T. Consensus-based cooperative formation control with collision avoidance for a multi-UAV system. In Proceedings of the American Control Conference, Portland, OR, USA, 4–6 June 2014.
11. Atn, G.M.; Stipanovi, D.M.; Voulgaris, P.G. Collision-free trajectory tracking while preserving connectivity in unicycle multi-agent systems. In Proceedings of the American Control Conference, Washington, DC, USA, 17–19 June 2013.
12. Tsankova, D.D.; Isapov, N. Potential field-based formation control in trajectory tracking and obstacle avoidance tasks. In Proceedings of the Intelligent Systems, Sofia, Bulgaria, 6–8 September 2012.
13. Hu, J.; Wang, L.; Hu, T. Autonomous Maneuver Decision Making of Dual-UAV Cooperative Air Combat Based on Deep Reinforcement Learning. *Electronics* **2022**, *11*, 467. [CrossRef]
14. Luo, Y.; Meng, G. Research on UAV Maneuver Decision-making Method Based on Markov Network. *J. Syst. Simul.* **2017**, *29*, 106–112.
15. Yang, Q.; Zhang, J.; Shi, G. Maneuver Decision of UAV in Short-Range Air Combat Based on Deep Reinforcement Learning. *IEEE Access* **2020**, *8*, 363–378. [CrossRef]
16. Li, Y.; Han, W.; Wang, Y. Deep Reinforcement Learning with Application to Air Confrontation Intelligent Decision-Making of Manned/Unmanned Aerial Vehicle Cooperative System. *IEEE Access* **2020**, *99*, 67887–67898. [CrossRef]
17. Wang, X.; Gu, Y.; Cheng, Y. Approximate Policy-Based Accelerated Deep Reinforcement Learning. *IEEE Trans. Neural Netw. Learn. Syst.* **2020**, *31*, 1820–1830. [CrossRef] [PubMed]
18. Hasselt, H.V.; Guez, A.; Silver, D. Deep Reinforcement Learning with Double Q-Learning. In Proceedings of the AAAI Conference on Artificial Intelligence, Canberra, Australia, 30 November–5 December2015.
19. Mnih, V.; Kavukcuoglu, K.; Silver, D. Playing Atari with deep reinforcement learning. *arXiv* **2013**, arXiv:1312.5602.
20. Silver, D.; Huang, A.; Maddison, C.J. Mastering the game of go with deep neural networks and the tree search. *Nature* **2016**, *529*, 484. [CrossRef]
21. Silver, D.; Lever, G.; Heess, N. Deterministic policy gradient algorithms. In Proceedings of the 31st International Conference on Machine Learning, Beijing, China, 21–26 June 2014.
22. Ioffe, S.; Szegedy, C. Batch normalization: Accelerating deep network training by reducing internal covariate shift. *arXiv* **2015**, arXiv:1502.03167.
23. Lillicrap, T.P.; Hunt, J.J.; Pritzel, A. Continuous control with deep reinforcement learning. *arXiv* **2015**, arXiv:1509.02971.
24. Wada, D.; Araujo-Estrada, S.A.; Windsor, S. Unmanned Aerial Vehicle Pitch Control Using Deep Reinforcement Learning with Discrete Actions in Wind Tunnel Test. *Aerospace* **2021**, *8*, 18. [CrossRef]
25. Haarnoja, T.; Zhou, A.; Abbeel, P. Soft Actor-Critic Algorithms and Applications. *arXiv* **2018**, arXiv:1812.05905.
26. Heess, N.; Silver, D.; Teh, Y.W. Actor-critic reinforcement learning with energy-based policies. In Proceedings of the Tenth European Workshop on Reinforcement Learning, Edinburgh, UK, 30 June–1 July 2012.
27. Schaul, T.; Quan, J.; Antonoglou, I. Prioritized experience replay. *arXiv* **2015**, arXiv:1511.05952.
28. Hu, Z.; Wan, K.; Gao, X.; Zhai, Y.; Wang, Q. Deep Reinforcement Learning Approach with Multiple Experience Pools for UAV's Autonomous Motion Planning in Complex Unknown Environments. *Sensors* **2020**, *20*, 1890. [CrossRef] [PubMed]
29. Neunert, M.; Abdolmaleki, A.; Wulfmeier, M. Continuous-Discrete Reinforcement Learning for Hybrid Control in Robotics. In Proceedings of the Conference on Robot Learning, Virtual Event, 30 October–1 November 2020.
30. Xiong, J.; Wang, Q.; Yang, Z. Parametrized Deep Q-Networks Learning: Reinforcement Learning with Discrete-Continuous Hybrid Action Space. *arXiv* **2018**, arXiv:1810.06394.
31. Anderson, M.R.; Robbins, A.C. Formation flight as a cooperative game. In Proceedings of the AIAA Guidance, Navigation, and Control Conference and Exhibit, Boston, MA, USA, 10–12 August 1998.

32. Kelley, H.J. Reduced-order modeling in aircraft mission analysis. *AIAA J.* **2015**, *9*, 349–350. [CrossRef]
33. Williams, P. Real-time computation of optimal three-dimensional aircraft trajectories including terrain-following. In Proceedings of the AIAA Guidance, Navigation, and Control Conference and Exhibit, Keystone, CO, USA, 24–26 August 2006.
34. Wang, X.; Guo, J.; Tang, S. Entry trajectory planning with terminal full states constraints and multiple geographic constraints. *Aerosp. Sci. Technol.* **2019**, *84*, 620–631. [CrossRef]
35. Snell, S.A.; Enns, D.F.; Garrard, W.L. Nonlinear inversion flight control for a supermaneuverable aircraft. In Proceedings of the AIAA Guidance, Navigation, and Control Conference and Exhibit, Portland, OR, USA, 20–22 August 1990.
36. Dukeman, G. Profile-Following Entry Guidance Using Linear Quadratic Regulator Theory. In Proceedings of the AIAA Guidance, Navigation, and Control Conference and Exhibit, Monterey, CA, USA, 5–8 August 2002.
37. Wen, Z.; Shu, T.; Hong, C. A simple reentry trajectory generation and tracking scheme for common aero vehicle. In Proceedings of the AIAA Guidance, Navigation, and Control Conference, Minneapolis, MN, USA, 13–16 August 2012.
38. Bryson, A.E.; Ho, Y. Applied Optimal Control. *Technometrics* **1979**, *21*, 3.

Review

A Review of Swarm Robotics in a NutShell

Muhammad Muzamal Shahzad [1,2], Zubair Saeed [3], Asima Akhtar [1], Hammad Munawar [2], Muhammad Haroon Yousaf [1,3,*], Naveed Khan Baloach [1,3] and Fawad Hussain [3]

[1] Swarm Robotics Lab (SRL), National Center of Robotics and Automation (NCRA), Taxila 47050, Pakistan; mmuzamil.shahzad@uettaxila.edu.pk (M.M.S.); asima.akhtar@uettaxila.edu.pk (A.A.); naveed.khan@uettaxila.edu.pk (N.K.B.)

[2] Department of Avionics Engineering, College of Aeronautical Engineering, National University of Science and Technology (NUST), Islamabad 44000, Pakistan; h.munawar@cae.nust.edu.pk

[3] Department of Computer Engineering, University of Engineering and Technology (UET), Taxila 47050, Pakistan; zubair.saeed@students.uettaxila.edu.pk (Z.S.); fawad.hussain@uettaxila.edu.pk (F.H.)

* Correspondence: haroon.yousaf@uettaxila.edu.pk

Abstract: A swarm of robots is the coordination of multiple robots that can perform a collective task and solve a problem more efficiently than a single robot. Over the last decade, this area of research has received significant interest from scientists due to its large field of applications in military or civil, including area exploration, target search and rescue, security and surveillance, agriculture, air defense, area coverage and real-time monitoring, providing wireless services, and delivery of goods. This research domain of collective behaviour draws inspiration from self-organizing systems in nature, such as honey bees, fish schools, social insects, bird flocks, and other social animals. By replicating the same set of interaction rules observed in these natural swarm systems, robot swarms can be created. The deployment of robot swarm or group of intelligent robots in a real-world scenario that can collectively perform a task or solve a problem is still a substantial research challenge. Swarm robots are differentiated from multi-agent robots by specific qualifying criteria, including the presence of at least three agents and the sharing of relative information such as altitude, position, and velocity among all agents. Each agent should be intelligent and follow the same set of interaction rules over the whole network. Also, the system's stability should not be affected by leaving or disconnecting an agent from a swarm. This survey illustrates swarm systems' basics and draws some projections from its history to its future. It discusses the important features of swarm robots, simulators, real-world applications, and future ideas.

Keywords: swarm intelligence; swarm behaviors; swarm robotics; industrial swarm; swarm robotics applications

Citation: Shahzad, M.M.; Saeed, Z.; Akhtar, A.; Munawar, H.; Yousaf, M.H.; Baloach, N.K.; Hussain, F. A Review of Swarm Robotics in a NutShell. *Drones* **2023**, *7*, 269. https://doi.org/10.3390/drones7040269

Academic Editors: Xiwang Dong, Mou Chen, Xiangke Wang and Fei Gao

Received: 10 March 2023
Revised: 9 April 2023
Accepted: 10 April 2023
Published: 14 April 2023

1. Introduction

A swarm of robots refers to the coordination of multiple individual entities, which traditionally operate without centralized control and instead rely on simple local behaviors to cooperate. Robot technology, particularly Unmanned Aerial Systems (UAS), is becoming more affordable, efficient, and is boosting the transmission capacity of robots as solutions to problems ranging from disaster relief to research mapping. Independent robots can perform tasks that need simple, ready to go solutions and a consistent real time approach. The autonomous robot can be a part of a robot swarm, if it fulfills at least three significant characteristics. These characteristics include the following: the minimum number of individual entities must be three or more, minimal or no human control, and cooperation between these robots based on a simple set of rules as depicted in Figure 1. Swarm robotics include a group of independent robots working collaboratively to complete a shared task without relying on any external infrastructure or a centralized control system/robot. Figure 2 illustrates how the fundamental concept of the swarm may be comprehended. In Figure 2a, the system is a robot swarm which consists of three autonomous agents that cooperate in response to the orders received from a single ground control station. Figure 2b indicates that the system is a sensor network rather than a swarm of robots. Each sensor is neither a robot nor an intelligent agent, and is solely responsible for providing data through readings without the capability of taking any actions. Figure 2c does not depict a robot swarm since a swarm necessitates more than two agents. Despite the robots working together towards a shared objective, each one has its own designated tasks to accomplish, which are directed by a separate operator. Figure 2d depicts a software system comprising multiple agents, which cannot be classified as a robot swarm as the agents are not autonomous robots, despite their collaboration on a shared hardware platform.

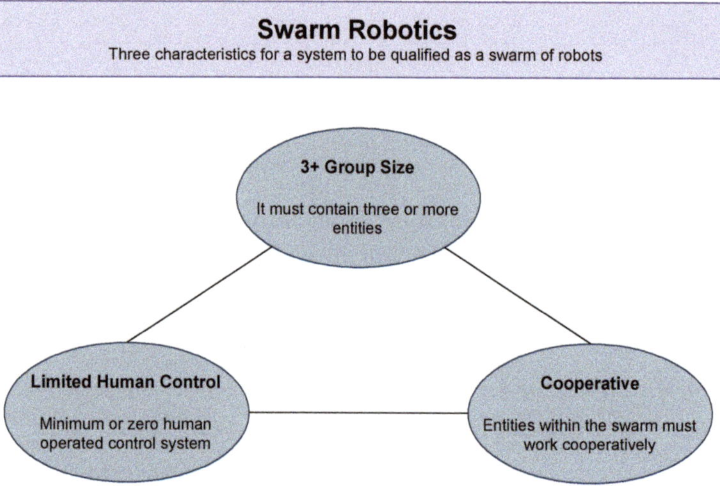

Figure 1. Basic Characteristics of Swarm Robotics.

Swarm robotics involves a group of robots that collaborate to address problems through the development of advantageous structures and behaviours that resemble those observed in nature, such as birds, fish, and bees. These robots, which can be either homogeneous or heterogeneous, form an intelligent network of a swarm, enabling individual robots to interact autonomously with each other and their environment by leveraging onboard communication, processing, and sensing capabilities. Such behaviours can be classified into four categories, namely *navigation , spatial organization, intelligent and precise decision-making*, and *miscellaneous* [1]. This study offers an in-depth analysis and mathematical comprehension of swarm intelligence algorithms. It also provides a comprehensive review of the evolution of swarm robotics from its inception to the present day and highlights the future ambitions of this field. Our aim is to present a broad overview of swarm robotics by exploring its history, current research, and future directions. The main contributions are as follows:

- To understand the fundamental difference between multi-agent and swarm of robots, along with the natural behaviours of a swarm.
- Multiple swarm intelligence algorithms derived from the natural set of rules and constraints for their transformation on multi-agent robots.
- Industrial and academic utilization of swarm robotics keeping in view the history and future perspectives.
- The objective is to address the research gap that exists between theoretical and industrial research in the field of swarm robotics. Theoretical research mainly involves simulating swarm behaviours using algorithms, while research in industrial settings are primarily focused on designing and developing hardware capable of executing

swarm behaviour. Therefore, it is imperative to deploy swarm algorithms using specific hardware that can accommodate swarm behaviour functionality.

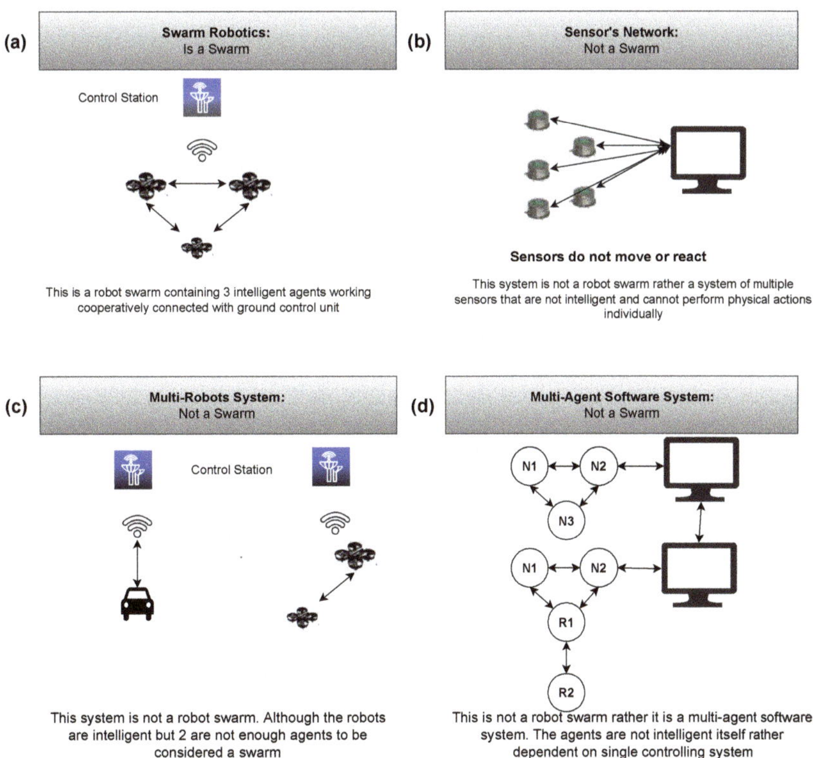

Figure 2. Comparison of Multi-agent Systems and Robot Swarm [2]. (**a**) depicts swarm robotics system, while (**b**–**d**) show non-swarm systems.

Figure 3 depicts the deployment of swarm behaviours in simulation and hardware, which is thoroughly explored in Section 2 of this article. The behaviours are simulated using existing and state-of-the-art swarm intelligence algorithms, as explained in Section 3 with mathematical reasoning. The simulation results demonstrate high accuracy in replicating natural animal behaviours. For the past two decades, the main research challenge in swarm robotics has been to develop multi-robot systems that are robust, flexible, fault-tolerant, and capable of incorporating self-organizing behaviours dynamically and by design. The swarm robotics field has evolved from algorithmic studies to mature academic, laboratory, and industrial-based solutions since the early 2000s. A comprehensive review of swarm robotics and its applications is presented in Table 1 and Section 4, respectively. Despite significant progress, cooperation and coordination in deploying the developed swarming algorithms among swarm robots remain limited [3]. Section 5 provides a brief overview of the era of swarm robotics, and the article concludes in the final section.

Table 1. Era of Swarm Robotics: Past, Present, and Future Perspectives.

1990–2000	The first robot tests show self-organization through indirect and local interactions, clearly inspired by swarm intelligence.	SW
2000–2005	The ability to generate swarms of robots that work together has now been expanded to a variety of additional tasks, including object handling, task allocation, and occupations that require significant teamwork to achieve.	SW
2002–2006	Swarm-bots is a project that shows how robot swarms self-assemble. Robots can construct pulling chains and massive constructions capable of transporting large loads and dealing with tough terrain.	HW and SW
2004–2008	The evolving swarm robotics technique was devised after the first demonstrations of autonomous assembly of robot swarms using evolutionary algorithms.	SW
2005–2009	For swarm robotics research, the first attempts at building standard swarm robotics platforms and small robots.	HW
2006–2010	Swarmanoid showed heterogeneous robot swarms made up of three different types of robots: flying, climbing, and ground-based robots for the first time.	HW and SW
2010–2015	Advanced autonomous design methods such as AutoMoDe, novelty search, design patterns, mean-field models, and optimal stochastic approaches are all employed in the creation of robot swarms.	SW
2016–2020	Decentralized solutions have been investigated and deployed as swarms of flying drones become available for investigation.	HW and SW
2020–2025	The first example of robot swarms that may self-learn suitable swarm behaviour in response to a specific set of challenges.	SW
2025–2030	Marine and deep-sea robotic swarms will be utilized for ecological monitoring, surveillance, and fishing,among other things.	HW
2030–2040	Small rover swarms will be utilized for the first mission to the Moon and Mars to expand the exploration area and showcase on-site construction capabilities.	HW
2040–2045	Soft-bodied robot swarms measuring in millimeters will be deployed to explore agricultural fields and aquatic areas to identify plastic usage and assist with pest control.	HW and SW
2035–2050	Clinical research with human volunteers will begin after nanoscale robot swarms have been shown for therapeutic objectives such as customized medication delivery.	HW and SW

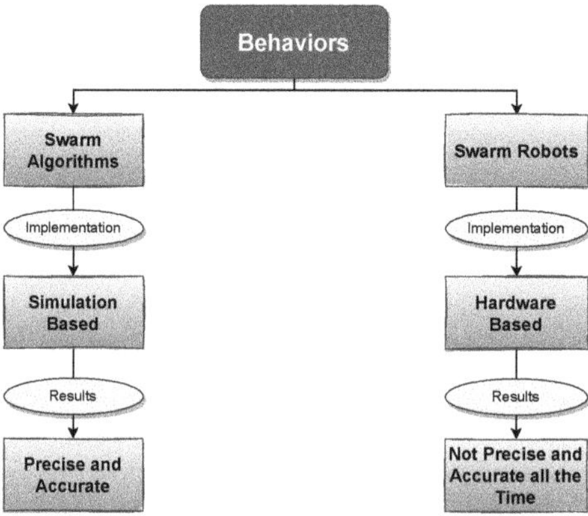

Figure 3. Swarming Behaviours' Deployment in Simulation and Hardware.

2. Swarm Robotics Fundamental Behaviours

Swarm algorithms are characterized by individual entities following local rules, resulting in the emergence of overall behaviour through swarm interactions. In swarm robotics, robots exhibit local behaviours based on a set of rules ranging from basic reactive mapping to complex local algorithms. These behaviours often involve interactions with the physical environment, such as other robots and surroundings [4]. The interaction process involves retrieving environmental values and subsequently processing them to drive the actuators in accordance with a set of instructions. This recurring process is referred to as the fundamental activity and persists until the desired state is attained. Figure 4 illustrates a summary of several naturally occurring behaviours that are further elaborated in the subsequent subsection.

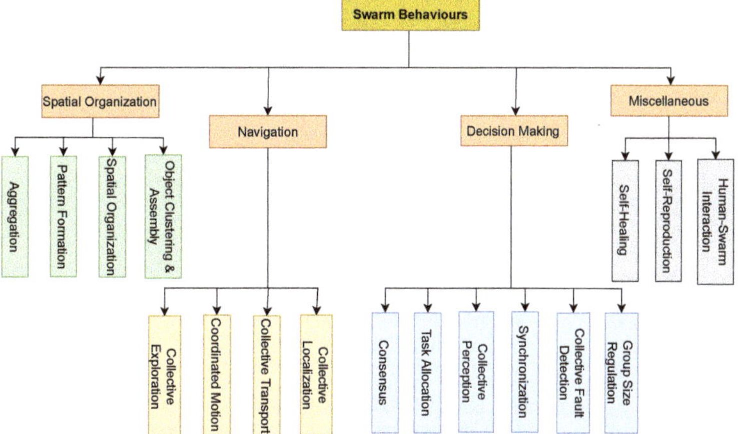

Figure 4. Swarm Behaviours [1].

2.1. Spatial Organization

These behaviours allow robots in a swarm to move around the environment and spatially arrange themselves around things.

Object Clustering and Assembly allow a swarm of robots to control geographically dispersed things. These are critical for construction processes. *Pattern Formation* organizes the robot swarm into a precise form. *Chain Formation* is a specific instance where robots construct a line to establish multi-hop communication between two places [1,5]. *Self-assembly* links robots to form structures. They can be connected physically or remotely via communication lines [1,6]. *Morphogenesis* is a specific instance in which the swarm grows into a predetermined form [1,7,8]. *Aggregation* pushes the individual robots to gather spatially in a certain location of the environment. This permits swarm members to get geographically near to one another for further interaction [1,9,10].

2.2. Navigation

These characteristics enable a swarm of numerous robots in the environment to move in unison. Thus, a group of robots move in harmony from one location to another or from a source to a final destination [1,11].

Collective Localization allows the swarm's robots to determine their location and orientation relative to one another by establishing a local coordinate system across the swarm [1]. In *Collective Transport*, a swarm of robots may collectively move things that are too heavy or massive for individual robots [1]. *Coordinated Motion* moves the swarm in a configuration that must have a well-defined shape or structure, such as a line, triangle, or arbitrary formation of robots, as in flocking [1]. *Collective Exploration* navigates the environment to examine things, monitor the environment, or create a robot-to-robot communication network [1,12].

2.3. Decision Making

This characteristic of swarm robotics facilitates collective decision-making for accomplishing specific tasks collaboratively. The *Group Size Regulation* feature empowers the swarm's robots to create groups of the required size, and if the swarm's size exceeds the required group size, it automatically divides into multiple groups or sub-swarms [1,13]. Additionally, the *Collective Fault Detection* feature detects individual robot shortcomings inside the swarm, enabling the identification of robots that deviate from the expected behaviour due to hardware or some algorithmic issues [1,14]. Furthermore, *Synchronization* aligns the frequency and phase of the swarm's oscillators, enabling the robots to share a common perception of time and execute tasks in synchrony. The *Collective Perception* feature aggregates the locally collected data from the swarm's robots into a comprehensive image. It allows the swarm to make collective decisions, such as accurately classifying objects, allocating a suitable percentage of robots to a given task, or determining the best solution to a global problem [1]. Moreover, the *Task Allocation* feature dynamically assigns emergent tasks to individual robots, aiming to maximize the overall performance of the swarm system. In cases where the robots possess diverse skill sets, the work can be assigned differently to further enhance the system's performance [1,15]. Finally, the *Consensus* feature allows the swarm of robots to converge on a single common point from multiple available options [1,16].

2.4. Miscellaneous

The swarm robots exhibit additional behaviours beyond the previously discussed categories. *Self-healing* behaviour allows the swarm to recover from individual robot failures, improving the swarm's reliability, resilience, and overall performance [1,17]. *Self-reproduction* enables a swarm of robots to add new robots/agents or replicate the patterns created by several individuals, thereby increasing the swarm's autonomy by eliminating the need for human intervention in the construction of additional robots. *Human-swarm Interaction* facilitates communication between humans and the swarm of robots, either remotely via a computer terminal or in a shared area using visual or auditory cues [1].

3. Swarm Intelligence Algorithms

Swarm Intelligence (SI) is a collective intelligence employed in various applications, including self-organized and decentralized systems [18]. Some examples are collective sorting, cooperative transportation, group foraging, and clustering. Self-organization and division of work are two essential notions in SI. The ability of robots to evolve into a proper pattern without external assistance is referred as *self-organization*. In contrast, division of labor refers to the simultaneous execution of multiple tasks by individual robots. It enables the swarm to execute a challenging task that requires individuals to collaborate. Genetic Algorithm (GA), Ant Colony Optimization (ACO), Particles Swarm Optimization (PSO), Differential Evolution (DE), Artificial Bee Colony (ABC), Glowworm Swarm Optimization (GSO), and Cuckoo Search Algorithm (CSA), are all examples of famous and currently used swarm intelligence algorithms.

3.1. Genetic Algorithm

Genetic Algorithms (GA) were introduced in 1975 by John Holland [19,20]. This type of algorithm mimics natural existing biological behaviours in order to evaluate the survival of the fittest. In a genetic algorithm (GA), a specified number of individuals, also known as members, comprise the population. Mathematical operators such as crossover, reproduction, and mutation are used to manipulate the genetic makeup of individuals. Based on these operators, the fitness value of each member is calculated and ranked accordingly. The previous population's traits, represented by chromosomes (or strings), are combined with new traits to generate a new population [21–24]. A GA algorithm with five basic steps is shown in Figure 5. The fitness function evaluates population members, which begins with an initial population that can be generated randomly or through a heuristic search. After the population members are assessed, the lowest-ranked chromosome is eliminated, and the remaining members are used for reproduction. The final step is mutation, in which the mutation operator modifies genes on a chromosome to ensure that every part of the problem space is explored. This process of evaluating and generating new populations continues until the best solution is found.

It has a vast area as an application, which includes, navigation and formation control [25], path planning [26], scheduling [27,28], machine learning [29], robotics [30,31], signal processing [32], business [33], mathematics [34], manufacturing [35] and routing [36].

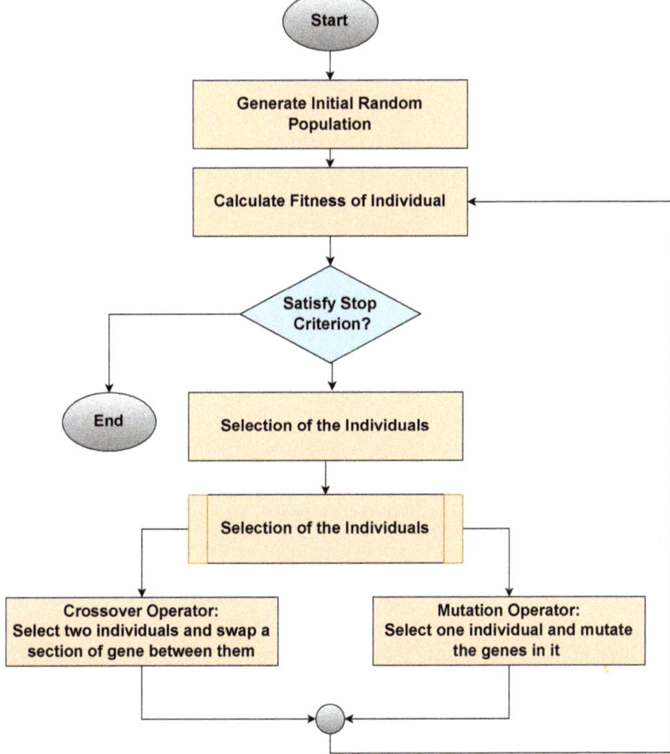

Figure 5. Flow Chart of Genetic Algorithm [37].

3.2. Ant Colony Optimization

Ant Colony Optimization (ACO) is a heuristic search-based algorithm that uses the ant colony system to solve problems. It was proposed by Marco Dorigo as part of his Ph.D. study in 1992 [38]. The four fundamental components of the ant-inspired foraging algorithm are the ant, pheromone, daemon action, and decentralized control. The *ant* acts as an imaginary agent which mimics the behaviour of exploitation and exploration processes in a search space and produces a chemical substance called *pheromones*. Its intensity varies with the passage of time due to the evaporation process and serves as a global memory for the ant's path of travel. *Daemon activity* is used to gather global data whereas, the decentralized control is used for the robustness of the ACO algorithm and to maintain flexibility within a dynamic environment. The Figure 6a–c show the initial, mid-range, and final outcomes of the ACO algorithm, respectively [38,39]. Figure 6a shows the initial random environment in which the agent (or ant) from the nest begins the process. When ants discover numerous viable paths from the nest to the source, they go through many iterations of execution, as shown in Figure 6b. The ant has chosen the shortest possible path, which contributes to the pheromone trail's high intensity. Equation (1) below is used as an initial step in determining the optimal solution to select the best node from the current search space.

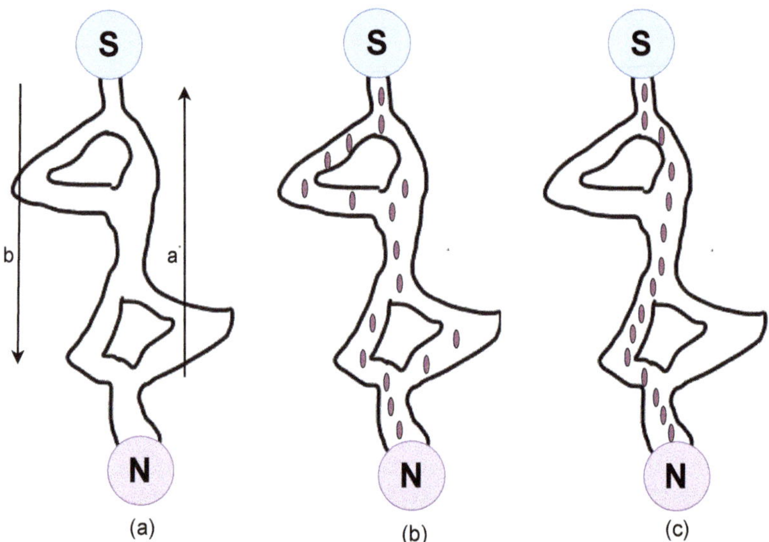

Figure 6. Nest and Food-Source have been shown by letters N and S, respectively. (**a**) depicts the early stages of the process, in which ants start to discover a passage between the nest and the source and lay their pheromones. (**b**) depicts the intermediate phase, in which the ants took all available pathways. (**c**) demonstrates that the majority of ants chose the road with the highest pheromone concentration [36].

$$p_{(n,m)}^{u}(t_o) = \frac{\left([\tau_{nm}(t_o)]^{\alpha} \cdot [\eta_{nm}]^{\beta}\right)}{\left(\sum_{u \in I_u}[\tau_{nm}(t_o)]^{\alpha} \cdot [\eta_{nm}]^{\beta}\right)} \tag{1}$$

The probability of travelling from node n to node m is $p_{(n,m)}$, I_u are the nodes to which the ant is permitted to go from node n, whereas $\eta_{(nm)}$ adds to visibility between nodes n and m and it indicates the quantity of un-evaporated pheromone between nodes at a time t_o. α and β in Equation (1) regulate the impact of $\tau_{nm}(t_o)$ and η_{nm}, where, if α is larger, the ant's searching behaviour is more pheromone-dependent, and if β is higher, then the ant's searching depends on its visibility or knowledge.

In order to deposit a pheromone, the following equation is used:

$$\Delta\tau_{nm}^{u}(t) = \begin{cases} \frac{Q}{L_u}(t) \\ 0 \end{cases} \tag{2}$$

Q is a constant, L is the cost of the ant's tour that represents the length of the created path, t is the iteration number and u shows a specific ant. Another key factor is pheromone evaporation rate, which shows exploration and exploitation behaviour of the ant. In Equation (3), s is the number of ants in the system and p is the pheromone evaporation rate or decay factor.

$$\tau_{nm}(t+1) = (1-p).\tau_{(n,m)(t)} + \sum_{k=1}^{s}[\Delta\tau_{n,m}^{u}(t)] \tag{3}$$

Compared to other heuristic-based approaches, ACO guarantees to converge, but the time required for it is uncertain and for better performance, the search space should be small [40,41]. Its applications include vehicle routing [42,43], network modelling problem [44,45], machine learning [46], path planning robots [47], path planning for Unmanned Aerial Vehicles (UAVs) [48], project management [49] and so on.

3.3. Particle Swarm Optimization

Kennedy and Eberhart invented Particle Swarm Optimization (PSO) in 1995, and it uses a simple method to encourage particles to explore optimal solutions [50]. It is based on flocking bird and schooling fish behaviours [51], by exhibiting three simple behaviours: separation, alignment, and cohesiveness. *Separation* is used to avoid congested local flockmates, *alignment* is the travelling of one flock-mate in the same average direction of the other flock-mates, and *cohesiveness* is the movement of flock-mates toward the average position. The PSO algorithm is as follows [50,52,53]:

$$v_{id}^{t+1} = v_{id}^{t} + c_1 \cdot \text{rand}(0,1) \cdot \left(p_{id}^{t} - x_{id}^{t}\right) + c_2 \cdot \text{rand}(0,1) \cdot \left(p_{gd}^{t} - x_{id}^{t}\right)$$
$$x_{id}^{t+1} = x_{id}^{t} + v_{id}^{t+1} \tag{4}$$

where v_{id}^{t} and x_{id}^{t} are particle velocity and position, whereas d is search space dimension, i represents particle index and t shows the iteration number. c_1 and c_2 depict the speed and regulating length of the swarm when it travels towards the optimal particle position. The optimal position attained by particle i is p_i and the best position found by neighbouring particles of i is p_g. The process of exploration ensues if either or both of the differences between the best of particle p_{id}^{t} and the previous position of particle x_{id}^{t} and between the population all-time best p_{gd}^{t} and the previous particle's position x_{id}^{t} are large. Similarly, the process of exploitation happens when both of these values are small. PSO has been demonstrated as an effective, robust, and stochastic optimization algorithm for high-dimensional spaces. The key parameters of PSO include the position of the agent in space, the number of particles, velocity, and the agent's neighbourhood [54–56].

The PSO algorithm begins by initializing the population, and the second step is to calculate the fitness of each particle. Whereas, the third step is followed by updating the individual and global best. In the fourth step velocity and neighbourhood of the particles are updated. Steps two to four keep repeating until the terminating condition is satisfied [51,54,57,58].

Figure 7 shows the working of the PSO method, where the particles are spread out in the first iteration to discover the best exploration. The best solution is identified in terms of neighbourhood topology, and each member's personal and global best particles are updated. As indicated in the figure, the convergence would be determined by attracting all particles towards the particle with the best solution.

PSO is simple to configure for efficient global search, has few parameters to set, is scale-insensitive, and parallelism for concurrent processing is also easy. Population size is one of the key factors that ensures precise and fast convergence for large population sizes [51,59]. Networking [60], power systems [61], signal processing [62], control systems [63], machine learning [64], and image processing [65–67] are some of the applications.

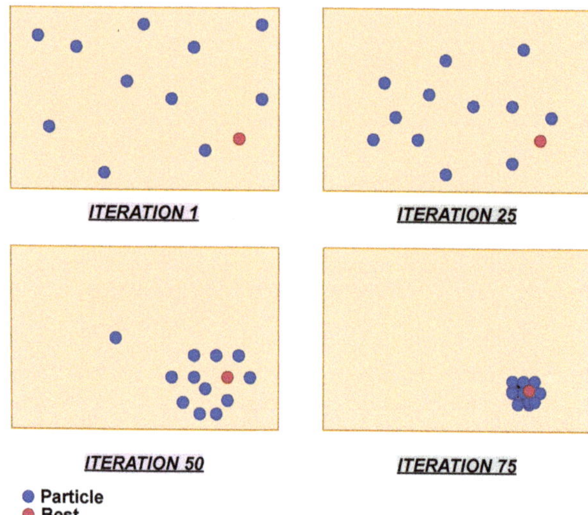

Figure 7. The operation of the PSO algorithm and its progress towards global optima as measured by iteration numbers [47].

3.4. Differential Evolution

Differential Evolution (DE) is similar to GA, using the same crossover, mutation, and selection operators. The fundamental difference between the two algorithms is that the DE utilizes the mutation operator while GA uses the crossover operator to produce a superior solution. Price and Storn first introduced it in 1997 [68]. DE repeatedly generated new populations using three properties: mutant vector, target vector, and trail vector explained in Figure 8. A crossover process between the target and mutant vectors produces the trailing vector. The mutant vector represents the mutation of the target vector, whereas the target vector represents the vector holding the search space solution [69,70]. The DE algorithm starts with population initialization and then evaluates the population to find the fittest members. The weighted difference between the two population vectors is added to the third vector to create new parameter vectors and this process is known as *mutation*. The vector is blended within the crossover to perform a final selection.

N parameter vector mutation is generated by using the following equation:

$$v_{j,N+1} = x_{l1,N} + F(x_{l2,N} - x_{l3,N}). \tag{5}$$

i shows the index of the 2D vector. x_{l1}, x_{l2}, and x_{l3}, are solution vectors selected randomly and the values of $l1$, $l2$ $l3$ and i should not be equal to each other. F is the scaling factor $\in [0,1]$, while, a crossover procedure is employed to improve the variety of the disconcerted parameter vectors. The parent and mutant vectors are combined in the following method to create a trial vector:

$$u_{i,G+1} = \begin{cases} v_{i,G+1} \text{ if } R_j \leq CR \\ x_{i,G} \text{ if } R_j > CR \end{cases} \tag{6}$$

where CR denotes the crossover constant. R_j denotes a random real number ϵ [0,1] while j depicts the resultant array's j^{th} component.

The primary distinction between DE and GA operations is that in DE, the probability of being selected as a parent is not based on fitness value. Increasing the population size can significantly improve DE performance.

DE can be found in a variety of fields, including, robot path planning [71,72] engineering [73], image processing domain [74], machine learning [75], and economics [76].

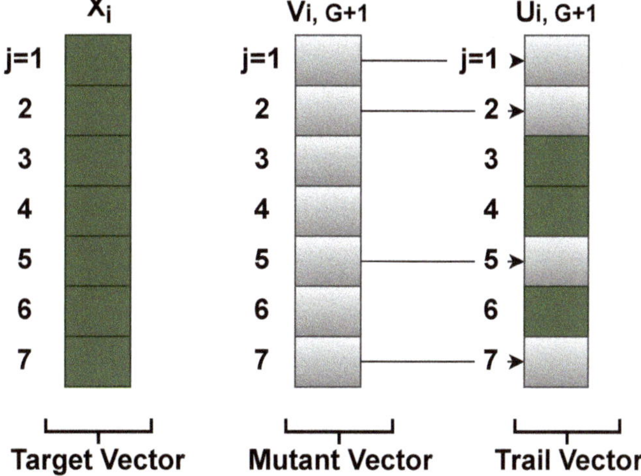

Figure 8. Demonstration of DE with a seven-vector dimension j. A target vector is a current approach; however, a mutant vector is also an alternative. After the crossover operation, the trailing vector is a new solution [55].

3.5. Artificial Bee Colony

Dervis Karaboga presented Artificial Bee Colony (ABC) as an important SI algorithm in 2005 [77]. Its performance is thoroughly examined in [78], which concluded that ABC outperforms other techniques. It is based on honey bees' intelligent behaviour in locating food and communicating information about that food with other bees. ABC is as straightforward as PSO, and DE [78], which divides artificial agents into three types: *employed*, *observer*, and *scout* bees. Each agent bee is given a particular task to finish the algorithm process. The employed bee concentrates and memorizes the food supply. The employed bee provides the observer bee with the information about the hive's food supply. The scout bee is on the lookout for new nectar and its sources. Figure 9 presents the algorithmic flow of the ABC. The ABC method's overall procedure and specifications of each step are explained below [77–79]:

Step 1. Initialization: Food sources, x_i, are initialized with $i = 1 \dots N$, where N is the number of scout bees in the population. l_i and u_i are the control parameters represent lower and upper limits, respectively. The following Equation (7) represents the initialization phase:

$$x_i = l_i + \text{rand}(0,1) * (u_i - l_i) \tag{7}$$

Step 2. Employed Bees: The search capacity for finding new neighbour food source v_i increases to accumulate more nectar around the neighbour food source x_i. Once they identify a nearby new food source supply, its profitability and fitness value are assessed. The following formula is used to define the new nearby food source:

$$v_i = x_i + \phi_i (x_i - x_j) \tag{8}$$

where x_j is a randomly selected food source. ϕ_i has random numbers of range between $[-a, a]$. After the profitability of the new source v_i is determined, a greedy selection is used between $\overrightarrow{x_i}$ and $\overrightarrow{v_i}$. The process of exploration occurs if $x_i - x_j$ is greater, otherwise exploitation happens. The fitness value $fit_i(\overrightarrow{x_i})$ is computed by the following Equation (9) and objective function with solution value x_i is $f_i(\overrightarrow{x_i})$.

$$fit_i(\overrightarrow{x_i}) = \begin{cases} \frac{1}{1+f_i(\overrightarrow{x_i})} & \text{if } f_i(\overrightarrow{x_i}) \geq 0 \\ 1 + abs(f_i(\overrightarrow{x_i})) & \text{if } f_i(\overrightarrow{x_i}) < 0 \end{cases} \tag{9}$$

Step 3. Onlooker Bees: After calculating the fitness value and by obtaining information from employed bees, a probability value p_i is computed by using Equation (10), and this value is then shared with the waiting bees in the hives for selecting food sources. These bees are known as onlooker bees.

$$p_i = \frac{fit_i(\overrightarrow{x_i})}{\sum_{i=1}^{SN} fit_i(\overrightarrow{x_i})} \tag{10}$$

Step 4. Scout Bees: Employed bees that cannot raise their fitness values after multiple repetitions become scout bees. These unemployed bees choose sources at random.

Step 5. Best Fitness: The best fitness value and the exact position with an associated value are memorized.

Step 6. Termination Checking Phase: The program terminates upon meeting the termination condition. If the termination condition may not be reached, the program goes back to step 2 and repeats the process until it is.

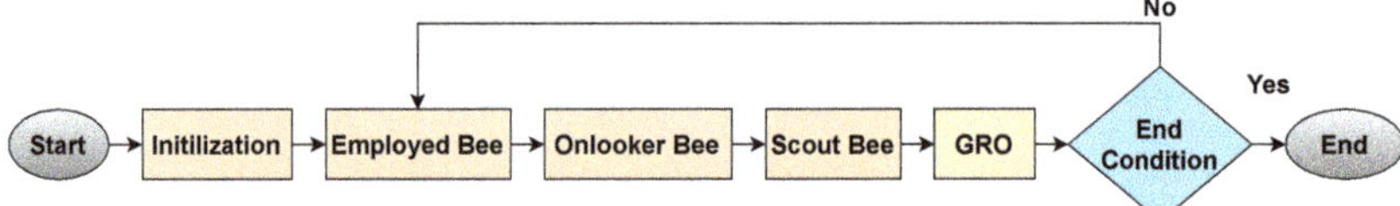

Figure 9. Flow Chat of ABC Algorithm.

Since ABC has only two control factors, colony size and maximum cycle number, it is straightforward to set up, robust and customize-able. It is also possible to add and remove bees without re-initializing the algorithm [80,81]. The disadvantage of ABC is that additional fitness tests for new parameters are required to increase the algorithm's overall performance. It is also slow when a large number of objective function evaluations are required [82]. Path planning for multi-UAVs [83], engineering design difficulties [84,85], networking [86], electronics [87], scheduling [87], and image processing [87] are some of the disciplines where it is used.

3.6. Glowworm Swarm Optimization

Glowworm Swarm Optimization (GSO) is a new SI based approach presented by Krishnanad and Ghose in 2005 [88,89] to optimize multimodal functions. In GSO, glowworms are real-life tangible creatures. There are three key parameters in a glowworm m condition at time t: a search space position $x_m(t)$, a luciferin level $l_m(t)$, and a neighbourhood range $r_m(t)$ [88–90]. These variables change over time, whereas the glowworms are distributed throughout the work area at random initially, and then the other settings are set using predetermined constants. It is similar to earlier algorithms, where three phases are continued until the termination condition is reached. The three steps of [88] are luciferin level update, glowworm migration, and neighbourhood range update. The fitness value of glowworm m's current position of luciferin level is updated by using the following equation:

$$l_m(t) = (1 - p) \cdot l_m(t - 1) + \gamma J(x_m(t) \tag{11}$$

where p is the luciferin evaporation factor and J represents the objective function. For position update in the search space, the following equation is used:

$$x_m(t) = x_m(t-1) + s\frac{(x_n(t-1) - x_m(t-1))}{||(x_n(t-1) - x_m(t-1))||} \tag{12}$$

where s is the step size, and $||.||$ is euclidean norm operator. Exploration and exploitation behaviours occur on the basis of x_n and x_m difference. Greater difference leads to exploration and smaller to exploitation behaviour.

If a glowworm has several neighbours to choose from, one is selected using the following probability equation and the glowworm m is the neighbour of glowworm n only if the distance between them is shorter than the neighbourhood range $r_m(t)$:

$$p_m(t) = \frac{l_m(t) - l_n(t)}{\sum_{k \in Ni(t)} l_k(t) - l_n(t)} \tag{13}$$

The following equation is used to compute the neighbourhood range:

$$r_m(t+1) = \min\{r_s, \max[0, r_m(t) + \beta(n_d - |n_m(t)|)]\} \tag{14}$$

r_s represents sensor range, n_d is the desired number of neighbours, $|n_m(t)|$ is several neighbours of the glowworm m at time t, and β is a model constant. The diagram below demonstrates two hypothetical scenarios in which agents developing methods result in distinct behaviours depending on the agents' placement in the search space and the accessible nearby agents. The glowworm's agents are represented by i, j, and k. Figure 10a signifies agent j's sensor range, whereas r_d^j denotes agent j's local-decision range. The same is true for i and k, where r_s^i and r_d^i, r_s^k and r_d^k respectively denote sensor range and local-decision range. It is applied in the circumstances where agent i is in the sensor range of agent j and k. Only agent j uses the input from agent i because the agents have different local decision domains. Glowworm agents are a, b, c, d, and e in Figure 10b. The glowworm agents are ranked 1, 2, 3, 4, and 5, depending on their luciferin values.

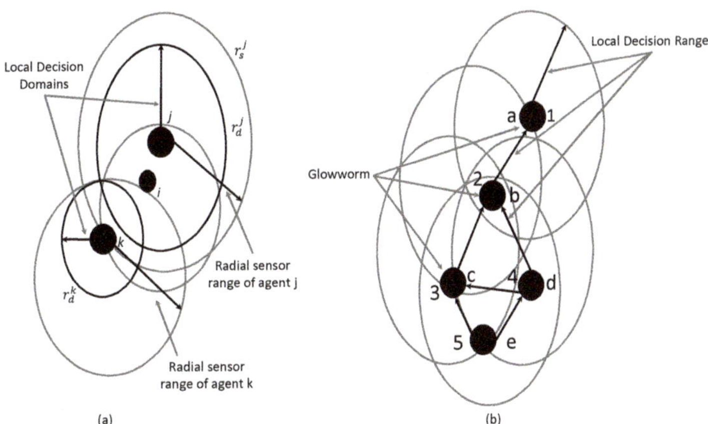

Figure 10. GSO in two different scenarios. The glowworm agents are a, b, c, d, e, f, i, j, and k. Three agents with varied sensor ranges and local-decision ranges are shown in (**a**). It demonstrates how agents gravitate towards agents with higher luciferin values when they are in the same local decision as another agent. Glowworm agent's rating is according to their luciferin levels, as shown in (**b**). Lower numbers indicate greater luciferin values and vice versa [67].

The following modifications can be considered to improve the performance of GSO. (i) To include all agents in the solution, consider increasing the number of neighbourhoods. When the best solution has been identified, all the agents can travel in the direction of the agent which has the best solution, because more agents will be within the optimal solution range and it will also increase the efficiency of exploitation; (ii) In the neighbourhood range, smallest possible number should be selected to increase the convergence rate of GSO. Since there are fewer calculations needed to estimate the probability and direction of the GSO's movement, this action may decrease the GSO's processing time.

GSO is useful in situations when only a small sensor range is required. It can detect many sources and can be used to resolve problems of numerical optimization [88–90]. It is also inaccurate and has a slow convergence rate [91,92]. 3-Dimensional path planning [93], self-organization based clustering scheme for UAVs [94], routing [95], swarm robotics [96], image processing [97], and localization [98,99] difficulties have all been solved using GSO.

3.7. Cuckoo Search Algorithm

Yang and Deb in 2009 proposed Cuckoo Search Algorithm (CSA) as one of the most current meta-heuristic techniques. The behavior of cuckoos, i.e., brood parasites, and the properties of Levy flights [100] inspired this algorithm. Three steps are followed throughout the implementation of this approach. First, in each repetition, each cuckoo lays one egg, and the nest in which the cuckoo lays its egg is chosen at random by the cuckoo. Quality eggs and nests are passed down from generation to generation in the second step. In the third step, the number of possible host nests are fixed, and a host bird uses probability $p_a \in$ [0, 1] to find a cuckoo egg. In other words, the host can either reject the egg or depart the nest and start over. These three major criteria are used to present the specifications of the acts taken in CSA. The following Levy flight equation is used to construct a new solution, $u(i+1)$ [100,101]:

$$u_m(i+1) = u_m(i) + \partial \oplus \text{Levy}(\beta) \tag{15}$$

$$Levy \sim s = t^{-1-\beta}(0 < \beta < 2) \tag{16}$$

The product $\oplus$ is an indication of multiplication, follows the same rules as entry-wise matrix multiplication, and ∂ is the step size and, in most circumstances, $\partial = 1$. The step size begins with a large value and gradually decreases until the last generation, allowing the population to converge on a solution, similar to the processes involved in reducing PSO linearly. Yang [102] introduces the additional component as follows:

$$u_m(i+1) = u_m(i) + \partial \oplus \text{Levy}(\beta) \sim 0.01\frac{s}{|v|^{1/\beta}}(u_n(i) - u_m(i)) \tag{17}$$

where s and v are selected using the normal distribution, which is defined as follows:

$$s \sim N\left(0, \sigma_s^2\right), v \sim N\left(0, \sigma_s^2\right) \tag{18}$$

where;

$$\sigma_u = \left\{ \frac{\left(\gamma(1+\beta)\sin\left(\frac{\pi\beta}{2}\right)\right)}{(\gamma[1+\beta]/2]\beta 2^{\frac{\beta-1}{2}}} \right\}^{1/\beta}, \sigma_v = 1 \tag{19}$$

γ is the standard gamma function [102]. Exploration happens when the difference between u_n and u_m is high, while exploitation occurs when the difference is minor.

Compared to other approaches, CSA offers the advantage of multi-model objectives and requires fewer parameters to fine-tune them. It is used in a variety of settings, including path planning for UAVs [103], neural networks [104], embedded systems [105], electromagnetics [106], economics [107], business [108], and the Traveling Salesman Problem (TSP) issue [109].

4. Applications of Swarm Robotics

Swarm robotics is an emerging area of research and development that has yet to gain significant industrial adoption. Still, academics have created a variety of platforms to test and analyze the algorithm. In [110], the authors mentioned that they are researching for future industrial platforms. Swarm robotics research (see Figure 11), and industrial efforts & products (see Figure 12), are the two areas of the survey which will be discussed later. Industrial projects and products are examples of deployment in a real-time scenario. The swarm robotics research platform assists researchers in demonstrating, verifying, and experimenting with swarming algorithms in a laboratory setting. The four categories for both platforms are terrestrial, aerial, aquatic, and extraterrestrial. Robotic vehicles include Unmanned Submarine Vehicles (UUV), Unmanned Aerial Vehicles (UAVs), Unmanned Surface Vehicles (USVs), and Unmanned Ground Vehicles (UGVs).

4.1. Research Platforms

This section includes the application from swarm algorithms to swarm robots. Advanced robotics research platforms, such as the *balboa* robot and others, exist but are not included in Figure 11 because, they are not designed to use in swarm applications.

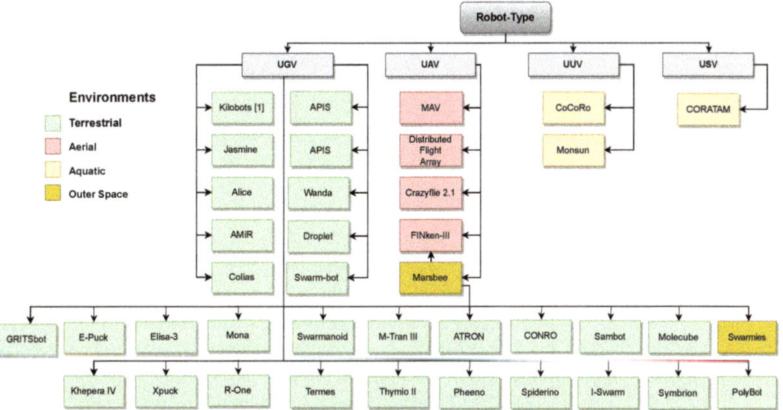

Figure 11. Classification of different research platforms for swarm robotics .

4.1.1. Terrestrial

The *kilobot* swarm is widely considered the best swarm of robots ever produced for educational and research purposes. They are little, measuring 33 mm in diameter. For propulsion, vibration motors are employed, and for communication, infrared light reflected from the ground is used. For swarming, 1024 robots are used and they are well-known for their capacity to self-assemble into various forms [111]. It is open-source and commercially accessible through K-Teams. *Jasmin*, an open-source platform, was created with a large-scale swarm investigation that required touch, proximity, distance, and color sensors. *Alice* [112] is another platform, with additional sensors, including a linear camera, increases the functionalities of swarming. Similarly, *AMiR* [113] and *Colias* [114] are open-source and commercially available swarm robots that provide a foundation for a number of research platforms. *Mona* is a commercial product as well as an open-source initiative. However, *R-One* may be used as a swarm robotics platform since it comes with a camera for ground-truth localization and software to connect all devices. The swarming platform *Elisa-3* incorporates an Arduino with eight infrared sensors, three accelerometers, and four ground sensors, all of which can be charged by a charging station and communicate through infrared or radio waves. The *Khepera IV* [115] was created for indoor use. K-Team is a tiny and unique swarming research platform with a linux core, color camera, WLAN, bluetooth, USB, accelerometer, loudspeakers, gyroscope, three RGB

LEDs, and it is also commercially available. The *GRITSbot* [116] is an open-source robot found at Georgia Tech's Robotarium in Atlanta. Researchers can utilize the resources by uploading code, performing experiments, and gathering data using Robotarium's remote access. As the size and quantity of these robots increase, more maintenance and usability aspects become crucial.

The *e-puck* and its successor, the *e-puck2*, are designed to make programming and controlling robot behavior simple for research and education. It includes an infrared proximity sensor, a CMOS camera, and a microphone. Both commercial and open-source versions are available. Its new edition, *Xpuck*, introduces new features, including aggregation of raw processing power, which is used in current mobile system-on-chip (SoC) devices with roughly two teraflops of processing power.

ArUco marker tracking in image processing computations is another example [117]. Similarly, *Thymio II* [118] swarm robots offer a range of sensors, including temperature, infrared distance, microphone, and accelerometer. Visual and text-based programming are also available. *Thymio II* is open-source and commercially available at Thymio, whereas *Pheeno* [119] is also a free and open-source swarm robotics platform for teaching and research. Custom modules with three degrees of freedom may be employed, and an IR sensor is used to communicate with the outside world. The open-source and locomotion-capable *Spiderino* [120] has six legs and has a hexapod toy-like design with an Arduino CPU, WLAN, and some reflected infrared sensors on a PCB.

I-Swarm (Intelligent Small-World Autonomous Robots for Tiny-Manipulation) is a swarming microrobot. Its sizes are $3 \times 3 \times 3$ mm, and it is solar-powered without a source. It travels by vibrating and communicates using infrared transceivers to establish a swarm of 1000 robots [121]. The prototype is on exhibit at the technology museum in Munich. The *Zooids* [122] human-computer interface is a novel type of HCI that handles interaction and presentation. It was built as a unique open-source robotics platform. Light patterns projected from an overhead projector regulate the swarming of *Zooids*. The APIS, or adaptable platform for an interactive swarm, comprises several components, i.e., the swarm's infrastructure and testing environment, software infrastructure, and simulation [123]. The focus is to experiment with human-swarm interaction. The platform uses an OLED display and a buzzer. With the help of the swarm, clean up the environment Wanda [124] is a robotics platform that might be useful. The authors have built the entire tool-chain from robot design and simulation to deployment. *Droplet* [125], a spherical robot that can organize itself into complex shapes with the help of vibration locomotion, is another ideal platform for education and study. The powered floor, which features alternating positive charge and ground stripes, has been used for both charging and communication between swarm robots. *Swarm-bots* [126,127] may automatically align themselves to various 3D shapes. Its design is open-source, and robots are made up of various insect-like shapes. They are built with low-cost, readily available components. They can adapt to any environment due to their self-assembling and self-organizing capabilities. The swarm can move heavy goods that would be too heavy for individual robots. *Swarmanoid* and its successor, are the first study of integrated design, development, and control of heterogeneous swarm robotics systems. It is open-source, and includes three types of autonomous robots. *Eye-bots* (UAVs that can stick to an interior ceiling), *Hand-bots* (UGVs that can climb), and *Footbots* (UGVs that can self-assemble) are among the varieties of UAVs that are developed [128]. Surprisingly, the *termes* robots [129] interact without the need for communication or GPS to build huge constructions using modular components. It is based on how termites construct their nests in nature, and they are block-carrying climbing robots that can also construct similar structures in unstructured situations. Other swarming platforms for research are *symbrion* and *replicator* [130]. They are two projects that are pretty much identical in terms of developing autonomous platforms for swarms. By physically connecting to each robot in the swarm, they may function individually or in a certain form and the goal was to devise a strategy for achieving robot organism evolvability. *PolyBots* [131] are self-configurable robots that can move in many ways. They have interchangeable object manipulation modules that may take on a variety of shapes depending on the situation, such as an earthworm for slithering over barriers or a spider for marching through hilly terrain. These robots are ideal for multitasking and usage in new areas. *M-TRAN I* [132], *M-TRAN II* [133], and *M-TRAN III* [134] are self-configurable robotics technologies. *ATRON* [135], *CONRO* [136], *sambot* [137], and *molecube* [138] are all open-source robotic systems and robots.

4.1.2. Aerial

Miniature and micro unmanned aerial vehicles (µUAVs) for swarming are affordable robots available for research and education [139] and Swetha et al. [140] both look into small-scale UAVs. Several off-the-shelf *Micro Air Vehicles* (µAVs) are available and famous in the gaming and commercial industries. Three rate gyroscopes and three accelerometers are used in UAVs developed for swarming robots in µAVs in [141], together with one ultrasonic sensor and four IR sensors. The Distributed Flight Array [142] is a popular platform used to construct *swarmanoid* [128] on it. Each UAV adds a single rotor to a big array. The module self-assembles into a multi-rotor system, in which all robots must exchange coordinates and local parameters for coordinated flying. *Crazyflies* [143], which are available commercially and open-source at Bitcraze, make use of a variety of sensors, including a high-precision pressure sensor, an accelerometer, a magnetometer, and a gyroscope. It can conduct experiments while minimizing the risk to humans because of its light weight of about 27 g. In *FINken-III* [144], is a powerful copter equipped with a better communication module (802.15.4) to communicate between ground station and other copters , and sensors like optical flow, infrared distance, and four sonar sensors.

4.1.3. Aquatic

The Collective Cognitive Robotics (CoCoRo) project has been developed with 41 heterogeneous Unmanned Underwater Vehicles (UUVs). Electric fields and sonar sensors are used to communicate, and the system applies to environmental monitoring, water pollution assessment in rivers and oceans, and global warming consequences. The *Monsun* [145] has two communication modes: a camera for identifying other swarm members and an underwater acoustic modem for transmitting data. *CORATAM* (Control of Aquatic Drones for Maritime Tasks) [146] has also been developed for swarms of USVs, with uses such as sea border patrols, marine life localization, and environmental monitoring. This open-source platform uses evolutionary computing to evaluate swarm methods [147].

4.1.4. Outer Space

NASA has developed swarmies to gather water, ice, and minerals on Mars. They have also established a swarmathon to aid academics in developing an ant-based swarm algorithm. In-situ Resource Utilization (ISRU) is the name given to this application. Twenty swarmies cover a distance of 42 km in around 8 h. Another NASA Innovation Advanced Concept (NIAC) program project aims to enrich knowledge on the Mars exploration swarm of Marsbees [148]. These have the size of the bumblebee for robotics flapping wing flyers. They can explore and discover themselves in an unfamiliar place. With NIAC financing, a flapping flyer with insect-like wings will be offered as a technical implementation.

4.2. Industrial Projects and Products

These include UAV, UGV, UUV, and USV swarm robots developed for industrial projects and products. The available robot with respective type has been shown in Figure 12.

4.2.1. Terrestrial

Agriculture is essential to a country's growth. Food demand is growing, but the output is still insufficient [149]. *SwarmBot 3.0* is being used to monitor fields autonomously using Unmanned Ground Vehicles (UGVs). Before beginning the specified task, this swarm collaborates via a centrally controlled timetable. The large area is automatically subdivided into smaller fields and then allocated to an individual robot in the swarm [150]. Their tasks include sowing, applying fertilizers to the assigned areas, harvesting, and irrigation which is the requirement of the agriculture sector. Another fascinating innovation from the Fendt firm is the *UGV Xaver*, which is used for seeding and is powered by a battery [151].

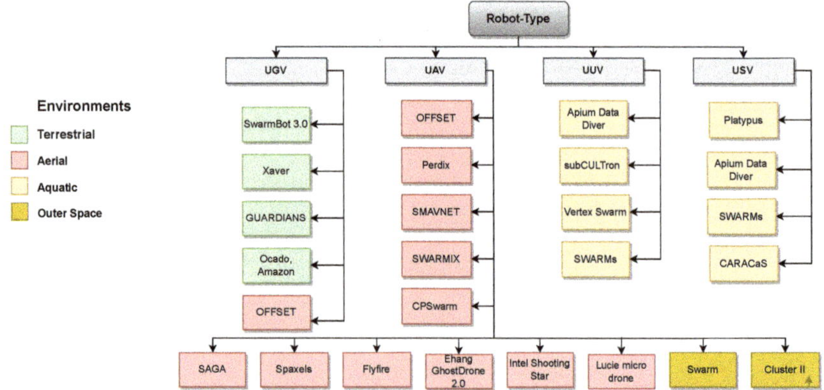

Figure 12. Classification of different industrial projects and products of swarm robotics.

The *GUARDIANS* (Group of Unmanned Assistant Robots Deployed in Aggregative Navigation by Scent) [152] has been used for emergency and rescue missions. They are used in places where human presence is prohibited or where the environment severely impairs human senses. This project assists in searching and warns against toxic chemicals using mobile communication links. They can form and navigate using potential fields and achieve the assigned task without explicit communication between the robots.

Another autonomous *Ocado* [153] warehouse has been developed that has a swarm of homogeneous cuboids and is being utilized for grocery orders and dispatching. A total of 1100 collaborative swarms of robots are used for the order and dispatch, where the workers put the customer order together. Robots are controlled from a central location by a cloud server, and data are exchanged via cellular technology between the robot and the cloud. Amazon [154], which employs Kiva, is the most prominent player in the swarming of robots in warehouses. An A* algorithm (with visual tags on the ground) searches for humans who assemble the customer's order. WLAN is used for the communication of robots, and dispatching is organized centrally. A low battery of robots is handled by the charging stations automatically. *Alibaba* [155] retailers are using a similar system for the autonomous order of goods and dispatching.

4.2.2. Aerial

The *OFFSET* (OFFensive Swarm Enabled Tactics) [156] projects are mostly deployed in military applications, although they can also be applied in other situations. This project aims to improve intra-city observations using UAVs and UGVs. These swarms of robots are capable of detecting hazards from the surroundings. *Perdix* [157], a military application swarm supported by the business. It is capable of performing its tasks without human piloting and has the ability to communicate with other drones to work collaboratively and achieve a common goal. These drones operate in a swarm of 20 or more and coordinate their actions to accomplish the desired outcome. *Pentagon* consisting of 103 drones, is another military application swarm. This swarm is not controlled by a single leader and can adapt to UAVs. They can fly in formation and make decisions as a group, making them useful for covert operations and targeted assassinations. The autonomous swarm is developed to install and manage WLAN network [158] as part of the *Swarming Micro Air Vehicle Network* (SMAVNET) project [159] in the emergency and rescue application sector. The project aims to gather rescue teams when disaster places have been explored and located. *SWARMIX* [160] is a similar search and rescue initiative in which a swarm of heterogeneous agents, such as humans, dogs, and Unmanned Aerial Vehicles (UAVs), create a swarm and engage in a search and eventually rescue operation.

Using a swarm of autonomous aerial vehicles, the Swarm robotics for Agricultural Applications (SAGA) project [161] seeks to do weed-spread monitoring and mapping. A swarm's fitness is decided by trade-off exploration and weed detection time in smart farming. Weeds and plants are detected and identified using a visual approach.

Nowadays, swarms are also providing entertainment in terms of light shows. The UAVs are equipped with colorful LEDs and perform the formation of different patterns accomplished by music to create a beautiful scene. In *Spaxels, Flyfire, Ehang, Intel* [162], and *Lucie micro*, 1000 Unmanned Aerial Vehicles (UAVs) are controlled from a central location and follow pre-programmed patterns.

4.2.3. Aquatic

Swarms are commonly used in aquatic environments to monitor the environment. *Platypus* [163] offers autonomous swarm robotics boats as USVs. They are utilized to keep track of water quality, produce a dense map of defined bodies beneath the surface, and stratify salinity and oxygen levels. *Apium Data Diver* is a prototype vehicle with a maximum depth of 100 m. It is meant for swarm operations on the surface and underwater, with temperature, pressure, and GPS among the sensors on board. It finds its application areas in defense, oceanography, hydrographic survey, and aquaculture. This type of swarm can be found in UUVs and USVs. It can accept high-level commands from a human operator and build a wide range of patterns [164]. *Hydromea's Vertex Swarm* is available in UUVs and can assess water quality in various places up to 300 m deep. It generates 3D data with great spatial and temporal resolution that is faster and more precise than manual approaches. The major purpose of the SWARMs (Smart Networking Underwater Robots in Cooperation Meshes) project [165] is to develop surface and underwater vehicles that can operate in maritime and offshore operations. It is responsible for designing and developing software and hardware components for the next generation of maritime vehicles, as well as assisting in the improvement of autonomy, robustness, cooperation, dependability, and cost-effectiveness. It uses offshore installations, chemical pollution monitoring, and plume tracking. Research focus lies on reliable underwater communication [166] and leveraging topology control [167].

The military has employed the CARACaS software kit, which is used in aquatic environment. NASA developed CARACaS (Control Architecture for Robotic Agent Command and Sensing), which has now been upgraded by ONR (Office of Naval Research) for autonomous Navy operations in the United States where USVs communicate with one another [168]. It enables USVs to choose their courses, protect assets in the navel, and intercept enemy boats as a group. In a demonstration at the James River in Virginia in 2014, CARACaS was installed on rigid-hulled boats and proved to be magnificent and successful [169]. Based on the discoveries of the CoCoRo, Submarine Cultures (SubCULTron) conduct long-term robotic exploration of unusual environmental niches. It is used on UUV robots to assess factors such as learning and self-sustainability.

4.2.4. Outer Space

Swarm was launched in 2013 and is made up of three identical spacecraft, two of which are side-by-side at 450 km and the third at 530 km above the ground. The mission of each satellite was to research the earth's magnetic field, and each was nine meters long [169]. Cluster II is a tetrahedral arrangement of four identical cylindrical spacecraft that was launched in 2000. It was initially capable of sending three-dimensional solar wind data on the earth's magnetosphere to investigate the sun's influence on the environment [170].

5. Swarm Robotics: Past, Present and Future Perspective

Social insects, fish schools, and bird flocks are examples of naturally self-organizing systems that display emergent collective behavior based on simple local knowledge [171,172]. Swarm robotics emerged as a branch of swarm intelligence, or the computational modeling of collective, self-organizing activity, which has yielded many successful optimization methods [173,174] that are now used in fields ranging from telecommunications [175] to crowd simulation, and prediction [176]. In contrast, swarm behavior in robots necessitate the installation of swarm intelligence algorithms on current robotic systems. Because of the expected ubiquity of autonomous robots in real-world applications and the challenge of allowing them to interact with one another and with their human users while avoiding the drawbacks of centralized control, swarm robotics research is gaining traction. Swarm robotics research will be crucial in addressing complex coordination problems in future robotics applications. It includes cooperative (i.e., robots working together to complete a common task) and semi-cooperative (i.e., self-interested robots benefiting from a globally efficient organization of activities, such as autonomous vehicles) scenarios. In the future, it will become a new and powerful tool in precision medicine, allowing for personalized therapies such as minimally invasive surgery or direct polytherapy delivery to malignant cells inside the human body [177,178]. Large numbers of robots with limited computation and communication capabilities, on the other hand, will push swarm robotics to its limits, necessitating the development of new conceptual tools in addition to tiny hardware or robotics devices [179].

In lab settings, robot swarms are shown using a small number of tiny robots [128,180]. Although technology advancements are pushing the bounds to ever-smaller sizes [177,181] and greater numbers [6,7], but the road to real-world applications remains lengthy and arduous. For example, *group scale*, from a few dozen to millions of people constituting the swarm and *physical scale*, from micro/nanorobots to massive terrestrial, aerial, and aquatic robots. Swarms that display prompt intervention and adaptability in a quickly changing environment to robots that work on months-long missions are examples of *temporal scale* (e.g., on a distant planet) from small-scale deployments to large-scale deployments and geographical scale. Previous, current, and future robotics achievements in terms of software, hardware, or a combination of the two are explained in the Table 1. Figure 13 shows the evolution of swarm robotics, to the best of our knowledge, from algorithmic research to the real-time best-performing swarm of robots.

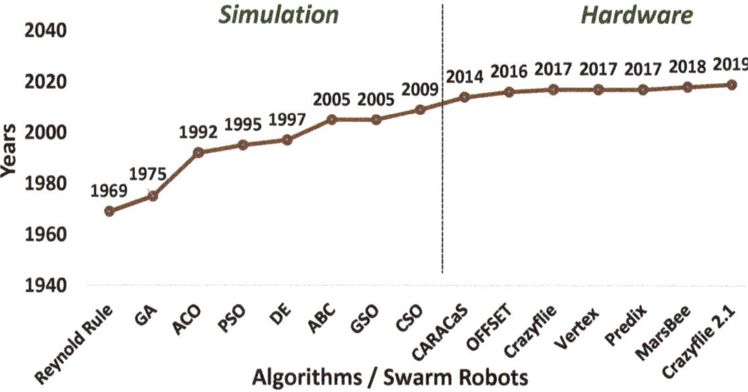

Figure 13. Evolution of swarm algorithms and swarm robotics.

6. Conclusions

Swarm robotics aims to develop simple, autonomous or self-governing robots that can cooperate to solve real-world problems collectively. Intelligent swarm algorithms are needed to enable the robots to interact autonomously and coordinate together without centralized control. The research on the swarm robotics domain started in the late 1900s, and the development work started in the early 2000s, gradually evolving the previous research and simulation work towards the actual real-world projection of swarm robotics. But there is a gap between theoretical and industrial research in swarm robotics. Theoretical research mainly focused on simulating swarm behaviours, while industrial research focuses on designing hardware that can execute swarm behaviour. Therefore, it is crucial to deploy swarm algorithms on hardware that can accommodate swarm behaviour functionality.

This article provides a comprehensive overview to new researchers of the swarm robotics field. It classifies the definition of swarm robots and identifies the difference between a multi-agent system and an actual swarm of agents. A detailed review of the swarm's most emerging swarm behaviors, and swarm intelligence algorithms is captured, keeping in view the limitation and the transformation towards the industrial application and development of the swarm robotic platform. In addition to the industrial application, this paper reviewed several research hardware platforms specifically designed to demonstrate or replicate any swarm behaviour. Finally, this paper concludes by reviving the era of swarm robotics from the past, present, and future projections with expected timelines of evolving the system and having real-world application, agnostic of swarm robotics platforms.

This article provides valuable insights for researchers in swarm robotics by highlighting various areas of research gaps, including algorithmic and hardware implementation. It emphasizes the importance of addressing these gaps to enable effective collaboration among robots. Researchers can bridge the gap between theoretical and industrial research in swarm robotics, leading to advancements in the field.

Author Contributions: M.M.S.: contributes to the main part of the research focused on the hardware including the swarm of robots, their limitations, future projections, and the overall organization of the paper structure. Z.S.: researched on swarm intelligence algorithms, swarm behaviors, and paper formatting. A.A.: review, rewriting, formatting and editing of the paper. H.M.: Supported in the review of industrial application, abstract, and conclusion. M.H.Y.: researched on the era of swarm robotics, allied applications and future projections. N.K.B. & F.H.: co-supervision of research work, paper reviewing, editing and structural guidance. All authors have read and agreed to the published version of the manuscript.

Funding: We acknowledge the Higher Education Commission (HEC) of Pakistan, for funding this project through a grant titled "Establishment of National Centre of Robotics and Automation (NCRA)" under Grant DF-1009-31.

Data Availability Statement: No new data were created or analyzed in this study. Data sharing is not applicable to this article.

Acknowledgments: The authors acknowledge the funding and support from National Centre of Robotics and Automation (NCRA) Pakistan for this research work.

Conflicts of Interest: The authors declare no conflict of interest.

References

1. Schranz, M.; Umlauft, M.; Sende, M.; Elmenreich, W. Swarm robotic behaviors and current applications. *Front. Robot. AI* **2020**, *7*, 36. [CrossRef] [PubMed]
2. Arnold, R.; Carey, K.; Abruzzo, B.; Korpela, C. What is a robot swarm: A definition for swarming robotics. In Proceedings of the 2019 IEEE 10th Annual Ubiquitous Computing, Electronics & Mobile Communication Conference (UEMCON), New York, NY, USA, 10–12 October 2019; pp. 74–81.
3. Dorigo, M.; Theraulaz, G.; Trianni, V. Reflections on the future of swarm robotics. *Sci. Robot.* **2020**, *5*, eabe4385. [CrossRef] [PubMed]
4. Floreano, D.; Mattiussi, C. *Bio-Inspired Artificial Intelligence*; MIT Press: Cambridge, MA, USA, 2008; Volume 5, pp. 335–396.

5. Maxim, P.M.; Spears, W.M.; Spears, D.F. Robotic chain formations. *IFAC Proc. Vol.* **2009**, *42*, 19–24. [CrossRef]
6. Rubenstein, M.; Cornejo, A.; Nagpal, R. Programmable self-assembly in a thousand-robot swarm . *Science* **2014**, *345*, 795–799. [CrossRef] [PubMed]
7. Slavkov, I.; Carrillo-Zapata, D.; Carranza, N.; Diego, X.; Jansson, F.; Kaorp, J.; Hauert, S.; Sharpe, J. Morphogenesis in robot swarms . *Sci. Robot.* **2018**, *3*, eaau9178. [CrossRef]
8. Carrillo, D.; Sharpe, J.; Winfield, A.F.; Giuggioli, L.; Hauert, S. Toward controllable morphogenesis in large robot swarms. *IEEE Robot. Autom. Lett.* **2019**, *4*, 3386–3393. [CrossRef]
9. Bayındır, L. A review of swarm robotics tasks. *Neurocomputing* **2016**, *172*, 292–321. [CrossRef]
10. Sion, A.; Reina, A.; Birattari, M.; Tuci, E. Controlling robot swarm aggregation through a minority of informed robots. In Proceedings of the Swarm Intelligence: 13th International Conference, ANTS 2022, Málaga, Spain, 2–4 November 2022; Springer: Cham, Switzerland, 2022; pp. 91–103.
11. Ducatelle, F.; Di Caro, G.A.; Förster, A.; Bonani, M.; Dorigo, M.; Magnenat, S.; Mondada, F.; O'Grady, R.; Pinciroli, C.; Rétornaz, P.; et al. Cooperative navigation in robotic swarms. *Swarm Intell.* **2013**, *8*, 1–33. [CrossRef]
12. Varga, M.; Bogdan, S.; Dragojević, M.; Miklić, D. Collective search and decision-making for target localization. *Math. Comput. Model. Dyn. Syst.* **2012**, *18*, 51–65. [CrossRef]
13. Firat, Z.; Ferrante, E.; Zakir, R.; Prasetyo, J.; Tuci, E. Group-size regulation in self-organized aggregation in robot swarms. In *Swarm Intelligence, Proceedings of the 12th International Conference, ANTS 2020, Barcelona, Spain, 26–28 October 2020;* Springer: Cham, Switzerland, 2020; pp. 315–323.
14. Christensen, A.L.; O'Grady, R.; Birattari, M.; Dorigo, M. Exogenous fault detection in a collective robotic task. In Proceedings of the Advances in Artificial Life: 9th European Conference, ECAL 2007, Lisbon, Portugal, 10–14 September 2007; Springer: Berlin/Heidelberg, Germany, 2007.
15. Alitappeh, R.J.; Jeddisaravi, K. Multi-robot exploration in task allocation problem. *Appl. Intell.* **2022**, *52*, 2189–211. [CrossRef]
16. Moussa, M.; Beltrame, G. On the robustness of consensus-based behaviors for robot swarms. *Swarm Intell.* **2020**, *14*, 205–231. [CrossRef]
17. Tang, W.; Zhang, C.; Zhong, Y.; Zhu, P.; Hu, Y.; Jiao, Z.; Wei, X.; Lu, G.; Wang, J.; Liang, Y.; et al. Customizing a self-healing soft pump for robot. *Nat. Commun.* **2021**, *12*, 2247. [CrossRef] [PubMed]
18. Bonabeau, E.; Dorigo, M.; Theraulaz, G.; Theraulaz, G. *Swarm Intelligence: From Natural to Artificial Systems* ; Oxford University Press: New York, NY, USA, 1999.
19. Goldberg, D.E. *Genetic Learning in Optimization, Search and Machine Learning;* Addisson Wesley: Boston, CA, USA, 1994.
20. Holland, J.H. Genetic algorithms. *Sci. Am.* **1992**, *267*, 66–73. [CrossRef]
21. Grefenstette, J.J. Navy Center for Applied Research in Artificial Intelligence Naval Research Laboratory, Washington, DC 20375-5000. *Mach. Learn. Proc. 1989* **2014**, *28*, 340.
22. Bhattacharjya, D.R. *Introduction to Genetic Algorithms;* IIT Guwahati: Guwahati, India, 1998.
23. Devooght, R. Multi-Objective Genetic Algorithm. 2010, 1–39. Available online: epb-physique.ulb.ac.be/IMG/pdf/devooght_2011.pdf (accessed on 1 January 2014).
24. Uzel, O.; Koc, E. Basic of Genetic Programming, Graduation Project I. 2012, 1–25. Available online: http://mcs.cankaya.edu.tr/proje/2012/guz/omer_erdem/Rapor.pdf (accessed on 3 January 2014).
25. de Oliveira, G.M.; Silva, R.G.; do Amaral, L.R.; Martins, L.G. An evolutionary-cooperative model based on cellular automata and genetic algorithms for the navigation of robots under formation control. In Proceedings of the 2018 7th Brazilian Conference on Intelligent Systems (BRACIS), Sao Paulo, Brazil, 22–25 October 2018; pp. 426–431.
26. Shao, J. Robot Path Planning Method Based on Genetic Algorithm. In *Journal of Physics: Conference Series;* IOP Publishing: Bristol, UK, 2021; Volume 1881, p. 022046.
27. Godinho, F.M.; Barco, C.F.; Tavares Neto, R.F. Using Genetic Algorithms to solve scheduling problems on flexible manufacturing systems (FMS): A literature survey, classification and analysis. *Flex. Serv. Manuf. J.* **2014**, *26*, 408–431. [CrossRef]
28. Cheng, C.; Yang, Z.; Xing, L.; Tan, Y. An improved genetic algorithm with local search for order acceptance and scheduling problems. In Proceedings of the 2013 IEEE Symposium on Computational Intelligence in Production and Logistics Systems (CIPLS), Singapore, 16–19 April 2013; pp. 115–122.
29. Sachdeva, J.; Kumar, V.; Gupta, I.; Khelwal, N.; Ahuja, C.K. Multiclass brain tumor classification using GA-SVM. In Proceedings of the 2011 Developments in E-systems Engineering, Dubai, United Arab Emirates, 6–8 December 2011; pp. 182–187.
30. Khuntia, A.K.; Choudhury, B.B.; Biswal, B.B.; Dash, K.K. A heuristics based multi-robot task allocation. In Proceedings of the 2011 IEEE Recent Advances in Intelligent Computational Systems, Trivandrum, India, 22–24 September 2011; pp. 407–410.
31. Yang, Q.; Yu, M.; Liu, S.; Chai, Z.M. Path planning of robotic fish based on genetic algorithm and modified dynamic programming. In Proceedings of the 2011 International Conference on Advanced Mechatronic Systems, Zhengzhou, China, 11–13 August 2011; pp. 419–424.
32. Kang, C.C.; Chuang, Y.J.; Tung, K.C.; Chao, C.C.; Tang, C.Y.; Peng, S.C.; Wong, D.S. A genetic algorithm-based boolean delay model of intracellular signal transduction in inflammation. *BMC Bioinform.* **2011**, *12*, 17. [CrossRef]
33. Foschini, L.; Tortonesi, M. Adaptive and business-driven service placement in federated Cloud computing environments. In Proceedings of the 2013 IFIP/IEEE International Symposium on Integrated Network Management (IM 2013), Ghent, Belgium, 27–31 May 2013; pp. 1245–1251.

34. Fu, H.H.; Li, Z.J.; Li, G.W.; Jin, X.T.; Zhu, P.H. Modeling and controlling of engineering ship based on genetic algorithm. In Proceedings of the 2012 International Conference on Modelling, Identification and Control, Wuhan, China, 24–26 June 2012; pp. 394–398.

35. Mahmudy, W.F.; Marian, R.M.; Luong, L.H. Optimization of part type selection and loading problem with alternative production plans in flexible manufacturing system using hybrid genetic algorithms-part 1: Modelling and representation. In Proceedings of the 2013 5th International Conference on Knowledge and Smart Technology (KST), Chonburi, Thailand, 31 January 2013–1 February 2013; pp. 75–80.

36. Jing, X.; Liu, Y.; Cao, W. A hybrid genetic algorithm for route optimization in multimodal transport . In Proceedings of the 2012 Fifth International Symposium on Computational Intelligence and Design, Hangzhou, China, 28–29 October 2012; pp. 261–264.

37. Murata, T.; Ishibuchi, H. MOGA: Multi-objective genetic algorithms. In Proceedings of the IEEE International Conference on Evolutionary Computation, Piscataway, NJ, USA, 29 November 1995; pp. 289–294.

38. Dorigo, M. Optimization, Learning and Natural Algorithms. Ph.D. Thesis, Politecnico di Milano, Milan, Italy, 1992.

39. Dorigo, M.; Birattari, M.; Stutzle, T. Ant colony optimization. *IEEE Comput. Intell. Mag.* **2006**, *1*, 28–39. [CrossRef]

40. Abreu, N.; Ajmal, M.; Kokkinogenis, Z.; Bozorg, B. *Ant Colony Optimization*; University of Porto: Porto, Portugal, 2011.

41. Selvi, V.; Umarani, R. Comparative analysis of ant colony and particle swarm optimization techniques. *Int. J. Comput. Appl.* **2010**, *5*, 1–6. [CrossRef]

42. Valdez, F.; Chaparro, I. Ant colony optimization for solving the TSP symetric with parallel processing. In Proceedings of the 2013 Joint IFSA World Congress and NAFIPS Annual Meeting (IFSA/NAFIPS), Edmonton, AB, Canada, 24–28 June 2013; pp. 1192–1196.

43. Tosun, U.; Dokeroglu, T.; Cosar, A. A robust island parallel genetic algorithm for the quadratic assignment problem. *Int. J. Prod. Res.* **2013**, *51*, 4117–4133. [CrossRef]

44. Yagmahan, B.; Yenisey, M.M. A multi-objective ant colony system algorithm for flow shop scheduling problem. *Expert Syst. Appl.* **2010**, *37*, 1361–1368. [CrossRef]

45. Abdelaziz, A.Y.; Elkhodary, S.M.; Osama, R.A. Distribution networks reconfiguration for loss reduction using the Hyper Cube Ant Colony Optimization. In Proceedings of the 2011 International Conference on Computer Engineering & Systems, Cairo, Egypt, 29 November–1 December 2011; pp. 79–84.

46. Kumar, S.B.; Myilsamy, G. Multi-target tracking in mobility sensor networks using Ant Colony Optimization. In Proceedings of the 2013 IEEE International Conference On Emerging Trends in Computing, Communication and Nanotechnology (ICECCN), Tirunelveli, India, 25–26 March 2013; pp. 350–354.

47. Agrawal, P.; Kaur, S.; Kaur, H.; Dhiman, A. Analysis and synthesis of an ant colony optimization technique for image edge detection . In Proceedings of the 2012 International Conference on Computing Sciences, Phagwara, India, 14–15 September 2012; pp. 127–131.

48. He, Y.; Zeng, Q.; Liu, J.; Xu, G.; Deng, X. Path planning for indoor UAV based on Ant Colony Optimization. In Proceedings of the 2013 25th Chinese Control and Decision Conference (CCDC), Guiyang, China, 25–27 May 2013; pp. 2919–2923.

49. Abdallah, H.; Emara, H.M.; Dorrah, H.T.; Bahgat, A. Using ant colony optimization algorithm for solving project management problems. *Expert Syst. Appl.* **2009**, *36*, 10004–10015. [CrossRef]

50. Kennedy, J.; Eberhart, R. Particle swarm optimization. In Proceedings of the ICNN'95-International Conference on Neural Networks, Perth, WA, Australia, 27 November–1 December 1995; Volume 4, pp. 1942–1948.

51. Poli, R.; Kennedy, J.; Blackwell, T. Particle swarm optimization: An overview. *Swarm Intell.* **2007**, *1*, 33–57. [CrossRef]

52. Del Valle, Y.; Venayagamoorthy, G.K.; Mohagheghi, S.; Hernandez, J.C.; Harley, R.G. Particle swarm optimization: Basic concepts, variants and applications in power systems. *IEEE Trans. Evol. Comput.* **2008**, *12*, 171–195. [CrossRef]

53. Shi, Y.; Eberhart, R. A modified particle swarm optimizer. In Proceedings of the 1998 IEEE International Conference on Evolutionary Computation Proceedings. IEEE World Congress on Computational Intelligence (Cat. No. 98TH8360), Anchorage, AK, USA, 4–9 May 1998; pp. 69–73.

54. Yan, X.; Wu, Q.; Liu, H.; Huang, W. An improved particle swarm optimization algorithm and its application. *Int. J. Comput. Sci. Issues* **2013**, *10*, 316.

55. Arumugam, M.S.; Rao, M.V.; Tan, A.W. A novel and effective particle swarm optimization like algorithm with extrapolation technique . *Appl. Soft Comput.* **2009**, *9*, 308–320. [CrossRef]

56. Kiranyaz, S.; Ince, T.; Yildirim, A.; Gabbouj, M. Fractional particle swarm optimization in multidimensional search space. *IEEE Trans. Syst. Man Cybern. Part B* **2009**, *40*, 298–319. [CrossRef] [PubMed]

57. Banks, A.; Vincent, J.; Anyakoha, C. A review of particle swarm optimization. Part I: Background and development. *Nat. Comput.* **2007**, *6*, 467–484. [CrossRef]

58. Banks, A.; Vincent, J.; Anyakoha, C. A review of particle swarm optimization. Part II: Hybridisation, combinatorial, multicriteria and constrained optimization, and indicative applications. *Nat. Comput.* **2008**, *7*, 109–124. [CrossRef]

59. Gong, D.; Lu, L.; Li, M. Robot path planning in uncertain environments based on particle swarm optimization. In Proceedings of the 2009 IEEE Congress on Evolutionary Computation, Trondheim, Norway, 18–21 May 2009; pp. 2127–2134.

60. Gong, M.; Cai, Q.; Chen, X.; Ma, L. Complex network clustering by multiobjective discrete particle swarm optimization based on decomposition. *IEEE Trans. Evol. Comput.* **2013**, *18*, 82–97. [CrossRef]

61. Sivakumar, P.; Grace, S.S.; Azeezur, R.A. Investigations on the impacts of uncertain wind power dispersion on power system stability and enhancement through PSO technique. In Proceedings of the 2013 International Conference on Energy Efficient Technologies for Sustainability, Nagercoil, India, 10–12 April 2013; pp. 1370–1375.
62. Li, F.; Li, D.; Wang, C.; Wang, Z. Network signal processing and intrusion detection by a hybrid model of LSSVM and PSO. In Proceedings of the 2013 15th IEEE International Conference on Communication Technology, Guilin, China, 17–19 November 2013; pp. 11–14.
63. Jun, Z.; Kanyu, Z. A Particle Swarm Optimization Approach for Optimal Design of PID Controller for Temperature Control in HVAC. In Proceedings of the 2011 Third International Conference on Measuring Technology and Mechatronics Automation, Shanghai, China, 6–7 January 2011; pp. 230–233.
64. Atyabi, A.; Luerssen, M.; Fitzgibbon, S.P.; Powers, D.M.W. PSO-Based Dimension Reduction of EEG Recordings: Implications for Subject Transfer in BCI. *Neurocomputing* **2013**, *119*, 319–331. [CrossRef]
65. Mohan, S.; Mahesh, T.R. Particle Swarm Optimization based Contrast Limited enhancement for mammogram images. In Proceedings of the 2013 7th International Conference on Intelligent Systems and Control (ISCO), Coimbatore, India, 4–5 January 2013; pp. 384–388.
66. Gorai, A.; Ghosh, A. Hue-preserving colour image enhancement using particle swarm optimization. In Proceedings of the 2011 IEEE Recent Advances in Intelligent Computational Systems, Trivandrum, India, 22–24 September 2011; pp. 563–568.
67. Na, L.; Yuanxiang, L. Image Restoration Using Improved Particle Swarm Optimization . In Proceedings of the 2011 International Conference on Network Computing and Information Security, Guilin, China, 14–15 May 2011; pp. 394–397.
68. Storn, R.; Price, K. Differential Evolution—A Simple and Efficient Heuristic for Global Optimization over Continuous Spaces. *J. Glob. Optim.* **1997**, *11*, 341–359. [CrossRef]
69. Price, K.; Storn, R.; Lampinen, J. *Differential Evolution: A Practical Approach to Global Optimization*; Springer: Berlin/Heidelberg, Germany, 2005; Volume 104, p. 542.
70. Wu, Y.; Lee, W.; Chien, C. Modified the Performance of Differential Evolution Algorithm with Dual Evolution Strategy. *Int. Conf. Mach. Learn. Comput.* **2011**, *3*, 57–63.
71. Muñoz, J.; López, B.; Quevedo, F.; Barber, R.; Garrido, S.; Moreno, L. Geometrically constrained path planning for robotic grasping with Differential Evolution and Fast Marching Square. *Robotica* **2023**, *41*, 414–432. [CrossRef]
72. Mahmoud, Z.S.; Powers, D.; Sammut, K.; Yazdani, A.M. Differential evolution for efficient AUV path planning in time variant uncertain underwater environment. *arXiv* **2016**, arXiv:1604.02523.
73. Dragoi, E.; Curteanu, S.; Vlad, D. Differential evolution applications in electromagnetics. In Proceedings of the 2012 International Conference and Exposition on Electrical and Power Engineering, Iasi, Romania, 25–27 October 2012; pp. 636–640.
74. Myeong-Chun, L.; Sung-Bae, C. Interactive differential evolution for image enhancement application in smart phone. In Proceedings of the IEEE Congress on Evolutionary Computation, Brisbane, QLD, Australia, 10–15 June 2012; pp. 1–6.
75. Yilmaz, A.R.; Yavuz, O.; Erkmen, B. Training multilayer perceptron using differential evolution algorithm for signature recognition application. In Proceedings of the 2013 21st Signal Processing and Communications Applications Conference (SIU), Haspolat, Turkey, 24–26 April 2013; pp. 1–4.
76. Chiou, J.; Chang, C.; Wang, C. Hybrid differential evolution for static economic dispatch. In Proceedings of the 2014 International Symposium on Computer, Consumer and Control, Taichung, Taiwan, 10–12 June 2014; pp. 950–953.
77. Karaboga, D. *An Idea Based on Honeybee Swarm for Numerica Optimization*; Technical Report TR06; Erciyes University: Talas/Kayseri, Türkiye, 2005.
78. Karaboga, D.; Basturk, B. A powerful and efficient algorithm for numeric function optimization: Artificial Bee Colony (ABC) algorithm. *J. Glob. Optim.* **2007**, *39*, 459–471. [CrossRef]
79. Karaboga, D. Artificial bee colony algorithm. *Scholarpedia* **2010**, *5*, 6915. [CrossRef]
80. Singh, A. An artificial bee colony algorithm for the leaf-constrained minimum spanning tree problem. *Appl. Soft Comput.* **2009**, *9*, 625–631. [CrossRef]
81. Abu-Mouti, F.S.; El-Hawary, M.E. Overview of Artificial Bee Colony (ABC) algorithm and its applications. In Proceedings of the 2012 IEEE International Systems Conference SysCon, Vancouver, BC, Canada, 19–22 March 2012; pp. 1–6.
82. Gerhardt, E.; Gomes, H.M. Artificial Bee Colony (ABC) algorithm for engineering optimization problems. In Proceedings of the International Conference on Engineering Optimization, Rio de Janeiro, Brazil, 1–5 July 2012; pp. 1–11.
83. Tan, L.; Shi, J.; Gao, J.; Wang, H.; Zhang, H.; Zhang, Y. Multi-UAV path planning based on IB-ABC with restricted planned arrival sequence. *Robotica* **2023**, *41*, 1244–1257. [CrossRef]
84. Sharma, T.K.; Pant, M. Golden search based artificial bee colony algorithm and its application to solve engineering design problems. In Proceedings of the 2012 Second International Conference on Advanced Computing & Communication Technologies, Rohtak, India, 7–8 January 2012; pp. 156–160.
85. Lee, C.H.; Park, J.Y.; Park, J.Y.; Han, S.Y. Application of artificial bee colony algorithm for structural topology optimization. In Proceedings of the 2012 8th International Conference on Natural Computation, Chongqing, China, 29–31 May 2012; pp. 1023–1025.
86. Lee, T.E.; Cheng, J.H.; Jiang, L.L. A new artificial bee colony based clustering method and its application to the business failure prediction. In Proceedings of the 2012 International Symposium on Computer, Consumer and Control, Taichung, Taiwan, 4–6 June 2012; pp. 72–75.

87. Karaboga, D.; Gorkemli, B.; Ozturk, C.; Karaboga, N. A comprehensive survey: Artificial bee colony (ABC) algorithm and applications. *Artif. Intell. Rev.* **2014**, *42*, 21–57. [CrossRef]
88. Krishnan, K.N.; Ghose, D. Glowworm swarm optimization for simultaneous capture of multiple local optima of multimodal functions. *Swarm Intell.* **2009**, *3*, 87–124. [CrossRef]
89. Krishnan, K.N.; Ghose, D. Glowworm swarm optimisation: A new method for optimising multi-modal functions. *Int. J. Comput. Intell. Stud.* **2009**, *1*, 93–119.
90. Krihnanand, K.N.; Amruth, P.; Guruprasad, M.H. Glowworm-inspired Robot Swarm for Simultaneous Taxis towards Multiple Radiation Sources. In Proceedings of the 2006 IEEE International Conference on Robotics and Automation, (ICRA), Orlando, FL, USA, 15–19 May 2006; pp. 958–963.
91. Zainal, N.; Zain, A.M.; Radzi, N.H.; Udin, A. Glowworm swarm optimization (GSO) algorithm for optimization problems: A state-of-the-art review. *Appl. Mech. Mater.* **2013**, *421*, 507–511. [CrossRef]
92. Zhang, Y.; Ma, X.; Miao, Y. Localization of multiple odor sources using modified glowworm swarm optimization with collective robots. In Proceedings of the 30th Chinese Control Conference, Yantai, China, 22–24 July 2011; pp. 1899–1904.
93. Goel, U.; Varshney, S.; Jain, A.; Maheshwari, S.; Shukla, A. Three dimensional path planning for UAVs in dynamic environment using glow-worm swarm optimization. *Procedia Comput. Sci.* **2018**, *133*, 230–239. [CrossRef]
94. Khan, A.; Aftab, F.; Zhang, Z. Self-organization based clustering scheme for FANETs using Glowworm Swarm Optimization. *Phys. Commun.* **2019**, *36*, 100769. [CrossRef]
95. He, D.X.; Zhu, H.Z.; Liu, G.-Q. Glowworm swarm optimization algorithm for solving multi-constrained QoS multicast routing problem. In Proceedings of the 2011 Seventh International Conference on Computational Intelligence and Security, Sanya, China, 3–4 December 2011; pp. 66–70.
96. Krishnan, K.N.; Ghose, D. Detection of multiple source locations using a glowworm metaphor with applications to collective robotics. In Proceedings of the 2005 IEEE Swarm Intelligence Symposium, Pasadena, CA, USA, 8–10 June 2005; pp. 84–91.
97. Senthilnath, J.; Omkar, S.N.; Mani, V.; Tejovanth, N.; Diwakar, P.G.; Shenoy, A. Multi-spectral satellite image classification using glowworm swarm optimization. In Proceedings of the 2011 IEEE International Geoscience and Remote Sensing Symposium, Vancouver, BC, Canada, 24–29 July 2011; pp. 47–50.
98. McGill, K.; Taylor, S. Comparing swarm algorithms for large scale multi-source localization. In Proceedings of the 2009 IEEE International Conference on Technologies for Practical Robot Applications, Woburn, MA, USA, 9–10 November 2009; pp. 48–54.
99. Menon, P.P.; Ghose, D. Simultaneous source localization and boundary mapping for contaminants. In Proceedings of the 2012 American Control Conference (ACC), Montreal, QC, Canada, 27–29 June 2012; pp. 4174–4179.
100. Yang, X.S.; Deb, S. Cuckoo search via Lévy flights. In Proceedings of the 2009 World Congress on Nature & Biologically Inspired Computing (NaBIC), Coimbatore, India, 9–11 December 2009; pp. 210–214.
101. Yang, X.S.; Deb, S. Engineering optimisation by cuckoo search. *Int. J. Math. Model. Numer. Optim.* **2010**, *1*, 330–343. [CrossRef]
102. Yang, X.S.; Deb, S. Multi-objective cuckoo search for design optimization. *Comput. Oper. Res.* **2013**, *40*, 1616–1624. [CrossRef]
103. Xie, C.; Zheng, H. Application of improved Cuckoo search algorithm to path planning unmanned aerial vehicle. In Proceedings of the Intelligent Computing Theories and Application: 12th International Conference, ICIC 2016, Lanzhou, China, 2–5 August 2016; Springer: Berlin/Heidelberg, Germany, 2016; pp. 722–729.
104. Chaowanawatee, K.; Heednacram, A. Implementation of cuckoo search in RBF neural network for flood forecasting. In Proceedings of the 2012 4th International Conference on Computational Intelligence, Communication Systems and Networks, Phuket, Thailand, 24–26 July 2012; pp. 22–26.
105. Kumar, A.; Chakarverty, S. Design optimization for reliable embedded system using Cuckoo Search. In Proceedings of the 2011 3rd International Conference on Electronics Computer Technology, Kanyakumari, India, 8–10 April 2011; Volume 1, pp. 264–268.
106. Khodier, M. Optimisation of antenna arrays using the cuckoo search algorithm. *IET Microwaves Antennas Propag.* **2013**, *7*, 458–464. [CrossRef]
107. Vo, D.N.; Schegner, P.; Ongsakul, W. Cuckoo search algorithm for non-convex economic dispatch. *Gener. Transm. Distrib.* **2013**, *7*, 645–654. [CrossRef]
108. Yang, X.S.; Deb, S.; Karamanoglu, M.; Xingshi, N. Cuckoo search for business optimization applications. In Proceedings of the 2012 National Conference on Computing and Communication Systems, Durgapur, India, 21–22 November 2012; pp. 1–5.
109. Jati, G.K.; Manurung, H.M.; Suyanto, S. Discrete cuckoo search for traveling salesman problem. In Proceedings of the 2012 7th International Conference on Computing and Convergence Technology (ICCCT), Seoul, Republic of Korea, 3–5 December 2012; pp. 993–997.
110. Sharkey, A.J. Swarm robotics and minimalism. *Connect. Sci.* **2007**, *19*, 245–260. [CrossRef]
111. Programmable Robot Swarms. Available online: https://wyss.harvard.edu/technology/programmable-robot-swarms/ (accessed on 6 November 2019).
112. Caprari, G.; Balmer, P.; Piguet, R.; Siegwart, R. The Autonomous Micro Robot "Alice": A platform for scientific and commercial applications. In Proceedings of the MHA'98, 1998 International Symposium on Micromechatronics and Human Science Creation of New Industry (Cat. No. 98TH8388), Nagoya, Japan, 25–28 November 1998; pp. 231–235.
113. Arvin, F.; Samsudin, K.; Ramli, A.R. Development of a miniature robot for swarm robotic application. *Int. J. Comput. Electr. Eng.* **2009**, *1*, 436–442. [CrossRef]

114. Arvin, F.; Murray, J.; Zhang, C.; Yue, S. Colias: An autonomous micro robot for swarm robotic applications. *Int. J. Adv. Robot. Syst.* **2014**, *11*, 113. [CrossRef]

115. Soares, J.M.; Navarro, I.; Martinoli, A. The Khepera IV mobile robot: Performance evaluation, sensory data and software toolbox. In Proceedings of the Robot 2015: Second Iberian Robotics Conference: Advances in Robotics, Lisbon, Portugal, 19–21 November 2015; Volume 1 , pp. 767–781.

116. Pickem, D.; Lee, M.; Egerstedt, M. The GRITSBot in its natural habitat-a multi-robot testbed. In Proceedings of the 2015 IEEE International Conference on Robotics and Automation (ICRA), Seattle, WA, USA, 26–30 May 2015; pp. 4062–4067.

117. Jones, S.; Studley, M.; Hauert, S.; Winfield, A.F. A two teraflop swarm. *Front. Robot. AI* **2018**, *5*, 11. [CrossRef]

118. Riedo, F.; Chevalier, M.; Magnenat, S.; Mondada, F. Thymio II, a robot that grows wiser with children. In Proceedings of the 2013 IEEE Workshop on Advanced Robotics and Its Social Impacts, Tokyo, Japan, 7–9 November 2013; pp. 187–193.

119. Wilson, S.; Gameros, R.; Sheely, M.; Lin, M.; Dover, K.; Gevorkyan, R.; Haberl, M.; Bertozzi, A.; Berman, S. Pheeno, a versatile swarm robotic research and education platform. *IEEE Robot. Autom. Lett.* **2016**, *1*, 884–891. [CrossRef]

120. Jdeed, M.; Zhevzhyk, S.; Steinkellner, F.; Elmenreich, W. Spiderino-a low-cost robot for swarm research and educational purposes. In Proceedings of the 2017 13th Workshop on Intelligent Solutions in Embedded Systems (WISES), Hamburg, Germany, 12–13 June 2017; pp. 35–39.

121. Seyfried, J.; Szymanski, M.; Bender, N.; Estana, R.; Thiel, M.; Wörn, H. The I-SWARM project: Intelligent small world autonomous robots for micro-manipulation. In Proceedings of the Swarm Robotics: SAB 2004 International Workshop, Santa Monica, CA, USA, 17 July 2004.

122. Le Goc, M.; Kim, L.H.; Parsaei, A.; Fekete, J.D.; Dragicevic, P.; Follmer, S. Zooids: Building blocks for swarm user interfaces. In Proceedings of the 29th Annual Symposium on User Interface Software and Technology, Tokyo, Japan, 16–19 October 2016; pp. 97–109.

123. Dhanaraj, N.; Hewitt, N.; Edmonds-Estes, C.; Jarman, R.; Seo, J.; Gunner, H. Adaptable platform for interactive swarm robotics (apis): A human-swarm interaction research testbed. In Proceedings of the 19th International Conference on Advanced Robotics (Belo Horizonte), Belo Horizonte, Brazil, 2–6 December 2019; pp. 720–726. [CrossRef]

124. Kettler, A.; Szymanski, M.; Wörn, H. The wanda robot and its development system for swarm algorithms. In *Advances in Autonomous Mini Robots, Proceedings of the 6-th AMiRE Symposium, Bielefeld, Germany, 23–25 May 2011*; Springer: Berlin/Heidelberg, Germany, 2012; pp. 133–146.

125. Klingner, J.; Kanakia, A.; Farrow, N.; Reishus, D.; Correll, N. A stick-slip omnidirectional powertrain for low-cost swarm robotics: Mechanism, calibration, and control. In Proceedings of the 2014 IEEE International Conference on Intelligent Robots and Systems, Chicago, IL, USA, 14–18 September 2014; pp. 846–851. [CrossRef]

126. Mondada, F.; Pettinaro, G.C.; Kwee, I.W.; Guignard, A.; Gambardella, L.M.; Floreano, D.; Nolfi, S.; Deneubourg, J.L.; Dorigo, M. *SWARM-BOT: A Swarm of Autonomous Mobile Robots with Self-Assembling Capabilities*; Technical Report; ETH-Zürich: Zürich, Switzerland, 2002.

127. Groß, R.; Bonani, M.; Mondada, F.; Dorigo, M. Autonomous self-assembly in swarm-bots. *IEEE Trans. Robot.* **2006**, *22*, 1115–1130.

128. Dorigo, M.; Floreano, D.; Gambardella, L.M.; Mondada, F.; Nolfi, S.; Baaboura, T.; Birattari, M.; Bonani, M.; Brambilla, M.; Brutschy, A.; et al. Swarmanoid: A novel concept for the study of heterogeneous robotic swarms . *IEEE Robot. Autom. Mag.* **2013**, *20*, 60–71. [CrossRef]

129. Petersen, K.H.; Nagpal, R.; Werfel, J.K. Termes: An autonomous robotic system for three-dimensional collective construction. In *Robotics: Science and Systems VII*; MIT Press: Cambridge, CA, USA, 2011 .

130. Kernbach, S.; Meister, E.; Schlachter, F.; Jebens, K.; Szymanski, M.; Liedke, J.; Laneri, D.; Winkler, L.; Schmickl, T.; Thenius, R.; et al. Symbiotic robot organisms: REPLICATOR and SYMBRION projects. In Proceedings of the 8th Workshop on Performance Metrics for Intelligent Systems, Gaithersburg, MD, USA, 19–21 August 2008; pp. 62–69.

131. Duff, D.; Yim, M.; Roufas, K. Evolution of polybot: A modular reconfigurable robot. In Proceedings of the Harmonic Drive Intelligent Symposium, Nagano, Japan, 19 November 2001.

132. Murata, S.; Yoshida, E.; Kamimura, A.; Kurokawa, H.; Tomita, K.; Kokaji, S. M-TRAN: Self-reconfigurable modular robotic system. *IEEE/ASME Trans. Mechatronics* **2002**, *7*, 431–441. [CrossRef]

133. Kurokawa, H.; Kamimura, A.; Yoshida, E.; Tomita, K.; Kokaji, S.; Murata, S. M-TRAN II: Metamorphosis from a four-legged walker to a caterpillar. In Proceedings of the 2003 IEEE/RSJ International Conference on Intelligent Robots and Systems (IROS 2003) (Cat. No. 03CH37453), Las Vegas, NV, USA, 27–31 October 2003; Volume 3, pp. 2454–2459.

134. Kurokawa, H.; Tomita, K.; Kamimura, A.; Kokaji, S.; Hasuo, T.; Murata, S. Distributed self-reconfiguration of M-TRAN III modular robotic system. *Int. J. Robot. Res.* **2008**, *27*, 373–386. [CrossRef]

135. Brandt, D.; Christensen, D.J.; Lund, H.H. ATRON robots: Versatility from self-reconfigurable modules. In Proceedings of the 2007 IEEE International Conference on Mechatronics and Automation, Harbin, China, 5–8 August 2007; pp. 26–32.

136. Castano, A.; Behar, A.; Will, P.M. The Conro modules for reconfigurable robots. *IEEE/ASME Trans. Mechatronics* **2002**, *7*, 403–409. [CrossRef]

137. Wei, H.; Cai, Y.; Li, H.; Li, D.; Wang, T. Sambot: A self-assembly modular robot for swarm robot. In Proceedings of the 2010 IEEE International Conference on Robotics and Automation, Anchorage, AK, USA, 3–7 May 2010; pp. 66–71.

138. Zykov, V.; Mytilinaios, E.; Desnoyer, M.; Lipson, H. Evolved and designed self-reproducing modular robotics. *IEEE Trans. Robot.* **2007**, *23*, 308–319. [CrossRef]

139. Cheng, J.; Cheng, W.; Nagpal, R. Robust and self-repairing formation control for swarms of mobile agents. In Proceedings of the Twentieth National Conference on Artificial Intelligence and the Seventeenth Innovative Applications of Artificial Intelligence Conference, Pittsburgh, PA, USA, 9–13 July 2005; Volume 5.

140. Swetha, S.; Anandan, R.; Kalaivani, K. An investigation of micro aerial vehicles (μAV). *Int. J. Eng. Technol.* **2018**, *7*, 174–177. [CrossRef]

141. Roberts, J.F.; Stirling, T.S.; Zufferey, J.; Floreano, D. Quadrotor using minimal sensing for autonomous indoor flight. In Proceedings of the European Micro Air Vehicle Conference and Flight Competition (Toulouse), Toulouse, France, 20 September 2007.

142. Oung, R.; D'Andrea, R. The distributed flight array. *Mechatronics* **2011**, *21*, 908–917. [CrossRef]

143. Preiss, J.A.; Honig, W.; Sukhatme, G.S.; Ayanian, N. Crazyswarm: A large nano-quadcopter swarm. In Proceedings of the 2017 IEEE International Conference on Robotics and Automation (ICRA), Singapore, 29 May–3 June 2017; pp. 3299–3304.

144. Heckert, A. Entwicklung eines Dynamischen Modells und Parameterschätzung für den FINken 3 Quadkopter. Bachelor's Thesis, Fakultät für Elektrotechnik und Informationstechnik, Otto-von-Guericke-Universitat Magdeburg, Magdeburg, Germany, 2016 .

145. Osterloh, C.; Pionteck, T.; Maehle, E. MONSUN II: A small and inexpensive AUV for underwater swarms. In Proceedings of the 7th German Conference on Robotics, Munich, Germany, 21–22 May 2012; pp. 1–6.

146. Christensen, A.L.; Duarte, M.; Postolache, O.; Sargento, S.; Oliveira, M.J.; Santana, P.; Nunes, L.; Velez, F.; Oliveira, S.M.; Sebastião, P.; et al. Design of communication and control for swarms of aquatic surface drones. *Des. Commun. Control. Swarms Aquat. Surf. Drones* **2015**, 548–555. [CrossRef]

147. Duarte, M.; Costa, V.; Gomes, J.; Rodrigues, T.; Silva, F.; Oliveira, S.M.; Christensen, A.L. Evolution of collective behaviors for a real swarm of aquatic surface robots. *PLoS ONE* **2016**, *11*, e0151834. [CrossRef] [PubMed]

148. Kang, C.K.; Fahimi, F.; Griffin, R.; Landrum, D.B.; Mesmer, B.; Zhang, G.; Lee, T.; Aono, H.; Pohly, J.; McCain, J.; et al. Marsbee-Swarm of Flapping Wing Flyers for Enhanced Mars Exploration. 2019. Available online: https://ntrs.nasa.gov/citations/201900 02496 (accessed on 1 March 2023) .

149. Tilman, D.; Balzer, C.; Hill, J.; Befort, B.L. Global food demand and the sustainable intensification of agriculture. *Proc. Natl. Acad. Sci. USA* **2011**, *108*, 20260–20264. [CrossRef] [PubMed]

150. Ball, D.; Ross, P.; English, A.; Patten, T.; Upcroft, B.; Fitch, R.; Sukkarieh, S.; Wyeth, G.; Corke, P. Robotics for sustainable broad-acre agriculture. In Proceedings of the Field and Service Robotics: Results of the 9th International Conference, Brisbane, Australia, 9–11 December 2013; Springer: Berlin/Heidelberg, Germany, 2015; pp. 439–453.

151. Blender, T.; Buchner, T.; Fernandez, B.; Pichlmaier, B.; Schlegel, C. Managing a mobile agricultural robot swarm for a seeding task. In Proceedings of the IECON 2016-42nd Annual Conference of the IEEE Industrial Electronics Society, Florence, Italy, 23–26 October 2016; pp. 6879–6886.

152. Saez-Pons, J.; Alboul, L.; Penders, J.; Nomdedeu, L. Multi-robot team formation control in the GUARDIANS project. *Ind. Robot. Int. J.* **2010**, *37*, 372–383. [CrossRef]

153. Telegraph, T. The Science Behind the Swarm of Robots Picking Your Grocery Order. Available online: https://www.telegraph.co. uk/technology/2018/06/04/science-behind-swarm-robots-picking-grocery-order/ (accessed on 2 August 2018).

154. Brown A. Rise of the Machines? Amazon's Army of More Than 100,000 Warehouse Robots Still Can't Replace Humans Because They Lack 'Common Sense'. Available online: http://www.dailymail.co.uk/sciencetech/article5808319/Amazon-100-000-warehouse-robots-company-insists-replacehumans.html (accessed on 2 August 2019).

155. Pickering, J. Take a Look Inside Alibaba's Smart Warehouse Where Robots do 70% of the Work. Available online: https://www.businessinsider.com/inside-alibaba-smart-warehouse-robots-70-per-cent-work-technologylogistics-2017-9?IR=T (accessed on 2 August 2019).

156. Chung, T.H. OFFensive Swarm-Enabled Tactics (OFFSET). Available online: https://www.darpa.mil/attachments/OFFSET_ProposersDay.pdf (accessed on 2 April 2019).

157. Mizokami, K. The Pentagon's Autonomous Swarming Drones Are the Most Unsettling Thing You'll See Today. Available online: https://www.popularmechanics.com/military/aviation/a24675/pentagon-autonomousswarming-drones/ (accessed on 2 April 2018).

158. Varga, M.; Basiri, M.; Heitz, G.; Floreano, D. Distributed formation control of fixed wing micro aerial vehicles for area coverage. In Proceedings of the 2015 IEEE/RSJ International Conference on Intelligent Robots and Systems (IROS), Hamburg, Germany, 28 September–2 October 2015; pp. 669–674.

159. Hauert, S.; Zufferey, J.C.; Floreano, D. Evolved swarming without positioning information: An application in aerial communication relay. *Auton. Robot.* **2009**, *26*, 21–32. [CrossRef]

160. Flushing, E.F.; Gambardella, L.M.; Di Caro, G.A. A mathematical programming approach to collaborative missions with heterogeneous teams. In Proceedings of the 2014 IEEE/RSJ International Conference on Intelligent Robots and Systems, Chicago, IL, USA, 14–18 September 2014; pp. 396–403.

161. Albani, D.; Manoni, T.; Arik, A.; Nardi, D.; Trianni, V. Field coverage for weed mapping: Toward experiments with a UAV swarm. In Proceedings of the Bio-inspired Information and Communication Technologies: 11th EAI International Conference, BICT 2019, Pittsburgh, PA, USA, 13–14 March 2019; Springer: Cham, Switzerland, 2019.

162. Barrett B. Intel Lights Up the Night with 500 'Shooting Star' Drones. Available online: https://www.wired.com/story/more-inclusive-fourth-of-july-drone-show/ (accessed on 2 August 2019).

163. Jeradi, A.; Raeissi, M.M.; Farinelli, A.; Brooks, N.; Scerri, P. Focused Exploration for Cooperative Robotic Watercraft. In Proceedings of the 2015 International Workshop on Artificial Intelligence and Robotics, Las Vegas, NV, USA, 9–10 May 2015; pp. 83–93.

164. Maccready, T. Multiscale vorticity from a swarm of drifters. In Proceedings of the 2015 IEEE/OES Eleveth Current, Waves and Turbulence Measurement (CWTM), St. Petersburg, FL, USA, 2–6 March 2015; pp. 1–6.

165. Real-Arce, D.A.; Morales, T.; Barrera, C.; Hernández, J.; Llinás, O. Smart and networking underwater robots in cooperation meshes: The swarms ECSEL: H2020 project. *Instrum. Viewp.* **2016**, *19*.

166. Rodríguez-Molina, J.; Bilbao, S.; Martínez, B.; Frasheri, M.; Cürüklü, B. An optimized, data distribution service-based solution for reliable data exchange among autonomous underwater vehicles. *Sensors* **2017**, *17*, 1802. [CrossRef]

167. Li, N.; Cürüklü, B.; Bastos, J.; Sucasas, V.; Fernandez, J.A.; Rodriguez, J. A probabilistic and highly efficient topology control algorithm for underwater cooperating AUV networks. *Sensors* **2017**, *17*, 1022. [CrossRef]

168. Smalley, D. Autonomous Swarmboats: New Missions, Safe Harbors. *Off. Nav. Res.* **2016**, *14*.

169. Hsu, J. US navy's drone boat swarm practices harbor defense. *IEEE Spectr.* **2016**, *19*.

170. Escoubet, C.P.; Fehringer, M.; Goldstein, M. Introduction the cluster mission. In *Proceeding Annual Geophysics*; Copernicus GmbH: Göttingen, Germany, 2001; Volume 19, pp. 1197–1200.

171. Sneyd, J.; Theraula, G.; Bonabeau, E.; Deneubourg, J.L.; Franks, N.R. *Self-Organization in Biological Systems*; Princeton University Press: Princeton, NJ, USA, 2003.

172. Sumpter, D.J. Collective animal behavior. In *Collective Animal Behavior*; Princeton University Press: Princeton, NJ, USA, 2010.

173. Dorigo, M. Ant colony optimization. *Scholarpedia* **2007**, *2*, 1461. [CrossRef]

174. Dorigo, M.; de Oca, M.A.; Engelbrecht, A. Particle swarm optimization. *Scholarpedia* **2008**, *3*, 1486. [CrossRef]

175. Di Caro, G.; Dorigo, M. AntNet: Distributed stigmergetic control for communications networks. *J. Artif. Intell. Res.* **1998**, *9*, 317–365. [CrossRef]

176. Moussaïd, M.; Helbing, D.; Theraulaz, G. How simple rules determine pedestrian behavior and crowd disasters. *Proc. Natl. Acad. Sci. USA* **2011**, *108*, 6884–6888. [CrossRef]

177. Jeon, S.; Kim, S.; Ha, S.; Lee, S.; Kim, E.; Kim, S.Y.; Park, S.H.; Jeon, J.H.; Kim, S.W.; Moon, C.; et al. Magnetically actuated microrobots as a platform for stem cell transplantation . *Sci. Robot.* **2019**, *4*, eaav4317. [CrossRef]

178. Dong, X.; Sitti, M. Controlling two-dimensional collective formation and cooperative behavior of magnetic microrobot swarms. *Int. J. Robot. Res.* **2020**, *39*, 617–638. [CrossRef]

179. Sitti, M. *Mobile Microrobotics*; MIT Press: Cambridge, MA, USA, 2017.

180. Nouyan, S.; Groß, R.; Bonani, M.; Mondada, F.; Dorigo, M. Teamwork in self-organized robot colonies. *IEEE Trans. Evol. Comput.* **2009**, *13*, 695–711. [CrossRef]

181. Xie, H.; Sun, M.; Fan, X.; Lin, Z.; Chen, W.; Wang, L.; Dong, L.; He, Q. Reconfigurable magnetic microrobot swarm: Multimode transformation, locomotion, and manipulation. *Sci. Robot.* **2019**, *4*, eaav8006. [CrossRef]

drones

Article

A Sampling-Based Distributed Exploration Method for UAV Cluster in Unknown Environments

Yue Wang [1,*], Xinpeng Li [1], Xing Zhuang [1], Fanyu Li [2] and Yutao Liang [2]

1 School of Mechatronical Engineering, Beijing Institute of Technology, Beijing 100081, China
2 School of Computer Science and Communication Engineering, Jiangsu University, Zhenjiang 212013, China
* Correspondence: jackwy@bit.edu.cn

Abstract: Rapidly completing the exploration and construction of unknown environments is an important task of a UAV cluster. However, the formulation of an online autonomous exploration strategy based on a real-time detection map is still a problem that needs to be discussed and optimized. In this paper, we propose a distributed unknown environment exploration framework for a UAV cluster that comprehensively considers the path and terminal state gain, which is called the Distributed Next-Best-Path and Terminal (DNBPT) method. This method calculates the gain by comprehensively calculating the new exploration grid brought by the exploration path and the guidance of the terminal state to the unexplored area to guide the UAV's next decision. We propose a suitable multistep selective sampling method and an improved Discrete Binary Particle Swarm Optimization algorithm for path optimization. The simulation results show that the DNBPT can realize rapid exploration under high coverage conditions in multiple scenes.

Keywords: exploration of unknown environment; UAV cluster; sampling and optimization; distributed path planning; particle swarm optimization

Citation: Wang, Y.; Li, X.; Zhuang, X.; Li, F.; Liang, Y. A Sampling-Based Distributed Exploration Method for UAV Cluster in Unknown Environments. *Drones* 2023, 7, 246. https://doi.org/10.3390/drones7040246

Academic Editors: Xiwang Dong, Mou Chen, Xiangke Wang and Fei Gao

Received: 2 March 2023
Revised: 20 March 2023
Accepted: 30 March 2023
Published: 1 April 2023

1. Introduction

At present, unmanned aerial vehicles (UAVs) are widely used to perform tasks in various environments, especially in complex and unknown scenes. One of the typical tasks is to explore an unknown environment, which is widely used for search [1,2], rescue [3,4], and dangerous area reconnaissance [5]. Unknown environment exploration means that UAVs or a UAV cluster can make decisions on their own actions in real-time by relying on their detection equipment under the condition that there is no prior environmental information to achieve a fully independent construction of highly saturated environmental information. Compared with other missions, unknown environment exploration lacks prior map information. It is crucial to set the autonomous strategy of the exploration action to complete the environmental construction of the whole region as soon as possible. The coordination in the cluster and avoidance of repeated exploration must also be considered.

The traditional exploration of unknown environments adopts the ploughing method for complete coverage path planning [6], but it only aims at specific conditions [7] without obstacles in the environment. When encountering obstacles, the ploughing method adopts a simple wall-following strategy [8] to avoid sudden obstacles in the path, which also has great limitations. Yamauchi initiated a frontier-based exploration strategy [9] and extended it to multiple robots [10], which is considered to be an important classical method for unknown environment exploration. The frontier is defined as the boundary between the unexplored grid and the explored grid while excluding the explored obstacle grid. The frontier-based method obtains exploration information by navigating the robot to the frontier grid. Many of the most advanced methods are based on frontier-based exploration [11–13]. Ref. [14] developed a frontier-selection strategy that minimizes the change in velocity necessary to reach it to achieve the high-speed movement of quadrotors.

Ref. [15] proposed a hierarchical planning framework based on frontier information (FIS), enabling a UAV cluster to quickly explore indoor environments.

The sampling-based method is another particularly effective method for exploration. The central idea is to calculate the information gain of the sampled state and select the best one to execute, which can adapt to various gain calculation forms and has strong flexibility. The sampling-based method is also combined well with the frontier-based method. For example, the SRT [16] algorithm drives the motion update of the robot through sampling in the sensor safety space and the selection of random exploration angles. The Next-Best-View (NBV) [17,18] is an exploration method introduced from 3D reconstruction and has become a widely used sample-based exploration method. Authors in [19] proposed the Receding Horizon Next-Best-View, combining NBV with a path planning algorithm similar to RRT [20] and RRT* [21], and obtained a fine effect in indoor exploration and reconstruction mapping. Authors in [22] proposed a UFO exploration method. Based on the rapidly updated map format called the UFO Map, the maximum information gain was not considered but adopted the nearest point with the information gain as the exploration decision, which achieved the effect of rapid exploration at a small cost of computing resources. In addition, exploration methods based on machine learning are also considered to have great potential, and many scholars are conducting relevant research [23–25]. However, its engineering applications for unknown environment exploration are still relatively few, and it performs poorly in the generalization ability to different environments, which still needs further exploration and research [26,27].

For large scenes, such as in underground garages or large factories, to ensure that the task can be completed quickly, the cluster is generally used. For the exploration of the environment of a robot cluster, the distributed cluster structure is considered to be better in this scenario [28]. It can not only avoid excessive pressure on central computing resources but also flexibly handle the impact of poor communication in the cluster or sudden failures [29–31] to minimize efficiency loss. However, for distributed clusters, designing the exploration strategy of each platform to avoid repeated exploration and complete the exploration quickly under the premise of cluster collision avoidance is still a difficult problem.

The above method seems to simply consider the state of the next step to calculate its gain, but the impact of its motion process on the exploration is considered to be negligible, especially when the sensor is limited by the field of view (FOV) or is in the area near the obstacle, and at the same time, dynamics should be considered to increase the efficiency. Therefore, this paper proposes a distributed exploration framework for unknown environments considering the path and terminal gain. In this framework, multiple exploration paths are obtained by considering the dynamic constraints of the state sequence, and the optimal path is obtained by using optimization methods. The evaluation factors include the energy loss of dynamics, the growth of map exploration in the path process, the benefits of the terminal state to the next exploration, and collision avoidance in the cluster. The paths are planned for a period of time in the future and take the frontmost path to implement until the exploration coverage of the entire cluster meets the requirements. To ensure the efficiency of online planning, a multistep selective optimal sampling method, a gain calculation method of path exploration, and an efficient improved Discrete Binary Particle Swarm Optimization (BPSO) algorithm are given. The results show the effectiveness and superiority of the algorithm in the exploration of unknown environments of a UAV cluster in multiple scenes.

The contributions of this paper are as follows:

1. A Distributed Next-Best-Path and Terminal framework for real-time path planning for UAV cluster unknown environment exploration.
2. A multistep selective sampling method for the initial generation of the exploration path with the calculation method of progress and terminal gain.
3. An improved Discrete Binary Particle Swarm Optimization algorithm to generate the best exploration path.

The remainder of this paper is organized as follows: Section 2 describes the problem of unknown environment exploration and introduces our distributed unknown environment exploration framework and the construction of the specific model. Section 3 discusses the multistep selective sampling method we propose and the improved BPSO algorithm of the optimal path solution. Section 4 presents the simulation to verify the algorithm performance. Section 5 gives a summary and introduces further work.

2. Framework and Model Establishment

2.1. System Framework

In an unknown environment, the UAV cluster detects and builds the map using its own sensors and independently plans the next path or action according to the real-time built map. In the exploration process, UAVs intercept the global map separately at the ground station and generate a local map for prediction and planning, while the new environmental data obtained in the movement are sent to the ground station for the global map update. The map is in the form of a grid map. Each grid has three states: unknown, known free, and known obstacle. Position and attitude messages can be obtained for planning cluster collision avoidance through communication.

For each individual in the UAV cluster, the critical factors for the decision behavior include two aspects: the new situation of the possible exploration map after the path is executed and the advantage of the terminal of the path for the next exploration action. Among them, the map change comes from the increment in the grid in the unknown state to the known state, which is calculated by the fast approximate method mentioned below. Considering the tendency of the terminal state to the unexplored area and referring to the generation method of the frontier, the frontier closest to the current position is generated from the map before the path is executed, and the terminal state closer to the frontier after the path is executed is considered the better terminal. Other factors that must be considered in cluster exploration include path obstacle conflict, path energy loss, and collision avoidance in the cluster, as shown on the right side of Figure 1.

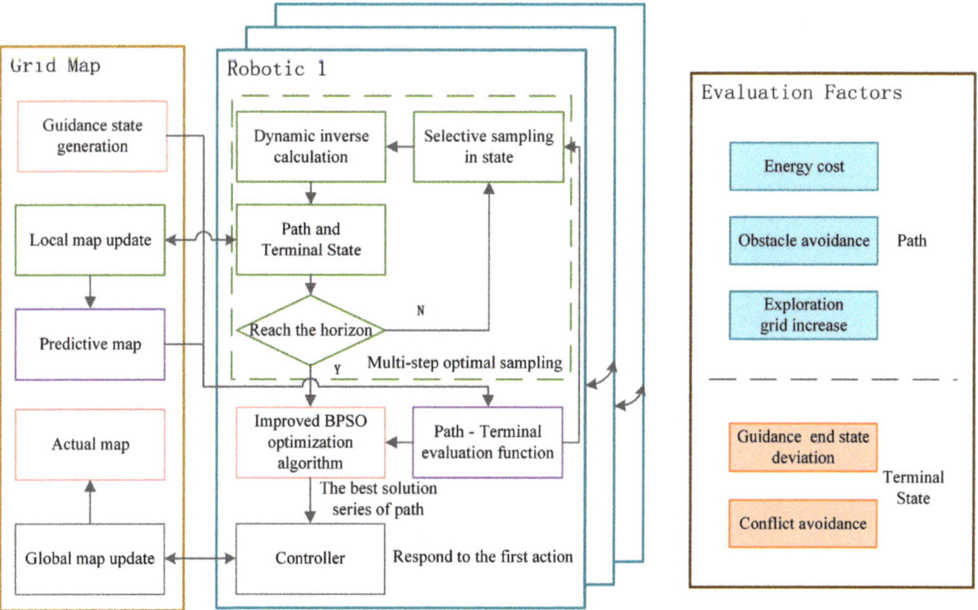

Figure 1. Distributed Next-Best-Path and Terminal exploration framework.

The left side of Figure 1 shows the DNBPT framework we propose. We set a prediction horizon as the time domain for each plan and plan the series paths of the UAV with a fixed length. We take a multistep sampling and preferential growth method, sample in a limited state space, and inversely calculate the dynamic path through the terminal state to obtain the path and terminal state. Considering the factors mentioned above to evaluate, we select a certain number of action sequences with high evaluation. Due to the nonoptimal solution by limited sampling, the series of paths obtained is used as the initial solution of the improved BPSO algorithm to further optimize and finally obtain the optimal path and terminal sequence. The controller responds to the first step of the optimal sequence to address sudden obstacles or other situations in the actual movement. In this framework, UAVs can join or leave the exploration mission freely without causing disorder in the whole system.

2.2. Exploration Model in Unknown Environments

The dynamics of the UAV have the property of differential flatness [32]. In the planning process, the flat output $s = \left[p_x, p_y, p_z, \varphi\right]^T$ is used as the planning quantity to reduce the planning dimension and improve the timeliness. p_x, p_y, p_z is the position of the UAV and φ is the yaw angle.

The obtained detection area is fitted with the local grid map to obtain the index of the grid map and bring it into the global map, as shown in Equation (1).

$$Grid(Map_{new}) = Grid(explored) \cup Grid(Map_{old}) \tag{1}$$

Assume that k is the length of the prediction horizon and $S_k = [s(t+1|t), \ldots, s(t+k|t)]$ is the state sequence in the prediction horizon as the input of the predictive map update and the evaluation function.

The UAV carries sensors with an observation field of view (FOV) to detect environmental information. In this paper, the sensor mapping algorithm is not considered, and it is assumed that the sensor can obtain the environment information within the angle range. The sensor can be equivalent to a sector, and its detection range within the prediction interval can identify the trajectory of the sensor's sector area driven by the k segment control for integration, as shown in Equation (2), where $G(°)$ represents the processing program of the region on the grid map and $N_{increase}$ represents the number of new grids.

$$N_{increase} = \sum_{i=1}^{k} G\left(\int_{l_i}^{l_{i+1}} sector(FOV)\, dl\right), \quad FOV \in (0, \pi] \tag{2}$$

Due to the complexity of the integral calculation and the grid form of the map, we propose a simplification to predict the update of the map and calculate the number of new detection grids, as shown in Figure 2. The sampling range and time interval are limited to ensure that the state falls in the previous detection sector. For a state S_i in the terminal state sequence S_k, we connect the two points P_{i-1} and Q_{i-1} of the previous sector and the two points P_i and Q_i of the current sector. A convex quadrilateral is formed, and each convex quadrilateral is connected. The obtained convex quadrilateral is placed into the local map to calculate the number of new exploration grids as the evaluation basis of the path-terminal sequence, as shown in Equation (3).

$$\tilde{N}_{increase} = G\left(\sum_{i=1}^{k} A_i(P_{i-1}Q_{i-1}P_iQ_i)\right) \tag{3}$$

This equivalent method improves the calculation efficiency to support online UAV planning.

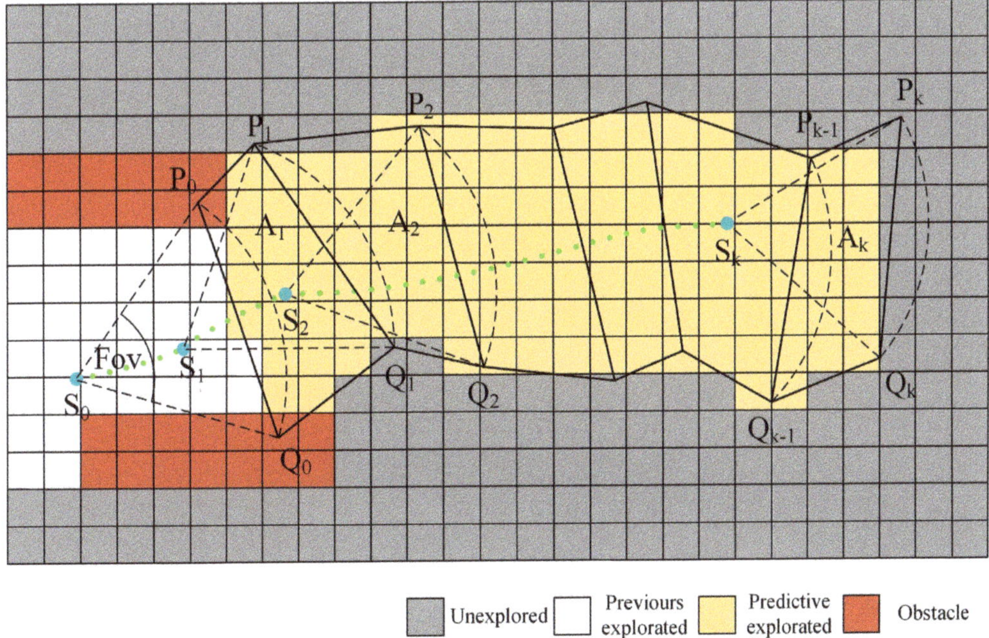

Figure 2. Detection model and approximate calculation of grid increase.

2.3. Construction of the Evaluation Function

In the process of exploring the unknown environment, the form of the solution is the terminal state sequence of the predicted horizon and the path between the state transitions. Its optimality evaluation includes two aspects: the path and the terminal state.

First, the path needs to be checked for obstacle conflicts, and it is necessary to ensure that the planned path of the cluster will not collide with the obstacles that have been explored. The path points in the process are extracted by interpolation, and the obstacle is checked in the grid map. The results are accumulated for the evaluation function value, such as Equation (4). Note that this is not applied in sampling but in the optimization algorithm.

$$J_0 = \begin{cases} -100, & \text{collision} \\ 0, & \text{collision free} \end{cases} \tag{4}$$

In the action space, under the constraint conditions, the average energy consumption of the action sequence is smallest, as shown in Equation (5), where dim (u) is the dimension of the action space.

$$J_1 = 1 - \frac{\sum_{i=1}^{k} \sum u_i^T \mathrm{diag}\left(\frac{1}{u_{max}^2}\right) u_i}{k \cdot \dim(u)} \tag{5}$$

To ensure the continuity and smoothness of the front and back actions, it is necessary to minimize the front and back deviation of each step in the action sequence, as shown in Equation (6):

$$J_2 = 1 - \frac{\sum_{i=0}^{k} \sum_{j=0}^{l} \frac{\|u_{j+1} - u_j\|}{\|u_{j+1}\|}}{k} \tag{6}$$

The exploration of unknown space during the UAV's movement brings gain. The calculation method of the number of new exploration grids is proposed above, and the evaluation function is shown in Equation (7). $N_{reference}$ refers to the number of exploration grids in the ideal state. The calculation method is shown in Equation (8), where r is the sensor detection distance, d is the fixed maximum distance for sampling, and rev is the map resolution (magnification).

$$J_3 = \frac{\widetilde{N}_{increase}}{N_{reference}} \tag{7}$$

$$N_{reference} = 2krd \sin\left(\frac{Fov}{2}\right) rev \tag{8}$$

Because the environment is unknown, the exploration gain under different paths may be similar. In the later stage of the exploration process, it may occur that the UAV's surrounding environment has been completely explored. It is easy to fall into the local optimum, causing invalid and repeated paths. Therefore, the guidance of other factors is needed to enable the UAV to move in the direction that the gain may increase. The rating function of this part is shown as Equation (9):

$$J_4 = 1 - e^{1 - \frac{\|s_0 - s_r\|}{\|s_k - s_r\|}} \tag{9}$$

where s_k is the state at moment k, s_0 is the initial state, and s_r is the reference guidance state. We use frontier coordinates and straight orientation as references in this paper, and the frontier is rapidly generated through the edge detection of OpenCV.

To improve the exploration efficiency of the cluster, the UAVs should be distributed as far as possible. Therefore, the evaluation function is designed for the terminal state, as in Equation (10), where r_s is the set safety distance and n is the number of UAVs in the cluster.

$$J_5 = \sum \frac{r_s}{(n-1) \cdot \|s_{robot}(p_x, p_y) - s_{other}(p_x, p_y)\|} \tag{10}$$

Based on the above evaluation factors, the total evaluation functions used in selective sampling and improved BPSO, respectively, are designed as Equations (11) and (12), respectively:

$$J_{sampling} = \sum_{j=1}^{5} \omega_j J_j \tag{11}$$

$$J_{iBPSO} = J_0 + \sum_{j=1}^{5} \omega_j J_j \tag{12}$$

where $\omega_j \epsilon [0, 1]$ is the weight value, which can be adjusted according to the actual situation, while $\sum_{j=1}^{5} \omega_j = 1$.

3. Method and Algorithm

3.1. Multistep Selective Sampling Algorithm

To make UAVs better adapt to unknown environments, planning is often performed in multiple steps. Planning the multipath within a certain planning horizon and executing the first segment of the optimal path are needed. In the calculation process, the number of samples will increase exponentially with the increase in the number of segments in the planning horizon, and the calculation cost is unaffordable. Therefore, we design a multistep selective sampling method. During each round of sampling, we sample the sequence with high current evaluation in the next step. The pseudocode of Algorithm 1 shows more details of the multistep selective sampling method.

An empty set $\mathcal{X}$ saves path and terminal sequences with lengths less than k, and an empty set $\mathcal{F}$ saves sequences with lengths equal to k. In the sampling space under the constraint conditions, m terminal states are randomly taken to be composed on the initial state as the initial sequence, and then the loop begins while the sequence gradually grows. In each loop, the best n solutions in the set $\mathcal{X}$ are selected for the next step of sampling.

Each solution also takes m states randomly in the updated sampling space based on the current state to make the sequence grow and update these sequences in $\mathcal{X}$. The sequence with a length of k is placed in $\mathcal{F}$ and will not be selected again for the next sampling. When the evaluation value of the n th better sequence in $\mathcal{F}$ is greater than the best evaluation value of the sequence in $\mathcal{X}$, the sampling process is finished.

After multistep selective sampling, n path and terminal state sequences are obtained as the basis of the next optimization algorithm.

Algorithm 1 Multistep selective sampling

Input: Grid Map, initial state x_0, sample space $\mathcal{U}$, other states x_{others}
Parameters: planning horizon k, number of samples m, n, safe distance r_s
Output: aggregate of terminal state sequence X_k^n with path
update $\mathcal{U}$
$\mathcal{X}, \mathcal{F} \leftarrow \varnothing$
sampling random m in $\mathcal{U}$, generate X_1^n, $\mathcal{X} \leftarrow X_1^n$
while $\mathcal{X}$ is not empty
    select the best m sequences with length $<k$
    update $\mathcal{U}$
    uniform sampling m in $\mathcal{U}$ based on the selected sequences $X_i^n, i \in [1, k-1]$
    $X_i^n \cdot X_{i+1}^n$
    update the evaluation value of X_i^n //according to the Equation (11)
    if $i < k$ **then** $\mathcal{X} \leftarrow X_i^n$
    else $\mathcal{F} \leftarrow X_k^n$
    if the nth best X in $\mathcal{F}$ better than the best X in $\mathcal{X}$ **then**
    break
select the best n in $\mathcal{X}$ as output

3.2. Improved Discrete Binary Particle Swarm Optimization Algorithm

To select the optimal path from the generated multiple paths, we propose an optimal path selection method based on improved BPSO. We propose a mutation strategy to increase the diversity of the population, which can help the particle swarm jump out of the local optimum trap. In addition, we introduce a contraction factor to ensure the convergence performance of the algorithm [33], which controls the final convergence of the system behavior and can effectively search different regions. This method can obtain high-quality solutions.

The velocity and position update formulas of the particle swarm after introducing the contraction factor is shown as Equations (13) and (14), respectively:

$$v_{id}^{t+1} = \lambda \left(v_{id}^t + c_1 r_1 \left(pbest_{id}^t - x_{id}^t \right) + c_2 r_2 \left(gbest_{id}^t - x_{id}^t \right) \right) \tag{13}$$

$$x_{id}^{t+1} = x_{id}^t + v_{id}^{t+1} \tag{14}$$

where λ represents the shrinkage factor, as shown in Equation (15), t represents the current iteration number, c_1 and c_2 are the learning factors [34–36], r_1 and r_2 are two random values uniformly distributed in [0, 1], and $pbest_{id}$ and $gbest_{id}$ represent the individual optimal position and the global optimal position of the particle, respectively.

$$\lambda = \frac{2}{\left| 2 - (c_1 + c_2) - \sqrt{(c_1 + c_2)^2 - 4 * (c_1 + c_2)} \right|} \tag{15}$$

Our mutation strategy introduces the idea of the dMOPSO [37] algorithm, and age is used to represent the number of times that the individual optimal position (*pbest*) of the current particle has not been updated continuously in the loop. When the local optimal position of the particle has not been updated for a long time, it means that the particle is likely to have fallen into the local optimal position, and it is necessary to perturb the particle. The specific perturbation example is shown in Figure 3.

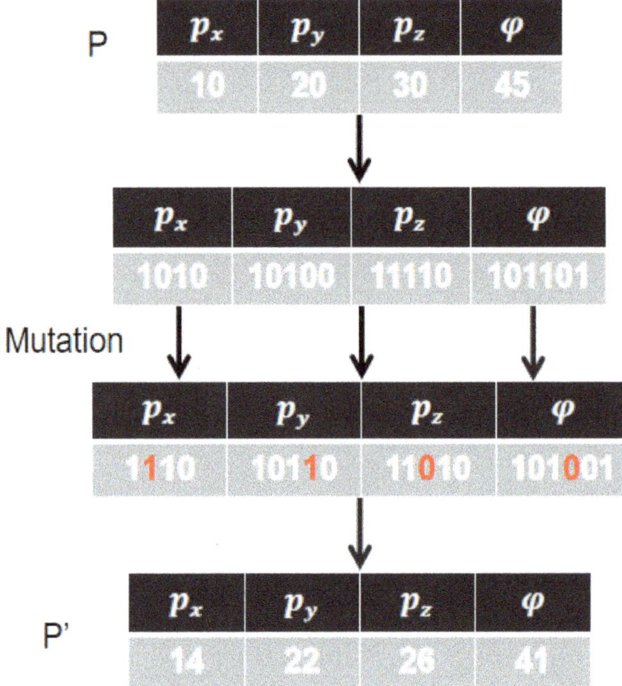

Figure 3. Perturbing the particles that have been in the local optimum for a long time.

The state of particles is converted into binary form. In each coordinate, a bit of the position is randomly selected for the mutation operation. After the mutation operation, the particle's age is reset to 0. If the particle's age does not reach the age threshold, only the particle's age is increased. The pseudocode of the mutation operation is shown in Algorithm 2:

Algorithm 2 Mutation strategy

Input: Pop(swarm), P_{best}, N(size of the population), Ta(threshold of age)
Output: NewPop(new swarm)
for $i = 1 : N$ **do**
    if age(P_i) > Ta **then**
        $P_i' \leftarrow$ Mutation(P_i)
        Fitness $\leftarrow$ CalculateFitness(P_i) // *according to the Equation (12)*
        if fitness(P_i') > fitness(P_i) **then**
$$P_i = P_i'$$
        age(P_i) $\leftarrow$ 0
    else
        age(P_i) $\leftarrow$ age(P_i) + 1
    end if
end for
return NewPop

The proposed improved BPSO algorithm is mainly divided into two stages. The first stage is the initialization stage. We encode and initialize the particle swarm according to the input path information, randomly initialize the speed and the age of the initialization particle, and finally calculate the fitness value of the particle. The second stage is the main loop stage. When the particle's age exceeds the age threshold, the mutation operation is performed. The loop is finished when the termination condition is met. The final returned G_{best} is the optimal sequence of terminals with paths. The pseudocode of the Algorithm 3 is:

Algorithm 3 Improved DBPSO

Input: original Pop, size of the population N, maximal generation number *maxgen*
Output: G_{best} (optimal sequence of terminal state with path)

$P \leftarrow$ InitializeParticles(N)
$Age \leftarrow$ InitializeAge(N)
Fitness $\leftarrow$ CalculateFitness(N)
While NCT (Number of current iterations) <= *maxgen* **do**
    for $i = 1 : N$ **do**
      $G_{best} \leftarrow$ SelectGbest(Pop)
        Pop $\leftarrow$ Updateparticles(N) //*according to the Equations (13)–(15)*
        NewPop $\leftarrow$ Mutation(P_i)
        Fitness $\leftarrow$ CalculateFitness(N) //*according to the Equation (12)*
        Pop $\leftarrow$ NewPop
      $G_{best} \leftarrow$ SelectGbest(Pop)
    end for
end while
return G_{best}

---

4. Simulation and Analysis

4.1. Simulation in Fixed-Obstacle Scenes

We design three indoor scenes with fixed obstacles of different sizes based on the interior of the building, and the number of UAVs in each scene is different. Due to the indoor scene, we assume that the UAV is flying at a fixed altitude. The sizes of the scene are 20 m long and 50 m wide, 50 m long and 50 m wide, and 100 m long and 100 m wide, respectively. The numbers of UAVs are 3, 4, and 5. For each scene, simulations of a fixed initial state and a random initial state are carried out. The initial states of all scenes are shown in Figures 4a, 5a, 6a, 7a, 8a and 9a. More detailed parameters about the scenes and algorithm are shown in Tables 1 and 2.

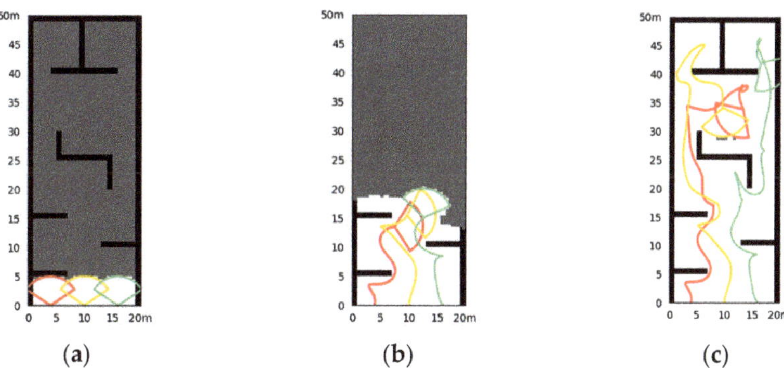

(a)   (b)   (c)

Figure 4. Simulation results of the fixed initial state in Scene I. (**a**) The initial state of the simulation and the obstacle; (**b**) the situation at a certain moment (10.8 s) in the exploration process; (**c**) the situation of exploration results.

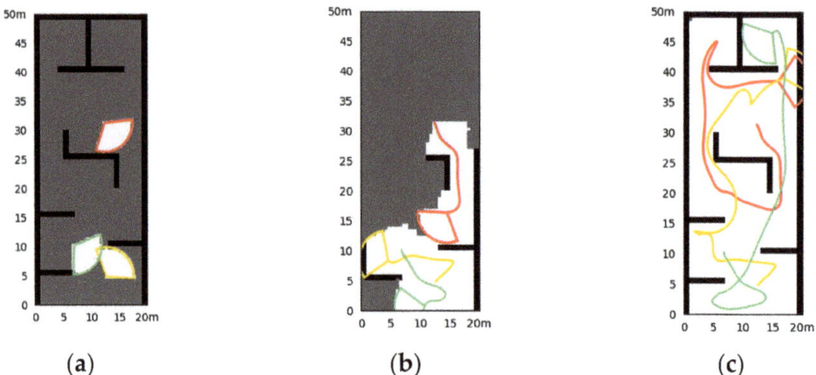

(a)   (b)   (c)

Figure 5. Simulation results of the random initial state in Scene I. (**a**) The initial state of the simulation and the obstacle; (**b**) the situation at a certain moment (8.4 s) in the exploration process; (**c**) the situation of exploration results.

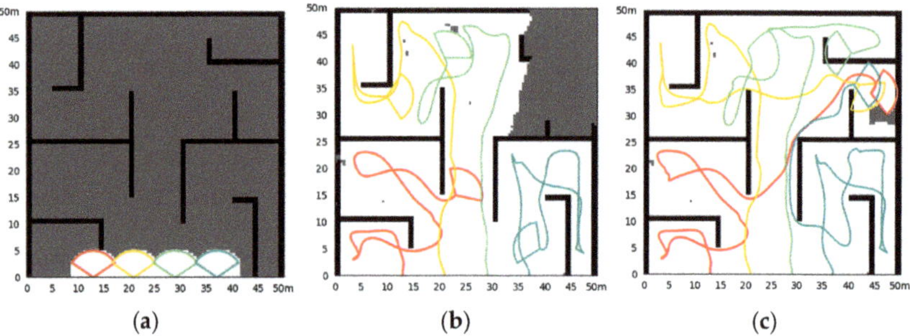

(a)   (b)   (c)

Figure 6. Simulation results of the fixed initial state in Scene II. (**a**) The initial state of the simulation and the obstacle; (**b**) the situation at a certain moment (46.8 s) in the exploration process; (**c**) the situation of exploration results.

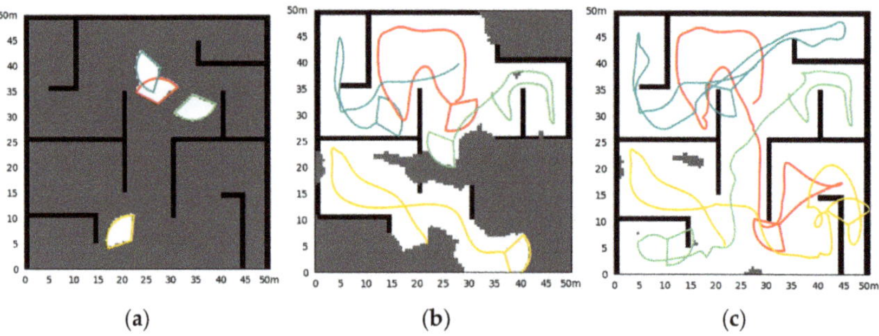

(a)   (b)   (c)

Figure 7. Simulation results of the random initial state in Scene II. (**a**) The initial state of the simulation and the obstacle; (**b**) the situation at a certain moment (34.0 s) in the exploration process; (**c**) the situation of exploration results.

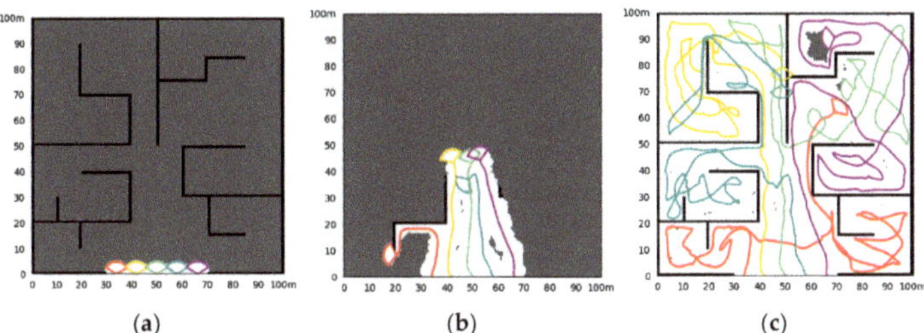

(a)   (b)   (c)

Figure 8. Simulation results of the fixed initial state in Scene III. (**a**) The initial state of the simulation and the obstacle; (**b**) the situation at a certain moment (20.0 s) in the exploration process; (**c**) the situation of exploration results.

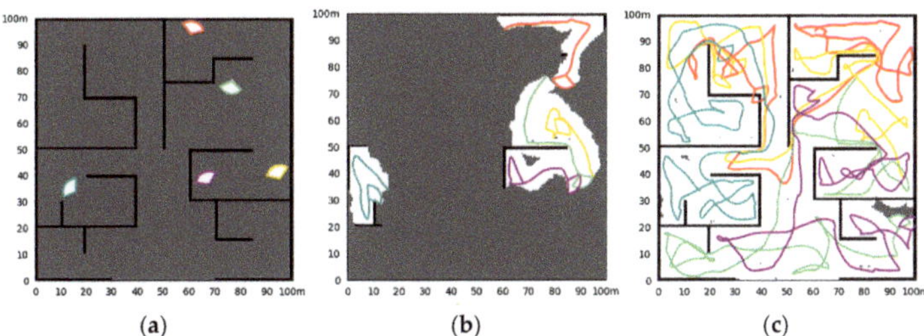

(a)   (b)   (c)

Figure 9. Simulation results of the random initial state in Scene III. (**a**) The initial state of the simulation and the obstacle; (**b**) the situation at a certain moment (25.6 s) in the exploration process; (**c**) the situation of exploration results.

Table 1. Parameter setting for three scenes.

Scene I
Map Parameters: map size: 20 m × 50 m, resolution: 0.25 m × 0.25 m
Initialization $(p_x, p_y, \varphi)^*$: UAV1:(4,0,90), UAV2:(10,0,90), UAV3:(16,0,90)
detection radius: 5 m, Fov: 104°, r_s: 3 m, max velocity: 2.5 m/s, end rate: 99.5%

Scene II
Map Parameters: map size: 50 m ×50 m, resolution: 0.4 m ×0.4 m
Initialization $(p_x, p_y, \varphi)^*$: UAV1:(13,0,90), UAV2:(21,0,90), UAV3:(29,0,90), UAV4:(37,0,90)
detection radius: 5 m, Fov: 104°, r_s: 3 m, max velocity: 2.5 m/s, end rate: 99%

Scene III
Map Parameters: map size: 100 m ×100 m, resolution: 0.5 m ×0.5 m
Initialization $(p_x, p_y, \varphi)^*$: UAV1:(34,0,90), UAV2:(42,0,90), UAV3:(50,0,90), UAV4:(58,0,90), UAV5:(66,0,90)
detection radius: 5 m, Fov: 104°, r_s: 3 m, max velocity: 2.5 m/s, end rate: 99%

* Only for the fixed initial state.

Table 2. Parameter setting for three algorithms.

Parameters	Value
predict horizon	$k = 5$
sample num	$m = 10, n = 50$
weight distribution	$\omega_1 = 0.1, \omega_2 = 0.1, \omega_3 = 0.5, \omega_4 = 0.2, \omega_5 = 0.1$
learning factor	$c_1, c_2 = 1.46$
threshold of age	$Ta = 3$
population size	$N = 50$
max number of generations	$maxgen = 100$
simulation step	$t = 0.2$ s

1. Simulation in Scene I

    Figure 4 shows the simulation results of three UAV explorations in Scene I, with fixed initial UAV states to start. The exploration takes 36.8 s. It can be seen that the cluster can realize well the exploration of the environment in a short time, and there are few repeated paths, unless it is necessary to leave the impasse that is surrounded by obstacles. Figure 5 shows the results of random initial states, and the exploration takes 34.6 s. We can see that the cluster can explore well in any initial state, with the same short time cost and fewer repeated paths.

2. Simulation in Scene II

    Figure 6 shows the simulation results of four UAV explorations in Scene II, with fixed initial UAV states to start. The exploration takes 62.0 s. Figure 7 shows the results of random initial states, and the exploration takes 63.2 s. In medium-size scenes, UAVs in the cluster avoid repeated exploration in the same area through a distributed strategy. The UAV can quickly return to exploring other areas after the exploration of corners or the impasse. In the random initial state, the UAV's performance is almost unaffected.

3. Simulation in Scene III

    Figure 8 shows the simulation results of four UAV explorations in Scene III, with fixed initial UAV states to start, and Figure 9 shows the situation of random initial states. Exploration takes 190.0 s and 214.8 s, respectively.

    With the expansion of the scale of the exploration scene and the increase in the complexity of the internal structure, the difficulty of cluster exploration is also increasing, and the UAVs show a complex movement. The random initial state brings uncertainty to the exploration process. Under the above factors, whether the exploration efficiency can be maintained is the key point of the exploration method. The proposed method can still maintain the complete exploration of the area in large scenes, and there is less repeated exploration. There may be a tendency for multiple UAVs to move in the same direction at the end of the exploration. This is because we do not allow the UAV to be idle, to achieve the fastest exploration speed. As the environment is unknown, it is difficult to define which UAV can reach the unexplored area faster, so we keep every UAV in the cluster continuously exploring until the area is fully explored. This ensures the shortest exploration time, but may bring a waste of energy for engineering applications. It can be adjusted according to the actual application, for example, using conditional judgments to make some UAVs idle.

4. Comparing the methods in three scenes

    Due to the randomness of the environment exploration process, we conduct 100 simulations for each situation (fixed initial state and random initial state in each scene) and count the time cost of exploration and compare it with the two classical methods, as shown in Table 3. The frontier-based method is a method with fixed results when the frontier generation, map, and initial state are fixed. The NBV and DNBPT methods have some randomness in the process for deeper exploration. The average exploration efficiency of a single UAV in all scenes is also compared, as shown in Figure 10.

    Compared with the frontier-based method and the NBV, the exploration efficiency of the proposed method in each scene has a great advance. In Scene I, the exploration efficiency is increased on average by approximately 85.7% and 34.4% with fixed initial states and 71.4% and 33.1% with random initial states, respectively. In Scene II, the efficiency is increased by approximately 107.0% and 36.0% with fixed initial states and 108.7% and 35.8% with random initial states, respectively. In Scene III, the efficiency is increased by approximately 122.1% and 36.6% with fixed initial states and by 124.4% and 33.9% with random initial states, respectively. For larger and more complex indoor scenes, the improvement effect of the proposed method is more obvious.

Table 3. Results and comparison of multiple simulation data samples in the fixed obstacle scenes.

			Exploration Time (s)			
	Initial	Method	Mean	Best	Worst	Std
Scene I	Fixed	Proposed	32.6	26.4	38.8	3.0
		Frontier-based	61.2	-	-	-
		NBV	44.3	36.8	62.6	6.0
	Random	Proposed	35.9	30.4	42.0	3.1
		Frontier-based	61.6	39.2	75.6	6.6
		NBV	47.8	39.6	68.6	6.9
Scene II	Fixed	Proposed	65.9	60.8	76.4	4.0
		Frontier-based	136.4	-	-	-
		NBV	89.5	74.2	107.0	6.7
	Random	Proposed	70.5	61.6	80.8	4.6
		Frontier-based	147.1	95.2	185.6	16.9
		NBV	95.7	80.8	112.4	7.4
Scene III	Fixed	Proposed	205.4	188.2	221.0	8.9
		Frontier-based	456.2	-	-	-
		NBV	280.5	247.4	302.6	14.7
	Random	Proposed	211.9	193.2	249.2	10.0
		Frontier-based	475.5	296.2	700.4	43.1
		NBV	283.9	251.8	305.8	14.6

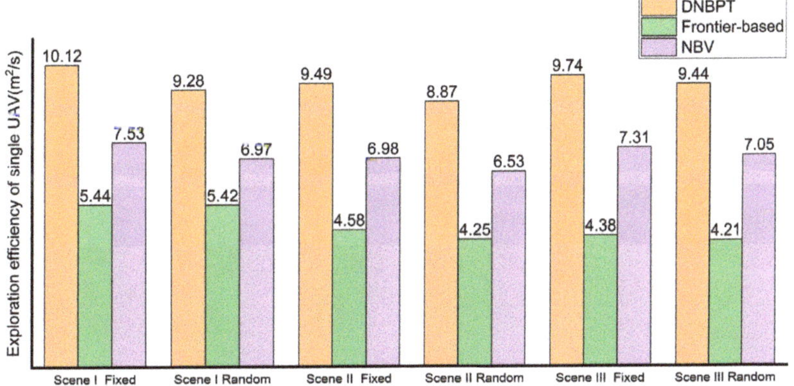

Figure 10. Comparison of the single UAV exploration efficiency of each method.

In the simulation process of algorithm comparison, it is found that the frontier-based method has a good effect in the early stage of exploration. However, in the end stage, due to its greedy strategy that tends to the nearest point, many omissions in the early stage need to be explored in reverse, resulting in a waste of efficiency. This becomes more obvious with increasing exploration rate requirements. The NBV method can carry out deeper exploration locally, but it loses the directional guidance of the global environment and produces repeated meaningless paths. The proposed method combines the advantages of the two methods, including deep local exploration and global guidance, to improve the exploration efficiency of UAV clusters and reduce repetitive paths.

This proves the effectiveness and superiority of the method in complex indoor scenes. In addition, the proposed method shows a stabler exploration efficiency in uncertain scenes and can complete the exploration quickly in any initial state.

4.2. Simulation in Random-Obstacle Scenes

To explore the applicability of our method in other scenes, we design a random obstacle scene for simulation. We simulate scenes with dense small obstacles such as trees, where obstacles are randomly generated and their size is limited, as shown in Figure 11. We also design the situations of a fixed initial state and a random initial state. The scene size is set to be 100 m long and 30 m wide, and four UAVs form a cluster. The number of obstacles in the scene is 40, and the maximum side length of obstacles is 3 m. The simulation is also designed for constant-altitude flight. The initial state (p_x, p_y, φ) of the UAV in fixed scenes is $(0, 3, 0)$, $(0, 11, 0)$, $(0, 19, 0)$, and $(0, 27, 0)$. The terminal condition for exploration is that the map exploration rate reaches 99%.

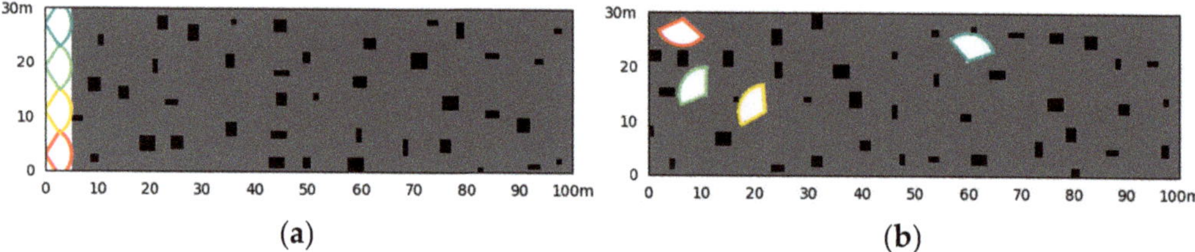

Figure 11. Scene with randomly dense small obstacles. (**a**) The situation of the fixed initial UAV states; (**b**) the situation of random initial UAV states.

Similarly, we conduct 100 simulations for a fixed initial state and a random initial state and compare them with other algorithms. The simulation results of once in each scene are shown in Figures 12 and 13, and the statistical data are shown in Table 4. Under the condition of fixed initial states, the cluster can complete the exploration with few repeated backtrackings and a high rate of coverage while crossing the obstacle area. The random initial state has an impact on the exploration, leading to more possible backtrack and repetitions, but it can still be handled well for reduction.

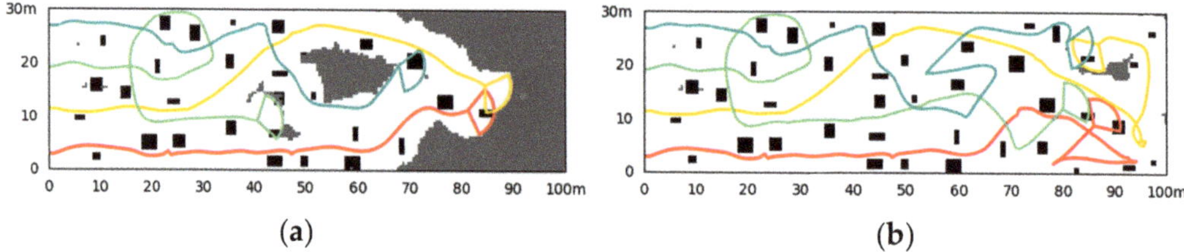

Figure 12. Simulation results of the fixed initial state. (**a**) The situation at a certain moment (41.2 s) in the exploration process; (**b**) the situation of exploration results (64.0 s).

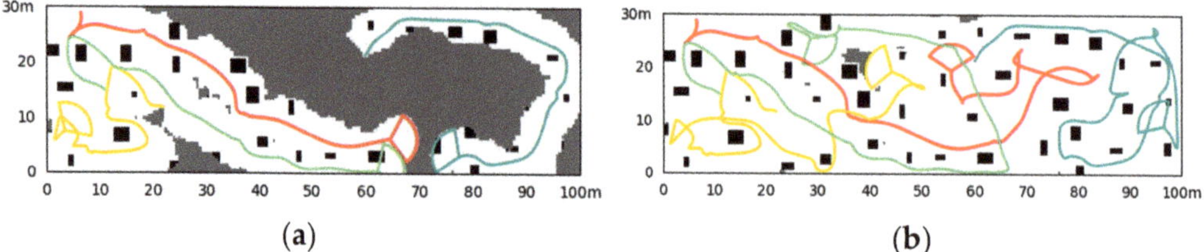

Figure 13. Simulation results of the random initial state. (**a**) The situation at a certain moment (36.0 s) in the exploration process; (**b**) the situation of exploration results (72.4 s).

Table 4. Results and comparison of multiple simulation data samples in random obstacle scene.

Initial	Method	Exploration Time (s)			
		Mean	Best	Worst	Std
	Proposed	73.7	60.0	82.8	6.1
Fixed	Frontier-based	101.3	90.2	131.4	9.7
	NBV	92.4	82.8	109.2	9.0
	Proposed	74.8	64.8	86.4	6.2
Random	Frontier-based	104.8	91.4	138.8	10.9
	NBV	93.3	80.4	114.6	8.3

Regarding the exploration of areas with dense small obstacles, compared with the frontier-based method and NBV, the exploration efficiency is increased on average by approximately 37.4% and 25.2% with fixed initial states and 40.1% and 24.8% with random initial states, respectively. The comparison proves the good performance in the environment with dense small obstacles.

5. Conclusions

In this paper, we propose a DNBPT method for UAV clusters to explore unknown environments. The gain is calculated by comprehensively considering the contribution of the path process and the terminal state to the exploration, and the optimal path is evaluated and selected by multistep optimal sampling and the improved BPSO algorithm. The simulation results show that this method has advantages in different types and sizes of scenes. In addition, this method has strong generality and can be transplanted to other robot platforms.

Author Contributions: Conceptualization, Y.W. and X.L.; methodology, Y.W. and X.L.; software, X.L.; validation, X.L. and X.Z.; formal analysis, X.L., X.Z., F.L. and Y.L.; investigation, F.L. and Y.L.; resources, Y.W.; writing—original draft preparation, X.L.; writing—review and editing, Y.W. and X.Z.; supervision, Y.W. All authors have read and agreed to the published version of the manuscript.

Funding: This research received no external funding.

Data Availability Statement: Not applicable.

Conflicts of Interest: The authors declare no conflict of interest.

References

1. Zheng, Y.-J.; Du, Y.-C.; Ling, H.-F.; Sheng, W.-G.; Chen, S.-Y. Evolutionary Collaborative Human-UAV Search for Escaped Criminals. *IEEE Trans. Evol. Comput.* **2019**, *24*, 217–231. [CrossRef]
2. Unal, G. Visual Target Detection and Tracking Based on Kalman Filter. *J. Aeronaut. Space Technol.* **2021**, *14*, 251–259.
3. Alotaibi, E.T.; Alqefari, S.S.; Koubaa, A. Lsar: Multi-uav Collaboration for Search and Rescue Missions. *IEEE Access* **2019**, *7*, 55817–55832. [CrossRef]
4. Yang, Y.; Xiong, X.; Yan, Y. UAV Formation Trajectory Planning Algorithms: A Review. *Drones* **2023**, *7*, 62. [CrossRef]
5. Chen, Y.; Dong, Q.; Shang, X.; Wu, Z.; Wang, J. Multi-UAV Autonomous Path Planning in Reconnaissance Missions Considering Incomplete Information: A Reinforcement Learning Method. *Drones* **2022**, *7*, 10. [CrossRef]
6. Chi, W.; Ding, Z.; Wang, J.; Chen, G.; Sun, L. A Generalized Voronoi Diagram-based Efficient Heuristic Path Planning Method for RRTs in Mobile Robots. *IEEE Trans. Ind. Electron.* **2021**, *69*, 4926–4937. [CrossRef]
7. Sun, Y.; Tan, Q.; Yan, C.; Chang, Y.; Xiang, X.; Zhou, H. Multi-UAV Coverage through Two-Step Auction in Dynamic Environments. *Drones* **2022**, *6*, 153. [CrossRef]
8. Cheng, L.; Yuan, Y. Adaptive Multi-player Pursuit–evasion Games with Unknown General Quadratic Objectives. *ISA Trans.* **2022**, *131*, 73–82. [CrossRef]
9. Elsisi, M. Improved Grey wolf Optimizer Based on Opposition and Quasi Learning Approaches for Optimization: Case Study Autonomous Cehicle Including Vision System. *Artif. Intell. Rev.* **2022**, *55*, 5597–5620. [CrossRef]
10. Batinovic, A.; Petrovic, T.; Ivanovic, A.; Petric, F.; Bogdan, S. A Multi-resolution Frontier-based Planner for Autonomous 3D Exploration. *IEEE Robot. Autom. Lett.* **2021**, *6*, 4528–4535. [CrossRef]
11. Gomez, C.; Hernandez, A.C.; Barber, R. Topological Frontier-based Exploration and Map-building Using Semantic Information. *Sensors* **2019**, *19*, 4595. [CrossRef] [PubMed]

12. Tang, C.; Sun, R.; Yu, S.; Chen, L.; Zheng, J. Autonomous Indoor Mobile Robot Exploration Based on Wavefront Algorithm. In *Intelligent Robotics and Applications: 12th International Conference, ICIRA 2019, Shenyang, China, 8–11 August 2019; Proceedings, Part V 12, 2019*; Springer: Berlin/Heidelberg, Germany, 2019; pp. 338–348.

13. Ahmad, S.; Mills, A.B.; Rush, E.R.; Frew, E.W.; Humbert, J.S. 3d Reactive Control and Frontier-based Exploration for Unstructured Environments. In Proceedings of the 2021 IEEE/RSJ International Conference on Intelligent Robots and Systems (IROS), Prague, Czech Republic, 27 September–1 October 2021; IEEE: Piscataway, NJ, USA, 2021; pp. 2289–2296.

14. Cieslewski, T.; Kaufmann, E.; Scaramuzza, D. Rapid Exploration with Multi-rotors: A Frontier Selection Method for High Speed Flight. In Proceedings of the 2017 IEEE/RSJ International Conference on Intelligent Robots and Systems (IROS), Vancouver, BC, Canada, 24–28 September 2017; IEEE: Piscataway, NJ, USA, 2017; pp. 2135–2142.

15. Zhou, B.; Zhang, Y.; Chen, X.; Shen, S. Fuel: Fast Uav Exploration Using Incremental Frontier Structure and Hierarchical Planning. *IEEE Robot. Autom. Lett.* **2021**, *6*, 779–786. [CrossRef]

16. Ashutosh, K.; Kumar, S.; Chaudhuri, S. 3d-nvs: A 3D Supervision Approach for Next View Selection. In Proceedings of the 2022 26th International Conference on Pattern Recognition (ICPR), Montreal, QC, Canada, 21–25 August 2022; IEEE: Piscataway, NJ, USA, 2022; pp. 3929–3936.

17. Kim, J.; Bonadies, S.; Lee, A.; Gadsden, S.A. A Cooperative Exploration Strategy with Efficient Backtracking for Mobile Robots. In Proceedings of the 2017 IEEE International Symposium on Robotics and Intelligent Sensors (IRIS), Ottawa, ON, Canada, 5–7 October 2017; IEEE: Piscataway, NJ, USA, 2017; pp. 104–110.

18. Palazzolo, E.; Stachniss, C. Effective Exploration for MAVs Based on the Expected Information Gain. *Drones* **2018**, *2*, 9. [CrossRef]

19. Li, J.; Li, C.; Chen, T.; Zhang, Y. Improved RRT Algorithm for AUV Target Search in Unknown 3D Environment. *J. Mar. Sci. Eng.* **2022**, *10*, 826. [CrossRef]

20. Vasquez-Gomez, J.I.; Troncoso, D.; Becerra, I.; Sucar, E.; Murrieta-Cid, R. Next-best-view Regression Using a 3D Convolutional Neural Network. *Mach. Vis. Appl.* **2021**, *32*, 1–14.

21. Duberg, D.; Jensfelt, P. Ufoexplorer: Fast and Scalable Sampling-based Exploration with a Graph-based Planning Structure. IEEE Robot. *Autom. Lett.* **2022**, *7*, 2487–2494. [CrossRef]

22. Wang, J.; Li, T.; Li, B.; Meng, M.Q.-H. GMR-RRT*: Sampling-based path Planning Using Gaussian Mixture Regression. *IEEE Trans. Intell. Veh.* **2022**, *7*, 690–700. [CrossRef]

23. Li, H.; Zhang, Q.; Zhao, D. Deep Reinforcement Learning-based Automatic Exploration for Navigation in Unknown Environment. *IEEE Trans. Neural Netw. Learn. Syst.* **2019**, *31*, 2064–2076. [CrossRef]

24. Ramezani Dooraki, A.; Lee, D.-J. An End-to-end Deep Reinforcement Learning-based Intelligent Agent Capable of Autonomous Exploration in Unknown Environments. *Sensors* **2018**, *18*, 3575. [CrossRef]

25. Niroui, F.; Zhang, K.; Kashino, Z.; Nejat, G. Deep Reinforcement Learning Robot for Search and Rescue Applications: Exploration in Unknown Cluttered Environments. *IEEE Robot. Autom. Lett.* **2019**, *4*, 610–617. [CrossRef]

26. Garaffa, L.C.; Basso, M.; Konzen, A.A.; de Freitas, E.P. Reinforcement Learning For Mobile Robotics Exploration: A Survey. *IEEE Trans. Neural Netw. Learn. Syst.* **2021**, 1–15. [CrossRef]

27. Xu, Y.; Yu, J.; Tang, J.; Qiu, J.; Wang, J.; Shen, Y.; Wang, Y.; Yang, H. In Explore-bench: Data Sets, Metrics and Evaluations for Frontier-based and Deep-reinforcement-learning-based Autonomous Exploration. In Proceedings of the 2022 International Conference on Robotics and Automation (ICRA), Philadelphia, PA, USA, 23–27 May 2022; IEEE: Piscataway, NJ, USA, 2022; pp. 6225–6231.

28. Elmokadem, T.; Savkin, A.V. Computationally-Efficient Distributed Algorithms of Navigation of Teams of Autonomous UAVs for 3D Coverage and Flocking. *Drones* **2021**, *5*, 124. [CrossRef]

29. Unal, G. Fuzzy Robust Fault Estimation Scheme for Fault Tolerant Flight Control Systems Based on Unknown Input Observer. *Aircr. Eng. Aerosp. Technol.* **2021**, *93*, 1624–1631. [CrossRef]

30. Kilic, U.; Unal, G. Aircraft Air Data System Fault Detection and Reconstruction Scheme Design. *Aircr. Eng. Aerosp. Technol.* **2021**, *93*, 1104–1114. [CrossRef]

31. Hong, Y.; Kim, S.; Kim, Y.; Cha, J. Quadrotor Path Planning Using A* Search Algorithm and Minimum Snap Trajectory Generation. *ETRI J.* **2021**, *43*, 1013–1023. [CrossRef]

32. Unal, G. Integrated Design of Fault-tolerant Control for Flight Control Systems Using Observer and Fuzzy logic. *Aircr. Eng. Aerosp. Technol.* **2021**, *93*, 723–732. [CrossRef]

33. Amoozegar, M.; Minaei-Bidgoli, B. Optimizing Multi-objective PSO Based Feature Selection Method Using a Feature Elitism Mechanism. *Expert Syst. Appl.* **2018**, *113*, 499–514. [CrossRef]

34. Yuan, Q.; Sun, R.; Du, X. Path Planning of Mobile Robots Based on an Improved Particle Swarm Optimization Algorithm. *Processes* **2022**, *11*, 26. [CrossRef]

35. Han, F.; Chen, W.-T.; Ling, Q.-H.; Han, H. Multi-objective Particle Swarm Optimization with Adaptive Strategies for Feature Selection. *Swarm Evol. Comput.* **2021**, *62*, 100847. [CrossRef]

36. Han, F.; Wang, T.; Ling, Q. An Improved Feature Selection Method Based on Angle-guided Multi-objective PSO and Feature-label Mutual Information. *Appl. Intell.* **2023**, *53*, 3545–3562. [CrossRef]
37. Shaikh, P.W.; El-Abd, M.; Khanafer, M.; Gao, K. A Review on Swarm Intelligence and Evolutionary Algorithms for Solving the Traffic Signal Control Problem. *IEEE Trans. Intell. Transp. Syst.* **2020**, *23*, 48–63. [CrossRef]

Article

Swarm Cooperative Navigation Using Centralized Training and Decentralized Execution

Rana Azzam [1,2,*], Igor Boiko [3] and Yahya Zweiri [1,4]

[1]   Aerospace Engineering Department, Khalifa University of Science and Technology,
      Abu Dhabi P.O. Box 127788, United Arab Emirates
[2]   Khalifa University Center for Autonomous Robotic Systems (KUCARS), Khalifa University of Science
      and Technology, Abu Dhabi P.O. Box 127788, United Arab Emirates
[3]   Electrical Engineering and Computer Science Department, Khalifa University of Science and Technology,
      Abu Dhabi P.O. Box 127788, United Arab Emirates
[4]   Advanced Research and Innovation Center (ARIC), Khalifa University of Science and Technology,
      Abu Dhabi P.O. Box 127788, United Arab Emirates
*   Correspondence: rana.azzam@ku.ac.ae

Abstract: The demand for autonomous UAV swarm operations has been on the rise following the success of UAVs in various challenging tasks. Yet conventional swarm control approaches are inadequate for coping with swarm scalability, computational requirements, and real-time performance. In this paper, we demonstrate the capability of emerging multi-agent reinforcement learning (MARL) approaches to successfully and efficiently make sequential decisions during UAV swarm collaborative tasks. We propose a scalable, real-time, MARL approach for UAV collaborative navigation where members of the swarm have to arrive at target locations at the same time. Centralized training and decentralized execution (CTDE) are used to achieve this, where a combination of negative and positive reinforcement is employed in the reward function. Curriculum learning is used to facilitate the sought performance, especially due to the high complexity of the problem which requires extensive exploration. A UAV model that highly resembles the respective physical platform is used for training the proposed framework to make training and testing realistic. The scalability of the platform to various swarm sizes, speeds, goal positions, environment dimensions, and UAV masses has been showcased in (1) a load drop-off scenario, and (2) UAV swarm formation without requiring any re-training or fine-tuning of the agents. The obtained simulation results have proven the effectiveness and generalizability of our proposed MARL framework for cooperative UAV navigation.

Keywords: UAV cooperative navigation; multi-agent reinforcement learning; autonomous decision making; centralized training and decentralized execution; curriculum learning

Citation: Azzam, R.; Boiko, I.; Zweiri, Y. Swarm Cooperative Navigation Using Centralized Training and Decentralized Execution. *Drones* **2023**, *7*, 193. https://doi.org/10.3390/drones7030193

Academic Editors: Xiwang Dong, Mou Chen, Xiangke Wang and Fei Gao

Received: 20 January 2023
Revised: 19 February 2023
Accepted: 22 February 2023
Published: 11 March 2023

1. Introduction

A UAV swarm is a cyber-physical system consisting of multiple, possibly heterogeneous, UAVs that cooperate to execute a particular mission. A significant amount of swarm applications involve making decisions on how the swarm members will maneuver to cooperatively achieve their objective, such as load delivery [1,2], area coverage [3], search and rescue [4], formation [5], path planning [6], and collision avoidance [7], among others. There are various benefits of deploying swarms of UAVs to carry out cooperative tasks as compared to a single agent, such as fault tolerance, task distribution, execution efficiency and effectiveness, and flexibility, to name a few. This has paved the way for further developments of swarms, particularly through artificial intelligence. As opposed to conventional approaches, learning-based decision-making involves less complex computations, requires neither prior nor global knowledge of the environment, and exhibits better scalability.

Deep reinforcement learning (DRL) [8] is a cutting-edge learning paradigm that accommodates sequential decision-making capabilities and has proved effective in a plethora of robotic applications. By interacting with the environment, a DRL agent controlling a robotic platform is able to learn a certain behavior through incentives and penalties provided by the environment as a result of certain actions decided by the agent. DRL can be extended to multiple agents through varying levels of centralization [9]. Centralized training and centralized execution (CTCE) is the most direct extension, where a single DRL agent is trained to control multiple platforms simultaneously. Although this approach exhibits high efficiency, it is computationally expensive, susceptible to failure upon communication loss, and hence is not robust. The second variant is the decentralized training and decentralized execution approach which is definitely more scalable and robust to a communication failure. This is attributed to the fact that a separate agent is trained to control every entity in the swarm which makes the approach less efficient and more computationally expensive. An alternative approach that combines the advantages of both levels of centralization is centralized training and decentralized execution (CTDE). In CTDE, agents are trained in a centralized manner and hence they exhibit collaborative behavior while maintaining the flexibility and scalability of the swarm. Various works in the literature have been carried out to employ multi-agent reinforcement learning (MARL) in various formulations to solve concurrent challenges concerning cooperative UAV applications. In the following section, a synopsis of the most recent related work on MARL-based UAV applications is presented.

1.1. Related Work

In reference [10], reinforcement learning-based path planning of muli-UAV systems is proposed using CTDE. A long short-term memory (LSTM) layer is used within a proximal policy optimization (PPO) agent, to facilitate making decisions based on current and past observations of the environment. Their reward function was designed as a weighted sum of the objectives that the agent is expected to achieve. Model validation was carried out in a simulated environment with three UAVs. By visualizing the reported results, the planned paths for the UAVs are not very smooth. This behavior may result due to various factors, such as oscillations in subsequent actions.

The work presented in [11] addresses the problem of flocking control of UAVs using a CTDE approach based on PPO. The approach aimed at maintaining a flocking behavior following the model suggested by Reynolds [12] and training was done using a simplified UAV model. The task was defined in such a way that the UAV swarm safely travels as fast as possible towards the goal with minimal distance to the swarm's spatial center. The reward formulation was in terms of the Euclidean distances to the goal, the obstacles, and the swarm center. The UAVs in this work are assumed to fly at different altitudes and fixed speeds. The former condition simplifies exploration by excluding swarm collisions from the experiences, and the latter limits the control of the agent to the heading of the UAV. Controlling the speed or the position of the UAV using the reinforcement learning agent allows for more flexibility and efficiency, yet makes exploration much more challenging. This approach also requires communication between the UAVs in the swarm members, which makes the approach susceptible to communication failure. Simulation results were demonstrated with swarms including up to ten UAVs.

In reference [13], a multi-agent UAV navigation approach was developed using an extension of the original multi-agent deep deterministic policy gradient (MADDPG) [14]. The experiences collected by the agent during training are assigned priorities. Based on these priorities, the experiences are sampled out of the buffer to update the trainable parameters of the neural networks that constitute the MADDPG agent. This means that better experiences have a higher chance of being selected to update the network. However, it is also important for the agent to learn about undesired behaviors since it is highly likely that the agent will encounter previously unseen experiences during real-time deployment.

Another CTDE multi-agent reinforcement learning approach was presented in [15] for the application of collision avoidance of homogeneous UAVs. A PPO agent was adopted to decide on the acceleration of the UAVs in the swarm to maintain safety by avoiding collisions. Every UAV is aware of the positions and velocities of all other UAVs in the environment. This condition might be challenging to achieve in real-world scenarios and may require strong communication if some UAVs are out of the observation range of others in the environment. To circumvent the scalability issue, the algorithm uses an LSTM that encodes the states of all the agents in the swarm into a fixed-size vector. Quantitative results report a high success rate, however, the smoothness of the generated UAV trajectories could be improved. In reference [16], a hierarchy of reinforcement learning agents was used to achieve a multi-objective UAV swarm suppression of an enemy air-defense (SEAD) mission. The top-level agent is concerned about pinpointing the target location to be attacked, while the lower-level agent makes decisions on how the swarm will cooperatively attack the target. Training the agents was done in a decentralized manner, without any experience sharing between the two agent levels.

The work proposed in [17] addresses fixed-wing UAV formation using a leader–follower approach through deep reinforcement learning. The leader UAV makes decisions on how to maneuver, while the others (the followers) try to maintain a certain formation by executing the control commands specified by the leader and communicating the resulting states back. The swarm is rewarded based on defined relative positions between the UAVs respective to a certain formation. The proposed algorithm requires communication between the swarm members and for that, the authors proposed a communication protocol to ensure every UAV has a communication link with at least one member in the swarm. However, any loss of communication would result in undesired formation since the followers rely completely on the leader. An improvement to the original PPO algorithm was proposed to encourage better exploration.

A MARL-based multi-UAV decision making approach was proposed in [18]. A simple UAV model was used to train a multi-agent UAV system for an air-combat mission. A gated recurrent unit and an attention mechanism were used in the decentralized actor and centralized critic networks, respectively, to train a policy that is robust to environmental complexities. The action space combined continuous and discrete actions to make decisions concerning the UAV motion and the combat activity, respectively.

Several other multi-UAV flocking and navigation approaches were proposed using centralized reinforcement learning, such as [19,20]. However, such approaches require communication between the UAVs, rely on global information about the environment, and may not be flexible in terms of the size of the swarm.

An interesting research direction in MARL is credit assignment. When a reinforcement learning agent interacts with the environment, it receives a single scalar value as a reward/penalty for its action(s). In the case of cooperative tasks, multiple agents perform the learning task by taking actions to optimize a single reward that represents them all. This setting introduces a new challenge to MARL, in which agents become "lazy" [21]. In other words, some agents may not perform well as everyone else in the team, and yet receive the same reward collectively. Researchers have proposed several learning [21,22] and non-learning [23–25] approaches to tackle this issue by assigning credit to each agent based on their contribution to the success of the collaborative task. The learning-based methods rely on training agent-specific critic networks in addition to the global critic network to assist with factorizing the global reward into values that reflect the actual contribution of each agent in the team. The other non-learning methods use no additional networks; rather, they employ a difference-reward of various formulations to compute the advantage of each agent's contribution to the collaborative outcome. For instance, the advantage function reflects the value of an agent's actions [23,24] or the agent's actions and observations [25]. Specifically, the approach in [25] implements a multi-agent collision avoidance approach using CTDE. Upon updating the network parameters, the advantage of each agent in the swarm is computed based on the contribution of their action and observation to the global state. The objects used to represent the UAVs in the swarm were defined using primitive kinematic equations, which are very simplistic and hard to transfer to reality. Furthermore, the action space is the heading of the UAV, where UAVs are assumed to fly at a fixed speed. This simplifies exploration, since the action space is bounded and varies along a single dimension.

1.2. Contributions

In this paper, a multi-agent reinforcement learning (MARL)-based cooperative navigation of a swarm of UAVs (as depicted in Figure 1) is developed through centralized training and decentralized execution (CTDE). Curriculum learning is used to facilitate and expedite convergence, in presence of various task complexities arising from partial environment observability, multi-agent training, and exploration in continuous state and action space scenarios. A reward function, combining positive and negative reinforcement, is formulated to encourage cooperative behavior and to ensure that agents achieve their individual goals simultaneously, although executed in a decentralized manner. The cooperative navigation approach is scaled-up to work with a large number of agents without requiring re-training or varying the number of agents during training, as opposed to the approaches in the literature, such as [25], where changing the swarm size requires retraining the agent since the observation space and hence the dimensions of the neural network inputs will differ. Scalability of the proposed approach was also achieved in terms of the swarm speed, and the size of the task environment. The generalizability of the proposed approach was demonstrated through a load delivery application, where the mass of the platform changes during the cooperative navigation task after the swarm drops off payloads (of variable mass per UAV). The swarm was shown to continue the task, and arrive at the final navigation goal (which is set during the task) at the same time, without fine-tuning the parameters of the MARL agent. Extensive testing of the proposed approach was carried out in simulations with varying swarm speeds, navigation goals, and environment sizes. Sample UAV swarm formation scenarios are also showcased and the convergence of the proposed approach is demonstrated.

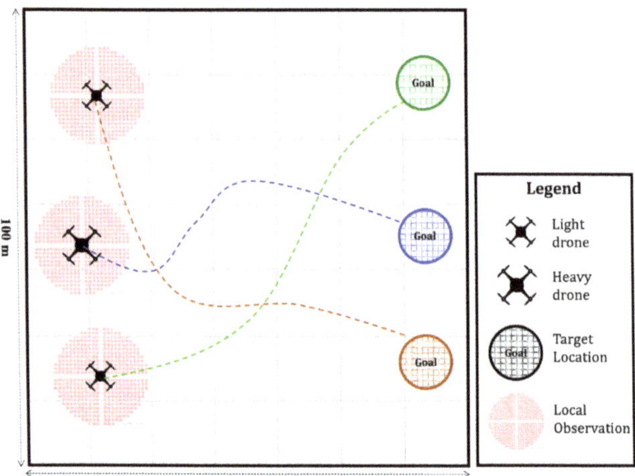

Figure 1. UAV swarm cooperative navigation.

In summary, the contributions of this paper are listed below:

- The development of a scalable, real-time, autonomous MARL-based collaborative navigation approach for a swarm of UAVs using centralized training and decentralized execution.
- The training of the proposed collaborative navigation approach based on a combination of curriculum learning and early stopping using a reward formulation that

encourages cooperative behavior during decentralized execution by means of positive reinforcement.

- Demonstration of the proposed collaborative navigation approach in a load delivery scenario and in swarm formation.
- Extensive testing of the proposed approach across various initial conditions, swarm sizes, UAV speeds, UAV loads, and environment sizes.

2. Methods

2.1. Task Description

The proposed MARL approach is designed for a cooperative navigation task in which a set of agents, in this case, UAVs, are expected to navigate to a set of locations in the task environment. Starting from an initial position, every UAV safely maneuvers to a specific target, within a certain period of time. UAVs are expected to simultaneously arrive at their target locations while operating independently in a decentralized manner. During execution, each agent will only obtain access to local observations within a certain range around the corresponding UAV. Cooperative behavior during decentralized execution is achievable because policies are obtained through centralized training, based on a reward formulation that encourages goal achievement at the individual and collaborative levels. Particularly, training is done with access to global observations collected by all members of the swarm and the parameters of the involved neural networks are updated based on the rewards pertaining to the collective swarm behavior. Collaborative navigation could be deployed in environments with various sizes, and consequently, the maximum allowable speeds may need to be adjusted based on the available space. In addition, the number of UAVs participating in the collaborative task varies based on the application. Flexibility and scalability of the swarm are essential and need to be accounted for in any swarm application.

2.2. Centralized Training and Decentralized Execution

Centralized training and decentralized execution (CTDE) [14] is an approach to MARL where the computational complexity is offloaded onto the training process rather than execution. A popular implementation of this approach is the centralized critic training and decentralized actor execution (as illustrated in Figure 2), which is an extension of the policy–gradient actor–critic model. Particularly, the critic network is trained offline, without constraints on real-time performance. The main purpose is to facilitate obtaining decentralized policies that could accomplish the cooperative multi-agent task through access to global information obtained by multiple agents during training, but not execution. In such a setting, every agent partially observes the environment and hence the problem could be modeled using an extension of Markov decision processes for multiple agents. This extension is referred to as a decentralized partially observable Markov decision process (Dec-POMDP) and is formulated as a five-tuple $(\mathcal{S}, \mathcal{O}, \mathcal{A}, R, \mathcal{T})$ encapsulating:

- State space ($\mathcal{S}$): the global setting of the environment including all the agents.
- Observation space ($\mathcal{O}$): the set of individual observations that agents perceive from the environment.
- Action space ($\mathcal{A}$): a set of actions that the agents execute in the environment.
- Reward (R): the incentives that agents receive upon acting in the environment.
- Transition function ($\mathcal{T}$): defines how agents transition from one state to another.

While operating, each agent attempts to maximize its expected return $\mathcal{R}$ from the ongoing task, as defined in (1).

$$\mathcal{R} = \sum_{t=0}^{T} \gamma^t R_{t+1} \tag{1}$$

where T is the time horizon, and γ is a discount factor that determines the importance of future rewards and falls in the range $[0, 1)$.

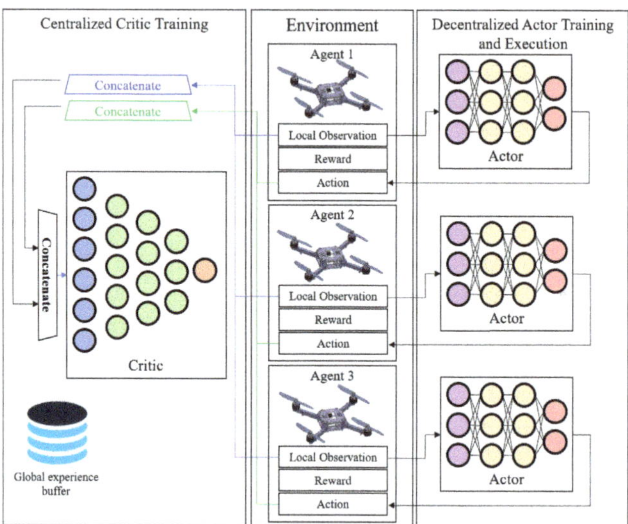

Figure 2. Overall centralized training and decentralized execution (CTDE) multi-agent reinforcement learning (MARL) framework for collaborative UAV swarm navigation.

In a policy gradient method where neural networks are used as policy estimators, the trainable parameters of the network are directly updated to maximize the objective of the optimization which in this case is the agent's total return. The update is carried out by taking steps in the direction of the gradient of the objective function. The objective function and its gradient for a deterministic policy are formulated in (2) and (3).

$$J(\theta) = \mathbb{E}_{s \sim p^\mu}[\mathcal{R}(s, a)] \tag{2}$$

$$\nabla_\theta J(\theta) = \mathbb{E}_{s \sim \mathcal{D}}[\nabla_\theta \mu_\theta(a|s) \nabla_a \mathcal{Q}^\mu(s, a)|_{a=\mu_\theta(s)}] \tag{3}$$

where $s \in \mathcal{S}$, p^μ is the state distribution, $a \in \mathcal{A}$ is an action, $\mathcal{D}$ is a set of transitions collected through experiences and stored in the experience buffer, and $\mathcal{Q}^\mu(s, a)$ is the action-value function associated with the deterministic policy μ.

For the case of CTDE-based MARL, the policy gradient algorithm could be extended to perform centralized critic training based on global observation ($\mathbf{x}$) of N agents, each following a policy μ_i with a set of trainable parameters θ_i, where $i \in 1, ..., N$. The updated formulation of the policy gradient is shown in (4).

$$\nabla_{\theta_i} J(\theta_i) = \mathbb{E}_{\mathbf{x}, a \sim \mathcal{D}}[\nabla_{\theta_i} \mu_i(a_i|o_i) \nabla_{a_i} \mathcal{Q}_i^\mu(\mathbf{x}, a_1, ..., a_N)|_{a_i=\mu_i(o_i)}] \tag{4}$$

where $o_i \in \mathbf{O}$ is the observation of agent i. Every transition in the experience buffer $\mathcal{D}$ in the multi-agent setting contains the current global state, the next global state, the individual actions per agent, and the corresponding rewards. In the current work, the deterministic policy and the corresponding value function and computed using neural networks, referred to as the actor and critic, respectively.

2.3. Proposed Model

2.3.1. Actor and Critic Architecture

An actor–critic agent is adopted to perform the cooperative navigation task. The critic is centralized and hence receives global input from the swarm, while the actor is decentralized where it processes local observations. Figure 3 shows a detailed description of the architecture of both the critic and actor. The critic consists of two input paths, one for the global state and the other for the swarm actions. The states are passed through seven hidden dense layers, activated using the rectified linear unit (ReLU), while the actions are passed through a single ReLU activated dense layer.

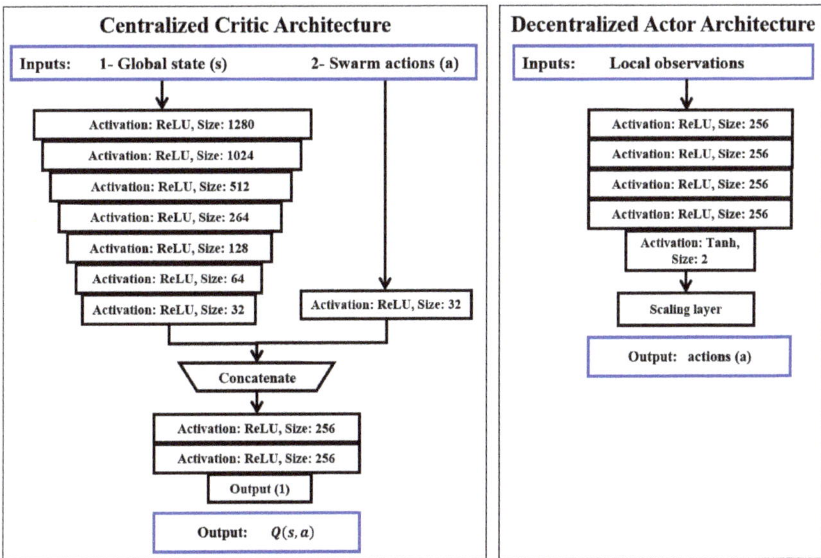

Figure 3. Architecture of the proposed actor and critic networks.

$$ReLU(x) = max(0, x) \tag{5}$$

The outputs of the two paths are then concatenated and passed into two ReLU activated dense layers. Finally, a single-neuron layer outputs the value of an action (a) taken in the state (s).

The actor-network, on the other hand, consists of four hidden dense layers activated using ReLU, followed by a two-neuron layer activated using hyperbolic-tan (tanh) to output actions in the range $[-1, 1]$.

$$tanh(x) = \frac{e^x - e^{-x}}{e^x + e^{-x}} \tag{6}$$

However, since it is sometimes desired to fly UAVs at higher speeds, particularly when the target locations are far apart, a scaling layer was used to set the maximum UAV speed per task.

It is worth noting that the actor-critic agent contains duplicate networks of the actor and critic, referred to as target actor and target critic, respectively. These networks are initialized to the same parameters as the actor and critic but are updated less frequently to achieve learning stability.

2.3.2. State Space and Action Space

At time step t, an agent i observes $o_i^t \in \mathcal{O}$ which represents its local surrounding; namely the relative distance to any other agent within the observation range, the speed of the observed neighbor, and the relative distance to the target location. It is worth noting that each agent is able to observe the environment up to certain spatial limits, and anything outside this range is not perceived by the agent and hence does not affect its decisions.

The action space used in the proposed approach is continuous and two-dimensional. More particularly, the actor-network outputs the reference velocities that will be passed to the UAV's low-level controller to guide each UAV in the swarm from its initial position to its target location. At the time t, the action generated for agent i is denoted as $a_t^i \in \mathcal{A}$.

2.3.3. Reward Formulation

Rewards are incentives that guide agent training to achieve a particular task, by praising actions taken towards the goal achievement and penalizing actions that hinder the completion of the task, such as collisions. For the cooperative navigation task, a major component of the reward function is concerned with reducing the Euclidean distance to the target location. This component was chosen to be continuous to ease exploration and facilitate convergence as defined in (7).

$$r_{euclidean} = -\sqrt{(x_{target_i} - x_{uav_i})^2 + (y_{target_i} - y_{uav_i})^2} \tag{7}$$

where (x_{uav_i}, y_{uav_i}) is the position of the UAV controlled by the ith agent, and $(x_{target_i}, y_{target_i})$ is the 2D target location set for this agent in the environment. It is assumed that UAVs fly at a fixed altitude.

To achieve cooperative behavior, a large positive reward ($r_{swarm\_goal} = 100$) was granted to the UAV swarm if *all* agents arrived at the goal position at the same time. During training, it was observed that reaching the goal position was frequently achievable by the agents individually at different time instances during the episode. However, staying at the target location was challenging. To that end, positive reinforcement was used to reward individual agents that arrive at the goal position ($r_{individual\_goal} = 10$). To maximize its own return, an agent will try to remain within the target area to collect as many rewards as possible. This component of the reward facilitated achieving the sought swarm objective, where all agents have to be at the target location at the same time. More specifically, the agent generates actions to reduce the speed of the UAV around the goal position. In case an action causes an agent to collide with other agents, a sparse negative penalty ($r_{collision} = -100$) is used to discourage this behavior.

The global reward associated with a set of actions taken in a certain state at time t is a weighted sum of these four components as indicated in (8).

$$R_t = \omega \sum_{i=1}^{N} r^i_{euclidean} + \sum_{i=1}^{N} r^i_{individual\_goal} + \sum_{i=1}^{N} r^i_{collision} + r_{swarm\_goal} \tag{8}$$

where N is the number of agents, and ω was set to 0.01 to scale down the value of the Euclidean distance since training was done in a 100×100 m^2 environment.

2.4. Curriculum Learning

Curriculum learning is a training strategy in which a neural network is gradually exposed to task complexity as originally proposed in [26]. The concept behind this strategy is inspired by nature, where humans progressively learn the skills they need over their lifespan. Curriculum learning has two major advantages: (1) it facilitates fast convergence, and (2) it helps achieve better local minima when solving non-convex optimization. In the context of neural networks, curriculum learning guides training toward convergence in a timely manner.

In this work, training the proposed MARL framework was carried out in stages, in a way that supports exploration. UAVs were placed in an environment and were expected to navigate to a target position that required them to maneuver along a single dimension. Given the decentralized nature of the actor training/execution, each UAV receives an action based on its current local observation. Consequently, in every training step, every member in the swarm contributes a different experience towards achieving a common goal. In view of the fact that the action space is continuous, this has expedited the exploration of the action space and has facilitated convergence towards the required cooperative goal. The UAVs are considered to have achieved the goal if they arrive in the vicinity of the target location up to a certain radius. This spatial threshold was set to a large value in the first training stage (50 m) then gradually reduced to 2 m in the following stages.

The complexity of the task was then increased to require UAV control in two dimensions in order to arrive at the goal position. Instead of starting the training over, the neural networks were initialized using the weights from the previous stage. This has significantly accelerated convergence toward achieving the swarm goal. Due to the high variance of the training process, early stopping was also adopted to terminate training after the model had converged for a few hundred episodes.

2.5. UAV Dynamics

The UAV multirotor model that is used to train the proposed approach highly resembles the dynamics of a physical multicoptor to facilitate transferability to real experiments at later stages of this work. The model encapsulates various nonlinear dynamics [27], namely, (1) nonlinear drag dynamics for which a linearized drag model [28] was used, as verified in [29,30], (2) nonlinear propulsion dynamics for which electronic speed controllers (ESCs) are used to linearly map ESC inputs to corresponding thrust, (3) nonlinearities arising from motor saturation which are avoided through operation strictly in the non-saturation regime, and (4) nonlinear kinematics caused by under actuation and gravity, which are linearized using a geometric tracking controller [31] and hence a feedback linearization controller is obtained.

The adopted altitude and attitude dynamics are shown in (9)–(11) and a summary of the used transfer functions and symbols is provided in Table 1.

$$G_{prop}(s) = \frac{K_{prop}e^{-\tau_{act}s}}{T_{prop}s + 1} \tag{9}$$

$$G_{att,alt}(s) = \frac{K_p}{s(T_1s + 1)} \tag{10}$$

$$G_{in}(s) = \frac{K_{eq}e^{-\tau_{in}s}}{s(T_{prop}s + 1)(T_1s + 1)} \tag{11}$$

Table 1. Linearized altitude and attitude dynamics.

Transfer Function	Type	Purpose	Symbols
(9)	First order plus time delay	Maps ESC inputs to force/torque output	K_{prop}: propulsion static gain τ_{act}: propulsion system delay T_{prop}: propulsion time constant
(10)	First order system with an integrator	Models attitude and altitude dynamics	T_1: time constant - drag dynamics K_p: system inertia
(11)	$G_{att,alt}(s)$ cascaded with $G_{prop}(s)$	Maps ESC commands to UAV attitude and altitude	$K_{eq} = K_p K_{prop}$ τ_{in}: total inner dynamics' delay

The work presented in [27] demonstrates the high resemblance of the UAV behavior in simulations and experiments using this model. The lateral motion dynamics of the UAV are adopted from [32] to describe the change in attitude in the direction of motion. The equations are listed below (12)–(13) and explained in Table 2.

$$G_{out}(s) = \frac{K_{eq}e^{-\tau_{out}s}}{s(T_2s + 1)} \tag{12}$$

$$G_{lat}(s) = \frac{K_{eq,l}e^{-(\tau_{in}+\tau_{out})s}}{s^2(T_{prop}s + 1)(T_1s + 1)(T_2s + 1)} \tag{13}$$

Table 2. Linearized lateral motion dynamics.

Transfer Function	Purpose	Symbols
(12)	Maps the multirotor's tilt angle to its lateral position	$K_{eq,l}$: overall lateral dynamics gain τ_{out}: lateral motion sensor delay T_2: lateral motion drag.
(13)	Maps ESC commands to UAV lateral position	-

The deep neural network and the modified relay feedback test (DNN-MRFT) identification approach [32] is used to experimentally identify the presented model parameters. First, a domain for the unknown time parameters is chosen for both the inner and lateral dynamic parameters as in [30,32], respectively. The selected domains are discretized to guarantee up to 10% performance sub-optimality. MRFT is then performed and the results are passed to the DNN which will select the best-suited model parameters. The corresponding controller parameters may then be obtained using the derivative-free Nelder–Mead simplex algorithm.

3. Results and Discussion

3.1. Model Training

The proposed model structure was developed using the TensorFlow [33] library on a Dell desktop, with Intel Xeon(R) W-2145 CPU @ 3.70 GHz × 16. The initial stage of training extended for 10,000 episodes, each consisting of a maximum of 3000 steps. Every step runs for 0.1 s, i.e., a new action is generated at the beginning of each step and the agent executes the action for the remaining time in that step. It is worth noting that during execution on a physical platform, the actor is capable of generating actions at 100 Hz by means of an Intel NUC onboard computer. The episode was selected to be long enough to allow sufficient exploration with various speeds in the task environment which spans 100×100 m^2. It is worth noting that the motion of the UAV swarm was restricted to the defined environment boundaries, where actions that lead to exiting the environment were ignored. In case the swarm goal is achieved or a collision occurs between the UAVs, the training episode is terminated. In subsequent training stages where the complexity of the task was increased, training was conducted with less exploration noise and a lower learning rate, and was suspended when convergence was observed. The training was repeated many times to ensure that the results are not affected by the initial random seed.

The plots depicted in Figure 4 show the cooperative navigation scenario on which the agent was trained. Three UAVs were guided through a 100×100 m^2 environment to stop at the same time at set locations. The maximum speed of the swarm in this scenario was 1 m/s which is extremely slow for the total traveled distance per UAV. Hence, the swarm arrived at their goal positions, which are 80 m away from the initial position in 2600 steps.

One of the common problems in the reinforcement learning literature is the oscillatory behavior in consecutive actions generated by a trained agent [34]. Such oscillations may result in undesired behavior and may lead to damaging the platform in case of aggressive maneuvers. While testing the trained model, this behavior was not encountered in any of the scenarios across various speeds, various locations, and initial conditions, as will be shown in the next sections. Consecutive reference velocities generated by the MARL agent gradually decrease upon approaching the goal. This has resulted in smooth flights for all the members of the swarm.

The behavior exhibited by the agents upon approaching the goal is essential to achieving the swarm goal, particularly with decentralized execution, since agents are required to be at the target locations at the same time. The sparse positive reward used to incentivize individual agents for reaching their goal locations has contributed to this behavior, especially when agents have to traverse variable distances, as will be seen in the following sections.

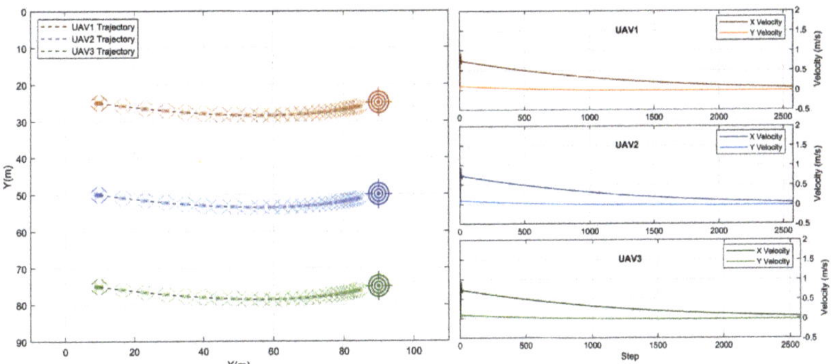

Figure 4. Result: cooperative navigation training scenario.

3.2. Testing with Variable Swarm Speeds

In this section, the same scenario presented in the previous section is used, however, the swarm speed was much higher than in the training scenario. The maximum speed per UAV was 12 m/s which is 12 times the speed in the previous section. The decentralized policies were still able to successfully achieve the swarm goal and the three UAVs arrived at their target locations at the same time, after gradually and smoothly slowing down near the set locations. The swarm was at the target locations in less than 150 steps, which is equivalent to 15 s. The trajectories and the corresponding UAV speeds at each time step are shown in Figure 5.

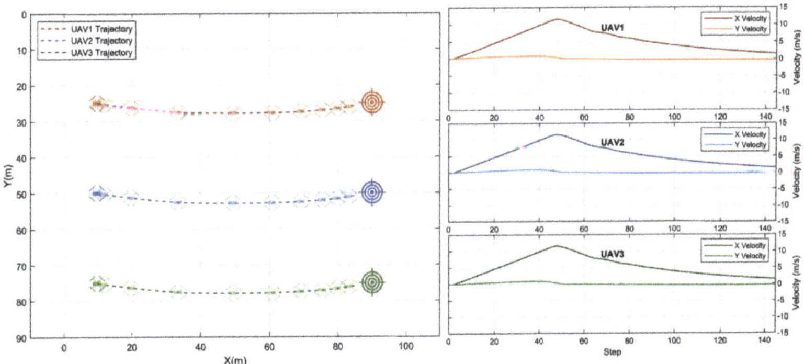

Figure 5. Result: cooperative navigation with high swarm speed.

3.3. Testing with Different Goal Positions

In this example, the agents were assigned target locations at variable distances from the UAVs' initial locations, in both dimensions (x, y). Figure 6 shows the scenario and the obtained results using the proposed approach. UAV1 (in orange) has to travel the longest distance, followed by UAV2 (in green), and lastly UAV3 whose target location is the closest. Because of the centralized training nature, the decentralized policies exhibit collaborative behavior and are able to effectively achieve the goal of the swarm. In order for the three UAVs to arrive at their goal locations at the same time, the agents generated reference velocities based on each UAV's distance from its target. Obviously, UAV1 was the fastest, followed by UAV3, and then UAV2. In the y dimension, the generated reference velocities were also different since the target locations were above, below, and along the initial location for UAV 1, 3, and 2 respectively. The maximum swarm speed was 8 m/s and the swarm goal was achieved after approximately 270 steps. The speed of each UAV was drastically reduced near the goal position.

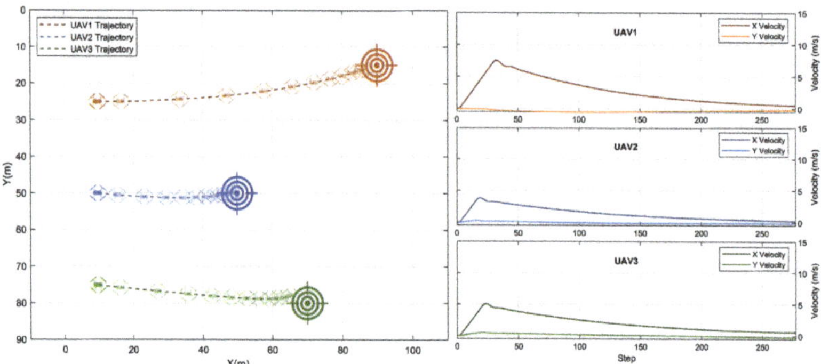

Figure 6. Result: cooperative navigation with different goal Positions.

3.4. Load Drop-Off Scenario in a Large Environment

In this section, we demonstrate a load drop-off scenario using the proposed MARL-based cooperative navigation framework. The scenario is demonstrated in a larger environment than that used for training, involves changing the goal position during operation, and requires the ability to handle the change in the platform mass to achieve successful cooperative navigation, as depicted in Figure 7.

Starting from their initial positions, every UAV is assumed to carry loads weighing 10%, 20%, and 30% of the platform mass, respectively. The three UAVs are expected to drop the load off simultaneously at locations 50 m, 70 m, and 90 m away from the initial positions. To achieve that, UAV3 commanded the highest reference velocity (approximately 8 m/s), while UAV1 traveled at the lowest speed among the other agents (approximately 5 m/s). The agents were able to drop their loads off simultaneously after about 28 s. Right then, the UAVs (with their reduced masses) were assigned updated target locations that are 120 m, 100 m, and 80 m apart from the drop-off locations of UAV 1, 2, and 3, respectively. It is worth noting that the UAVs were not completely stopped at the drop-off location. To arrive at the new target locations at the same time, the maximum speed for UAV1 was 10 m/s, while UAV2 and UAV3 traveled at lower speeds. All three UAVs arrived at the new target locations simultaneously and gradually slowed down in the target vicinity.

The results obtained in this test have proven the scalability of the proposed approach to a larger environment, its ability to handle changes in the platform mass in-flight, and its ability to cope with dynamic target locations during the mission.

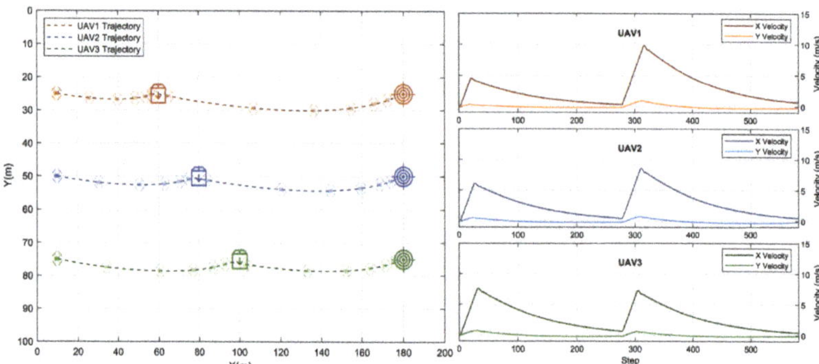

Figure 7. Result: load drop-off scenario in a large environment.

3.5. Testing with Variable Swarm Sizes

The scenarios presented here show how the developed MARL cooperative navigation framework can be used for any number of UAVs in the swarm, given that it was trained to work for three only. In its original design, every agent observes its own distance to the goal, its own speed, and its relative distance to the other members of the swarm and their velocities if they fall within the observation range. Since the input to dense neural networks has to be of fixed size, the size of the observation vector of each agent was set to always fit the states of the two neighboring members of the swarm. In case they were out of the observation range, the corresponding values in the observation vector are set to zero. To make that work for a large swarm, we have added a function to check for the closest two neighbors to every agent in the swarm during operation then included their states in the observation vector of that agent. This has added flexibility to the number of allowable UAVs in the swarm and facilitated testing with larger numbers of UAVs without requiring retraining of the MARL agent. All agents demonstrated collaborative behavior and were able to achieve the swarm goal collectively.

The results illustrated in Figure 8 show an example scenario where six UAVs have maneuvered into a triangular formation starting from their initial positions where they were lined up at $y = 10$. The target locations were set at various distances in x and y dimensions. At any time instance, every UAV may observe the closest two members in the swarm. The velocity plots in the same figure show how each decentralized agent generated different reference velocities depending on the relative distance between the UAV and its corresponding target, to allow all UAVs to achieve their goals at the same time.

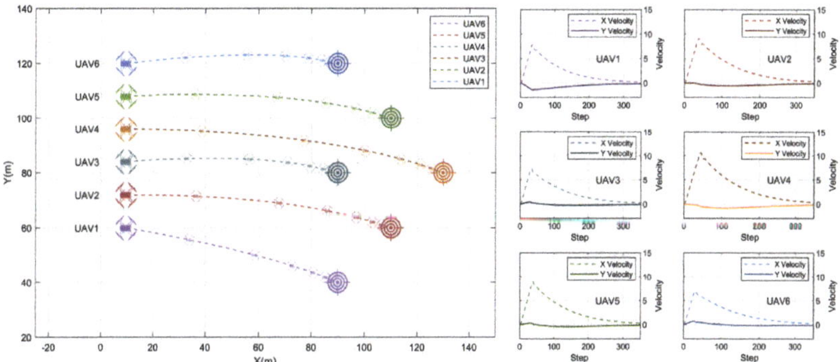

Figure 8. Result: swarm formation example.

Figure 9 demonstrates another formation scenario where ten UAVs were assigned colinear target locations starting from opposite sides in the environment. In this example, the MARL agent was responsible for generating the magnitudes of the reference velocities and an external function was used to decide the direction of the velocity based on each agent's relative position to its target. All ten UAVs were able to be within 2 m of the set target locations at the same time and all the flights show a high level of smoothness. The cooperative task was completed in 34 s.

3.6. Action Smoothness

In this section, a test with a swarm of 50 UAVs was conducted to demonstrate the ability of the proposed approach to generate smooth actions across consecutive steps. Every UAV started from a different position in the environment and was assigned a target location at a different distance than the other members in the swarm. This test shows the flexibility and the generalizability of the proposed approach and proves that oscillatory behavior is not encountered over a large range of states and actions. The trajectories followed by the UAVs are depicted in Figure 10 and the corresponding actions in the x direction are shown in Figure 11. The same test was repeated with much lower velocities and oscillations were not encountered at all.

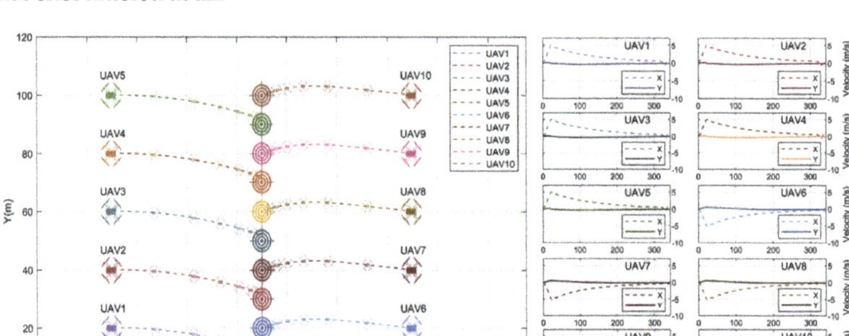

Figure 9. Result: swarm formation in two directions.

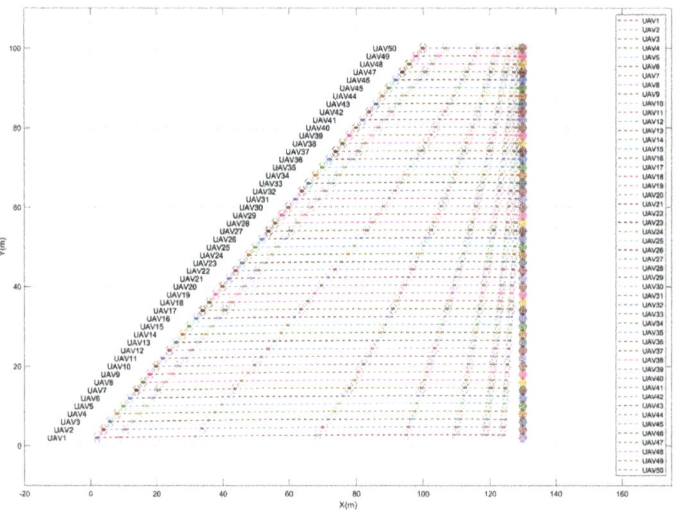

Figure 10. Result: collaborative navigation of a swarm of 50 UAVs.

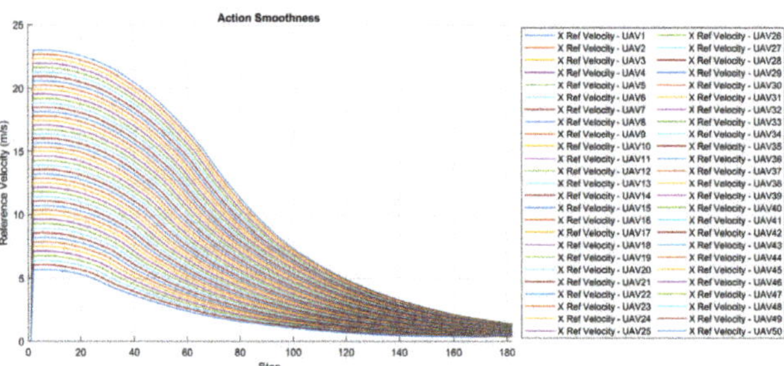

Figure 11. Result: action smoothness with a swarm of 50 UAVs.

3.7. Training Convergence

Training the proposed approach was carried out in various stages starting from simple tasks to more difficult ones. During training, the performance of the model was evaluated based on the episodic reward, as well as the performance of the agents in the environment. Once the desired behavior was achieved by the proposed framework, training the model in that stage was halted (a training strategy referred to as early stopping). Afterward, the complexity of the problem was increased and the model retrained, where the neural networks were initialized into the values obtained in the previous stage. In later training stages, the learning rates of both the actor and critic are reduced to benefit more from what the model has already learned earlier. Figure 12 demonstrates the convergence of the model in the final training stages.

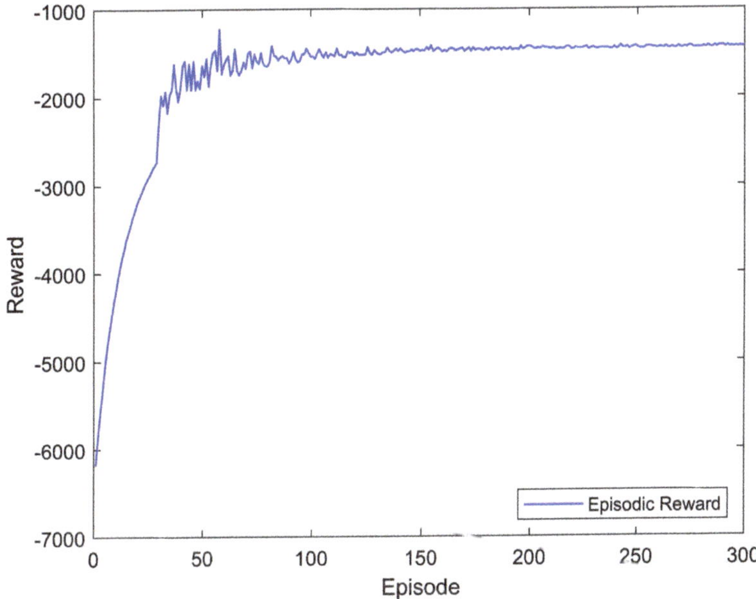

Figure 12. Proposed MARL training convergence in final stages of curriculum learning.

The adopted training strategy has expedited converge to a policy that exhibits cooperative behavior although is executed in a decentralized manner. In addition, the actions generated by the policy resulted in smooth maneuvers as demonstrated in all the test results.

Without curriculum learning and early stopping, the convergence of the model is much more challenging due to the large environment size, continuous action and state spaces, multi-agent setting, limited observability, and the instability of the environment in presence of multiple dynamic entities at the same time. Figure 13a,b show examples of unstable training of the same model if exploration is performed in one shot. It is worth noting that positive rewards were achieved in these cases because of the positive component of the reward formulation that an agent receives when it arrives at its target location. Larger positive episodic rewards mean that one or two agents were at their target locations accumulating the positive rewards while waiting for the remaining two or one agent, respectively, to arrive at their target location. The latter agents in such cases would be exploring a different area in the environment and hence the episode was not terminated, until the specified number of steps ended. Furthermore, Figure 13c shows the episodic rewards of the same model with credit assignment as proposed in [25], where decentralized agents do not use the global reward to update their parameters, but rather a reward value that reflects their contribution to the success of the task. Every episode may extend to 3000 steps if no collisions between the UAVs happened. It is worth noting that the highest reward values, in this case, were achieved due to the early termination of the training episode, and hence the negative penalty corresponding to the Euclidean distance to the target location was not accumulated for a long time. A sample testing scenario with credit assignment where the collaborative navigation task was not achieved is shown in Figure 14. These examples demonstrate the importance of curriculum learning to the training convergence and task achievement.

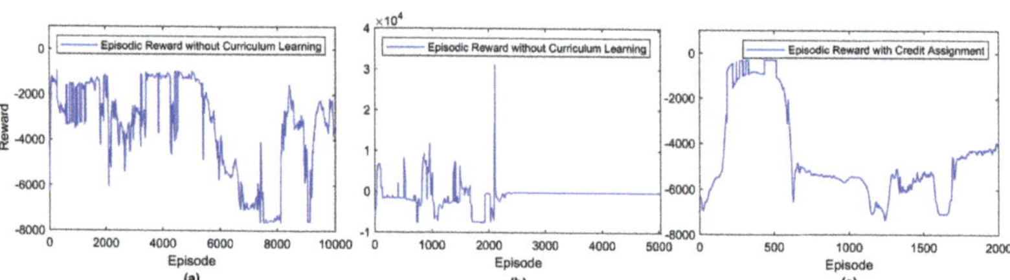

Figure 13. (**a**,**b**): Sample learning curves by training the model without curriculum learning or early stopping, (**c**) Episodic reward with credit assignment as proposed in [25] without curriculum learning or early stopping.

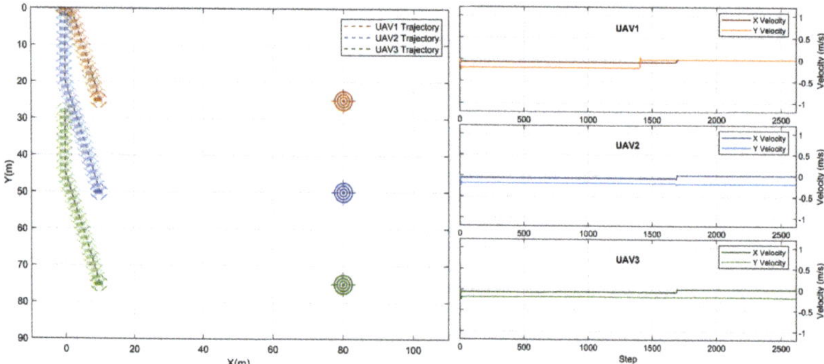

Figure 14. Result: collaborative navigation with credit assignment.

In the future, the proposed MARL-based cooperative navigation approach will be tested in real-world experiments. It is anticipated that the policy will transfer well to the physical platforms in real environments. This was demonstrated in our previous work [35], in which a single agent was trained to perform a goal oriented task and the transferability to reality was seamless without any model retraining or finetuning. The same UAV model was used for training the current approach, and hence we conjecture that no additional tuning is required for simulation to reality transfer.

3.8. Centralized Collaborative Navigation

In this section, a centralized DDPG agent was trained to perform collaborative navigation in exactly the same settings as our proposed approach. The architecture of the actor and critic networks and the reward formulation were not altered. However, in the centralized approach, one actor network is used to generate the actions for all the UAVs in the swarm at the same time. The centralized agent was trained for more than 2 M steps but convergence was not achieved. The resulting behavior of the swarm after training is shown in Figure 15. One of the UAVs left the environment, while the other two collided. In addition, the actions demonstrate variations throughout the episode as opposed to the actions generated by our proposed approach which demonstrate much higher smoothness. Extensive exploration is yet needed for the centralized agent to achieve the sought performance. It is also worth noting that changes to the size of the swarm would require retraining the actor since the size of the output vector will change. Our proposed approach is more flexible since the decentralized nature of execution circumvents this problem.

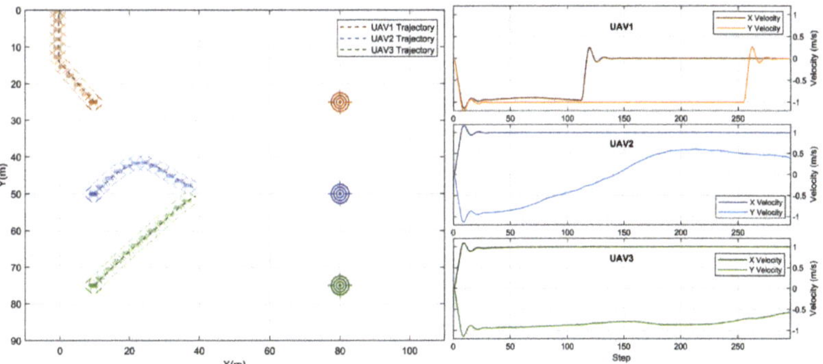

Figure 15. Result: centralized collaborative navigation.

4. Conclusions

In this paper, a multi-agent reinforcement learning-based swarm cooperative navigation framework was proposed. The centralized training and decentralized execution approach was adopted with a reward formulation combining negative and positive reinforcement. The training was carried out using a high-fidelity UAV model to facilitate simulation to reality transfer. In order to achieve the desired behavior and reduce the complexity of exploration, training was performed in multiple stages where the difficulty of the swarm goal was gradually increased. The proposed framework was extensively tested in simulated scenarios which vary from the one used for training, and demonstrated remarkable performance. It generalized well to larger environment sizes, a large number of UAVs in the swarm, high speeds, various UAV masses, variable goal positions, and changes to the target locations in flight. The effectiveness and scalability of the multi-agent reinforcement based UAV collaborative navigation were demonstrated through load dropoff and UAV formation scenarios. The training convergence of the proposed framework was demonstrated and the importance of curriculum learning was highlighted by analyzing the stability of the learning-in-one-shot of the same framework and another variant that uses credit assignment.

In the future, the proposed framework will be tested in real experiments and the complexity of the swarm goal will be increased to make the environment more challenging. In addition, navigation in 3D will be investigated to improve collision avoidance flexibility in presence of obstacles that could be avoided by flying at varying altitudes.

Author Contributions: Conceptualization: R.A., Y.Z. and I.B.; methodology, testing, validation, writing, original draft preparation, visualization: R.A.; review and editing, supervision, project administration, funding acquisition: Y.Z. and I.B. All authors have read and agreed to the published version of the manuscript.

Funding: This research was funded by Khalifa University Grant CIRA-2020-082.

Data Availability Statement: No new data were created or analyzed in this study. Data sharing is not applicable to this article.

Conflicts of Interest: The authors declare no conflict of interest.

References

1. Cavone, G.; Epicoco, N.; Carli, R.; Del Zotti, A.; Paulo Ribeiro Pereira, J.; Dotoli, M. Parcel Delivery with Drones: Multi-criteria Analysis of Trendy System Architectures. In Proceedings of the 29th Mediterranean Conference on Control and Automation (MED), Bari, Italy, 22–25 June 2021; pp. 693–698. [CrossRef]
2. Saunders, J.; Saeedi, S.; Li, W. Autonomous Aerial Delivery Vehicles, a Survey of Techniques on how Aerial Package Delivery is Achieved. *arXiv* **2021**, arXiv:2110.02429.
3. Li, M.; Richards, A.; Sooriyabandara, M. Asynchronous Reliability-Aware Multi-UAV Coverage Path Planning. In Proceedings of the 2021 IEEE International Conference on Robotics and Automation (ICRA), Xi'an, China, 30 May–5 June 2021; pp. 10023–10029. [CrossRef]
4. Alotaibi, E.T.; Alqefari, S.S.; Koubaa, A. LSAR: Multi-UAV Collaboration for Search and Rescue Missions. *IEEE Access* **2019**, *7*, 55817–55832. [CrossRef]
5. Jiang, Y.; Bai, T.; Wang, Y. Formation Control Algorithm of Multi-UAVs Based on Alliance. *Drones* **2022**, *6*, 431. [CrossRef]
6. Abichandani, P.; Lobo, D.; Muralidharan, M.; Runk, N.; McIntyre, W.; Bucci, D.; Benson, H. Distributed Motion Planning for Multiple Quadrotors in Presence of Wind Gusts. *Drones* **2023**, *7*, 58. [CrossRef]
7. Huang, Y.; Tang, J.; Lao, S. Cooperative Multi-UAV Collision Avoidance Based on a Complex Network. *Appl. Sci.* **2019**, *9*, 3943. [CrossRef]
8. Plaat, A. Deep Reinforcement Learning. *arXiv* **2022**, arXiv:2201.02135.
9. Zhang, K.; Yang, Z.; Basar, T. Multi-Agent Reinforcement Learning: A Selective Overview of Theories and Algorithms. *arXiv* **2019**, arXiv:1911.10635.
10. Chen, Y.; Dong, Q.; Shang, X.; Wu, Z.; Wang, J. Multi-UAV Autonomous Path Planning in Reconnaissance Missions Considering Incomplete Information: A Reinforcement Learning Method. *Drones* **2023**, *7*, 10. [CrossRef]
11. Yan, P.; Bai, C.; Zheng, H.; Guo, J. Flocking Control of UAV Swarms with Deep Reinforcement Learning Approach. In Proceedings of the 2020 3rd International Conference on Unmanned Systems (ICUS), Harbin, China, 27–28 November 2020; pp. 592–599. [CrossRef]
12. Reynolds, C.W. Flocks, Herds and Schools: A Distributed Behavioral Model. *SIGGRAPH Comput. Graph.* **1987**, *21*, 25–34. [CrossRef]
13. Wu, D.; Wan, K.; Tang, J.; Gao, X.; Zhai, Y.; Qi, Z. An Improved Method towards Multi-UAV Autonomous Navigation Using Deep Reinforcement Learning. In Proceedings of the 2022 7th International Conference on Control and Robotics Engineering (ICCRE), Beijing, China, 15–17 April 2022; pp. 96–101. [CrossRef]
14. Lowe, R.; Wu, Y.; Tamar, A.; Harb, J.; Abbeel, P.; Mordatch, I. Multi-Agent Actor-Critic for Mixed Cooperative-Competitive Environments. *arXiv* **2017**, arXiv:1706.02275.
15. Thumiger, N.; Deghat, M. A Multi-Agent Deep Reinforcement Learning Approach for Practical Decentralized UAV Collision Avoidance. *IEEE Control. Syst. Lett.* **2022**, *6*, 2174–2179. [CrossRef]
16. Yue, L.; Yang, R.; Zuo, J.; Zhang, Y.; Li, Q.; Zhang, Y. Unmanned Aerial Vehicle Swarm Cooperative Decision-Making for SEAD Mission: A Hierarchical Multiagent Reinforcement Learning Approach. *IEEE Access* **2022**, *10*, 92177–92191. [CrossRef]
17. Xu, D.; Guo, Y.; Yu, Z.; Wang, Z.; Lan, R.; Zhao, R.; Xie, X.; Long, H. PPO-Exp: Keeping Fixed-Wing UAV Formation with Deep Reinforcement Learning. *Drones* **2023**, *7*, 28. [CrossRef]
18. Li, S.; Jia, Y.; Yang, F.; Qin, Q.; Gao, H.; Zhou, Y. Collaborative Decision-Making Method for Multi-UAV Based on Multiagent Reinforcement Learning. *IEEE Access* **2022**, *10*, 91385–91396. [CrossRef]
19. Wang, W.; Wang, L.; Wu, J.; Tao, X.; Wu, H. Oracle-Guided Deep Reinforcement Learning for Large-Scale Multi-UAVs Flocking and Navigation. *IEEE Trans. Veh. Technol.* **2022**, *71*, 10280–10292. [CrossRef]
20. Shen, G.; Lei, L.; Li, Z.; Cai, S.; Zhang, L.; Cao, P.; Liu, X. Deep Reinforcement Learning for Flocking Motion of Multi-UAV Systems: Learn From a Digital Twin. *IEEE Internet Things J.* **2022**, *9*, 11141–11153. [CrossRef]
21. Sunehag, P.; Lever, G.; Gruslys, A.; Czarnecki, W.M.; Zambaldi, V.F.; Jaderberg, M.; Lanctot, M.; Sonnerat, N.; Leibo, J.Z.; Tuyls, K.; et al. Value-Decomposition Networks For Cooperative Multi-Agent Learning. *arXiv* **2017**, arXiv:1706.05296.
22. Feng, L.; Xie, Y.; Liu, B.; Wang, S. Multi-Level Credit Assignment for Cooperative Multi-Agent Reinforcement Learning. *Appl. Sci.* **2022**, *12*, 6938. [CrossRef]
23. Foerster, J.N.; Farquhar, G.; Afouras, T.; Nardelli, N.; Whiteson, S. Counterfactual Multi-Agent Policy Gradients. *arXiv* **2017**, arXiv:1705.08926.
24. Li, J.; Kuang, K.; Wang, B.; Liu, F.; Chen, L.; Wu, F.; Xiao, J. Shapley Counterfactual Credits for Multi-Agent Reinforcement Learning. In Proceedings of the 27th ACM SIGKDD Conference on Knowledge Discovery & Data Mining, Online Conference, 14–18 August 2021; ACM: New York, NY, USA, 2021. [CrossRef]
25. Huang, S.; Zhang, H.; Huang, Z. Multi-UAV Collision Avoidance Using Multi-Agent Reinforcement Learning with Counterfactual Credit Assignment. *arXiv* **2022**, arXiv:2204.08594. https://doi.org/10.48550/ARXIV.2204.08594.
26. Bengio, Y.; Louradour, J.; Collobert, R.; Weston, J. Curriculum Learning. In Proceedings of the 26th Annual International Conference on Machine Learning, Montreal, QC, Canada, 14–18 June 2009; Association for Computing Machinery: New York, NY, USA, 2009; Volume ICML '09, pp. 41–48. [CrossRef]
27. AlKayas, A.Y.; Chehadeh, M.; Ayyad, A.; Zweiri, Y. Systematic Online Tuning of Multirotor UAVs for Accurate Trajectory Tracking Under Wind Disturbances and In-Flight Dynamics Changes. *IEEE Access* **2022**, *10*, 6798–6813. [CrossRef]

28. Pounds, P.; Mahony, R.; Corke, P. Modelling and control of a large quadrotor robot. *Control. Eng. Pract.* **2010**, *18*, 691–699. [CrossRef]

29. Chehadeh, M.S.; Boiko, I. Design of rules for in-flight non-parametric tuning of PID controllers for unmanned aerial vehicles. *J. Frankl. Inst.* **2019**, *356*, 474–491. [CrossRef]

30. Ayyad, A.; Chehadeh, M.; Awad, M.I.; Zweiri, Y. Real-Time System Identification Using Deep Learning for Linear Processes With Application to Unmanned Aerial Vehicles. *IEEE Access* **2020**, *8*, 122539–122553. [CrossRef]

31. Lee, T.; Leok, M.; McClamroch, N.H. Geometric tracking control of a quadrotor UAV on SE (3). In Proceedings of the 49th IEEE Conference on Decision and Control (CDC), Atlanta, GA, USA, 15–17 December 2010; IEEE: PIscatway, NJ, USA, 2010; pp. 5420–5425.

32. Ayyad, A.; Chehadeh, M.; Silva, P.H.; Wahbah, M.; Hay, O.A.; Boiko, I.; Zweiri, Y. Multirotors From Takeoff to Real-Time Full Identification Using the Modified Relay Feedback Test and Deep Neural Networks. *IEEE Trans. Control. Syst. Technol.* **2021**, 1–17. [CrossRef]

33. Abadi, M.; Agarwal, A.; Barham, P.; Brevdo, E.; Chen, Z.; Citro, C.; Corrado, G.S.; Davis, A.; Dean, J.; Devin, M.; et al. TensorFlow: Large-Scale Machine Learning on Heterogeneous Systems, 2015. Available online: https://www.tensorflow.org (accessed on 15 September 2022).

34. Ibarz, J.; Tan, J.; Finn, C.; Kalakrishnan, M.; Pastor, P.; Levine, S. How to Train Your Robot with Deep Reinforcement Learning; Lessons We've Learned. *arXiv* **2021**, arXiv:2102.02915.

35. Azzam, R.; Chehadeh, M.; Hay, O.A.; Boiko, I.; Zweiri, Y. Learning to Navigate Through Reinforcement Across the Sim2Real Gap. *arXiv* **2022**, arXiv:20138960. https://doi.org/10.36227/techrxiv.20138960.v1.

Article

Multi-UAV Formation Control in Complex Conditions Based on Improved Consistency Algorithm

Canhui Tao [1], Ru Zhang [2,*], Zhiping Song [1], Baoshou Wang [1] and Yang Jin [3]

[1]   China Ship Scientific Research Center, Wuxi 214082, China
[2]   School of Automation Science and Electrical Engineering, Beihang University, Beijing 100191, China
[3]   Shanghai Electro-Mechanical Engineering Institute, Shanghai 201109, China
*   Correspondence: zrrleo@buaa.edu.cn

Abstract: Formation control is a prerequisite for the formation to complete specified tasks safely and efficiently. Considering non-symmetrical communication interference and network congestion, this article aims to design a control protocol by studying the formation model with communication delay and switching topology. Based on the requirements during the flight and the features of the motion model, the three-degrees-of-freedom kinematics equation of the UAV is given by using the autopilot model of longitudinal and lateral decoupling. Acceleration, velocity, and angular velocity constraints in all directions are defined according to the requirements of flight performance and maneuverability. The control protocol is adjusted according to the constraints. The results show that the improved control protocol can quickly converge the UAV formation state to the specified value and maintain the specified formation with communication delay and switching topology.

Keywords: formation control; consistency theory; communication delay; constraints

Citation: Tao, C.; Zhang, R.; Song, Z.; Wang, B.; Jin, Y. Multi-UAV Formation Control in Complex Conditions Based on Improved Consistency Algorithm. *Drones* **2023**, 7, 185. https://doi.org/10.3390/drones7030185

Academic Editors: Xiwang Dong, Mou Chen, Xiangke Wang and Fei Gao

Received: 30 December 2022
Revised: 14 February 2023
Accepted: 2 March 2023
Published: 7 March 2023

1. Introduction

Due to their low cost, strong maneuverability, and wide application range, UAVs have good application prospects whether in the civilian or military fields [1,2]. With the complexity and diversification of mission requirements, the low efficiency of a single UAV has emerged. In order to solve this problem, in addition to improving the function and utility of a single UAV, UAV formation flight has also become a research focus [3]. Formation flight means that drones can fly in an expected formation. When the environment or tasks change, the formation can be changed according to the requirements [4]. The technology of UAV formation has broad development and application prospects, and using drones to fly in the expected formation can allow for the completion of more complex tasks and significant improvement in the efficiency of tasks [5].

The studies on multi-UAV formation focus on formation control, formation reconfiguration, real-time path planning, etc. Formation control is the basis and focus of formation flight. Commonly used formation control methods mainly include the leader−follower method, virtual structure method, and behavior control method. The virtual structure method can simplify the assignment of tasks with high accuracy. The disadvantages are that it is difficult to perform fault-tolerant processing and requires a lot of communication. The most mature traditional formation control method is the leader−follow method [6–9]. The leader−follower method simplifies the control of the multi-UAV model [10]. However, it still has certain limitations, for example, its tracking error will be propagated backward step by step and thus be amplified. Other methods are combined with the leader−follower method to solve the problems above [11–13]. Every aircraft receives the same information, namely the trajectory of the virtual leader in the virtual−leader method. The advantages of the virtual structure method are that it simplifies the description and assignment of tasks, and has high formation control accuracy. The disadvantages are that it is difficult to perform fault-tolerant processing and requires a large amount of communication as a centralized control method. The behavior-based method, which is based on the information obtained from the sensor to determine the responses of the UAVs, has strong robustness and flexibility, but cannot achieve accurate formation maintenance. The work of [14] studies multi-UAV formation by applying the behavior control method.

REN indicates that the above three formation control methods can be unified under the framework of the consistency theory; formation control based on the consistency theory can overcome some shortcomings of these traditional methods [15]. The formation control method based on the consistency theory is such that every agent can realize large-scale and distributed formation control through the communication between neighboring UAVs under a certain communication network without centralized coordination [16]. The impact of interaction models on the coherence of collective decision-making is discussed in [17–19].

Formation control methods based on the consistency theory have yielded some valuable research results in recent years [20–23]. The work of [24] studied the problem of time-varying formation control under the constraint of communication delay and designed a consistent control method that can deal with communication delay. The work of [25] studied the consensus formation control method based on time-varying communication topology. The work of [26] considered the existence of random communication noise and information packet loss constraints in the network and adopted the polygon method of information exchange based on the consistency theory to realize formation control. The work of [27] studied the cooperative formation control problem of the multi-aircraft system based on the consistency theory with a fixed connectivity of the network topology.

Control laws based on consistency are often adopted to solve the problems of multi-UAV formation, and the maneuvering performance and flight performance of UAVs will impose restrictions on the control variables and flight states in different ways. In addition, communication between aircraft is often affected by factors such as transmission speed and network congestion, resulting in communication delay; due to communication interference and complex terrain, the multi-aircraft system network topology changes. Therefore, the research on multi-UAV formation considering communication constraints and flight constraints has important value.

Aiming at the problems above, the main contributions of this paper are as follows: (1) This paper adopts the three-degrees-of-freedom kinematics model of a drone which is based on autopilot, and the lateral heading autopilot and the longitudinal autopilot are decoupled. (2) This paper proposes an improved basic consistency algorithm. During the flying process of drones, the communication constraints, such as topology switching and non-symmetrical communication delay, are considered to design the consistency algorithm. (3) In addition to communication constraints, mobility constraints are also considered to improve the consistency algorithm. Compared with other existing methods based on the consistency algorithm, the improved method considers the formation control in complex conditions. The communication constraints and flight constraints are both considered. The communication constraints include the communication delay and switching topology, and the flight performance and maneuverability constraints include the speed, acceleration, and heading angular velocity of the UAV. The improved algorithm can not only achieve multi-UAV formation control when topology switching and communication delay exist, but it also satisfies the constraints of UAV maneuverability and flight performance.

This article is organized as follows: The three-degrees-of-freedom kinematics model of a drone which is based on autopilot is adopted, and the lateral heading autopilot and the longitudinal autopilot are decoupled in Section 2. Section 3 proposes an improved consistency algorithm that is effective with topology switching and communication delay. The minimum adjustment is used to adjust for flight constraints. Then, the convergence proof of the improved consistency algorithm is given. Section 4 discusses the simulation. The results show that the consistency control protocol proposed can meet the mobility requirements with communication delay and switching topology.

2. UAV Dynamics Modeling and Consistency Algorithm

This section first establishes the coordinate system, describes the formation, and then gives the kinematics model. The consensus algorithm is presented to prepare for the subsequent proposed multi-UAV control protocol with switching topology and communication delay. Finally, the control protocol is adjusted considering the constraints of flight status and maneuverability.

2.1. UAV Formation Description

Firstly, a coordinate system is created to express the position of the UAVs. The UAV is considered a mass point. To describe the movement state of the drone, we use the ground coordinate system. On the horizontal plane, the origin O can be arbitrarily selected.

There are two ways to describe the three-dimensional plane of the UAV formation, the $l - \psi$ method and the $l - l$ method in [28]. In this paper, the method $l - l$ is selected. The positional relationship between UAVs can be described by the relative positional relationship matrix R_x, R_y, R_z.

$$
\begin{cases}
R_x = \begin{bmatrix} x_{11} & x_{12} & \cdots & x_{1n} \\ x_{21} & x_{22} & \cdots & x_{2n} \\ \vdots & \vdots & \vdots & \vdots \\ x_{n1} & x_{n2} & \cdots & x_{nn} \end{bmatrix} \\[2mm]
R_y = \begin{bmatrix} y_{11} & y_{12} & \cdots & y_{1n} \\ y_{21} & y_{22} & \cdots & y_{2n} \\ \vdots & \vdots & \vdots & \vdots \\ y_{n1} & y_{n2} & \cdots & y_{nn} \end{bmatrix} \\[2mm]
R_z = \begin{bmatrix} z_{11} & z_{12} & \cdots & z_{1n} \\ z_{21} & z_{22} & \cdots & z_{2n} \\ \vdots & \vdots & \vdots & \vdots \\ z_{n1} & z_{n2} & \cdots & z_{nn} \end{bmatrix}
\end{cases}
\tag{1}
$$

where $(x_{ij}, y_{ij}, z_{ij})(i, j = 1, \cdots, n)$ describes the difference in position between two drones. $x_{ii} = y_{ii} = z_{ii} = 0$.

The conditions when multi-UAV forms a stable, desired formation are as follows:

$$
\begin{cases}
\left| x_i - x_j \right| \to x_{ij} \\
\left| y_i - y_j \right| \to y_{ij} \\
\left| z_i - z_j \right| \to z_{ij} \\
\left| v_i - v_j \right| \to 0
\end{cases}
\tag{2}
$$

where x_i, y_i, z_i are coordinates for UAVs. v_i is velocity.

2.2. UAV Kinematics Model

In the UAV formation, the three-degrees-of-freedom kinematics model with autopilot is usually adopted. The longitudinal and lateral movements of the basic kinematic equations of UAV formation control are coupled. The work of [29] decouples the lateral heading autopilot and the longitudinal autopilot and obtains a kinematic model of lateral and longitudinal separation. The motion model of the UAV#i is given by Equation (3):

$$
\begin{cases}
\dot{x}_i = v_i \cos \theta_i \\
\dot{y}_i = v_i \sin \theta_i \\
\dot{\theta}_i = \omega_i \\
\dot{v}_i = \frac{1}{\tau_v}(v_{ci} - v_i) \\
\dot{\theta}_i = \frac{1}{\tau_\theta}(\theta_{ci} - \theta_i) \\
\ddot{z}_i = -\frac{1}{\tau_z}\dot{z}_i + \frac{1}{\tau_z}(z_{ci} - z_i)
\end{cases}
\tag{3}
$$

where v_i is the velocity of the aircraft on the XOY-plane; θ_i is the heading; ω_i is the course angular velocity; $\dot{z}_i$ is the climb rate; $\ddot{z}_i$ is the climb acceleration; τ_v is the speed corresponding to the flight state constant; τ_θ is the flight state constant corresponding to the heading angle; v_{ci} is the speed reference input for the UAV autopilot; θ_{ci} is the course angle reference input of the UAV autopilot; and z_{ci} is the altitude reference input for the UAV autopilot.

In Equation (3), the relationship of θ_i, v_i and the velocity component along the OX-axis and the OY-axis is:

$$\begin{cases} \tan \theta_i = \frac{v_{yi}}{v_{xi}} \\ v_i = \sqrt{v_{xi}^2 + v_{yi}^2} \end{cases} \tag{4}$$

where v_{xi}, v_{yi} are the velocity component.

The dynamic equation with the autonomous driver can be converted into Equation (5):

$$\begin{cases} \dot{x}_i = v_{xi} \\ \dot{y}_i = v_{yi} \\ \dot{z}_i = v_{zi} \\ \dot{v}_{xi} = \frac{1}{\tau_v}\left(v_{xi}^c - v_{xi}\right) \\ \dot{v}_{yi} = \frac{1}{\tau_v}\left(v_{yi}^c - v_{yi}\right) \\ \ddot{z}_i = -\frac{1}{\tau_z}\dot{z}_i + \frac{1}{\tau_z}\left(z_i^c - z_i\right) \end{cases} \tag{5}$$

where v_{zi} is the speed of the drone along the OZ-axis.

The speed, acceleration, and heading angular velocity of the UAV must be changed within a certain range:

$$\begin{cases} v_i \in \left(v_{\min}, v_{\max}\right) \\ \dot{v}_i \in \left(a_{\min}, a_{\max}\right) \\ \dot{\theta}_i \in \left(\omega_{\min}, \omega_{\max}\right) \\ \dot{z}_i \in \left(\dot{z}_{\min}, \dot{z}_{\max}\right) \\ \ddot{z}_i \in \left(\ddot{z}_{\min}, \ddot{z}_{\max}\right) \end{cases} \tag{6}$$

2.3. The Basic Principle of Consensus Algorithm

For any vehicle, its motion states are described by differential equations:

$$\begin{cases} \dot{\xi}_i(t) = \zeta_i(t) \\ \dot{\zeta}_i(t) = u_i(t) \end{cases} \tag{7}$$

where $\xi_i \in \mathbb{R}^n$ is the coordinate vector of the drone; $\zeta_i \in \mathbb{R}^n$ is the speed vector; $u_i(t) \in \mathbb{R}^n$ is the control input vector.

In [30], the basic consensus algorithm given by Equation (7) is:

$$u_i(t) = -\sum_{j=1}^{n} a_{ij}\left[\left(\xi_i(t) - \xi_j(t)\right) + \alpha\left(\zeta_i(t) - \zeta_j(t)\right)\right] \tag{8}$$

where $\alpha > 0$; a_{ij} is an element of the matrix A_n; matrix A_n is the communication topology; and i, j are two different voyages. If UAV#j can send messages to UAV#i, then $a_{ij} = 1$, else $a_{ij} = 0$.

For UAVs, the information exchange topology is G_n, the element l_{ij} in the Laplace matrix L_n is defined as:

$$l_{ij} = \begin{cases} -a_{ij} & if \quad i \neq j \\ \sum_{j=1, i \neq j,}^{n} a_{ij} & if \quad i = j \end{cases} \tag{9}$$

We assume that the topology of ξ_i, ζ_i is consistent during the flight of the UAVs; then, the condition for the consensus algorithm to converge is:

Lemma 1 [31] *If G_n has a directed spanning tree, $\alpha > \bar{\alpha}$, the state of the UAV formation can be asymptotically consistent. If the $n-1$ non-zero eigenvalues of $-L_n$ are negative, then $\bar{\alpha} = 0$, otherwise:*

$$\bar{\alpha} = \max_{\forall \mathrm{Im}(\eta_i)>0, \mathrm{Re}(\eta_i)<0} \sqrt{\frac{2}{|\eta_i| \cos\left(\arctan \frac{\mathrm{Im}(\eta_i)}{-\mathrm{Re}(\eta_i)}\right)}} \tag{10}$$

The lemma shows that, when the state is able to converge according to the consensus algorithm, then, for any initial state such as $x_i(0)$ and $v_i(0)$, when $t \to \infty$, there are $|x_i(t) - x_j(t)| \to 0$ and $|v_i(t) - v_j(t)| \to 0$.

It is necessary to set a reasonable communication topology and α value so that the state of the drone converges to the same level.

3. Improved Consistency Algorithm

The above Equation (8) does not consider the network communication delay and network topology switching in the formation flight of UAVs, nor does it consider the constraints of UAV maneuverability and flight performance; therefore, the consensus algorithm needs to be improved for the actual flight of UAVs.

Firstly, for the situation of non-symmetrical communication delay and topology switching in formation flight, we design the consensus control protocol for the formation flight. Then, the designed control protocol is modified to make itself and the corresponding state output meet the constraints of UAV maneuvering and flight performance.

3.1. Formation State Control

This section studies the consensus control protocol in the case of joint connectivity communication topology.

The state of the dynamic equation of UAV#i is shown in Equation (7).

If the formation protocol can ensure that the states of the UAVs meet the conditions: $[\xi_i - \xi_j] \to r_{ij}$ and $\zeta_i \to \zeta_i \to \zeta^*$, ($r_{ij}$ is the expected difference in position between two drones, and $r_{ij} = -r_{ji}$, ζ^* is the desired speed vector). This shows that the control algorithm can make the multiple UAVs form our expected formation and move forward according to the expected flight speed finally.

The work of [31] gives a control protocol that can make the multi-aircraft system form the desired formation and achieve a given speed, but only for fixed communication topology, and does not consider the communication delay of the system.

This section refers to the control protocol idea of [15] for the multi-aircraft formation flight control system with non-symmetrical communication delay and a communication topology map that is jointly connected. The formation protocol for the UAV is Equation (11):

$$\begin{aligned} u_i(t) &= \sum_{j \in N_i(t)} a_{ij}(t) \left\{ k_1 \left[\xi_j(t-\tau_{ij}) - \xi_i(t-\tau_{ii}) - r_{ji} \right] + k_2 \left[\zeta_j(t-\tau_{ij}) - \zeta_i(t-\tau_{ii}) \right] \right\} \\ &+ \dot{\zeta}^* - k_3(\zeta_i(t) - \zeta^*) \end{aligned} \tag{11}$$

u_{xi}, u_{yi}, u_{zi} are shown in Equations (12)–(14), as follows:

$$\left\{ \begin{aligned} v_{xci} &= v_{xi} + \tau_v u_{xi} \\ u_{xi} &= \sum_{V_j \in N_i(t)} a_{ij}(t) \left\{ k_1 [x_j - x_i - \bar{x}_{ji}] + \frac{2}{k_2}[v_{xj} - v_{xi}] \right\} + \dot{v}^*_x - k_3^x(v_{xi} - v_x^*) \end{aligned} \right. \tag{12}$$

$$\left\{ \begin{aligned} v_{yci} &= v_{yi} + \tau_v u_{yi} \\ u_{yi} &= \sum_{V_j \in N_i(t)} a_{ij}(t) \left\{ k_1 \left[y_j - y_i - \bar{y}_{ji} \right] + \frac{2}{k_2}[v_{yj} - v_{yi}] \right\} + \dot{v}^*_y - k_3\left(v_{yi} - v_y^*\right) \end{aligned} \right. \tag{13}$$

$$\begin{cases} z_i^c = z_i + \frac{\tau_z}{\tau_z}\dot{z}_i + \tau_z u_{zi} \\ u_{zi} = \sum_{V_j \in N_i(t)} a_{ij}(t)\left\{ k_1\left[z_j - z_i - \overline{z}_{ji}\right] + \frac{2}{k_2}\left[v_{zj} - v_{zi}\right]\right\} + \dot{v}^*_z - k_3\left(v_{yi} - v_y^*\right) \end{cases} \tag{14}$$

where τ_{ii} is the time change in the UAV#i itself, a type of latency which is caused by measurements or calculations; τ_{ij} represents the time delay for UAV#j to receive the state information from UAV#i; $N_i(t)$ is a collection of neighbors of node i, and $k_1, k_2, k_3 > 0, k_3 = k_1 k_2$.

Suppose there are M numbers of different time delay in total, it is expressed as $\tau_m(t) \in \left\{\tau_{ii}(t), \tau_{ij}(t), i, j \in \updownarrow\right\}, m = 1, 2, 3, \cdots M$, and satisfies Assumption 1.

Assumption 1. *For specific normal values $h_m > 0, d_m > 0$, time-varying delay time $\tau_m(t)$, $m = 1, 2, 3, \cdots, M$ satisfies $0 \leq \tau_m \leq h_m$ and $\dot{\tau}_m \leq d_m \leq 1$.*

When the network topology is switched and there is a delay in communication, this control protocol can realize the coordinated flight of multiple UAVs.

3.2. Formation Control Protocol Adjustment under Constraints

Section 3.1 does not consider the constraints of Equation (6) when designing the formation control protocol, so the generated control commands and corresponding flight states may not meet the requirements of UAV maneuverability and flight performance. This section proposes a strategy called minimum adjustment to adjust the formation control protocol in Section 2.1 so that both itself and the corresponding state output meet the constraints of UAV maneuvering and flight performance.

The control command u_{xi}, u_{yi} is adjusted in the XOY-plane, so that it satisfies the constraints of velocity v_i, acceleration $\dot{v}_i$, and heading angular velocity $\dot{\varphi}_i$. Then, the values of u_{xi}, u_{yi} are fixed and the value of the control instruction u_{zi} is adjusted to meet the constraints of the OZ-axis direction of the climbing speed $\dot{z}_i$ and the climbing acceleration $\ddot{z}_i$.

Then, u_{xi}, u_{yi} are adjusted in Equations (12) and (13), then the related constraints of speed v_i, acceleration $\dot{v}_i$, and heading angular velocity $\dot{\varphi}_i$, $v_i(t + \Delta t)$ can be obtained through the current flight status:

$$\begin{cases} \alpha_i(t) = \sqrt{u_{xi}^2(t) + u_{yi}^2(t)} \\ v_i(t + \Delta t) = v_i(t) + \alpha_i(t)\Delta t \end{cases} \tag{15}$$

If $v_i(t + \Delta t)$ does not satisfy the constraint $v_i(t + \Delta t) \in (v_{\min}, v_{\max})$, the following variables can be defined as follows:

$$\begin{cases} \alpha'_{\min,i}(t) = \frac{v_{\min} - v_i(t)}{\Delta t} \\ \alpha'_{\max,i}(t) = \frac{v_{\max} - v_i(t)}{\Delta t} \end{cases} \tag{16}$$

where $\alpha'_{\min,i}(t)$ and $\alpha'_{\max,i}(t)$ are the accelerations of the UAV#i at time t.

When the speeds are $v_{\min}, v_{\max}$ at $t + \Delta t$, $\alpha_i(t) \in \left[\alpha'_{\min,i}(t), \alpha'_{\max,i}(t)\right]$. $a'_{\min,i}(t)$ is compared with $a_{\min}$ and $a'_{\max,i}(t)$ is compared with $a_{\max}$, respectively, to obtain the updated constraints of acceleration:

$$\begin{cases} a^{new}_{\min,i}(t) = \max\left(a_{\min}, a'_{\min,i}(t)\right) \\ a^{new}_{\max,i}(t) = \min\left(a_{\max}, a'_{\max,i}(t)\right) \end{cases} \tag{17}$$

where $a^{new}_{\max,i}(t)$ is the upper limit of $a_i(t)$, and $a^{new}_{\min,i}(t)$ is the lower limit of $a_i(t)$.

Equation (17) actually includes constraints on the $v_i(t + \Delta t)$. As long as the acceleration $a_i(t)$ of the UAV# i satisfies Equation (17), the two constraints on the acceleration and velocity in the XOY-plane can be satisfied at the same time. If $v_i(t + \Delta t) \in (v_{\min}, v_{\max})$, then the values of $a_{\min}$ and $a_{\max}$ do not need to be updated by Equation (16). Constraint $a_i \in \left[a^{new}_{\min,i}, a^{new}_{\max,i}\right]$ is used to adjust u_{xi}, u_{yi} so that it satisfies the constraints.

The adjustment of $u_{xi}(t), u_{yi}(t)$ needs to be carried out synchronously, and the influence on the original acceleration value should be as small as possible. The following adjustments can be made to $u_{xi}(t), u_{yi}(t)$:

$$\begin{cases} u'_{xi}(t) = a^{new}_{max,i}(t)\frac{u_{xi}(t)}{a_i(t)}, u'_{yi}(t) = a^{new}_{max,i}(t)\frac{u_{yi}(t)}{a_i(t)} \\ \quad if \quad a_i(t) > a^{new}_{max,i}(t) \\ u'_{xi}(t) = a^{new}_{min,i}(t)\frac{u_{xi}(t)}{a_i(t)}, u'_{yi}(t) = a^{new}_{min,i}(t)\frac{u_{yi}(t)}{a_i(t)} \\ \quad if \quad a_i(t) < a^{new}_{min,i}(t) \end{cases} \tag{18}$$

Then, the values of u'_{xi}, u'_{yi} meet the constraints of acceleration and speed after the above adjustments, and the adjustment range is the smallest.

After that, the heading angular velocity $\dot{\theta}_i(t)$ constraint is processed, and $u'_{xi}(t)u'_{yi}(t)$ will be adjusted in the next step.

According to the constraints of $\dot{\theta}_i \in (\omega_{min}, \omega_{max})$, the allowable value range of the heading angle θ_i at the next sampling time can be obtained as:

$$\begin{cases} \theta_{min,i}(t+\Delta t) = \theta_i(t) + \omega_{min}\Delta t \\ \theta_{max,i}(t+\Delta t) = \theta_i(t) + \omega_{max}\Delta t \end{cases} \tag{19}$$

The heading angle $\theta_i(t+\Delta t)$ at the next sampling time is:

$$\theta_i(t+\Delta t) = \arctan\frac{v_{yi}(t) + u'_{yi}(t)\Delta t}{v_{xi}(t) + u'_{xi}(t)\Delta t} \tag{20}$$

where $v_{xi}(t)$ and $v_{yi}(t)$ are the speeds of the drone at time t.

If $\theta_i(t+\Delta t) \notin [\theta_{min,i}, \theta_{max,i}]$, then $u'_{xi}(t), u'_{yi}(t)$ should be adjusted by Equations (21) and (22).

$$\begin{cases} \frac{v_{yi}(t)+u''_{yi}(t)\Delta t}{v_{xi}(t)+u''_{xi}(t)\Delta t} = \tan(\theta_{max,i}(t+\Delta t)) \\ u''^2_{xi}(t) + u''^2_{yi}(t) = a'^2_i(t) \\ \theta_i(t+\Delta t) > \theta_{max,i}(t+\Delta t) \end{cases} \tag{21}$$

$$\begin{cases} \frac{v_{yi}(t)+u''_{yi}(t)\Delta t}{v_{xi}(t)+u''_{xi}(t)\Delta t} = \tan(\theta_{min,i}(t+\Delta t)) \\ u''^2_{xi}(t) + u''^2_{yi}(t) = a'^2_i(t) \\ \theta_i(t+\Delta t) < \theta_{min,i}(t+\Delta t) \end{cases} \tag{22}$$

where Equations (21) and (22) are binary quadratic equations; usually, there are two different sets of solutions, denoted as $u''_{xi1}(t), u''_{yi1}(t)$ and $u''_{xi2}(t), u''_{yi2}(t)$. Because both sets of solutions satisfy the constraints of the heading angular velocity $\dot{\varphi}_i$, it is necessary to further confirm which set is finally selected as the result according to the "minimum adjustment" strategy.

Let $\gamma'_{ai}(t) = \arctan\frac{u'_{yi}(t)}{u'_{xi}(t)}$ represent the direction of the acceleration $a'_i(t)$; the adjusted values should not only keep the value of $a'_i(t)$ unchanged, but also the direction of $a'_i(t)$ should change minimally.

$\gamma''_{ai1}(t) = \arctan\frac{u''_{yi1}(t)}{u''_{xi1}(t)}$ and $\gamma''_{ai2}(t) = \arctan\frac{u''_{yi2}(t)}{u''_{xi2}(t)}$ represent the directions of the UAV#i's acceleration in the XOY-plane after adjustment by Equations (21) or (22).

Among these two sets of solutions, the set of solutions corresponding to $\min\left(\left|\gamma''_{ai1}(t) - \gamma'_{ai}(t)\right|, \left|\gamma''_{ai2}(t) - \gamma'_{ai}(t)\right|\right)$ is selected as the values required.

The above procedure makes minimal adjustments to $u_{xi}(t), u_{yi}(t)$ and satisfies the constraints of the XOY-plane.

For the constraints in the direction of the OZ-axis, the climbing rate $\dot{z}_i(t)$ and climb acceleration $\ddot{z}_i(t)$ of the drone are limited and $u_{zi}(t)$ is adjusted. First, the climbing rate of the next sampling moment $\dot{z}_i(t + \Delta t)$ is calculated through the current time of the drone flight status as $\dot{z}_i(t + \Delta t) = \dot{z}_i(t) + u_{zi}(t)\Delta t$.

If $\dot{z}_i(t + \Delta t) \notin [\dot{z}_{\min}, \dot{z}_{\max}]$, then the updating constraints of the climbing acceleration $\ddot{z}_i(t)$ are shown in the following formula:

$$\begin{cases} \ddot{z}'_{\min,i}(t) = \frac{\dot{z}_{\min} - \dot{z}_i(t)}{\Delta t} \\ \ddot{z}'_{\max,i}(t) = \frac{\dot{z}_{\max} - \dot{z}_i(t)}{\Delta t} \end{cases} \tag{23}$$

where $\ddot{z}'_{\min,i}(t)$ and $\ddot{z}'_{\max,i}(t)$ are the lower limit and upper limit of the constraints after the climbing rate constraints are converted to the climbing acceleration at time t, respectively. If $\dot{z}_{\min} \leq \dot{z}_i(t + \Delta t) \leq \dot{z}_{\max}$, then $\ddot{z}'_{\min,i}(t)$ and $\ddot{z}'_{\max,i}(t)$ do not need to be updated.

Through Equation (24), the constraints on the climbing rate are also converted into the constraint on the climbing acceleration at time t. Then, $\ddot{z}'_{\min,i}(t)$, $\ddot{z}_{\min}$, $\ddot{z}'_{\max,i}(t)$, and $\ddot{z}_{\max}$ are compared, and the updated rising acceleration constraint conditions is determined as:

$$\begin{cases} \ddot{z}^{new}_{\min,i}(t) = \max\left(\ddot{z}_{\min}, \ddot{z}'_{\min,i}(t)\right) \\ \ddot{z}^{new}_{\max,i}(t) = \min\left(\ddot{z}_{\max}, \ddot{z}'_{\max,i}(t)\right) \end{cases} \tag{24}$$

where $\ddot{z}^{new}_{\min,i}(t)$ and $\ddot{z}^{new}_{\max,i}(t)$ are the final lower limit and upper limit values of the climbing acceleration, respectively, after the climbing rate and climbing acceleration constraints have been considered.

Finally, we limit the current climb acceleration $u_{zi}(t)$ to the allowable range:

$$u'_{zi}(t) = \max\left(\ddot{z}^{new}_{\min,i}(t), \min\left(u_{zi}(t), \ddot{z}^{new}_{\max,i}(t)\right)\right) \tag{25}$$

where $u'_{zi}(t)$ is the adjusted climbing acceleration. From Equation (25), when $u_{zi}(t) \in \left[\ddot{z}^{new}_{\min,i}(t), \ddot{z}^{new}_{\max,i}(t)\right]$, we have $u'_{zi}(t) = u_{zi}(t)$; when $u_{zi}(t) < \ddot{z}^{new}_{\min,i}(t)$, we have $u'_{zi}(t) = \ddot{z}^{new}_{\min,i}(t)$; when $u_{zi}(t) > \ddot{z}^{new}_{\max,i}(t)$, we have $u'_{zi}(t) = \ddot{z}^{new}_{\max,i}(t)$.

3.3. Convergence Proof of Improved Consistency Algorithm

Let $\bar{\bar{\xi}}_i = \xi_i - \xi_0 - r_i, \bar{\zeta}_i = \zeta_i - \zeta^*$, then Equation (11) can be transformed into:

$$\begin{aligned} u_i(t) = \sum_{s_j \in N_i(t)} a_{ij}(t)\Big\{ &k_1\Big[\big(\bar{\bar{\xi}}_j(t - \tau_{ij}(t))\big) - \big(\bar{\bar{\xi}}_i(t - \tau_{ii}(t))\big)\Big] \\ &+ \frac{2}{k_2}\Big[\bar{\zeta}_j(t - \tau_{ij}(t)) - \bar{\zeta}_i(t - \tau_{ii}(t))\Big] + \dot{\zeta}^* - k_3\bar{\zeta}_i(t) \end{aligned} \tag{26}$$

If $\hat{\zeta}_i(t) = 2\bar{\zeta}_i(t)/k_1 k_2 + \bar{\bar{\xi}}_i(t), \varepsilon(t) = [\bar{\bar{\xi}}_1(t), \hat{\zeta}_i(t), \cdots, \bar{\bar{\xi}}_n(t), \hat{\zeta}_n(t)]^T$, then:

$$B = \begin{bmatrix} -k_3/2 & k_3/2 \\ k_3/2 & k_3/2 \end{bmatrix}, Q = \begin{bmatrix} 0 & 0 \\ 0 & 2/k_2 \end{bmatrix}$$

According to Equation (26), the closed-loop dynamic equation is Equation (27):

$$\dot{\varepsilon}(t) = (I_n \otimes B)\varepsilon(t) - \sum_{m=1}^{M} (L_{\sigma m} \otimes Q)\varepsilon(t - \tau_m) \tag{27}$$

In fact, if we have $\lim\limits_{t \to +\infty} \varepsilon(t) = 0$, then $\lim\limits_{t \to +\infty} \xi_j(t) - \xi_i(t) = r_{ji}$, $\lim\limits_{t \to +\infty} \zeta(t) = \zeta^*$.

Next, we show that the above closed-loop control system can achieve $\lim\limits_{t \to +\infty} \varepsilon(t) = 0$.

Referring to the definition of switching topology, it is assumed that the time-invariant topology $\overline{G}_\sigma$ in a certain sub-interval $[t_{k_b}, t_{k_{b+1}})$ has $q(q \geq 1)$ the numbers of connected parts, and its corresponding node set is denoted by $\psi_{k_j}^1, \psi_{k_j}^2, \cdots, \psi_{k_j}^{d_\sigma}$, and f is the node number in $\psi_{k_j}^i$. Then, a permutation matrix $P_\sigma \in \mathbb{R}^{n \times n}$ is obtained by satisfying:

$$P_\sigma^T L_\sigma P_\sigma = diag\left\{ L_\sigma^1, L_\sigma^2, \cdots, L_\sigma^{d_\sigma} \right\}$$
$$P_\sigma^T L_{\sigma m} P_\sigma = diag\left\{ L_{\sigma m}^1, L_{\sigma m}^2, \cdots, L_{\sigma m}^q \right\} \tag{28}$$

$$\varepsilon^T(t)(P_\sigma \otimes I_2) = \left[\varepsilon_\sigma^{1\ T}, \varepsilon_\sigma^{2\ T}, \cdots, \varepsilon_\sigma^{q\ T} \right] \tag{29}$$

where $L_\sigma^i \in \mathbb{R}^{f_\sigma^i \times f_\sigma^i}$ is the Laplacian matrix which corresponds to the part which is connected, and $L_{\sigma m}^i \in \mathbb{R}^{f \times f}$, $L_\sigma^i = \sum_{m=1}^M L_{\sigma m}^i$. Therefore, in $[t_{k_b}, t_{k_{b+1}})$, it can be broken down into q numbers of subsystems:

$$\dot{\varepsilon}_\sigma^i(t) = (I_f \otimes B)\varepsilon_\sigma^i(t) - \sum_{m=1}^M (L_{\sigma m}^i \otimes Q)\varepsilon_\sigma^i(t - \tau_m), i = 1, 2, \cdots, q \tag{30}$$

where $\varepsilon_\sigma^i(t) = [\varepsilon_{\sigma 1}^i(t), \cdots \varepsilon_{\sigma 2 f_\sigma^i}^i(t)] \in \mathbb{R}^{2 f_\sigma^i}$.

Lemma 2. [32] *If there is $D_n = n I_n - \mathbf{1}\mathbf{1}^T$, then there must be an orthogonal matrix $U_n \in \mathbb{R}^{n \times n}$ which makes $U_n^T D U_n = diag\{n I_{n-1}, 0\}$, where the last column U_n is $1/\sqrt{n}$. We give a matrix $D \in \mathbb{R}^{n \times n}$ and make it satisfy $\mathbf{1}^T D = \mathbf{0}$ and $D\mathbf{1} = \mathbf{0}$, then $U_n^T D U_n = diag\left\{ \overline{U}_n^T D \overline{U}_n, 0 \right\}$.*

Lemma 3. [33] *For any function of actual cable vector $x(t) \in R^n$, any function of cable scalar $\tau(t) \in [0, a]$, and any constant matrix $\mathbf{0} < H = H^T \in R^{n \times n}$, there is:*

$$\frac{1}{a}[x(t) - x(t - \tau(t))]^T H[x(t) - x(t - \tau(t))] \leq \int_{t-\tau(t)}^t \dot{x}^T(s) H \dot{x}(s) ds, t \geq 0 \tag{31}$$

where $a > 0$.

Theorem 1. *Considering a multi-UAV system with non-uniform time delay and switching topology, in any sub-interval $[t_{r_b}, t_{r_{b+1}})$, if $\gamma > 0$, and $F_\sigma^i \in \mathbb{R}^{f \times f}, i = 1, 2, \cdots, q$, there is:*

$$F_\sigma^{i\ T} \Xi_\sigma^i F_\sigma^i < 0 \tag{32}$$

then there is $\lim\limits_{t \to \infty} \xi_j(t) - \xi_i(t) = r_{ji}, \lim\limits_{t \to \infty} \zeta_i(t) = \zeta^*$.

where $F_\sigma^i = diag\left\{ U_{2f}, I_{2Mf} \right\}$, and the definition of U_{2f} is as shown in Lemma 2.

where $\Xi_\sigma^i = \begin{bmatrix} \Xi_{11} & \Xi_{12} \\ \Xi_{12}^T & \Xi_{22} \end{bmatrix}$, and

$$\Xi_{11} = 2\gamma\left(I_f \otimes B\right) + \sum_{m=1}^M h_m \left(I_f \otimes B\right)^T \left(I_f \otimes B\right) - \sum_{m=1}^M \frac{1 - d_m}{h_m} I_{2f}$$

$$\Xi_{12} = \left[-\gamma(L_{\sigma 1}^i \otimes Q) + \frac{1-d_1}{h_1} I_{2f} - \sum_{m=1}^M h_m \left(I_f \otimes B\right)^T (L_{\sigma 1}^i \otimes Q), \cdots \right.$$
$$\left. -\gamma(L_{\sigma M}^i \otimes Q) + \frac{1-d_M}{h_M} I_{2f} - \sum_{m=1}^M h_m \left(I_f \otimes B\right)^T (L_{\sigma M}^i \otimes Q) \right]$$

$$\Xi_{22} = \left[-diag\left\{ \frac{1-d_1}{h_1} I_{2f}, \frac{1-d_2}{h_2} I_{2f}, \cdots \frac{1-d_M}{h_M} I_{2f} \right\} + \right.$$
$$\left. \sum_{m=1}^M h_m \ [(L_{\sigma 1}^i \otimes Q), \cdots (L_{\sigma M}^i \otimes Q)]^T [(L_{\sigma 1}^i \otimes Q), \cdots (L_{\sigma M}^i \otimes Q)] \right.$$

Proof of Theorem 1. *The Lyapunov−Krasovskii function for Equation (11) can be defined following:*

$$V(t) = \gamma \varepsilon^T(t)\varepsilon(t) + \sum_{m=1}^{M} \int_{-\tau_m}^{0} \int_{t+a}^{t} \dot{\varepsilon}^T(s)\dot{\varepsilon}(s)\,ds\,da, \gamma > 0 \tag{33}$$

$\dot{V}(t)$ can be calculated as:

$$\begin{aligned}
\dot{V}(t) &= 2\gamma\varepsilon^T(t)\dot{\varepsilon}(t) + \sum_{m=1}^{M}\tau_m\dot{\varepsilon}^T(t)\dot{\varepsilon}(t) - \sum_{m=1}^{M}(1-\dot{\tau}_m)\int_{t-\tau_m}^{t}\dot{\varepsilon}^T(s)\dot{\varepsilon}(s)\,ds \\
&= 2\gamma\varepsilon^T(t)\left[\left(I_f \otimes B\right)\varepsilon(t)\right] - 2\gamma\varepsilon^T(t)\sum_{m=1}^{M}\left[(L_{\sigma m}^i \otimes Q)\varepsilon(t-\tau_m)\right] + \\
&\quad \sum_{m=1}^{M}\tau_m\dot{\varepsilon}^T(t)\dot{\varepsilon}(t) - \sum_{m=1}^{M}(1-\dot{\tau}_m)\int_{t-\tau_m}^{t}\dot{\varepsilon}^T(s)\dot{\varepsilon}(s)\,ds
\end{aligned} \tag{34}$$

According to Equation (33) and Assumption 1, $\dot{V}(t)$ is changed to the following form:

$$\begin{aligned}
\dot{V}(t) &\leq \sum_{i=1}^{q}\left\{2\gamma{\varepsilon_\sigma^i}^T(t)\left[\left(I_f \otimes B\right)\varepsilon_\sigma^i(t) - 2\gamma{\varepsilon_\sigma^i}^T(t)\sum_{m=1}^{M}\left[(L_{\sigma m}^i \otimes Q)\varepsilon_\sigma^i(t-\tau_m)\right]\right]\right. \\
&\quad \left. + \sum_{m=1}^{M}h_m{\dot{\varepsilon}_\sigma^i}^T(t)\dot{\varepsilon}_\sigma^i(t) - \sum_{m=1}^{M}(1-d_m)\int_{t-\tau_m}^{t}\dot{\varepsilon}^T(s)\dot{\varepsilon}(s)\,ds\right\}
\end{aligned} \tag{35}$$

According to Lemma 3, the following can be obtained:

$$\begin{aligned}
\dot{V}(t) &\leq \sum_{i=1}^{q}\left\{2\gamma{\varepsilon_\sigma^i}^T(t)\left[\left(I_f \otimes B\right)\varepsilon_\sigma^i(t) - 2\gamma\varepsilon_\sigma^{iT}(t)\sum_{m=1}^{M}\left[(L_{\sigma m}^i \otimes Q)\varepsilon_\sigma^i(t-\tau_m)\right]\right] + \sum_{m=1}^{M}h_m\dot{\varepsilon}_\sigma^{iT}(t)\dot{\varepsilon}_\sigma^i(t)\right. \\
&\quad \left. - \sum_{m=1}^{M}\frac{1-d_m}{h_m}\left[\dot{\varepsilon}_\sigma^{iT}(t)\dot{\varepsilon}_\sigma^i(t) - \dot{\varepsilon}_\sigma^{iT}(t)\dot{\varepsilon}_\sigma^i(t-\tau_m) - \varepsilon_\sigma^{iT}(t-\tau_m)\varepsilon_\sigma^i(t) + \varepsilon_\sigma^{iT}(t-\tau_m)\varepsilon_\sigma^i(t-\tau_m)\right]\right\} \\
&= \sum_{i=1}^{q}\delta_i^T\Xi_\sigma^i\delta_i
\end{aligned}$$

where $\delta_i = \left[{\varepsilon_\sigma^i}^T(t), {\varepsilon_{\sigma 1}^i}^T(t-\tau_1), {\varepsilon_{\sigma 2}^i}^T(t-\tau_2), \cdots {\varepsilon_{\sigma M}^i}^T(t-\tau_M)\right]$.

$\Xi_\sigma^i = {\Xi_\sigma^i}^T$, and $\Xi_\sigma^i\left[\mathbf{1}_{2f}^T, \mathbf{0}_{2Mf}^T\right]^T = 0$. According to Lemma 2, we can conclude

$\Xi_\sigma^i \leq 0$, when ${F_\sigma^i}^T\Xi_\sigma^i F_\sigma^i < 0, rank(\Xi_\sigma^i) = 2(M+1)f - 1$.

Let $\eta = \left[{\varepsilon_\sigma^i}^T(t) - h\mathbf{1}^T, {\varepsilon_{\sigma 1}^i}^T(t), {\varepsilon_{\sigma 2}^i}^T(t), \cdots {\varepsilon_{\sigma M}^i}^T(t)\right], h > 0$, then $\Xi_\sigma^i(\delta_i - \eta) = 0$, and we can obtain

$$\delta_i^T\Xi_\sigma^i\delta_i = \eta^T\Xi_\sigma^i\eta \leq \lambda\|\eta\|^2 \leq \lambda\left[\left\|\varepsilon_\sigma^i(t) - h\mathbf{1}\right\|^2 + \sum_{m=1}^{M}\sum_{K=1}^{f}\left(\varepsilon_{\sigma mk}^i\right)^2(t)\right] \tag{36}$$

where $\|\cdot\|$ is the Standard European norm and $\lambda < 0$ represents the maximum non-zero eigenvalue of Ξ_σ^i.

Therefore, $\dot{V}(t) \leq \lambda\sum_{i=1}^{q}\left[\left\|\varepsilon_\sigma^i(t) - h\mathbf{1}\right\|^2 + \sum_{m=1}^{M}\sum_{K=1}^{f}\left(\varepsilon_{\sigma mk}^i\right)^2(t)\right] \leq 0$.

Through the above analysis, Equation (11) is stable, and $\lim_{t\to+\infty}V(t) = 0$. Then, we can obtain $\lim_{t\to+\infty}\varepsilon(t) = 0$, then $\lim_{t\to+\infty}\overline{\xi}_i(t) = 0$, $\lim_{t\to+\infty}\hat{\zeta}_i(t) = 0$, and we can have $\lim_{t\to+\infty}\xi_j(t) - \xi_i(t) = r_{ji}$, $\lim_{t\to+\infty}\zeta_i(t) = \zeta^*$. That is, under the action of the control protocol of Equation (30), the drones can eventually form the specific formation at an expected velocity. $\square$

4. Simulation and Results

The effectiveness of the control protocol designed is verified by simulation. This section verifies the improved control protocol of the existing constraints, indicating the effectiveness of the strategies proposed. We assume that the formation has a non-symmetrical communication delay and has the jointly connected topologies in the example below.

We assume that the formation consists of six UAVs. The topology structure and the formation that we expect are shown in Figures 1 and 2.

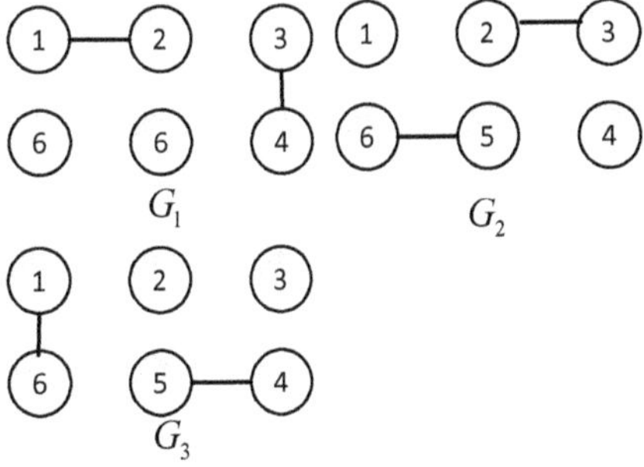

Figure 1. Communication topology.

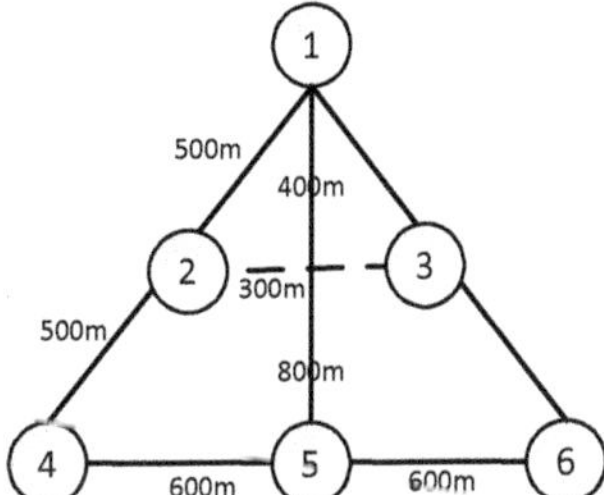

Figure 2. Expected formation.

The communication topology is switched in the order of (G_1, G_2, G_3, G_1), and the weight of each connected edge is 1. Assuming that there are three different time delays in the system as $\tau_1(t), \tau_2(t), \tau_3(t)$, for $\forall i \neq j$, then $\tau_{ii}(t) = \tau_{jj}(t) = \tau_1(t)$; $\tau_{12}(t) = \tau_{23}(t) = \tau_{34}(t) = \tau_{45}(t) = \tau_{56}(t) = \tau_{61}(t) = \tau_2(t)$; $\tau_{21}(t) = \tau_{32}(t) = \tau_{43}(t) = \tau_{54}(t) = \tau_{65}(t) = \tau_{16}(t) = \tau_3(t)$.

The time delays satisfy $0 \leq \tau_1(t) \leq 0.01, 0 \leq \tau_2(t) \leq 0.07, 0 \leq \tau_3(t) \leq 0.08$. The initial state of the six UAVs and the parameters setting are listed in Tables 1 and 2.

Table 1. The initial state of the six UAVs.

Number	1	2	3	4	5	6
x_i/m	20	60	10	90	43	60
y_i/m	66	56	96	56	86	86
z_i/m	50	10	40	330	350	240
$v_i/(m.s^{-1})$	15	35	55	75	65	90
$\theta_i/(\circ)$	36	-36	45	-45	-20	45
$\dot{z}/(m.s^{-1})$	4	3	2	1	5	3

Table 2. Parameter settings.

Parameter	$v_{\min}/\left(m.s^{-1}\right)$	$v_{\max}/\left(m.s^{-1}\right)$	$a_{\min}/g$	$a_{\max}/g$	$\dot{z}_{\min}/\left(m.s^{-1}\right)$	$\dot{z}_{\max}/\left(m.s^{-1}\right)$
Value	10	600	-5	5	-30	30
Parameter	$\ddot{z}_{\min}/\left(m.s^{-1}\right)$	$\ddot{z}_{\max}/\left(m.s^{-1}\right)$	$\omega_{\min}/\left(rad.s^{-1}\right)$	$\omega_{\max}/\left(rad.s^{-1}\right)$	$\zeta^{*}(m/s)$	$z^{*}(m)$
Value	-5	5	$-\pi/2$	$\pi/2$	50	300
Parameter	k_1	k_2	k_3	τ_v	τ_z	$\tau_{\dot{z}}$
Value	0.6	1.1	0.66	10	0.3	0.3

Under the improved control protocol, the position curves, speed curves, course angle curves, and expected formation of the six UAVs are shown in Figure 3.

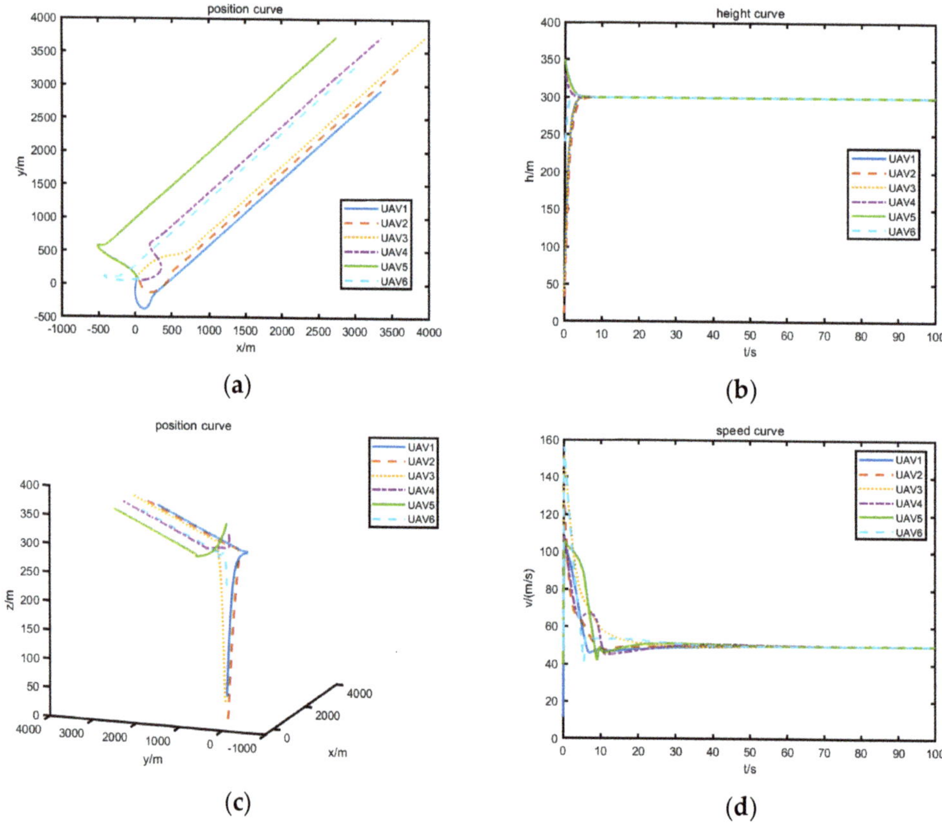

Figure 3. *Cont.*

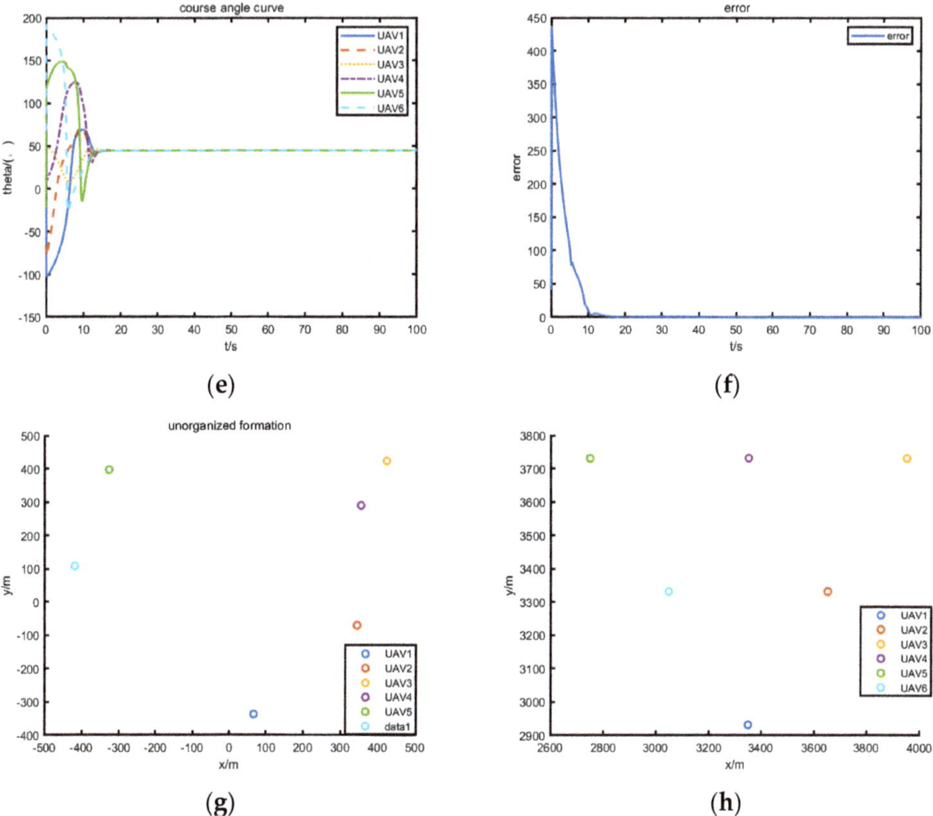

Figure 3. States of UAV formation. (**a**) Position curves (*XOY*-plane). (**b**) Height curves. (**c**) Position curves. (**d**) Speed curves. (**e**) Course angle curves. (**f**) Error. (**g**) Unorganized formation (5 s). (**h**) Final formation.

The figures show that, under the improved formation control protocol, the six drones can achieve the expected formation with the expected speed under the complex conditions of communication constraints and dynamic constraints; the composite error of the formation is 0, as shown in Figure 3f. This indicates that the formation control protocol is effective for UAV formation in the conditions of non-symmetrical communication delay and topology switching.

When drones form a stable formation, assuming that the formation needs to be changed during flight, the control protocol is still valid. The simulation results are shown in Figure 4.

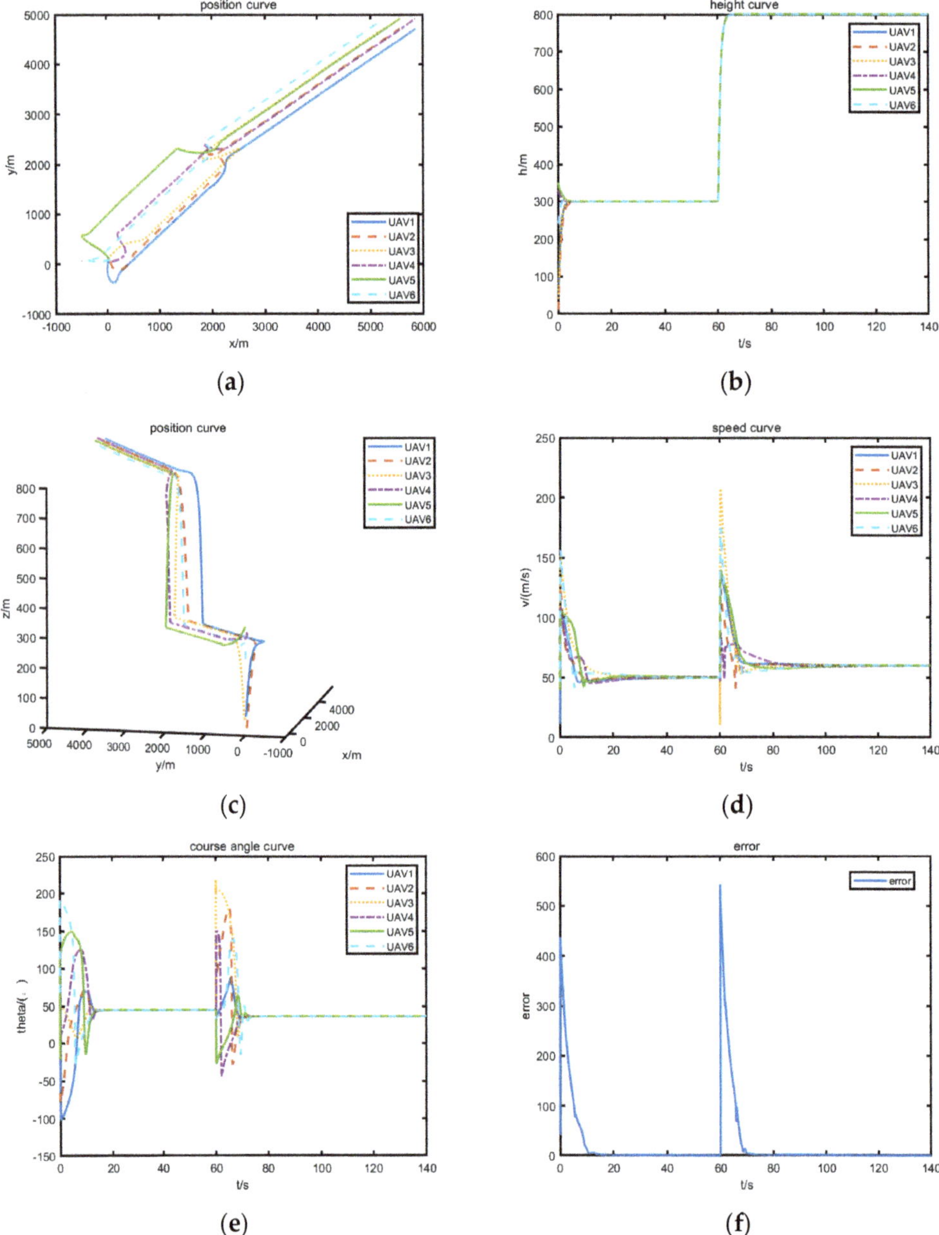

Figure 4. *Cont.*

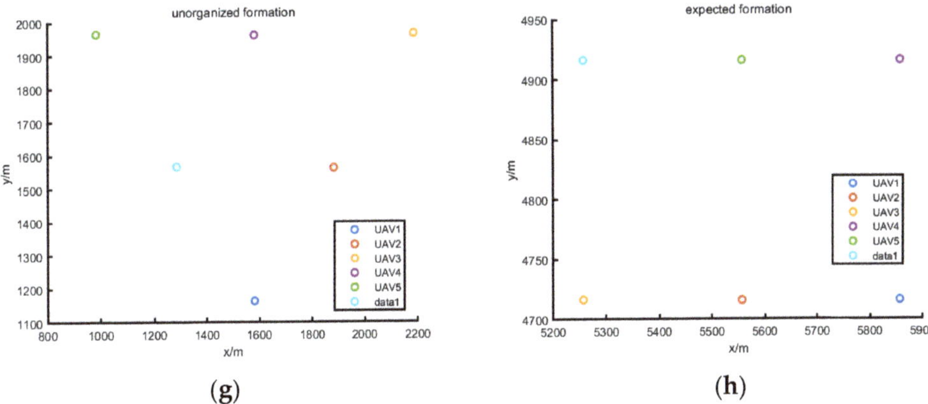

Figure 4. States of UAV formation. (**a**) Position curves (*XOY*-plane). (**b**) Height curves. (**c**) Position curves. (**d**) Speed curves. (**e**) Course angle curves (**f**) Error. (**g**) Formation (50 s). (**h**) Final formation.

The figures show that, under the formation control protocol, the six drones can achieve the expected formation with the expected speed. When the formation needs to be changed, under the control protocol, the new formation is formed. The drones can fly with the new expected speed and the designed formation control protocol is still valid. The results indicate that the improved control protocol is widely used.

5. Conclusions

This article studies the problem of formation control based on the consistency theory. This article focuses on the research of drone formation, thus ignores the gesture control of the drone. The three-degrees-of-freedom kinematics equation of the UAV is given by using the autopilot model of longitudinal and lateral decoupling. Considering the communication interference and network congestion, this paper designs the control protocol by studying the formation model with non-symmetrical communication delay and switching topology. Acceleration, velocity, and angular velocity constraints in all directions are defined according to the requirements of flight performance and maneuverability. Both communication and mobility constraints are considered in this paper. The improved control protocol is adjusted according to the constraints. The results show that the improved control protocol is effective and can quickly converge the UAV formation state to the specified value and can maintain the specified formation with communication delay and switching topology.

Author Contributions: Conceptualization, C.T. and R.Z.; Methodology, R.Z.; Software, R.Z. and Z.S.; Validation, Y.J.; Formal analysis, R.Z., B.W. and Y.J.; Investigation, B.W.; Resources, C.T.; Writing–original draft, C.T., Z.S. and Y.J.; Writing–review & editing, Z.S. All authors have read and agreed to the published version of the manuscript.

Funding: This research received no external funding.

Data Availability Statement: No new data were created.

Conflicts of Interest: The authors declare no conflict of interest.

References

1. Zhang, G.J.; Yan, H. Application and future development of unmanned aerial vehicle in forestry. *J. For. Eng.* **2019**, *4*, 8–16. (In Chinese)
2. Dai, W.; Zhang, H.T. Application analysis of UAV in future naval warfare field. *J. Ordnance Equip. Eng.* **2018**, *39*, 21–24. (In Chinese)

3. Han, L.; Ren, Z. Research on cooperative control method and application for multiple unmanned aerial vehicles. *Navig. Position. Timing* **2018**, *5*, 1–7. (In Chinese)

4. Zong, Q.; Wang, D.; Shao, S.; Zhang, B.; Han, Y. Research status and development of multi UAV coordinated formation flight control. *J. Harbin Inst. Technol.* **2017**, *49*, 1–14.

5. Pacher, M.; D'Azzo, J.J.; Proud, A.W. Tight Formation Flight Control. *J. Guid. Control Dyn.* **2001**, *24*, 246–254. [CrossRef]

6. Han, Q.; Zhang, A.; Wang, X.; Cao, R. Leader-following consensus of multi-agent system with multiple leaders under formation control. In Proceedings of the 2020 IEEE 3rd International Conference of Safe Production and Informatization (IICSPI), Chongqing, China, 28–30 November 2020; pp. 380–384.

7. Challa, V.R.; Ratnoo, A. Analysis of UAV Kinematic Constraints for Rigid Formation Flying. In Proceedings of the AIAA Guidance, Navigation, and Control Conference, San Diego, CA, USA, 4–8 January 2016.

8. Luo, D.; Zhou, T.; Wu, S. Obstacle Avoidance and Formation Regrouping Strategy and Control for UAV Formation Flight. In Proceedings of the 10th IEEE International Conference on Control and Automation, Hangzhou, China, 12–14 June 2013; pp. 1921–1926.

9. Vazquez Trejo, J.A.; Adam Medina, M.; Vázquez Trejo, J.A. Robust Observer-based Leader-Following Consensus for Multi-agent Systems. *IEEE Lat. Am. Trans.* **2021**, *11*, 1949–1958. [CrossRef]

10. Cam, P.G.; Napolitano, M.R.; Brad, S.; Perhinschi, M.G. Design of Control Laws for Maneuvered Formation Flight. In Proceedings of the IEEE Proceedings of the American Control Conference, Boston, MA, USA, 30 June–2 July 2004; pp. 2344–2349.

11. Jin, X.Z.; Wang, S.F.; Yang, G.H.; Ye, D. Robust adaptive hierarchical insensitive tracking control of a class of leader-follower agents. *Inf. Sci.* **2017**, *406*, 234–247. [CrossRef]

12. Jin, X.; Haddad, W.M. An adaptive control architecture for leader-follower multiagent systems with stochastic disturbances and sensor and actuator attacks. *Int. J. Control* **2019**, *92*, 2561–2570. [CrossRef]

13. Ebegbulem, J.; Guay, M. Distributed control of multiagent systems over unknown communication networks using extremum seeking. *J. Process Control* **2017**, *59*, 37–48. [CrossRef]

14. Yun, X.; Alptekin, G.; Albayrak, O. Line and Circle Formation of Distributed Physical Mobile Robots. *J. Robot. Syst.* **1997**, *14*, 63–76. [CrossRef]

15. Ren, W. Consensus strategies for cooperative control of vehicle formations. *IET Control Theory Appl.* **2007**, *1*, 505–512. [CrossRef]

16. Gulzar, M.M.; Rizvi, S.T.H. Multi-agent cooperative control consensus: A comparative review. *Electronics* **2018**, *7*, 22. [CrossRef]

17. Aust, T.; Talamali, M.S.; Dorigo, M.; Hamann, H.; Reina, A. The hidden benefits of limited communication and slow sensing in collective monitoring of dynamic environments. In Swarm Intelligence, Proceedings of the 13th International Conference, ANTS 2022, Málaga, Spain, 2–4 November 2022; Springer: Cham, Switzerland, 2022.

18. Rausch, I.; Khaluf, Y.; Simoens, P. Collective decision-making on triadic graphs. In *Complex Networks XI*; Springer: Cham, Switzerland, 2020; pp. 119–130.

19. Simoens, P.; Khaluf, Y. Adaptive foraging in dynamic environments using scale-free interaction networks Rausch. *Front. Robot. AI* **2020**, *7*, 86.

20. Wei, Q.Q.; Liu, R.; Zhang, H. High-order consistency formation control method for quadrotor UAV. *Electro.-Opt. Control* **2019**, *26*, 1–5.

21. Dong, X.W.; Yu, B.C.; Shi, Z.Y. Time-varying formation control for unmanned aerial vehicles: Theories and applications. *IEEE Trans. Control Syst. Technol.* **2014**, *23*, 340–348. [CrossRef]

22. Zeng, Z.W. Research on Multi-Agent Collaborative Control Based on Nonlinear Quantitative and Structural Constraints. Ph.D. Thesis, National University of Defense Technology, Changsha, China, 2017.

23. Zhao, J.; Xu, M.; Wang, Z. Dynamic Control Design for UAV Formation Tracking with Time Delays. In Proceedings of the IEEE 2nd International Conference on Electronic Technology, Communication and Information (ICETCI), Changchun, China, 27–29 May 2022.

24. He, L.L.; Zhang, J.Q. Time-varying formation control for UAV swarm with directed interaction topology and communication delay. *J. Beijing Univ. Aeronaut. Astronaut.* **2020**, *46*, 314–323. (In Chinese)

25. Dong, X.W.; Zhou, Y.; REN, Z. Time-Varying formation tracking for Second-Order Multi-Agent systems subjected to switching topologies with application to quadrotor formation flying. *IEEE Trans. Electron.* **2017**, *64*, 5014–5024. [CrossRef]

26. Joongbo, S.; Chaeik, K. Controller Design for UAV Formation Flight Using Consensus based Decentralized Approach. In Proceedings of the AIAA Aerospace Conference, Seattle, WA, USA, 6–9 April 2009.

27. Kuriki, Y.; Namerikawa, T. Consensus-based cooperative control for geometric configuration of UAVs flying in formation. In Proceedings of the IEEE SICE Annual Conference 2013, Nagoya, Japan, 14–17 September 2013; pp. 1237–1242.

28. Sun, F.; Turkoglu, K. Nonlinear consensus strategies for multi-agent networks under switching topologies. *Real-Time Receding Horiz. Approach. Aerosp. Sci. Technol.* **2019**, *87*, 323–330.

29. Desai, J.P.; Ostrowski, J.P.; Kumar, V. Modeling and Control of Formations of Nonholonomic Mobile Robots. *IEEE Trans. Robot. Autom.* **2002**, *17*, 905–908. [CrossRef]

30. Wang, X.M. *UAV Formation Flight Technology*; Northwestern Polytechnical University Press: Xian, China, 2015; pp. 149–232.

31. Ren, W. Consensus based formation control strategies for multi-vehicles systems. In Proceedings of the 2006 American Control Conference, Minneapolis, MN, USA, 14–16 June 2006.

32. Peng, L.; Jia, Y. Consensus of a Class of Second-Order Multi-Agent Systems with Time-Delay and Jointly-Connected Topologies. *IEEE Trans. Autom. Control* **2010**, *55*, 778–784. [CrossRef]
33. Peng, L.; Jia, Y. Multi-agent consensus with diverse time-delays and jointly-connected topologies. *Automatic* **2011**, *47*, 848–856.

 *drones*

Article

A Novel Semidefinite Programming-based UAV 3D Localization Algorithm with Gray Wolf Optimization

Zhijia Li [1], Xuewen Xia [1] and Yonghang Yan [1,2,*]

[1] School of Computer and Information Engineering, Henan University, Kaifeng 475004, China
[2] Henan Province Engineering Research Center of Spatial Information Processing, Henan University, Kaifeng 475004, China
* Correspondence: yanyonghang@henu.edu.cn

Abstract: The unmanned aerial vehicle (UAV) network has gained vigorous evolution in recent decades by virtue of its advanced nature, and UAV-based localization techniques have been extensively applied in a variety of fields. In most applications, the data captured by a UAV are only useful when associated with its geographic position. Efficient and low-cost positioning is of great significance for the development of UAV-aided technology. In this paper, we investigate an effective three-dimensional (3D) localization approach for multiple UAVs and propose a flipping ambiguity avoidance optimization algorithm. Specifically, beacon UAVs take charge of gaining global coordinates and collecting distance measurements from GPS-denied UAVs. We adopt a semidefinite programming (SDP)-based approach to estimate the global position of the target UAVs. Furthermore, when high noise interference causes missing distance pairs and measurement errors, an improved gray wolf optimization (I-GWO) algorithm is utilized to improve the positioning accuracy. Simulation results show that the proposed approach is superior to a number of alternative approaches.

Keywords: unmanned aerial vehicles; three-dimensional localization; semidefinite programming; flipping ambiguity avoidance; gray wolf optimization

Citation: Li, Z.; Xia, X.; Yan, Y. A Novel Semidefinite Programming-based UAV 3D Localization Algorithm with Gray Wolf Optimization. *Drones* **2023**, *7*, 113. https://doi.org/10.3390/drones7020113

Academic Editor: Diego González-Aguilera

Received: 9 January 2023
Revised: 29 January 2023
Accepted: 6 February 2023
Published: 7 February 2023

1. Introduction

The flying ad hoc network (FANET) is evolving at a tremendous rate and emerging technologies, and applications based on UAVs oriented toward 6G have received considerable attention [1,2]. UAV networks as a kind of FANET have promising applications in both military and civilian areas (e.g., urban fire emergency rescue, forest wildfire monitoring, enemy aircraft reconnaissance and air shows, etc.). UAV networks have many advantages over traditional cellular networks (CN) and mobile ad hoc networks (MANETs). With higher altitudes and wider coverage capabilities, the signal of line of sight (LoS) from UAVs is effectively utilized by ground terminals and users. As a result, UAVs are regarded as airborne base stations to service mobile users and ground stations by using air-to-ground (A2G) channels [3,4]. In addition, UAVs with characteristics such as small size and high mobility can be flexibly deployed. Therefore, rapid on-demand services can be provided in hazardous and harsh environments. However, due to high mobility and dynamics, UAVs suffer from end-to-end transmission delays because of frequent topology changes and disconnections of communication links in UAV networks. As a result, it is critical to know the precise, low-latency position information of each UAV [5].

Localization is an invaluable area for researchers in surveillance, path planning, wireless communication, and UAV networks [6,7]. In FANETs, most procedures (topology control, position-aware routing, etc.) require knowledge of the precise physical location of each UAV, because many actions and observations are implemented based on position. The technology of UAVs becomes meaningless if the two cannot be correlated. UAV localization is therefore a critical technology that requires in-depth exploration. Among the existing localization methods, the global positioning system (GPS) is widely used. However, there are limitations (e.g., it requires higher costs and/or is not achievable to be deployed in all scenarios). Moreover, since GPS requires light of sight (LoS) from a satellite, it cannot provide a reliable localization scheme when the LoS is obstructed by obstacles. Therefore, many UAV localization schemes that are non-GPS based have been proposed [8–10].

Distance-based UAV localization schemes are the focus of much current research, which defines network localization as the problem of determining the physical coordinates of unknown nodes given the anchors of coordinates and distance pairs between unknown nodes and anchors. Target node localization using anchor information and pairwise distance measurements among nodes is usually formulated as an optimization problem with quadratic-constrained quadratic programming (QCQP). Usually, the problem is NP-hard and nonconvex. Semidefinite programming (SDP) relaxation (a novel idea proposed by researchers) has been applied to solve this issue. The basic principle is to first change the nonconvexity of the problem, and the SDP relaxation causes the constraints of the original problem to be relaxed to a semipositive definite form, which transforms the initial problem into a convex optimization problem. The SDP-based problem is then solved by common convex optimization techniques. However, in the actual scene, the measurement value will be polluted by noise, which makes the measurement result inaccurate, and reduces the positioning precision. For decades, computational intelligence (CI)-based methods have been employed to improve the localization efficiency and accuracy. In [11], the numerical solution is carried out by using a rank-relaxed approach based on SDP, and the results are upgraded using the orthogonal Procrustes technique. The experimental results suggest that the method is feasible when DOA measurements are noisy. Arafat et al. proposed a bounding box model and applied it in swarm intelligent localization and hybrid gray wolf optimization (HGWO) localization algorithms to narrow the particle search space and estimate the location of UAVs, respectively [12,13].

The goal of this paper is to design an efficient localization optimization approach based on inter-UAV distance measurement in a 3D dynamic scene. SDP is the typical method for solving distance-based localization in general wireless communication networks. SDP and graph theory-based localization method for sensor networks was proposed as early as in [14], and its core philosophy is to derive error bounds by adding regularization terms to the SDP. Instead, our proposal is to transform the localization problem into a maximum likelihood estimation (MLE) problem with decreasing distance errors and to solve it by relaxing the bound through a composite algorithm called SDP + RLT. Furthermore, in order to improve localization accuracy at high noise levels, the I-GWO algorithm is inspired to refine the SDP + RLT results. Simulation results verify that our scheme outperforms existing localization schemes. Our contributions and innovations in this paper are summarized as follows:

- A system model for the design and analysis of 3D UAV localization is developed. We consider distance measurement-based UAV position estimation as an objective optimization problem with quadratic constraints and formulate it as a maximum likelihood estimation (MLE) problem. A localization model for the design of UAVs in a 3D moving scene is developed. We consider the distance-based UAV position estimation as an objective optimization problem with quadratic constraints and formulate it as an MLE problem.
- The SDP and RLT relaxation constraints are established based on the distance constraints of the localization problem, and the solvability and tightness of the proposed composite algorithm SDP + RLT are analyzed.
- In addition, our solution is extended to the case of noisy distance measurement errors and loss, and an I-GWO algorithm is proposed, which greatly improves localization accuracy. Finally, we validate the excellence of the proposed scheme by comparing multiple sets of experimental results.

The remaining sections are as follows. Related work on UAV localization is introduced in the next section. Section 3 introduces the network setting and problem formulation.

Section 4 proposes a localization scheme based on the combination of SDP and RLT relaxation. Section 5 presents a bionic optimization algorithm I-GWO. Numerical simulations are given in Section 6. Section 7 concludes this paper.

2. Related Works

The problem of determining the position of a UAV is called localization. According to the technical means, the data collected by UAVs to achieve localization vary. We will classify and describe UAV localization methods from the following aspects.

First, it can be divided into relative localization (RL) and absolute localization (AL) according to the type of measurement. AL is usually implemented based on GPS technology, but its performance is vulnerable to environmental factors. The technique of acquiring the topological shape of a network through relative distance or angle measurements between nodes is called RL. Guo et al. [15] estimated the relative positions of UAVs based on graph theory and UWB RCM in a GPS-rejected environment, thus enabling distributed formation control. Autopilot and guidance laws designed for fixed-wing UAVs without a priori knowledge were studied in [16,17]. Secondly, there are centralized and distributed algorithms based on the computational framework. The centralized approach is suitable for static networks, as it requires the collection of all measurements and the calculation of estimates at the fusion center, which requires powerful computational and communication capabilities. Location estimation that iteratively extends to the overall framework based on local information is called distributed localization, which has the advantage of load balancing and efficiency of the network. Two different distributed algorithms are described in [18,19]. Thirdly, we categorize them according to the use of anchors or not. Zhang et al. propose a unique anchorless localization algorithm that uses a combined distance and angle technique to establish a local coordinate system (LCS) and then estimates the relative position from a fixed coordinate system [20]. This method effectively avoids reflection blur. However, the high cost of angle measurements has resulted in few practical applications. A new idea of using a mobile single anchor to locate target nodes was proposed in [21] that employed the computational intelligence-based H-best particle swarm (HPSO) algorithm. Liu et al. [22] proposed a distributed UAV relative positioning framework and used the SDP method to obtain the global topology of the UAV cluster. The lower bounds of CRLB with and without anchors were then analyzed separately. In addition, the localization is classified according to the type of calculation (distance or hop count). An MDV-hop algorithm based on the DV-hop algorithm is proposed in [23] for locating wireless sensor nodes. Techniques based on distance information typically include SDP, least squares (LS), gradient descent (GD), and multidimensional scaling (MDS) [24–26]. Among these, SDP methods have the advantage of significantly better localization accuracy than other closed-form solutions. However, it is less resistant to interference in large-scale networks. Zou et al. [27] proposed a method that uses the initial estimation of the SDP to iterate the position and velocity of target nodes to improve the localization accuracy. A distributed gradient algorithm based on Barzilai–Borwein steps was applied to distributed distance measurements, which enabled node localization accuracy to meet expectations [28]. Figure 1 shows the UAV localization algorithms under different classifications. The problem of determining the location of a UAV node is called localization. Usually, distance-based UAV localization requires the estimation of absolute position (relative to a local or global frame of reference) from partial relative measurements between UAVs. That is, each UAV can measure relative positions from a set of neighboring UAVs, and then infer the absolute position of all UAVs from this information. UAV localization algorithms can be classified from different perspectives.

We denote $(\cdot)^T$ and $(\cdot)^{-1}$ as the transpose and inverse of a vector (or matrix). The rank of A is denoted as $Rank(A)$. a_i stands for the ith element of a. $||\cdot||$ is the Euclidean norm. 1_k and 0_k are the all-one and all-zero vectors of length k. I_k is the $k \times k$ identity matrix. $A \succcurlyeq B$ signify that $A - B$ is positive semidefinite.

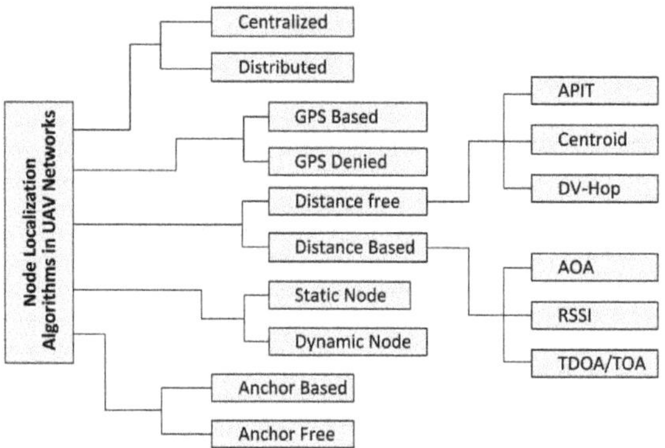

Figure 1. Node localization algorithms in UAV networks.

3. Problem Formulation

Consider a scenario of dynamic swarms of UAVs. The 3D network space $\mathbb{R}^3$ consists of m beacon UAVs (denoted as anchors and coordinates are known) and n target UAVs (positions are unknown). Let $N_a = \{1, 2, \ldots, m\}$ denote the set of anchors, and let $N_u = \{m + 1, m + 2, \ldots, m + n\}$ denote the set of target UAVs, where $N_R = N_a + N_u$. There exists a clock offset δ_i between the UAV$_i \in N_R$ and the standard time. The UAVs have certain communication capabilities that allow them to range with neighbor UAVs. The coordinate of the anchor UAVs and target UAVs is defined as $x_i = [x_i, y_i, z_i]^T \in \mathbb{R}^3$ and $a_k = [x_k, y_k, z_k]^T \in \mathbb{R}^3$, respectively.

It is assumed that the distance between UAVs is obtained by the signal time of arrival (TOA). If the distance of UAV i and j is within communication range R, their distance can be expressed by

$$d_{ij} = \left\| x_i - x_j \right\|_2, \forall i, j \in N_R \tag{1}$$

Furthermore, we regard the distance as symmetric, i.e., $d_{ij} = d_{ji}$. We denote T as the time duration of the signal $r_{ij}(t)$ from jth UAV to the ith UAV, given by

$$r_{ij}(t) = \alpha_{ij} s_j(t - \tau_{ij}) + n_{ij}(t), t \in [0, T] \tag{2}$$

where α_{ij} denotes amplitude, $s_j(t)$ is the known waveform, τ_{ij} is the transmit delay, and $n_{ij}(t)$ represents Gaussian noise.

Among the actual problems of UAV network localization, due to the limitation of communication distance, not all the measurement pairs of distance are known. Therefore, the pair distances of UAV/UAV pairs and UAV/anchor pairs are denoted as $(i, j) \in \mathcal{N}$ and $(k, j) \in \mathcal{M}$, respectively. Define d_{ij} as the true distance between target UAVs and $\hat{d}_{kj}$ as the true distance between the target UAV and anchor UAV. The measurement noise is modeled to obey Gaussian distribution, and thus the corresponding distance measurement of UAV is

$$\begin{aligned} d_{ij} &= d_{ij} + \varepsilon_{ij}, \forall (i, j) \in \mathcal{N} \\ \hat{d}_{kj} &= \hat{d}_{kj} + \gamma_{ij}, \forall (k, j) \in \mathcal{M} \end{aligned} \tag{3}$$

where d_{ij} and $\hat{d}_{kj}$ are the measurements between UAV i and UAV j with Gaussian noise $\varepsilon_{ij} \sim N(0, \sigma_{ij}^2)$, $\gamma_{ij} \sim N(0, \sigma_{kj}^2)$, and N represents normal random variables with mean 0 and variance σ^2, as well as those that are independent.

The final goal is to estimate the position of a large-scale UAV network in a dynamic scene using distance measurements. Figure 2 depicts the network model of the UAV network localization problem in a dynamic scenario. The nomenclature of used terms is provided in Table 1.

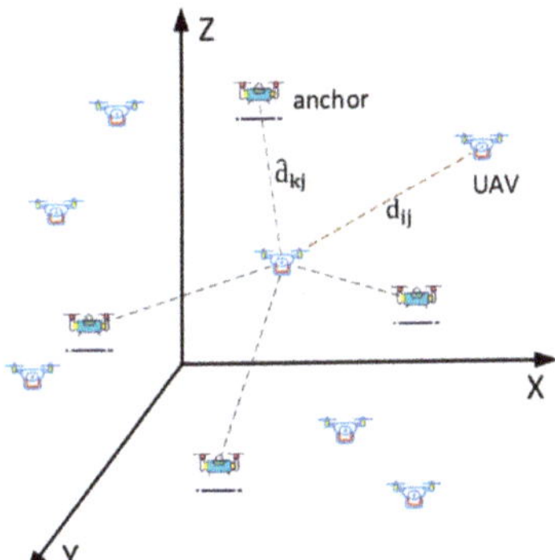

Figure 2. Localization model of UAVs in 3D space, in which green UAV and blue UAV represent anchors and target nodes, respectively.

Table 1. Nomenclature of used terms in this paper.

Symbol	Definition
Na, Nu, N_R	Set of anchors and target UAVs, where $N_R = Na + Nu$
x_i, a_k	Coordinate of the anchor UAVs and target UAVs
$d_{ij}, \hat{d}_{kj}$	Distance between target UAV and UAV (or anchor)
$\mathcal{N}, \mathcal{M}$	Set of distance pairs of UAV/UAV and UAV/anchor
$r_{ij}, \alpha_{ij}, s_j, n_{ij}$	Amplitude, waveform, and noise of UAV signal
$\varepsilon_{ij}, \gamma_{ij}$	Gaussian noise of distance measurements
X, Y, Z	X is UAV position matrix and $Y = X^T X$, $Z = [I_3\ X; X^T\ Y]$
l, u	Lower and upper bounds of variables in X
ϕ_i, ϕ_j	Variables greater than 1 and less than e
U	Population of gray wolves
$a, r1, r2$	Convergence factor and two random numbers of $[0,1]$

In the above figure, the distances between UAVs are represented by the corresponding edges; thus, the UAV network localization problem is

$$\begin{aligned}
Find \quad & x_1, x_2, \ldots, x_n \in \mathbb{R}^3 \\
s.t. \quad & \left\| x_j - x_i \right\|^2 = d_{ji}^2, \forall (i,j) \in \mathcal{N} \\
& \left\| x_j - a_k \right\|^2 = \hat{d}_{kj}^2, \forall (k,j) \in \mathcal{M}
\end{aligned} \tag{4}$$

In general, the above problem is considered as a nonconvex problem (NP-hard), which is tough to solve. Global optimization techniques have been widely used, such as MDS, SDP, and nonlinear least squares (NLS). Projection and dimensionality reduction are the main ideas of the MDS technique. NLS is usually resolved by direct derivatives, gradient descent, and quasi-Newton methods. Semidefinite relaxation (SDR) is a computationally efficient approximation to quadratically constrained quadratic programming (QCQP), and the UAV localization problem proposed in this article is classified as QCQP. The QCQP is approximated by semidefinite programming (SDP), in which reliable and efficient algorithms have been studied by previous scholars.

In this paper, a novel idea for solving distance-based UAV localization has been designed. We first transform the UAVs position estimation problem into an MLE problem that reduces the error between the true and the estimated value of the position, and then create relaxation constraints and derive an answer for the MLE problem via the composite localization algorithm SDP + RLT.

4. Proposed Localization Solution

4.1. System Model

Based on the above problem formulation, a UAV network consists of N_u target UAVs and N_a anchor UAVs. The distance measurement of UAVs is obtained by communicating with their neighbor UAVs. We combine the 3D coordinates of all target UAVs into a matrix $X = [x_1, x_2, \ldots, x_n] \in \mathbb{R}^{3 \times n}$, which needs to be settled. Due to the influence of UAV mobility and noise during distance measurement, there exists an error between the measurements and the accurate distance of UAVs. In order to account for noisy distance information, the idea of maximum likelihood estimation (MLE) is utilized to set up optimization problems that minimize the expected error in the UAV position estimation [29]. Therefore, an MLE problem of UAV positioning can be expressed as

$$\widetilde{X} = \underset{X}{\arg\min} \left\{ \sum_{(i,j) \in \mathcal{N}} w_{ij} \left| \|x_i - x_j\|^2 - d_{ij}^2 \right| + \sum_{(k,j) \in \mathcal{M}} w_{kj} \left| \|x_j - a_k\|^2 - \tilde{d}_{kj}^2 \right| \right\} \tag{5}$$

where $w_{ij} > 0$ are weights. The weights in Equation (5) are important to achieve precise positioning. We tend to give higher weights to distance measures with high confidence, which facilitates higher-quality estimator error reduction.

The position error is defined as

$$\Sigma = \|\widetilde{x} - x\|2 \tag{6}$$

We further model the UAV localization problem as

$$\begin{aligned}
Find \quad & X \in \mathbb{R}^{3 \times n}, Y \in \mathbb{R}^{n \times n} \\
s.t. \quad & e_{ij}^T X^T X e_{ij} = d_{i,j}^2, \forall (i,j) \in \mathcal{N} \\
& (a_k; e_j)^T \begin{pmatrix} I_3 & X \\ X^T & Y \end{pmatrix} (a_k; e_j) = \tilde{d}_{kj}^2, \forall (k,j) \in \mathcal{M} \\
& Y = X^T X
\end{aligned} \tag{7}$$

Here, $e_{ij} \in \mathbb{R}^n$ is the vector with 1 at the ith position, -1 at the jth position, and zero everywhere else; $e_j \in \mathbb{R}^n$ is the vector of all zeros except an -1 at the jth position; I_3 is the 3×3 identity matrix.

In the SDP approach, the constraint relaxation is $Y \succeq X^T X$, which means $Y - X^T X \succeq 0$, the constraint is equivalent to the following linear matrix inequality:

$$Z = \begin{pmatrix} I_3 & X \\ X^T & Y \end{pmatrix} \succeq 0 \tag{8}$$

Then formulate the relaxation problem as a standard SDP problem, that is, to find the symmetric matrix $Z \in \mathbb{R}^{(3+n) \times (3+n)}$ such that:

$$\begin{aligned}
\underset{Z}{maximize} \quad & 0 \\
s.t. \quad & Z_{1:3,1:3} = I_3 \\
& (0; e_j)(0; e_j)^T \cdot Z = d_{i,j}^2, \forall (i,j) \in \mathcal{N} \\
& (a_k; e_j)(a_k; e_j)^T \cdot Z = \tilde{d}_{kj}^2, \forall (k,j) \in \mathcal{M} \\
& Z \succeq 0
\end{aligned} \tag{9}$$

where $Z_{1:3}$ is the upper three-dimensional master submatrix of Z. After the SDP normalization of the localization problem, the problem solvability will be analyzed. Then, whether UAVs can be uniquely localized is considered, and the proof is given in the next section.

4.1.1. SDP Solvability Analysis

We now analyze the solvability of Problem (9) through SDP pairwise relaxation theory. The rank of any possible feasible solution matrix of (9) is highlighted in [30] as being at least 3. Assume that (9) has a feasible solution. This happens when d_{kj} and d_{ij} denote the accurate values of the position $\widetilde{X} = [\widetilde{x}_1, \widetilde{x}_2, \cdots, \widetilde{x}_n]$. Then, $\widetilde{Z} = Q^T Q$, where $Q = (I_3; \widetilde{X})^T$ is a feasible solution to (9). Since the original is feasible, the dual must have a minimum value of 0. The dual of the SDP relaxation is represented below

$$minimize \qquad I_3 \cdot C + \sum_{(i,j) \in \mathcal{N}} u_{ij} d_{ij}^2 + \sum_{(i,j) \in \mathcal{M}} v_{ij} \tilde{d}_{kj}^2$$

$$s.t. \qquad \begin{pmatrix} C & 0 \\ 0 & 0 \end{pmatrix} + \sum_{(i,j) \in \mathcal{N}} u_{ij}(0; e_{ij})(0; e_{ij})^T + \sum_{(i,j) \in \mathcal{M}} v_{ij}(a_k; e_j)(a_k; e_j)^T \succcurlyeq 0$$

Let U be a $(3 + n)$-dimensional dual relaxation matrix, that is

$$U = \begin{pmatrix} C & 0 \\ 0 & 0 \end{pmatrix} + \sum_{(i,j) \in \varepsilon_{uu}} u_{ij}(0; e_{ij})(0; e_{ij})^T + \sum_{(i,j) \in \varepsilon_{au}} v_{ij}(a_k; e_j)(a_k; e_j)^T$$

According to the dual theory of SDP, the matrices Z and U satisfy $Rank(\widetilde{Z}) \geq 3$ and $Rank(\widetilde{U}) \leq n$. The following theory can be derived.

Theory 1. *If the optimal dual relaxation matrix has rank n, then each solution of (4) has Rank 3. That is, Problems (4) and (9) are equivalent, and (4) can be solved in polynomial time as an SDP.*

For the UAV localization problem, the SDP relaxation in (9) can find an accurate solution to Problem (7). Note that localization is unique only in some cases where the network map is generally rigid, while nodes are locatable when the graph is globally rigid. The solvability of the UAV localization problem will be investigated by graph theory below.

The issue of the network location problem based on relative ranging measurement pairs of distance is actually a general distance geometry and graphical implementation problem. Suppose a set of wireless nodes are placed in a Euclidean space $\mathbb{R}^d$ and they communicate with each other to form a certain topology. This topology can be abstracted as an undirected connectivity graph $\mathcal{G}_N$, which contains vertices set $\mathcal{V} = \{1, 2, \dots, n\}$ and a set of edges ε. The vertices and edges represent the nodes and communication links (distance measurements) in the communication network, respectively. Thus, the goal of network localization derives the location information of all unknown nodes based on known anchor information x_j, topological connectivity information of the graph $\mathcal{G}_N$, and distance information d_{ij}. The proof process is omitted here. See [30] for details, and the following theorem is valid.

Theorem 1. *The network localization problem is just considered solvable if there exists a set of positions $\{x_{m+1}, \dots, x_n\}$ in $\mathbb{R}^d$ corresponds to the given data $(\mathcal{G}_N, \{x_1, x_2, \dots, x_m\}$ and $d_{ij})$.*

Theorem 1 provides sufficient conditions for the UAV network 3D localization problem of particular relevance in the case of global rigidity. Having solved the solvability, whether localization is uniquely solvable is discussed below.

4.1.2. Unique Solvability Analysis

The optimal value in Problem (9) is 0 if the distance measurements of UAVs are exact. However, the sufficient number of distance information is also a significant feature of the challenge to achieve unknown node localization. Biswas et al. [29] have discussed this problem in detail and provided a unique solvability condition. We repeat it here in terms of localization in 3D space. Here, the matrix Z has $3n + n(n + 1)/2$ unknown variables. Hence, there are at least $3n + n(n + 1)/2$ linear equations in the constraint. In addition, in case these equations are linearly independent, then there is a unique solution to Z. Therefore, we have the following proposal: If there are $3n + n(n + 1)/2$ distance pairs in 3D localization, each distance pair has an exact distance metric. Then, Problem (7) has a unique feasible solution:

$$\overline{Z} = \begin{pmatrix} I_3 & \overline{X} \\ \overline{X}^T & \overline{Y} \end{pmatrix} \tag{10}$$

Now, it could be said that it equals the true position vectors of the target UAVs. That is, the SDP relaxation solves the original problem exactly.

Theorem 2. *If Problem (7) has a unique feasible solution* $\overline{X}$ *in* $\mathbb{R}^{3 \times n}$ *and no* x_j *($j = 1, 2, \cdots, n$) in* $\mathbb{R}^h$ *, then (5) is uniquely locatable, where $h > 3$ (excluding the case of attaching all zeros to $\overline{X}$), such that*

$$\begin{aligned} \left\| x_j - x_i \right\|^2 &= d_{ji}^2, \forall (i, j) \in \mathcal{N} \\ \left\| x_j - (a_k; 0) \right\|^2 &= \hat{d}_{kj}^2, \forall (k, j) \in \mathcal{M} \end{aligned} \tag{11}$$

The latter condition in Theorem 2 states that the problem cannot be confined to a higher-dimensional space. In this space, the anchors are increased to $(a_k; 0) \in \mathbb{R}^h, j = 1, \ldots, M$. The results in [29] mean that the relaxation problem (4) solves (2) exactly when the problem could be uniquely localized.

Based on the above SDP uniqueness analysis, we learn that it is extremely crucial that the localization problem can be solved uniquely, especially for A2G communication. For unique localization in 2D and 3D space, at least three points and four points are required, respectively, and they are not collinear or on the same plane.

Figure 3 shows a typical network model for 3D localization. In these models, at least four anchor nodes are required that need to locate the unknown node, and all four anchor nodes cannot be on the same line or the same UAV on the same line. Figure 3a illustrates the case where four nodes can locate an unknown node E with anchors (A, B, C, D). The second case shown in Figure 3b shows that two potential positions (E, F) can be calculated for nodes (A, B, C, D) in the same plane. In Figure 3c, four anchors are on the same line and the exact position cannot be found because all anchor positions are on the same line, and $(\alpha_1, \alpha_2, \ldots, \alpha_n)$ may be the correct position.

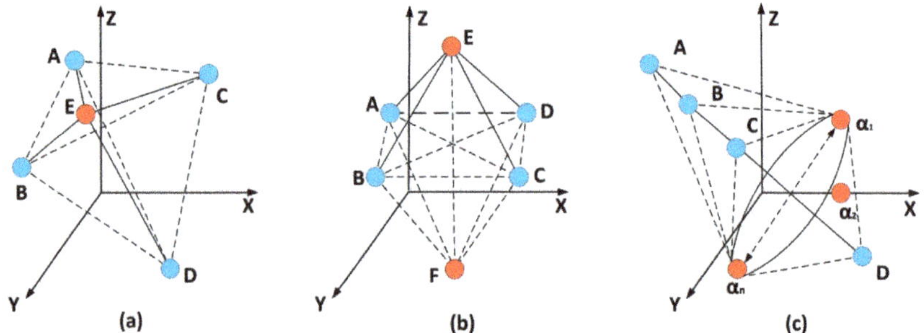

Figure 3. A network localization model in 3D space. The blue points represents the anchor nodes, and the red point represents the unknown node to be located.

4.2. SDP Plus RLT Relaxation Scheme

RLT is an effective approach for continuous and discrete nonconvex or QCQP issues [31]. As a linear programming method, the researched problem exhibits a nonconvexity when there are second-order terms in the objective function and constraints. RLT can transform the problem into a solvable convex-optimal form by introducing new variables containing constraints to replace the constraints of the original problem in the reformulation phase. Anstreicher et al. [31] demonstrated that for typical values of the original variables, the semidefiniteness constraint removes a large part of this feasible region. It has been shown in [32] that using RLT plus SDP constraints can achieve better results than one of the two. In this article, we introduce RTL into SDP to solve the distance-based UAV localization problem.

First, determine the boundaries of each coordinate in the position matrix X of the UAV target nodes The upper bounds are denoted as $u = [u_1, u_2, \cdots, u_n]$, and the lower bounds are denoted as $l = [l_1, l_2, \cdots, l_n]$. Next, the boundary constraints are determined. With two variables $\phi_i, \phi_j \in X$, four constraints are established $\phi_i - l_i \geq 0$, $u_i - \phi_i \geq 0$, $\phi_j - l_j \geq 0$, $u_j - \phi_j \geq 0$. Multiplying the new constraints containing ϕ_i and ϕ_j, at the same time using Y_{ij} instead $\phi_i \phi_j$, we obtain

$$\begin{cases} Y_{ij} - l_i \varphi_j - l_j \varphi_i \geq -l_i l_j \\ Y_{ij} - u_i \varphi_j - u_j \varphi_i \geq -u_i u_j \\ Y_{ij} - l_i \varphi_j - u_j \varphi_i \geq -l_i u_j \\ Y_{ij} - l_j \varphi_i - u_i \varphi_j \geq -l_j u_i \end{cases} \tag{12}$$

It is clear that $Y_{ij} = Y_{ji}$, hence the last two constraints are equivalent.

$$\begin{cases} Y_{ij} - l_i \phi_j - l_j \phi_i \geq -l_i l_j \\ Y_{ij} - u_i \phi_j - u_j \phi_i \geq -u_i u_j \\ Y_{ij} - l_i \phi_j - u_j \phi_i \leq -l_i u_j \end{cases} \tag{13}$$

Next, after adding the SDP constraint and RLT, (9) can be reformulated as

$$\begin{aligned}
\underset{Z}{maxmize} \quad & 0 \\
s.t. \quad & Z_{1:3,1:3} = I_3 \\
& (0; e_j)(0; e_j)^T \cdot Z = d_{ij}^2, \forall(i,j) \in \mathcal{N} \\
& (a_k; e_j)(a_k; e_j)^T \cdot Z = \hat{d}_{kj}^2, \forall(k,j) \in \mathcal{M} \\
& Y_{ij} - l_i \phi_j - l_j \phi_i \geq -l_i l_j \\
& Y_{ij} - u_i \phi_j - u_j \phi_i \geq -u_i u_j, \forall i,j = 1,\ldots,n \\
& Y_{ij} - l_i \phi_j - u_j \phi_i \leq -l_i u_j \\
& Z \succeq 0
\end{aligned} \tag{14}$$

The Tightness Analysis

Tightness analysis is performed to assess the impact of our proposed solution when it comes to UAV localization performance. We concentrate on the variety of the feasible regions of Y_{ij} when attaching SDP to it.

The RLT relaxation does not change the affine transformation of the initial constraint factors. We suppose that $l = 0$ and $u = e$ without loss of generality. We consider variables ϕ_i, ϕ_j, and assuming $i = 1, j = 2$ and $1 < \phi_i < \phi_j < e$. Then, the RLT constraints on Y_{11}, Y_{22}, and Y_{12} become

$$\begin{aligned}
0 \leq Y_{11} \leq \phi_1, \\
0 \leq Y_{22} \leq \phi_2, \\
0 \leq Y_{12} \leq \phi_1.
\end{aligned} \tag{15}$$

Next, we exert the SDP constraint on the RLT. The SDP can be written as follows

$$\begin{pmatrix} 1 & \phi_1 & \phi_2 \\ \phi_1 & Y_{11} & Y_{12} \\ \phi_2 & Y_{12} & Y_{22} \end{pmatrix} \succeq 0 \tag{16}$$

For the specified values of x_1 and x_2, comparing the feasible region of $Y_{11}Y_{22}$ before and after adding (15), the equivalent constraint can be derived as follows.

$$
\begin{cases}
Y_{11} \geq \phi_1^2, \\
Y_{11} \geq \phi_1^2, \\
Y_{11} \leq \phi_1\phi_2 + \sqrt{\left(Y_{11} - \phi_1^2\right)\left(Y_{22} - \phi_2^2\right)}, \\
Y_{11} \geq \phi_1\phi_2 - \sqrt{\left(Y_{11} - \phi_1^2\right)\left(Y_{22} - \phi_2^2\right)}.
\end{cases}
\tag{17}
$$

It is straightforward that applying the SDP to RLT improves the lower bounds on Y_{11} and Y_{22}, but has no effect on the upper bounds. Thus, combining the RLT constraint on the SDP tightens the feasible region of Problem (13). Similarly, the feasible domain of (16) is tighter than the feasible domain of (9). Therefore, it can be demonstrated that our approach can speed up the localization convergence rate. Algorithm 1 shows the SDP + RLT method for UAV localization.

Algorithm 1 SDP + RLT Method for UAV Localization

Input: $a_k, m, n, \mathcal{N}, \mathcal{M}, \hat{d}_{k,j}, d_{i,j}, R, l, u, \epsilon$.
Output: positions of unknown UAV nodes $x_1, x_2, \dots, x_n$.
1: **begin**
/*Initialization*/
2: Initialize the Euclidean distance matrix $Z = [I_3, X; X^T, Y]$, where $Y = X^T X$.
3:   $Z = \text{Symmetrize }(Z)$
4:   **cvx begin**
5:     *minimize* $\text{norm}(X, 2)$
6:     s.t.

$$
\begin{aligned}
&Z \succcurlyeq 0 \\
&Y_{ij} - l_i\phi_j - l_j\phi_i \geq -l_i l_j \\
&Y_{ij} - u_i\phi_j - u_j\phi_i \geq -u_i u_j \\
&Y_{ij} - l_i\phi_j - u_j\phi_i \leq -l_i u_j
\end{aligned}
$$

7:      **for** each UAV $i \leftarrow 1$ to n **do**
8:       update Z by solving Problem (14).
9:      **end for**
10:     **if** $Z \leq 0$ **then**
11:      break;
12:     **end if**
13: **cvx end**
14: **end**

---

5. Bionic Optimization Algorithm

In this section, we develop an improved strategy called I-GWO based on the classical gray wolf optimization (GWO) to optimize the UAV localization results in the presence of noise interference, such that the localization results have smaller errors and higher accuracy.

5.1. Motivation

As seen above, the exact resolution of the localization problem can be gained using SDP + RLT when the distance information between UAVs is accurate. However, the localization problem suffers from measurement noise, and there is no scheme to satisfy the constraints in (7). Bioinspired optimization algorithms have attracted the attention of researchers who have attempted to solve UAV localization problems by employing them because of their high precision and low complexity. Previous studies have proposed several bionic-based localization algorithms. Arafat et al. [12] proposed an improved PSO algorithm, which uses a grouping approach to achieve fast convergence. Raguraman et al. [33] proposed a dimension-based hybrid algorithm (HDPSO) to reduce localization errors.

Motivated by the bionic optimization algorithm, a bionic optimization algorithm I-GWO is proposed in this paper for improving localization accuracy. Since the localization results of UAVs are subject to errors when applying the SDP + RLT technique to distance measurements containing noise, we consider the localization results of SDP + RLT as input and then apply the I-GWO algorithm to improve the localization accuracy.

The localization of target UAVs can be regarded as an error optimization issue. Let $\theta_i = [\theta_{xi}, \theta_{yi}, \theta_{zi}]^T$ be the optimized position estimate of the unknown UAVs. The distance between anchor a_k and θ_i is $\widetilde{d}_{ki}$. Consequently, the objective function of the localization optimization is denoted as

$$f(\theta_i, a_k) = \frac{1}{K} \sum_{K=1}^{K} \sqrt{\left(\hat{d}_{ki} - \widetilde{d}_{ki}\right)^2} \tag{18}$$

where K is the number of anchors whose location is close to the target node (usually smaller than R). $\hat{d}_{ki}$ is the noise-measured distance between the target drone and anchor, and $\widetilde{d}_{ki}$ is the distance between them after calculation of optimization, defined as

$$\widetilde{d}_{ki} = \sqrt{\left(\theta_{x_i} - a_{x_k}\right)^2 + \left(\theta_{y_i} - a_{y_k}\right)^2 + \left(\theta_{z_i} - a_{z_k}\right)^2} \tag{19}$$

5.2. I-GWO Algorithm

The GWO algorithm is inspired by the unique hierarchy and hunting patterns of wolves in nature. In the algorithm, the gray wolf population is divided into four classes, including alpha, beta, delta, and omega (hereinafter will be omitted as α, β, δ, and ω) in order from high to low. The position of each wolf is regarded as a potential solution for the prey position (target node). The α wolf leads the wolves to proceed with the search, encirclement, and hunting activities for the prey. In order to imitate the social hierarchy in the design of I-GWO, the location of α wolf is taken as the best solution (optimal fitness of individuals), β as the sub-optimal solution, δ as the third optimal solution, and the remaining candidate solution is named ω. Within the prey searching phase, the movement of lower-ranked gray wolves in the population is based on the top three ranks, and the wolves' ranks are constantly updated. When the algorithm reaches the maximum iterations, the leader of the wolf population (α wolf) is considered to have caught the prey, at which case the hunt is finished. Studies show that the I-GWO algorithm can dramatically reduce the probability of being premature and falling into a local optimum.

Define $U = [U_1, U_2, \ldots, U_w]$ as the population of gray wolves and w denotes the number of wolves in the population. The ith wolf is denoted as $U_i = [U_i{}^1, U_i{}^2, U_i{}^3]^T$ in the 3D situation. In the encircling prey phase, the gray wolf position update is defined as

$$U(t + 1) = U_p(t) - A \cdot D \tag{20}$$

$$D = |C \cdot U_p(t) - U(t)| \tag{21}$$

where t denotes the iteration, U_p denotes the location of the prey, and D expresses the distance vector between the individual wolf and prey. A and C are constantly varying coefficients.

$$A = 2a \cdot r_1 - a, \quad C = 2r_2 \tag{22}$$

where r_1 and r_2 are random numbers of [0,1], and a is the convergence factor (decreases from $2 \to 0$ with the t).

$$a = 2 - \sqrt{t^2 / (t_{max} + t)(t_{max} - t)} \tag{23}$$

It can be seen from the equation that the gray wolf group moves toward α wolf (who is nearest to the prey as the leader). The wolf group's movement is motivated by its position U_i and the random vector C. The step length is determined by D and A. When $|A| > 1$, it is far from the target and shows stronger global seeking ability; On the contrary, if $|A| < 1$, it approaches the prey and shows stronger local search ability.

During the hunting stage, gray wolves are assumed to identify the prey's position (which is set to the α wolf position at the first algorithm loop) and encircle it. However, actually, the desired position of prey is unknown during the whole optimization phase. Usually, it is assumed that the top three rank wolves (e.g., α, β, and δ) have a better knowledge of the prey's potential location. Therefore, we saved these three solutions so far and asked other gray wolves (including ω) to update their positions with the iteration based on the first three best-searched positions. Here, the following equation is proposed.

$$
\begin{aligned}
U_1(t+1) &= U_\alpha(t) - A_\alpha |C_\alpha \cdot U_\alpha(t) - U(t)| \\
U_2(t+1) &= U_\beta(t) - A_\beta |C_\beta \cdot U_\beta(t) - U(t)| \\
U_3(t+1) &= U_\delta(t) - A_\delta |C_\delta \cdot U_\delta(t) - U(t)|
\end{aligned}
\tag{24}
$$

$$
U_i(t+1) = \frac{1}{3} \sum_{j=1,2,3} U_j(t+1)
\tag{25}
$$

The step length and direction of wth individuals wolf in the group toward α, β, and δ wolves are defined by (24), and the final position of ω and other wolves is defined by (25). In order to achieve higher localization accuracy of prey (e.g., target UAV), the proposed algorithm I-GWO has the following improvements.

(i) On the one hand, the particle initialization is refined. Since we have already conducted a preliminary estimation of the UAV target node position by the SDP method in the preliminary phase of localization, the α wolf position is assigned by the result of SDP + RLT. We generate a regular polyhedron with the position of the α wolf as the center and the R_w as the radius. Then, its individual vertices are assumed to be the initial positions of β wolf, δ wolf, ω wolf, and other wolves.

(ii) On the other hand, to further enhance the exploration capability of the I-GWO algorithm, the wolves' positions are updated according to the mean value of the random weight sum of α, β, and δ wolves, and the position updating is modified as

$$
U_i(t+1) = \frac{1}{3} \sum_{j=1,2,3} q_j U_j(t+1)
\tag{26}
$$

where q_i is the dynamic weight coefficient generated by random numbers of [0,1], and $q_1 + q_2 + q_3 = 1$. The update rule of I-GWO is described in Figure 4, and the optimization procedure of I-GWO is exhibited in Algorithm 2.

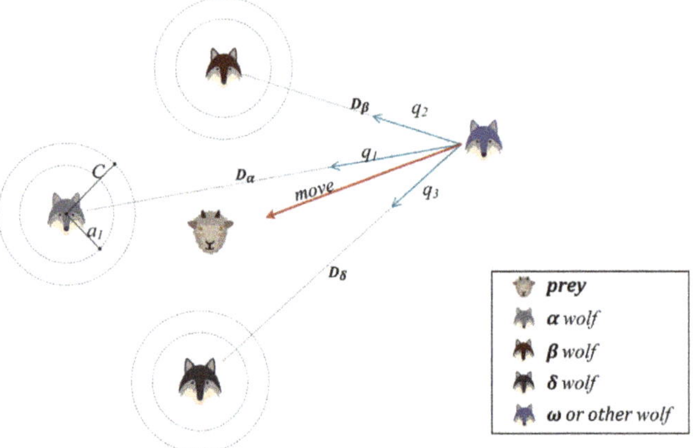

Figure 4. Location update rule in I-GWO.

Algorithm 2 I-GWO of UAV Localization Optimization

Input: Preliminary results X_i of SDP + RLT, population size U, max iterations t_{max}, dimension d, and coefficients $r_1, r_2, q_1, q_2,$ and q_3.
Output: Optimal position of the unknown UAVs.
/*Initialization*/
1: Initialize the gray wolf pack $U = [U_1, U_2, \ldots, U_w]$ using X_i.
2: Initialize the GWO parameters (a, A, C).
3: Initialize the fitness value $(U_\alpha, U_\beta, U_\delta)$.
/*Computation*/
4: **while** $(t \leq t_{max})$ **do**
5:    **for** each wolf w = 1: W **do**
6:      Update the current search agent position using (18)
7:    **end for**
/*I-GWO loop*/
8:    **for** each wolf w = 1: W **do**
9:      Evaluate the fitness value and update $(U_\alpha, U_\beta, U_\delta)$
10:      Obtain the variable a based on (23).
11:    **end for**
12:    **for** each wolf w = 1: W **do**
13:      Calculate A and C based on (19)
14:      Update the position of wolf w by (24)(25)
15:    **end for**
16:    $w = w + 1$
17:    $t = t + 1$
18: **end while**
19: Terminate the process and output the optimal position by U_α
20: **end**

---

6. Simulation

The performance of the proposed algorithms will be evaluated through MATLAB simulator simulations when solving UAVs that cannot be localized or uniquely localized because of limited measurements. First, the hybrid algorithm (named SDP + RLT) is compared with the SDP + O algorithm by Russell et al. [11], the MDS method [18], and the least squares (LS) algorithm [24]. Second, the performance of the I-GWO algorithm is evaluated under noisy measurements and compared with PSO algorithms [12], HPSO algorithms [21], and HGWO algorithms proposed by Arafat et al. [13].

6.1. Performance of Proposed Method

Let 50 target UAVs be randomly placed within the simulated area $[-300m, 300m] \times [-300m, 300m]$. The altitudes of the UAVs are in the range $[100m, 600m]$, and 6 anchor UAVs are evenly placed. The UAV motions follow the Gaussian random walk model. In addition, we set $R = 300m$, then the distance measurements can be made with the UAVs within its communication range. For the noisy range measurement, the *Gauss error* with fixed standard deviation is used. To be closer to the real scenario, it is also assumed that the distance pairs between UAVs will be lost randomly by 1%. The localization error (LE) is defined as the deviation of the estimated position of a UAV node from the actual position.

$$
\begin{aligned}
LE_i &= \sqrt{\left(X_{ei} - X_{ri}\right)^2} \\
&= \sqrt{\left(x_{ei} - x_{ri}\right)^2 + \left(y_{ei} - y_{ri}\right)^2 + \left(z_{ei} - z_{ri}\right)^2}
\end{aligned}
\tag{27}
$$

where the 3D coordinates of the estimation and actual value of the target nodes are denoted as $X_e = (x_e, y_e, z_e)^T$ and $X_r = (x_r, y_r, z_r)^T$, respectively. We evaluate the UAV localization performance using the root mean square error (RMSE) as the evaluation metric, which is averaged over 100 runs.

$$\mathrm{RMSE} = \sqrt{\frac{1}{n}\sum_{i=1}^{n}\left(X_{ei}-X_{ri}\right)^2} = \sqrt{\frac{1}{n}\cdot\sum_{i=1}^{n}LE_i} \tag{28}$$

where X_{ri} denotes the real position matrix of the target UAV node, and X_{ei} denotes the matrix of estimated positions. NMSE is given by the ratio of the matrix $X - \widehat{X}$ to the F-norm of X. In addition, we evaluate the localization accuracy by referring to the mean absolute error (MAE) of the nodes, which is also taken as the average of 100 runs.

$$\mathrm{MAE} = \frac{1}{n\times d}\sum_{i=1}^{n}\sum_{j=1}^{d}\left|X_{e_{i,j}}-X_{r_{i,j}}\right| \tag{29}$$

where $X_{e_{i,j}}, X_{r_{i,j}}$, denote the estimated and true positions of UAV, respectively.

6.2. Simulation Results and Discussion

Figure 5 depicts the accuracy of different localization algorithms for the target node under the measurement noise ratio (noise ratio is 10%) and loss ratio (loss ratio is 1%). Figure 5a shows the LS-based localization effect, where the blue line represents the deviation between the true and estimated positions. Since LS localization is carried out iteratively, the localization deviation of previous nodes will affect the localization of subsequent nodes, so the overall RMSE of localization is large. Figure 5b,c shows MDS localization and SPD + O algorithm, in which 70% of nodes can estimate their positions well, while the rest of the nodes have big deviations, and their MAEs are 4.236 m and 4.013 m, respectively. Figure 5d shows that the proposed localization method SDP + RLT has an MAE of 2.082, which is only half of the SDP + O algorithm. In conclusion, the performance of the SDP + RLT algorithm can be demonstrated by other existing localization algorithms.

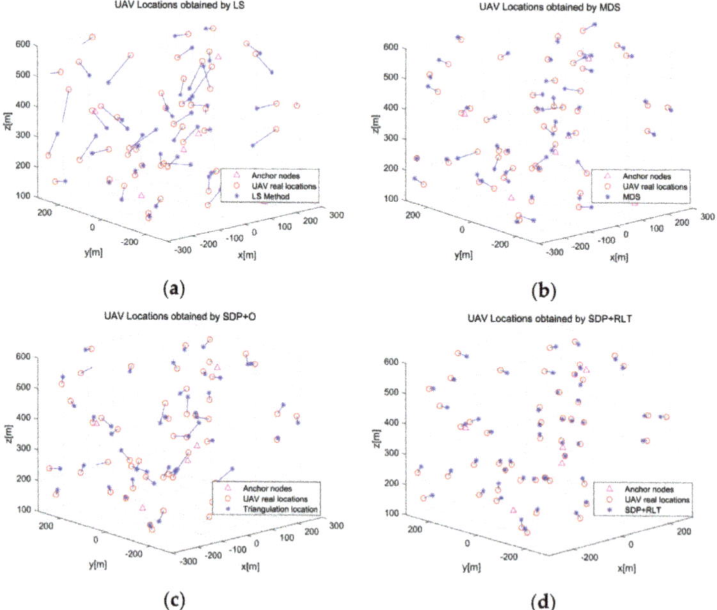

Figure 5. Comparison of LS, MDS, SDP + O, and SDP + RLT localization results in the same scenario with $m = 6$, $n = 50$ and $R = 300$. The "Δ" represents the anchor node, the "○" represents the actual position of the UAV, and the "*" represents the estimated position.

We further demonstrate the localization performance. Figure 6 and Table 2 describe the RMSE of different algorithms when the number of target UAVs is increased with noise ratio = 10%, distance loss ratio = 1%, and anchors = 6. Figure 6 clearly demonstrates that the localization performance of our proposed algorithm incrementally improves by increasing the number of UAVs, which is attributed to the fact that more neighbors are available. In addition, the SDP + RLT algorithm outperforms the MDS by about 1.5 orders of magnitude in terms of RMSE and by about 2 orders of magnitude over the LS and SDP + O algorithms. However, the SDP + RLT algorithm costs more time since it uses the "CVX toolbox" to semidefinite constraints on the distance matrix during the run. Table 2 provides accurate numerical statistics. In future research, we consider applying a distributed method to reduce the localization time.

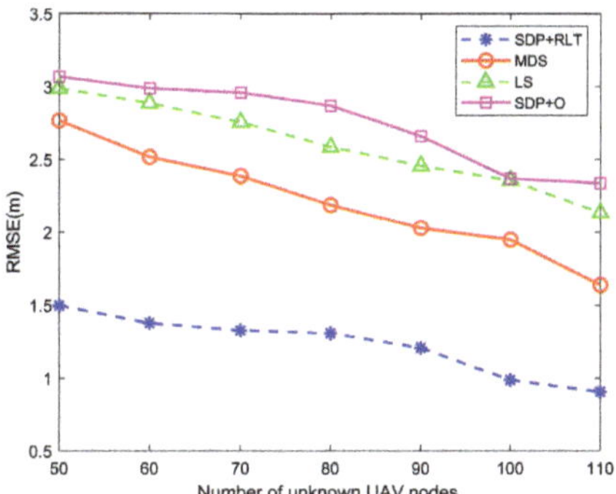

Figure 6. Comparison of RMSE vs. UAV numbers.

Table 2. Number of UAVs as a function of RMSE for different localization algorithms.

Localization Algorithm/RMSE	Numbers of Unknown UAV Nodes						
	50	60	70	80	90	100	110
LS	2.9912	2.8923	2.7633	2.5908	2.4648	2.3567	2.1435
MDS	2.7732	2.5243	2.3957	2.1894	2.0362	1.9533	1.6415
SDP + O	3.0796	2.9923	2.9617	2.8796	2.6695	2.3709	2.3474
SDP + RLT	1.5057	1.3809	1.3385	1.3196	1.2061	0.9921	0.9107

Figure 7 shows that the localization accuracy of the LS, MDS, and SDP + O algorithms improves substantially when anchors increase while the target UAVs are fixed at 50. The RMSE improvement is about 200%, which indicates that they are highly dependent on the anchors and therefore not suitable for large-scale network applications. With increasing anchors, the RMSE of the SDP + RLT algorithm improves by about 50%, so we infer that our algorithm is more scalable and robust in the face of network expansion. In addition, the relatively small number of anchors used implies a lower GPS cost.

Next, the noise factor during distance measurement was evaluated by considering the effects of realistic factors (geography, weather conditions, or UAV's own maneuverability, etc.). Figure 8 shows that the RMSE of all the algorithms increases when the noise rate increases. However, the SDP + RLT algorithm outperforms the others algorithms in terms of interference immunity. It is notable that the RMSE is equal to 0 when the noise rate is 0. This also justifies the conclusion in Theorem 2 that the SDP method can achieve unique accurate localization.

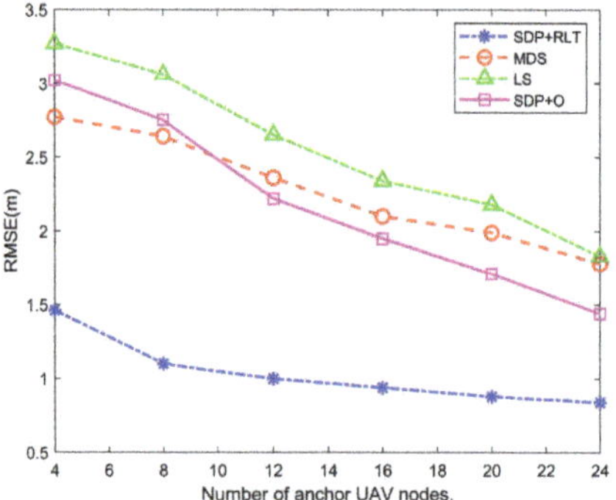

Figure 7. Comparison of RMSE vs. anchor numbers.

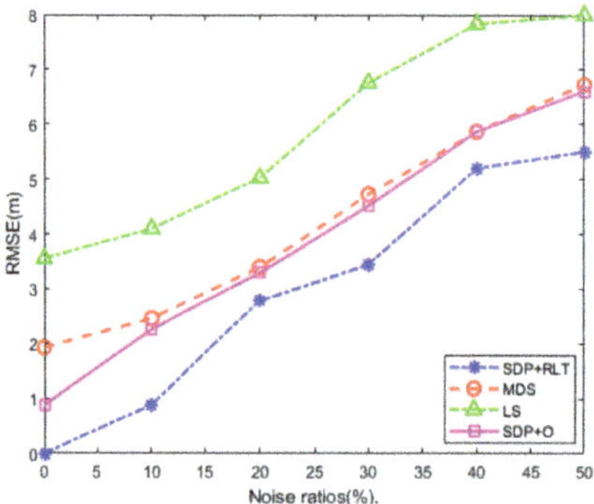

Figure 8. Comparison of RMSE vs. noise ratios.

Finally, the improvement of positioning accuracy of different optimization algorithms is demonstrated through comparative experiments. Figure 9 shows the comparison of different affine optimization algorithms applied for the initial localization results of SDP + RLT. Table 3 details the numerical results of the different optimization algorithms. It means that the convergence speed of the I-GWO algorithm is faster than that of PSO, HPSO, and HGWO, and the RMSE is smaller. Therefore, based on the localization results of the SDP + RLT method, the I-GWO algorithm can achieve the goal of high-precision localization of UAVs.

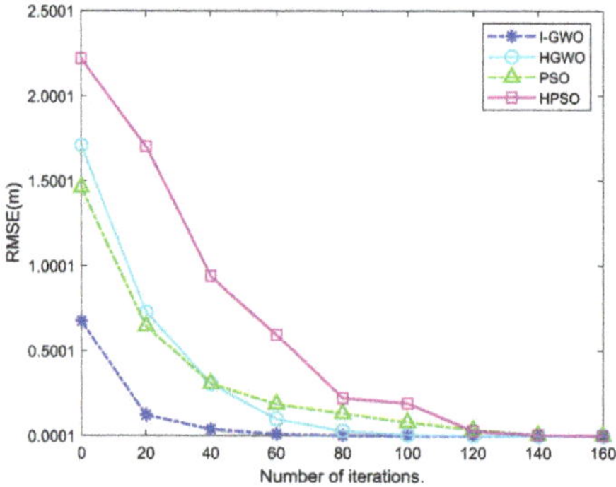

Figure 9. Comparison of RMSE vs. iterations.

Table 3. Number of iterations as a function of RMSE for different optimization algorithms.

Optimization Algorithm/RMSE	Iterations								
	0	20	40	60	80	100	120	140	160
PSO	1.4602	0.6434	0.3061	0.1853	0.1302	0.0794	0.0337	0.0036	0.0015
HPSO	1.7132	0.7266	0.3068	0.0988	0.0287	0.005	0.0027	5.56×10^{-5}	7.35×10^{-7}
HGWO	2.2296	1.7022	0.9412	0.5926	0.2205	0.1899	0.0274	0.0038	0.0006
I-GWO	0.6757	0.1209	0.0395	0.0096	0.0061	0.0021	3.21×10^{-6}	2.57×10^{-8}	1.82×10^{-11}

7. Conclusions

In this article, SDP + RLT and I-GWO algorithms have been proposed for solving the localization problem in 3D cyberspace for swarms of UAVs using only distance measurements and a few GPS-equipped UAVs, which are essential for cooperative UAV swarm flying. Then, the solvability and unique solvability have been analyzed theoretically. Further, the I-GWO is proposed to improve the localization accuracy for distance measurements that contain noise. Through simulation results exhibited in MATLAB, we compare the performance of SDP + RLT with other advanced localization methods (i.e., LS, MDS, and SDP + O) and demonstrate its efficiency in terms of RMSE, measurement error, and scalability. I-GWO is compared with the classical optimization algorithms PSO, HPSO, and HGWO. The RMSE and MAE results prove the superiority of the algorithms. Our study still has limitations in terms of computational speed since the proposed approach is based on SDP, where semidefinite constraints consume more time when the target UAVs increase. Hence the proposed method is more suitable for centralized UAV routing schemes. Future work includes the introduction of distributed methods for efficient and high-precision localization of UAV swarms in mobile scenarios to reduce the localization delay. In addition, finding the optimal trajectory of a single mobile anchor UAV to achieve localization is also a future research trend.

Author Contributions: Methodology, Z.L.; Software, X.X.; Validation, Z.L.; Formal analysis, X.X. and Y.Y.; Investigation, Z.L. and Y.Y.; Resources, X.X.; Writing—original draft, Z.L.; Supervision, Y.Y. All authors have read and agreed to the published version of the manuscript.

Funding: This research was funded by science and technology research project of the Henan province grant number 222102240014.

Institutional Review Board Statement: Not applicable.

Data Availability Statement: The data used to support the findings of this study are included in the article.

Conflicts of Interest: The authors declare that there are no conflict of interest regarding the publication of this paper.

References

1. Ashish, S.; Prakash, J. Future FANET with application and enabling techniques: Anatomization and sustainability issues. *Comput. Sci. Rev.* **2021**, *39*, 100359.
2. Tang, F.; Kawamoto, Y.; Kato, N.; Liu, J. Future intelligent and secure vehicular network toward 6G: Machine-learning approaches. *Proc. IEEE* **2020**, *108*, 292–307. [CrossRef]
3. Orfanus, D.; De Freitas, E.; Eliassen, F. Self-Organization as a Supporting Paradigm for Military UAV Relay Networks. *IEEE Commun. Lett.* **2016**, *20*, 804–807. [CrossRef]
4. Mozaffari, M.; Saad, W.; Bennis, M.; Debbah, M. Unmanned Aerial Vehicle with under Laid Device-to-Device Communications: Performance and Tradeoffs. IEEE Trans. *Wirel. Commun.* **2016**, *15*, 3949–3963.
5. Guo, K.; Li, X.; Xie, L. Simultaneous cooperative relative localization and distributed formation control for multiple UAVs. *Sci. China Inf. Sci.* **2020**, *63*, 119201. [CrossRef]
6. Zheng, Z.; Sun, J.; Wang, W.-Q.; Yang, H. Classification and localization of mixed near-field and far-field sources using mixed-order statistics. *Signal Process.* **2018**, *143*, 134–139. [CrossRef]
7. Li, W.; Jelfs, B.; Kealy, A.; Wang, X.; Moran, B. Cooperative Localization Using Distance Measurements for Mobile Nodes. *Sensors* **2021**, *21*, 1507. [CrossRef]

8. Xu, C.; Wang, Z.; Wang, Y.; Wang, Z.; Yu, L. Three passive TDOA-AOA receivers-based flying-UAV positioning in extreme environments. *IEEE Sens. J.* **2020**, *20*, 9589–9595. [CrossRef]
9. Kang, Y.; Wang, Q.; Wang, J.; Chen, R. A high-accuracy TOA-based localization method without time synchronization in a three-dimensional space. *IEEE Trans. Ind. Inform.* **2018**, *15*, 173–182. [CrossRef]
10. Mur-Artal, R.; Montiel, J.M.M.; Tardos, J.D. ORB-SLAM: A versatile and accurate monocular SLAM system. *IEEE Trans. Robot.* **2015**, *31*, 1147–1163. [CrossRef]
11. Russell, J.S.; Ye, M.; Anderson, B.D.O.; Hmam, H.; Sarunic, P. Cooperative localization of a GPS-denied UAV using direction-of-arrival measurements. *IEEE Trans. Aerosp. Electron. Syst.* **2019**, *56*, 1966–1978. [CrossRef]
12. Arafat, M.Y.; Moh, S. Localization and clustering based on swarm intelligence in UAV networks for emergency communications. *IEEE Internet Things J.* **2019**, *6*, 8958–8976. [CrossRef]
13. Yeasir, A.M.; Moh, S. Bio-inspired approaches for energy-efficient localization and clustering in UAV networks for monitoring wildfires in remote areas. *IEEE Access* **2021**, *9*, 18649–18669.
14. Biswas, P.; Liang, T.-C.; Toh, K.-C.; Ye, Y.; Wang, T.C. Semidefinite programming approaches for sensor network localization with noisy distance measurements. *IEEE Trans. Autom. Sci. Eng.* **2006**, *3*, 360–371. [CrossRef]
15. Guo, K.; Li, X.; Xie, L. Ultra-wideband and odometry-based cooperative relative localization with application to multi-uav formation control. *IEEE Trans. Cybern.* **2019**, *50*, 2590–2603. [CrossRef] [PubMed]
16. Baldi, S.; Roy, S.; Yang, K. Towards adaptive autopilots for fixed-wing unmanned aerial vehicles. In Proceedings of the 2020 59th IEEE Conference on Decision and Control (CDC), Jeju, Republic of Korea, 14–18 December 2020; pp. 4724–4729. [CrossRef]
17. Wang, X.; Roy, S.; Farì, S.; Baldi, S. Adaptive Vector Field Guidance Without a Priori Knowledge of Course Dynamics and Wind. *IEEE/ASME Trans. Mechatron.* **2022**, *27*, 4597–4607. [CrossRef]
18. Shang, Y.; Ruml, W. Improved MDS-based localization. In Proceedings of the IEEE INFOCOM 2004, Hong Kong, China, 7–11 March 2004; Volume 4, pp. 2640–2651.
19. Wan, C.; You, S.; Jing, G.; Dai, R. A distributed algorithm for sensor network localization with limited measurements of relative distance. In Proceedings of the 2019 American Control Conference (ACC), Philadelphia, PA, USA, 10–12 July 2019; pp. 4677–4682.
20. Zhang, Y.; Chen, Y.; Liu, Y. Towards unique and anchor-free localization for wireless sensor networks. *Wirel. Pers. Commun.* **2012**, *63*, 261–278. [CrossRef]
21. Parulpreet, S.; Khosla, A.; Kumar, A.; Khosla, M. Optimized localization of target nodes using single mobile anchor node in wireless sensor network. *AEU-Int. J. Electron. Commun.* **2018**, *91*, 55–65.
22. Liu, Y.; Wang, Y.; Wang, J.; Shen, Y. Distributed 3D relative localization of UAVs. *IEEE Trans. Veh. Technol.* **2020**, *69*, 11756–11770. [CrossRef]
23. Wang, W.; Yang, Y.; Liu, W.; Wang, L. PMDV-hop: An effective range-free 3D localization scheme based on the particle swarm optimization in wireless sensor network. *KSII Trans. Internet Inf. Syst. (TIIS)* **2018**, *12*, 61–80.
24. Nguyen, T.V.; Jeong, Y.; Shin, H.; Win, M.Z. Least square cooperative localization. *IEEE Trans. Veh. Technol.* **2015**, *64*, 1318–1330. [CrossRef]
25. Wang, G.; Li, Y.; Ansari, N. A semidefinite relaxation method for source localization using TDOA and FDOA measurements. *IEEE Trans. Veh. Technol.* **2013**, *62*, 853–862. [CrossRef]
26. Shang, Y.; Ruml, W.; Zhang, Y.; Fromherz, M. Localization from connectivity in sensor networks. *IEEE Trans. Parallel Distrib. Syst.* **2004**, *15*, 961–974. [CrossRef]
27. Zou, Y.; Liu, H.; Wan, Q. An Iterative Method for Moving Target Localization Using TDOA and FDOA Measurements. *IEEE Access* **2018**, *6*, 2746–2754. [CrossRef]
28. Calafiore Giuseppe, C.; Carlone, L.; Wei, M. A distributed technique for localization of agent formations from relative range measurements. *IEEE Trans. Syst. Man Cybern.-Part A Syst. Hum.* **2012**, *42*, 1065–1076. [CrossRef]
29. Pratik, B.; Lian, T.-C.; Wang, T.; Ye, Y. Semidefinite programming based algorithms for sensor network localization. *ACM Trans. Sens. Netw. (TOSN)* **2006**, *2*, 188–220.
30. Man-Cho, A.; Ye, Y. Theory of semidefinite programming for sensor network localization. *Math. Program.* **2007**, *109*, 367–384.
31. Anstreicher, K.M. Semidefinite programming versus the reformulation-linearization technique for nonconvex quadratically constrained quadratic programming. *J. Glob. Optim.* **2009**, *43*, 471–484. [CrossRef]
32. Zheng, Z.; Zhang, H.; Wang, W.; So, H.C. Source localization using TDOA and FDOA measurements based on semidefinite programming and reformulation linearization. *J. Frankl. Inst.* **2019**, *356*, 11817–11838. [CrossRef]
33. Raguraman, P.; Ramasundaram, M.; Balakrishnan, V. Localization in wireless sensor networks: A dimension based pruning approach in 3D environments. *Appl. Soft Comput.* **2018**, *68*, 219–232. [CrossRef]

 *drones*

Article

PPO-Exp: Keeping Fixed-Wing UAV Formation with Deep Reinforcement Learning

Dan Xu [1], Yunxiao Guo [2], Zhongyi Yu [3], Zhenfeng Wang [2], Rongze Lan [2], Runhao Zhao [1], Xinjia Xie [4,*] and Han Long [2,*]

1 College of System Engineering, National University of Defense Technology, Changsha 410073, China
2 College of Sciences, National University of Defense Technology, Changsha 410073, China
3 College of Advanced Interdisciplinary Studies, National University of Defense Technology, Changsha 410073, China
4 College of Computer Science, National University of Defense Technology, Changsha 410073, China
* Correspondence: xiexinjia97@nudt.edu.cn (X.X.); longhan@nudt.edu.cn (H.L.)

Abstract: Flocking for fixed-Wing Unmanned Aerial Vehicles (UAVs) is an extremely complex challenge due to fixed-wing UAV's control problem and the system's coordinate difficulty. Recently, flocking approaches based on reinforcement learning have attracted attention. However, current methods also require that each UAV makes the decision decentralized, which increases the cost and computation of the whole UAV system. This paper researches a low-cost UAV formation system consisting of one leader (equipped with the intelligence chip) with five followers (without the intelligence chip), and proposes a centralized collision-free formation-keeping method. The communication in the whole process is considered and the protocol is designed by minimizing the communication cost. In addition, an analysis of the Proximal Policy Optimization (PPO) algorithm is provided; the paper derives the estimation error bound, and reveals the relationship between the bound and exploration. To encourage the agent to balance their exploration and estimation error bound, a version of PPO named PPO-Exploration (PPO-Exp) is proposed. It can adjust the clip constraint parameter and make the exploration mechanism more flexible. The results of the experiments show that PPO-Exp performs better than the current algorithms in these tasks.

Keywords: fixed-wing UAV; formation keeping; reinforcement learning

Citation: Xu, D.; Guo, Y.; Yu, Z.; Wang, Z.; Lan, R.; Zhao, R.; Xie, X.; Long, H. PPO-Exp: Keeping Fixed-Wing UAV Formation with Deep Reinforcement Learning. *Drones* 2023, 7, 28. https://doi.org/10.3390/drones7010028

Academic Editor: Oleg Yakimenko

Received: 27 November 2022
Revised: 24 December 2022
Accepted: 28 December 2022
Published: 31 December 2022

1. Introduction

In recent years, unmanned aerial vehicles (UAVs) have been widely used in military and civil fields, such as in tracking [1], surveillance [2], delivery [3], and communication [4]. Due to the inherent defects, such as fewer platform functions and a light payload, it is difficult for a single UAV to perform diversified tasks in complex environments [5]. The cooperative formation composed of multiple UAVs can effectively compensate for the lack of performance and has many advantages in performing combat tasks. Thus, the formation control of UAVs has become a hot topic and attracted much attention [6,7].

Traditional solutions are usually based on accurate models of the platform and disturbance, such as model predictive control [8] and consistency theory [9]. This paper [10] proposed a group-based hierarchical flocking control approach, which did not need the global information of the UAV swarms. The study in [11] researched the mission-oriented miniature fixed-wing UAV flocking problem and proposed an architecture that decomposes the complex problem; it was the first work that successfully integrated the formation flight, target recognition, and tracking missions into simply an architecture. However, due to the influence of environmental disruption, these methods are difficult to accurately model [12]. This seriously limits the application scope of traditional analysis methods. Therefore, with the emergence of machine learning (ML), the reinforcement learning (RL) [13,14] method to solve the above problem has received increasing attention [15]. RL applies to decision-making control problems in unknown environments and has achieved successful applications in the robotics field [16–18].

At present, some works have integrated RL into the formation coordination control problem solution and preliminarily verified the feasibility and effectiveness in the simulation environment. Most existing schemes use the particle agent model for the rotary-wing UAV. The researchers [19] first researched RL in coordinated control, and applied the Q-learning algorithm and potential field force method to learn the aggregation strategy. After that, ref. [20] proposed a multi-agent self-organizing system based on a Q-learning algorithm. Ref. [21] investigates second-order multi-agent flocking systems and proposed a single critic reinforcement learning approach. The study in [22] proposes a UAV formation coordination control method based on the Deep Deterministic Policy Gradient algorithm, which enabled UAVs to perform navigation tasks in a completely decentralized manner in a large-scale complex environment.

Different from rotary-wing UAVs, the formation coordination control of fixed-wing UAVs is more complex and more vulnerable to environmental disturbance; therefore, different control strategies are required [23]. The Dyna-Q(λ) and Q-flocking algorithm are proposed [24,25] for solving the discrete state & action space fixed-wing UAV flocking problem under complex noise environments with deep reinforcement learning. To deal with the continuous space, ref. [26,27] proposed a fixed-wing UAV flocking method in continuous spaces based on deep RL with the actor–critic model. The learned policy can be directly transferred to the semi-physical simulation. Ref. [28] focused on the nonlinear attitude control problem and devised a proof-of-concept controller using proximal policy optimization.

However, the above methods also assume that UAVs fly with different attitudes, so the interaction (collision) between the followers can be ignored, and the followers in the above methods are seen as independent. Under the independent condition, these single-agent reinforcement learning algorithms can be effective due to the stationary environment [29]. However, in real tasks, even when the attitude is different, the collision still may happen when the attitude difference is not significant, and the UAVs adjust their roll angles.

In real tasks, the followers can interact with each other, and it is also common for them to collide in some scenarios, such as the identical attitude flocking task. However, this scenario is rarely studied. Ref. [30] proposed a collision-free multi-agent flocking approach MA2D3QN by using the local situation map to generate the collision risk map. The experimental results demonstrate that it can reduce the collision rate. The followers' reward function in MA2D3QN is only related to the leader and itself; however, other followers can also provide some information. This indicates that the method did not fully consider the interaction between the followers.

However, MA2D3QN did not demonstrate the ability to manage the non-stationary multi-agent environment [29], and the experiments also show collision judgments with high computation. With the number of UAVs rising, the computation time also increases. Furthermore, some problems in the above methods on fixed-wing UAVs have not been adequately solved, such as the generalization aspect and communication protocol; the most concerning problem is the minimum cost of the formation.

To consider the communication protocol of the formation, this paper takes the maximum communication distance between the UAVs into consideration, with a minimum cost communication protocol to guide the UAVs to send the message in the formation-keeping process. Under this protocol, the centralized training method for the UAVs is designed; only the leader needs to equip the intelligence chip. The main contributions of this work are as follows:

1. Research the formation keeping task with continuous space through reinforcement learning, and building the RL formation-keeping environment with OpenAI gym, and constructing the reward function for the task.
2. Design the communication protocol for the UAVs' formation with one leader who can make decisions intelligently and five followers who receive the decisions from

    the leader. The protocol is feasible even when the UAVs are far away from each other. Under this protocol, the followers and leader can communicate at a low cost.

3.   Analyze the PPO-Clip algorithm, give the estimation error bound of its surrogate, and elaborate on the relationship between the bound and hyperparameter ε: the higher ε, the more exploration, the larger the bound.

4.   Propose a variation of PPO-clip: PPO-Exp. The PPO-Exp separates the exploration reward and regular reward in the task of formation keeping, and estimates the advantage function from them, respectively. The adaptive mechanism is used to adjust ε to balance the estimation error bound and exploration. The experiments demonstrate this mechanism with effectiveness for improving performance.

This paper is organized as follows. The first section introduced the current research on UAV flocking. Section 3 describes the background of the formation-keeping task and introduces reinforcement learning briefly. In Section 4, the formation-keeping environment is constructed, and the reward of the formation process is designed. Section 5 discusses the dilemma between the estimation error bound and exploration ability of PPO-Clip, and proposes PPO-Exp to balance the dilemma. Section 6 shows the experimental setup and results. Section 7 provides the conclusions of the paper.

2. Related Work

This section reviews current research about fixed-wing UAV flocking and formation-keeping approaches with deep reinforcement learning. According to the training architecture, this paper divides the current methods into the following two categories: centralized and decentralized. The difference between the two categories is as follows:

The centralized methods utilize the leader and all the followers' states in the training model, and the obtained optimal policy can control all of the followers so that they flock to the leader. The decentralized methods only use one follower and the leader's state to train the policy, and the obtained optimal policy could only control one follower. If there are several followers in the task, the policy and intelligence chip should be deployed on all of the followers.

2.1. Decentralized Approach

The paper [24] proposed a reinforcement learning flocking approach Dyna-Q(λ) to flock the fixed-wing UAV under the stochastic environment. To learn a model in the complex environment, the authors used Q(λ) [31] and Dyna architecture to train each fixed-wing follower to follow the leader, and combined internal models to deal with the influence of the stochastic environment. In [25], the authors further proposed Q-Flocking, which is a model-free and variable learning parameter algorithm based on Q-learning. Compared to Dyna-Q(λ), Q-Flocking removed the internal models and proved it could also converge to the solutions. For simplification, Q-Flocking and Dyna-Q(λ) also require that the state and action spaces are discrete, which is inappropriate. In [26], the authors first developed a DRL-based approach for the continuous state and action spaces fixed-wing UAV flocking. The proposed method is based on the Continuous Actor-Critic Learning Automation(CACLA) algorithm [32], with the experience replay technique embedded to improve the training efficiency. Ref. [33] considered a more complex flocking scenario, where the enemy threat is considered in the dynamic environment. To learn the optimal control policies, the authors use the situation assessment module to transfer the state of UAVs to the situation map stack. Then, the stack is input into the proposed Dueling Double Deep Q-network(D3QN) algorithm to update the policies until convergence. Ref. [34] proposed the Multi-Agent PPO algorithm to decentralize learning in the two–group fixed-wing UAV swarms dog fight control. To accelerate the learning speed, the classical rewarding scheme is added to the resource baseline, which could reduce the state and action spaces.

The advantage of decentralized methods is that these methods could be deployed on the distribution UAV systems, which could extend to the large-scale UAV formation. The disadvantage of the centralized methods is as follows:

- These methods also require all of the followers to be equipped with intelligence chips, which increase the costs.
- These methods do not consider the collision and communication problem, due to the use of only local information.

The decentralized approaches also assume that UAVs fly at different heights, and then the collision problem could be ignored. However, in real-world applications, the collision problem must be considered [30].

2.2. Centralized Approach

Ref. [35] studied the collision avoidance fixed-wing UAV flocking problem. To manage collision among the UAVs, the authors proposed the PS-CACER algorithm, which receives the global information of UAV swarms through the plug-n-play embedding module. Ref. [30] proposed a collision-free approach by transferring the global state information to the local situation map and constructing the collision risk function for training. To improve the training efficiency, the reference-point-based action selection technique is proposed to assist the UAVs' decisions.

The advantages of the centralized methods are as follows:

- These methods could reduce the cost of the formation. Under the centralized architecture, the formation system only requires the leader to equip the intelligence chip. The followers only need to send their state information to the leader and receive the feedback commands.
- These methods could consider collision avoidance and communication in the formation due to their use of global information.

The disadvantage of the centralized method is the dependence on the leader. Ref. [36] pointed out that the defect or jamming of the leader causes failure in the whole formation system.

When the number of UAVs increases or the tasks are complex, the centralized methods face the dimension curse and lack of learning ability problems. A popular approach is learning the complex tasks with a hierarchical method [37,38], which divides the complex tasks into several sub-tasks and uses the centralized method to optimize the hierarchies. The hierarchical reinforcement learning approaches are applied in the quadrotors swarm system [37,38], but are rarely used in fixed-wing UAV systems.

Even when using global information in training, the current centralized approaches fail to consider communication in the formation. Compared to current centralized approaches, the approach proposed in this paper considers the communication in formation, and provides the communication protocol. Through the communication protocol, the formation system could be considered as one leader with an intelligence chip and five followers without intelligence chips; the leader collects the followers' information, with a centralized train on the intelligence chip. The followers receive the command from the leader through this protocol and execute.

3. Background

This section will introduce the kinematic model of the fixed-wing UAV, restate the formation keeping problem, and briefly introduce reinforcement learning.

3.1. Problem Description

The formation task can be described as follows: At the beginning, the formation is orderly (shown in Figure 1), which is a common formation designed in [39]). The goal of the task is to reach the target area (the green circle area) with the formation in as orderly a way as possible; when the leader enters the target area, the mission is complete.

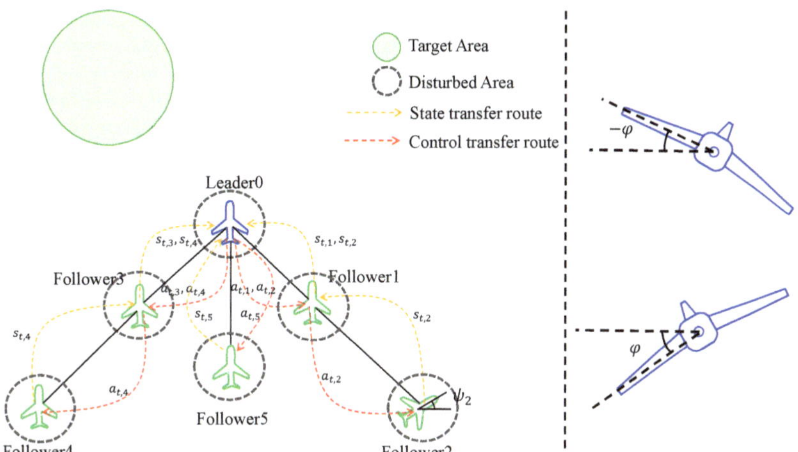

Figure 1. Left: The Leader–Follower formation topology structure and the task schematic diagram. **Right**: The action of UAV.

During the task, assume the UAVs are flying at a fixed attitude; then, each UAV in the formation can also be described as a six-degree of freedom (6DoF) dynamic model. However, analyzing the six-degree model directly is very complex; it will increase the space scale and make control more difficult. The 6DoF model can be simplified to the 4DoF model; to compensate for the loss incurred during this simplification, random noise is introduced into the model [27], and the dynamic equations of *i*th UAV in the formation can be written as follows:

$$\dot{\zeta}_i = \frac{\mathrm{d}}{\mathrm{d}t}\begin{bmatrix} x_i \\ y_i \\ \psi_i \\ \varphi_i \end{bmatrix} = \begin{bmatrix} v_i \cos\psi_i + \eta_{x_i} \\ v_i \sin\psi_i + \eta_{y_i} \\ -(\alpha_g/v_i)\tan\varphi_i + \eta_{\psi_i} \\ f(\varphi_i, \varphi_{i,d}) \end{bmatrix} \tag{1}$$

where $(x_i, y_i) \in \mathbb{R}^2$ is the planar position, and $\psi_i \in \mathbb{R}^1$, $\varphi_i \in \mathbb{R}^1$ represent the heading and roll angle, respectively, (see Figure 1). The v_i is the velocity, and α_g is the gravity acceleration. The random noise values $\eta_{x_i}, \eta_{y_i}, \eta_{\varphi_i}, \eta_{\psi_i}$ are the normal distributions, its means are $\mu_{x_i}, \mu_{y_i}, \mu_{\varphi_i}, \mu_{\psi_i}$, and its variances are $\sigma_{x_i}^2, \sigma_{y_i}^2, \sigma_{\varphi_i}^2, \sigma_{\psi_i}^2$, respectively, (the gray dotted circles in Figure 1 show the area of influence, of random factors); they represent the random factors introduced by simplification and environment noise.

A simple control strategy can make the formation satisfactory when the environment's noise is low. However, under a strong inference environment, such as one with strong turbulence, the random factors will be apparent, leading the formation to maintain the complexity of the task. If no effective control is provided, the formation will break up quickly, (this is demonstrated in Figure 2), and a crash may happen.

Furthermore, even though there is an effective control policy for the formation, the coupling between the control and communication protocol can also be an unsolved challenge. Because the communication range of UAVs is limited, if the UAV wants to know others' states, it has to wait for other UAVs out of range to send state information to UAVs it can communicate with, which in turn send state information to it. If no harmonic protocol is applied in the formation control, the asynchronous and nonstationary elements will be introduced into the formation control, making the control strategy more complex.

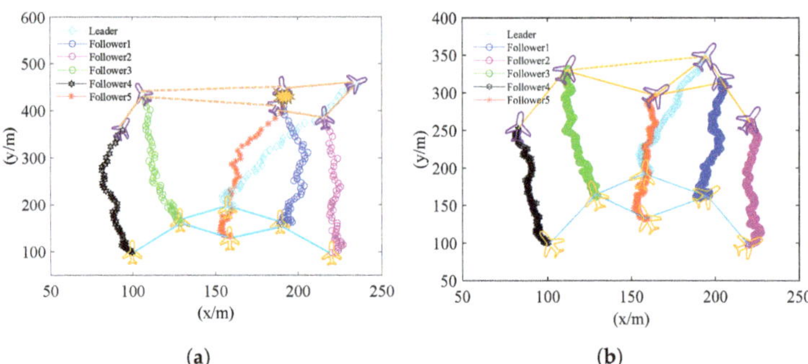

Figure 2. (a): The ablation experiment result of environment noise: track of formation with no control; **(b)**: The ablation experiment result of exploration balance point: PPO-Clip with $\varepsilon = 0.05$.

3.2. Reinforcement Learning

In the last part, the solution of differential Equation (1) can be represented as the current dynamic parameters adding the integral items by difference equation methods such as the Runge–Kutta method. So, the UAV formation control can be modeled as a Markov Decision Process(MDP), which refers to the decision process that satisfies the Markov property.

The MDP also can be described as the tuple $(\mathcal{S}, \mathcal{A}, \mathcal{P}, r, \gamma)$. $\mathcal{S}$ represents the state space, $\mathcal{A}$ represents the action space, and $\mathcal{P} : \mathcal{S} \times \mathcal{A} \times \mathcal{S} \to \mathcal{R}$ is the transition probability. The reward function is $r : \mathcal{S} \times \mathcal{A} \to \mathcal{R}$, and $\gamma \in (0, 1)$ is the discount factor, which leads the agent to pay more attention to the current reward.

Reinforcement learning can solve the MDP well to maximize the discounted return, as follows: $R_t = \sum_{t=0}^{\infty} \gamma^t r(s_t)$. The main approaches of RL are divided into the following three categories: value-based, model-based, and policy-based. The policy-based methods have been developed and widely used in various tasks in recent years. These methods directly optimize the value function by the policy gradient:

$$\nabla \mathcal{J}_\pi(\theta) = \mathbb{E}_{\pi_\theta} \left[\nabla_\theta \sum_{t=0}^{T} \log \pi_\theta(s_t, a_t) A_\pi \right] \tag{2}$$

where A_π is the advantage function that is equal to the state-action value function, and the the state value function is subtracted, as follows:

$$A_\pi(S_t, a_t) = E_\pi \left[\sum_{k=0}^{\infty} \gamma^k r_{t+k} | S_t = s, a_t = a \right] - E_\pi \left[\sum_{k=0}^{\infty} \gamma^k r_{t+k} | S_t = s \right] \tag{3}$$

PPO (Proximal Policy Optimization) is one of the most famous policy gradient methods in continuous state and action space [40]. In policy gradient descent, PPO updates the following equation at each update epoch :

$$\mathcal{L}^{\text{Clip}, \theta} = \mathbb{E}_{\pi_{\theta_{old}}} \left[\min \left(r_t(\theta) A_{\pi_{\theta_{old}}}, clip(r_t(\theta), 1 - \varepsilon, 1 + \varepsilon) A_{\pi_{\theta_{old}}} \right) \right] \tag{4}$$

However, using the constant clip coefficient ε, the PPO also proved its lack of exploration ability and difficulty in convergence. Therefore, designing an efficient dynamic mechanism to adjust ε and ensure greater exploration and faster convergence is also challenging.

4. Formation Environment

This section constructs the fixed-wing UAV formation-keeping environment, the formation topology, communication and control protocols, and collision. Communication loss is also considered in the environment through the reward design.

4.1. State and Action Spaces

In the course of the formation task, based on the 4DoF Equation (1), it is modified to a more realistic control environment. For the ith UAV, assume the thrust of the UAV is controllable, and it will generate a linear acceleration $\alpha_{v_i} = \dot{v}_i$. Moreover, assume the torque of the roll angle is controllable too, and add the roll angle acceleration $\alpha_{\varphi_i} = \dot{w}_i = \ddot{\varphi}_i$ into the dynamic equations. Finally, the dynamic equations of ith UAV can be modified as follows:

$$\dot{\zeta}_i = \frac{\mathrm{d}}{\mathrm{d}t} \begin{bmatrix} x_i \\ y_i \\ \psi_i \\ \varphi_i \\ v_i \\ w_i \end{bmatrix} = \begin{bmatrix} v_i \cos \psi_i + \alpha_{v_i} \cos(\psi_i)t + \eta_{x_i} \\ v_i \sin \psi_i + \alpha_{v_i} \sin(\psi_i)t + \eta_{y_i} \\ -(\alpha_g/v_i)\tan \varphi_i + \eta_{\psi_i} \\ w_i + \eta_{w_i} \\ \alpha_{v_i} \\ \alpha_{\varphi_i} \end{bmatrix} \tag{5}$$

To control the UAVs, linear acceleration and roll angle acceleration are input. For control, we have the dynamic model of ith UAV:

$$\ddot{\zeta}_i = \frac{\mathrm{d}}{\mathrm{d}t} \begin{bmatrix} \dot{x}_i \\ \dot{y}_i \\ \dot{\psi}_i \\ \dot{\varphi}_i \end{bmatrix} = \begin{bmatrix} \alpha_{v_i} \cos \psi_i \\ \alpha_{v_i} \sin \psi_i \\ \frac{-\alpha_g f(\varphi_i, \varphi_{i,d})}{v_i \cos^2 \varphi_i} + \frac{\alpha_{v_i} \alpha_g \tan \varphi_i}{v_i^2} \\ \alpha_{\varphi_i} \end{bmatrix} \tag{6}$$

The state and action spaces for existing methods in UAVs controlled by reinforcement learning are often discrete, but in the real world, the state space is continuous and changes continuously as time goes on. Therefore, combining the analysis of the previous dynamics, we define the state tuple of the ith UAV as $\zeta_i := (x_i, y_i, \psi_i, \varphi_i, v_i, w_i)$. The planar position $(x_i, y_i) \in \mathbb{R}^2$, heading $\psi_i \in S^1$, roll angle $\varphi_i \in S^1$, line and angle velocity $v, w \in \mathbb{R}$ are determined by solving the differential Equation (5).

In the action space, although the engine can produce fixed thrust, the real thrust acting on the UAVs in the nonuniform atmospheric environment is not of the same value as the engine product. So, we define the action space by $a_i := (\alpha_{v_i}, \alpha_{\varphi_i})$. Assume the UAVs can also produce the same acceleration in positive and negative directions, where we have $\alpha_{v_i} \in [-\alpha_{v_i max}, \alpha_{v_i max}]$, and $\alpha_{\varphi_i} \in [-\alpha_{\varphi_i max}, \alpha_{\varphi_i max}]$. The action will influence $\dot{\zeta}_i$ through Equation (6), and then influence the $\dot{\zeta}_i$ indirectly.

After defining the individual state and action of the UAV, we define the formation system state and action by sticking to the individual state (action) as a vector. Define the state of system $\boldsymbol{\zeta} := [\zeta_1, \cdots, \zeta_6]$, and the action of system $\boldsymbol{a} := [a_1, \cdots, a_6]$.

4.2. Communication and Control Protocol

To ensure the UAV formation consumes less energy in the information send and receive process, and ensure the reinforcement learning method can be helpful in the task, the communication and control protocol for the UAV formation will be provided in this part.

As is shown in Figure 1, the formation is of a Leader–Follower structure; in terms of hardware, all the UAVs are equipped with gyroscopes and accelerometers to monitor their action and state parameters. Only the leader has the "brain" chip that can make decisions intelligently; the followers only have the chips that can receive the control command signals, take the command action and send the state signals.

To describe this relationship, the graph model is introduced. Use the communication graph $\mathcal{G}_t$ to describe the communication ability of the formation at time t [39]:

$$\mathcal{G}_t = (6, \mathcal{V}_t, \mathcal{E}_t) \tag{7}$$

where $\mathcal{V}_t = \{v_1, \cdots, v_6\}$ is the set of nodes that represent UAVs, the $\mathcal{E}_t$ represent the arc set at time t, e.g., $e_{i,j} \in \mathcal{E}_t$ denote an arc from node i to node j, which means the UAV i can communicate with UAV j directly at time t. The adjacent matrix $\mathcal{A}_t = \{a_{i,j}\}$ of graph $\mathcal{G}_t$ is used to describe the communicated situation of formation in real-time, e.g., at the initial time, the adjacent matrix is as follows:

$$\mathcal{A}_0 = \begin{bmatrix} 0 & 1 & 0 & 1 & 0 & 1 \\ 1 & 0 & 1 & 0 & 0 & 0 \\ 0 & 1 & 0 & 0 & 0 & 0 \\ 1 & 0 & 0 & 0 & 1 & 0 \\ 0 & 0 & 0 & 1 & 0 & 0 \\ 1 & 0 & 0 & 0 & 0 & 0 \end{bmatrix} \tag{8}$$

The adjacent matrix is symmetric, and its element a_{ij} indicates the communication situation of UAV i and UAV j. If $a_{ij} = 1$, then $a_{ji} = 1$, and the ith and jth UAVs can share their state, and the control command can be sent from i to j or j to i. The adjacent matrix is updated in real-time. If the distance between two UAVs is greater than the communication limited distance d_{com}, the corresponding elements of the adjacent matrix will be 0.

Additionally, at the initial time, the formation is connected, and the connected component $\mathcal{W}$ is 1. l If the UAVs want to keep in communication with all the others, the graph $\mathcal{G}$ should only have one connected component. In the graph model, this condition could be transferred to $\mathcal{G}$. The methods that judge whether an undirected graph is connected include union-find disjoint sets, DFS, and BFS [41]. So, after DFS or BFS, the task fails when the connected component number $\mathcal{W}$ of graph $\mathcal{G}$ is more than 1. When the formation works, $\mathcal{W}$ should be 1.

When the formation works, the protocol should be active to support the UAVs communicating with each other. The communication protocol's primary purpose is to send all the UAVs' states to the leader for the decision; the control protocol sends the action command to all the UAVs. When the formation is as orderly as it was at first, the information only needs to obey the transfer route (shown in Figure 1), so the whole formation can be controlled well. However, when the noise disturbs the position of UAVs, it makes the connection between the UAVs that are not connected at the initial time. It breaks the connection between the UAVs that are connected at the initial time. To handle the chaos brought about by the noise, a communication and control protocol is shown in Figure 3.

In Figure 3a the communication protocol is shown, where the block in ith row represents the communication priority of the corresponding UAV. For the priority, the bigger the number, the higher the priority. Priority 1, 2 determines the order of communication. If the priority is 0, both parties have no communication probability. i.e., when the leader0 and follower3 are within the communication range of follower5, the follower5 will send the information to leader0 instead of follower3.

The protocol is designed based on the communication object: to send all the followers' state information to the leader to support the decision. So, the principle of the protocol is to give the followers closer to the leader higher priority, such as followers 1, 3 and 5.

Figure 3b has a similar meaning to the control protocol. The target of the control protocol sends the control information to all the UAVs. The control protocol motivates the leader to send the control information to the followers that connects as much as the followers. Therefore, leader0, and follower1, 2, and 5 have priority 2 because they can connect with up to 2 other followers.

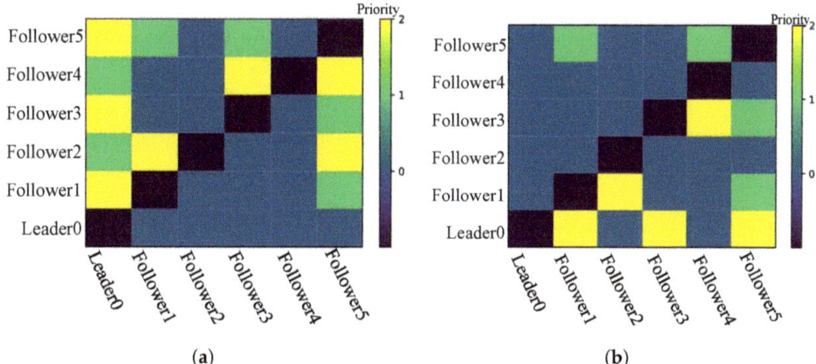

(a)   (b)

Figure 3. (a): The communication protocol of the UAVs formation; **(b)**: The control protocol of the UAVs formation.

4.3. Reward Scheme

The goal of the formation-keeping task is to reach the target area and ensure the formation is as orderly as possible. At first, the orderliness of the formation is of primary concern. So, some geometric parameters are defined to describe the formation. The followers in the formation can be divided into two categories, one is on an oblique line with the leader, like followers 3 and 4, and another is on a straight line with the leader. Only follower 5 belongs to this category. The linear between the leader and the position where the follower should be located is called the baseline (see the back lines in Figure 4). Then, it is easy to know the first category followers have a baseline with a slope, and the second follower's baseline does not. For the follower i, the length of the initial baseline is l_i, and the initial slope is $k_i = \tan\theta$ (the first category).

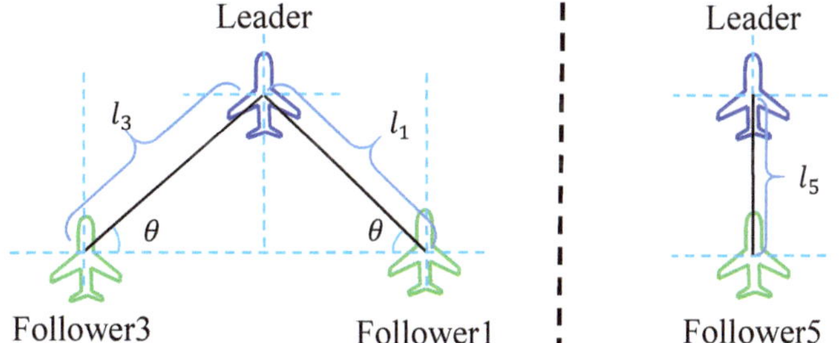

Figure 4. The communication and control protocol under the topology of the formation.

To make sure the UAV agent can return to the position that makes formation more orderly, for ith UAV, the formation reward is designed as follows:

$$R_{f,i} = -\max\{dis_{a,i}, |dis_{b,i} - l_i|\} \tag{9}$$

where $dis_{a,i}$ represents the distance between the follower i and the baseline along the vertical line of the baseline, and $dis_{b,i}$ represents the distance between the leader and follower i along the baseline. The formation reward is $R_{f,i}$.

When the UAVs belong to the first category follower (e.g., follower 3), the distance $dis_{a,3}$ can be calculated by the following formula:

$$dis_{a,3} = \frac{|x_3 \tan\theta - y_3 + (y_0 - x_0 \tan\theta)|}{\sqrt{1 + \tan\theta}} \tag{10}$$

Followers 2 and 4 have the same dis_a as in the above equation. The distance dis_b also can be obtained with the following formula:

$$dis_{b,3} = (x_3 - x_0)\cos\theta + (y_3 - y_0)\sin\theta + l_3 \tag{11}$$

For the second category follower (follower 5), it is easy to know that the reward can be represented as the following simple formation:

$$R_{f,5} = -\max\{|x_0 - x_5|, |y_0 - y_5 - l_5|\} \tag{12}$$

Furthermore, the main target of UAVs formation is to reach the target area, which is a circle with center coordinates (x_{tar}, y_{tar}) and radius r_{tar}. To encourage the formation to reach the target area, a sparse reward is designed as the destination reward:

$$R_d = \begin{cases} 0, \sqrt{(x_0 - x_{tar})^2 + (y_0 - y_{tar})^2} \le r_{tar} \\ 10{,}000, otherwise \end{cases} \tag{13}$$

We only calculate the distance of the leader. Only when the formation reaches the target area do the UAVs receive this sparse reward, and the learning process will halt. It leads to the UAVs not only needing to take minor actions to ensure that the orderly formation is not disorganized by the disturbance, but also needing to adjust direction to reach the target area. From the reward design view, UAV agents need to try different actions to discover and obtain a sparse signal. To accelerate the learning, the exploration rewards, as described in the literature [42], are designed as the incentive reward:

$$R_{e,i} = -max\{|x_i - x_{tar}|, |y_i - y_{tar}|\} \tag{14}$$

When the formation is closer to the target area, it will receive a higher exploration reward, leading the UAV agent to learn to reach the target area.

Meanwhile, some UAVs are too close and crash together, or they are too far and cease communicating with each other. In that case, the formation will suffer permanent destruction, and the task will halt.

Setting the minimum distance for crashes makes it easy to obtain the halt condition of UAV crashes. Then, the penalty should be added to avoid the above situation. This penalty is designed as a formal sparse reward as follows:

$$R_p = \begin{cases} -10{,}000, d_{i,j} \le d_{cra}, \forall i, j = 0, 1, \cdots 5 \\ -10{,}000, \mathcal{W} > 1 \\ 0, others \end{cases} \tag{15}$$

where the $d_{.,j}$ represents the minimum distance between the jth UAV and another five UAVs: $d_{.,j} = \min_i\{d_{i,j}\}, \forall i = 1, \cdots, 6, i \ne j$. The lowest communication distance is d_{com}, once the minimum distance $d_{.,j}$ less than d_{com}, the jth UAV will lose the communication ability with other UAVs. In addition, d_{cra} is the crash distance; as long as the distance between two UAVs is less than this, the two UAVs might crash.

Finally, the reward of the formation system at time T can be represented as the sum of the following reward function:

$$R(T) = \sum_{i=1}^{6} \left[R_{f,i}(T) + R_{e,i}(T)\right] + R_d(T) + R_p(T) \tag{16}$$

5. PPO-Exp

PPO is one of the most popular deep reinforcement learning algorithms in continuous tasks that achieved outstanding performance. The PPO embedded the Actor–Critic algorithm, which uses a deep neural network as an Actor for policy generation, and another deep neural network as a Critic for policy estimating. The structure of PPO can be seen in

Figure 5; the Actor interacts with the environment, collects the trajectories: $\{s_t, a_t, r_t, s_{t+1}\}$ and stores them in the buffer, then it uses the buffer and the value function estimated by the Critic to optimize the Actor network's hyperparameter according to following surrogate:

$$\mathcal{L}_t^{\text{Clip},\theta} = \begin{cases} (1+\varepsilon)A_{\pi_{\theta_{t-1}}}; A_{\pi_{\theta_{t-1}}} > 0, r_t > 1 + \varepsilon \\ (1-\varepsilon)A_{\pi_{\theta_{t-1}}}; A_{\pi_{\theta_{t-1}}} < 0, r_t < 1 - \varepsilon \\ r_t \cdot A_{\pi_{\theta_{t-1}}}; otherwise \end{cases} \tag{17}$$

where the $A_{\pi_{\theta_{t-1}}}$ is the advantage function defined in Equation (3). The Critic network's hyperparameter ϕ is updated by minimizing the following MSE error:

$$\mathcal{L}_t^{\text{Clip},\phi} = \sum_t \left(y_t - Q_\phi(s_t, a_t)\right)^2 \tag{18}$$

$$y_t = r_t + \gamma \cdot Q_\phi(s_{t+1}, \pi_{\theta_{old}}(s_{t+1})) \tag{19}$$

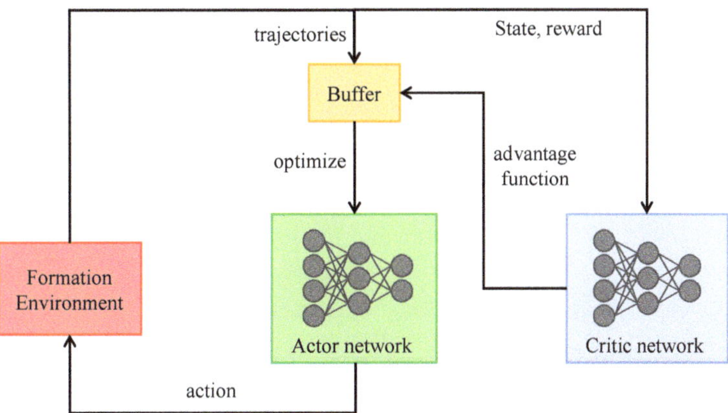

Figure 5. The structure of PPO with experience replay.

The gradient of Equations (17) and (18) is computed and used to update the hyperparameters θ and ϕ until they converge or reach maximum steps. In surrogate (17), the PPO restricted the difference between new and old policy by using the clip trick to restrain the ratio $r_t = \frac{\pi_\theta(s_t, a_t)}{\pi_{\theta_{old}}(s_t, a_t)}$. It could be considered a constraint on updated policy; under it, the ratio should satisfy the following constraint: $1 - \varepsilon \leq r_t \leq 1 + \varepsilon$. Then, the updated policy is restricted as follows:

$$\frac{|\pi_\theta(s_t, a_t) - \pi_{\theta_{old}}(s_t, a_t)|}{\pi_{\theta_{old}}(s_t, a_t)} \leq \varepsilon \tag{20}$$

The coefficient ε is also a constant in the range $(0, 1)$ in PPO-Clip; from the inequality (20), it can be seen that the relative deviation is bound between $\pi_{\theta_{old}}$ and π_θ. When this deviation is under ε, as the increase in r_t is observed, the $\mathcal{L}_t^{\text{Clip},\theta}$ increase as well, but when the deviation exceeds ε, even if the r_t is increases, the $\mathcal{L}_t^{\text{Clip},\theta}$ maintains its value. It shows the exploration within the constraint ε; however, when the relative difference is beyond ε, the exploration is not encouraged by clipping the result to $(1+\varepsilon)A_{\pi_{\theta_{old}}}$. Figure 6 shows the surrogate of PPO-Clip in different ε. The large ε could encourage the agent to explore more and accept more policies. However, enlarging ε will lead to the estimated error of the surrogate. The PPO-Clip is the off-policy algorithm. The data generated by the old policy will be used as new policy updates. When $\|r_t - 1\| \leq \varepsilon$, the estimated error bound of $\mathcal{L}^{\text{Clip},\theta}$ will increase as ε increases. For convenience, denote the following assumption:

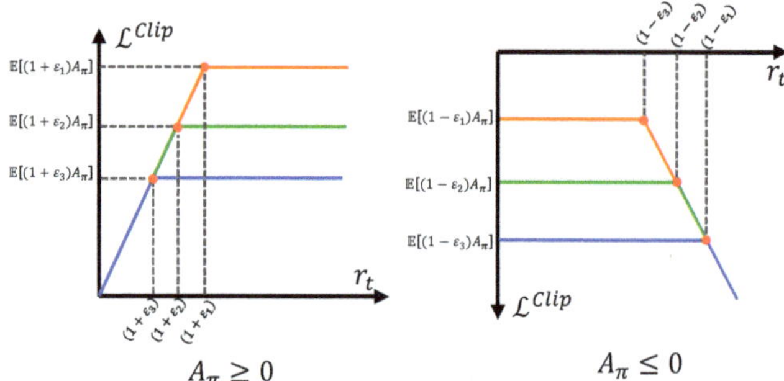

Figure 6. The surrogate of PPO-Clip in different ε. The relationship of different ε: $\varepsilon_3 > \varepsilon_2 > \varepsilon_1$.

Assumption 1. *In the previous t timestep of policy update, the ratio r_k satisfies $\|r_k - 1\| \leq \varepsilon$, $\forall k = 1, \cdots, t$.*

Under Assumption 1, the following Lemma is given for auxiliary proof of the error bound:

Lemma 1. *Under Assumption 1, the difference of state distribution resulting from the policy satisfies the following inequality:*

$$\|\rho^{\pi_{\theta_t}} - \rho^{\pi_{\theta_{t-1}}}\| \leq \frac{\varepsilon \cdot \gamma}{1 - \gamma} \tag{21}$$

Proof. The distribution ρ^{π_θ} can be rewritten as [43]:

$$\rho^{\pi_\theta} = (1 - \gamma) \sum_{k=0}^{\infty} \gamma^k \cdot d_{\pi_\theta}^k \tag{22}$$

where $d_{\pi_\theta}^k$ is the distribution resulting from π_θ at k timestep. Using the Markov property, $\forall s' \in \mathcal{S}$, the $d_{\pi_\theta}^k(s')$ could be decompose as follows:

$$d_{\pi_\theta}^k(s') = \sum_{s,a} d_{\pi_\theta}^{k-1}(s) \cdot \pi_\theta(a|s) \cdot P(s'|s,a) \tag{23}$$

Using the decomposition, the following equation holds:

$$\begin{aligned}
d_{\pi_{\theta_t}}^k(s') - d_{\pi_{\theta_{t-1}}}^k(s') &= \sum_{s,a} \left[d_{\pi_{\theta_t}}^{k-1}(s) \cdot \pi_{\theta_t}(a|s) - d_{\pi_{\theta_{t-1}}}^{k-1}(s)\pi_{\theta_{t-1}}(a|s) \right] P(s'|s,a) \\
&= \sum_{s,a} \left[d_{\pi_{\theta_t}}^{k-1}(s) \cdot \pi_{\theta_t}(a|s) - d_{\pi_{\theta_{t-1}}}^{k-1}(s)\pi_{\theta_{t-1}}(a|s) + d_{\pi_{\theta_{t-1}}}^{k-1}(s)\pi_{\theta_{t-1}}(a|s) - d_{\pi_{\theta_{t-1}}}^{k-1}(s)\pi_{\theta_{t-1}}(a|s) \right] \\
&\quad \cdot P(s'|s,a) \\
&= \sum_{s,a} \left[\pi_{\theta_t}(a|s) - \pi_{\theta_{t-1}}(a|s) \right] \cdot d_{\pi_{\theta_{t-1}}}^{k-1}(s) \cdot P(s'|s,a) \\
&\quad + \sum_{s,a} \left[d_{\pi_{\theta_{t-1}}}^{k-1}(s) - d_{\pi_{\theta_{t-1}}}^{k-1}(s) \right] \cdot \pi_{\theta_{t-1}}(a|s) \cdot P(s'|s,a)
\end{aligned} \tag{24}$$

Using the triangle inequality, the following equation hold:

$$\sum_{s,a} \|\pi_{\theta_t}(a|s) - \pi_{\theta_{t-1}}(a|s)\| \cdot d_{\pi_{\theta_t}}^{k-1}(s)P(s'|s,a) + \sum_{s,a} \|d_{\pi_{\theta_t}}^{k-1}(s) - d_{\pi_{\theta_{t-1}}}^{k-1}(s)\| \cdot \pi_{\theta_{t-1}}(a|s) \cdot P(s'|s,a)$$

$$\geq \| \sum_{s,a} \left[\pi_{\theta_t}(a|s) - \pi_{\theta_{t-1}}(a|s)\right] \cdot d_{\pi_{\theta_{t-1}}}^{k-1}(s) \cdot P(s'|s,a) \tag{25}$$

$$+ \sum_{s,a} \left[d_{\pi_{\theta_{t-1}}}^{k-1}(s) - d_{\pi_{\theta_{t-1}}}^{k-1}(s)\right] \cdot \pi_{\theta_{t-1}}(a|s) \cdot P(s'|s,a)\| = \|d_{\pi_{\theta_t}}^{k}(s') - d_{\pi_{\theta_{t-1}}}^{k}(s')\|$$

Sum up the inequality (26) to calculate the expectation on s':

$$
\begin{aligned}
&\|d_{\pi_{\theta_t}}^{k} - d_{\pi_{\theta_{t-1}}}^{k}\| \\
&= \sum_{s'} \|d_{\pi_{\theta_t}}^{k}(s') - d_{\pi_{\theta_{t-1}}}^{k}(s')\| \leq \sum_{s,a} \|\pi_{\theta_t}(a|s) - \pi_{\theta_{t-1}}(a|s)\| \cdot d_{\pi_{\theta_t}}^{k-1}(s) \sum_{s'} P(s'|s,a) \\
&\quad + \sum_{s,a} \|d_{\pi_{\theta_t}}^{k-1}(s) - d_{\pi_{\theta_{t-1}}}^{k-1}(s)\| \cdot \pi_{\theta_{t-1}}(a|s) \cdot \sum_{s'} P(s'|s,a) \\
&= \|\pi_{\theta_t}(a|s) - \pi_{\theta_{t-1}}(a|s)\| + \|d_{\pi_{\theta_t}}^{k-1} - d_{\pi_{\theta_{t-1}}}^{k-1}\| \\
&\leq \| \frac{\pi_{\theta_t}(a|s) - \pi_{\theta_{t-1}}(a|s)}{\pi_{\theta_{t-1}}(a|s)}\| \cdot \|\pi_{\theta_{t-1}}(a|s)\| + \|d_{\pi_{\theta_t}}^{k-1} - d_{\pi_{\theta_{t-1}}}^{k-1}\| \\
&\leq \| \frac{\pi_{\theta_t}(a|s) - \pi_{\theta_{t-1}}(a|s)}{\pi_{\theta_{t-1}}(a|s)}\| + \|d_{\pi_{\theta_t}}^{k-1} - d_{\pi_{\theta_{t-1}}}^{k-1}\| \leq \varepsilon + \|d_{\pi_{\theta_t}}^{k-1} - d_{\pi_{\theta_{t-1}}}^{k-1}\| \\
&\leq 2\varepsilon + \|d_{\pi_{\theta_t}}^{k-2} - d_{\pi_{\theta_{t-1}}}^{k-2}\| \leq k\varepsilon
\end{aligned}
\tag{26}
$$

Using Equation (22), the following equation holds:

$$\|\rho^{\pi_{\theta_t}} - \rho^{\pi_{\theta_{t-1}}}\| \leq \frac{1}{\gamma - 1} \sum_{k=0}^{\infty} \gamma^k \|d_{\pi_{\theta_t}}^{k} - d_{\pi_{\theta_{t-1}}}^{k}\|$$

$$\leq \frac{1}{\gamma - 1} \sum_{k=0}^{\infty} \gamma^k \cdot k \cdot \varepsilon = \frac{\varepsilon \cdot \gamma}{1 - \gamma} \tag{27}$$

$\square$

Using this Lemma, the estimation error of the PPO-Clip could be obtained:

Theorem 1. *Under the Assumption 1, the estimation error of PPO-Clip is satisfied:*

$$Err\left[\mathcal{L}^{Clip,\theta}\right] = Err\left[\mathbb{E}_{\pi_{\theta_{old}}}\left[\frac{\pi_\theta}{\pi_{\theta_{old}}} A_{\pi_{\theta_{old}}}\right]\right] \leq \frac{\varepsilon \cdot \gamma}{1 - \gamma} \mathbb{E}_{s \sim Unif_S, a \sim \pi_\theta}[A_\pi(s,a)] \tag{28}$$

Proof. When $\|r_t - 1\| \leq \varepsilon$, the surrogate of the PPO-Clip will be degraded [40]:

$$\mathcal{L}^{Clip,\theta} = \mathbb{E}_{\pi_{\theta_{old}}}\left[\frac{\pi_\theta}{\pi_{\theta_{old}}} A_{\pi_{\theta_{old}}}\right], \|\frac{\pi_\theta - \pi_{\theta_{old}}}{\pi_{\theta_{old}}}\| \leq \varepsilon \tag{29}$$

The above surrogate is the importance sampling estimator of the objective of the new policy [44]:

$$\mathbb{E}_{\pi_{\theta_{old}}}\left[\frac{\pi_\theta}{\pi_{\theta_{old}}} A_\pi(s, \pi_{\theta_{old}}(s))\right] \approx \mathbb{E}_{\pi_\theta}[A_\pi(s, \pi_\theta(s))] \tag{30}$$

However, the estimator uses the data generated by $\pi_{\theta_{old}}$, and the state distribution of $\mathcal{L}^{Clip,\theta}$ is derived from $\rho^{\pi_{\theta_{old}}}$. Therefore, the estimation error is satisfied:

$$\text{Err}\left[\mathbb{E}_{\pi_{\theta_{old}}}\left[\frac{\pi_\theta}{\pi_{\theta_{old}}}A_\pi(s,\pi_{\theta_{old}}(s))\right]\right] = \left\|\mathbb{E}_{\pi_{\theta_{old}}}\left[\frac{\pi_\theta}{\pi_{\theta_{old}}}A_\pi(s,\pi_{\theta_{old}}(s))\right] - \mathbb{E}_{\pi_\theta}[A_\pi(s,\pi_\theta(s))]\right\|$$

$$= \left\|\int_s \rho^{\pi_{\theta_{old}}}(s)\int_{a\sim\pi_{\theta_{old}}}\frac{\pi_\theta(a|s)}{\pi_{\theta_{old}}(a|s)}A_\pi(s,a)dads - \int_s \rho^{\pi_\theta}(s)\int_{a\sim\pi_\theta}A_\pi(s,a)dads\right\|$$

$$\leq \int_s \left\|\rho^{\pi_{\theta_{old}}}(s)-\rho^{\pi_\theta}(s)\right\|\int_{a\sim\pi_\theta}\|A_\pi(s,a)\|dads \tag{31}$$

Consider the positive advantage situation and expand the integral of a; the following equation will hold:

$$\text{Err}\left[\mathbb{E}_{\pi_{\theta_{old}}}\left[\frac{\pi_\theta}{\pi_{\theta_{old}}}A_\pi(s,\pi_{\theta_{old}}(s))\right]\right] \leq \int_s \left\|\rho^{\pi_{\theta_{old}}}(s)-\rho^{\pi_\theta}(s)\right\|\int_a \pi_\theta(a|s)A_\pi(s,a)dads \tag{32}$$

Using the conclusion of Lemma 1, the following error bound could be obtained:

$$\text{Err}\left[\mathbb{E}_{\pi_{\theta_{old}}}\left[\frac{\pi_\theta}{\pi_{\theta_{old}}}A_\pi(s,\pi_{\theta_{old}}(s))\right]\right] \leq \int_s \frac{\varepsilon\cdot\gamma}{1-\gamma}\int_a \pi_\theta(a|s)A_\pi(s,a)dads$$

$$= \int_s \frac{\varepsilon\cdot\gamma}{1-\gamma}\cdot|\mathcal{S}|\cdot\frac{1}{|\mathcal{S}|}\int_a \pi_\theta(a|s)A_\pi(s,a)dads$$

$$= \frac{\varepsilon\cdot\gamma}{1-\gamma}\cdot|\mathcal{S}|\cdot\int_s \frac{1}{|\mathcal{S}|}\int_a \pi_\theta(a|s)A_\pi(s,a)dads$$

$$= \frac{\varepsilon\cdot\gamma}{1-\gamma}\cdot|\mathcal{S}|\cdot\mathbb{E}_{s\sim\text{Unif}_\mathcal{S},a\sim\pi_\theta}[A_\pi(s,a)] \tag{33}$$

where the $\text{Unif}_\mathcal{S}$ represents the uniform distribution of the state. $\square$

Theorem 1 confirms the positive relationship between the estimation error and ε. By using it, a more clear conclusion could be obtained:

Remark 1. *In PPO-Clip, the high ε could enhance the exploration but will result in a high estimation error bound of the surrogate; the low ε could decrease the error bound but will restrict the exploration.*

Therefore, to deal with the exploration and estimation error problems mentioned in Remark 1, this paper considers making the ε adaptive in different situations. The last part designed the sparse reward R_d, and the exploration reward R_e is designed as the incentive reward. The agent should explore more in the task to receive a high-level R_d and R_e. So, when these rewards are too low, the agent should release the restriction on r_t to encourage the exploration. When these rewards are high and stable, the restriction on r_t increases to ensure the estimation of the surrogate is accurate.

So, the exploration advantage function $A_\pi^{exp}(s_t,a_t)$ can be used to represent the advantage function that is estimated by R_d and R_e, which can reflect the exploration ability of the agent:

$$A_\pi^{exp}(s_t,a_t) = E_\pi\left[\sum_{k=0}^\infty \gamma^k(R_d(t+k)+\sum_{i=1}^6 R_{e,i}(t+k))|S_t=s,a_t=a\right] -$$

$$E_\pi\left[\sum_{k=0}^\infty \gamma^k(R_d(t+k)+\sum_{i=1}^6 R_{e,i}(t+k))|S_t=s\right] \tag{34}$$

According to the exploration function, an exploration PPO algorithm is proposed with an adaptive clip parameter ε. When the exploration advantage function is lower than last time, to improve the exploration ability, ε will be enlarged. Otherwise, the ε will be reduced, restraining the updated policy in a trust region. To sum up, the adaptive mechanism is designed as follows:

$$\varepsilon(t) = \begin{cases} \varepsilon(t-1) - clip(\dfrac{A^{exp}_{\pi_{\theta_t}} - A^{exp}_{\pi_{\theta_{t-1}}}}{A^{exp}_{\pi_{\theta_{t-1}}}}, 0, \dfrac{\varepsilon(t-1)}{2}); \ A^{exp}_{\pi_{\theta_t}} - A^{exp}_{\pi_{\theta_{t-1}}} > 0 \\[3ex] \varepsilon(t-1) + clip(\dfrac{A^{exp}_{\pi_{\theta_t}} - A^{exp}_{\pi_{\theta_{t-1}}}}{A^{exp}_{\pi_{\theta_{t-1}}}}, 0, \dfrac{\varepsilon(t-1)}{2}); \ A^{exp}_{\pi_{\theta_t}} - A^{exp}_{\pi_{\theta_{t-1}}} < 0 \\[3ex] \varepsilon(t-1); otherwise \end{cases} \tag{35}$$

The clip function in the above equations is to restrict the adaptive mechanism and avoid the ε being abnormal. Through the variation of the exploration advantage function, the exploration-based adaptive ε mechanism is proposed. When simply replacing the constant ε with the adaptive ε, the PPO will be PPO-Explorationε(PPO-Exp). With the restriction of old policies, new policies will be adjusted automatically. The surrogate of the PPO-Exp is as follows:

$$\mathcal{L}^{Exp,\theta} = \mathbb{E}_{\pi_{\theta_{old}}} \left[\min \left(r_t(\theta) A_{\pi_{\theta_{old}}}, clip(r_t(\theta), 1 - \varepsilon(t), 1 + \varepsilon(t)) A_{\pi_{\theta_{old}}} \right) \right] \tag{36}$$

The Algorithm of PPO-Exp in the formation environment could be seen in Algorithm 1. The exploration and estimation error problem in PPO-Exp could be adapted without delay, and the following Proposition will give the exploration range and the estimation error decrease rate in different situations:

Algorithm 1 PPO-Exploration ε with formation keeping task.

Initialize π_0, ϕ_0.
for $i = 0, 1, 2, \ldots N$ **do**
    for $t = 1, \cdots, T$ **do**
        The leader0 collects state information $\{s_{t,i} | i = 1, \cdots, 5\}$ through the communication protocol (Figure 3a)
        Run policy π_θ, obtain the action $\{a_{t,i} | i = 0, 1, \cdots, 5\}$, and send them using the control protocol (Figure 3b).
        The leader and followers execute the action commands and receive a reward as follows: $(R_f(t), R_e(t), R_d(t), R_p(t))$
        Store (s_t, a_t, s_{t+1}, R_t) at the buffer.
    end for
  Transitions data from buffer, and estimate $\hat{A}_{\pi_{\theta_t}}$, $\hat{A}^{exp}_{\pi_{\theta_t}}$, respectively.
    if $\hat{A}^{exp}_{\pi_{\theta_t}} - \hat{A}^{exp}_{\pi_{\theta_{t-1}}} > 0$ **then**
$$\varepsilon(t) = \varepsilon(t-1) - clip(\frac{\hat{A}^{exp}_{\pi_{\theta_t}} - \hat{A}^{exp}_{\pi_{\theta_{t-1}}}}{\hat{A}^{exp}_{\pi_{\theta_{t-1}}}}, 0, \frac{\varepsilon(t-1)}{2})$$
    end if
    if $\hat{A}^{exp}_{\pi_{\theta_t}} - \hat{A}^{exp}_{\pi_{\theta_{t-1}}} < 0$ **then**
$$\varepsilon(t) = \varepsilon(t-1) + clip(\frac{\hat{A}^{exp}_{\pi_{\theta_t}} - \hat{A}^{exp}_{\pi_{\theta_{t-1}}}}{\hat{A}^{exp}_{\pi_{\theta_{t-1}}}}, 0, \frac{\varepsilon(t-1)}{2})$$
    end if
    for $j = 1, \cdots, M$ **do**
        $\hat{\mathcal{L}}_\theta = \sum_{t=1}^{T} \min(r_t \cdot \hat{A}_{\pi_{\theta_t}}, clip(1 - \varepsilon, 1 + \varepsilon, r) \hat{A}_{\pi_{\theta_t}})$
        Update θ by SGD or Adam.
    end for
  Update critic network parameter ϕ_t by minimizing:
  $\sum_{k=1}^{T}(\sum_{t'>k} \gamma^{t'-t} R_t - V_\phi(s_t))^2$
end for

Proposition 1. *In PPO-Exp, when $\hat{A}^{exp}_{\pi_{\theta_t}} - \hat{A}^{exp}_{\pi_{\theta_{t-1}}} < 0$, the exploration range of next policy will be expanded to $\frac{\|\pi_{\theta_t} - \pi_{\theta_{t-1}}\|}{\|\pi_{\theta_{t-1}}\|} \leq \varepsilon + \frac{\hat{A}^{exp}_t - \hat{A}^{exp}_{t-1}}{\hat{A}^{exp}_{t-1}} \leq \frac{3\varepsilon(t-1)}{2}$; when $\hat{A}^{exp}_{\pi_{\theta_t}} - \hat{A}^{exp}_{\pi_{\theta_{t-1}}} > 0$, in next update, the error bound of the surrogate will decrease to $O(\frac{\varepsilon(t-1)}{2})$.*

Proof. When $\hat{A}^{exp}_{\pi_{\theta_t}} - \hat{A}^{exp}_{\pi_{\theta_{t-1}}} < 0$, according to Equation (35), it is easy to see the next policy will be expanded to $\frac{\|\pi_{\theta_t} - \pi_{\theta_{t-1}}\|}{\|\pi_{\theta_{t-1}}\|} \leq \varepsilon + clip(\frac{\hat{A}^{exp}_t - \hat{A}^{exp}_{t-1}}{\hat{A}^{exp}_{t-1}}, 0, \frac{\varepsilon(t-1)}{2})$. Then, the following inequality will hold:

$$0 \leq clip(\frac{\hat{A}^{exp}_t - \hat{A}^{exp}_{t-1}}{\hat{A}^{exp}_{t-1}}, 0, \frac{\varepsilon(t-1)}{2}) \leq \frac{\varepsilon(t-1)}{2} \tag{37}$$

So, the following inequality is held:

$$\frac{\|\pi_{\theta_t} - \pi_{\theta_{t-1}}\|}{\|\pi_{\theta_{t-1}}\|} \leq \varepsilon + \frac{\varepsilon(t-1)}{2} = \frac{3\varepsilon(t-1)}{2} \tag{38}$$

When $\hat{A}^{exp}_{\pi_{\theta_t}} - \hat{A}^{exp}_{\pi_{\theta_{t-1}}} > 0$, and Assumption 1 is satisfied, it is obvious that the conclusion of Theorem 1 could be used in PPO-Exp. So, using Equation (35) and Theorem 1, the PPO-Exp's decrease rate of the bound is as follows:

$$\begin{aligned}
\Delta\text{Err}\left[\mathcal{L}^{\text{Exp}}_\theta\right] &= \text{Err}\left[\mathbb{E}_{\pi_{\theta_{t-1}}}\left[\frac{\pi_{\theta_t}}{\pi_{\theta_{t-1}}} A_{\pi_{\theta_{t-1}}}\right]\right] - \text{Err}\left[\mathbb{E}_{\pi_{\theta_{t-2}}}\left[\frac{\pi_{\theta_{t-1}}}{\pi_{\theta_{t-2}}} A_{\pi_{\theta_{t-2}}}\right]\right] \\
&\leq \gamma\frac{\varepsilon(t) - \varepsilon(t-1)}{1-\gamma} \cdot |\mathcal{S}| \cdot \left\|\mathbb{E}_{s\sim\text{Unif}_\mathcal{S},a\sim\pi_{\theta_t}}[A_\pi(s,a)] - \mathbb{E}_{s\sim\text{Unif}_\mathcal{S},a\sim\pi_{\theta_{t-1}}}[A_\pi(s,a)]\right\| \\
&\leq \gamma\frac{\varepsilon(t) - \varepsilon(t-1)}{1-\gamma} \cdot |\mathcal{S}| \cdot \Gamma \\
&\leq \gamma\frac{(\varepsilon(t-1) + clip(\frac{\hat{A}^{exp}_t - \hat{A}^{exp}_{t-1}}{\hat{A}^{exp}_{t-1}}, 0, \frac{\varepsilon(t-1)}{2})) - \varepsilon(t-1)}{1-\gamma} \cdot |\mathcal{S}| \cdot \Gamma \\
&\leq \gamma\frac{\frac{3\varepsilon(t-1)}{2}}{1-\gamma} \cdot |\mathcal{S}| \cdot \Gamma = O(\frac{\varepsilon(t-1)}{2})
\end{aligned} \tag{39}$$

where the Γ is the upper bound of advantage:

$$\Gamma = \max_{\forall t}\left\|\mathbb{E}_{s\sim\text{Unif}_\mathcal{S},a\sim\pi_{\theta_t}}[A_\pi(s,a)] - \mathbb{E}_{s\sim\text{Unif}_\mathcal{S},a\sim\pi_{\theta_{t-1}}}[A_\pi(s,a)]\right\| \tag{40}$$

$\square$

Proposition 1 indicates that the PPO-Exp could encourage the agents to adjust the exploration in different situations. The next section will validate it through numerical experiments.

6. Numerical Experiments

This section compares the PPO-Exp with four common reinforcement learning algorithms (PPO-Clip, PPO-KL, TD3, DDPG) in the formation-keeping task, and compared the performace of PPO-Exp and PPO-Clip in the formation changing task and obstacle avoidance task.

6.1. Experimental Setup

In terms of hardware, all the experiments are completed on the Windows 10 (64-bit) operating system, Intel(R) Core i7 processor, 16 GB memory, and 4 GB video memory. As for software, OpenAI-gym [45] is used to design the reinforcement learning environment and the physics rulers of the UAVs' formation.

The formation task is modeled on the OpenAI gym environment. See Figure 1; the position of the leader and followers can be seen in Table 1. The formation is updated by the dynamic equations solved by the difference method per 0.5 s per time mesh grid. The environment noises are set as $N(0,1)$ default. The target area is designed as a circle at $(200, 400)$ with a radius of 40.

Table 1. The initial position of UAVs' formation.

	Leader0	Follower1	Follower2	Follower3	Follower4	Follower5
Position X	160	190	220	130	100	160
Position Y	190	160	100	160	100	130

6.2. Experiments on PPO-Exploration ε

The following famous continuous space RL algorithms are explored in this section: TD3, DDPG, PPO-KL, and PPO-clip; they are compared to the proposed method under the formation-keeping task.

- PPO-Clip [40]: Proximal Policy Optimization with Clip(PPO-Clip) function.
- PPO-KL [40]: Proximal Policy Optimization with KL-divergence(PPO-KL) constrain.
- DDPG [46]: Deep Deterministic Policy Gradient(DDPG) algorithm, which is a continuous action deep reinforcement learning algorithm that uses Actor–Critic architecture. In DDPG, the deterministic policy gradient is used to update the Actor parameter.
- TD3 [47]: Twin Delayed Deep Deterministic (TD3) policy gradient algorithm, which is a variant of DDPG. The TD3 introduced the delaying policy updates mechanism and the double network architecture to manage the per-update error and overestimation bias in DDPG.

The main hyperparameters of the contrast experiment are shown in Table 2. The blank area in the above table means the algorithm does not include this parameter.

Table 2. The main hyperparameters of the algorithm used in the experiment.

Parameter Name	TD3	DDPG	PPO-KL	PPO-Clip	PPO-Exp
γ	0.9	0.9	0.9	0.9	0.9
A_{LR}	0.00005	0.00005	0.00005	0.00005	0.00005
C_{LR}	0.0002	0.0002	0.0002	0.0002	0.0002
Batch	32	32	32	32	32
A_{US}			10	10	10
C_{US}			10	10	10
EPS			10^{-8}	10^{-8}	10^{-8}
$D_{KL}(target)$			0.01		
λ			0.5		
ε_{clip}			0.1	0.1	0.1
τ_{DDPG}		0.01			
VAR_{DDPG}		3			
Explore Step	500				
dim_{HIDDEN}	32				

Set the episode length be 200; the results of PPO-Exploration ε and other comparing algorithms are shown in Figure 7a. As the learning curves indicated, the PPO series methods achieved better performance; in all variations of PPO, the PPO-Exp has the best performance. It is validated that the adaptive mechanism based on exploration makes sense during policy updating. Figure 7b shows the change of ε; the series $\varepsilon(t)$ is stationary, and varies around 0.05, although the initial value is 0.1, which means 0.05 is the balance point between exploration and exploitation found by PPO-Exp. Meanwhile, the episode reward curve of PPO-Exp is higher than PPO-Clip's, validating the idea that exploration from PPO-Exp is efficient.

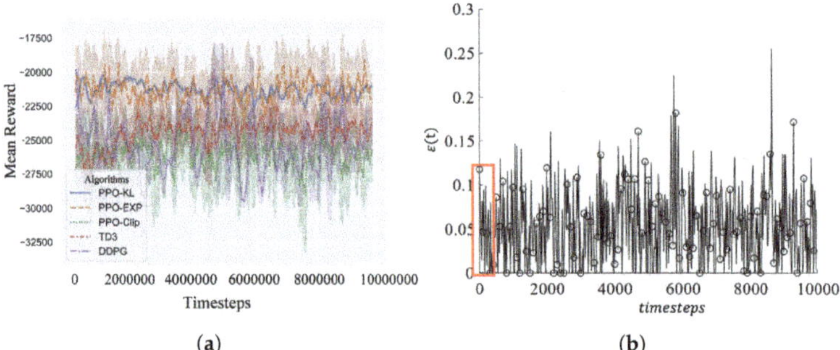

Figure 7. (**a**): Learning curves of TD3, DDPG, PPO-KL, PPO-Clip, and PPO-Exp; (**b**): The on of ε of PPO-Exploration ε during the training process.

6.3. Experiments on Formation Keeping

Only the learning curve was unable to declare whether the algorithm works well, so the trained PPO-Exp is used to perform 200s; the formation track can be seen in Figure 8. In this way, there is only a slight distortion in the formation, indicating that PPO-Exp can perform better in real tasks than PPO-Clip.

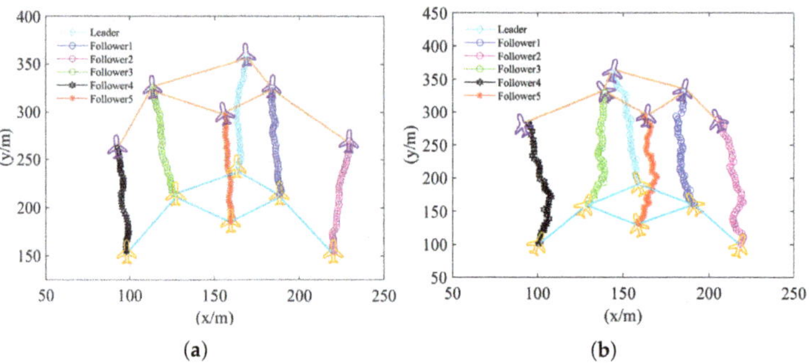

Figure 8. (**a**): The flight track of formation that is controlled by trained PPO-Exp ; (**b**): The flight track of formation that is controlled by trained PPO-Clip.

Furthermore, to evaluate the results, we plotted the heading ψ and the velocity v during 200 s in Figure 9. Figure 9a shows that followers 1, 4, and 5 are approaching gradually as time goes on. Followers 2, 3 and the leader, have no such trend to converge gradually; however, all the heading deviations are no more than $10°$. In Figure 9b, the velocity of each UAV is shown. The velocities of followers 1, 3, 4, and 5 diverge a little and then converge. Corresponding to Figure 9a, followers 1, 4, and 5 are closer in terms of the value of velocity and heading; the leader and follower 2 are far away from these followers, but the velocity difference is not more than 1.5 m/s as well. This inspired us to design the reward based on the velocity and heading.

To illustrate the influence of environmental noise on formation keeping, the results show the formation track with no control in Figure 2a. To verify that the proposed centralized method saves time, this section further compares the decentralized version of PPO-Exp: PPO-Exp-Dec, which, similar to MAPPO, needs all six UAV agents to learn the control policy at the same time.

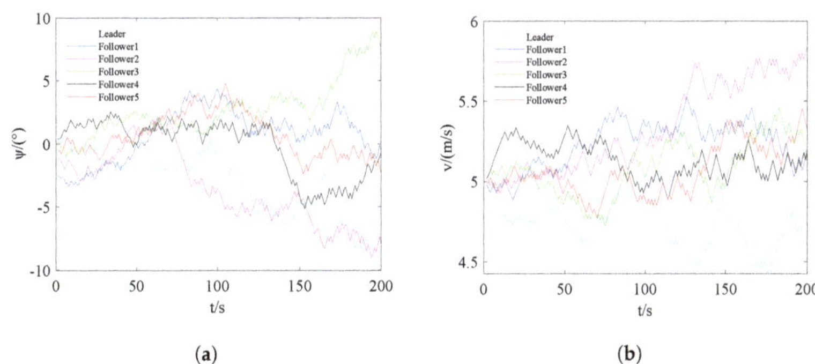

Figure 9. (**a**) The test results in the heading angle of PPO-Exp; (**b**) The test results in the velocity of PPO-Exp.

To validate that the protocol can reduce the communication cost and avoid placing the UAVs out of the communication range, this section also compares the protocol-free version: PPO-exp-pro. The results can be seen in Table 3. Γ represents the episode reward, T represents time per episode, r_{col} and r_{fai} represents the collision rate and failure to communicate rate, respectively.

Table 3. The experimental results in different algorithms.

Algorithm	Γ	T	$r_{coll}(\%)$	$r_{fail}(\%)$
PPO-exp	$-19{,}197.2 \pm 1307.4$	2.19 ± 0.04	0.93 ± 0.01	0.32 ± 0.02
PPO-exp-dec	$-20{,}374.7 \pm 1926.4$	10.06 ± 0.08	1.01 ± 0.02	0.35 ± 0.01
PPO-exp-pro	$-23{,}001.3 \pm 2507.2$	2.43 ± 0.03	0.98 ± 0.03	12.48 ± 1.76
PPO-clip	$-20{,}305.7 \pm 1588.6$	2.14 ± 0.06	0.97 ± 0.02	0.94 ± 0.03
Greedy	$-39{,}074.5 \pm 3806.5$	1.15 ± 0.04	12.32 ± 1.32	10.56 ± 0.65

To further verify the effectiveness of the proposed method, ablation experiments are performed (see Figures 2a,b and 8b). Figure 8b shows the trained PPO-clip without the exploration mechanism. Although there is no UAV crash, the leader and follower3 are very close, and the formation is not as orderly as the PPO-Exp. Figure 2a shows the result of no action taken, where the UAVs will crash, and the formation will break up. Figure 2b shows the trained PPO-clip with $\varepsilon = 0.05$, which is the balance point in the PPO-Exp. However, the experimental result shows it performs worse; there is one follower that loses communication with leader, and one follower almost crashes with the leader. The result illustrates that the PPO-Exp with adaptive ε is better than the PPO-Clip with a good ε. In summary, the ablation experiments also indicated that PPO-Exp performs better than other algorithms in terms of learning curves and the real-task.

6.4. Experiment on More Complex Tasks

To further show the efficiency of PPO-Exp in fixed-wing UAV formation keeping, this part design two more complex scenarios: formation changing and obstacle avoidance task, the UAV formation perform 120 s on each task. This part mainly compared the performance of PPO-Exp and PPO-Clip on these tasks.

The goal for the formation changing task is changing the formation shown in Figure 1 to the vertical formation. The vertical formation also expects the differences between leader and followers are as small as possible in coordinates on the x-axis. For guiding the followers to change the formation, this paper utilizes the absolute difference value of x

coordinates to modify the flocking reward. The modified flocking reward (9) and (12) could be represented as follows:

$$R_{f,i} = \|x_0 - x_i\|, \forall i = 1, \cdots, 5 \tag{41}$$

Then the total reward (47) can be rewritten as follows:

$$R(T) = \sum_{i=0}^{5}[\|x_0(T) - x_i(T)\| + R_{e,i}(T)] + R_d(T) + R_p(T) \tag{42}$$

where the $x_0(T), x_i(T)$ represent the x coordinates of leader and ith follower at time T, respectively. To encourage the UAV system to take more exploration on forming new formation, the flocking reward is added to the exploration advantage function:

$$A_\pi^{exp}(S_t, a_t) = E_\pi\left[\sum_{k=0}^{\infty}\gamma^k(R_d(t+k) + \sum_{i=0}^{5}[\|x_0(T) - x_i(T)\| + R_{e,i}(t+k))]|S_t = s, a_t = a\right] -$$

$$E_\pi\left[\sum_{k=0}^{\infty}\gamma^k(R_d(t+k) + \sum_{i=0}^{5}[\|x_0(T) - x_i(T)\| + R_{e,i}(t+k))]|S_t = s\right] \tag{43}$$

Training the task with PPO-Exp and PPO-Clip, the training parameters are kept as same as in the previous part except episode length. After training, the test result of PPO-Exp is shown in Figure 10a, and the PPO-Clip is shown in Figure 10b. To evaluate the performance, this paper draws the plots of the x coordinates and timesteps of the leader and followers in Figure 10c,d. The closer the x coordinates of followers to that of the leader, the better the performance will be. The x coordinates of followers in (c) converge to the leader faster than (d), representing that PPO-Exp can change vertical formation faster than PPO-Clip.

To further evaluate the formed vertical formation. Denote the terminal time as t_{ter}, calculate the average difference between the followers and leader in x coordinates in the last ten timesteps, and denote the result as δ_x, which can be represented as follows:

$$\delta_x = \frac{1}{5}\sum_{i=1}^{5}\sum_{t>t_{ter}-10}\|x_0(t) - x_i(t)\| \tag{44}$$

The low δ_x indicates the follower is close to the leader in x coordinates. In PPO-Clip, the calculated $\delta_x \approx 95.383$, but in PPO-Exp, the calculated $\delta_x \approx 43.816$, which is nearly half of the PPO-Clip.

Compared to the control strategy in formation keeping, the followers in formation changing tasks perform good cooperation. All followers maneuver orderly to the position where the leader's x-coordinate is located. To avoid the UAVs collide each other, the followers decided to move to different positions on the y-axis. The followers take different maneuvers depending on their initial position to reach the position. e.g., follower 4, in the initial time, is far away from the leader in x-coordinates. For follower 4, a collision avoidance path is moving to the tail of the newly formed formation. Therefore, the follower4 achieves a large angle arc maneuver and moves to the tail of the formed vertical formation.

The target of the obstacle avoidance task is to reach the target area and avoid crashing into the obstacle. This paper considers a circle area on the plane as an obstacle. Denote the coordinates of the obstacle center is (x_{obs}, y_{obs}), and the radius is r_{obs}. A simple approach to consider this situation is to add a penalty on the formation system reward when the UAVs crash on the obstacle, the penalty effect. The penalty for crashing into the obstacle is denoted as follows:

$$R_{o,i} = \begin{cases} 0, \sqrt{(x_i - x_{obs})^2 + (y_i - y_{obs})^2} \leq r_{obs} \\ -10,000, otherwise \end{cases} \tag{45}$$

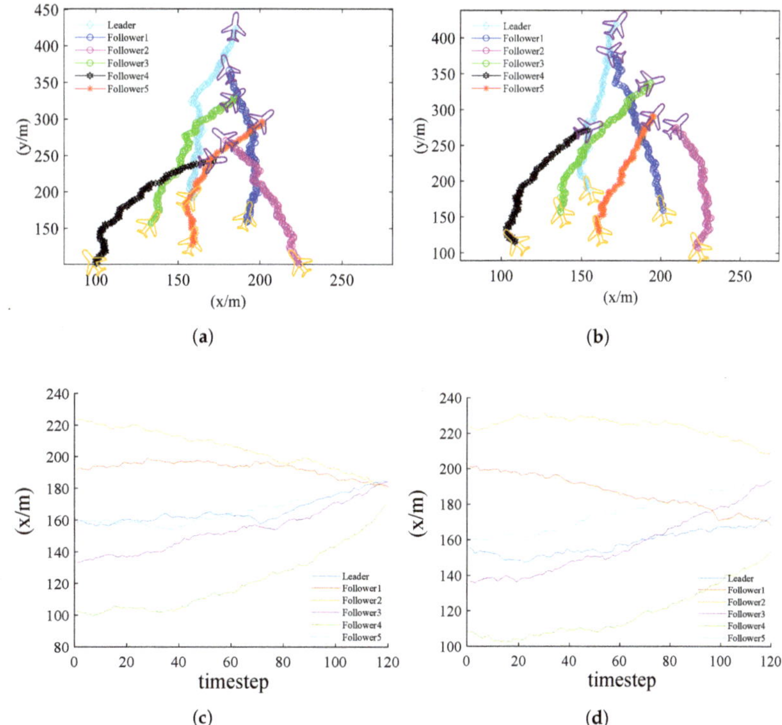

Figure 10. (**a**): The performance of vertical formation changing task by PPO-Exp; (**b**): The performance of vertical formation changing task by PPO-Clip (**c**): The x coordinate of formation system in PPO-Exp; (**d**): The x coordinate of formation system in PPO-Clip

Similar to the exploration reward $R_{e,i}$, to

$$R_{e,i}^{obs} = min\{|x_i - x_{obs}|, |y_i - y_{obs}|\} \tag{46}$$

Then the total reward (47) can be rewritten as follows:

$$R(T) = \sum_{i=0}^{5} \left[R_{f,i}(T) + R_{e,i}(T) + R_{e,i}^{obs}(T) + R_{o,i}(T) \right] + R_d(T) + R_p(T) \tag{47}$$

To encourage the UAV system to take more exploration on avoid obstacle, the exploration reward in avoid obstacle $R_{r,i}^{obs}$ is added to the exploration advantage function:

$$A_\pi^{exp}(S_t, a_t) = E_\pi \left[\sum_{k=0}^{\infty} \gamma^k (R_d(t+k) + \sum_{i=0}^{5} \left[R_{f,i}(T) + R_{e,i}(t+k) + R_{e,i}^{obs}(T)) \right] | S_t = s, a_t = a \right] -$$

$$E_\pi \left[\sum_{k=0}^{\infty} \gamma^k (R_d(t+k) + \sum_{i=0}^{5} [R_{f,i}(T) + R_{e,i}(t+k) + R_{e,i}^{obs}(T))] | S_t = s \right] \tag{48}$$

Training the obstacle to avoid task with PPO-Exp and PPO-Clip, the training parameters are kept as same as the previous part except episode length. After training with PPO-Exp and PPO-Clip, the test results of obstacle avoid task are shown in Figure 11a, and the results of PPO-Clip can be seen in Figure 11b. A follower in the formation trained by PPO-Clip crashed on the obstacle at 94 timesteps. The formation trained by PPO-Exp performed the arc maneuvers and avoided the obstacle. PPO-Exp performs better than PPO-Clip because it can explore more policies to reach the target area and discover a good path to avoid obstacles. However, the PPO-Clip still tries to reach the target area straight.

Compared to the formation keeping task without obstacles, the obstacle scenario requires the formation system to explore more to avoid the obstacle. Therefore, in this scenario, compared to the fixed ε PPO-Clip, the PPO-Exp shows better performance because it could adjust their ε to balance exploration and estimation error. Then the PPO-Exp explored the large-angle arc maneuvers and performed them to avoid the obstacle.

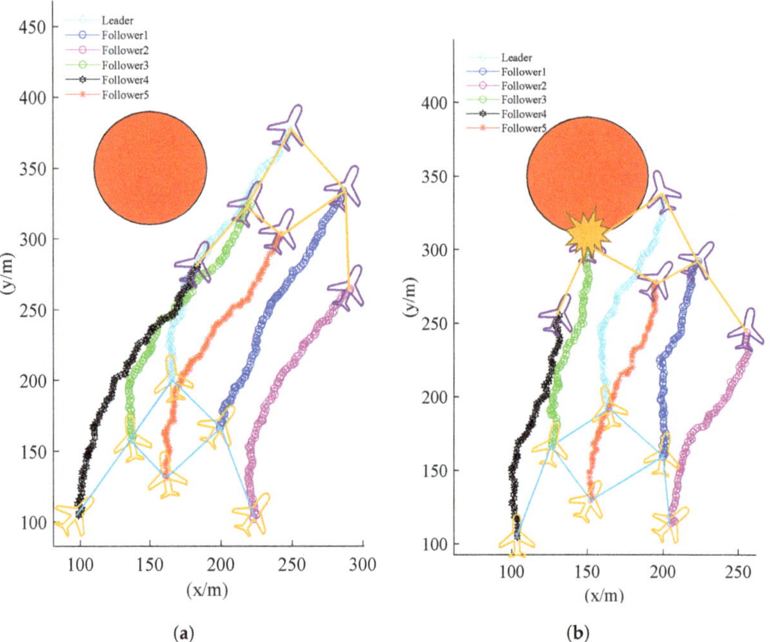

Figure 11. (**a**): The performance of formation keeping with obstacle avoid task by PPO-Exp; (**b**): The performance of formation keeping with obstacle avoid task by PPO-Clip.

7. Conclusions

This paper studies a flocking scenario consistent with one leader (with an intelligence chip) and several followers(without an intelligence chip). The reinforcement learning environment is constructed (continuous action and state space) with an OpenAI gym, and the reward is designed as a regular part and an exploration part. A low-communication cost protocol is provided to ensure the UAVs can communicate the state and action information between leader and followers. In addition, a variation of Proximal Policy Optimization is proposed to balance the dilemma between the estimation error bound and the exploration ability of PPO. The proposed method can help UAVs adjust the explore strategy, and the experiments demonstrate it has better performance than the current algorithms such as PPO-KL, PPO-clip, and DDPG.

Author Contributions: Conceptualization, D.X. and H.L.; Methodology, D.X., Y.G.; Supervision, D.X. and H.L.; Software, Y.G., Z.Y., Z.W., R.L., R.Z.; Formal analysis, Y.G. and H.L.; Writing—original draft, Y.G., R.Z.; Validation, Z.Y., Z.W., R.L., R.Z.; Visualization, Z.Y., Z.W.; Funding acquisition, D.X.; Resources, R.L.; Investigation, X.X., Data curation, X.X., Writing—review and editing, X.X. All authors have read and agreed to the published version of the manuscript.

Funding: This research received no external funding.

Data Availability Statement: Not applicable.

Conflicts of Interest: The authors declare no conflict of interest.

References

1. Zhou, W.; Li, J.; Zhang, Q. Joint Communication and Action Learning in Multi-Target Tracking of UAV Swarms with Deep Reinforcement Learning. *Drones* **2022**, *6*, 339. [CrossRef]
2. Tian, S.; Wen, X.; Wei, B.; Wu, G. Cooperatively Routing a Truck and Multiple Drones for Target Surveillance. *Sensors* **2022**, *22*, 2909. [CrossRef] [PubMed]
3. Wu, G.; Fan, M.; Shi, J.; Feng, Y. Reinforcement Learning based Truck-and-Drone Coordinated Delivery. *IEEE Trans. Artif. Intell.* **2021**. [CrossRef]
4. Gupta, L.; Jain, R.; Vaszkun, G. Survey of important issues in uav communication networks. *IEEE Commun. Surv. Tutor.* **2015**, *18*, 1123–1152. [CrossRef]
5. Wu, Q.; Zeng, Y.; Zhang R. Joint trajectory and communication design for multi-uav enabled wireless networks. *IEEE Trans. Wirel. Commun.* **2018**, *17*, 2109–2121. [CrossRef]
6. Eisenbeiss, H. A mini unmanned aerial vehicle (uav): System overview and image acquisition. *Int. Arch. Photogramm. Remote Sens. Spat. Inf. Sci.* **2004**, *36*, 1–7. Available online: https://www.isprs.org/proceedings/XXXVI/5-W1/papers/11.pdf (accessed on 29 November 2022).
7. Wang, Y.; Xing, L.; Chen, Y.; Zhao, X.; Huang, K. Self-organized UAV swarm path planning based on multi-objective optimization. *J. Command. Control* **2021**, *7*, 257–268. [CrossRef]
8. Kuriki, Y.; Namerikawa, T. Formation control with collision avoidance for a multi-uav system using decentralized mpc and consensus-based control. *SICE J. Control Meas. Syst. Integr.* **2015**, *8*, 285–294. [CrossRef]
9. Saif, O.; Fantoni, I.; Zavala-Río A. Distributed integral control of multiple uavs: Precise flocking and navigation. *IET Contr. Theory Appl.* **2019**, *13*, 2008–2017. [CrossRef]
10. Chen, H.; Wang, X. Formation flight of fixed-wing UAV swarms: A group-based hierarchical approach. *Chin. J. Aeronaut.* **2021**, *34*, 504–515. [CrossRef]
11. Liu, Z.; Wang, X.; Shen, L.; Zhao, S.; Cong, Y.; Li, J.; Yin, D.; Jia, S.; Xiang, X. Mission-Oriented Miniature Fixed-Wing UAV Swarms: A Multilayered and Distributed Architecture. *IEEE Trans. Syst. Man Cybern. Syst.* **2022**, *1*, 2168–2216. [CrossRef]
12. Koch, W.; Mancuso, R.; West, R.; Bestavros, A. Reinforcement learning for uav attitude control. *ACM Trans. Cyber-Phys. Syst.* **2019**, *3*, 1–21. [CrossRef]
13. Kaelbling, L.; Littman, M.; Moore, A. Reinforcement learning: A survey. *J. Artif. Intell. Res.* **1996**, *4*, 237–285. 10.1613/jair.301. [CrossRef]
14. Li, Y. Deep reinforcement learning: An overview. *arXiv* **2017**, arXiv:1701.07274. Available online: https://arxiv.org/pdf/1701.07274.pdf (accessed on 29 November 2022).
15. Huy, P.; Hung, L.; David, S. Autonomous uav navigation using reinforcement learning. *arXiv* **2018**, arXiv:1801.05086. Available online: https://arxiv.org/pdf/1801.05086.pdf (accessed on 29 November 2022).
16. Gullapalli, V.; Franklin, J.; Benbrahim, H. Acquiring robot skills via reinforcement learning. *IEEE Control Syst. Mag.* **1994**, *14*, 13–24. [CrossRef]
17. Huang, J.; Mo, Z.; Zhang, Z.; Chen, Y. Behavioral control task supervisor with memory based on reinforcement learning for human—Multi-robot coordination systems. *Front. Inf. Technol. Electron. Eng.* **2022**, *23*, 1174–1188. FITEE.2100280. [CrossRef]
18. Zhang, F.; Leitner, J.; Milford, M.; Upcroft, B.; Corke, P. Towards vision-based deep reinforcement learning for robotic motion control. *arXiv* **2017**, arXiv:1511.03791. Available online: https://arxiv.org/pdf/1511.03791.pdf (accessed on 29 November 2022).
19. Tomimasu, M.; Morihiro, K.; Nishimura, H. A reinforcement learning scheme of adaptive flocking behavior. In Proceedings of the 10th International Symposium on Artificial Life and Robotics (AROB), Oita, Japan, 4–6 February 2005.
20. Morihiro, K.; Isokawa, T.; Nishimura, H.; Matsui, N. Characteristics of flocking behavior model by reinforcement learning scheme. In Proceedings of the 2006 SICE-ICASE International Joint Conference, Busan, Republic of Korea, 18–21 October 2006. [CrossRef]
21. Shao, W.; Chen, Y.; Huang, J. Optimized Formation Control for a Class of Second-order Multi-agent Systems based on Single Critic Reinforcement Learning Method. In Proceedings of the 2021 IEEE International Conference on Networking, Sensing and Control (ICNSC), Xiamen, China, 3–5 December 2021; pp. 1–6. [CrossRef]
22. Wang, C.; Wang, J.; Zhang, X. A deep reinforcement learning approach to flocking and navigation of uavs in large-scale complex environments. In Proceedings of the 2018 IEEE Global Conference on Signal and Information Processing (GlobalSIP), Anaheim, CA, USA, 26–28 November 2018. [CrossRef]
23. Beard, R.; Kingston, D.; Quigley, M.; Snyder, D.; Christiansen, R.; Johnson, W.; McLain, T.; Goodrich, M. Autonomous vehicle technologies for small fixed-wing uavs. *J. Aerosp. Comput. Inf. Commun.* **2005**, *2*, 92–108. [CrossRef]
24. Hung, S.; Givigi, S.; Noureldin, A. A dyna-q (lambda) approach to flocking with fixed-wing uavs in a stochastic environment. In Proceedings of the 2015 IEEE International Conference on Systems, Man, and Cybernetics(SMC), Hong Kong, China, 9–12 October 2015. [CrossRef]

25. Hung, S.; Givigi, S. A Q-learning approach to flocking with UAVs in a stochastic environment. *IEEE Trans. Cybern.* **2016**, *47*, 186–197. [CrossRef]

26. Yan, C.; Xiang, X.; Wang, C. Fixed-wing uavs flocking in continuous spaces: A deep reinforcement learning approach. *Robot. Auton. Syst.* **2020**, *131*, 103594. [CrossRef]

27. Wang, C.; Yan, C.; Xiang, X.; Zhou, H. A continuous actor-critic reinforcement learning approach to flocking with fixed-wing UAVs. In Proceedings of the 2019 Asian Conference on Machine Learning(ACML), Nagoya, Japan, 17–19 November 2019. Available online: http://proceedings.mlr.press/v101/wang19a/wang19a.pdf (accessed on 29 November 2022).

28. Bøhn, E.; Coates, E.; Moe, E.; Johansen, T.A. Deep reinforcement learning attitude control of fixed-wing uavs using proximal policy optimization. In Proceedings of the 2019 International Conference on Unmanned Aircraft Systems (ICUAS), Atlanta, GA, USA, 11–14 June 2019. [CrossRef]

29. Hernandez, P.; Kaisers, M.; Baarslag, T.; de Cote, E.M. A survey of learning in multiagent environments: Dealing with non-stationarity. *arXiv* **2017**, arXiv:1707.09183. Available online: https://arxiv.org/pdf/1707.09183.pdf (accessed on 29 November 2022).

30. Yan, C.; Wang, C.; Xiang, X.; Lan, Z.; Jiang, Y. Deep reinforcement learning of collision-free flocking policies for multiple fixed-wing uavs using local situation maps. *IEEE Trans. Ind. Inform.* **2021**, *18*, 1260–1270. [CrossRef]

31. Peng, J.; Williams, R. Incremental multi-step Q-learning. *Mach. Learn.* **1996**, *22*, 283–290.:1018076709321. [CrossRef]

32. Hasselt, H.; Marco, W. Reinforcement Learning in Continuous Action Spaces. In Proceedings of the 2007 IEEE International Symposium on Approximate Dynamic Programming and Reinforcement Learning, Honolulu, HI, USA, 1–5 April 2007; pp. 272–279. [CrossRef]

33. Wang, C.; Wu, L.; Yan, C.; Wang, Z.; Long, H.; Yu, C. Coactive design of explainable agent-based task planning and deep reinforcement learning for human-UAVs teamwork. *Chin. J. Aeronaut.* **2020**, *33*, 2930–2945. [CrossRef]

34. Zhao, Z.; Rao, Y.; Long, H.; Sun, X.; Liu, Z. Resource Baseline MAPPO for Multi-UAV Dog Fighting. In Proceedings of the 2021 International Conference on Autonomous Unmanned Systems (ICAUS), Changsha, China, 24–26 September 2021._327. [CrossRef]

35. Yan, C.; Xiang, X.; Wang, C.; Lan, Z. Flocking and Collision Avoidance for a Dynamic Squad of Fixed-Wing UAVs Using Deep Reinforcement Learning. In Proceedings of the 2021 IEEE/RSJ International Conference on Intelligent Robots and Systems (IROS), Prague, Czech Republic, 27 September–1 October 2021; pp. 4738-4744. [CrossRef]

36. Song, Y.; Choi, J.; Oh, H.; Lee, M.; Lim, S.; Lee, J. Improvement of Decentralized Flocking Flight Efficiency of Fixed-wing UAVs Using Inactive Agents. In Proceedings of the AIAA Scitech 2019 Forum, San Diego, CA, USA, 7–11 January 2019.

37. Yan, Y.; Wang, H.; Chen, X. Collaborative Path Planning based on MAXQ Hierarchical Reinforcement Learning for Manned/Unmanned Aerial Vehicles. In Proceedings of the 39th Chinese Control Conference (CCC), Shenyang, China, 27–29 July 2020; pp. 4837–4842. [CrossRef]

38. Ren, T.; Niu, J.; Liu, X.; Hu, Z.; Xu, M.; Guizani, M. Enabling Efficient Scheduling in Large-Scale UAV-Assisted Mobile-Edge Computing via Hierarchical Reinforcement Learning. *IEEE Internet Things J.* **2021**, *9*, 7095–7109. [CrossRef]

39. Yang, H.; Jiang, B.; Zhang, Y. Fault-tolerant shortest connection topology design for formation control. *Int. J. Control Autom. Syst.* **2014**, *12*, 29–36. [CrossRef]

40. Schulman, J.; Wolski, F.; Dhariwal, P.; Radford, A.; Klimov, O. Proximal policy optimization algorithms. *arXiv* **2017**, arXiv:1707.06347. Available online: https://arxiv.org/pdf/1707.06347.pdf (accessed on 29 November 2022).

41. Banerjee, N.; Chakraborty, S.; Raman, V.; Satti, S.R. Space efficient linear time algorithms for bfs, dfs and applications. *Theory Comput. Syst.* **2018**, *62*, 1736–1762. [CrossRef]

42. Bansal, T.; Pachocki, J.; Sidor, S.; Sutskever, I.; Mordatch, I. Emergent Complexity via Multi-Agent Competition. *arXiv* **2017**, arXiv:1710.03748.

43. Sutton, R.; Barto, A. *Reinforcement Learning: An Introduction*, 2nd ed.; MIT Press: Cambridge MA, USA, 2018.

44. Schulman, J.; Levine, S.; Abbeel, P.; Jordan, M.; Moritz, P. Trust Region Policy Optimization. In Proceeding of the 2015 International Conference on Machine Learning(ICML), Lille, France, 6–11 July 2015; pp. 1889–1897.

45. Brockman, G.; Cheung, V.; Pettersson, L.; Schneider, J.; Schulman, J.; Tang, J.; Zaremba, W. Openai gym. *arXiv* **2016**, arXiv:1606.01540. Available online: https://arxiv.org/pdf/1606.01540.pdf (accessed on 29 November 2022).

46. Lillicrap, T.; Hunt, J.; Pritzel, A.; Heess, N.; Erez, T.; Tassa, Y.; Wierstra, D. Continuous control with deep reinforcement learning. In Proceeding of the 4th International Conference on Learning Representations (ICLR), San Juan, Puerto Rico, 2–4 May 2016; pp. 1582–1591.

47. Fujimoto, S.; Herke, H.; David, M. Addressing Function Approximation Error in Actor-Critic Methods. In Proceeding of the 2018 International Conference on Machine Learning (ICML), Stockholm, Sweden, 10–15 July 2018; pp. 1582–1591. Available online: http://proceedings.mlr.press/v80/fujimoto18a/fujimoto18a.pdf (accessed on 29 November 2022).

Article

Onboard Distributed Trajectory Planning through Intelligent Search for Multi-UAV Cooperative Flight

Kunfeng Lu, Ruiguang Hu *, Zheng Yao and Huixia Wang

National Key Laboratory of Science and Technology on Aerospace Intelligent Control, Beijing Aerospace Automatic Control Institute, Beijing 100854, China
* Correspondence: rghu258@163.com

Abstract: Trajectory planning and obstacle avoidance play essential roles in the cooperative flight of multiple unmanned aerial vehicles (UAVs). In this paper, a unified framework for onboard distributed trajectory planning is proposed, which takes full advantage of intelligent discrete and continuous search algorithms. Firstly, the Monte Carlo tree search (MCTS) is used as the task allocation algorithm to solve the cooperative obstacle avoidance problem. Taking the task allocation decisions as the constraint, knowledge-based particle swarm optimization (Know-PSO) is used as the optimization algorithm to solve the onboard distributed cooperative trajectory planning problem. Simulation results demonstrate that the proposed intelligent MCTS-PSO search framework is effective and flexible for multiple UAVs to conduct the cooperative trajectory planning and obstacle avoidance. Further, it has been applied in practical experiments and achieved promising results.

Keywords: multiple UAVs; trajectory planning; task allocation; obstacle avoidance; intelligent search; Monte Carlo tree search; knowledge-based particle swarm optimization

Citation: Lu, K.; Hu, R.; Yao, Z.; Wang, H. Onboard Distributed Trajectory Planning through Intelligent Search for Multi-UAV Cooperative Flight. *Drones* **2023**, *7*, 16. https://doi.org/10.3390/drones7010016

Academic Editors: Xiwang Dong, Mou Chen, Xiangke Wang and Fei Gao

Received: 21 November 2022
Revised: 22 December 2022
Accepted: 22 December 2022
Published: 26 December 2022

1. Introduction

Unmanned aerial vehicles (UAVs) have been extensively used in many areas, such as surveying [1–4], military surveillance [5–8], disaster rescue [9,10], etc. Although a single UAV can conduct the above-mentioned tasks, some disadvantages, including energy and others, have placed severe limitations on its applications. That is the reason why cooperative multiple UAVs have been further developed and utilized, which can overcome the disadvantages of single UAVs and accomplish tasks more robustly and intelligently with less time consumption [11].

Trajectory planning plays an essential role in the cooperative flight of multiple UAVs, which can be seen as continuous search algorithms mathematically. Commonly with some environmental constraints and UAVs' own constraints, the trajectories that minimize the cost function are seen as the best ones. Trajectory planning has been widely researched for many years, and two classes of trajectory planning methods have been developed. The first one is those traditional algorithms, including dynamic planning, the Voronoi diagram [12,13], the Dijkstra algorithm [14,15], and the A* algorithm [16]. The other class is swarm intelligence algorithms, such as the artificial bee colony (ABC) algorithm [17], the ant colony optimization (ACO) algorithm [18], and especially, the particle swarm optimization (PSO) algorithm [19–21]. In the literature [22], the trajectories of multiple UAVs were firstly given out by an offline simulation software program and saved in a text file. Then, the trajectories of all UAVs were transferred into the ground station software and displayed on a map. Finally, the trajectories were downloaded to the UAVs through the ground stations, and the UAVs completed the flight. The offline operation as in the literature [22] is now the mainstream execution style for cooperative trajectory planning.

Obstacle avoidance plays another essential role in the cooperative flight of multiple UAVs, for which the artificial potential field (APF) method is widely used [23,24]. To some extent, cooperative obstacle avoidance of multiple UAVs can be seen as a task allocation problem [25], in which multiple UAVs have to choose one side or the other to pass through the threat area. The Hungarian algorithm [22] and the auction algorithm [26,27] are two mainstream algorithms for the task allocation of multiple UAVs. In the literature [28], a two-step auction mechanism was first proposed to select the optimal action. Then, an obstacle avoidance mechanism was designed by defining several heuristic rules. Finally, a reverse auction mechanism was developed to balance the workload between multiple UAVs.

Although a large amount of effort has been devoted to addressing the trajectory planning and task allocation problems for multiple UAVs, there is still some room for further improvement. For instance, traditional task allocation algorithms often require that the input dimension must be equal to the output dimension, namely, the allocation matrix has identical row and column numbers, which severely restricts the application for more general task allocation problems. In addition, the efficiency and effectiveness of trajectory planning algorithms for multiple UAVs are not satisfactory in some circumstances. Meanwhile, the trajectory planning and task allocation for multiple UAVs are conducted offline in some literature, which cannot meet the requirements in complex and intense environments, where multiple UAVs must respond to commands rationally and quickly.

Taking the abovementioned issues into consideration, a unified framework for onboard distributed trajectory planning is proposed in this paper, which takes full advantage of intelligent discrete and continuous search algorithms. The main contributions of this paper are as follows:

(1) The Monte Carlo tree search (MCTS) is used as a task allocation algorithm to conduct obstacle avoidance, which does not require the equality of the row and column numbers of the allocation matrix. Further, the obstacle avoidance for multiple UAVs takes the energy constraint into account.
(2) Knowledge-based particle swarm optimization (Know-PSO) is used as the optimization algorithm to solve the onboard distributed cooperative trajectory planning problem, in which the motion energies of a few good particles are used to improve the velocities of those bad particles, and the information of the individual worst particles and global worst particle are also used. Furthermore, the interaction among multiple UAVs is utilized to avoid conflicts.
(3) The decisions of MCTS are taken as constraints for Know-PSO to form a unified framework for onboard distributed trajectory planning.
(4) The method proposed in this paper has been verified by actual flights and achieved good practical results.

The remainder of this paper is organized as follows: Section 2 formulates the trajectory planning and obstacle avoidance problem. Section 3 presents the proposed intelligent MCTS-PSO search framework. In Section 4, a series of simulations and actual experiments were conducted to evaluate the performance of MCTS-PSO. Finally, Section 5 concludes the paper and presents the future direction for the next work.

2. Mathematical Model

Cooperative trajectory planning driven by obstacle avoidance for multiple UAVs can be presented in the following mathematical models:

(1) Task allocation to obstacle avoidance: As shown in Figure 1, m UAVs are configured to go to the target area to conduct some important operations, in which the threat area must be avoided. Regarding each UAV, there are two choices for it to avoid the threat area, going through one side of the threat area or the other. Consequently, the cooperative obstacle avoidance problem can be translated into the task allocation problem, for which there are m decisions that have to be made. Generally, suppose

there are n choices for each UAV, then the mathematical model for task allocation is as below:

$$\max \sum_{j=1}^{n} w_j \left[1 - \prod_{i=1}^{m} (1 - e_{ij})^{x_{ij}} \right]$$
$$s.t. \begin{cases} \sum_{j=1}^{n} x_{ij} = 1, \ i = 1, 2, \ldots, m \\ x_{ij} = 0 \ or \ 1, \ i = 1, 2, \ldots, m, j = 1, 2, \ldots n \end{cases} \tag{1}$$

where w_j is the threat value of j choice, e_{ij} is the capacity evaluation of the i UAV to pass through the j choice and can be seen as the energy constraints, x_{ij} is the final decision, whether the i UAV passes through the j choice or not. The MCTS is used to solve the model to obtain x_{ij}. All x_{ij} compose the task allocation matrix, which is a 0–1 matrix and has m rows and n columns. Each row has only one, which means that the corresponding UAV can only make one choice.

(2) Cooperative trajectory planning driven by obstacle avoidance: The goal of cooperative trajectory planning is to minimize the total distances of m UAVs from the start area to the target area; in the meantime, m UAVs must avoid the threat area and not collide with each other. The mathematical model is as below:

$$\min \sum_{i=1}^{m} \sum_{j=1}^{n} l_{ij} x_{ij}$$
$$s.t. \begin{cases} dm_{i,k} > d_s, \ i, k = 1, 2, \ldots, m, i \neq k \\ x_{ij} = 0 \ or \ 1, \ i = 1, 2, \ldots, m, j = 1, 2, \ldots n \end{cases} \tag{2}$$

where l_{ij} is the length of a trajectory that the i UAV pass through the j choice, $dm_{i,k}$ is the margin distance between trajectories of two different UAVs, and d_s is the safe distance between two adjacent UAVs. x_{ij} is the final decision, whether the i UAV passes through the j choice or not as in formula (1), which is solved by the MCTS method.

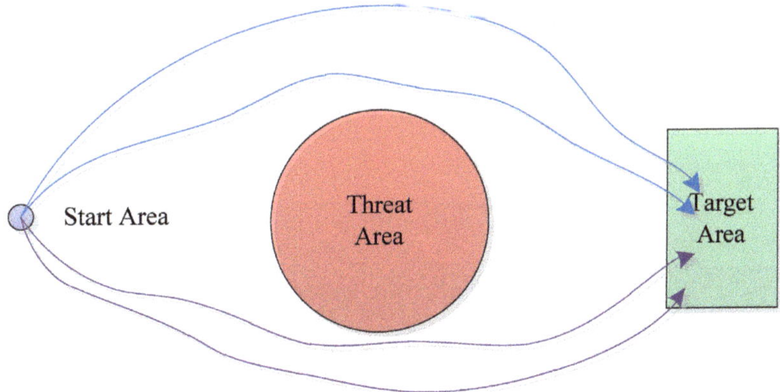

Figure 1. Representative scene of cooperative obstacle avoidance and trajectory planning.

3. MCTS-PSO Framework for Onboard Distributed Trajectory Planning

In this section, we elaborate on the cooperative trajectory planning driven by obstacle avoidance, which can be conducted in an onboard distributed mode. Firstly, MCTS is used as a task allocation algorithm to conduct obstacle avoidance, taking the energy constraint into consideration; secondly, the decisions of MCTS are taken as constraints for Know-PSO to conduct onboard distributed trajectory planning for multiple UAVs. The schematic diagram of the MCTS-PSO framework is shown in Figure 2, and the Pseudo-code of the MCTS-PSO Algorithm 1 is:

Algorithm 1: MCTS-PSO framework

Input: UAVs number m,
        choices number n,
        start position P_s and target position P_t,
        Threat area center P_c and radius r,
        the safe distance d_s,
Output: best trajectories
1: for $i \leftarrow 1$ to m do
2: for $j \leftarrow 1$ to n do
3: Evaluate the threat values w_j;
4: Evaluate the capacity values e_{ij};
5: end for
6: end for
7: Use MCTS to solve formula (1) to get decisions x_{ij};
8: Use Know-PSO to generate a trajectory for one UAV i;
9: for $k \leftarrow 1$ to m do
10: if $k == i$
11: continue;
12: else
13: Use Know-PSO to generate a trajectory for UAV k with d_s;
14: Check whether $dm_{i,k}$ is larger than d_s or not;
15: end if
16: end for
17: Return m best trajectories;
    The resulting m best trajectories obtained after the above steps is the optimal
    solution for onboard distributed cooperative trajectory planning;
end

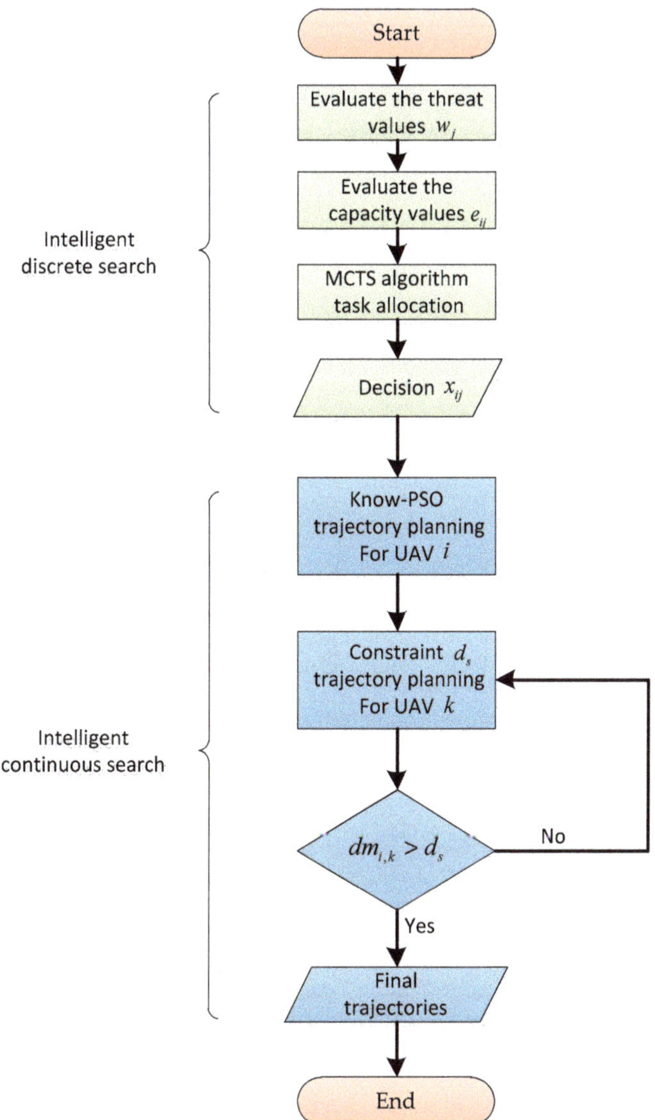

Figure 2. The schematic diagram of the MCTS-PSO framework.

3.1. MCTS Task Allocation for Multiple UAVs

Task allocation plays an essential role in the cooperative flight of multiple UAVs. In this paper, MCTS, as an intelligent discrete search algorithm, is used to conduct task allocation. MCTS does not require the equality of the row and column numbers of the allocation matrix, beyond the traditional mainstream Hungarian algorithm and auction algorithm. The flow chart of MCTS task allocation is shown in Figure 3, whose specific steps are presented as follows:

Step 1: Input UAV number, choices number, threat values, UAV capacities, and iteration number. The UAV capacities can be seen as the energy constraints:

$$e_{ij} = \frac{1}{\alpha * dist(UAV_i, choice_j)} \tag{3}$$

where α is the energy coefficient.

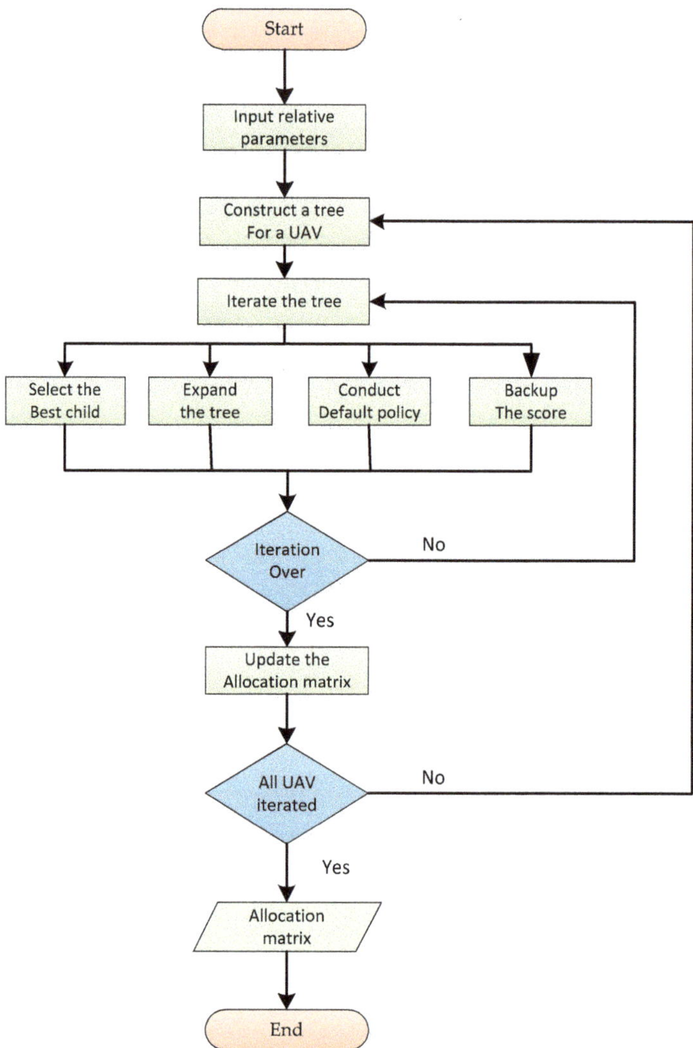

Figure 3. The flow chart of MCTS task allocation.

Step 2: Construct a new search tree and initialize the root node.
Step 3: Iterate the search tree until the iteration number:

 (1) Select the best child layer by layer to find a leaf node;
 (2) Expand the tree from the leaf node;
 (3) Conduct the default policy;
 (4) Back up the score and update nodes' attributes.

Step 4: Update the allocation matrix.

Step 5: Repeat steps 2–4 for all UAVs.

The resulting 0–1 matrix obtained after the above steps is the optimal solution for multiple UAV task allocation problems.

Pseudo-code of the MCTS task allocation Algorithm 2 is:

Algorithm 2: MCTS task allocation

Input: UAVs number m,
       choices number n,
       threat values w,
       capacities e,
       IterNum,
Output: Allocation matrix *AlloMx*
1: for $i \leftarrow 1$ to m do
2: Create a new tree with *root* node and initialize *root*:
3: $root.N \leftarrow 0$, $root.Q \leftarrow 0$;
4: for $j \leftarrow 1$ to *IterNum* do
5: node $p \leftarrow root$;
6: $AloMx\_copy \leftarrow AloMx$;
7: $i\_temp \leftarrow i$;
8: while(True)
9: if p is leaf
10: break;
11: end if
12: find the best child of p and its index *ind*;
13: $p \leftarrow best\ child\ of\ p$;
14: $AloMx\_copy[i\_temp][ind] \leftarrow 1$;
15: $i\_temp \leftarrow i\_temp + 1$;
16: end while
17: if $i\_temp \sim= m$
18: Expand the node p;
19: end if
20: Conduct the default policy for *AlloMx_copy* and get the score;
21: Back up the score;
22: end for
23: find the best child of *root* and its index *ind_best*;
24: $AloMx[i][ind\_best] \leftarrow 1$;
25: end for
26: Return *AlloMx*;
    The resulting 0–1 matrix obtained after the above steps is the optimal solution for task allocation;
end

3.2. Onboard Distributed Cooperative Trajectory Planning for Multiple UAVs

Trajectory planning plays another essential role in the cooperative flight of multiple UAVs. PSO, as an intelligent continuous search algorithm, has been widely used to conduct trajectory planning. Nevertheless, the standard PSO algorithm has some limitations such as premature convergence. In this paper, knowledge-based particle swarm optimization (Know-PSO) is proposed as the optimization algorithm to solve the onboard distributed cooperative trajectory planning problem, in which the motion energies of a few good particles are used to improve the velocities of those bad particles, and the information of the individual worst particles and global worst particle is also used. Furthermore, the interaction among multiple UAVs is utilized to avoid conflicts.

The well-known standard PSO algorithm is

$$\begin{cases} V_i^{k+1} = \omega V_i^k + c_1 r_1 (p_{best,i}^k - x_i^k) + c_2 r_2 (g_{best}^k - x_i^k) \\ x_i^{k+1} = x_i^k + V_i^{k+1} \end{cases} \tag{4}$$

Practically, it has been found that the information of individual worst particles and global worst particle are also beneficial, which are introduced as

$$\begin{cases} V_i^{k+1} = \omega V_i^k + c_1 r_1 (p_{best,i}^k - x_i^k) + c_2 r_2 (g_{best}^k - x_i^k) \\ \qquad\quad -c_3 r_3 (p_{worst,i}^k - x_i^k) - c_4 r_4 (g_{worst}^k - x_i^k) \\ x_i^{k+1} = x_i^k + V_i^{k+1} \end{cases} \tag{5}$$

Further, the motion energies of a few good particles are used to improve the velocities of those bad particles, which is defined as

$$E^{k,m_1} = \sum_{i=1}^{m_1} \left(V_i^k\right)^T \left(V_i^k\right) \tag{6}$$

The energy loss is

$$\Delta E^{k,m_1} = E^{k,m_1} - E^{k-1,m_1} \tag{7}$$

Consequently, the updated equation for bad particles is

$$\begin{cases} V_i^{k+1} = \omega V_i^k + c_1 r_1 (p_{best,i}^k - x_i^k) + c_2 r_2 (g_{best}^k - x_i^k) \\ \qquad\quad -c_3 r_3 (p_{worst,i}^k - x_i^k) - c_4 r_4 (g_{worst}^k - x_i^k) + \Delta E^{k,m_1} / m_1 \\ x_i^{k+1} = x_i^k + V_i^{k+1} \end{cases} \tag{8}$$

The decisions of MCTS are taken as constraints for Know-PSO to form a unified framework for onboard distributed trajectory planning. Specifically,

$$cost(x_{ij}) = \begin{cases} 0, & x_{ij} = 1 \\ 1000, & x_{ij} \neq 0 \end{cases} \tag{9}$$

The flow chart of Know-PSO for onboard distributed cooperative trajectory planning is shown in Figure 4, whose specific steps are presented as follows:

Step 1: Input particle number, point number, start position and target position, threat area center and radius, task allocation matrix, iteration number, UAV number, the safe distance, and the max velocity.

Step 2: Initialize the particles and best values.

Step 3: Iterate the particles until the iteration number:

    (1)   Compute the cost of particles;
    (2)   Update the best values;
    (3)   Update the velocities;
    (4)   Update the particles.

Step 4: Generate one best trajectory.

Step 5: Repeat steps 2–4 for all UAVs considering the safe distance between them.

The resulting best trajectories obtained after the above steps is the optimal solution for onboard distributed cooperative trajectory planning.

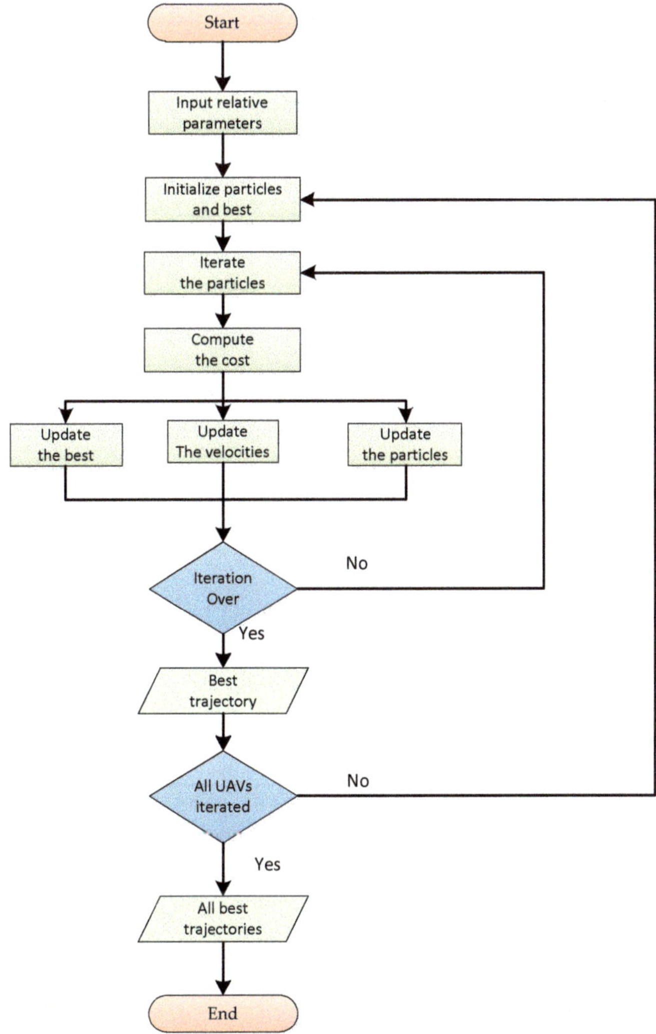

Figure 4. The flow chart of Know-PSO cooperative trajectory planning.

Pseudo-code of the cooperative trajectory planning Algorithm 3 is:

Algorithm 3: Onboard distributed cooperative trajectory planning

---

Input: particle number mp,
    point number n,
    start position P_S and target position P_t,
    Threat area center P_c and radius r,
    $AlloMx$,
    $IterNum$,
    UAVs number m,
    the safe distance d_s,
    V_{max},
Output: best trajectories
1: $particles \leftarrow rand(mp, n)$;
2: $particlesBest \leftarrow zeros(mp, n)$;
3: $globalBest \leftarrow zeros(1, n)$;
4: $particlesV \leftarrow rand(mp, n) * 2V_{max} - V_{max}$
5: for $iter \leftarrow 1$ to $IterNum$ do
6: for $i \leftarrow 1$ to mp do
7: Compute the $cost$ of $particles[i, :]$ with decision $AlloMx$ and P_c, r;
8: if $cost$ is descending
9: $particlesBest[i, :] \leftarrow particles[i, :]$;
10: $globalBest \leftarrow particles[i, :]$;
11: end if
12: for $j \leftarrow 1$ to n do
13: Update $particlesV[i, j]$ according Formulas (5) and (8);
14: Adjust $particlesV[i, j]$ into $[-V_{max}, V_{max}]$;
15: $particles[i, j] \leftarrow particles[i, j] + particlesV[i, j]$;
16: end for
17: end for
18: end for
19: Here, we got the best trajectory for one UAV.
20: for $i \leftarrow 2$ to m do
21: Repeat 1~21 considering the safe distance d_s;
22: end for
23: Return m best trajectories;
    The resulting m best trajectories obtained after the above steps is the optimal
    solution for onboard distributed cooperative trajectory planning;
end

---

4. Experiments and Analysis

This section demonstrates the performance of the MCTS-PSO framework by conducting a series of experiments.

A cooperative processor was deployed for each UAV. The processor had a four-core CPU, whose main frequency was 1.5G Hz and AI computational power was 20 TOPS(int8).

The experiments were all conducted in open environments, including plains and mountainous areas. The environmental temperature was generally higher than $-20\,^{\circ}\mathrm{C}$.

Wind has a very important influence on flight in open environments. According to all experiments, UAVs could be controlled stably if the wind velocity was smaller than 15 m/s; otherwise, there would be some accidents.

Firstly, a flight of two drones was conducted, whose trajectories were shown in Figure 5, of which the red circle was a threat area. It can be seen that the two drones avoided the threat area successfully and passed it by one side, respectively. The task allocation matrix of them was shown in Table 1.

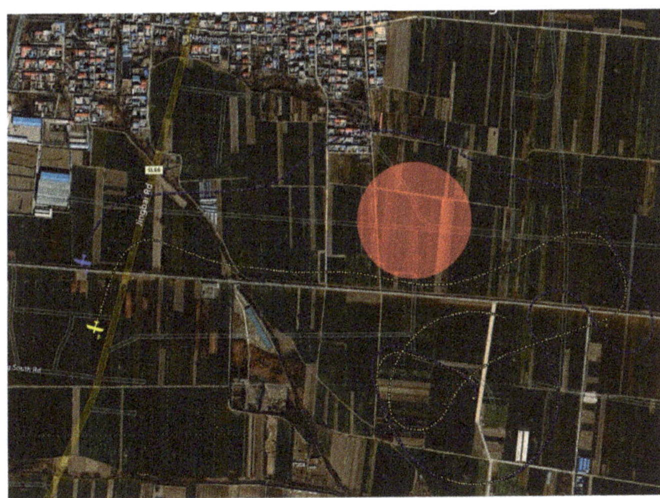

Figure 5. MCTS-PSO framework for two drones.

Table 1. Task allocation matrix for two drones in Figure 5.

	Top Channel	Bottom Channel
UAV1	1	0
UAV2	0	1

Secondly, a flight of three drones was conducted, whose trajectories were shown in Figure 6, of which the red circle was a threat area. It can be seen that the three drones avoided the threat area successfully. The task allocation matrix of them was shown in Table 2.

Thirdly, a flight of four drones was conducted in another environment, whose trajectories were shown in Figure 7, of which the red circle was a threat area. It can be seen that the four drones avoided the threat area successfully. The task allocation matrix of them was shown in Table 3.

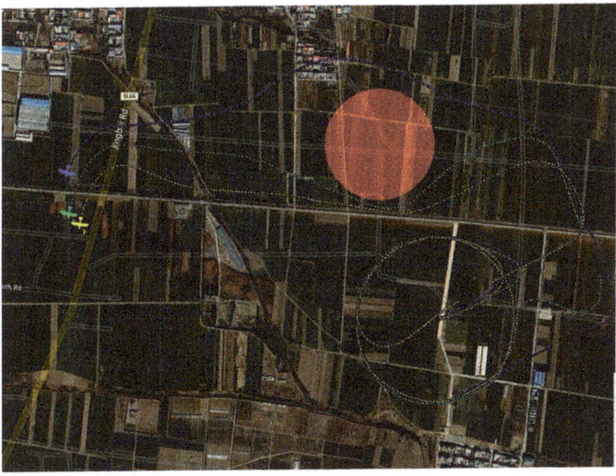

Figure 6. MCTS-PSO framework for three drones.

Table 2. Task allocation matrix for three drones in Figure 6.

	Top Channel	Bottom Channel
UAV1	1	0
UAV2	0	1
UAV3	0	1

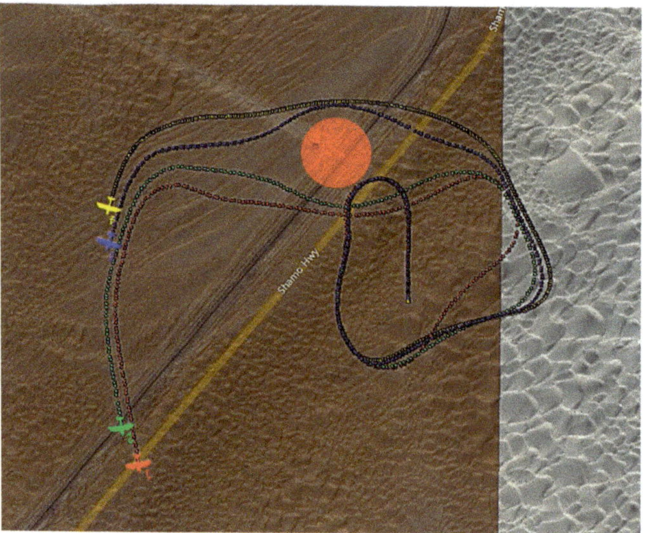

Figure 7. MCTS-PSO framework for four drones.

Table 3. Task allocation matrix for four drones in Figure 7.

	Top Channel	Bottom Channel
UAV1	1	0
UAV2	1	0
UAV3	0	1
UAV4	0	1

It was shown in these experiments, when the number of UAVs changed, the trajectories were slightly adjusted automatically. For example, comparing Figures 5 and 6, when the green UAV joined in, the trajectory of the yellow UAV was automatically adjusted to leave some space for the green one. This reflected the intelligent onboard adjustment ability of the algorithm.

Cooperative flight in an open environment was shown in Figure 8, in which five quadcopters were used. It must be noted that the MCTS-PSO framework can also be applied to other kinds of UAVs, such as fixed-wing UAVs.

Figure 8. Cooperative flying drones in open environments.

The further computational time of the proposed distributed framework and common centralized framework were shown in Table 4.

Table 4. Computational time.

	Distributed Framework	Centralized Framework
2 UAVs	0.72s	2.2s
3 UAVs	0.73s	3.1s
4 UAVs	0.73s	4.2s

It is quite clear that the proposed distributed framework was more efficient than a common centralized framework, especially when there were more UAVs, because each UAV planned its trajectory using its own processor.

The distance between the start area and the target area was about 1.4 Km. Note that the start area was where the proposed distributed framework conducted cooperative trajectory planning, not the location where the UAVs were launched. The total distances were shown in Table 5.

Table 5. Total distances.

	Total Distances
2 UAVs	3.2 Km
3 UAVs	4.9 Km
4 UAVs	6.6 Km

In open environments, other factors could also influence the real flight trajectories, such as maneuverability, wind, etc. Moreover, PSO variants could not always generate the best trajectories. These were the reasons why the trajectories in Figures 5–7 were not very close to the threat area.

Generally, the effectiveness, scalability, and adaptability of our framework were verified through the quantitative experiments with different UAV numbers. Moreover, it realized the avoidance of conflicts between multiple UAVs. All experiments demonstrated that the MCTS-PSO framework could be applied in dynamic and complex environments.

5. Conclusions

In this paper, the unified MCTS-PSO framework for onboard distributed trajectory planning is proposed, which takes full advantage of intelligent discrete and continuous search algorithms. The effectiveness, scalability, and adaptability of our framework have been verified through a series of experiments with different UAV numbers. Moreover, the proposed framework can also be applied in other similar swarm systems.

In future work, large-scale UAVs will be tested with the MCTS-PSO framework. Further, other intelligent algorithms such as multi-agent reinforcement learning will be introduced into the framework. In addition, we would like to evaluate our framework in intense confrontation applications.

Author Contributions: Conceptualization, R.H.; methodology, R.H. and K.L.; software, R.H. and Z.Y.; validation, K.L.; formal analysis, H.W.; investigation, H.W.; resources, R.H.; data curation, K.L.; writing—original draft preparation, K.L. and R.H.; writing—review and editing, Z.Y. and K.L.; visualization, K.L.; supervision, R.H. All authors have read and agreed to the published version of the manuscript.

Funding: This research received no external funding.

Data Availability Statement: Not applicable.

Acknowledgments: The authors would like to thank the National Key Laboratory of Science and Technology on Aerospace Intelligent Control, Beijing, China, for the resources provided by them.

Conflicts of Interest: The authors declare no conflict of interest.

References

1. Huang, R.; Zhou, H.; Liu, T.; Sheng, H. Multi-UAV collaboration to survey Tibetan antelopes in Hoh Xil. *Drones* **2022**, *6*, 196. [CrossRef]
2. Bollard, B.; Doshi, A.; Gilbert, N.; Poirot, C.; Gillman, L. Drone technology for monitoring protected areas in remote and fragile environments. *Drones* **2022**, *6*, 42. [CrossRef]
3. Ball, Z.; Odonkor, P.; Chowdhury, S. A swarm-intelligence approach to oil spill mapping using unmanned aerial vehicles. In Proceedings of the AIAA Information Systems—AIAA Infotech@ Aerospace, Grapevine, TX, USA, 9–13 January 2017.
4. Alevizos, E.; Oikonomou, D.; Argyriou, A.V.; Alexakis, D.D. Fusion of drone-Based RGB and multi-spectral imagery for shallow water bathymetry inversion. *Remote Sens.* **2022**, *14*, 1127. [CrossRef]
5. Tong, H.; Andi, T.; Huan, Z.; Dengwu, X.; Lei, X. Multiple UAV cooperative path planning based on LASSA method. *Syst. Eng. Electron.* **2022**, *44*, 233–241.
6. Pannozzi, P.; Valavanis, K.P.; Rutherford, M.J.; Guglieri, G.; Scanavino, M.; Quagliotti, F. Urban monitoring of smart communities using UAS. In Proceedings of the 2019 International Conference on Unmanned Aircraft Systems, Atlanta, GA, USA, 11–14 June 2019.
7. Ni, J.; Tang, G.; Mo, Z.; Cao, W.; Yang, S.X. An improved potential game theory based method for multi-UAV cooperative search. *IEEE Access* **2020**, *8*, 47787–47796. [CrossRef]
8. Zhang, H.; Sun, M.; Li, Q.; Liu, L.; Liu, M.; Ji, Y. An empirical study of multi-scale object detection in high resolution UAV images. *Neurocomputing* **2021**, *421*, 173–182. [CrossRef]
9. Lin, L.; Goodrich, M.A. UAV intelligent path planning for wilderness search and rescue. In Proceedings of the 2009 IEEE/RSJ International Conference on Intelligent Robots and Systems, St. Louis, MO, USA, 10–15 October 2009.
10. Kohlbrecher, S.; Kunz, F.; Koert, D.; Rose, C.; Manns, P.; Daun, K.; Schubert, J.; Stumpf, A.; von Stryk, O. Towards highly reliable autonomy for urban search and rescue robots. In Proceedings of the Robot Soccer World Cup, Berlin/Heidelberg, Germany, 12 May 2015.
11. McGuire, K.N.; DeWagter, C.; Tuyls, K.; Kappen, H.J.; de Croon, G.C. Minimal navigation solution for a swarm of tiny flying robots to explore an unknown environment. *Sci. Robot.* **2019**, *4*, 971. [CrossRef]
12. Junlan, N.; Qingjie, Z.; Yanfen, W. UAV path planning based on weighted-Voronoi diagram. *Flight Dyn.* **2015**, *33*, 339–343.
13. Chen, X.; Chen, X. The UAV dynamic path planning algorithm research based on Voronoi diagram. In Proceedings of the 26th Chinese Control and Decision Conference (2014 CCDC), Changsha, China, 31 May–2 June 2014.
14. Singh, Y.; Sharma, S.; Sutton, R.; Hatton, D.C. Towards use of Dijkstra algorithm for optimal navigation of an unmanned surface vehicle in a real-time marine environment with results from artificial potential field. *TransNav Int. J. Mar. Navig. Saf. Sea Transp.* **2018**, *12*, 125–131. [CrossRef]
15. Cheng, N.; Liu, Z.; Li, Y. Path planning algorithm of Dijkstra-based intelligent aircraft under multiple constraints. *Xibei Gongye Daxue Xuebao/J. Northwest. Polytech. Univ.* **2020**, *38*, 1284–1290. [CrossRef]
16. Mandloi, D.; Arya, R.; Verma, A.K. Unmanned aerial vehicle path planning based on A * algorithm and its variants in 3d environment. *Int. J. Syst. Assur. Eng. Manag.* **2021**, *12*, 990–1000. [CrossRef]
17. Zhou, J.; Yao, X. Multi-objective hybrid artificial bee colony algorithm enhanced with Lévy flight and self-adaption for cloud manufacturing service composition. *Appl. Intell.* **2017**, *47*, 721–742. [CrossRef]

18. Xianqiang, L.I.; Rong, M.A.; Zhang, S. Improved design of ant colony algorithm and its application in path planning. *Acta Aeronaut. Astronaut. Sin.* **2020**, *41*, 724381.
19. Jiang, T.; Li, J.; Huang, K. Longitudinal parameter identification of a small unmanned aerial vehicle based on modified particle swarm optimization. *Chin. J. Aeronaut.* **2015**, *28*, 865–873. [CrossRef]
20. Liu, Y.; Zhang, X.; Guan, X.; Delahaye, D. Adaptive sensitivity decision based path planning algorithm for unmanned aerial vehicle with improved particle swarm optimization. *Aerosp. Sci. Technol.* **2016**, *58*, 92–102. [CrossRef]
21. Wenzhao, S.; Naigang, C.; Bei, H.; Xiaogang, W.; Yuliang, B. Multiple UAV cooperative path planning based on PSO-HJ method. *J. Chin. Inert. Technol.* **2020**, *28*, 122–128.
22. Zhang, J.; Sheng, H.; Chen, Q.; Zhou, H.; Yin, B.; Li, J.; Li, M. A Four-dimensional space-time automatic obstacle avoidance trajectory planning method for multi-UAV cooperative formation flight. *Drones* **2022**, *6*, 192. [CrossRef]
23. Chen, Y.B.; Luo, G.C.; Mei, Y.S.; Yu, J.Q.; Su, X.L. UAV path planning using artificial potential field method updated by optimal control theory. *Int. J. Syst. Sci.* **2016**, *47*, 1407–1420. [CrossRef]
24. Sudhakara, P.; Ganapathy, V.; Priyadharshini, B.; Sundaran, K. Obstacle avoidance and navigation planning of a wheeled mobile robot using amended artificial potential field method. *Procedia Comput. Sci.* **2018**, *133*, 998–1004. [CrossRef]
25. Khamis, A.; Hussein, A.; Elmogy, A. Multi-robot task allocation: A review of the state-of-the-art. *Coop. Robot. Sens. Netw.* **2015**, *8*, 31–51.
26. Gao, G.Q.; Xin, B. A-STC: Auction-based spanning tree coverage algorithm for motion planning of cooperative robots. *Front. Inf.Technol. Electron. Eng.* **2019**, *20*, 18–31. [CrossRef]
27. Khan, A.S.; Chen, G.; Rahulamathavan, Y.; Zheng, G.; Assadhan, B.; Lambotharan, S. Trusted UAV network coverage using blockchain, machine learning, and auction mechanisms. *IEEE Access* **2020**, *8*, 118219–118234. [CrossRef]
28. Sun, Y.; Tan, Q.; Yan, C.; Chang, Y.; Xiang, X.; Zhou, H. Multi-UAV Coverage through Two-Step Auction in Dynamic Environments. *Drones* **2022**, *6*, 153. [CrossRef]

Article

A Lightweight Uav Swarm Detection Method Integrated Attention Mechanism

Chuanyun Wang [1], Linlin Meng [1], Qian Gao [1,*], Jingjing Wang [2], Tian Wang [3], Xiaona Liu [4], Furui Du [5], Linlin Wang [1] and Ershen Wang [6]

[1] College of Artificial Intelligence, Shenyang Aerospace University, Shenyang 110136, China
[2] China Academic of Electronics and Information Technology, Beijing 100041, China
[3] Institute of Artificial Intelligence, Beihang University, Beijing 100191, China
[4] Yantai Science and Technology Innovation Promotion Center, Yantai 264003, China
[5] State Key Laboratory of Automatic Control Technology for Mining and Metallurgy Process, Beijing 102628, China
[6] School of Electronic and Information Engineering, Shenyang Aerospace University, Shenyang 110136, China
* Correspondence: gaoqian@buaa.edu.cn

Abstract: Aiming at the problems of low detection accuracy and large computing resource consumption of existing Unmanned Aerial Vehicle (UAV) detection algorithms for anti-UAV, this paper proposes a lightweight UAV swarm detection method based on You Only Look Once Version X (YOLOX). This method uses depthwise separable convolution to simplify and optimize the network, and greatly simplifies the total parameters, while the accuracy is only partially reduced. Meanwhile, a Squeeze-and-Extraction (SE) module is introduced into the backbone to improve the model's ability to extract features; the introduction of a Convolutional Block Attention Module (CBAM) in the feature fusion network makes the network pay more attention to important features and suppress unnecessary features. Furthermore, Distance-IoU (DIoU) is used to replace Intersection over Union (IoU) to calculate the regression loss for model optimization, and data augmentation technology is used to expand the dataset to achieve a better detection effect. The experimental results show that the mean Average Precision (mAP) of the proposed method reaches 82.32%, approximately 2% higher than the baseline model, while the number of parameters is only about 1/10th of that of YOLOX-S, with the size of 3.85 MB. The proposed approach is, thus, a lightweight model with high detection accuracy and suitable for various edge computing devices.

Keywords: object detection; Unmanned Aerial Vehicle (UAV) swarm; lightweight model; attention mechanism; data augment

Citation: Wang, C.; Meng, L.; Gao, Q.; Wang, J.; Wang, T.; Liu, X.; Du, F.; Wang, L.; Wang, E. A Lightweight Uav Swarm Detection Method Integrated Attention Mechanism. *Drones* **2023**, *7*, 13. https://doi.org/10.3390/drones7010013

Academic Editor: Carlos Tavares Calafate

Received: 27 November 2022
Revised: 22 December 2022
Accepted: 23 December 2022
Published: 25 December 2022

1. Introduction

In recent years, with the rapid development and wide application of Unmanned Aerial Vehicle (UAV) technology in civil and military fields, there has been a tremendous escalation in the development of applications using UAV swarms. Currently, the main research effort in this context is directed toward developing unmanned aerial systems for UAV cooperation, multi-UAV autonomous navigation, and UAV pursuit-evasion problems [1].

In modern wars, where UAVs are widely used, the technical requirements for anti-UAV technologies are becoming increasingly significant [2]. However, existing anti-UAV technologies are not enough to effectively deal with the suppression of UAV swarms [3]. For a small number of UAVs, countermeasures such as physical capture, navigation deception, seizing control and physical destruction can be used. But it is difficult to cope with a large number of UAVs once they gather together to form a UAV swarm. It is imperative to be able to detect the incoming UAV swarm from a long distance in time and then carry out scale estimation, target tracking, and other operations.

Therefore, the development of UAV swarm target detection and tracking, etc., is the premise and key to achieving comprehensive awareness, scientific decision-making, and active response in battlefield situations. Because anti-UAV swarm systems have high requirements for the accuracy and speed of object location and tracking methods and radar detection, passive location, and other methods experience significant interference from other signal clutter, resulting in false detections or missing detection problems. Therefore, using computer vision technology to detect and track the UAV swarm has significant research value.

This paper focuses on UAV target detection for anti-UAV systems. Specifically, under our proposed method, once the UAV swarm is detected, detectors are rapidly deployed on the ground to obtain video, and quickly and accurately detect the target to facilitate subsequent countermeasures.

Most of the existing object detection algorithms consider the object scale to be of medium size, while a low-flying UAV accounts for a very small proportion of the image, and there is little available texture information. It is difficult to extract useful features, especially against a complex background, and, thus, it is easy to mistakenly detect or miss the UAV target. Therefore, in order to improve the capability of UAV swarm detection in different scales and complex scenes, and meet the application requirements in resource-constrained situations, such as in terms of computing power and storage space, this paper proposes a lightweight UAV swarm detection method that integrates an attention mechanism. Data augmentation technology is applied to expand the dataset to improve the diversity of the training set. In addition, depthwise separable convolution [4] is used to compress the main structure of the network, with the aim of building a model that meets the accuracy requirements and takes up as little computing resources as possible.

We train and test based on the UAVSwarm dataset [5], and the experimental results show that the mAP value of the proposed method reaches 82.32%, while the number of parameters is only about 1/10th of that of the YOLOX-S model and the model size is only 3.85 Mb. Under the same experimental conditions, compared with other YOLO series lightweight models, the detection accuracy of the proposed method is 15.59%, 15.41%, 1.78%, 0.58%, and 1.82% higher than MobileNetv3-yolov4, GhostNet-Yolov4, YOLOv4-Tiny, YOLOX-Tiny, and YOLOX-Nano models, respectively. At the same time, the total network parameters and model size are excellent.

The main innovations of this paper are as follows:

(1) The depthwise separable convolution method is used to compress the model, and a nano network is constructed to achieve the lightweight UAV swarm detection network.

(2) A Squeeze-and-Extraction (SE) module [6] is introduced into the backbone to improve the network's ability to extract object features. The introduction of a Convolutional Block Attention Module (CBAM) [7] in the feature fusion network makes the network pay more attention to important features and suppress unnecessary features.

(3) During the training process, Distance-IoU (DIoU) [8] is used instead of Intersection over Union (IoU) to calculate the regression loss, which is beneficial for model optimization. At the same time, Mosaic [9] and Mixup [10] data augmentation technologies are used to expand the dataset to achieve a better detection effect.

2. Related Work

Swarm intelligence algorithms play an extremely important role in multiple UAV collaborations such as collision avoidance, task assignment, path planning, and formation reconfiguration. Object detection is an important computer vision task. Traditional object detection methods are mostly based on manual feature construction [11], which has weak generalization ability and takes up large computing resources. In recent years, with the vigorous development and wide application of deep learning technology in various fields, algorithms based on deep learning have been widely studied by researchers.

2.1. Related Work for UAV Swarm

Currently, lots of researchers pay attention to the development of UAV systems for UAV cooperation, multi-UAV autonomous navigation, and UAV pursuit–evasion problems. For successful communication among collaborating UAVs in a swarm, Cheriguene, Y. et al. [12] proposed COCOMA, an energy-efficient multicast routing protocol for UAV swarms. This method builds a multicast tree that can convey data from a single source to the swarm's UAVs in order to pick the shortest distance between UAVs, optimize total network energy consumption, and extend the network lifetime. Tzoumas, G. et al. [13] newly developed a control algorithm called dynamic space partition (DSP) for a swarm system consisting of high payload UAVs to monitor large areas for firefighting operations. Sastre, C. et al. [14] proposed and validated different algorithms to optimize the take-off time of drones belonging to a swarm, and the experiments proved that the proposed algorithms provide a robust solution within a reasonable time frame. Sastre, C. et al. [15] proposed a collision-free take-off strategy for UAV swarms. Experimental results show that the proposed method can significantly improve time efficiency and keep the risk of collision at zero.

2.2. Related Work for UAV Detection

Object detection algorithms based on deep learning are mainly divided into two categories: two-stage and one-stage detectors. In the former approach, first, a region proposal network is used to estimate a candidate object bounding box. Then, in the second stage, the network extracts features from each candidate box and performs classification and bounding box regression. In this manner, several methods such as R-CNN [16], Fast R-CNN [17], and Faster R-CNN [18] have been proposed. The latter object detector uses a single deep neural network with a regression strategy to directly classify and detect objects. It should be noted that, in this approach, the process of region proposal is avoided. In this manner, several methods such as the You Only Look Once (YOLO) series, Single Shot Detector (SSD) [19], and RetinaNet [20] have been proposed.

With the rapid development of computer vision technology, researchers have carried out a lot of research on image-based UAV detection algorithms. Hu Y. et al. [21] introduced an algorithm based on YOLOv3 into UAV object detection for the first time. In the prediction process, the last four scale feature maps are adopted to conduct multi-scale prediction to enrich the texture and contour information. At the same time, the size of the UAV in four scales feature maps is calculated according to input data, and then the number of anchor boxes is also adjusted. This approach improves the accuracy of small object detection while ensuring speed. Sun H. et al. [22] proposed a UAV detection network named TIB-Net, integrating a structure called cyclic pathway into the existing efficient method Extremely Tiny Face Detector (EXTD) to enhance the capability of the model to extract effective features of small objects. Furthermore, they integrated a spatial attention module into the backbone network to better locate small-size UAVs and further improve detection performance. Ma J. et al. [23] integrated the attention mechanism module into the PP-YOLO detection algorithm and introduced the Mish activation function to eliminate the gradient disappearance problem in the back-propagation process, which significantly improved the detection accuracy. Yavariabdi A. et al. [24] proposed a multi-UAV detection network named FastUAV-NET based on YOLOv3-tiny that can be used for embedded platforms. By increasing the depth and width of the backbone network, local and global features are extracted from the input video stream, providing higher detection accuracy and saving computing time. Liu B. et al. [25] replaced the backbone with a lightweight network Efficient-lite based on YOLOv5s to reduce the number of parameters of the model, introduced adaptive spatial feature fusion technology to balance the loss of accuracy caused by simplifying the network model, and, finally, introduced a constraint of angle into the original regression loss function to avoid the mismatch between the prediction frame and the real frame orientation during the training process in order to improve the speed of network convergence. Wang C. et al. [26] used the Se-ResNet as a feature extraction network by introducing the SeNet attention mechanism into the backbone to improve the correlation between feature channels and enhance the features of the target to solve the problem of UAVs in low-altitude airspace being submerged in complex background clutter. The differences between the proposed method and the existing UAV detection frameworks are shown in Table 1.

2.3. Related Work for Lightweight Network

Although R-CNN, YOLO, and SSD series algorithms have excellent performance in object detection, they generally have high computational complexity and large model volume, which makes them unable to fully meet the application requirements in resource-constrained situations such as limited computing power, storage space, and/or power consumption [27]. With the development of intelligent mobile devices toward marginalization and mobility, various lightweight object detection algorithms have been developed successively. The goal is to keep good detection performance on devices with low hardware conditions to adapt to the development trend of intelligent devices. The basic idea of MobileNet [4,28,29], based on depthwise separable convolution, is to use depthwise convolution to replace the filter in traditional convolution for feature extraction, and use point convolution instead of filter to combine features while reducing the number of parameters and amount of computation. ShuffleNet [30,31] was published by Zhang X. et al. in 2018, the core of which is to reduce the computation of a large number of point convolutions in MobileNet by using the strategy of combining group convolution and channel shuffle. Tan M. et al. [32] proposed MnasNet in 2019. Its core innovation lies in the proposed multi-objective optimization function and the decomposed hierarchical search space, which correspond to the optimization accuracy and reasoning delay, and improve the diversity between different layers. Han K. et al. [33] proposed GhostNet in 2020. First, they used less convolution to check the input for conventional convolution, obtaining the output features with fewer channels as the internal feature map. Then, they linearly transformed each channel of the internal feature map to obtain its corresponding Ghost feature map. Finally, they connected the internal feature map with the Ghost feature map to obtain the final GhostNet convolution output feature. Xiong Y. et al. [34] proposed a lightweight object detection network called MobileDets in 2021, based on a Neural Architecture Search (NAS) network architecture for object detection tasks, and achieved the state-of-the-art in mobile accelerators.

Table 1. A comparative overview of UAV detection methods.

Method	Detection Strategy	Backbone	Dataset
Hu Y. et al. [14]	YOLOv3-based	DarkNet-53	self-built
Sun H. et al. [15]	TIB-Net	EXTD	self-built
Ma J. et al. [16]	PP-YOLO	ResNet50-vd	Drone-vs-Bird, TIB-Net, and self-built
Yavariabdi A. et al. [17]	FastUAV-NET	Inception module	self-built
Liu B. et al. [18]	YOLOv5-based	Efficientlite	self-built
Wang C. et al. [19]	Se-ResNet	ResNet-18	Drone-vs-Bird and LaSOT
Proposed Method	YOLOX-based	CSPDarkNet	UAVSwarm [33]

3. Materials and Methods

3.1. Overview

The object detection model can generally be abstracted into backbone, neck, and head networks, as shown in Figure 1. The backbone network performs feature extraction, the neck network performs multi-scale feature fusion on the feature layer obtained by the backbone network, and the head network performs classification and regression analysis.

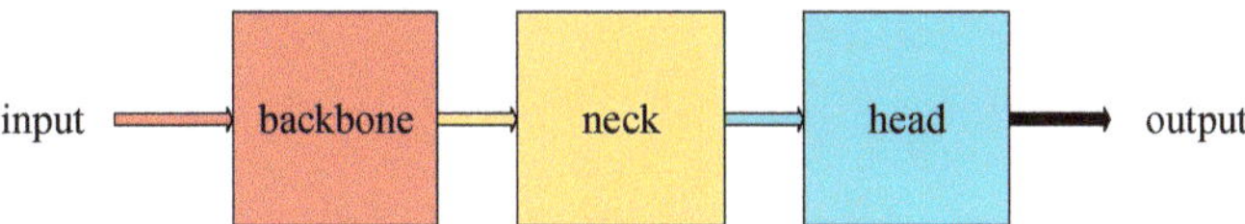

Figure 1. Overall structure of object detection model.

3.2. Backbone Network

CSPDarkNet [35] is used as the backbone of our UAV swarm detection model, consisting mainly of convolution layers and a CSP structure, as shown in Table 2. First, a 640 × 640 RGB three-channel image is input into the network, and the image size and the number of channels are adjusted through Focus. Then, four stacked Resblock body modules are used for feature extraction. In the last Resblock body module, the image is processed through the SPP module; that is, the max pooling operation with different kernel sizes is used for feature extraction to improve the receptive field of the network. The final output of CSPDarkNet is the feature maps of the 2nd, 3rd, and 4th Resblock body modules, with the shapes of 80 × 80 × 256, 40 × 40 × 512, and 20 × 20 × 1024, respectively.

Table 2. The structure of CSPDarkNet.

Module	Structure	Output	Valid Output
Inputs	Inputs	640 × 640 × 3	
Focus	Focus	320 × 320 × 12	
Conv2D-BN-SiLU	Conv2D-BN-SiLU	320 × 320 × 64	
Resblock body1	Conv2D-BN-SiLU	160 × 160 × 128	
	CSPLayer	160 × 160 × 128	
Resblock body2	Conv2D-BN-SiLU	80 × 80 × 256	Output 1
	CSPLayer	80 × 80 × 256	
Resblock body3	Conv2D-BN-SiLU	40 × 40 × 512	Output 2
	CSPLayer	40 × 40 × 512	
Resblock body4	Conv2D-BN-SiLU	20 × 20 × 1024	Output 3
	SPPBottleneck	20 × 20 × 1024	
	CSPLayer	20 × 20 × 1024	

In the Resblock body module, the CSPLayer is similar to the residual structure. The input first passes through convolutional layers with a kernel size of 1×1 and 3×3 for n times, then the result and the original input are concatenated as output, as shown in Figure 2a. In order to further improve the detection of UAV swarm targets, referring to the MobileNet V3 model [29], a Squeeze-and-Extraction (SE) module [6] is introduced into the CSPLayer structure, as shown in Figure 2b. The UAV detection model uses SiLU as the activation function. SiLU has the characteristics of no upper bound, with a lower bound and smooth and non-monotone functions. It can converge faster during training and its formula is as follows:

$$SiLU(x) = x \cdot sigmoid(x), \tag{1}$$

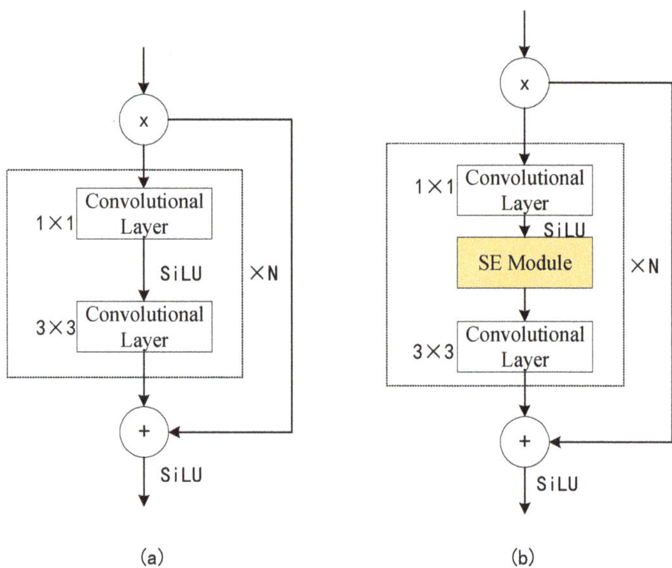

Figure 2. (a) Original CSPLayer structure (b) CSPLayer structure with SE module.

Specifically, the SE module generates different weight coefficients for each channel by using the correlation between feature channels, multiplies them with the previous features, and adds them to the original features to enhance the features. As shown in Figure 3, the detailed process of the SE attention mechanism is as follows: First, the extracted feature $X \in \mathbb{R}^{H' \times W' \times C'}$ is mapped to $U \in \mathbb{R}^{H \times W \times C}$ through the conversion function F_{tr}. Then, the global information of each channel is represented with a channel characteristic description value through global average pooling $F_{sq}(\cdot)$; and then the channel characteristic description value is adaptively calibrated by $F_{ex}(\cdot, W)$ to make the weight value more accurate. Finally, the enhanced feature is obtained by multiplying the weight value and the original feature through $F_{scale}(\cdot, \cdot)$.

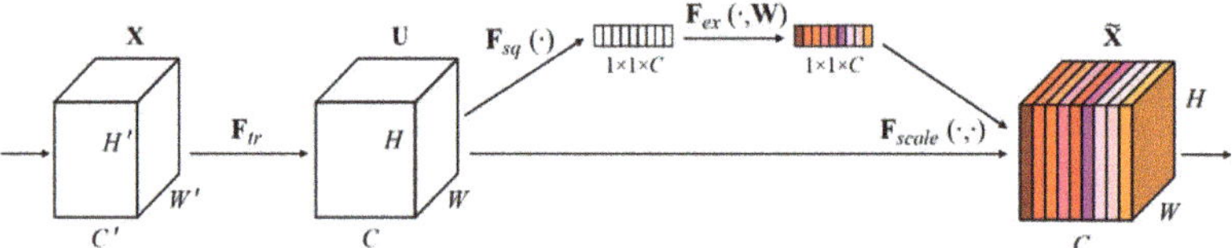

Figure 3. Diagrammatic sketch of the SE module.

3.3. Neck Network

The three feature layers obtained by CSPDarkNet are sent to the neck network for enhanced feature extraction and feature fusion. The neck network of the UAV swarm detection model is constructed based on the Path Aggregation Network (PANet) [36]. The input feature map is resized through a convolution layer and then fused through up- and down-sampling operations. The specific network structure is shown in Figure 4.

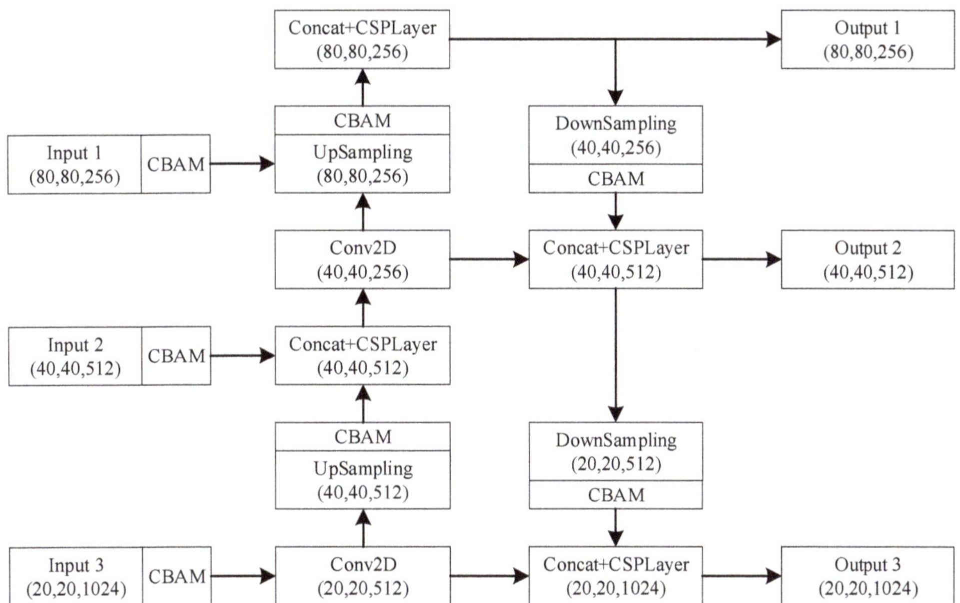

Figure 4. The structure of the neck network.

In order to improve the detection performance of the model for UAVs, a Convolutional Block Attention Module (CBAM) [7] is firstly applied to the three feature maps obtained by the backbone network, and then sent to the neck network for feature fusion. The CBAM is also applied after each up-sampling and down-sampling operation in the PANet. CBAM is a simple and effective attention module for feedforward convolutional neural networks. It combines the two dimensions of channel and spatial features. When the feature map is input, it first goes through the Channel Attention Module (CAM) and then Spatial Attention Module (SAM). The calculation formula is (2):

$$\begin{cases} F_c = M_c(F) \otimes F \\ F_a = M_s(F_c) \otimes F_c' \end{cases} \tag{2}$$

As shown in Figure 5, in the CAM, for the input feature map $F \in \mathbb{R}^{C \times H \times W}$, first, Global Average Pooling (GAP) and Global Maximum Pooling (GMP) operations are performed based on the width and height of the input feature map, and then they are processed by a shared neural network Multilayer Perceptron (MLP), respectively. The two processed results are added together, and a one-dimensional channel attention vector $M_c \in \mathbb{R}^{C \times 1 \times 1}$ is obtained through the Sigmoid function. Finally, the feature map F_c with channel weights is generated by multiplying the channel attention vector M_c and the feature map F. In the SAM module, for the feature map F_c obtained by CAM, a pooling operation is performed, and then the 7×7 convolution and Sigmoid function to obtain a two-dimensional spatial attention vector $M_s \in \mathbb{R}^{1 \times H \times W}$. Finally, F_c and M_s are multiplied to obtain the final feature map F_a.

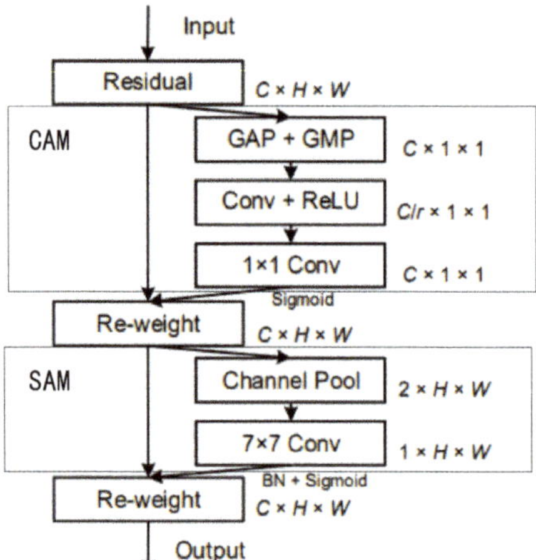

Figure 5. Diagrammatic sketch of the CBAM.

3.4. Head Network

After feature fusion and enhanced feature extraction are completed, the head network conducts classification and regression analysis on the three feature layers of different scales, and finally outputs the recognition results. Its network structure is shown in Figure 6. The head network of the UAV detection model proposed in this paper has two convolution branches [35], one of which is used to achieve object classification and output object categories. The other branch is used to judge whether the object in the feature point exists and regress the coordinates of the bounding box. Thus, for each feature layer, three prediction results can be obtained:

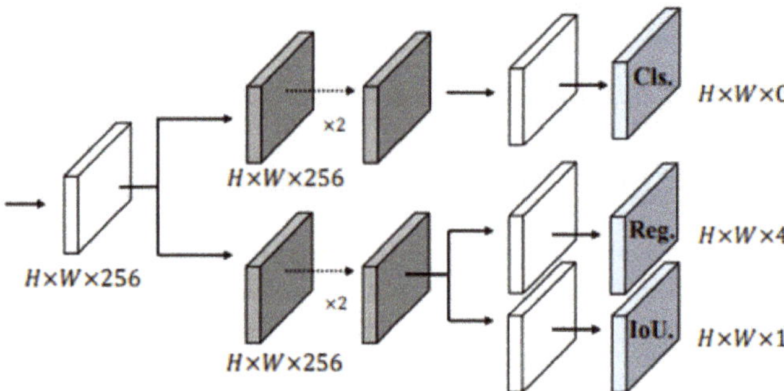

Figure 6. The structure of the head network.

(1) Reg (h, w, 4): The position information of the target is predicted. The four parameters are x, y, w, and h, where x and y are the coordinates of the center point of the prediction box, and w and h are the width and height.

(2) Obj (h, w, 1): This is used to judge whether the prediction box is a foreground or a background. After being processed by the Sigmoid function, it provides the confidence of the object contained in each prediction box. The closer the confidence is to 1, the greater the probability of the existence of a target.

(3) Cls (h, w, num_classes): Determine what type of object, each object is, give each type of object a score, and obtain the confidence level after the sigmoid function processing.

The above three prediction results are stacked, and the prediction result of each feature layer is (h, w, 4+1+num_classes). The first four parameters of the last dimension are regression parameters of each feature point and the fifth parameter is used to judge whether each feature point contains an object, and the last num_classes parameter is used to judge the category of the object contained in each feature point.

3.5. Lightweight Model

The essence of a lightweight model is to solve the limitations of storage space and energy consumption on the performance of traditional neural networks on equipment with low-performance hardware. Aiming at the problem that the traditional deep convolution neural network consumes a large amount of computing resources, this paper pays more attention to how to reduce the complexity of the model and the amount of computation, while improving the accuracy of object detection. Considering that the depthwise separable convolution method [4] can effectively compress the model size while retaining the ability of feature extraction, this paper uses it to simplify and optimize the UAV swarm detection network.

A standard convolution both filters and combines inputs into a new set of outputs in one step. Depthwise separable convolution splits this into two layers, for filtering and combining. While minimizing the loss of accuracy, this approach can greatly simplify the network parameters and reduce the amount of calculation. The depth-separable convolution operation divides the traditional convolution operation into two steps: depthwise convolution and pointwise convolution. The depthwise convolution applies a single filter to each input channel. The pointwise convolution then applies a 1×1 convolution to combine the outputs of the depthwise convolution. The standard convolution operation and the depthwise separable convolution operation are shown in Figure 7a,b, respectively. The depthwise separable convolution is used to replace the traditional convolution in the UAV detection network while reducing the network parameters and computation.

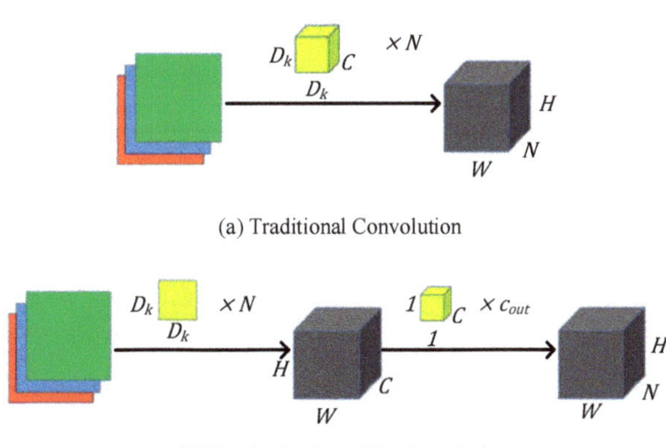

Figure 7. Comparison of the two convolution methods.

The total number of convolution kernel parameters and the total amount of convolution operations are analyzed to determine the amount of internal product operations. If we assume that N groups of convolutions, having the same kernel size, $D_k \times D_k \times C$, are taken to check the input image for convolution and that the required feature map size is $W \times H \times N$, then the quantity of parameters and operation required by the two methods are shown in Table 3. The depthwise separable convolution method can compress the network size and reduce the amount of computation. In the process of obtaining a fixed-size feature map using convolution kernels of the same width and height, by expressing convolution as a two-step process of filtering and combining, we get a reduction in necessary computations of $1/N + 1/D_k{}^2$, which is the key to achieving light weight.

Table 3. Comparison of the two convolution methods.

Method	Parameters	Computational Amount	Ratio
standard convolution	$D_k \times D_k \times C \times N$	$D_k \times D_k \times C \times N \times W \times H$	1
depthwise separable convolution	$D_k \times D_k \times C + C \times N$	$D_k \times D_k \times C \times W \times H + C \times N \times W \times H$	$\frac{1}{N} + \frac{1}{D_k^2}$

3.6. Model Training

In the training process of the UAV swarm detection model, Distance-IoU (DIoU) [8] is used instead of Intersection over Union (IoU) to calculate regression loss, which is beneficial for model optimization. In addition, Mosaic [9] and Mixup [10] data augmentation technologies are used to expand the dataset to achieve better UAV detection.

3.6.1. Data Augmentation

Data augmentation is a means of expanding the dataset in computer vision. The approach enhances the image data to compensate for the problem of insufficient training dataset images and achieve the purpose of expanding the training data. As the UAV swarm dataset UAVSwarm used for the experiment has few training samples and repetitive scenes, the Mosaic and Mixup algorithms are used in the image data preprocessing process in order to increase the diversity of training samples and enrich the background of the target, to avoid, as far as possible, the network falling into overfitting during the training process and improve the recognition accuracy and generalization ability of the network model.

The enhancement effect of the Mosaic algorithm is shown in Figure 8. First, the four images are randomly cut, scaled, and rotated, and then they are spliced into a new image as the input image for model training. It should be noted that the image during processing contains the coordinate information of the bounding box of the target, so the new image obtained also contains the coordinate information of the bounding box of the UAV. The advantage of this is that, on the one hand, the size of the object in the picture is reduced to meet the requirements for small object detection accuracy, and, on the other hand, the complexity of the background is increased, so that the UAV swarm detection model has better robustness toward complex backgrounds.

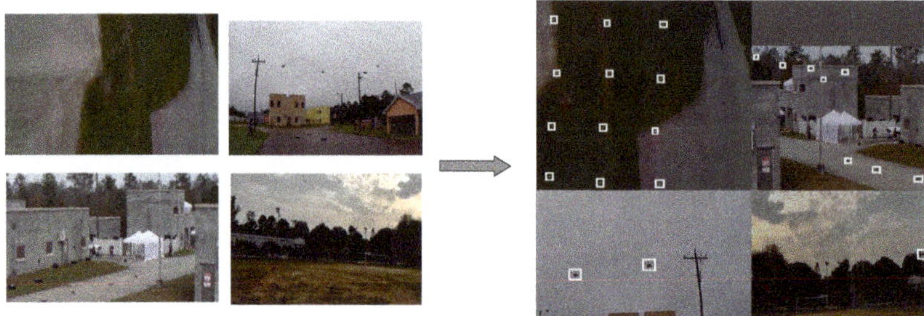

Figure 8. The enhancement effect of Mosaic.

The Mixup algorithm was originally used for image classification tasks. The core idea is to randomly select two images from each batch and mix them up to generate new images in a certain proportion. The Mixup algorithm is more lightweight than Mosaic, requiring only minimal computational overhead, and can significantly improve the operation speed of the model. Its mathematical expression is as follows:

$$\begin{cases} \tilde{x} = \lambda x_i + (1 - \lambda)x_j \\ \tilde{y} = \lambda y_i + (1 - \lambda)y_j \, , \\ \qquad \lambda \epsilon [0,\ 1] \end{cases} \tag{3}$$

where (x_i, y_i) and (x_j, y_j) are two randomly selected samples and their corresponding labels, $(\tilde{x}, \tilde{y})$ are the newly generated samples and their corresponding labels that will be used to train the neural network model, and λ is a fusion coefficient. It can be seen from Formula (3) that Mixup essentially fuses two samples through a fusion coefficient. The enhancement effect of the Mixup algorithm is shown in Figure 9.

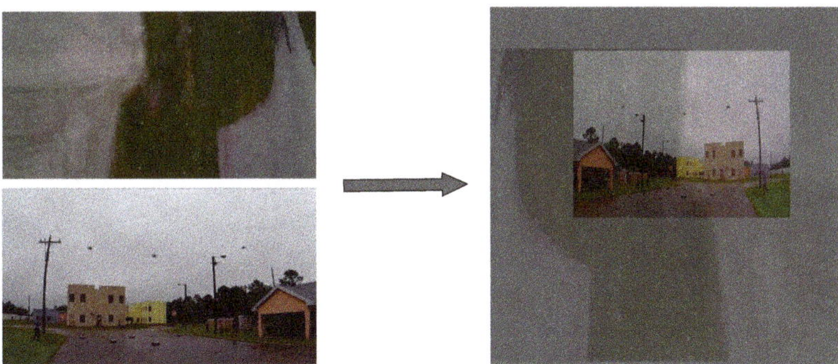

Figure 9. The enhancement effect of Mixup.

3.6.2. Loss Function

The goal of network training is to reduce the loss function and make the prediction box close to the ground truth box to obtain a more robust model. The loss function of object detection needs to indicate the proximity between the prediction box and the ground truth, whether the prediction box contains the target to be detected, and whether the object category in the prediction box is true. As predicted by the head network, the loss function consists of three parts, which are given by Formula (4):

$$Loss = Loss_{Reg} + Loss_{Obj} + Loss_{Cls} \, , \tag{4}$$

(1) Regression loss ($Loss_{Reg}$) is the loss of position error between the prediction box and the ground truth. The x, y, w, and h parameters predicted by the model can locate the position of the prediction box, and the loss is calculated based on the DIoU of the ground truth and the prediction box. Figure 10 shows the principle for DIoU to calculate regression loss, and the corresponding calculation Formula is (5):

$$DIoU = IoU - \frac{\rho^2\left(b, b^{gt}\right)}{c^2} = IoU - \frac{d^2}{c^2} \, , \tag{5}$$

where b represents the parameter of the center coordinate of the prediction box and b^{gt} represents the parameter of the center coordinate of the ground truth; d is the distance between the center point of the prediction box and the ground truth; and c represents the diagonal length of the maximum bounding rectangle of the union of the prediction box and ground truth. IoU measures the intersection ratio between the prediction box and the ground truth. However, if there is no intersection between them, the result of IoU will always be 0. When one of the two boxes is inside the other, if the size of the box remains unchanged, the calculated IoU value will not change, which will make the model difficult to optimize. If DIoU is used to calculate regression loss, this problem can be effectively solved and a good measurement effect can be obtained.

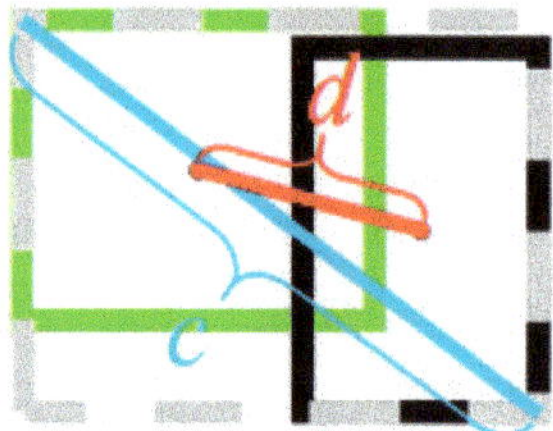

Figure 10. Schematic diagram of DIoU.

(2) Object loss ($Loss_{Obj}$) is to determine whether there is an object in the predicted box, which is a binary classification problem. According to the result predicted by the head network, whether the target is included can be known, while the feature points corresponding to all ground truths are positive samples, and the remaining feature points are negative samples. The Binary Cross-Entropy loss is calculated according to the prediction results of whether the positive and negative samples include the target.

(3) Classification loss ($Loss_{Cls}$) is applied to reflect the error in object classification. According to the feature points predicted by the model, the predicted category results of the feature points are extracted, and then the Binary Cross-Entropy loss is calculated according to the category of the ground truth and prediction results.

4. Experimental Results and Analysis

The proposed UAV swarm target detection model is constructed based on the deep learning framework Pytorch. The size of the input images needs to be adjusted to 640×640, and the number of input images in each batch is set to 12 during the training process; a total of 100 epochs are trained without using a pre-training weight. Furthermore, the Mosaic and Mixup data augment algorithms are used for the first 70 epochs and canceled for the last 30 epochs. The gradient descent optimization strategy adopts the SGD optimizer, and the initial learning rate is set to 0.01. In this experiment, the mean Average Precision (mAP), the number of parameters, model size, latency, and Frame Per Second (FPS) are used as measurement metrics of the experimental results. We train and test on a computer equipped with dual Intel Xeon E5 2.40GHz CPUs, a single NVIDIA GTX 1080TI GPU, and 32 GB RAM.

4.1. UAVSwarm Dataset

Wang C. et al. [5] collected 72 UAV image sequences and manually annotated them, creating a new UAV swarm dataset named UAVSwarm for UAV multi-object detection and tracking. This dataset contains 12,598 images in total, of which 23 are included in the images with the largest number of UAVs, 36 image sequences (6844 images) are included in the training set, and the remaining 36 sequences (5754 images) are included in the test set. The dataset we used largely excludes all objects (flocks of birds, etc.) except UAVs but some scenes are complex, leading to the UAVs being blocked. Figure 11 shows some images and annotation information of the dataset.

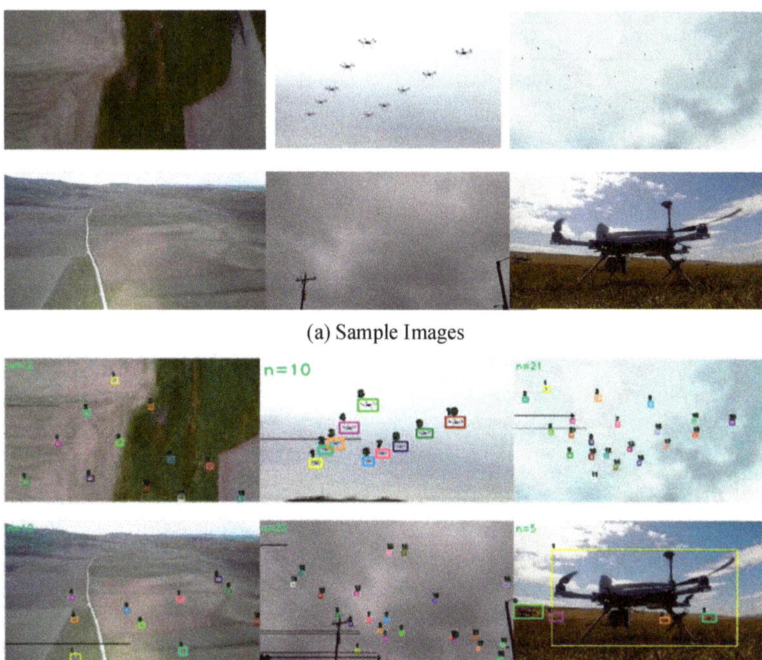

(a) Sample Images

(b) Sample Images with Labels

Figure 11. Samples of the UAVSwarm dataset.

4.2. Ablation Experiment

Firstly, the structure of YOLOX is simplified and optimized by depthwise separable convolution to build a nano network. In order to verify the effectiveness of the lightweight module, this paper uses the same training strategy to train three lightweight YOLOX models, namely, YOLOX-S, YOLOX-Tiny, and YOLOX-Nano, and tests them on the same test set to analyze their performance differences. It can be seen from Table 4 that the UAV detection accuracy of the three differently scaled YOLOX models toward the test set is more than 80%. As far as the network accuracy and scale of YOLOX of the same series are concerned, the results of this experiment are consistent with the general law of the object detection network—that is, the more layers and parameters of the convolutional neural network, the stronger its feature extraction and generalization ability, and the higher its recognition accuracy. When the DIoU threshold score is set to 0.5, the mAP scores of the three networks are largely the same, but the size of the model and the total number of parameters greatly differ Among them, the model size and the total number of parameters of nano network are about 1/10th of those of the version S network. This shows that under the same hardware conditions, the lightweight nano model can process more input images and reduce the equipment cost on the premise of meeting the accuracy requirement. Therefore, YOLOX-Nano is selected as the baseline model for research in this paper.

Table 4. Performance comparison of YOLOX models with different scales.

Model	mAP@0.5 (%)	Params ($\times 10^6$)	Model Size/Mb
YOLOX-S	82.12	8.94	34.30
YOLOX-Tiny	81.74	5.03	13.70
YOLOX-Nano	80.50	0.89	3.70

Table 5 shows that the mAP score of the proposed model on the UAVSwarm test set is 82.32%, which is about 2% higher than that of the baseline model, while the total number of parameters and model size are only about 40 Kb higher. It can also be found that the introduction of the SE and CBAM modules and the improvement of the loss functions have brought about an increase in mAP compared with the baseline model, which proves the effectiveness of the above three modules.

Table 5. Ablation experiment based on YOLOX-Nano.

SE	CBAM	Loss	mAP@0.5 (%)	Params ($\times 10^6$)	Model Size/Mb
-	-	-	80.50	0.89	3.70
✓	-	-	81.25	0.92	3.78
-	✓	-	80.62	0.92	3.84
-	-	✓	82.14	0.89	3.70
✓	✓	✓	82.32	0.93	3.85

The Loss curves of the different network models during the training process are shown in Figure 12. The abscissa and ordinate are the Epoch and Loss values, respectively. It can be seen that in the training process, the convergence of different models of the YOLOX series is similar. The loss of the training set decreases rapidly in the early stage. With the increase in epoch, the loss value gradually decreases and tends to be stable. Finally, the loss value of the proposed network model is the lowest, which proves that the training strategy and parameter settings are reasonable and effective for improving the model detection accuracy.

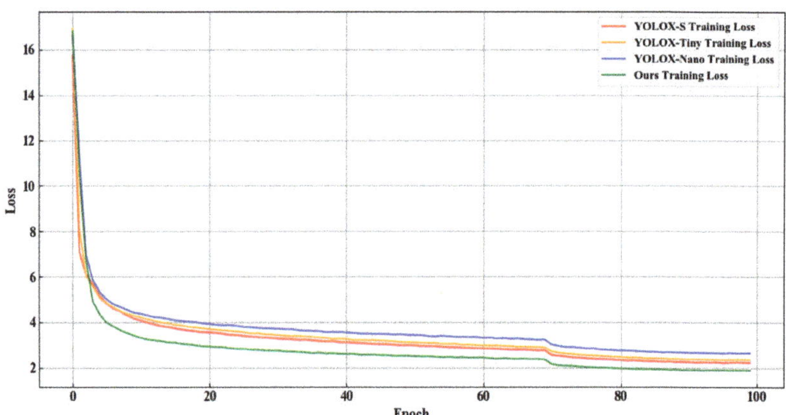

Figure 12. Loss curves of the different networks.

For the UAVSwarm dataset, as a typical small object, the network model proposed in this paper has improved the Precision and Recall indicators of the baseline model by 0.63% and 1.84%, respectively, when the threshold scores are both 0.5, as shown in Figure 13. The improvement of Recall shows that the optimization strategy we used can effectively increase the learning effect of the model on the target of positive foreground samples.

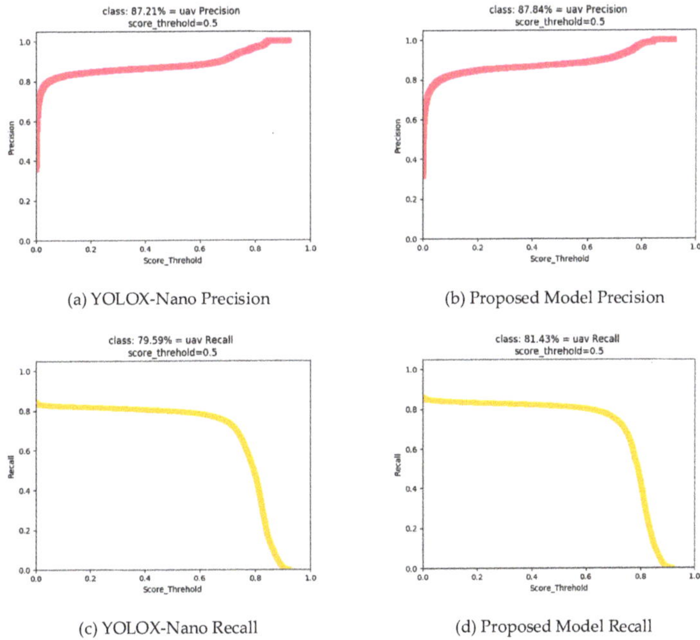

(a) YOLOX-Nano Precision

(b) Proposed Model Precision

(c) YOLOX-Nano Recall

(d) Proposed Model Recall

Figure 13. Precision and Recall of UAV detection.

4.3. Comparison Experiment

In order to objectively reflect the performance of the UAV swarm object detection network proposed in this paper, this study also uses the same settings to train other lightweight YOLO models and conducts a comparative analysis. The comparison results are shown in Table 6. It can be seen that under the same test set, the mAP value of the proposed UAV detection model has reached 82.32%, which is 15.59%, 15.41%, 1.78%, 0.58%, and 1.82% higher than MobileNetv3-yolov4, GhostNet-YOLOv4, YOLOv4-Tiny, YOLOX-Tiny, and YOLOX-Nano, respectively. At the same time, the total network parameters and model size are optimal.

Table 6. Comparison experiment of different networks.

Method	mAP@0.5 (%)	Params ($\times 10^6$)	Model Size/Mb	Latency/ms	FPS
GhostNet-YOLOv4	66.91	11.00	42.40	83	12
MobileNetv3-YOLOv4	66.73	11.30	53.70	90	11
YOLOv4-Tiny	80.54	5.87	22.40	45	22
YOLOX-Tiny	81.74	5.03	13.70	55	18
YOLOX-Nano	80.50	0.89	3.70	71	14
Ours	82.32	0.93	3.85	71	14

The experimental results that are obtained on the computational time are tabulated in Table 6. It should be said that, due to limitations in the hardware platform, our experiment did not achieve the effect in the original paper, but we believe that the proposed method will have lower latency and higher processing speed under the condition of higher computing power. Table 6 illustrates that, compared with the baseline model YOLOX-Nano, the proposed model achieves higher recognition accuracy under approximately the same inference time. Thus, our model can also meet the requirements of being real-time and is suitable for applications that require low latency. However, the YOLOX-Tiny and YOLOv4-Tiny networks are quicker than the proposed model, as they run 18 FPS and 22 FPS, respectively. Even though the proposed architecture is slower than these two models, significantly, it provides higher detection accuracy. Therefore, the proposed model is a lightweight model with high detection accuracy and suitable for various edge computing devices.

In order to verify the detection effect of the UAV detection model proposed in this paper in actual scenes, some images in the dataset are selected for detection, and the comparison diagram of the detection effect is shown in Figure 14. As shown in Figure 14a, in an environment with a simple background, most models can successfully identify UAVs. The proposed method can detect more small and distant objects than the baseline model YOLOX-Nano. As shown in Figure 14b, when the background is complex or UAVs are densely distributed, there are many undetected phenomena with the other models. The algorithm in this paper shows better detection performance and can accurately detect UAVs when occlusion occurs among objects. As shown in Figure 14c, when the image resolution is low, the algorithm in this paper can still accurately detect UAVs with complex backgrounds and provide higher confidence scores. Through comparison, it can be demonstrated that our method has improved detection accuracy and confidence, which shows that it can satisfy the requirements of being lightweight and providing higher accuracy, meeting the requirements of industrial applications.

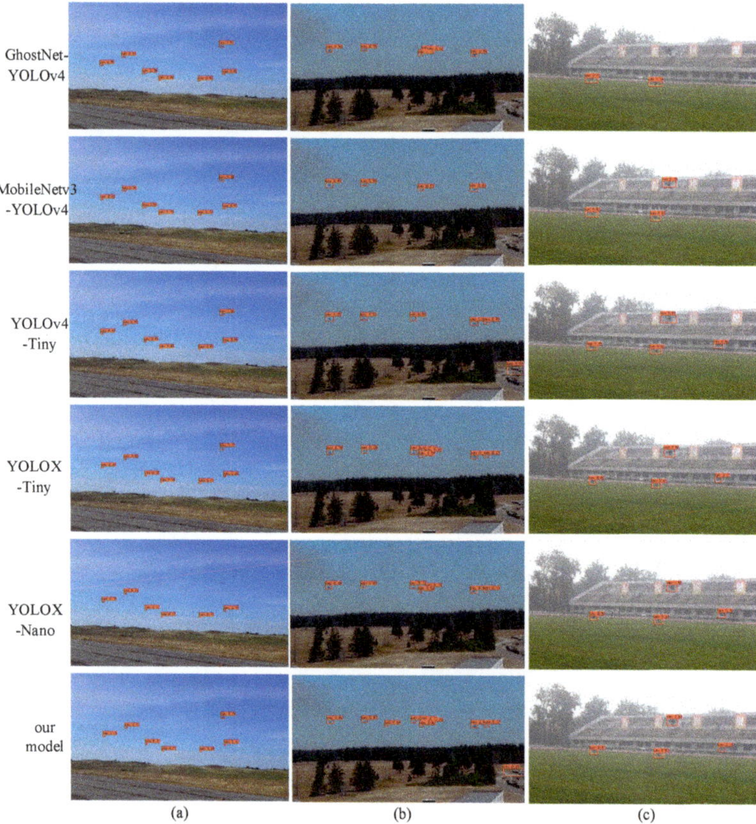

Figure 14. Comparison of detection results of different models.

In addition to the construction of an anti-UAV system, the algorithm proposed in this paper can also be applied as a swarm intelligence algorithm to achieve UAV swarm formation, multi-UAV cooperation, etc. By deploying lightweight object detection methods, UAVs in the swarm can quickly and accurately obtain the position and status of other partners, so that they can adjust themselves in time according to the swarm intelligence algorithm. Swarms of UAVs may have enhanced performance during performing some missions where having coordination among multiple UAVs may enable broader mission coverage and provide more efficient operating performance. Moreover, the proposed method will greatly help to improve the performance of UAV swarm systems including total energy, average end-to-end delay, packets delivery ratio, and throughput. According to the analysis, the application of the lightweight model provided by this paper may reduce the total energy demand and average end-to-end delay of the UAV swarm system, while increasing the packets delivery ratio and throughput of the system. This means that the UAV swarm system may be able to achieve higher data transmission efficiency at a lower cost, thus, better performing tasks.

5. Conclusions

This paper proposes a lightweight UAV swarm detection model integrating an attention mechanism. First, the structure of the network is simplified and optimized by using the depthwise separable convolution method, which greatly reduces the total number of parameters of the network. Then, a SE module is introduced into the backbone network to improve the model's ability to extract object features; the introduction of a CBAM in the feature fusion network makes the network pay more attention to important features and suppress unnecessary features. Finally, in the training process, a loss function based on DIoU can better describe the overlapping information and make the regression faster and more accurate. In addition, two data augmentation technologies are used to expand the UAVSwarm dataset to achieve better UAV detection. The proposed model is a lightweight model with high detection accuracy and only 3.85 MB in size, which is suitable for embedded devices and mobile terminals. In conclusion, the real-time performance and accuracy of the UAV swarm detection model proposed in this paper meet the requirements of rapid detection of UAVs in real environments, which has practical significance for the construction of anti-UAV systems. In our future work, we will continue to study and optimize the improvement strategy, so that it can achieve better recognition accuracy and real-time performance under the premise of minimizing the complexity of the model.

Author Contributions: Conceptualization, C.W.; methodology, C.W.; software, L.M.; validation, L.M., Q.G. and J.W.; formal analysis, Q.G., X.L. and F.D.; investigation, L.M.; resources, C.W.; data curation, L.M.; writing—original draft preparation, L.M.; writing—review and editing, C.W., Q.G. and T.W.; visualization, L.M., L.W. and E.W.; supervision, C.W.; project administration, C.W.; funding acquisition, C.W. All authors have read and agreed to the published version of the manuscript.

Funding: This research was supported by the National Natural Science Foundation of China (Grant No. 61703287, 62173237, and 61972016), the Scientific Research Program of Liaoning Provincial Education Department of China (Grant Nos. LJKZ0218, LJKMZ20220556, and JYT2020045), the Young and Middle-aged Science and Technology Innovation Talents Project of Shenyang of China (Grant No. RC210401), the Doctoral Scientific Research Foundation of Shenyang Aerospace University (Grant No. 22YB03).

Data Availability Statement: Not applicable.

Conflicts of Interest: The authors declare no conflict of interest.

References

1. Tang, J.; Duan, H.; Lao, S. Swarm intelligence algorithms for multiple unmanned aerial vehicles collaboration: A comprehensive review. *Artif. Intell. Rev.* **2022**, 1–33. [CrossRef]
2. Zhang, D.; Cheng, Y.; Lin, Q.; Yu, G.F.; Xiao, L.W. Key Technologies and Development Trend of UAV Swarm Operation. *China New Telecommun.* **2022**, *24*, 56–58.

3. Cai, J.; Wang, F.; Yang, B. Exploration of UAV cluster defense technology. *Aerodyn. Missile J.* **2020**, *12*, 5.
4. Andrew, G.; Zhu, M.; Chen, B.; Kalenichenko, D.; Wang, W.; Weyand, T.; Andreetto, M.; Adam, H. MobileNets: Efficient Convolutional Neural Networks for Mobile Vision Applications. *arXiV* **2017**, arXiv:1704.04861.
5. Wang, C.; Su, Y.; Wang, J.; Wang, T.; Gao, Q. UAVSwarm Dataset: An Unmanned Aerial Vehicle Swarm Dataset for Multiple Object Tracking. *Remote Sens.* **2022**, *14*, 2601. [CrossRef]
6. Hu, J.; Shen, L.; Albanie, S.; Sun, G.; Wu, E. Squeeze-and-Excitation Networks. *IEEE T. Pattern Anal.* **2020**, *42*, 2011–2023. [CrossRef] [PubMed]
7. Woo, S.; Park, J.; Lee, J.Y.; Kweon, I. CBAM: Convolutional Block Attention Module. In *European Conference on Computer Vision*; Springer: Cham, Switzerland, 2018; pp. 3–19.
8. Zheng, Z.; Wang, P.; Liu, W.; Li, J.; Ye, R.; Ren, D. Distance-IoU Loss: Faster and Better Learning for Bounding Box Regression. In Proceedings of the AAAI Conference on Artificial Intelligence, New York, NY, USA, 7–12 February 2020; pp. 12993–13000.
9. Bochkovskiy, A.; Wang, C.-Y.; Liao, H. YOLOv4: Optimal Speed and Accuracy of Object Detection. *arXiv* **2020**, arXiv:2004.10934.
10. Zhang, H.; Cisse, M.; Dauphin, Y.N.; David, L.P. mixup: Beyond Empirical Risk Minimization. *arXiv* **2017**, arXiv:1710.09412.
11. Zou, Z.; Shi, Z.; Guo, Y.; Ye, J. Object Detection in 20 Years: A Survey. *arXiv* **2019**, arXiv:1905.05055v2.
12. Cheriguene, Y.; Bousbaa, F.Z.; Kerrache, C.A.; Djellikh, S.; Lagraa, N.; Lahby, M.; Lakas, A. COCOMA: A resource-optimized cooperative UAVs communication protocol for surveillance and monitoring applications. *Wireless Netw.* **2022**, 1–17. [CrossRef]
13. Tzoumas, G.; Pitonakova, L.; Salinas, L.; Scales, C.; Richardson, T.; Hauert, S. Wildfire detection in large-scale environments using force-based control for swarms of UAVs. *Swarm Intell.* **2022**, 1–27. [CrossRef]
14. Sastre, C.; Wubben, J.; Calafate, C.T.; Cano, J.C.; Manzoni, P. Safe and Efficient Take-Off of VTOL UAV Swarms. *Electronics* **2022**, *11*, 1128. [CrossRef]
15. Sastre, C.; Wubben, J.; Calafate, C.T.; Cano, J.C.; Manzoni, P. Collision-free swarm take-off based on trajectory analysis and UAV grouping. In Proceedings of the 2022 IEEE 23rd International Symposium on a World of Wireless, Mobile and Multimedia Networks (WoWMoM), Belfast, UK, 9 August 2022; pp. 477–482.
16. Girshick, R.; Donahue, J.; Darrell, T.; Malik, J. Rich Feature Hierarchies for Accurate Object Detection and Semantic Segmentation. In Proceedings of the 2014 IEEE Conference on Computer Vision and Pattern Recognition, Columbus, OH, USA, 23–28 June 2014; pp. 580–587.
17. Girshick, R. Fast R-CNN. In Proceedings of the 2015 IEEE International Conference on Computer Vision (ICCV), Santiago, Chile, 7–13 December 2015; pp. 1440–1448.
18. Ren, S.; He, K.; Girshick, R.; Sun, J. Faster R-CNN: Towards real-time object detection with region proposal networks. *IEEE Trans. Pattern Anal. Mach. Intell.* **2017**, *39*, 1137–1149. [CrossRef] [PubMed]
19. Liu, W.; Anguelov, D.; Erhan, D.; Szegedy, C.; Cheng, Y.; Alexander, C.B. SSD: Single Shot MultiBox Detector. In *European Conference on Computer Vision*; Springer: Cham, Switzerland, 2016; pp. 21–37.
20. Lin, T.; Goyal, P.; Girshick, R.; He, K.; Dollar, P. Focal Loss for Dense Object Detection. In Proceedings of the 2017 IEEE International Conference on Computer Vision (ICCV), Venice, Italy, 22–29 October 2017; pp. 2980–2988.
21. Hu, Y.; Wu, X.; Zheng, G.; Liu, X. Object Detection of UAV for Anti-UAV Based on Improved YOLO v3. In Proceedings of the 2019 Chinese Control Conference (CCC), Guangzhou, China, 27–30 July 2019; pp. 8386–8390.
22. Sun, H.; Yang, G.; Shen, J.; Liang, D.; Liu, N.; Zhou, H. TIB-Net: Drone Detection Network with Tiny Iterative Backbone. *IEEE Access* **2020**, *8*, 130697–130707. [CrossRef]
23. Ma, J.; Yao, Z.; Xu, C.; Chen, S. Multi-UAV real-time tracking algorithm based on improved PP-YOLO and Deep-SORT. *J. Comput. Appl.* **2022**, *42*, 2885–2892.
24. Yavariabdi, A.; Kusetogullari, H.; Celik, T.; Cicek, H. FastUAV-NET: A Multi-UAV Detection Algorithm for Embedded Platforms. *Electronics* **2021**, *10*, 724. [CrossRef]
25. Liu, B.; Luo, H. An Improved Yolov5 for Multi-Rotor UAV Detection. *Electronics* **2022**, *11*, 2330. [CrossRef]
26. Wang, C.; Shi, Z.; Meng, L.; Wang, J.; Wang, T.; Gao, Q.; Wang, E. Anti-Occlusion UAV Tracking Algorithm with a Low-Altitude Complex Background by Integrating Attention Mechanism. *Drones* **2022**, *6*, 149. [CrossRef]
27. Yang, Y.; Liao, Y.; Lin, C.; Ni, S.; Wu, Z. A Survey of Object Detection Algorithms for Lightweight Convolutional Neural Networks. *Ship Electron. Eng.* **2021**, *41*, 31–36.
28. Sandler, M.; Howard, A.; Zhu, M.; Zhmoginov, A.; Chen, L. MobileNetV2: Inverted Residuals and Linear Bottlenecks. In Proceedings of the 2018 IEEE/CVF Conference on Computer Vision and Pattern Recognition, Salt Lake City, UT, USA, 18–23 June 2018; pp. 4510–4520.
29. Howard, A.; Sandler, M.; Chu, G.; Chen, L.C.; Chen, B.; Tan, M.; Wang, W.; Zhu, Y.; Pang, R.; Vasudevan, V.; et al. Searching for MobileNetV3. In Proceedings of the 2019 IEEE/CVF International Conference on Computer Vision (ICCV), Seoul, Republic of Korea, 27 October–2 November 2019; pp. 1314–1324.
30. Zhang, X.; Zhou, X.; Lin, M.; Sun, J. ShuffleNet: An Extremely Efficient Convolutional Neural Network for Mobile Devices. In Proceedings of the 2018 IEEE/CVF Conference on Computer Vision and Pattern Recognition, Salt Lake City, UT, USA, 18–23 June 2018; pp. 6848–6856.
31. Ma, N.; Zhang, X.; Zheng, H.-T.; Sun, J. ShuffleNet V2: Practical Guidelines for Efficient CNN Architecture Design. In *European Conference on Computer Vision*; Springer: Cham, Switzerland, 2018; pp. 122–138.

32. Tan, M.; Chen, B.; Pang, R.; Vasudevan, V.; Vasudevan, V.; Sandler, M.; Howard, A.; Le, Q.V. MnasNet: Platform-Aware Neural Architecture Search for Mobile. In Proceedings of the 2019 IEEE/CVF Conference on Computer Vision and Pattern Recognition (CVPR), Long Beach, CA, USA, 15–20 June 2019; pp. 2815–2823.
33. Han, K.; Wang, Y.; Tian, Q.; Guo, J.; Xu, C.; Xu, C. GhostNet: More Features from Cheap Operations. In Proceedings of the 2020 IEEE/CVF Conference on Computer Vision and Pattern Recognition (CVPR), Seattle, WA, USA, 13–19 June 2020; pp. 1577–1586.
34. Xiong, Y.; Liu, H.; Gupta, S.; Akin, B.; Bender, G.; Wang, Y.; Kindermans, P.J.; Tan, M.; Singh, V.; Chen, B. MobileDets: Searching for Object Detection Architectures for Mobile Accelerators. In Proceedings of the 2021 IEEE/CVF Conference on Computer Vision and Pattern Recognition (CVPR), Nashville, TN, USA, 20–25 June 2021; pp. 3824–3833.
35. Ge, Z.; Liu, S.; Wang, F.; Li, Z.; Sun, J. YOLOX: Exceeding YOLO Series in 2021. *arXiv* **2021**, arXiv:2107.08430.
36. Liu, S.; Qin, H.; Shi, J.; Jia, J. Path Aggregation Network for Instance Segmentation. In Proceedings of the 2018 IEEE/CVF Conference on Computer Vision and Pattern Recognition, Salt Lake City, UT, USA, 18–23 June 2018; pp. 8759–8768.

Article

Multi-UAV Autonomous Path Planning in Reconnaissance Missions Considering Incomplete Information: A Reinforcement Learning Method

Yu Chen [1], Qi Dong [1,2,*], Xiaozhou Shang [2], Zhenyu Wu [3] and Jinyu Wang [4]

[1] Institute of Advanced Technology, University of Science and Technology of China, Hefei 230026, China
[2] China Academy of Electronics and Information Technology, Beijing 100049, China
[3] School of Information and Electronics, Beijing Institute of Technology, Beijing 100081, China
[4] School of Information and Communication Engineering, Beijing University of Posts and Telecommunications, Beijing 100876, China
* Correspondence: dongqiouc@126.com

Abstract: Unmanned aerial vehicles (UAVs) are important in reconnaissance missions because of their flexibility and convenience. Vitally, UAVs are capable of autonomous navigation, which means they can be used to plan safe paths to target positions in dangerous surroundings. Traditional path-planning algorithms do not perform well when the environmental state is dynamic and partially observable. It is difficult for a UAV to make the correct decision with incomplete information. In this study, we proposed a multi-UAV path planning algorithm based on multi-agent reinforcement learning which entails the adoption of centralized training–decentralized execution architecture to coordinate all the UAVs. Additionally, we introduced a hidden state of the recurrent neural network to utilize the historical observation information. To solve the multi-objective optimization problem, We designed a joint reward function to guide UAVs to learn optimal policies under the multiple constraints. The results demonstrate that by using our method, we were able to solve the problem of incomplete information and low efficiency caused by partial observations and sparse rewards in reinforcement learning, and we realized kdiff multi-UAV cooperative autonomous path planning in unknown environment.

Keywords: multi-UAV; path planning; incomplete information; multi-objective, reinforcement learning

Citation: Chen, Y.; Dong, Q.; Shang, X.; Wu, Z.; Wang, J. Multi-UAV Autonomous Path Planning in Reconnaissance Missions Considering Incomplete Information: A Reinforcement Learning Method. *Drones* **2023**, *7*, 10. https://doi.org/10.3390/drones7010010

Academic Editors: Mou Chen, Xiwang Dong, Xiangke Wang and Fei Gao

Received: 30 November 2022
Revised: 14 December 2022
Accepted: 17 December 2022
Published: 23 December 2022

1. Introduction

Multi-UAV perform well in complex tasks because of their robustness and high efficiency [1]. When multi-UAV perform reconnaissance tasks cooperatively in an unknown environment, they have to perceive the environment through their own sensors and plan the optimal path online according to the current environmental state to reach the target points safely. It is important for UAVs to be capable of autonomous navigation in complex and unknown environments. Moreover, a greater coordination is needed between all UAVs. Thus, we have to consider we can guide multi-UAV to achieve a common goal.

Multi-UAV path planning can be considered as a Multi-Agent Path Planning (MAPF) problem [2], which is a model used to find the optimal path for multi-agents from the starting positions to destinations without conflicts. In fact, MAPF is a relatively complex joint objective optimization problem. The state space of this problem grows exponentially with the number of agents, and it has been proved to be an NP-hard problem [3]. In the reconnaissance tasks, multi-UAV not only have to avoid dangerous areas and reach the target points safely, but they must also cover a larger area in a shorter time. However, the time cost and coverage area are in conflict, as these are multi-objective optimization problems, and we must make a trade-off between two or more conflicting goals to enable optimal decision making. It is impossible to find a solution that can achieve the optimal performance of all objectives; therefore, for the multi-objective optimization problem, we usually use a set of non-inferior solutions called the "Pareto solution set" [4].

Most of the previous research regarding multi-UAV path planning has focused on intelligent optimization algorithms, such as evolutionary algorithms [5] including Particle Swarm Optimization(PSO) [6–8]. Shao et al. [9] proposed a more accurate and faster PSO algorithm to effectively improve the convergence speed and solution optimality, and the proposed PSO was successfully used in UAV formation path planning under terrain, threat, and collision avoidance constraints. Evan et al. [10] proposed a PSO algorithm for use in navigating in an unknown environment, which was able to reach a pre-defined goal and become collision-free. Ajeil et al. [11] proposed a hybridized PSO-modified which was shown to minimize the distance and follow path smoothness criteria to form an optimized path. Evolutionary algorithms based on swarm intelligence can iteratively search for local optimal solutions, but this method is difficult to expand to online and real-time optimization due to its limited speed, and it is not suitable for use in reconnaissance tasks.

In recent years, with the rapid development of Deep Reinforcement Learning (DRL), its powerful representation and learning capabilities have enabled it to perform well in decision-making problems [12]; therefore, researchers are beginning to explore the application of reinforcement learning in multi-UAV path planning and navigation [13–15]. Compared with traditional algorithms, reinforcement learning performs better when the environment is unknown and dynamic. Moreover, the inference speed and generalization of reinforcement learning are advantages in real-time decision-making tasks.

In our research, the perception abilities of multi-UAV were limited, and only partial observations of the environment were made, meaning that it was difficult for the multi-UAV to make the optimal decisions when global states were lacking because the state transitions were unknown. The action of each UAV could change the environment's state, especially in a learning-based algorithm, such as reinforcement learning, the incomplete information will lead to poor efficiency and convergence. Moreover, it is vital to design training architecture to coordinate multi-UAVs to achieve a common goal. It is unwise to adopt a completely distributed training architecture to solve MAPF problems because of the high complexity. The same applies to multi-objective optimization problems. Lowe, firstly, proposed a framework of centralized training with decentralized execution [16], allowing extra information to be used in policies to make training easier. This framework has been proved to be capable of handling collaborative problems, such as multi-agent path planning. For instance, Jose et al. [17] proposed a DRL model with a centralized training and decentralized execution paradigm to solve vehicles routing problem, which was shown to be able to produce near-optimal solutions through cooperative actions. Marc et al. [18] adopted a distributed multi-agent variable framework to solve conflicts between UAVs, and also to train agents using centralized learning. Wang et al. [19] adopted a centralized training and decentralized executing framework to enable dynamic routing, introducing a counterfactual baseline scheme to improve the convergence speed. Moreover, the reward function of reinforcement learning should be reviewed in light of multi-objective optimization problems. On the one hand, a reward function that is too simple maybe cause "Reward Hacking" [20] and exacerbate the difficulties of policy learning due to incomplete information. On the other hand, a reward function that is too complex will lead the worse generalization. The most commonly used solution is to design a reward to satisfy multiple objectives of different weights according to the prior knowledge, in which a multi-objective optimization problem will be changed into a singe-objective optimization problem. In fact, this solution is near-optimal. Li [21] proposed an end-to-end framework for use in solving multi-objective optimization problems using deep reinforcement learning. Xu [22] proposed prediction-guided multi-objective reinforcement learning for use in solving continuous robot control problems. In multi-UAV path planning, some constraints, such as time cost, security, and coverage, must be considered.

To solve these problems, we proposed an improved multi-agent reinforcement learning algorithm based on centralized training and decentralized execution architecture. The policy learning algorithm is proximal policy optimization (PPO) [23]. It is a model-free reinforcement learning algorithm, which can adapt to a dynamic environment and provide good generalization. The critic network of PPO is used to coordinate all the UAVs to maximize team returns through centralized training by receiving joint observations, and the actor network of PPO is used to output actions. We also added a recurrent neural network to the actor–critic network to gather the historical information from the hidden state of the recurrent neural network [24], which solves the problem of incomplete information caused by partial observations. In addition, we designed a joint reward function to guide multi-UAV to learn optimal policies. When the training stage is completed, each UAV can execute an action based on its local observations in the reference stage. The contributions of our research are as follows:

1. We solved the problem caused by multi-UAV path planning with incomplete information through reinforcement learning based on the centralized training and decentralized execution architecture. We deeply explored the reasons why centralized training and decentralized execution architecture improves model performance, and we explained the benefits of centralized training compared to fully distributed methods.

2. When designing the reward function, we decomposed the multi-objective optimization problem into multiple sub-problems based on the idea of decomposition, solving the multi-objective optimization problem through reinforcement learning.

Experiments show that by using our method, the performance was significantly improved compared with baselines, and we demonstrated the high application value of reinforcement learning in multi-UAV path planning. In the execution stage, our method could be used to plan paths online, far exceeding the speed of heuristic algorithms. Section 2 introduces the backgrounds of our research. Section 3 describes our methodologies in details. Section 4 introduces the experimentation setup and results. We provide a conclusion in Section 5.

2. Background

2.1. Problem Description

When multi-UAV perform reconnaissance missions, they need to make real-time decisions based on current state information, and a collision-free path to reach the target points must be planned. In addition, time cost and coverage need to be considered. Therefore, multi-UAV autonomous path planning is a online decision-making problem under the constraints of incomplete information. It has three characteristics: distributed decision-making, partial observation and multi-objective optimization. Multi-UAV autonomous path planning is considered to be a fully cooperative task. The objective of all the participants in such a task is to obtain the maximum team returns. Therefore, we could establish a multi-agent real-time sequential decision-making model by the Decentralized Partially Observable Markov Decision Process (Dec-POMDP) theoretical framework [25].

Dec-POMDP is a general model used to solve multi-agent objective optimization problems in cooperative environments, which generalizes the Partially Observable Markov Decision Process (POMDP) to multi-agent environments. It allows for the distributed control of multiple agents which may not be able to observe global states of environment. In every step, each agent chooses an action based on local observations (all agents in parallel), and then obtains its own reward from the environment, and all of agents cooperate to obtain common long-term benefits and maximize returns. Generally, a Dec-POMDP model is described by a tuple: $\langle I, S, \{A_i\}, T, R, \{\Omega_i\}, O, h \rangle$

- I is a set of N agents.
- S is a set of states of the environment, and S_0 is the initial state.
- $\{A\}$ is a set of actions for the agents. It is an action tuple $A_1, A_2, \ldots, A_i$.
- T is the state transition probability function $P(S'|S, A)$.
- R is the reward when agents take actions $\{A\}$ in state S, it depends on all the agents.
- $\{\Omega\}$ is a set of observations for the agents.
- O is a table of the observation probabilities, where $O(o_1, o_2, \ldots, o_i|S', A)$ is the probability that $(o_1, o_2, \ldots, o_i)$ are observed by all the agents, respectively.
- h is the maximum number of steps in an episode which is called "horizon".

However, the complexity of the optimal solution of this distributed model is

$$O\left[\left(|\mathcal{A}|^{\frac{|o|^h-1}{|o|-1}}\right)^n\right] \tag{1}$$

which is double exponential [26]; it is hard to compute directly, and reinforcement learning is usually used to obtain the approximate solution.

2.2. Actor–Critic Algorithm

In reinforcement learning, an agent interacts with the environment continuously to optimize the policy through the feedback (reward) given by the environment. Reinforcement learning is mainly divided into value-based methods and policy-based methods. A policy-gradient algorithm can easily select the appropriate action in the continuous action space, while value-based algorithm cannot. However, the limitation of the policy-gradient algorithm is its poor learning efficiency. Therefore, researchers proposed a method that combines the policy-gradient and value-based algorithms, called the actor–critic algorithm [27]. The architecture of actor–critic is shown in Figure 1. Actor–critic uses a value-based network and policy-based network as the critic network and the actor network, respectively. The critic network can realize single-step updates to overcome the poor learning efficiency, and the actor network outputs actions according to the current observation, while the critic network can judge whether the current action is good or bad, which can lead the actor network to output a better action. Currently, the algorithms based on the actor–critic framework, such as DDPG, PPO, and A3C, are very popular.

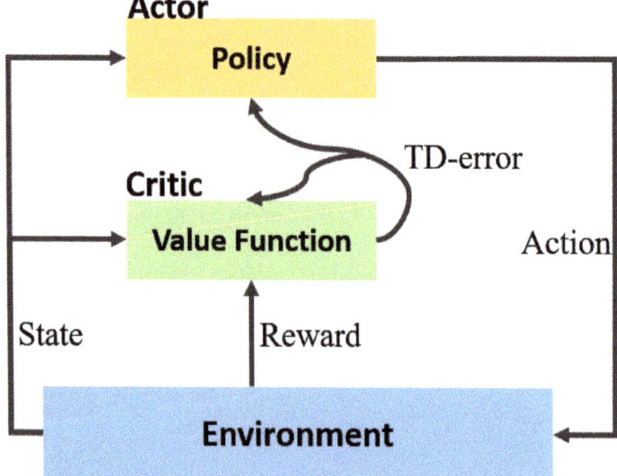

Figure 1. Actor–critic algorithm.

2.3. Centralized Training and Decentralized Execution Architecture

When there are multiple agents in a completely cooperative environment, we can establish a Dec-POMDP framework. Reinforcement learning is a great method to seek the optimal solution of a model. However, if we directly use single-agent reinforcement learning algorithms to train agents independently, it is hard to converge them, because the actions of each agent will change the environment, meaning the environment will be unstable for each agent and lead to learning difficulties. The MADDPG trains multi-agents through a centralized critic network, providing a good solution for the training of multi-agent systems. As shown in Figure 2, the input of the critic network is the joint observation of all of the agents in the environment in the training stage, and the actor network only inputs its own local observations and output actions according to the observations in the inference stage. This architecture enables each agent's actions in the environment to be observed by other agents, ensuring the stability of the environment. Therefore, multi-agent reinforcement learning is mostly based on centralized training and decentralized execution (CTDE) architecture, such as COMA [28].

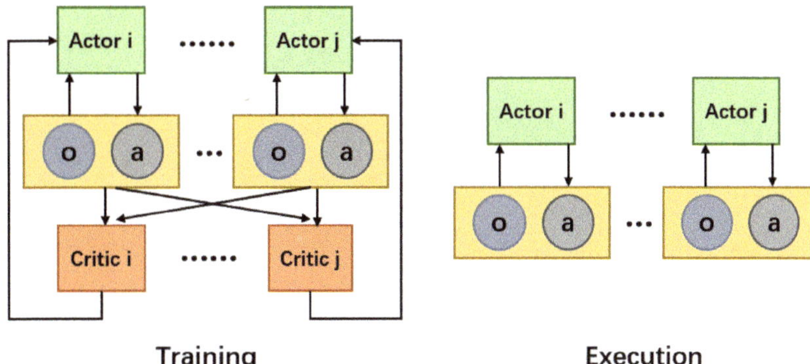

Figure 2. Centralized training (**left**) and decentralized execution (**right**).

3. Methodology

3.1. Proximal Policy Optimization with CTDE

We choose proximal policy optimization (PPO) to guide UAVs in learning policies. The PPO algorithm is based on the actor–critic architecture, which can more effectively achieve continuous control in high-dimensional space, and it is also an on-policy reinforcement learning algorithm. The learning approach of PPO is policy gradient. However, the policy-gradient algorithm is unstable, and this makes it difficult to choose an appropriate steps. If the difference between the old policy and new policy is too great during the training process, it is not conducive to learning. Using the PPO algorithm, a new objective function was proposed which can be updated in small batches in multiple training steps, which solves the problem of steps being difficult to determine in the policy-gradient algorithm. The algorithm takes into account the difference between an old network and an new network when updating parameters. In order to avoid the difference being too great, a clip is introduced to limit:

$$\nabla \overline{\mathcal{R}}(\tau) = E_{\tau \sim \pi\theta(\tau)}[A^{\pi}(s_t, a_t)\nabla log p_\theta(a_t|s_t)] \tag{2}$$

$$A^{\pi}(s, a) = Q^{\pi}(s, a) - V^{\pi}(s) \tag{3}$$

$$\mathbb{E}_{x \sim p}[f(x)] = \int f(x)p(x)dx = \int f(x)\frac{p(x)}{q(x)}q(x)dx = \mathbb{E}_{x \sim q}\left[f(x)\frac{p(x)}{q(x)}\right] \tag{4}$$

$$\mathcal{L}^{CLIP}(\theta) = \hat{E}_t \left\{ min \left[\left(\frac{\pi_\theta(a_t|s_t)}{\pi_{\theta_{old}}(a_t|s_t)} \right) \hat{A}_t, clip \left(\frac{\pi_\theta(a_t|s_t)}{\pi_{\theta_{old}}(a_t|s_t)}, 1 - \epsilon, 1 + \epsilon \right) \hat{A}_t \right] \right\} \tag{5}$$

where A is an advantage function, indicating the return of the action a in the current state.

Based on the PPO algorithm, we adopted the training method of the CTDE architecture and designed a multi-agent PPO algorithm in a multi-agent environment. Compared with the single-agent environment, the critic network's input is the joint observation of multi-UAV, which is equivalent to a central controller, each drone can obtain more information. The actor network is updated to maximize the objective:

$$\mathcal{L}(\theta) = \frac{1}{Bn} \sum_{i=1}^{B} \sum_{k=1}^{n} \left[min \left(r_{\theta,i}^k \mathcal{A}_i^k, clip \left(r_{\theta,i}^k, 1 - \epsilon, 1 + \epsilon \right) \mathcal{A}_i^k \right) + \sigma * \mathcal{S}_\pi \right] \tag{6}$$

where $r_{\theta,i}^k = \frac{\pi_\theta(a_i^k|o_i^k)}{\pi_{\theta_{old}}(a_i^k|o_i^k)}$. The critic network is updated to minimize the value loss:

$$\begin{aligned} \mathcal{L}(\phi) = \frac{1}{Bn} \sum_{i=1}^{B} \sum_{k=1}^{n} max \Bigg[& \left(V_\phi \left(s_i^k \right) - \hat{\mathcal{R}}_i \right)^2, \\ & \left(clip \left(V_\phi \left(s_i^k \right), V_{\phi_{old}} \left(s_i^k \right) - \epsilon, V_{\phi_{old}} \left(s_i^k \right) + \epsilon \right) - \hat{\mathcal{R}}_i \right)^2 \Bigg] \end{aligned} \tag{7}$$

The weights of two networks are updated in every episodes, the process is shown in Figure 3.

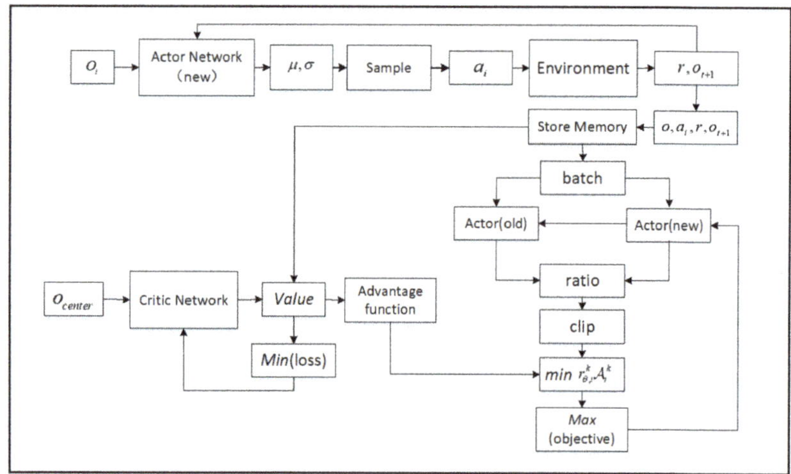

Figure 3. The weights of actor–critic network are updating in every episodes.

3.2. Adding RNN Layer For Incomplete Information

One of the difficulties within multi-UAV autonomous path-planning tasks is partial observation, which leads to limited information being obtained by UAVs. A solution to this is the utilization of the previous state to avoid falling into a local optimum.

Recurrent neural networks can memorize the previous information and apply it to the calculation of the current output. The nodes between the hidden layers are connected, and the input of the hidden layer includes the current input and the previous output, as well as the output of the hidden layer at the moment. The study of deep recurrent Q-learning (DRQN) was the first to combine an RNN with reinforcement learning [29]. As shown in Figure 4, DRQN essentially turns one of the linear layers of DQN into an RNN layer. Due to the addition of RNN, DRQN has short-term memory, and it can achieve similar scores to DQN in "Atari Games" without frame stack technology. Taking inspiration from this idea, we added the RNN layer to the PPO network, and we used the RNN layer to process historical information to solve the problem of incomplete information in the training process.

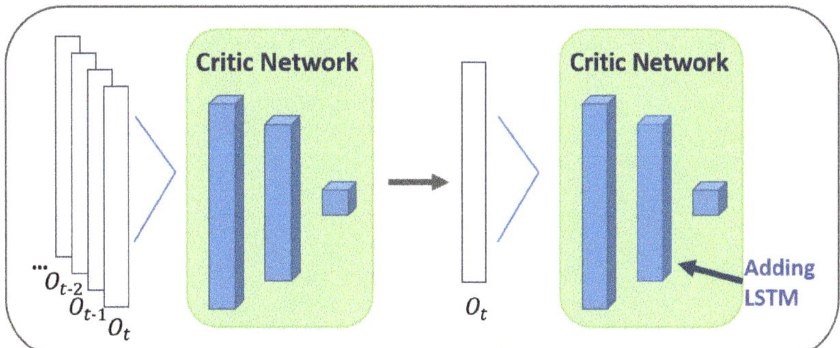

Figure 4. Adding LSTM layer have the same effect compared with frame stack, and reduce the dimension of input.

However, RNNs suffer from short-term memory. If a sequence is too long, it is difficult to transfer information from an earlier time step to a later time step. During back propagation, the gradient easily vanishes. Long Short-Term Memory (LSTM) is a variant of the RNN. It can select the information to be remembered or forgotten through the gate mechanism. As shown in Figure 5, the forget gate determines which relevant information in the previous step needs to be retained; the input gate determines which information in the current input is important and needs to be added; the output gate determines what the next hidden state should be. These "gates" can keep the important information in the sequence and discard the useless information, preventing the gradient from vanishing.

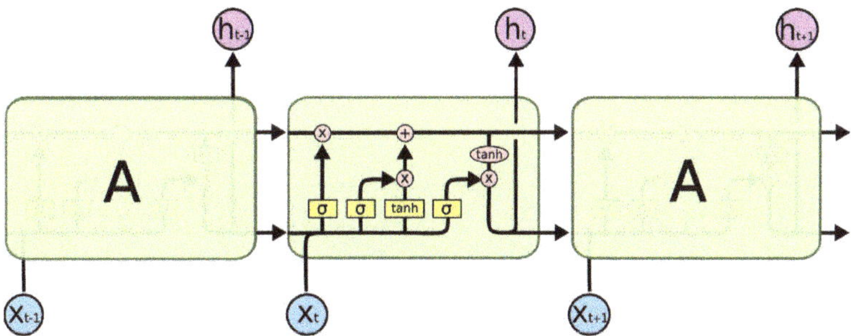

Figure 5. LSTM structure: at timestep t, X_t is input, C_t is cell state, and h_t is hidden state.

$$\begin{cases} f_t &= \sigma\left(W_f \bullet [h_{t-1}, x_t] + b_f\right) \\ i_t &= \sigma(W_i \bullet [h_{t-1}, x_t] + b_i) \\ \widetilde{C}_t &= \tanh(W_c \bullet [h_{t-1}, x_t] + b_c) \\ C_t &= f_t * C_{t-1} + i_t * \widetilde{C}_t \\ o_t &= \sigma(W_o \bullet [h_{t-1}, x_t] + b_o) \\ h_t &= o_t * \tanh(C_t) \end{cases} \tag{8}$$

where C_t is the cell state, and h_t is the current hidden state. Therefore, we added LSTM layers to both the actor and critic networks of PPO. After adding the LSTM layer, the historical information was remembered by updating the cell state and hidden state at each timestep T, and the problem of incomplete information caused by the local observations was alleviated.

Batches of τ are used to update the parameters of actor and critic networks to maximize $\mathcal{L}(\theta)$ and minimize $\mathcal{L}(\phi)$ through gradient descent. $\tau = [s_t, o_t, a_t, r_t, s_{t+1}, o_{t+1}, a_{t+1} \ldots]$ Add an LSTM layer to the network, two elements $h_{t,\pi}$ and $h_{t,V}$ are added to τ, which changed into $[s_t, o_t, h_{t,\pi}, h_{t,V}, a_t, r_t, s_{t+1}, o_{t+1}, h_{t+1,\pi}, h_{t+1,V}, a_{t+1} \ldots]$, $h_{t,\pi}, h_{t,V}$ are the hidden state of timestep t in an LSTM layer of the actor network and critic network.

$$\mathcal{L}(\theta) = \frac{1}{Bn} \sum_{i=1}^{B} \sum_{k=1}^{n} \left[min\left(r_{\theta,i}^k \mathcal{A}_i^k, clip\left(r_{\theta,i}^k, 1 - \epsilon, 1 + \epsilon \right) \mathcal{A}_i^k \right) + \sigma * \mathcal{S}_\pi \right] \tag{9}$$

Moreover, we sought to establish the different roles of the critic network and actor network in the multi-agent reinforcement learning algorithm based on the CTDE architecture. We believe that the critic network acts as a central controller to process all observation information, meaning that adding an RNN layer to the critic network can, theoretically, greatly enhance the performance of the model in partially observable environments. The actor network acts as a policy network for each agent, and adding the RNN layer to the actor network has less of an effect on its performance. The experimental results prove our analysis, it also confirms that the CTDE architecture is effective in this task.

3.3. Multi-Objective Joint Optimization

Multi-UAV autonomous path planning is a multi-objective optimization. The time cost, coverage area, and security of multi-UAV systems need to be considered in Figure 6. Multi-objective optimization is the optimal selection of decision variables in a discrete decision space.

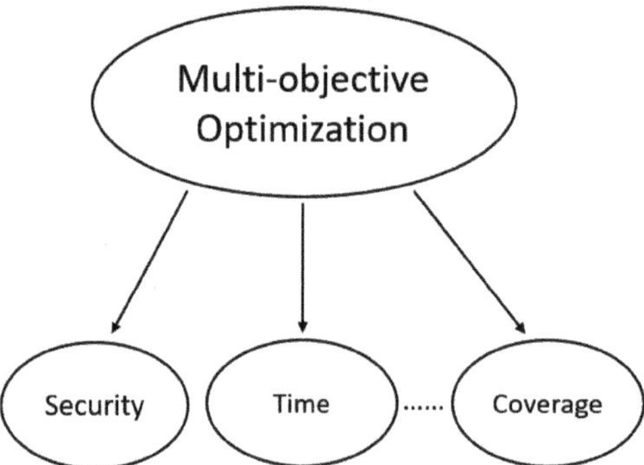

Figure 6. Multi-objective optimization seeks optimal solutions under the constraints of security, time, and coverage.

The mathematical expression is as follows:

$$max_\pi F(\pi) = max_\pi [f_1(\pi), f_2(\pi), \ldots, f_m(\pi)] \tag{10}$$

where m is the number of the objectives, and π is the policy.

This is very similar to the "action selection" of reinforcement learning, and the "offline training, online decision-making" characteristic of deep reinforcement learning make it possible for an online, real-time solution of a multi-objective optimization problem to be achieved. Therefore, deep reinforcement learning methods are a good choice when used to solve traditional multi-objective optimization problems, and the learning-based model has a good generalizability.

We designed the reward function based on multi-objective optimization and combined it with prior knowledge regarding navigation, decomposing the multi-objective optimization problem into multiple sub-problems. We shaped a joint reward function by considering the constraints of security, time, and coverage. There are several ways for multi-UAV to obtain feedback.

$$r_{total} = \alpha * r_{timecost} + \beta * r_{security} + \gamma * r_{coverage} \tag{11}$$

where $r_{security} = \sum |distance(UAV_i - UAV_j)| - |distance(UAV_i - target_i)|$, which guides the multi-UAV to reach the target points and avoid each other. The purpose of this design is to achieve larger coverage by distributing all of the UAVs, ensuring that the drone does not collide with other drones or obstacles. $r_{coverage} = \sum new\ area_{UAVi}$, it means the UAV will be rewarded if new areas are explored; this reward encourages multi-UAV to explore an environment, not simply reach the required points. $r_{timecost}$ guides the drone to reach the target point with the shortest possible number of steps. These three different rewards constitute the reward function that guides the multi-UAV autonomous path planning under constraints.

Usually, we give these rewards different weights to change a multi-objective problem into single-objective problem and to seek a solution. The advantage of this design is that the aggressiveness of the agent's learning strategy can be changed amending intended meaning has been retrained the manually set rules, but this is a near-optimal solution under the constraint. A multi-objective optimization problem can be solved through multi-objective reinforcement learning, as shown in Figure 7.

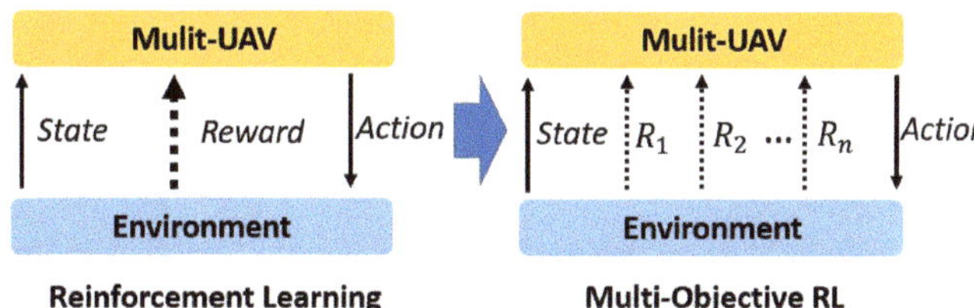

Figure 7. The difference between reinforcement learning and multi-objective reinforcement learning.

A set of solutions called "Pareto front" can represent the optimal solutions in all of the different weights. The differences between two methods are as shown in Figure 8.

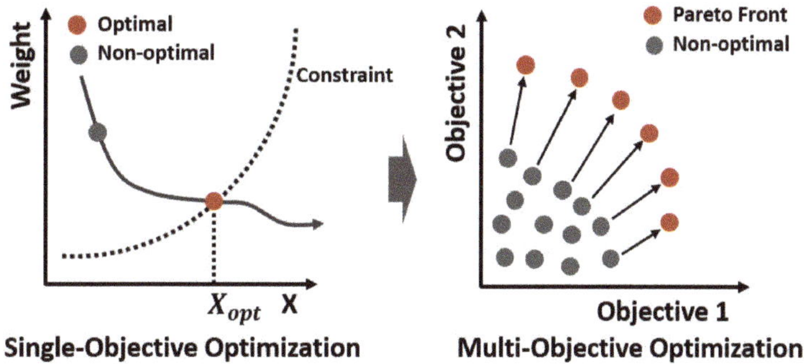

Figure 8. Multi-objective optimization: obtaining the optimal solution under the constraints of security, time, and coverage.

A multi-objective gradient optimizes the policy to maximize the weight–sum reward, where w is the weight of every objective, meaning that the policy gradient has changed.

$$\mathcal{J}(\theta,\omega) = \omega^T F(\pi) = \sum_{i=1}^{m} \omega_i f_i(\pi) = \sum_{i=1}^{m} \omega_i J_i^{\pi} \tag{12}$$

$$\begin{aligned}
\nabla_\theta \mathcal{J}(\theta,\omega) &= \sum_{i=1}^{m} \omega_i \nabla_\theta J_i(\theta) \\
&= \mathbb{E}\left[\sum_{t=0}^{T} \omega^T A^{\pi}(s_t, a_t) \nabla_{\pi\theta}(a_t \mid s_t) \right] \\
&= \mathbb{E}\left[\sum_{t=0}^{T} A_\omega^{\pi}(s_t, a_t) \nabla_{\pi\theta}(a_t \mid s_t) \right]
\end{aligned} \tag{13}$$

4. Experiment

4.1. Experimental Setup

We built a simulation platform based on Unreal Engine 4 to support quad-rotor dynamics simulation. In this platform, we created a scenario to simulate multi-UAV reconnaissance missions, as shown in Figure 9. The reconnaissance area was 2 km × 2 km, and the scenario contained four movable anti-drone devices. Once the drone entered the coverage area of these devices, it would be destroyed. The perception radius of a drone was 200 m × 200 m, and all of the drones communicated with each other by default.

Figure 9. Multi-UAV simulation platform.

Three UAVs started from the starting points and planed a collision-free path to three target points online. The area covered by all of the UAVs was the final total coverage area, and the total path length of the UAVs was the path cost. Figure 10 shows that three UAVs started from different points, and there were four threat areas in the environment. By default, the drones could only perceive dangerous areas within their capability radius.

In this simulation platform, low-level and high-level commands were used to control the motion of a UAV. As shown in Figure 11, in order to simulate a real flight, we choose to control the motion of the UAV through the underlying control method. The policy network outputs (*pitch, roll, yaw_rate, throttle*, and *duration*) a five-dimension vector in every step, where *pitch, roll*, and *yaw_rate* controlled the attitude and direction of a UAV, and *throttle* and *duration* made the UAV to accelerate or decelerate for a period of time.

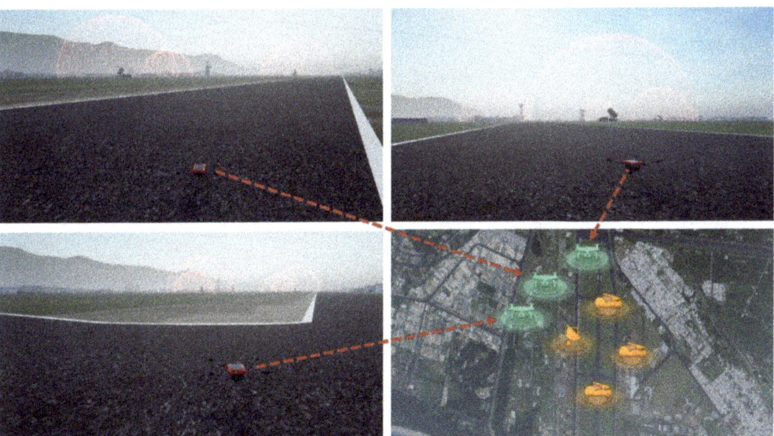

Figure 10. Initial state when multi-UAV perform a reconnaissance task.

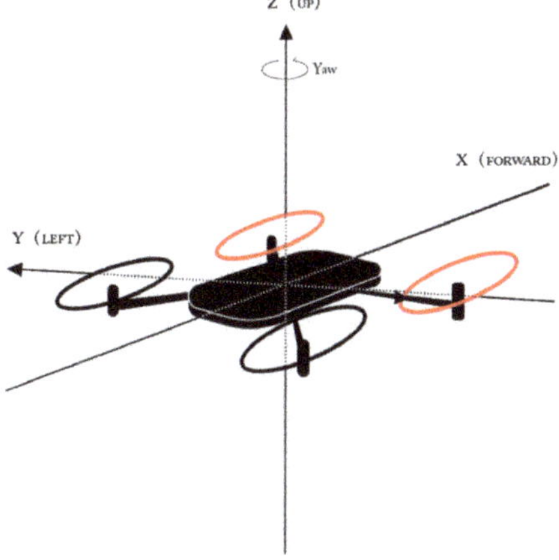

Figure 11. Kinematics of a quad-rotor.

4.2. Network Architecture

We used PyTorch to build a three layers neural network for the actor and critic networks of PPO, respectively. We used a centralized training and decentralized execution architecture to coordinate all of the UAVs; the intuitive difference between centralized training and independent training is the input of the value network. In this experiment, we connected the local observations of all of the UAVs into a high-dimensional vector as the joint observations, and then input the value network, called O_{center}, and the actor network input was the observation O_i of each UAV itself as shown in Figure 12. We set up four control experiments to compare the performance between CTDE and independent training in this task. The first and third layers of the networks were fully connected layers, and the second layer was an LSTM layer. In order to validate whether adding lstm was effective, we use the same network architecture to build a network without an LSTM layer as a comparison with the specific aim of verifying which one of critic and value was more dependent on historical information, thus confirming the role of CTDE architecture.

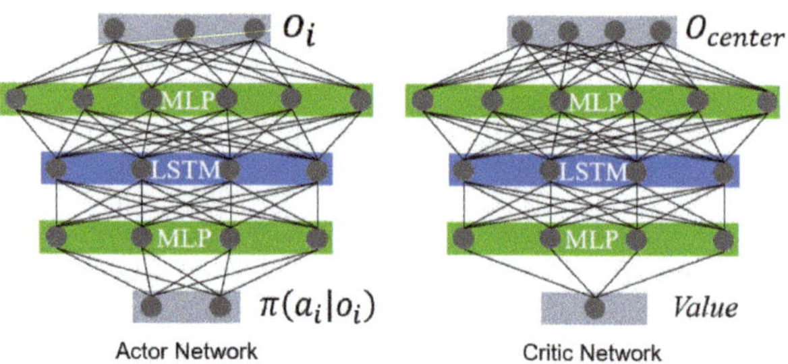

Figure 12. Actor network and critic network, and all of the agents that share the networks.

PPO outputs random policies, meaning that the outputs of the actor network are μ, σ, which are the expectation and variance of a Gaussian distribution, and the output action is randomly sampled by this Gaussian distribution. In the experiment, all of the agents shared common networks parameters as shown in Table 1.

Table 1. Network parameter table.

Episode	Episode length	Rollout thread	Clip	Discount	Entropy coefficient
625	200	16	0.2	0.99	0.1
Buffer size	**Batch size**	**FC layer dim**	**RNN hidden dim**	**Activation**	**Optimizer**
500	32	128	64	Relu	Adam

4.3. Results

After 1,000,000 steps of training, by analyzing the experimental results, we came to the conclusion that the PPO algorithm based on the centralized training decentralized execution architecture performed better compared to independent training in multi-UAV autonomous path planning tasks. As the results show in Figure 13, it is difficult for a completely independent and distributed training method to perform well in multi-UAV tasks. The adoption of CTDE architecture obviously and significantly improved the performance, the reward became positive, and the performance was even improved when the number of UAVs was larger. This proves that CTDE architecture is effective in such distributed tasks. A center controller can coordinate all of the UAVs. However, it does not indicate the number of UAVs, which can be unlimited. In fact, we found when there was more than six UAVs, the center controller could not effectively handle it, due to the dimension of joint observation being too high.

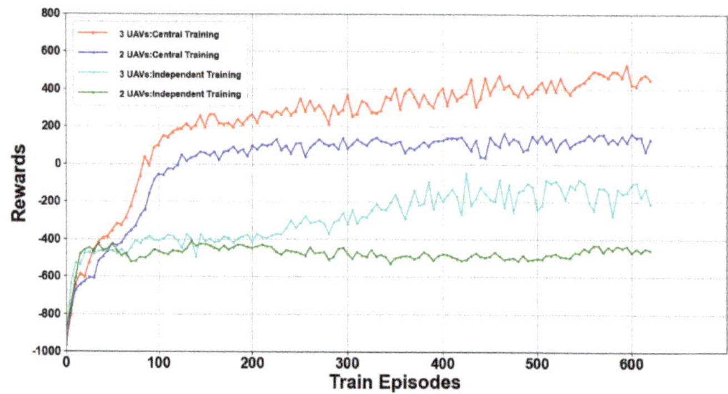

Figure 13. Comparison of the CTDE and independent architecture.

In addition, we carried out a set of control experiments to verify whether adding the RNN layer could solve the problem of multi-UAV learning difficulties with incomplete information. According to the experimental results in Figure 14, we found that adding the RNN layer to both the actor and critic networks significantly improved the performance of the model. Adding the RNN layer to the critic network also achieved practically the same effect, with the convergence speed being slower. The method of only adding the RNN layer to the actor network did not significantly improve the model performance, and it failed to solve the problem caused by partial observations of multi-UAVs. This result also verified our analysis: in the CTDE architecture, the critic network is the central controller, it coordinates all of the UAVs to complete common goals through the input of joint observations. The addition of the RNN layer to the critic network is effective, and the problem of incomplete information is solved through the hidden state.

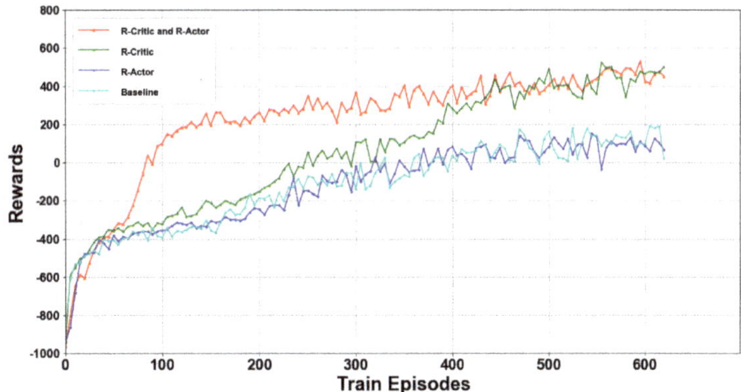

Figure 14. After the addition of the LSTM layer, better performance in an average reward was achieved.

In order to solve the decision-making problem with incomplete information, we chose CTDE architecture and added RNN layer to utilize historical information. In model-free reinforcement learning algorithms, reward represent an important evaluation criterion. Similarly, value loss, policy loss, and action entropy are also key components to evaluating algorithm performance. Value loss evaluates a value output of critic network and determines whether the prediction is accurate, and the action entropy reflects the randomness of the actor network strategy output. Here, we hoped that the action entropy would be larger enough to facilitate adequate exploration. The experimental results prove that our algorithm significantly improved the performance. As shown in Figure 15, after about 300 episodes, the loss function begin to stabilize, and the rapid convergence of value loss also showed that the value predicted by the critic network was more accurate. Similarly, after adding an LSTM layer, the critic loss was decreased, and the policy entropy value descended smoothly, which was a good performance and meant that the agents did not fall into a local optimum. We did not want this value to descend too rapidly or too slowly. Policy entropy is the variance of the output actions, and a smooth curve shows that multi-UAVs have learned a stable policy after a sufficient exploration, because exploration is indispensable in reinforcement learning.

Moreover, we found that adding a LSTM layer greatly improved the performance of the algorithm. After adding an LSTM, the average reward is significantly increased, which proved that the agent could make more correct decisions, and the policy entropy and critic loss converge faster, which shows that our method for adding an LSTM to the network effectively utilized historical information. The parameters of the model were continuously updated during the training phase.

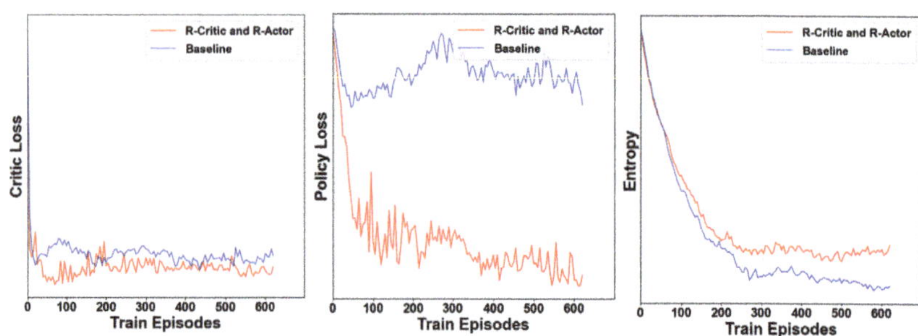

Figure 15. Critic loss, policy loss, and entropy.

Reinforcement learning is much faster than traditional swarm intelligence algorithms in the execution phase, and it is suitable for real-time decision-making tasks, as shown in Figure 16. Once the training stage was completed, the weight parameters of the networks were frozen during the execution phase. When we performed the navigation task with the trained model, the total average reward of our method was higher and stable.

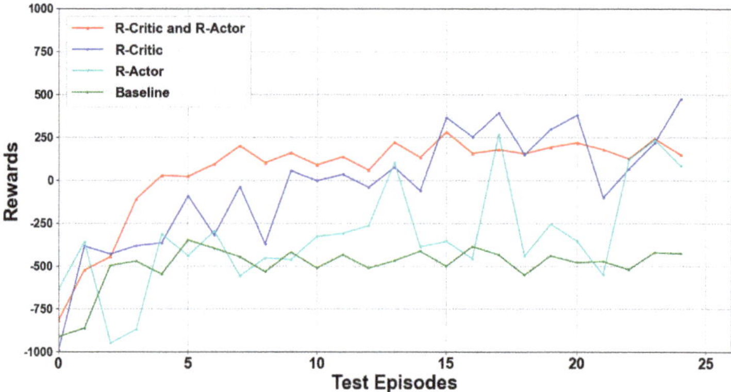

Figure 16. Our improved algorithm performs better in test.

In the simulation platform, multi-UAV realized path planning online by the pre-model in the inference stage as shown in Figure 17. We found that our method performs well under the constraints of security, time, and coverage. As shown in Table 2, compared with the state-of-the-art particle swarms optimization algorithms, our method has a better performance in many aspects especially the speed of reference. Reinforcement learning shows the powerful ability in real-time path planning task.

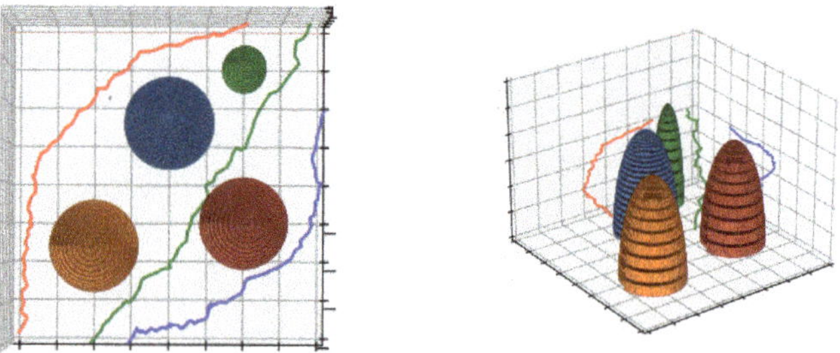

Figure 17. Multi-UAV path planning in three-dimensional environment through reinforcement learning.

Table 2. Performance comparison between our method and other state-of-the-art algorithms.

	Time Cost	Coverage	Success Rate	Reference Speed
Our method	92	57.9%	92.7%	0.126 s
RL baseline	93	54.5%	90.1%	0.115 s
DPSO	97	41.2%	65.2%	1.35 s
GAPSO	95	43.7%	62.1%	1.16 s

5. Conclusions

In this study, we proposed a multi-UAV autonomous path planning algorithm based on model-free reinforcement learning, which is able to adapt to dynamic environments. It was shown that the algorithm coordinates all of the UAVs through centralized training, which effectively lessens the difficulty of training distributed systems. When the training stage is completed, each UAV can make optimal decisions based on its own observations. We also introduced an RNN to remember historical information and prevent the model from falling into the local optimum due to incomplete information caused by partial observations. Finally, we designed a joint reward function to cooperatively guide the UAVs. Our experiments performs well in this type of task. Considering its communication capabilities in the real world, we plan to constrain the communication range and communication frequency between UAVs in follow-up research. The authors of [30,31] have contributed new ideas regarding the security of UAV communication. We believe this algorithm can be deployed to real drone swarms.

Author Contributions: Conceptualization, Y.C. and Q.D.; methodology, Y.C.; software, X.S.; validation, Y.C., Z.W. and J.W.; resources, Z.W.; data curation, J.W. All authors have read and agreed to the published version of the manuscript.

Funding: The project was supported by Open Fund of Anhui Province Key Laboratory of Cyberspace Security Situation Awareness and Evaluation under Grant CSSAE-2021-003 and Grant CSSAE-2021-004.

Data Availability Statement: Not applicable.

Conflicts of Interest: The authors declare no conflict of interest.

References

1. Mohiuddin, A.; Tarek, T.; Zweiri, Y.; Gan, D. A survey of single and multi-UAV aerial manipulation. *Unmanned Syst.* **2020**, *8*, 119–147. [CrossRef]
2. Stern, R. Multi-agent path finding–An overview. In *Artificial Intelligence*; Springer: Berlin/Heidelberg, Germany, 2019; pp. 96–115.
3. Ma, H.; Wagner, G.; Felner, A.; Li, J.; Kumar, T.; Koenig, S. Multi-agent path finding with deadlines. *arXiv* **2018**, arXiv:1806.04216.
4. Ngatchou, P.; Zarei, A.; El-Sharkawi, A. Pareto multi objective optimization. In Proceedings of the 13th international Conference on, Intelligent Systems Application to Power Systems, Arlington, VA, USA, 6–10 November 2005; pp. 84–91.
5. Konak, A.; Coit, D.W.; Smith, A.E. Multi-objective optimization using genetic algorithms: A tutorial. *Reliab. Eng. Syst. Saf.* **2006**, *91*, 992–1007. [CrossRef]
6. Tang, J.; Duan, H.; Lao, S. Swarm intelligence algorithms for multiple unmanned aerial vehicles collaboration: A comprehensive review. In *Artificial Intelligence Review*; Springer: Berlin/Heidelberg, Germany, 2022; pp. 1–33.
7. Wang, Y.; Bai, P.; Liang, X.; Wang, W.; Zhang, J.; Fu, Q. Reconnaissance mission conducted by UAV swarms based on distributed PSO path planning algorithms. *IEEE Access* **2019**, *7*, 105086–105099. [CrossRef]
8. Wan, Y.; Zhong, Y.; Ma, A.; Zhang, L. An Accurate UAV 3-D Path Planning Method for Disaster Emergency Response Based on an Improved Multiobjective Swarm Intelligence Algorithm. *IEEE Trans. Cybern.* **2022**. [CrossRef] [PubMed]
9. Shao, S.; Peng, Y.; He, C.; Du, Y. Efficient path planning for UAV formation via comprehensively improved particle swarm optimization. *ISA Trans.* **2020**, *97*, 415–430. [CrossRef] [PubMed]
10. Krell, E.; Sheta, A.; Balasubramanian, A.P.R.; King, S.A. Collision-free autonomous robot navigation in unknown environments utilizing PSO for path planning. *J. Artif. Intell. Soft Comput. Res.* **2019**, *9*, 267–282 [CrossRef]
11. Ajeil, F.H.; Ibraheem, I.K.; Sahib, M.A.; Humaidi, A.J. Multi-objective path planning of an autonomous mobile robot using hybrid PSO-MFB optimization algorithm. *Appl. Soft Comput.* **2020**, *89*, 106076. [CrossRef]
12. Mnih, V.; Kavukcuoglu, K.; Silver, D.; Graves, A.; Antonoglou, I.; Wierstra, D.; Riedmiller, M. Playing atari with deep reinforcement learning. *arXiv* **2013**, arXiv:1312.5602.

13. Luo, W.; Tang, Q.; Fu, C.; Eberhard, P. Deep-sarsa based multi-UAV path planning and obstacle avoidance in a dynamic environment. In *Proceedings of the International Conference on Swarm Intelligence*; Springer: Berlin/Heidelberg, Germany, 2018; pp. 102–111.
14. Liu, C.H.; Ma, X.; Gao, X.; Tang, J. Distributed energy-efficient multi-UAV navigation for long-term communication coverage by deep reinforcement learning. *IEEE Trans. Mob. Comput.* **2019**, *19*, 1274–1285. [CrossRef]
15. Bayerlein, H.; Theile, M.; Caccamo, M.; Gesbert, D. Multi-UAV path planning for wireless data harvesting with deep reinforcement learning. *IEEE Open J. Commun. Soc.* **2021**, *2*, 1171–1187. [CrossRef]
16. Lowe, R.; Wu, Y.I.; Tamar, A.; Harb, J.; Pieter Abbeel, O.; Mordatch, I. Multi-agent actor-critic for mixed cooperative-competitive environments. In Proceedings of the Advances in Neural Information Processing Systems, Long Beach, CA, USA, 4–9 December 2017; Volume 30.
17. Vera, J.M.; Abad, A.G. Deep reinforcement learning for routing a heterogeneous fleet of vehicles. In Proceedings of the 2019 IEEE Latin American Conference on Computational Intelligence (LA-CCI), Guayaquil, Ecuador, 11–15 November 2019; pp. 1–6.
18. Brittain, M.; Wei, P. Autonomous separation assurance in an high-density en route sector: A deep multi-agent reinforcement learning approach. In Proceedings of the 2019 IEEE Intelligent Transportation Systems Conference (ITSC), Auckland, NZ, USA, 27–30 October 2019; pp. 3256–3262.
19. Wang, Z.; Yao, H.; Mai, T.; Xiong, Z.; Yu, F.R. Cooperative Reinforcement Learning Aided Dynamic Routing in UAV Swarm Networks. In Proceedings of the ICC 2022-IEEE International Conference on Communications, Seoul, Republic of Korea, 16–20 May 2022; pp. 1–6.
20. Ibarz, B.; Leike, J.; Pohlen, T.; Irving, G.; Legg, S.; Amodei, D. Reward learning from human preferences and demonstrations in atari. In Proceedings of the Advances in Neural Information Processing Systems, Montreal, Canada, 3–8 December 2018; Volume 31.
21. Li, K.; Zhang, T.; Wang, R. Deep reinforcement learning for multiobjective optimization. *IEEE Trans. Cybern.* **2020**, *51*, 3103–3114. [CrossRef] [PubMed]
22. Xu, J.; Tian, Y.; Ma, P.; Rus, D.; Sueda, S.; Matusik, W. Prediction-guided multi-objective reinforcement learning for continuous robot control. In Proceedings of the International Conference on Machine Learning, Virtual, 13–18 July 2020; pp. 10607–10616.
23. Schulman, J.; Wolski, F.; Dhariwal, P.; Radford, A.; Klimov, O. Proximal policy optimization algorithms. *arXiv* **2017**, arXiv:1707.06347.
24. Hochreiter, S.; Schmidhuber, J. Long short-term memory. *Neural Comput.* **1997**, *9*, 1735–1780. [CrossRef] [PubMed]
25. Amato, C.; Konidaris, G.; Cruz, G.; Maynor, C.A.; How, J.P.; Kaelbling, L.P. Planning for decentralized control of multiple robots under uncertainty. In Proceedings of the 2015 IEEE international conference on robotics and automation (ICRA), Seattle, WA, USA, 25–30 May 2015; pp. 1241–1248.
26. Bernstein, D.S.; Givan, R.; Immerman, N.; Zilberstein, S. The complexity of decentralized control of Markov decision processes. *Math. Oper. Res.* **2002**, *27*, 819–840. [CrossRef]
27. Konda, V.; Tsitsiklis, J. Actor-critic algorithms. In Proceedings of the Advances in Neural Information Processing Systems, Denver, CO, USA, 29 November–4 December 1999; Volume 12.
28. Foerster, J.; Farquhar, G.; Afouras, T.; Nardelli, N.; Whiteson, S. Counterfactual multi-agent policy gradients. In Proceedings of the AAAI Conference on Artificial Intelligence, New Orleans, LA, USA, 2–7 February 2018; Volume 32.
29. Hausknecht, M.; Stone, P. Deep recurrent q-learning for partially observable mdps. In Proceedings of the 2015 AAAI Fall Symposium Series, Arlington, VA, USA, 12–14 November 2015.
30. Krichen, M.; Adoni, W.Y.H.; Mihoub, A.; Alzahrani, M.Y.; Nahhal, T. Security Challenges for Drone Communications: Possible Threats, Attacks and Countermeasures. In Proceedings of the 2022 2nd International Conference of Smart Systems and Emerging Technologies (SMARTTECH), Riyadh, Saudi Arabia, 9–11 May 2022; pp. 184–189.
31. Alrayes, F.S.; Alotaibi, S.S.; Alissa, K.A.; Maashi, M.; Alhogail, A.; Alotaibi, N.; Mohsen, H.; Motwakel, A. Artificial Intelligence-Based Secure Communication and Classification for Drone-Enabled Emergency Monitoring Systems. *Drones* **2022**, *6*, 222. [CrossRef]

Article

A Multi-Agent System Using Decentralized Decision-Making Techniques for Area Surveillance and Intruder Monitoring

Niki Patrinopoulou [1,*], Ioannis Daramouskas [1,2], Dimitrios Meimetis [1], Vaios Lappas [3] and Vassilios Kostopoulos [1]

[1] Applied Mechanics Lab, University of Patras, 26504 Patras, Greece
[2] Computer Technology Institute and Press "Diophantus, N. Kazantzaki Str., University Campus, 26504 Patras, Greece
[3] Department of Aerospace Science & Technology, National Kapodistrian University of Athens, 10563 Athens, Greece
* Correspondence: n.patrinopoulou@upnet.gr

Abstract: A decentralized swarm of quadcopters designed for monitoring an open area and detecting intruders is proposed. The system is designed to be scalable and robust. The most important aspect of the system is the swarm intelligent decision-making process that was developed. The rest of the algorithms essential for the system to be completed are also described. The designed algorithms were developed using ROS and tested with SITL simulations in the GAZEBO environment. The proposed approach was tested against two other similar surveilling swarms and one approach using static cameras. The addition of the real-time decision-making capability offers the swarm a clear advantage over similar systems, as depicted in the simulation results.

Keywords: real-time decision making; decentralized monitoring; swarm surveillance algorithm; autonomous quadcopters; swarm intelligence

Citation: Patrinopoulou, N.; Daramouskas, I.; Meimetis, D.; Lappas, V.; Kostopoulos, V. A Multi-Agent System Using Decentralized Decision-Making Techniques for Area Surveillance and Intruder Monitoring. *Drones* **2022**, *6*, 357. https://doi.org/10.3390/drones6110357

Academic Editors: Mou Chen, Xiwang Dong, Xiangke Wang and Fei Gao

Received: 20 October 2022
Accepted: 12 November 2022
Published: 16 November 2022

Publisher's Note: MDPI stays neutral with regard to jurisdictional claims in published maps and institutional affiliations.

1. Introduction

The decision-making capability is an important attribute, essential for designing autonomous and intelligent systems. Agent-based real-time decision-making based on the data collected by the swarm is proven that can increase the efficiency of the solution and remain robust to dynamic changes and uncertainties. The aim of this work is to examine the efficiency of a decision-making algorithm for swarms compared with other methods, where the decision-making is not existing, and evaluate the methods with a series of metrics in six different scenarios ensuring that the swarm can operate autonomously and safely regarding the inter-agent collisions.

We present a scalable and robust swarm, designed for surveilling a specific area and tracking intruders. The concept is based on that when the swarm starts its operations, it does not have any knowledge about whether intruders exist or not in the monitored area. The intruders spawn at random places in the world during initialization and then there is a fixed time window in which new intruders spawn in the world. The main algorithm behind the swarm's operation is a stochastic optimization-based decision-making algorithm, responsible for selecting the next task of each agent from a large total of options. The selection criteria are designed so that decision- making is optimized in a system level rather than in an agent level, since we consider that global optimization provides better results for our system. The algorithms needed to support the operation of the swarm are described as implemented.

The key findings of our work are that a swarm with key components such as Task allocation, Collision Avoidance, V2V communications, and V2G communications can perform precisely and robustly a series of tasks in contrast to swarms with no cognitive intelligence, as proven by our experiments. We can observe that when the swarm activates the decentralized decision-making, the effectiveness of the system is increased significantly, as measured by a group of metrics.

Section 2 contains a brief state-of-the-art review focusing on decision-making and task allocation algorithms. Section 3.1 introduces the UAVs and sensors used in the system. The rest of Section 3 presents the developed algorithms and the proposed system architecture. Section 3.7 depicts the tools used to implement and simulate the designed system. In Sections 3.8 and 3.9, the behavior of the simulated intruders and the parameters of the experiment scenarios are presented accordingly. In Section 4, the metrics used to assess the algorithm and the experiment results are provided, where in Section 5 we present the key findings of our work. Finally, in Section 6 we present the conclusions we made conducting these series of experiments.

2. Related Work

The state-of-the-art presents a plethora of different approaches to the use of decision-making in the task allocation problem. According to [1], the multi-robot task allocation problem is an example of a Discrete Fair Division Problem, as an Optimal Assignment Problem, an ALLIANCE Efficiency Problem or a Multiple Traveling Salesman Problem. The methods to solve the multi-robot task allocation problem can be categorized to be auction based, game theory based, optimization based, learning based and hybrid, as they are listed below.

Auction based: In this type of approach, tasks are offered via auctions, the agents can bid for tasks and the agent with the higher bid is assigned the corresponding task. Each agent bids a value representing the gain of the utility function in case the agent gets assigned that task. The utility function is designed based on the criteria of each problem and takes as inputs the agent's current state, the task's description, and the local environment perception of the agent [2]. The auctioneer might be a central agent, or as it is more common, the auction could be held in a decentralized manner, such as in [3]. The authors in [4] address the task allocation problem for multiple vehicles using the consensus-based auction algorithm (CBAA) and the consensus-based bundle algorithm (CBBA), which is a modification of the first one to be applied in multi-vehicle problems. The Contract Net Protocol (CNP) presented in [5] was the first negotiation platform used in task allocation problems and constitutes the base for numerous task allocation algorithms. CNP was tested in [6] using a variety of simulation environments to solve the task allocation problem for multiple robots. The authors concluded that because of the interdependency of the tasks in a multi-robot task allocation problem, the original CNP approach does not solve the problem sufficiently.

Game theory based: Game theory-based approaches describe the strategic interactions between the players of a game. Each decision-making agent is considered a player and the game strategy of a player consists of the tasks that the player chose. When the task allocation solution proposed has been optimized globally, all the players will stop changing their strategies, since the optimal outcome has been reached; that condition is called Nash equilibrium. In [7], the authors present several applications of game-theoretic approaches to UAV swarms. Authors in [8] proposed a decentralized game theory-based approach for single-agent and multi-agent task assignment for detecting and neutralizing targets by UAVs. In their scenario, UAVs might not be aware of the strategies of other UAVs and a Nash Equilibrium is difficult to achieve. Instead, they used a correlated equilibrium.

Optimization based; Optimization algorithms focus on finding a solution from a set of possible solutions, so that the solution's cost is minimized, or the solution's profit is maximized depending on the specific problem's criteria. Optimization techniques can be distinguished into deterministic or stochastic methods. Deterministic methods always produce the same results for equal inputs, while stochastic methods produce with high probability similar results for equal inputs. Probably the most famous deterministic optimization method used for task allocation is the Hungarian Algorithm (HA) [9]. The HA attempts to solve the General Assignment Problem (GAP) in polynomial time by maximizing the weights of a bipartite graph. In [10], the authors approach the task allocation problem as a Vehicle Routing Problem (VRP) in order to solve a multi-agent collaborative route planning problem. In this case, HA is employed after it has been modified to be able to consider constrains, and a detour resolution stage has been added.

A subcategory of stochastic algorithms with great interest for us is the metaheuristics methods which include evolutionary algorithms, bio-inspired algorithms, swarm intelligence, etc. [11] presents the Modified Distributed Bees Algorithm (MDBA), a decentralized swarm intelligence approach for dynamic task allocation, which shows great results when compared with the state-of-the-art auction-based and swarm intelligence algorithms. In [12], three different algorithms are presented inspired from Swarm-GAP, a swarm intelligence, heuristic method for the GAP. Authors in [13] use a genetic algorithm (GA) optimization for decentralized and dynamic task assignment between UAV agents. The task assignment includes an order optimization stage, using GA optimization, for ordering the tasks from a single-agent point of view and a communications and negotiation stage for reallocating tasks between neighboring agents.

Learning based: A commonly used learning-based method is reinforcement learning, a machine learning subcategory. Reinforcement learning algorithms adjust their parameters based on the data gathered from their experiences, to achieve better behaviors. Q-learning is a model free reinforcement learning method, which describes the environment as a Markov Decision Process (MDP). In [14], a Q-learning implementation for the dynamic task allocation is presented, while the adaptability of Q-learning to uncertainties is showcased in [15], where it is used for multi-robot task allocation for the fire-disaster response.

Hybrid: Hybrid approaches combine some of the methods listed above to solve the task allocation problem. In [16], the authors study the Service Agent Transport Problem (SATP), a problem in the family of task-schedule planning problems, using a Mixed-integer linear programming (MILP) of the optimization-based category and an auction-based approach. [17] proposes an improved CNP technique for solving the problem of task allocation for multi-agent systems (MAS), combining CNP with an ant colony algorithm using the dynamic response threshold model and the pheromone model for the communication between agents. [18] uses a CBBA-based approach, combined with the Ant Colony System (ACS) algorithm and a greedy-based strategy to solve the problem of task allocation for multiple robots' unmanned search and rescue missions.

The multi-agent surveillance and multi-target monitoring and tracking problem has been studied by several researchers, and a variety of decision-making techniques have been proposed. A gradient model for optimizing target searching based on beliefs regarding the target's location is presented in [19,20]. They propose a decentralized architecture for the implementation of their algorithm, in which it is assumed that the agents' belief is globally known across the system, and each agent optimizes its own actions based on the global belief. Authors in [21] present a decentralized approach, in which UAV agents are organized in local teams, in which the target estimations are communicated. A particle filter is used to track the targets and the estimations are approximated as Gaussian Mixtures using the expectation-maximization algorithm. The leaders of the local teams are responsible for dynamically assigning regions to the team members. A system of UAVs and ground sensors is studied in [22] for surveillance applications. Targets are detected from both the ground and aerial sensors and UAVs are assigned targets based on a decision-making methodology, so that a multi-attribute utility function is maximized. Partially Observable Markov Decision Processes (POMDPs) have been proposed to model surveillance missions to deal with uncertainties. A methodology to use POMDPs in a scalable and decentralized system is presented in [23], based on a role-based auctioning method. In [24], an integrated decentralized POMDP model is presented to model the multi-target finding problem in GPS-denied environments with high uncertainty.

3. Materials and Methods

3.1. Drone Characteristics

3.1.1. Drone Kinematic Model

The vehicle used in our tests is a simple quadcopter, shown in Figure 1, that can be controlled by linear velocity commands in the x, y and z axis. The yaw of the vehicle remains constant with small variations at its initial value, yaw = 0. For all the experiments we assume a constant flight altitude is used.

Figure 1. The iris drone as it is visualized in the GAZEBO simulator.

3.1.2. Sensors

The camera of the agent is directed vertically downwards, as presented in Figure 2. The camera's field of view (FOV) for every given moment is a rectangle defined by its height, width, and center. The center of the rectangle coincides with the position of the drone, while the height and width are given by the Equations (1) and (2).

$$height_{fov} = 2 \times altitude \times \tan\left(\frac{fov\ vertical\ angle}{2}\right) \tag{1}$$

$$width_{fov} = 2 \times altitude \times \tan\left(\frac{fov\ horizontal\ angle}{2}\right) \tag{2}$$

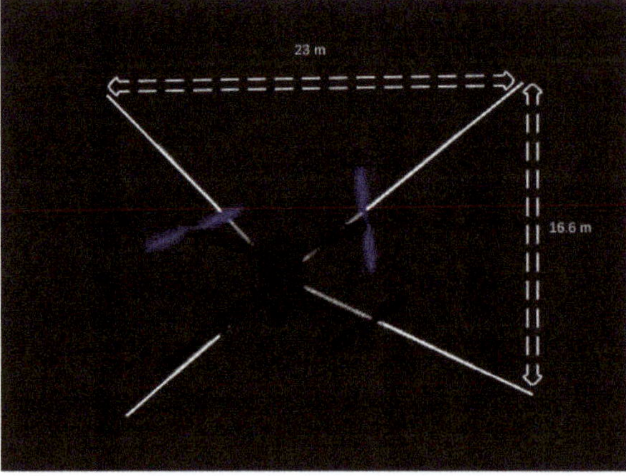

Figure 2. The field of view of the iris drone with downwards oriented optical camera. The dimensions of the field of view in this figure are measured for a flight altitude of 20 m.

In our case, the flight altitude of the drones was predefined to 20 m for all the simulations and the camera in use has fov vertical angle $= 0.785$ rad and fov horizontal angle $= 1.047$ rad.

3.2. System Overview

The system is described as a surveillance system with decentralized decision-making capabilities and a central entity acting as a single point of truth. Each agent runs the same code separately and can make its individual decisions. Before each decision is made the agent asks from the central entity to provide him with information about the map/world. That information is gathered in the central entity as each agent sends the data that he is collecting. For every agent an identification number, unique in the swarm, is allocated. Figure 3 shows the main data exchange between the agents and the central entity.

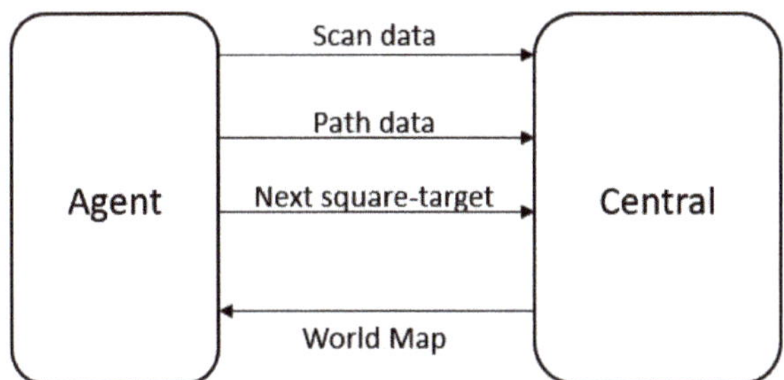

Figure 3. Central entity and agent data exchange.

The messages exchanged between each agent and the central entity are listed below:

- From an agent to the central entity:
- Scan data: The scan data message includes the identification number of the agent, the number of intruders caught in the square, the number of the intruders detected but not caught while scanning and the 2-D coordinates of the square scanned. The message is sent from the agent to the central entity every time that the agent transitions from the "Scan" mode to the "Go to" mode.
- Path data: The path data message includes the identification number of the agent and a list of the intruders that were detected and not caught while moving from the previous target to the next. The message is sent from the agent to the central entity every time the agent transitions from the "Go to" mode to the "Scan" mode, since that is when the agent has completed its path to the new target. Moreover, the path data message will be sent if in the process of following an intruder, another intruder gets detected.
- Next-square target: The next target message includes the identification number of the agent and the 2-D coordinates of the next target that the agent selected. That message is sent from the agent to the central entity every time the agent decides on a next target.
- From the central entity to an agent:
- World map: The world map message is a 2-D matrix with the information about the world, as described in Section 3.3.
- The agents' behavior consists of three different modes:
- Scan: "Scan" mode is activated when the agent is in the boundaries of its square-target. The agent delineates a zig-zag coverage pattern to surveille the whole square-target and check for intruders in that square. If an intruder is detected, then the agent will transit to "Follow intruder" mode. The algorithm used is described in detail in Section 3.4.
- Go to: In this mode, the agent has decided on the next square-target and it moves towards the target in a straight line connecting its current position and the vertex of the square-target that is closer to the current position.

- Follow intruder: Independent of the previous mode, when an intruder is detected the agent changes to "Follow intruder" mode. If the agent is already following an intruder, it will keep following the previous intruder and when the intruder is caught the agent will follow the new intruder if the new intruder is still in the agent's detection range (in the FOV of the agent), otherwise the agent will change to "Go to" mode and move towards the next square-target. In the case that another agent is in a distance that allows him to detect the intruder as well, the agent will drop the "Follow intruder" mode with a probability of 0.1. That characteristic is added to avoid agent congestion over a specific intruder or small group of intruders. The drop probability used may seem too small, but we need to consider that the algorithm runs in a ROS node with a frequency of 5 Hz, so for every second each agent in that situation has a probability of 0.5 to drop the mode.

The agents' modes and the trigger mechanisms for transitioning between modes are summed up in Figure 4.

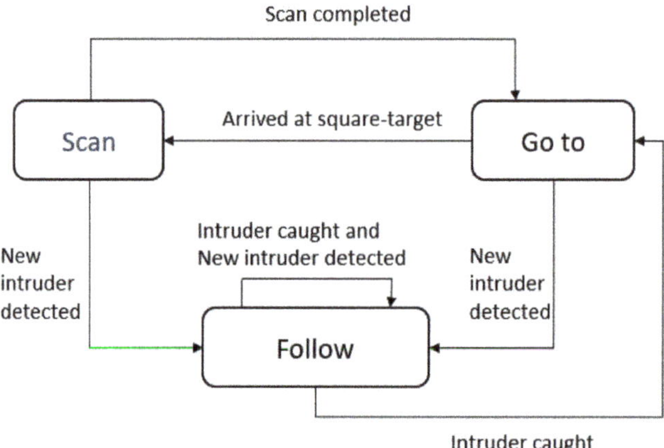

Figure 4. Agents' mode sequence and change triggers. The three modes of the agent "Scan", "Go to" and "Follow" are visualized as rectangles and the transitions between the modes are arrows, explaining the type of the cause that triggered the transition.

3.3. World Representation

The world is treated as a 2-D grid of $n \times n$ size, which consists of equal sized squares. A similar approach to discretize the area search problem has been introduced in [25,26]. Each square corresponds to one task and each task can be assigned to one agent at any given moment. Each agent is responsible for one task and only when that task is completed or dropped, is when the agent can select a different task. If an agent has selected a task, the central entity flags the square corresponding to that task, so that no other agent is able to select the same task. If two or more agents select the same task simultaneously then the central entity is responsible to inform one of them through a message asking to change their task and repeat the selection process.

The central entity initializes a 2-D matrix containing the grid's information. The matrix is updated by the central entity based on the data that are received from the agents. When an agent needs to select its next task considering the world information, the agent receives the grid matrix from the central entity. Each node of the matrix includes the following information:

- Time of last visit: that contains the time stamp of the last time that the corresponding node was scanned by an agent.
- Probability: that expresses the estimated probability of finding an uncaught intruder in that node. The probability is calculated based on the number of intruders that were

detected and not caught in that node and in its neighboring nodes. The probability value p_i is initialized at 0.1 for all the nodes (Equation (3)). When a square-target is selected by an agent, its corresponding node's probability takes a negative value so that no other agent selects that square-target until the current agent has completed its task (Equation (4)). The probability is repaired to its non-negative value when scanning is completed. When scan or path data are received, the probability updates as described at Equations (6)–(10).

Initialize all square probabilities to 0.1:

$$p_i = 0.1 \; \forall \, i \in Grid \tag{3}$$

When a next target message is received for square i as the target assigned to an agent:

$$p_i = p_i - 100 \tag{4}$$

When a scan message is received after scanning square i:

$$If \; p_i < 0 \; : \; p_i = p_i + 100 \tag{5}$$

- If no intruders were detected in the square i after a full scan:

$$p_i = 0.1 \tag{6}$$

- If N intruders were detected and not caught in the square i:

$$Find \; the \; neighborhood \; n_i \; of \; i \tag{7}$$

$$\forall \, square \; j \in n_i \; : \; v_j = \frac{1}{1 + e^{-(d_{max} - d_j)}} \tag{8}$$

$$\forall \, square \; j \in n_i \; : \; p_j = p_j + N * \frac{v_j}{\sum v_k} \tag{9}$$

$$If \; p_i > 1 : p_i = 1 \; \forall \, i \in Grid \tag{10}$$

The value v_j is computed for every square separately and it is dependent on its distance d_i from the center, since intruders tend to move towards the center and the probability of their next move to be in a square closer to the center has a higher probability. Where d_{max} is the maximum distance computed from the neighborhood to the target (in our experiments the center of the map). Before it is added to the probability of the square, the value v_j is divided by the sum of all values v_j calculated for the neighborhood so that $\sum_{k \in n_i} \frac{v_j}{\sum v_k} = 1$, where n_i is the neighborhood of square i. The size of the neighborhood depends on the speed of the intruders and the size of the squares. In our implementation, the neighborhood consisted of only the squares adjacent to the square i, creating a neighborhood of nine squares (3×3 square neighborhood), containing the square i.

3.4. Coverage Algorithm

The objective of the coverage path planning algorithms is to compute a path that crosses over all points of an area of interest while avoiding obstacles [27]. As mentioned above, each square of the grid corresponds to an agent's task. The task to be implemented is for the agent to scan the whole area of the square using a coverage algorithm. Since the main objective of the system is to detect intruders, the scanning is dropped if an intruder is detected, in which case the agent starts following the intruder, activating the "Follow intruder" mode. If no intruder is detected, the task is completed when the area of the square has been scanned.

The scan mode is activated only after the "Go to" mode and the event that triggers that transition is the arrival of the agent at one of the corners of the square to be scanned. Since the FOV of the agent is considered to be a rectangle, the agent does not actually have to be on the edges of the square for them to be scanned. We assume a rectangle smaller than the square and with the same center (the inner rectangle as presented in Figure 5). The height and width of the rectangle depends on the height and the width of the field of view accordingly and is given by Equations (11) and (12).

$$height_{rectangle} = edge_{square} - 2 \times \frac{2}{6} \times height_{fov} \tag{11}$$

$$width_{rectangle} = edge_{square} - 2 \times \frac{2}{6} \times width_{fov} \tag{12}$$

where the $height_{rectangle}$ and the $width_{rectangle}$ represent the height and the width accordingly of the inner rectangle, the $edge_{square}$ is the length of the edge of each square-target and the $height_{fov}$ and $width_{fov}$ are the height and width of the Field Of View of the agents.

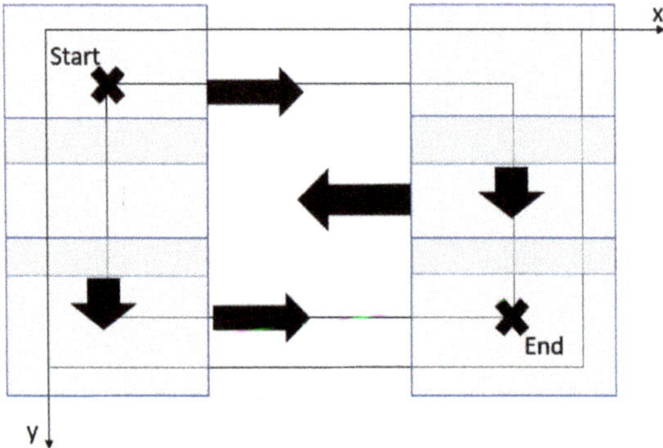

Figure 5. Scanning movement: The agent starts at the up-left corner of the inner rectangle. Then, the agent moves to the right along the x axis until it reaches the right edge of the inner rectangle. After, it moves downwards along the y axis for a distance equal to two thirds of the height of the FOV. The agent continues its movement, moving to the left along the x axis until it reaches the left edge of the inner rectangle. Finally, the agent repeats its downwards movement until it reaches the down side of the inner rectangle and it moves to the right until it reaches the down-right corner of the inner rectangle.

The agent moves in the boundaries of the inner rectangle, drawing a zig-zag shaped route. The scanning movement starts with a repeating shift on the x axis until the right-side or left-side (depending on the starting corner) boundary is reached and continues with a shift at the y axis for $\frac{2}{3} \times height_{fov}$. The sequence of shifts is repeated with the direction of the shift on the x axis to be inverted for each repetition until the upper-side or downer-side (depending on the starting corner) is reached. When the movement is completed, the agent has visited all the corners of the inner rectangle, and by doing so, it has scanned the whole area of the square.

3.5. Swarm Intelligence—Decision Making

The most important part of the system is the agents' ability of decision-making to select their next square-target. That is handled by a stochastic algorithm, partially inspired from the ant colony pheromone deposition [28] idea. The decision-making process is activated when an agent has completed a task and it needs to choose the next square-target as its task. To make its decision, it uses the world information provided by the central entity as a 2-D matrix, containing the probability and time of the last visit of all the square-targets of the grid. The decision-making process is depicted in Figure 6. The agent first decides if it will stay in its current neighborhood or travel to another neighborhood of the map. That decision is not deterministic, and the agent chooses its current neighborhood with a probability of 0.7, the center neighborhood with probability of 0.06 or a random square-target with probability of 0.24. The ability to travel across the map instead of staying in neighboring squares is added to force the agents to move around the map; this helps to escape local minima by exploring areas of the map that have not been explored recently or detect intruders during the flight and add more information to the world's matrix. After the agent decides the neighborhood of its next square-target, it needs to select the exact square-target. It computes the margin of every square of the neighborhood based on the Equation (13):

$$margin_i = probability_i^7 \times \frac{\left(time\ now - time\ of\ last\ visit\right)^3}{600} \tag{13}$$

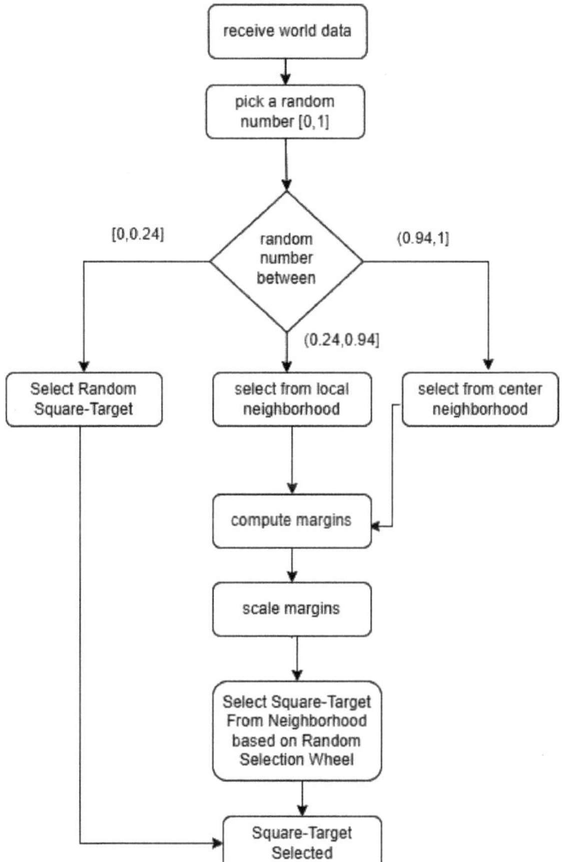

Figure 6. Flowchart of the proposed decision-making algorithm for the selection of the next square-target.

The sum of all the margins of the neighborhood gives the margin$_{sum}$:

$$margin_{sum} = \sum_{i\ \in\ neighborhood} margin_i \tag{14}$$

Each margin computed is divided by the margin$_{sum}$ to compute the probability of selecting each square-target.

$$selection\ prob_i = \frac{margin_i}{margin_{sum}} \tag{15}$$

Finally, the next square-target is selected in a non-deterministic manner and each square-target has a probability selection prob$_i$ to be selected. After the agent selects its next target, it informs the central entity by sending a "Next square-target" message containing its identification number and its selected target.

For the random selection based on probabilities, a simple wheel selection algorithm similar to the one proposed in [29] was developed. The algorithm is presented in Algorithm 1.

Algorithm 1. Random selection wheel

1: Choose a random number p in the range [0, 1]
2: Create a list prob_list containing all the probabilities
3: Initialize i as 0
4: Set prob as prob= prob_list[i]
5: If prob <= p
6:    The i element is selected, and the algorithm is terminated
7: Else
8:      p = p-prob
9:      i++
10:   Repeat from step 4

The decision-making algorithm uses the idea of pheromones and evaporation introduced in the ACS, which in our case is implemented by saving the time of the last visit of each square. The agent's decision is based on how recently the square that it is considering on selecting was visited. In that way, a square that has been scanned recently and hence has higher probability of not having intruders has a lower probability to be picked by the agent. It is clear that in our case the existence of pheromones acts as a suspending factor on visiting an area, which is in contrast to the way that the pheromones are used in the ant colony as described in [28], where the existence of pheromones increases the probability of an agent to visit the area.

The probability of finding intruders in a square can also be described as an attractive pheromone, which does not obey the evaporation phenomenon. The intruder-related pheromone only increases until the agent scans the corresponding square, and if no intruders are detected it is decreased to its initialization value of 0.1.

We should note here that in the scenario under study the behavior of one intruder is independent on the behavior of the rest of them. Under that assumption, it is not valid to use the information of an intruder that has been caught to predict the behavior of the rest of them. So, the probability of finding an intruder in a square is computed using information regarding only intruders that were detected, but they were not caught. It would be prudent to say that if the behavior of each intruder influences the rest of the intruders, the data concerning the intruders that have been caught would also be useful in determining the probability of finding an intruder in a specific area.

3.6. Collision Avoidance

The most crucial block when dealing with swarms is to ensure that each agent can perform autonomously with safety. Hence, a collision avoidance algorithm is needed to ensure that the agents do not collide on each other. In the literature, a variety of methods exists with many different characteristics and capabilities. A potential field method [30] was selected both for guiding the agents to a point of interest and for preventing inter-agent collisions. The implemented collision avoidance method is decentralized and it requires for every agent to be aware of the position of the other agents in a distance shorter or equal to 7 m by utilizing V2V communication. The collision avoidance block is enabled only at the "Go to" and "Intruder following" modes. In the "Scan" mode, no conflicts occur, since only one agent could be in the "Scan" mode on a particular square-target at every moment. If two or more agents either in the "Go to" or in the "Intruder following" mode detect a collision in their path, they all act to ensure deconfliction. If one or more agents not in the "Scan" mode detect a possible collision with an agent in the "Scan" mode, the agents that are not in "Scan" mode deconflict while the scanning agent continues its route.

In the "Go to" and the "Intruder following" modes, the objective is similar; navigate to a specific point of interest while avoiding collisions with other agents. The difference between the modes is the type of the point of interest, which is a constant point in the case of the "Go to" mode and a moving ground target in the case of the "Intruder following" mode. Thus, the calculation of the movement commands is conducted in the same way in both modes.

The computed desired velocity of each agent is the sum of attractive velocity and repulsive velocity. The attractive velocity is caused by an attractive force acting on the agent and causing it to move towards the point of interest. The repulsive velocity is caused by a repulsive force acting between agents, which is responsible for not allowing agents to come too close, preventing the possibility of a collision.

The attractive velocity is analyzed at $v\_attr_{i,x}$ and $v\_attr_{i,y}$ as shown in Equations (16) and (17) and it is dependent on the distance from the target. The coordinates of the target are given as a 2-D point ($goal_{i,x}$, $goal_{i,y}$), as is the position of the agent i ($position_{i,x}$, $position_{i,y}$).

$$v\_attr_{i,x} = \begin{cases} 2 \times \frac{goal_{i,x}-position_{i,x}}{|goal_{i,x}-position_{i,x}|}, & if \ |goal_{i,x} - position_{i,x}| \geq 2 \\ goal_{i,x} - position_{i,x}, & if \ |goal_{i,x} - position_{i,x}| < 2 \end{cases} \quad (16)$$

$$v\_attr_{i,y} = \begin{cases} 2 \times \frac{goal_{i,y}-position_{i,y}}{|goal_{i,y}-position_{i,y}|}, & if \ |goal_{i,y} - position_{i,y}| \geq 2 \\ goal_{i,y} - position_{i,y}, & if \ |goal_{i,y} - position_{i,y}| < 2 \end{cases} \quad (17)$$

The repulsive velocity is also analyzed at $v\_rep_{i,x}$ and $v\_rep_{i,y}$ and it is calculated from Equations (18) and (19), where the position of another agent j in the detection distance of 7 m is defined as ($position_{j,x}$, $position_{j,y}$), and $distance_{i,j}$ is the Euclidean distance between the two agents.

$$v\_rep_{i,x} = \begin{cases} \sum_{j} -2 \times \frac{position_{j,x}-position_{i,x}}{distance_{i,j}} & \forall j : 2 < distance_{i,j} \leq 7 \\ \sum_{j} -2 \times \frac{position_{j,x}-position_{i,x}}{|position_{j,x}-position_{i,x}|} & \forall j : distance_{i,j} \leq 2 \end{cases} \quad (18)$$

$$v\_rep_{i,y} = \begin{cases} \sum_{j} -2 \times \frac{position_{j,y}-position_{i,y}}{distance_{i,j}} & \forall j : 2 < distance_{i,j} \leq 7 \\ \sum_{j} -2 \times \frac{position_{j,y}-position_{i,y}}{|position_{j,y}-position_{i,y}|} & \forall j : distance_{i,j} \leq 2 \end{cases} \quad (19)$$

The overall desired velocity is expressed in the x, y axes as $v_{i,x}$ and $v_{i,y}$ for each agent i, and it is computed from Equations (20) and (21).

$$v_{i,x} = v\_attr_{i,x} + v\_rep_{i,x} \quad (20)$$

$$v_{i,x} = v\_attr_{i,y} + v\_rep_{i,y} \quad (21)$$

The computed velocity here is the desired velocity of the agent and it is sent to the autopilot, who is responsible for achieving it in a robust and efficient manner. That provides us with the freedom of not having to ensure the continuity of the velocity functions. If the velocities computed here were fed directly to the motors, the continuity of the velocity functions should be ensured, either by computing the velocity indirectly via computing the attraction or repulsion forces, or by adding a maximum velocity change step.

One of the main problems caused by the potential fields family of algorithms is the existence of local minimum that cause the agents to immobilize before they reach their goal [31]. Local minima could be resolved with three approaches: Local Minimum Removal, Local Minimum Avoidance and Local Minimum Escape (LME) [32]. Since the environment that we are working in does not contain any static obstacles, the agents could fall into local minimum caused only by the existence of other agents nearby. We choose to resolve local minimum using a local minimum escape method. In the LME approaches, the agents reach a local minimum and then an escape mechanism is triggered to resolve it.

The local minimum detection and resolution is implemented in a decentralized manner by each agent separately. After the agent has computed its desired velocity, it checks if he is trapped in a local minimum. If the agent's desired velocity is equal to zero (using a threshold near zero) and his attractive velocity does not equal to zero, then the agent is considered trapped. At that point, the agent assumes that all the other agents from which the agent is currently deconflicting are also trapped in the same local minimum. The agent computes the average position of all agents trapped in the same local minimum.

$$position_{local_{minimum}} = \frac{\sum_{i=0}^{n\_trapped} position_i}{n\_trapped} \tag{22}$$

where n_trapped is the number of the agents trapped in that local minimum and i belongs in the set of agents trapped in that local minimum. Each agent i performs a circular motion around the $position_{local_{minimum}}$ in an anti-clockwise direction with a constant speed. The agent recomputes its desired velocity in every time step and it continues with the circular motion until it is no longer trapped, in which case it continues with its path.

3.7. Implementation—Simulation

To validate our algorithms and the effectiveness of our system, we performed a series of experiments in simulated worlds. To make our swarm more realistic and applicable to real world scenarios, we decided to use the famous robotics framework ROS [33]. Using the ROS architecture capabilities, we can add to our system all the desirable aspects for every block we described. The nodes were developed at C++ and python and the ROS version used was ROS melodic. The simulations were conducted using the GAZEBO 7 physics engine [34], where the PX4 autopilot [35,36] was used to control the drones and the selected vehicle was the iris quadcopter, as provided by the PX4.

The central entity is managed by a python script that creates a ROS node is named the central_node, while a ROS node named drone_node was developed in C++ to control the agents. For each agent, an instance of the drone_node runs, given different values for each node. The essential data for each drone_node instance initialization are: the identification number, and the x and y cartesian coordinates of the corresponding agent's spawn position. The drone_node instances also send control commands with the desired velocity in the x, y and z axis to the PX4 autopilot. The intruders are managed by a python script, which creates a ROS node named intruders_node. The intruders_node is responsible for spawning them and moving them, as described in Section 3.8, and keeping logs of the metrics presented under Section 4.2. All of the components described communicate with each other by exchanging messages (publish or subscribe) to specific ROS topics. For the communication of the node developed by our team, special message types were developed to include the exact types of variables needed.

Figure 7 presents the overall system architecture of the implementation of a swarm containing two agents only for demonstration purposes. The figure has been produced from the rqt_graph ROS tool. The nodes are represented by eclipses, while the arrows connecting them represent the topics which they use to exchange messages. The gazebo and gazebo_gui nodes are related to the simulation and the simulation's graphical user interface. The uav0/mavros and uav1/mavros nodes' purpose is to transfer information between the ROS environment and the autopilot [37]. The MAVROS package [38] enables the data exchange between ROS nodes and autopilots equipped with the MAVLink communication protocol [39]. The nodes central_node, drone_node0 and drone_node1 were implemented by our team.

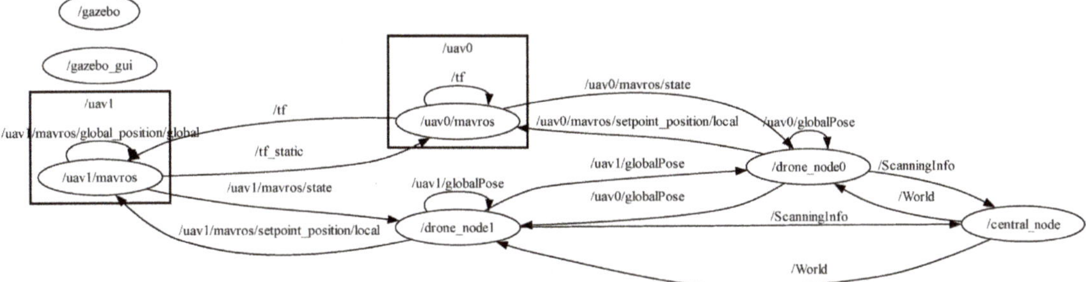

Figure 7. The rqt graph with the ROS nodes. The rqt graph includes two agents and the central entity for simplicity. The ROS nodes are represented by ellipses while the ROS topics used for message exchange between the nodes are the arrows connecting them. The /uav0/mavros and /uav1/mavros nodes are created from the mavros ROS package to enable the communication of the ROS nodes with the drones' firmware.

3.8. Intruders' Behavior

In this section, we will present the intruders' behavior. An intruder in our simulations can be ground moving objects (either people or robots with constant speed and smaller in amplitude to the drone's speed). An intruder's goal is to reach the center of the world and stay there for 10 s. The attributes defining the behavior of the simulated intruders are summarized here:

- Spawn positions: It is assumed that the world was not being surveilled before the simulation starts, so at the beginning of the simulation, five intruders are spawned at random positions through the world. After that, the intruders are spawned only at the edges of the world, randomly distributed along the four edges of the boundaries of the world.
- Spawn time: Spawn time is defined as the time interval between the spawn of two consequential spawning groups of intruders after the simulation starts. In our simulation, that value was constant and equal to 10 s and the size of the spawning group was set to two intruders, so every 10 s, two more intruders were spawned in the simulation.
- Movement type: The intruders' goal is to reach the target, so each intruder's average movement is on a straight line starting from its spawn position and ending at the target. To recreate a more realistic movement pattern, a stochastic element is added to the constant velocity movement. For every four steps that the intruders make, three of them are the right direction and one of them is in a random direction. After reaching the target, the intruders stay over it for 10 s before they complete their mission. If an intruder completes its mission, it is removed from the simulation.
- The intruders are simulated as non-dimensional points with holonomic movement. Since the intruders are assumed to be non-dimensional, inter-intruder collision is not considered.
- An intruder is considered caught after it has been tracked by an agent for a predefined tracking time. When an intruder is caught, it is removed from the simulation and the metrics related to the caught intruder are saved.
- An intruder is considered alive from its spawn time until it is caught, or it reaches the target.
- An intruder is detected from an agent, if the intruder is in the FOV of the agent's camera.

3.9. Scenario

Six scenarios were designed to test the performance of the algorithm. Each scenario has a different world size and swarm size to evaluate the scalability of the algorithm. The parameters to describe each scenario are listed below:

- World size: the size of the simulated world.
- Grid size: the size of the grid applied in the world.
- Square size: the size of the individual square of the grid depends on the size of the world and the size of the grid and is calculated based on the Equation (23).

$$square\ size = \frac{world\ size}{grid\ size} \tag{23}$$

- Swarm size: the number of the agents of the swarm.
- Environment type: an empty environment was selected with no static obstacles that would cause collision risks and visibility constraints.
- Simulation duration: the duration of the simulation remained constant for all three scenarios at 33 min in real time simulation.
- Intruders spawned: the total amount of intruders spawned during the simulation; that value is constant at 401 intruders for all the scenarios and experiments that were conducted.
- Intruders' average speed: That is computed by dividing the average time that the intruders need to reach the target by the average distance between their spawn position and the target.
- Target: The target is defined as the center of the world.
- Intruder tracking time: That is defined as the duration of time that an agent needs to track an intruder for the intruder to be considered caught. That was set to 10 s for all scenarios.
- Density of agents: That is defined as the number of agents of the swarm divided by the world area.

Table 1 summarizes the different parameters used between the different scenarios. Two sets of scenarios were designed, such that the density of the agents is maintained constant for all scenarios of the set. The size of the surveilled area, the swarm size and the speed of the intruders was changed in every scenario. The speed of the intruders changed proportionally to the area size to maintain the time of the intruders' life constant and test the algorithms in increasingly difficult scenarios.

Table 1. Scenarios' parameters.

Set 1	Scenario 1	Scenario 2	Scenario 3
World size	100 m × 100 m	140 m × 140 m	200 m × 200 m
Grid size	10 × 10	14 × 14	20 × 20
Swarm size	4	8	16
Intruders' speed	0.28 m.s^{-1}	0.39 m.s^{-1}	0.56 m.s^{-1}
Set 2	Scenario 1	Scenario 2	Scenario 3
World size	150 m × 150 m	210 m × 210 m	300 m × 300 m
Grid size	15 × 15	21 × 21	30 × 30
Swarm size	4	8	16
Intruders' speed	0.42 m.s^{-1}	0.59 m.s^{-1}	0.84 m.s^{-1}

In each scenario of the same set, the world size, number of agents and speed of the intruders is increased proportionally, aiming to examine the scalability of our system.

4. Results

In this section, the results from all the experiments conducted are presented.

4.1. Collision Avoidance

A separate scenario was designed for testing the collision avoidance algorithm developed. The scenario is simplified to focus on the collision avoidance. Each agent was given a specific destination point, so that several conflicts would occur in different or in the same position for multiple agents.

Figures 8 and 9 show the results of a collision avoidance simulation test using four agents. The agents are spawned simultaneously at the vertices of a rhombus and are assigned to go to the opposite vertex. All four of the agents detect the collision and deconflict. Figure 8 presents the trajectories of the four agents, while they conduct their individual mission and avoid collision with the other three agents. The trajectory of each agent is slightly altered to ensure a collision-free path, but the added cost of the path is not significant, considering that the agents replanned in real-time.

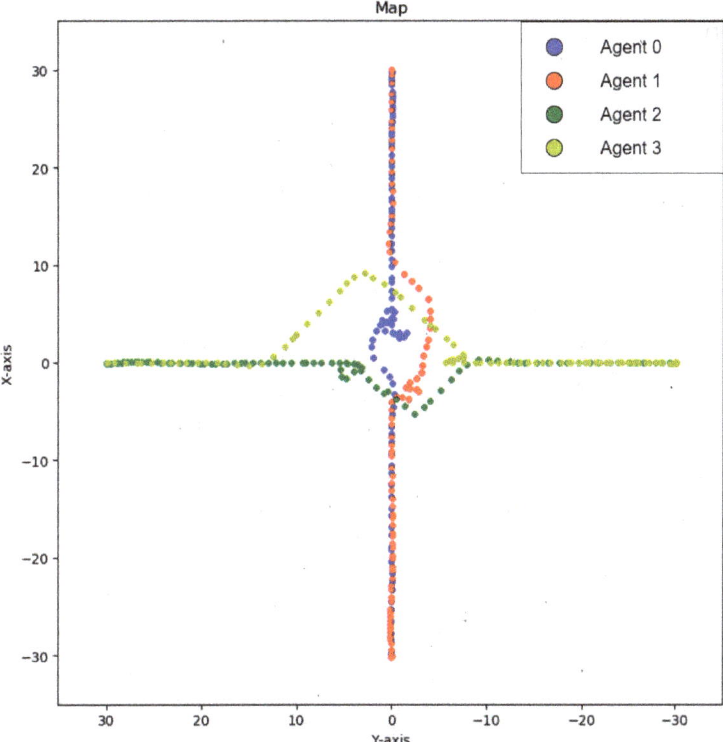

Figure 8. The agents' paths during the collision avoidance experiment. For this experiment, four agents were used and spawned simultaneously at the vertices of a rhombus. The agents were tasked to travel to the opposite vertex while using collision avoidance to ensure a safe flight. As expected, their paths intersected at the center and they adjusted their velocities to avoid collision.

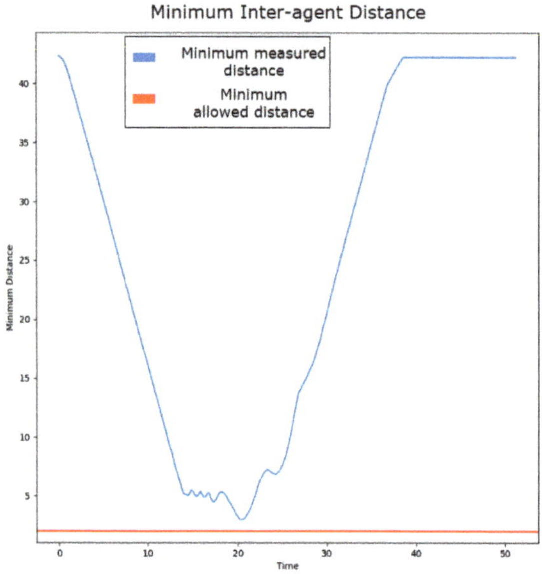

Figure 9. Minimum inter-agent distance in every moment where the red horizontal line is the 2 m distance boundary, the minimum allowed inter-agent distance. The graph is based on the same experiment that is presented in Figure 8.

Figure 9 is a diagram of the minimum inter-agent distance for every time step. The minimum measured inter-agent distance decreases significantly around the time value of 20 s, since the agents were in the center area deconflicting at that time, but it remains higher than the minimum allowed inter-agent distance, which for safety precautions was set to 2 m in our experiments.

4.2. Metrics

We propose a set of metrics that can be used to quantify the efficiency of our proposed algorithm regarding the detection of intruders and the area coverage to assess the decision-making process.

- Intruder-related metrics:
- Number of intruders caught: The sum of the intruders that the agents caught during the simulation run.
- Number of intruders reached the target: The sum of the intruders that reached the target during the simulation run.
- Average time of intruder's life: The average alive time of all the intruders during the simulation independently if the intruder was alive or not at the end of the simulation, measured in seconds.
- Average time of intruder's life for caught intruders: The average alive time of the intruders which were caught during the experiment, measured in seconds.
- Average time of intruder's life for reached intruders: The average alive time of the intruders that successfully reached the target, measured in seconds.
- Decision metric: The decision metric is the average time interval between two successive decisions of one agent. It is measured in seconds.
- Coverage metric: The coverage metric is defined as the percentage of the world that has been covered by the swarm. That metric is initialized every $t_{coverage}$ seconds, where $t_{coverage}$ was set to $t_{coverage} = 180$ s for our simulations. That metric is an indication of how effectively the area of interest is covered, but it is of less importance than the intruder's metrics in our case. We can easily understand that this metric

ensures us about the correct functionality of the decision-making process. Figure 10 shows an example of the coverage metric.

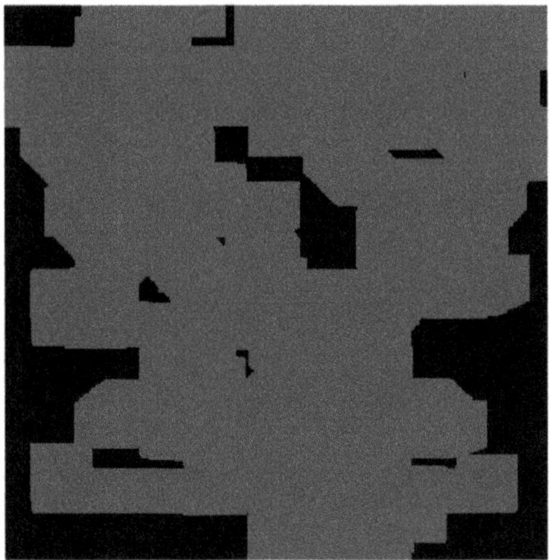

Figure 10. Coverage example with 10 agents, 78.02% coverage. The grey area depicts the coverage that the agents succeeded as a group in 180 s.

4.3. Competing Algorithms

Three competing surveillance methods were developed and implemented to compare their results with our method.

- Map division: The area of interest is divided into n rectangles, where n is the number of the agents of the swarm. Each agent undertakes the surveillance of one of the rectangles. The first action of each agent is to compute their rectangle and to move to it. After that, each agent changes to mode "Scan" and starts scanning the rectangle using zig-zag-like coverage. If the agent detects an intruder, it changes to "follow intruder" mode. When the intruder is caught, the agent carries on with scanning if the agent is in the boundaries of its rectangle. Otherwise, the agent changes to the "Go to" mode until it is in the boundaries of its rectangle and then changes to "scan" mode. Collision detection and avoidance is only activated if the agent is out of the boundaries of its rectangle since the rectangles do not overlap and there is no risk of collision when all the agents are the boundaries of their own rectangle. Algorithm 2 is used to divide the map into squares by setting the number of columns, nc, and rows, nr.

Algorithm 2. Map division

1: Set n the number of drones in the swarm
2: If the square root of n is an integer
3:     root = nc = nr = $\sqrt{n}$
4: Else
5:     nc = round ($\sqrt{n}$)
6:     nr = 1
7:     While nc > 0 and n% round(root) ! = 0
8:        nc = round (*root*)
9:        nr = $\frac{n}{nc}$
10:      root = root -—1

After the number of rows and columns is computed, each drone calculates the vertices of its square based on its ID, the world size, the coordinates of the center of the world and the computed number of rows and columns.

- Random decision: In this scenario, the agent's modes are the same as in our proposed algorithm, but the swarm intelligence has been removed. The agents do not make decisions based on the world information and the central entity does not exist. The agents select the next square-target at random each time.
- Static cameras: In this scenario, the agents take off and hover statically over a specific predefined position, different for each agent acting as static cameras. They are not allowed to follow intruders. Figure 11 presents the configuration of the static cameras for each scenario.

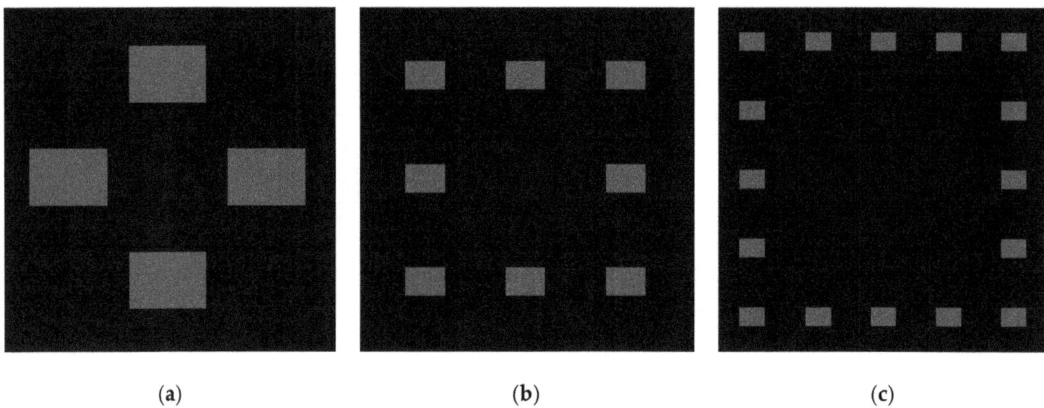

(a)   (b)   (c)

Figure 11. Positioning of the static cameras for scenarios 1, 2 and 3 accordingly at (**a–c**). The gray rectangles represent the field of view of the agents.

4.4. Experiment Results

This section includes the experimental results of the simulations conducted to assess the efficiency of our proposed algorithm and to compare the results with the competing algorithms. Each experiment was run five times and the results were averaged to be presented here. The number of intruders reached the target and the number of intruder-caught metrics are the most indicative of all the metrics used to assess the algorithms, since preventing the intruders from reaching the target is the main objective of the system.

In Figure 12, the results are presented for our first group of tests, where we maintain a UAV density of 25 square-targets per UAV. To keep the density constant, the area is increased linearly with the number of UAV agents. On the first graph of Figure 12, the results for 4 UAVs indicate that our decision-making algorithm outperforms all other algorithms, by letting just 10 intruders to reach their target. The random decision algorithm and map division algorithm perform closely to each other with 35 and 40 intruders reaching the target, respectively, and lastly, the static camera approach failed to catch most of the intruders, as 328 reached their target. We can observe that the proposed algorithm performs almost 350% better for the number of intruders reaching the target metric than the second best, which is the random decision.

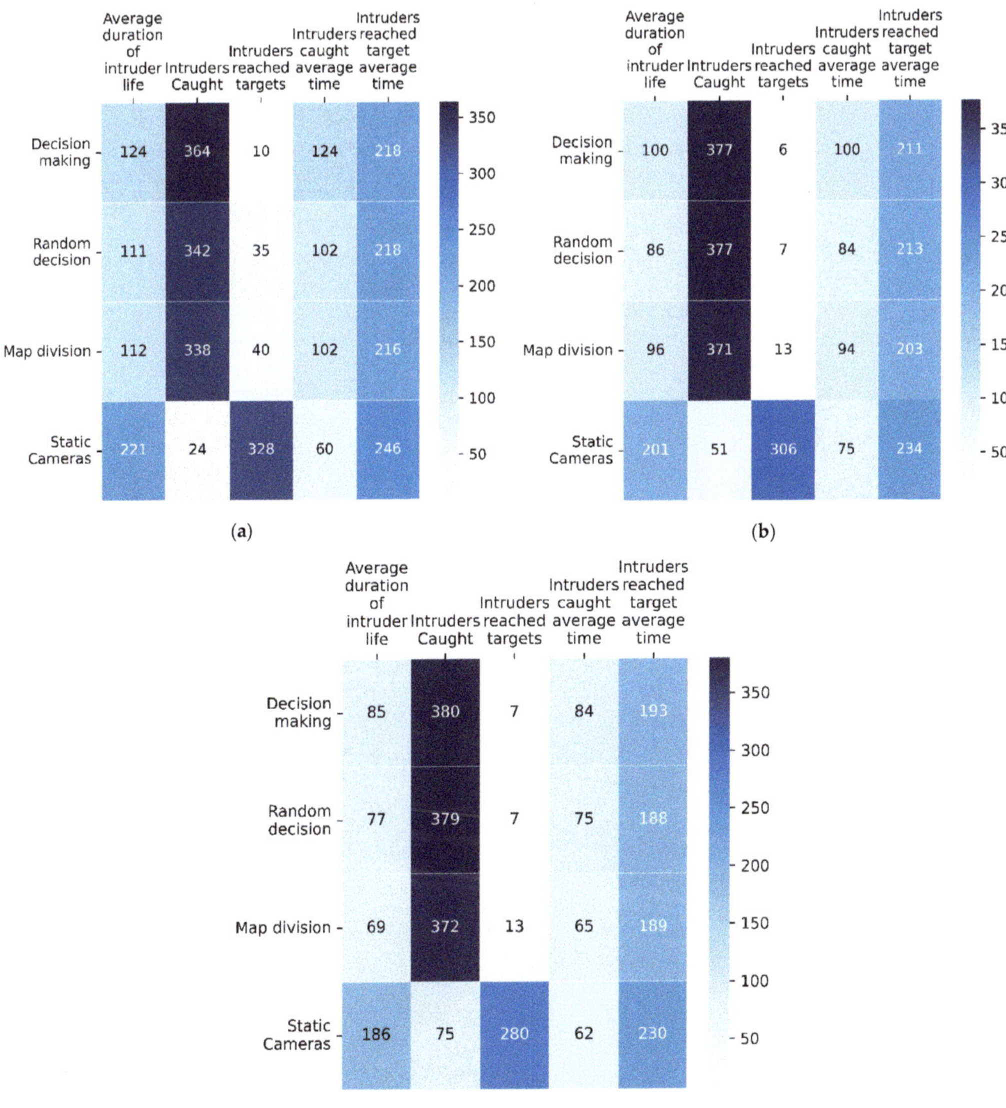

Figure 12. Intruder metrics for the first set of experiments. The three scenarios of set 1 correspond to (**a–c**) accordingly. (**a**) Scenario 1 of Set 1. A total of 4 UAVs for a world of 100 m × 100 m. (**b**) Scenario 2 of Set 1. A total of 8 UAVs for a world of 140 m × 140 m. (**c**) Scenario 3 of Set 1. A total of 16 UAVs for a world of 200 m × 200 m.

Our decision-making algorithm was able to catch 364 intruders, 22 more than the random decision algorithm and 26 more than the map division approach, by allocating resources in intruders' clusters, mostly close to the map center, where intruders converge. This in return increased the average alive time of caught intruders to 124 s, 22 more versus both the random decision and map division approaches. In this scenario, the system is stressed due to the low number of UAVs in comparison to the number of intruders, which results to most of the time being spent following intruders instead of actively searching.

When 8 and 16 UAVs are used as shown in the second and third graphs of Figure 12, we see that the decision-making algorithm performs similarly to the random decision one, with the map division approach performing a bit worse. The similar performance of the first two algorithms is explained by the low density of 25 square-targets per UAV, which in return minimizes the benefits of decision-making since a random approach still has a high chance of finding intruders. In all tests, static cameras proved inefficient and map division fell behind likely due to the inability of the system to migrate resources to hotspots.

In Figure 13, results are presented for the second experimental set, while we maintain a UAV density of 56 square-targets per UAV, more than twice higher than in set 1. In the first graph of Figure 13, the results for scenario 1 of set 2 are presented for four UAVs. The decision-making algorithm outperforms the three competing algorithms, but the performance is still rather poor, letting 33 intruders reach their target. The random decision algorithm and map division algorithm perform closely with 90 and 84 intruders reaching the target, respectively, and lastly, the static camera approach failed to catch most of the intruders, as 319 reached their target. The decision-making algorithm was able to catch 337 intruders, 48 more than the random decision and map division algorithm, which performed equally in this metric, while static cameras caught only 36 intruders. The problem described in the previous set of scenarios when four UAV agents are involved, is furtherly amplified by the increase in map size to achieve 56 square-targets per UAV. The average alive time of the caught intruders is 155 s, 42 more versus the random decision and 23 more versus the map division approach. These critical metrics show the worst performance than the first group of tests, attributed to the increased map size while still using four UAV agents.

When 8 and 16 UAV agents are used, as shown in the second and third graphs of Figure 13, the benefits of decision making are clearer when compared to other approaches as the higher amount of squares per UAV agent allows for a significant chance of a random decision being wrong. When 8 UAVs are involved, 20 intruders reached their target using the decision-making algorithm, 46 for random decision, and 43 for map division, which performed once again roughly equally. Static cameras once more proved to be significantly worst in these tests, as 302 intruders reached their targets. The decision-making system caught 355 intruders, 26 more when compared to random decision and 28 more when compared to map division. The trend continues for 16 UAVs with decision making having a large lead, catching 350 intruders, and missing just 24 intruders. In this case, the random decision proved better than map division, as 40 intruders reached their goal and 337 were caught, while the results were 58 and 321, respectively, for map division. Map division underperforms, likely due to the inability of the system to migrate resources to hotspots.

In all of the experiments presented above, the intruder speed was increased proportionally to the world's dimensions in an attempt to keep the difficulty equal in that regard. In Figure 15, the performance results of an extra scenario are presented for the case when 16 UAVs are deployed and 56 square-targets are assigned to each UAV, such as in the case of the scenario 3 of set 2. In this experiment, the speed of intruders was not adjusted to the world's dimensions, and it had the value of 0.28 m.s^{-1}. Intruders were not able to reach their target for the decision-making, random decision and map division approaches, and the average duration of their life is comparable for the three approaches. The excellent performance of the three approaches was probably caused by the long life-time required for an intruder to reach the target in this scenario. It seems that the increase in the world size would create a severe advantage for all approaches, and the results would not give a clear comparison between the approaches. Based on those results, an adjustable intruders' speed has been selected for all the experiments presented above.

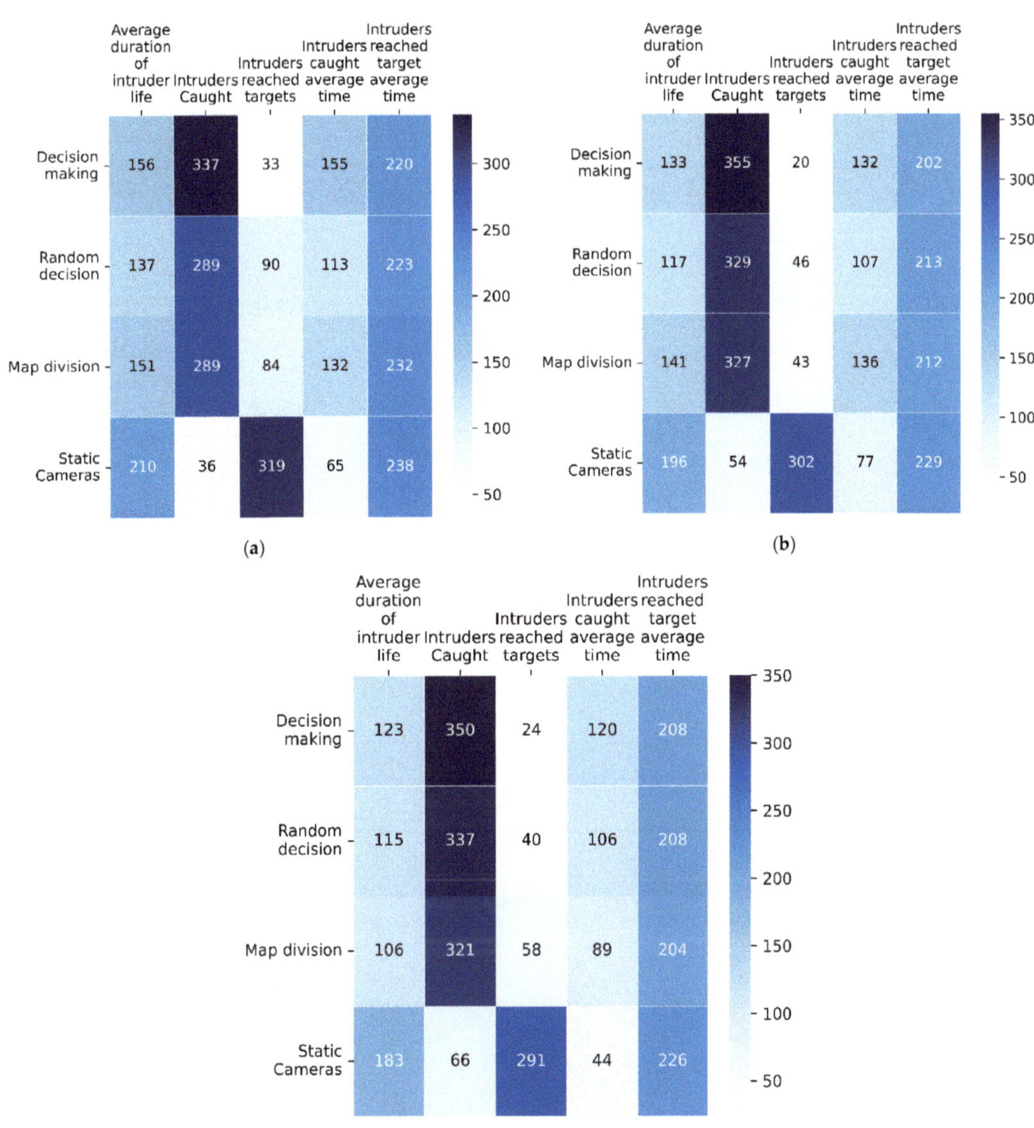

Figure 13. Intruder metrics for the second set of experiments. The three scenarios of set 2 correspond to (**a**–**c**), accordingly. (**a**) Scenario 1 of Set 2. A total of 4 UAVs for a world of 150 m × 150 m. (**b**) Scenario 2 of Set 2. A total of 8 UAVs for a world of 210 m × 210 m. (**c**) Scenario 3 of Set 2. A total of 16 UAVs for a world of 300 m × 300 m.

Figure 14 focuses on the number of intruders that reached the target for the different scenarios of each set. Plot (a) shows that the number of intruders to reach the target is relatively stable across the scenarios of set 1 and maintained in low values for the decision-making approach. That indicates that the system's performance fits the specific density used in set 1 of one UAV agent per 25 square-targets. The random decision and map division approaches demonstrate similar results to the decision-making approach as the size of the swarm increases, indicating that the number of UAVs is enough for monitoring the given area, even for systems with no decision-making capabilities. Plot (b) presents the same metric for the second experimental set. In this case, the density of UAVs per square-target is lower and the advantage of using agents capable of decision-making is clearer, as the proposed decision-making approach outperforms the three competing approaches.

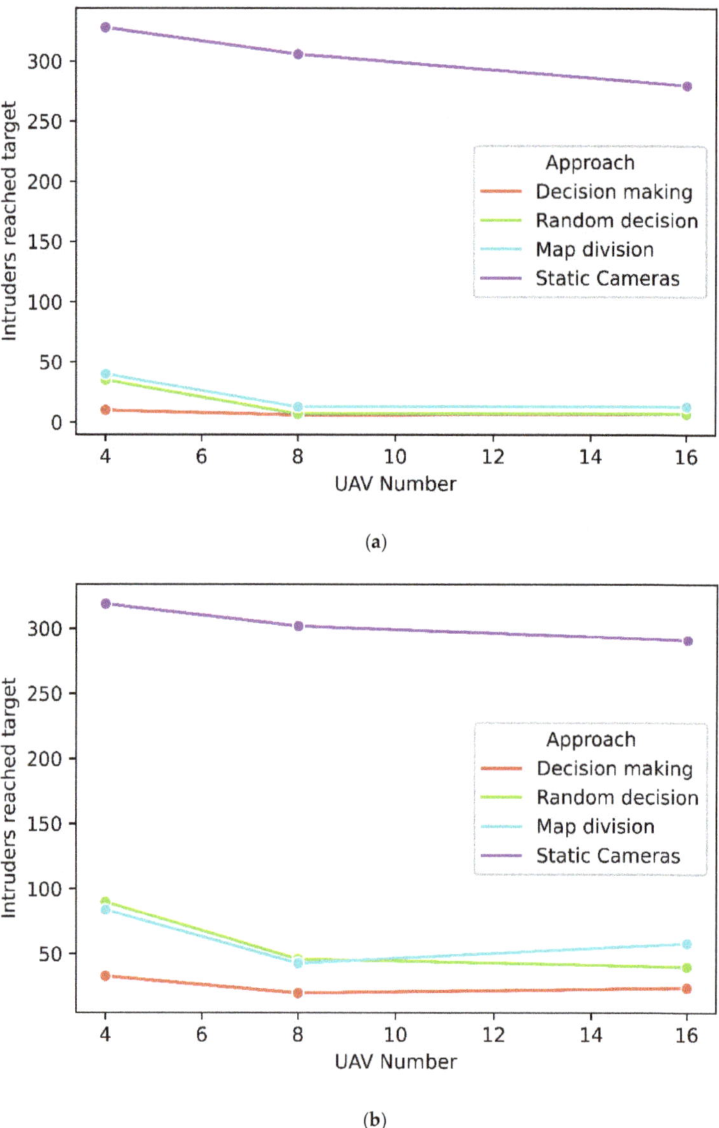

Figure 14. The intruders reached target metric depending on the number of UAVs for the four approaches. Graphs (**a**,**b**) correspond to the experimental sets 1 and 2 accordingly. (**a**) Number of intruders to reach target for scenarios of set 1. (**b**) Number of intruders to reach target for scenarios of set 2.

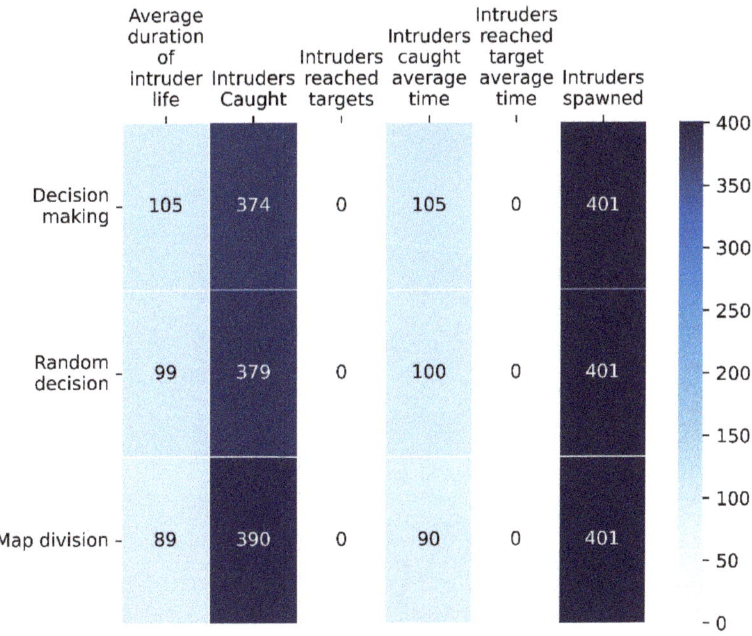

Figure 15. Intruders' metrics for a scenario of 16 UAVs in a 300 m × 300 m world, with the intruder speed at 0.28 m.s^{-1}.

Table 2 sums up the decision-making metric average results for the six scenarios. The time interval between two subsequent square-target selection is shorter for the decision-making algorithm than for the random decision that is explained because the random decision allows the agents to travel across the map in each decision, while the decision-making algorithm urges agents to stay in their neighborhoods with a large probability. By maintaining the decision-making metric small, the system will have a quicker reaction to new intruder data.

Table 2. The average of the decision metric in seconds for all the experiments and scenarios for the decision making and the random decision algorithms.

Set 1	Scenario 1	Scenario 2	Scenario 3
Decision making	40.9 s	33.8 s	38.7 s
Random decision	71.2 s	63.4 s	66.7 s
Set2	Scenario 1	Scenario 2	Scenario 3
Decision making	46.7 s	44.4 s	53.8 s
Random decision	75.4 s	75.8 s	93.6 s

Table 3 presents the average of the coverage metric for each scenario and implementation. It is noticeable that the random decision implementation offers a larger area coverage for each scenario. Before extracting any conclusions concerning the efficiency of the algorithms based on that metric, it should be noted that larger area coverage does not result to more efficient area coverage. The reason behind the lower area coverage provided by the decision-making algorithm is that agents tend to cluster over areas with high intruder density, which enables the detection of a larger amount of intruders.

Table 3. The average of the coverage metric for all the experiments and scenarios for the four competing algorithms.

Set 1	Decision Making	Random Decision	Map Division	Static Cameras
Scenario 1	68.41%	83.14%	73.89%	15.29%
Scenario 2	76.02%	87.2%	80.1%	15.59%
Scenario 3	82.64%	89.82%	87.23%	15.28%
Set 2	Decision making	Random decision	Map division	Static Cameras
Scenario 1	42.7%	62.32%	38.78%	6.8%
Scenario 2	53.87%	68.95%	40.79%	6.9%
Scenario 3	59.34%	67.41%	53.1%	6.8%

We can see in Figure 16 that the swarm manages to cover a big size of the area to be surveilled and is not biased in the selection of the next grid by selecting only certain areas of the world, resulting in the even distribution of the selection across the map based on the collected information. It is clear that of the two sequential coverage measurements in Figure 16a,b, that the swarm covers all the map and does not show preference to specific areas.

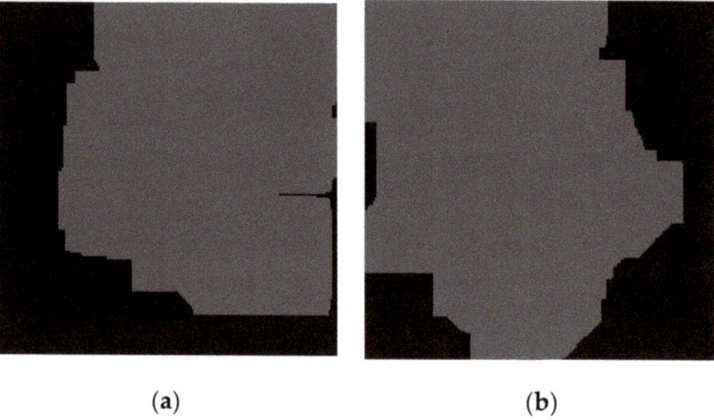

(a)   (b)

Figure 16. The coverage metric results for the decision-making algorithm in scenario 1 of set 1. (**a**,**b**) are two sequential measurements of the coverage figure.

Even though mostly two of the proposed metrics (the Number of intruders caught, and the Number of intruders reached the target) are used for the efficiency assessment of the algorithms, the rest of the metrics are of importance as well. All the proposed metrics are good indicators of how well tuned the decision-making algorithm is. It is a subject of further research to determine the exact equations to compute all the algorithm's parameters based on those metrics.

5. Discussion

Decision making is a crucial ability for autonomous systems and especially UAV swarms. It is an open-research area with most researchers in the field focusing on developing the theoretical background of the decision-making algorithms, while we propose a new optimization based, stochastic algorithm for real time decision making, and we describe the whole system implementation after testing it in SITL simulations. The literate review presented in Section 2 shows that there are multiple methods to approach the task allocation problem, offering a variety of solutions that provide different architectures and benefits. The proposed UAV swarm shows great scalability results, is considerate regarding the communication bandwidth, and reacts quickly to dynamic changes and uncertainties. Our system's nature is adaptable to information gathered from the environment and it dynamically reacts, facilitating global optimization. The decision-making algorithm has been designed to be decentralized and scalable ensuring fault tolerance through the operation of the system if UAV agents of the system suffer failures. Moreover, it is designed as a surveillance system for defense purposes of a friendly area, but it can be adapted to be used in multiple fields such as research missions in the research and rescue field, wildlife tracking missions, and wildfire monitoring missions. The algorithm can easily be modified to be optimized depending on the specific behavior of each intruder, or any other type of agent/object that the system is interested in observing and monitoring.

6. Conclusions

We present a system consisting of multiple UAV agents, designed for area surveillance and intruder monitoring. In addition to the state-of-the-art decentralized decision-making algorithm that is proposed, the supportive algorithms were also designed and implemented. The system was originally fine-tuned for a scenario with a swarm of four agents and a world size of 100 m × 100 m (scenario 1 of set 1). The results for this scenario are 363.6 intruders caught over the 401 intruders introduced in the world for our decision-making algorithm and 342.2, 337.6 and 24.4 accordingly for the random decision, map division and static cameras implementations. The average value of the intruders reaching the target for this scenario is 9.8 for our decision-making algorithm and 35.4, 39.4 and 328 accordingly for the random decision, map division and static cameras implementations. Overall, the system was tested in two experimental sets, maintaining a constant density of UAV agents per monitored area across the set. Each set included three scenarios, varying in the size of the swarm, the size of the world, and the intruders' speed. In all six scenarios, the proposed algorithm demonstrates superior results to the three competing systems. The proposed approach demonstrated comparable results across the three experiments of each set indicating that the UAV density is a stronger factor in the system's performance than the size of the monitoring area. That shows the scalability characteristic of the system. One exception to the stable performance of the system was identified for the first scenario of the second set, in which the system seems to have reached its limits, as the number of the intruders and their relatively high speed caused agents to chase intruders for most of their operational time, and the decision-making algorithm demonstrated a lower performance. As a result, we conclude that the existence of cognitive intelligence in a swarm is crucial and produces much higher situational awareness as opposed to the cases where the swarm is selfish and each agent act on his own without utilizing any shared information. The overall system was tested in real time simulations and demonstrated an improvement up to 350% when compared with similar systems that lacked the decision-making ability. Though the proposed decision-making algorithm was designed to be decentralized, the presented communication scheme of this work requires communication with a central agent, as the necessary processing power of the described central agent is very low, and that processing load may be allocated to the agents. Future work shall include the implementation of a decentralized communication layer for the world map data.

The key contribution of the present paper is the description of a decentralized decision-making algorithm designed for area monitoring and intruder tracking by a swarm of UAVs. The overall system was implemented to support the testing of the algorithm, including collision avoidance and area coverage algorithms. The system was developed in ROS and simulated in GAZEBO with swarms of up to 16 quadcopters. Experiments of this study included intruders incapable of planning to avoid UAV agents. Future research shall focus on adding strategy to the intruders' behavior and more elaborate models of estimating the intruders' near-future locations. It is of interest to investigate how the system will perform when faced with smarter intruders upgraded with self and group strategies to achieve their goal of reaching the target. We believe that the system's performance can be enhanced by the addition of alternative stochastic models describing the probability of an intruder's presence, especially in the case of intruders capable of strategic planning and collaboration. Finally, future research will also include the development of object detection, target tracking and localization techniques for detecting and following the intruders. This will allow us to study the uncertainties added during intruder detection and localization and may demonstrate some of the limitations of the system.

Author Contributions: Conceptualization, N.P., I.D., V.L. and V.K.; methodology, N.P. and I.D.; software, N.P., I.D. and D.M.; validation, N.P., I.D. and V.L.; formal analysis, N.P. and I.D.; investigation, N.P. and I.D.; resources, I.D. and V.L.; data curation, N.P. and D.M.; writing—original draft preparation, N.P.; writing—review and editing, N.P., I.D. and V.L.; visualization, D.M.; supervision, V.L. and V.K.; project administration, V.L. and V.K.; funding acquisition, V.L. and V.K. All authors have read and agreed to the published version of the manuscript.

Funding: This research is based upon work supported by the Air Force Office of Scientific Research under award number FA9550-19-1-7032 (Real-time Decision Making for Autonomous Systems using AI Methods).

Data Availability Statement: Not applicable.

Conflicts of Interest: The authors declare no conflict of interest. The funders had no role in the design of the study; in the collection, analyses, or interpretation of data; in the writing of the manuscript; or in the decision to publish the results.

References

1. Khamis, A.; Hussein, A.; Elmogy, A. Multi-robot Task Allocation: A Review of the State-of-the-Art. In *Cooperative Robots and Sensor Networks 2015*; Koubaa, A., Dios, J., Eds.; Springer International Publishing: Midtown Manhattan, NY, USA, 2015; pp. 31–51. [CrossRef]
2. Dias, M.B.; Zlot, R.; Kalra, N.; Stentz, A. Market-Based Multirobot Coordination: A Survey and Analysis. *Proc. IEEE* **2006**, *94*, 1257–1270. [CrossRef]
3. Turner, J.; Meng, Q.; Schaefer, G.; Whitbrook, A.; Soltoggio, A. Distributed Task Rescheduling With Time Constraints for the Optimization of Total Task Allocations in a Multirobot System. *IEEE Trans. Cybern.* **2017**, *48*, 2583–2597. [CrossRef] [PubMed]
4. Choi, H.-L.; Brunet, L.; How, J.P. Consensus-Based Decentralized Auctions for Robust Task Allocation. *IEEE Trans. Robot.* **2009**, *25*, 912–926. [CrossRef]
5. Smith, R.G. The Contract Net Protocol: High-Level Communication and Control in a Distributed Problem Solver. *IEEE Trans. Comput.* **1980**, *C-29*, 1104–1113. [CrossRef]
6. Liekna, A.; Lavendelis, E.; Grabovskis, A. Experimental Analysis of Contract NET Protocol in Multi-Robot Task Allocation. *Appl. Comput. Syst.* **2012**, *13*, 6–14. [CrossRef]
7. Mkiramweni, M.; Yang, C.; Li, J.; Han, Z. Game-Theoretic Approaches for Wireless Communications with Unmanned Aerial Vehicles. *IEEE Wirel. Commun.* **2018**, *25*, 104–112. [CrossRef]
8. Bardhan, R.; Bera, T.; Sundaram, S. A decentralized game theoretic approach for team formation and task assignment by autonomous unmanned aerial vehicles. In Proceedings of the 2017 International Conference on Unmanned Aircraft Systems (ICUAS), Miami, FL, USA, 13–16 June 2017. [CrossRef]
9. Kuhn, H.W. The Hungarian Method for the Assignment Problem. *Nav. Res. Logist. Q.* **1955**, *2*, 83–97. [CrossRef]
10. Yoon, S.; Kim, J. Efficient multi-agent task allocation for collaborative route planning with multiple unmanned vehicles. *IFAC-PapersOnLine* **2017**, *50*, 3580–3585. [CrossRef]
11. Tkach, I.; Jevtić, A.; Nof, S.Y.; Edan, Y. A Modified Distributed Bees Algorithm for Multi-Sensor Task Allocation. *Sensors* **2018**, *18*, 759. [CrossRef] [PubMed]
12. Schwarzrock, J.; Zacarias, I.; Bazzan, A.L.; Fernandes, R.Q.D.A.; Moreira, L.H.; de Freitas, E.P. Solving task allocation problem in multi Unmanned Aerial Vehicles. *Eng. Appl. Artif. Intell.* **2018**, *72*, 10–20. [CrossRef]
13. Hyun-Jin, C.; You-Dan, K.; Hyoun-Jin, K. Genetic Algorithm Based Decentralized Task Assignment for Multiple Unmanned Aerial Vehicles in Dynamic Environments. *Int. J. Aeronaut. Space Sci.* **2011**, *12*, 163–174. [CrossRef]
14. Noureddine, D.B.; Gharbi, A.; Ahmed, S.B. Multi-agent Deep Reinforcement Learning for Task Allocation in Dynamic Environment. In Proceedings of the 12th International Conference on Software Technologies (ICSOFT), Madrid, Spain, 24–26 July 2017. [CrossRef]
15. Tian, Y.-T.T.; Yang, M.; Qi, X.-Y.; Yang, Y.-M. Multi-robot task allocation for fire-disaster response based on reinforcement learning. In Proceedings of the International Conference on Machine Learning and Cybernetics, Baoding, China, 12–15 July 2009. [CrossRef]
16. Bays, M.J.; Wettergren, T.A. Partially-Decoupled Service Agent—Transport Agent Task Allocation and Scheduling. *J. Intell. Robot. Syst.* **2018**, *94*, 423–437. [CrossRef]
17. Zhang, J.; Wang, G.; Song, Y. Task Assignment of the Improved Contract Net Protocol under a Multi-Agent System. *Algorithms* **2019**, *12*, 70. [CrossRef]
18. Zitouni, F.; Harous, S.; Maamri, R. A Distributed Approach to the Multi-Robot Task Allocation Problem Using the Consensus-Based Bundle Algorithm and Ant Colony System. *IEEE Access* **2020**, *8*, 27479–27494. [CrossRef]

19. Gan, S.K.; Sukkarieh, S. Multi-UAV Target Search using Explicit Decentralized Gradient-Based Negotiation. In Proceedings of the 2011 IEEE International Conference on Robotics and Automation, Shanghai, China, 9–13 May 2011. [CrossRef]

20. Lanillos, P.; Gan, S.K.; Besada-Portas, E.; Pajares, G.; Sukkarieh, S. Multi-UAV target search using decentralized gradient-based negotiation with expected observation. *Inf. Sci.* **2014**, *282*, 92–110. [CrossRef]

21. Adamey, E.; Ozguner, U. A decentralized approach for multi-UAV multitarget tracking and surveillance. In Proceedings of the SPIE Defense, Security, and Sensing, Baltimore, MD, USA, 23–27 April 2012.

22. De Freitas, E.P.; Heimfarth, T.; Ferreira, A.M.; Pereira, C.E.; Wagner, F.R.; Larsson, T. Decentralized Task Distribution among Cooperative UAVs in Surveillance Systems Applications. In Proceedings of the 2010 Seventh International Conference on Wireless On-demand Network Systems and Services (WONS), Kranjska Gora, Slovenia, 3–5 February 2010. [CrossRef]

23. Capitan, J.; Merino, L.; Ollero, A. Decentralized Cooperation of Multiple UAS for Multi-target Surveillance under Uncertainties. In Proceedings of the 2014 International Conference on Unmanned Aircraft Systems (ICUAS), Orlando, FL, USA, 27–30 May 2014. [CrossRef]

24. Zhu, X.; Vanegas, F.; Gonzalez, F. Decentralised Multi-UAV Cooperative Searching Multi-Target in Cluttered and GPS-Denied Environments. In Proceedings of the 2022 IEEE Aerospace Conference (AERO), Big Sky, MT, USA, 5–12 March 2022. [CrossRef]

25. Zhang, Y.-Z.; Li, J.-W.; Hu, B.; Zhang, J.-D. An improved PSO algorithm for solving multi-UAV cooperative reconnaissance task decision-making problem. In Proceedings of the 2016 IEEE International Conference on Aircraft Utility Systems (AUS), Beijing, China, 10–12 October 2016. [CrossRef]

26. Venugopalan, T.; Subramanian, K.; Sundaram, S. Multi-UAV Task Allocation: A Team-Based Approach. In Proceedings of the 2015 IEEE Symposium Series on Computational Intelligence, Cape Town, South Africa, 7–10 December 2015. [CrossRef]

27. Galceran, E.; Carreras, M. A survey on coverage path planning for robotics. *Robot. Auton. Syst.* **2013**, *61*, 1258–1276. [CrossRef]

28. Colorni, A.; Dorigo, M.; Maniezzo, V. Distributed Optimization by Ant Colonies. In Proceedings of the ECAL91—European Conference on Artificial Life, Paris, France, 11–13 December 1991.

29. Goldberg, D. *Genetic Algorithms in Search Optimization and Machine Learning*; Addison-Wesley: Boston, MA, USA, 1989.

30. Khatib, O. Real-time obstacle avoidance for manipulators and mobile robots. In Proceedings of the 1985 IEEE International Conference on Robotics and Automation, St. Louis, MO, USA, 25–28 March 1985. [CrossRef]

31. Park, M.G.; Jeon, J.H.; Lee, M.C. Obstacle avoidance for mobile robots using artificial potential field approach with simulated annealing. In Proceedings of the ISIE 2001 IEEE International Symposium on Industrial Electronics Proceedings, Pusan, Republic of Korea, 12–16 June 2001. [CrossRef]

32. Doria, N.S.F.; Freire, E.O.; Basilio, J.C. An algorithm inspired by the deterministic annealing approach to avoid local minima in artificial potential fields. In Proceedings of the 2013 16th International Conference on Advanced Robotics (ICAR), Montevideo, Uruguay, 25–29 November 2013. [CrossRef]

33. Quigley, M.; Conley, K.; Gerkey, B.; Faust, J.; Foote, T.; Leibs, J.; Wheeler, R.; Ng, A. ROS: An open-source Robot Operating System. In Proceedings of the ICRA Workshop on Open Source Software, Kobe, Japan, 12–17 May 2009.

34. Koenig, N.; Howard, A. Design and Use Paradigms for Gazebo, An Open-Source Multi-Robot Simulator. In Proceedings of the Proceedings 01 2004 IEEE/RSJ International Conference on Intelligent Robots and Systems, Sendai, Japan, 28 September–2 October 2004. [CrossRef]

35. PX4, PX4 Project. Available online: http://px4.io (accessed on 14 November 2022).

36. Meier, L.; Honegger, D.; Pollefeys, M. PX4: A Node-Based Multithreaded Open Source Robotics Framework. In Proceedings of the 2015 IEEE International Conference on Robotics and Automation (ICRA), Seattle, WA, USA, 25–30 May 2015. [CrossRef]

37. Introduction—PX4 User Guide. Available online: https://docs.px4.io/master/en/ros/mavros_installation.html (accessed on 14 November 2022).

38. Mavros—ROS Wiki. Available online: http://wiki.ros.org/mavros (accessed on 14 November 2022).

39. Mavlink—ROS Wiki. Available online: http://wiki.ros.org/mavlink (accessed on 14 November 2022).

 **drones**

Review

Towards Resilient UAV Swarms—A Breakdown of Resiliency Requirements in UAV Swarms

Abhishek Phadke [1,2,*] and F. Antonio Medrano [1,2]

1   Conrad Blucher Institute for Surveying and Science, Texas A&M University-Corpus Christi, Corpus Christi, TX 78412, USA
2   Department of Computing Sciences, Texas A&M University-Corpus Christi, Corpus Christi, TX 78412, USA
*   Correspondence: aphadke@islander.tamucc.edu

Abstract: UAVs have rapidly become prevalent in applications related to surveillance, military operations, and disaster relief. Their low cost, operational flexibility, and unmanned capabilities make them ideal for accomplishing tasks in areas deemed dangerous for humans to enter. They can also accomplish previous high-cost and labor-intensive tasks, such as land surveying, in a faster and cheaper manner. Researchers studying UAV applications have realized that a swarm of UAVs working collaboratively on tasks can achieve better results. The dynamic work environment of UAVs makes controlling the vehicles a challenge. This is magnified by using multiple agents in a swarm. Resiliency is a broad concept that effectively defines how well a system handles disruptions in its normal functioning. The task of building resilient swarms has been attempted by researchers for the past decade. However, research on current trends shows gaps in swarm designs that make evaluating the resiliency of such swarms less than ideal. The authors believe that a complete well-defined system built from the ground up is the solution. This survey evaluates existing literature on resilient multi-UAV systems and lays down the groundwork for how best to develop a truly resilient system.

Keywords: UAV; swarm; resiliency; multi agent system

Citation: Phadke, A.; Medrano, F.A. Towards Resilient UAV Swarms— A Breakdown of Resiliency Requirements in UAV Swarms. *Drones* **2022**, *6*, 340. https://doi.org/10.3390/ drones6110340

Academic Editors: Xiwang Dong, Mou Chen, Xiangke Wang and Fei Gao

Received: 10 October 2022
Accepted: 1 November 2022
Published: 3 November 2022

Publisher's Note: MDPI stays neutral with regard to jurisdictional claims in published maps and institutional affiliations.

1. Introduction

Deploying multiple agents as part of a larger swarm has its advantages. Cooperative actions by several robots are a wide application domain [1]. Several possible advantages can be visualized particularly in the case of unmanned aerial vehicles (UAVs). A swarm of UAVs can search an area quicker than a single UAV making multiple passes over the same area. Higher-level approaches, such as search grid decomposition for individual agents, are more easily accomplished when multiple agents exist. Smaller size UAVs carry limited equipment to reduce equipment power consumption and reduce overall aircraft weight. It is possible to equip different agents in a swarm with different sensors. The result will be richer data streams that will be generated once the different sensor data is combined. Similar experiments can be envisioned where a swarm of UAV agents work at different altitudes in order to survey ground subjects, thereby providing multiple perspectives on the target. Such improvements in results by swarm agents are particularly useful considering the highly dynamic environments in which UAVs operate. Situations on the battlefield may already have changed by the time a single UAV makes a pass over the area and then moves on to cover other areas, and then returns. Similar effects are noticed while measuring large-scale phenomena such as red tide growth [2] or fish shoals [3]. Sensitive incidents such as a search-and-rescue (SAR) mission may require multiple agents to be deployed. An area may be too large for a single UAV to cover, and more agents improve the probability that a victim can be found quickly.

A multi-vehicle system can be described as effective, efficient, flexible, and exhibit higher tolerance to faults than a single agent [4]. This makes it more viable to have a swarm of UAVs attempt a particular task. However, the challenging environment they work in makes creating resilient UAV swarms a challenge. Successful UAV swarm implementations have demonstrated exceptional ability in performing tasks in various fields such as agriculture [5], natural resource surveying of water, soil, wildlife [6], search-and-rescue operations [7,8], and the military [9].

Unanticipated events such as inclement weather, intrusion from enemy agents, collision with foreign bodies or other swarm agents, loss of communication, or bugs in controlling schemes and software are just some of the events that may impede swarm function. Oftentimes, current multi-agent systems are interdependent to a high degree, making the loss of even a single agent disastrous for the swarm as a whole and its mission progress. However, failures can come in many different forms, both internal and external. Communication, Navigation, and Surveillance (CNS) failures [10] are categorized as internal failures, while weather and obstacles are external events. This study is part of an ongoing effort to improve resiliency in UAV swarms. To implement resiliency in swarms, we first need to conceptualize it into behavior responses that can then be implemented. Most modern systems exist and work in dynamic environments that are unpredictable in terms of their properties, composition, or behavior. Moreover, they have dependencies on input streams, power sources, and networks. Woods D.D in [11] perfectly condenses resilient behavior into four concepts.

1. Resilience as a rebound
2. Resilience as robustness
3. Resilience as graceful extensibility
4. Resilience as sustained adaptability

An UAV swarm is a perfect example of a cyber-physical system that works in areas that require it to exhibit resilient behavior. Networked, interdependent, and with limited capacity, UAV swarms are extremely susceptible to cascading system failures. It would be futile if an UAV swarm were to be brought down by the first disruption it faces. The remainder of the paper is arranged in the following manner. Subsections to this section discuss challenges faced while building resilient swarms and current research trends, Section 2 discusses every identified UAV swarm component as well as the various modules within them, Section 3 opens up directions for future research, while Section 4 presents conclusions based on the study and discussed statements. We define acronyms as they appear as well as provide a table of acronyms at the end of this article since some acronyms are repeated much later than their initial appearances.

1.1. Challenges to Building Resilient Swarms

UAV swarms have many challenges, such as the fact that the environment in which they operate is dynamic and unpredictable. During mission progress, external forces and internal incidents might affect critical hardware or software components of the swarm, leading to a reduction in performance, and perhaps an inability to perform a particular task such as record, fly, or navigate. Most application-specific swarm deployments require that the swarm be modified accordingly, and in situ adaptations are not uncommon [12]. Deploying and controlling even single UAV agent in the field has proven to be a challenge, and system complexity increases as the number of agents in the swarm increase. Mission needs might require specific sensors within the swarm but deploying such requirements to all agents in the swarm increases costs substantially. Madani in [13] describes resiliency engineering as the process of making a system capable enough to withstand disruptions. "Failure, in this context, is simply the absence of this ability, when needed" [13].

A halt in system function or a reduction in performance followed by the inability of the system to bounce back to the desired state is an indication of a non-resilient system [14–16]. Additionally, system intelligence must be capable of monitoring risk profiles for various components to actively avoid potential disruptions before they occur. In current UAV resiliency research, the authors have noted a stark lack of consideration of all UAV system components. Many publications mention this as beyond the scope of research, choosing instead to focus on one component. UAV swarms, on the other hand, require a ground-up integration of resilient behavior in all components to build a truly resilient system. The components here are defined as the many different areas of UAV operations. The authors here present a graphical visualization of these components in Section 1.3. Resilience is a multifaceted capability [15] and therefore cannot be summarized by modifying some aspects of system engineering while omitting the rest. We have observed that there are three major omissions in many existing resilience integration methodologies:

1. All UAV components are not considered for resiliency incorporation
2. All disruptions in their operational space are not considered.
3. Resiliency concepts developed for individual agents are attempted to be scaled and applied to a swarm.

Existing research does not address comprehensive resiliency requirements, instead choosing to focus on a few modules. Simulations of such results might not fully consider the dynamics of additional disruptions affecting system states. Article [16] discusses resiliency situations wherein they highlight how hard-coded action-policy lists are often unsuitable for environments they are not designed for. MANET (Mobile Ad-Hoc Network) topologies in a survey conducted by them such as distributed optimization and relay chaining address network problems in a specific context only. Labeled as the 'no-free lunch dilemma' [17], this is described as how a method will often address a problem well but may fail in previously unencountered scenarios. Article [18] reviews the research by Macek et al. in [19] where solutions for robot navigation consider only single agents and their safety, but fail to address collision-free navigation for robot swarms. This is our main argument in resilient system design. Methods for individual agents may not scale well for multiagent systems (MAS). Critical flaws in system design are not addressed in simulation scenarios that measure system behavior. Article [20] describes their SHARKS protocol and states that its motivation is security and resiliency. However, agents that run the security protocol alone cannot be labeled resilient. Additional parameters should be conceived for the purpose. Article [21] designs a control and optimization strategy for SAR missions, but takes certain steps such as optimizing inputs onboard the UAVs rather than on-ground control to prevent communication issues. They state resilient network design to be beyond the scope of their study, and its application for the deployment of multi-agents in smart cities may have other disruptions that can cause a swarm to go down. A cyberattack or network jam is not the only cause for a swarm to cease function. Moreover, the SHARKS protocol describes only circling stationary targets, and thus generalized assumptions on target mobility cannot be made, since targets may have the ability to go mobile. Target tracking capabilities require extensive algorithms for path planning, and optimized resource allocation, accounting for the loss of communication capability while following targets in low signal strength areas. A change in control design is needed for collision avoidance during dynamic formation changes during target tracking and damaged agent recovery. It is difficult to realize the large-scale aerial performance of UAV swarms because this process involves complex multi-UAV recovery scheduling, path planning, rendezvous, and acquisition problems [22]. A survey conducted by [23] on routing protocols reports how most developments focus on performance and not security. Unsecure protocols and networks, however resilient to connectivity loss, are not an indication of a resilient swarm.

1.2. Analysis of Current Research Trends

To gain an understanding of the trends in current research, it was first necessary to construct a database of related studies. The two most popular research databases were used, Google Scholar and Scopus. IEEE and MDPI databases were also examined. However, since the results from both are indexed by Google Scholar, mentioning it gets precedence. While Google Scholar provides higher numbers for the i and h indices due to its wider reach and slightly different citation trawl method, Scopus is more ordered and allows better access to articles. The primary keywords used in the search process were UAV, RESILIENCY, and MAS. Keywords were used to search the title and abstracts of articles.

A time constraint of the past 10 years (2011–2021) was applied to search for articles plus current work through June 2022. More than 1100 hits were generated on relevant research. These were manually filtered to remove duplicates and incorrect entries. Incorrect entries included resiliency recommendations for electric grids, outer space, and nonrelated cyber-physical systems. An analysis of the remaining 258 papers provided us with seven different sections that formed the UAV swarm components classified in Figure 1. All research was then categorized individually into these seven components with cross-links established for papers that had a combination of or covered multiple modules. The definition for combination modules was defined as research making progress in multiple subcategories, such as algorithms for heterogenous agents performing balanced path planning [24] or management architectures for drone service platforms that provide mission planning and resource handling for agents [25].

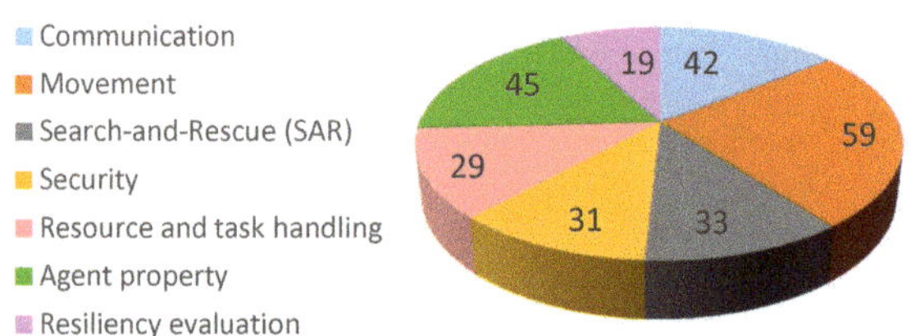

Figure 1. Examined research categorized into UAV swarm components.

An analysis of the final data set provided the number of articles that covered each component for resiliency integration as a major part of their research proposal. Figure 1 shows the seven components and the number of research articles in each one.

Figure 2 shows the above research categorized by the number of components they each cover. More than 50% of the research examined covered just one component as its target for resiliency integration while excluding the others. Less than 15% of the research covered three components of the recognized components for UAV swarm operations for resiliency incorporation. Research work such as swarm management and control policies were excluded from this count as it did not directly relate to any unique component. Thus, an error of ±3% is assumed.

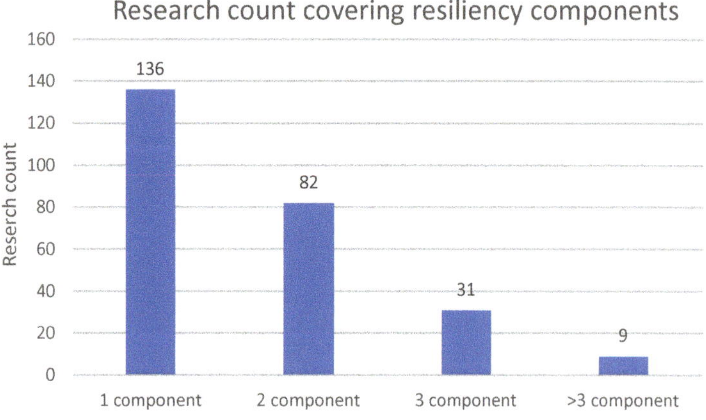

Figure 2. Research articles classified by the number of recognized UAV swarm components they cover.

A generalized search using Dimensions.ai [26] for related research in the past decade shows an increased uptrend. Figure 3 shows the number of research articles excluding book chapters that were published each year on UAV swarm resiliency from the year 2011. While research on swarm resiliency is certainly on the rise, the work is primarily focused on certain components, while others are excluded or cited as beyond the research scope. This study examines the resilience of UAV swarms from a broader perspective and recognizes the need for system-wide integration of resilient characteristics in operations.

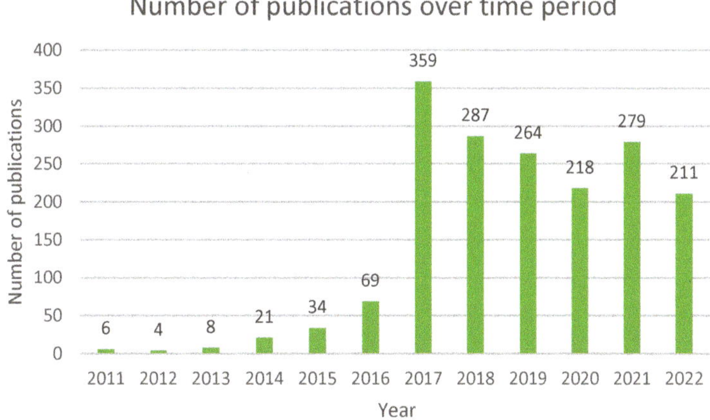

Figure 3. Number of publications on UAV swarms from 2011 to June 2022.

1.3. Scope and Contributions

This study examines UAV swarm resiliency research and creates a categorization of the various modules in swarm operations that require an integration of resiliency principles. The drawback of current resiliency research is that no single study addresses all types of factors that contribute to disruptions in UAV flights. An analysis of literature as shown in Section 1.2 supports author views. Existing research on UAV resiliency deals with some of the modules classified in this study. Other modules are recognized, but simply excluded for brevity, or recognized as beyond the scope of research. In this study, the authors have made efforts to identify every aspect of swarm operations and to categorize them. To our knowledge, this is among the very few studies that comprehensively address the resiliency issue by recognizing swarm components into modules and studying how resiliency features can be incorporated into them. Related existing research is discussed in detail. The state-of-the-art for all individual modules is discussed, and current challenges are highlighted.

Figure 4 categorizes UAV swarm operations into components and modules. The main components identified are Communication, Movement, SAR, Security, Resource and Task Handling, Agent Properties and Resilience Evaluation. Every component has subcategories (i.e., modules) that further extend operations. There is a recognized need for resilience engineering in all components and subcategories. Although it has been recognized that each module needs resilient incorporation, the module functions are interdependent. Failure of one can lead to a cascaded stop in operations. Every module is discussed as an important aspect of building resilience in UAVs. The major research in these modules is examined with their current implications and future possibilities. This study presents a classification-based review of resiliency research and lays the foundation for our current work on building comprehensive resilient swarms.

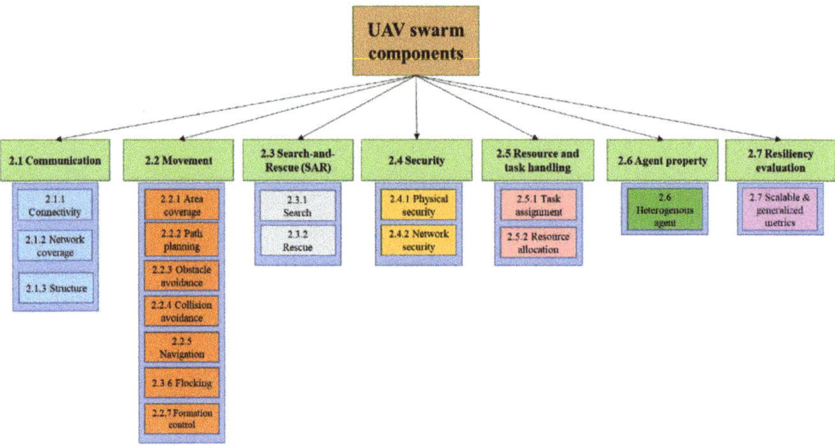

Figure 4. A categorization of UAV swarm components and modules.

To create a novel study, we also examined recent surveys and discussions (2019–2022) on UAV swarms. This list provides an idea of areas where UAV swarm research has been well covered, and where research gaps remain. Table 1 below cites a list of review literature on UAV swarms that was examined. The research focus section provides a general trendline of study directions and indicates that a comprehensive study that provides an overview of all UAV swarm components and modules was thus far not carried out.

Table 1. An examination of other related review and survey literature of UAV swarm components sorted by their publication year.

Reference	Research Focus	Published
[27]	A review of linear and model-based nonlinear controllers for UAV swarms as well as general swarm characteristics	2019
[28]	A review of ML-based techniques to improve UAV-based communication modules such as resource management and security	2019
[23]	A survey of swarm communication architectures and routing protocols and their open research challenges	2020
[29]	A categorization of swarm behaviors into organization, navigation, decision, and miscellaneous modules	2020
[30]	Studying issues in UAV swarm communications: Power issues, routing protocols, and quality of service along with open research challenges	2020
[31]	A survey of application-specific scenarios in internet of things (IoT) by using swarm intelligence	2020
[32]	A survey of UAV swarm intelligence based on hierarchical frameworks	2020
[33]	A study of the path planning problem for interception of mobile targets using an UAV swarm	2021
[34]	Examining dynamic task allocation methods for UAV swarms	2021
[35]	AI-backed routing protocols used for UAV swarms with an emphasis on dynamic topology properties	2022
[36]	A broad discussion on UAV, with an overview of swarms, agent characteristics, and applications	2022
[37]	A discussion on high-level swarm autonomy and the use of cellular networks for swarm communication	2022
[38]	A review of four main AI-based path planning methods in UAV swarms	2022
[39]	A summary of swarm intelligence techniques for Multi UAV collaboration	2022

2. Resilient UAV Swarm Components and Modules

This section reviews previous literature on all of the recognized components and modules in Figure 4. A tabular summary of the discussions is also provided in Table 2 below.

Table 2. A summary major research focus for each module in the seven recognized components of UAV swarm resilience.

Parent Component	Module	Major Focus
2.1 Communication	2.1.1 Connectivity	Connectivity maintenance
	2.1.2 Network coverage	Efficient coverage of an area with strongly interconnected agents
	2.1.3 Network structure	Types of network topologies
2.2 Movement	2.2.1 Area coverage	Optimized area coverage by agents
	2.2.2 Path planning	Path planning protocols for agents
	2.2.3 Obstacle avoidance	Protocols to avoid agent interaction with environmental obstacles
	2.2.4 Collision avoidance	Protocols to avoid agent interactions with other agents in the same swarm
	2.2.5 Navigation	Navigation and localization for agents
	2.2.6 Flocking	Flocking dynamics for agent swarms
	2.2.7 Formation control	Formation control for agent swarms
2.3 Search and Rescue (SAR)	2.3.1 Search	Searching for lost swarm agents
	2.3.2 Rescue	Rescue and connectivity of located agents
2.4 Security	2.4.1 Physical security	Ensuring physical security for swarm agents
	2.4.2 Network security	Network security and intrusion detection of swarm networks
2.5 Resource and task handling	2.5.1 Task assignment	Task assignment protocols for agents in a swarm
	2.5.2 Resource allocation	Resource allocation and assignment policies for swarms
2.6 Agent property	2.6 Heterogenous agents	The inclusion of heterogenous agents in swarms
2.7 Resiliency evaluation	2.7 Scalable and generalized metrics	Development of scalable and generalized metrics for evaluating swarm resilience

2.1. Communication

The main modules of the communication component that need to be addressed are connectivity, network coverage, structure, and characteristics. Each is a vital part of the communication process required by agents in the swarm to maintain contact with the base and each other. Important functionalities such as data transfer and action control take place through the communication pipeline. Keeping complete communication is often the first step towards resilient systems. Communication issues include communication delays between swarm agents with one another or with external entities, such as ground control [40]. Swarm agents may fail to communicate with each other due to a variety of reasons. Some agents might stray out of the communication area as a result of path planning and navigational actions. In such cases, the swarm as a whole must be flexible enough to select the optimal agent deployment area by considering communication equipment limitations. Communication at some point might be disconnected completely. This can be due to failure of communication equipment or loss of critical swarm agents responsible for handling connections. Certain UAV task algorithms might overwhelm agent computational capacities to the point that they become unresponsive and reduce the system to a standstill. In-path obstacles might also result in a temporary loss of communication with the swarm. Ongoing research on communication such formation control using ad hoc networks [40] identifies issues and proposes solutions. If some agents in the swarm become disconnected, flexible formation control can restructure swarm positions to bring back agents within the connectivity sphere. Transmission delays can also be offset using formation switching to alternate topology to position swarm agents closer to broadcast handling agents. Passive beacons installed on the ground can help recover agents from failure by guiding them to failsafe points. Section 2.6 discusses the addition of heterogeneous agents in a swarm as a means of increasing operational resiliency. Ground vehicles can assist in providing emergency communication to aerial swarms and vice versa, as well as perform functions such as visual detection of navigation beacons to coordinate transmission to aerial swarms.

2.1.1. Connectivity

This section deals with maintaining connectivity between the swarm and communication systems at ground control, ground beacons, and the user. The communication path from ground control to UAV agents in operational space has numerous vulnerabilities. Swarms might go off course due to winds or might have to change path due to sudden obstacles. This might affect the range of communication links used such as Wi-Fi and radio. Additionally, obstacles might also block transmissions resulting in delay or loss. Swarm agents with limited fuel capacity have additional problems. Attempts at reconnection and prolonged communication at low signal strength might deplete power reserves more quickly, reducing flight range on the mission. Adaptable connectivity protocols are needed in this scenario [41]. Here, a hierarchical topology is described, where a master drone controls a fleet of lower-level drones that can fly the search area. The master drone acts as a data pathway to the control center. Single link topological frameworks such as this always have the issue of data pathways failing. Since these are also agents that are exposed to operational space dynamics, there is a probability of them failing as well. By assigning different area restrictions to master and low-level drones, the study ensures that there is a persistent data pathway between ground control and end agents. Figure 5 shows a high-altitude fixed wing aircraft that has the equipment necessary to connect it to a ground-based communication system. A low altitude leader-follower quadcopter swarm is connected to the fixed wing aircraft. The swarm is able to communicate with ground-based stations located at a considerable distance using the fixed wing. However, there is still a probability that the master drone fails due to obstacles. Relay based connectivity maintenance [42] uses a similar communication link that uses relay and articulation UAVs to connect surveillance UAV agents in the swarm to a ground station.

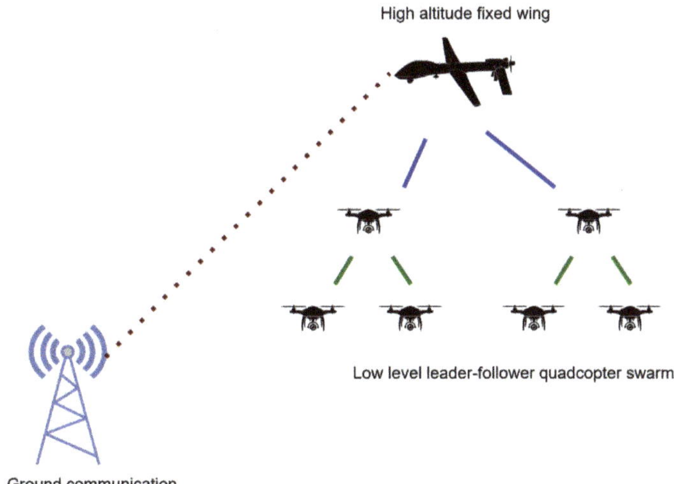

Figure 5. A high-altitude fixed-wing aircraft providing connectivity support to a low-altitude quadcopter swarm arranged in a multi-level hierarchical tree topology.

Multi-hop communications are key to such models where relays are used to read a wider range. While [42] designates articulation agents to balance mission with network connectivity, it brings all nodes within the same operational space. Without an excellent swarm defense and onboard intrusion detection system (IDS) in place, enemy agents can focus their attack on mission-critical nodes and target them first to bring down swarm operations. Spreading relay nodes over a wider range of topography can bring down the ROF (rate of failure). For example, if the aircraft was flown at a sufficiently low flight altitude, ground-based relay stations could be used. Figure 6 shows a multi-hop data pathway selection process that can occur between a source agent and a destination agent to relay messages. The routing protocol governing this transmission has to consider various factors such as agent locations, energy required, transmission time, etc.

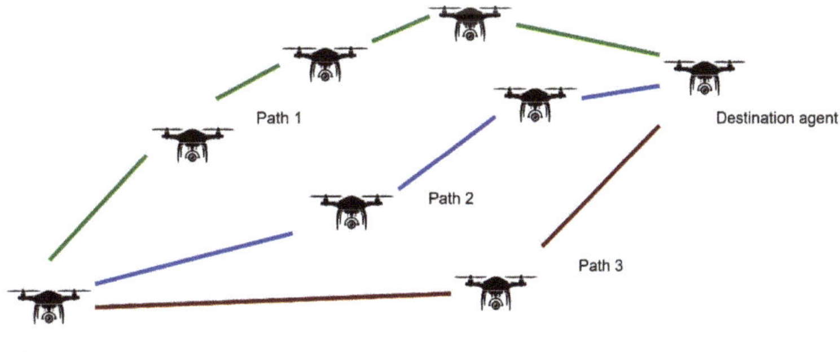

Figure 6. A multi-hop data pathway used by a source agent to communicate with the destination agent.

Routing protocols govern data transmission between network nodes. Hereby a swarm of UAV agents is capable of forming an ad-hoc MANET, FANET (flying ad-hoc network), or VANET (vehicular ad hoc network) depending on the topology and study in question. UAV networks are more advanced than those mentioned above due to faster flight speeds and energy constraints. High mobility and dynamic distribution of agents make the design of reliable communication links difficult [43,44]. The existing challenges to resilient protocol design are furthered by considering the nature of the UAV agent and the environment.

Routing protocols depending on their design can be position-based, topology-based, and cluster-based. A survey conducted in [9] summarizes several design considerations for routing protocols such as topology, flight formation, latency, collision avoidance, and external disturbances. Essentially, these factors are related to the resilience modules presented in this study. Although they apply to agent communication, the same factors can be scaled and considered as issues when designing a resilient UAV swarm. Routing frameworks such as the one proposed in [45] claim resiliency in communication by using software-defined networking (SDN) controllers. Connectivity graphs can be created by using available channels and relaying information for a particular area. However, this often requires knowledge of operational space beforehand. If that is available, shortest-path algorithms determine paths that allow agents in a swarm to be separated over a wide geographical area to prevent jamming by an adversarial entity that seeks to block communication waves in one area. Smart selection methods capable of switching data delivery schemes are required. The four basic schemes among agents are unicast–between two agents, broadcast–from one agent to all other agents, multicast–from one agent to several agents, and geocast–from one agent to some agents in the swarm selected due to absolute location or relative positioning.

Another challenge with routing protocols is that most utilize unicast routing. Multicast protocols assume dense and slow-moving node structures. Efforts to increase efficiency in protocol design, such as in [46], design a multicast protocol based on predicting the trajectory of agents. By selecting a node with higher priority metrics, which has a high chance of encountering an agent that needs to forward information, messages can be relayed faster. The calculated priority metric was dependent on start time and duration of the encounter with the transmitting agent. A priority encounter graph was constructed for this purpose. By equipping multiple modes of transmission depending on the position density and velocity of agents, routing protocols can act efficiently and guarantee constant connectivity and message delivery assurance despite disruptions.

2.1.2. Network Coverage

Optimal coverage of an area by a swarm of UAVs falls under the connectivity domain. Network coverage involves using the agents in a swarm to blanket a particular area such that all of the area is covered during application-specific purposes such as surveillance, while at the same time maintaining a standard quality of connection with other agents in the swarm. There is a tradeoff between coverage and connectivity that is recognized, and metrics such as those proposed in [47] study it. Multiple UAV swarms have applications in delivering persistent surveillance. However, features such as dynamic target tracking can lead agents astray and out of connectivity limits, which can lead to the loss of data and agents. Simulations on the exploratory problem using MAS show that to accomplish covering a certain area, particularly in cases where high rewards are offered to explore it, an agent might stray from the swarm thereby sacrificing connection with other agents or the base while doing so. Optimized models designed to balance connectivity and coverage such as those designed by [48] study the effects of network density on coverage and performance. Swarms with a smaller number of agents thus need high-level models, balanced with increased computational power, to handle more frequent formation control than swarms with a higher number of agents. A model parameter that requires agents to remain near each other might prevent agents from spreading out thereby restricting coverage. Bio-inspired ant colony optimization algorithms try to imitate ant behavior for communication and pathways based on pheromones. Chaotic ant colony optimization for coverage-connectivity [47] uses repulsive pheromones to visit unexplored areas. Alternatively, they were also used to avoid collisions between agents. With this approach, less work exists on balancing connectivity and coverage decision processes for swarms.

These forces can be calibrated to consider network strength. Thus, areas that might weaken connections between swarm components are assigned higher values to discourage single agents from attempting exploratory tasks. Intelligent algorithms can however assign a group of agents to visit high force areas such that the entire network shifts in that particular direction thereby maintaining network strength.

2.1.3. Network Structure and Characteristics

Network structure and characteristics play a major role in defining how resilient an UAV swarm will be to disruptions. While security components protect swarms from malicious intent, inherent network topographic structures, organization policies, and characteristics such as jamming resilience and energy efficiency can reduce other disruption margins and aid in strong connectivity [49]. While network structures typically define how a network is built with regards to nodes and resources, topology, and hierarchies for decision making, network characteristics deal with any intrinsic abilities that may be built into its structures such as jamming resilience, energy efficiency, and self-healing capabilities. This section examines both structure and characteristics in detail. Modified network topologies can have a substantial impact on how a network responds to a disruption. The most conventional of these is the hierarchical agent topology where some agents have a higher-level designation and act as leaders of a swarm. These agents may differ in their nature or hardware, leading to swarms being heterogeneous, perhaps even simply by their designation.

Article [41] created a similar topology in which agents are masters or followers. Each follower drone is an explorer whereas master drones are data collectors and major communication nodes. The master drones assign exploration areas to the follower drones. One master is responsible for several follower drones. The duties of the master are to maintain a set distance and signal strength with ground control at all times. Such tree topology implementations are often the first step in developing structured network systems that may increase swarm resiliency in a stepwise manner. Additional modifications to these networks are possible such as the addition of heterogenous agents as discussed in Section 2.6. Master drones are often high-altitude fixed-wing aircraft that control numerous low-altitude quadcopters.

Additional network layer modifications can also influence vital factors of communication such as the transmission power. Article [50] proposes the addition of a new layer between the data link and network layers. Each agent node has the ability to adjust its transmission power for sending data while maintaining stable connections with other nodes. The additional layer has decision constraints to prune unnecessary links and set the transmission power based on the nearby agent range. The proposed network has energy efficiency properties due to fine control of data transmission power. Depending on the number of nodes in the network, the target area can be divided accordingly, and an appropriate number of agents can be assigned to cover it. Article [51] proposes an advanced model that can differentiate between an agent in standby mode or active flight. The amount of energy consumed by an agent in active flight vs. being in standby mode is considerable. A model inspired by [52] splits the power required by the agents into flight power and standby power. Splitting energy demands allows for finer control over consumption as well as other functions such as task assignments and agent callback/deployment functions. Models that consider the state of agents and their energy consumption can be upgraded to allow for additional functionality. An agent in active flight may consume more power while performing certain tasks, allowing other idle agents to be on standby. This agent will have a lower availability of flight power "e" than standby agents. During task assignment, the control scheme can decide to recall this agent to redeploy a new agent or use the new energy value as a constraint in determining task allocations to the swarm. While fine control over energy consumption such as in [50] may not always enable superior performance, the inclusion of energy-efficient features in all areas of agent control may extend swarm flight time considerably. Conventional ad-hoc UAV networks do not have such flexibility, and all nodes transmit at the same power level for a fixed transmission range. However, transmitting at power levels higher than required can reduce network efficiency and impact flight time.

Solutions such as this can also solve the malicious attack problems discussed in Section 2.1.3 where nodes are compromised and forced to transmit packets at lower transmission powers to misdirect the attack or to cause latency issues. However, flexible power transmission control can allow neighboring authentic agents to efficiently capture these data packets and ensure that they reach the destination, making such attacks fail. The role of blockchain implementations as a security upgrade to UAV swarm networks has been discussed briefly in Section 2.4.2. However, their integration into network structure can also have added benefits towards building resilient networks. An agile implementation [53] defines an UAV network that is decentralized by using SDN and blockchain technology. It has the usual application, control, and data planes. The data plane is present on the UAV whereas the other two are at ground control.

The location of network planes is important. In [53] for example, the ground control location of the control plane gives it a global view of the network that other routing protocols lack. All agents in the network are aware of the swarm status. Control stations act as the deployment of the blockchain. Multiple ground stations share and synchronize data over the blockchain. An added advantage to this is the fact that the swarm has up-to-date data on the shortest routes and destination status anywhere they go, as long as they are in the range of a single ground station.

Swarm agents that act as the main creators of data on the data plane also add information to the blockchain. Task scheduling and flow table information are maintained and updated using smart contracts. Article [54] studies application-specific functionality of drones for hybrid cellular network deployment. Studies show that prolonged use of UAVs, particularly for drones as a service (DAS), has been in limited deployment due to factors such as limited flight time, collision avoidance, and onboard hardware capacity. Examples include the use of drones to provide emergency communications or vehicular micro-cloud services [55]. Modified network systems are needed to improve agent performance, and this study proposes a swarming strategy of agents based on continuously monitoring swarm geometry and the environment. Langevin dynamics were used to study agent interactions. Self-organizing characteristics were based on crystallization of molecular structures in condensed matter physics. However, self-organization occurs in chemical, robotic, cognitive, and natural systems as well. By allowing network nodes to develop a self-imposed hierarchy based on external or internal factors, an order can be established from an initially disordered system. By using external stimuli, or in this case demand requests by activated end users, matching supply movements can be created.

Application-specific research such as [56], which provides solutions for a disaster-resistant network by using UAV deployment, can also be used to make UAV swarms inherently robust. Although the problem addressed is mainly UAV deployment and formation to provide network services, these solutions can be applied to address network coverage problems to create ad hoc resilient UAV networks.

The proposed approach has the deployed UAVs act as aerial base stations (ABS) and estimates the position of user equipment in a disaster-struck region by using their uplink signals. User and ABS locations go to a central location where the signal-to-interference plus noise ratio values are calculated. Users are assigned to a particular ABS according to the max user per ABS limit and a clustering algorithm deploys ABS to final locations. Similar approaches can be used where agents can act as nodes in a network topology to which a set number of other swarm agents connect. Once connected, by using range limits of onboard communication equipment as a weight parameter, agents can be deployed to create a multi-node UAV net where every agent in the swarm maintains a stable connection with a specified number of swarm mates. Further upgrades to such networks such as energy efficiency and jam resilience can be introduced. Jamming resilience is typically used to defend against attacks that seek to disrupt routing protocols by compromising selective nodes. Data sent through these nodes are then captured and analyzed for additional information on how to compromise the network. Such attacks can be executed by corrupting a valid node in the network or by the addition of a new foreign agent-based or ground node that the swarm assumes is a part of the network. Every drone is essentially a node in the FANET that has localization capabilities. Link quality is measured using metrics such as RSSI (received signal strength indication) and SINR (signal to interference plus noise ratio). Additional schemes for traffic load and spatial distance are used to create a 'jamroute' protocol that has source nodes search their real-time changing routing tables for the shortest routes to the destination node [57]. In the case that the node is busy, a broadcast request is sent that has a packet of information such as source node, destination node, and max traffic loads. Each node has unique identifiers that associate it with the network, thus making it difficult to add malicious nodes to the network as a means of attack. Additional proposed schemes for link quality and traffic load prevent attacks such as a limited transmission power attack where the compromised node drops a data packet by transmitting it with reduced power. Protocols those such as in [57] prevent the loss of data packets by creating multiple forwarding paths. Even if one node attempts sabotage maneuvers, other paths ensure all packets reach the destination node.

Based on the characteristics of the routing table, the protocols can be further categorized. In static routing protocols, once created, the routing table cannot be updated. However, this is not suitable for dynamic formations and large swarms. Article [57] discusses other protocols, such as proactive tables, where tables are updated periodically. Hybrid routing protocols combine proactive and on-demand routing. Fixed network topologies are often an issue when it comes to improving ad-hoc networks. Ad-hoc networks are necessary to supplement higher costs and range-limited communication such as cellular communication and Bluetooth. Fixed topology networks do not scale for ad-hoc connections very well. Dynamic topology reconstruction protocol (DTRP) models such as the one in [58] consider changes in UAV network topology. The DTRP uses a master node and a worker node. The master controls the internal swarm network, and the worker node is responsible for the transmission of sensing and flight information. The master node transmits initialization messages to all follower nodes in the area. The worker nodes then change the node information of the message from that of the master to their own and forward the message to other workers that are out of direct reach of the master node. If a set time elapses and an acknowledgement message is not received from the worker node, the master node is converted to the lowest node in the link table. Minting link information tables such as these assist in verifying the position of each node on the map, as well as information control. The return messages contain location information, which when overlaid on a map pinpoint the location of each node in an area and also areas where no nodes are present. Therefore, such areas can be serviced by redeploying some agents in the swarm to them.

2.2. Movement

This component covers the decision process that involves the movement of the swarm agents in operational space, including flocking, optimized area coverage, path planning, and obstacle avoidance. As such, these are the physical behaviors that a swarm might exhibit during its operation. Major modules of the movement component are recognized to be area coverage, path planning, obstacle avoidance, collision avoidance, navigation, formation control, and flocking.

2.2.1. Area Coverage

While Section 2.1.2 discussed tradeoffs between network coverage and connectivity, this section discusses area coverage as a part of the movement component. These problems are often intermixed with each other. The coverage problem determines the success and probability of when the area will be completely scanned. It is often defined as increasing coverage while managing trajectories and disruptions. Article [59] defines swarm coverage as a process to cover a selected region. Swarm area coverage is an important decision process for swarm systems. Area coverage is often utilized in application-specific scenarios to cover a region of interest (ROI) with a swarm of agents, particularly as a movement problem, with less regard for connectivity maintenance and more focus on completing a particular goal. A similar problem framework is discussed in [60]. "Chaos-enhanced mobility models for multilevel swarms of UAVs" mentions how area coverage is an original problem by describing it as "focusing on the mobility management of a swarm of autonomous UAVs to maximize the coverage of a squared geographical area". The resilient component differentiates this from a simple path-planning objective. The two main objectives are for the swarm to cover a given area as well as counter any unpredictable disruptions that occur during the process. Considering the target area properties is an important input for decision-making models for swarm operations. Network coverage and area coverage differ in terms of how ROI is used. While network coverage focuses on making sure that agents in the area are always optimally connected, area coverage addresses the coverage problem and describes the actions taken by a swarm of agents to cover the ROI. The target space may differ in terrain and ground cover, as well as the presence of water bodies, wind channels, and stationary or dynamic obstacles. Current issues with area coverage lie mostly in the dimensional space in which they are tested. Most existing simulations portray coverage in a two-dimensional space. In [60], altitude information is not considered. Formation deployment controls may require agents flying through constricted spaces to fly in tighter formations thus requiring agents to vary in flight altitude rather than sweep area. This is especially prevalent when reaching consensus after obstacle navigation and post-deployment primary formations. The control methodology in which agents cover an area can be categorized into different types and [61] offers a naming convention for them as follows. Static coverage is a standard agent deployment method in which an UAV examines a particular spot for its target. Several such agents examine individual spots. Barrier coverage forms a perimeter that can detect the entry of any object through the barrier. These are usually deployed on security-specific applications. Sweeping coverage is the name the authors have given to the dynamic deployments of agents which can change formation as they move through the area. This is the standard procedure followed across any SAR protocols. Several coverage models have been described such as imposing grid cells on the ROI to ensure that every cell is covered or dividing the area into small bits that are assigned to the agent's area decomposition [62,63] and sweep motions [64,65]. Article [59] mentions two primary methods for cooperating coverage, centralized, and distributed decision-making. They propose a self-organized decision-making approach for the problem modeled in velocity space. The approach is divided into perception, decision, and actions. Multiple UAVs coordinate with each other for sharing position, velocity, and obstacle information. Their decision uses a reciprocal coverage method that creates collision-free optimal spaces. The swarm coverage decision model considers the above parameters with collision avoidance with other agents, obstacle avoidance, and optimal velocity decision in each iteration. This is carried out using the Monte Carlo method for velocity-finding in confirmed space.

A different obstacle characteristic is considered by [66] while solving the area coverage problem. Not all obstacles have the same threat levels, nor are all of them equal in terms of dimensions and nature. Additionally, energy constraints on UAVs during coverage problems have often not been addressed in complex scenarios. They propose a two-step auction framework for energy-constrained UAVs in a given area. The agents evaluate the threat levels of each area cell referred to in their paper as a module and bid in an auction for the UAV to come to it. If two bidding prices match, energy loss is also considered. The UAV determines the winning module. The second step is the obstacle avoidance strategy for any obstacles that the UAV might face while traveling to the winning module. In the case that an obstacle is unreachable by flying over it, the second-best module is selected. Additional constraints for sleep mode and two UAV bid clashes are also designed. Such strategies are also viable alternatives to reward-based ones where agents are rewarded for considering a particular area. Additional parameters such as energy considerations can thus be programmed.

2.2.2. Path Planning

Efficient path planning for multi-agent systems is a prevalent challenge in swarm development. Algorithms such as Particle Swarm Optimization (PSO) are developed to find near-optimal solutions. Multiple iterations on a solution may provide better results. Several biologically inspired heuristic algorithms have been proposed for path planning. Bio-inspired algorithms have found remarkable success in the movement development of multi-agent systems as they exist predominately in the animal world. The development of such algorithms drew its inspiration from group behaviors in fish shoals, bees, and ants [67]. Article [68] proposes a modified fruit fly optimization algorithm. The original fruit fly swarming is inspired by fruit flies making their way to ripening fruit. The modified algorithm divides the swarm into smaller subswarms. By allowing flexible search parameters there is a shift from a global search to a local search as the mission timeline progresses. This expands the search space considerably. The FOA (Fruit fly optimization algorithm) uses two search parameters, sight, and olfactory. While the visual search is a greedy search, the olfactory search is a directional search that examines the greatest concentration of the target in a cell and proceeds in areas with increased concentrations. Both processes are repeated until the termination stage is reached near the target destination. UAV inputs can be used to program essential paths for individual agents. Paths to the destination can be calculated by using terrain data, weather, and network signals as inputs. Based on pre-established network availability from base to destination, each agent can move on a path that has the best-preestablished parameters such as SINR. Secondary inputs such as terrain can be analyzed using the visual input.

Distributed path planning for multi-UAVs works similarly, ref. [67] where SAI (Surveillance Area Importance) values for each cell are analyzed and a connection is established that shares each UAV's location. The leader agent checks the area based on past SAI values, and uncovered areas are subdivided. Individual trajectories for each agent are generated. SAI is an intrinsic value generated and defined by [67] based on the probability of outside agents entering a restricted airspace. Since this is an application-specific development, such values are required. For general purpose use, however, the above-suggested values such as network strengths can be used to establish the grid importance. An alternative approach used by [69] uses external threat models to create channels through which agents can travel. The weather threat model measures wind and rain states and the transmission tower model calculates a safe path some distance away from transmission towers to prevent their electromagnetic waves from disrupting agent navigation systems. An upland threat model measures the terrain below the agents to maintain optimal distance between aircraft and the ground. An adaptive genetic algorithm controls the path generation schema.

Modeling individual threats as functions that act as inputs to learning algorithms is an efficient approach to creating low-cost shortest route solutions. By allowing models to scale as per a particular environment, threat functions that are not present can be eliminated thereby allowing quicker convergence on solutions. For example, the transmission tower threat function will not be used in an area that does not have them, thereby reducing overall model complexity.

While most research generically uses the term collision avoidance and categorizes together all impacts of UAV agents with other agents or in-path obstacles, this study classifies them separately. Obstacle avoidance deals with any static or dynamic obstacles that might be present in the flight path. Examples include geographic terrain, buildings, trees, birds, etc. Collision avoidance deals specifically with avoiding impacts with other UAV agents in the same swarm. Each is discussed separately in the following sections.

2.2.3. Obstacle Avoidance

Movement systems of UAV swarms for path planning, obstacle avoidance, and general navigation require comprehensive integration with each other to provide acceptable performance. Article [18] discusses a unique system that develops drone reflexes to maximize agent safety. They utilize a dynamic evolutionary algorithm to create drone routes, and a reinforcement-based learning algorithm to use system state data and create feedback loops. Drone reflexes are labeled as reactive actions performed by agents to prevent collisions with sudden in-path obstacles. Case studies such as this highlight the differences in modules for obstacle avoidance and path planning while also demonstrating the need for these modules to work closely together. Quick computation of alternate routes is required once a collision is foreseen by the learning algorithms. Moreover, path planning has a sub-module that calculates the trajectory progress of all agents in the swarm. While quick solution methods to reach optimal performance in collision avoidance dictate that the trajectories developed do not intersect, this is not an ideal solution. Real-world scenarios often require agent trajectories to overlap during application-specific functions such as target search. In such cases, the trajectory planner also has the responsibility of near real-time tracking of all agent progress concerning their defined trajectories. Any overlapping paths must not have agents present at the point of intersection. A wait-and-go action process must be implemented where an agent waits in hover mode while the other passes the point of intersection. Alternative solutions include introducing an altitude adjustment component. Two agents may follow intersecting trajectories without waiting if they are separated by a safe flight altitude.

The autonomous navigation system in [18] uses an offline component to generate the shortest paths between start and end points based on standard information. The online component is a dynamic monitoring system that utilizes a feedback mechanism to detect changes in swarm reconfiguration and suggest reorganization of swarm routes when necessary. Prediction features use monitoring data to predict drone movement and collision. Additional modules for safe area computation and reflex computation are present. The reflex module outputs reflexes for drones to avoid in-flight collisions.

The weighing mechanism in the hierarchical methodology in [70] is assigned levels based on the distance from the master UAV for each sub-swarm. Obstacle avoidance is inspired by occasion and behavior. If it is determined that the intended movement of the swarm and the obstacle point positions satisfy constraints that make it impossible to avoid, the master UAV then attempts to move in a direction tangent to the obstacle. The flocking control uses a hierarchical model that labels master UAVs as information UAVS. These decision-making agents act as the center for a flock of ordinary UAVs thus forming a sub-swarm. The objective of the study is to ensure that ordinary agents follow the information agent. Thus, by controlling the actions of the information UAV for obstacle avoidance and using flocking control on ordinary UAVs, the sub-swarm is shown to successfully avoid obstacles.

2.2.4. Collision Avoidance

The collision avoidance problem has been recognized by research as being separate from obstacle avoidance. While research such as [70] uses cooperative formation control to avoid obstacles, ref. [71] uses predictive state space to generate collision free trajectories for agents. The most prevalent way among current research of avoiding collisions is by using agent state information and generating artificial potential fields. The mission-based collision avoidance protocol (MBCAP) in [72] uses a similar strategy. Every agent broadcasts its current and intended next stage position. A collision detection process receives them and checks for coordinate matches of agents. In the case that a match is found, a stop signal is sent to those agents. Upgrades such as agent priority are also included where each agent has a priority level, where in the case of a collision threat, the lower priority agent gets a stop command while the higher priority agent can pass.

Additionally, altitude adjustments are often used to avoid mishaps. On a collision alert, the two agents can adjust their altitude for one to pass over the other. Article [73] uses a modified trajectory modification such as the one described above to climb and descend in order to provide vertical separation at the closest approach points. The global aim is to have the least number of modifications overall. However, this necessitates a higher-level integration into swarm motion to avoid collision with other agents in the same altitude range. Higher-level decision models can be computationally expensive and take more time to reach the decision stage. The second method is the artificial potential field method and several functions such as the flat-h function and hyper quadratic surface functions for the shape of potential fields have been studied [74–76]. The method considers each swarm agent as a charged particle and can set a potential field based on current configurations. The two forces, attractive and repulsive, decide how surrounding agents behave. Potential field models generally have higher computational costs than using location-based methods but have the advantage of lesser offset and error rates than image-based detection and avoidance of other agents. Moreover, using image sensing equipment for agent avoidance may employ equipment that can otherwise be used for mission purposes. Potential field generation and location tables can be handled at ground level and on unit, respectively. Potential field functions set artificial barriers around an agent that another agent recognizes and cannot cross. While these can be dynamic during the flight timeline, most current studies discuss a static volume for the generated field. This is comparable to the generation of a miniature no-fly zone around every agent. Article [71] divides the air space into cubic grid cells as they are easier to use when calculating agent positions. The rolling optimization algorithm they use estimates agent collision and offers updated trajectories to avoid them. It uses heading direction concerning current positions and the distance between two UAV agents as constraints for improved trajectory generation. Here, the distance between UAV agents is calculated with protected spheres, which are state spaces around agents with a radius greater than the safety distance. If two spheres were to overlap, the UAV agents would collide.

Similar approaches construct dynamic collision avoidance zones around aircraft in the horizontal [77,78] and vertical planes. These are used to detect flight conflict trends. The construction of both planes to detect a collision is simulated in [79]. While their particular method was applied toward a pursuit–evader scenario, similar scales can be applied to swarm agents to detect and actively avoid other agents nearby.

2.2.5. Navigation and Localization

Localization and navigation are two important properties required during swarm movement, as the two are closely interlinked. A navigation path cannot be established if agents are unable to sense their position relative to important points on the map. Localization is necessary for successful swarm navigation, as most multi-agent system applications require that each agent has a level of awareness regarding its surroundings as well as an ability to discover its location with respect to the local environment. Global navigation satellite system (GNSS) modules on agents can handle the problem, however not all UAV models are equipped with the necessary hardware. Low-cost agents often do not have GNSS modules due to additional costs. Other constraints such as aircraft weight and fuel constraints also prevent upgrades. Swarms equipped with GNSS may not work well in indoor environments, varied geographical features, or underwater due to weak GNSS signals. Additional methods need to be implemented especially if the swarm includes heterogeneous agents that are spread out over varied operational spaces, for example, an underwater submersible together with an UAV. While the addition of such agents has resilient implications, it mandates the need for upgraded localization methods [80].

Additionally, as [9] mentions, even GNSS module inputs may be insufficient for localization with smaller update intervals. Differential or assisted GNSS solutions were examined and provided far more accurate results. Article [81] provides a combination solution that uses GNSS, an inertial measurement unit, a baro-altimeter, and range radio measurements if GNSS is available. A fused reading is obtained by integrating all four sources. In case GNSS is unavailable due to any factor, the platform switches to an alternate solution based on the other three. While this may not be as accurate, it provides an estimated relative position estimate as well as reduces overall swarm drift that occurs in an IMU-only type system. The study also progressively explores some members of the swarm being resilient to GNSS interference, an observation can that can be utilized in the development of heterogenous swarms.

Localization methods can be of different types depending on how they function. There are methods that do not rely on the range and methods that do. Article [80] classifies them as range-free and range based. Range-free methods are more localized using techniques such as nearest neighbor information or node information [82]. Range-free solutions may be faster and computationally easier to process, however at the expense of reduced accuracy [82–86]. Alternate solutions exist such as using vision systems equipped on agents to estimate location. Image matching though is typically more suitable for indoor environments due to computational costs and time delays. Moreover, persistent localization may be a challenge as this might require vision sensors to be up to par with capture requirements. Article [87] mentions an indoor localization process using image capture. Passive beacons such as navigational pads can be assigned coordinates on a cartesian plane, and all agents include image processing algorithms designed to recognize the mission pads. A similar approach is used by DJI Tello drones [88] by capturing and recognizing quick response (QR) codes. However, the range of camera sensors and their directional position limit widespread adoption.

These methods use the vision sensors installed on the agents. Other methods may use an externally located camera along with image processing to detect the agents and estimate their location [89]. Such approaches have widely been used in indoor localization techniques. The drawback with using these systems outside as opposed to indoors is that it is much more difficult and processing-intensive to implement in an unknown environment.

Article [90] conducted an experiment where a cooperative path planning strategy was implemented between agents that have GNSS coverage and with others that do not. The agents with nominal or more GNSS coverage stay in the visual line of sight of agents that lack it, thereby guiding them through an obstacle filled environment. Range-based solutions [91–93] use actual distance values for unknown nodes, but additional hardware is required on the nodes in exchange for more accurate results and equipment cost may be an added factor.

Article [80] proposes a backtracking search algorithm (BSA) with multi-hop localization that improves resilience if reference nodes are not recognized correctly. The BSA is an evolutionary optimization algorithm that uses past solutions to find solutions with better fitness. The proposed approach here uses each unknown node to first estimate its distance from a preset point using multi-hop signals. A min-max method obtains coarse values for node location. The last stage takes multiple such contributions of the estimated node locations from the other nodes and provides a refined approach. Additional measures such as a confidence factor evaluate accuracy and can be programmed in multiple ways. The range of neighbors whose contributions are considered can be expanded or multiple iterations of coarse readings can be taken to establish finer mean values. Agents in a swarm act as nodes that broadcast their routing information periodically. Other nodes receive this information, and every node maintains a table of other agents. Neighboring nodes are given prevalence in terms of their receiving data packets. Prediction algorithms use the routing table to predict the location of agents at every turn to estimate the location of the nodes. Agents in sectors present in the direction of the destination node are naturally given preference [94].

2.2.6. Flocking

Flocking is a group behavior where all individuals in a large group exhibit consensus in terms of movement, navigation, and obstacle avoidance. Local rules dominate swarm decisions [95].

To realize flocking, all members must follow a similar trajectory often defined by higher-level agents in the swarm. Article [96] provides the groundwork that assists in classifying flocking in separate sub-modules from formation control and collision avoidance. While flocking and formation control are both sub-modules for swarm control, they should be categorized independently. Flocking dynamics are not particular in terms of the final goal [97], but rather focus on substages in swarm movement such as before and after an obstacle is avoided. Formation control may specify higher-level dynamics and control over agent hierarchy and the decision process that assists in creating swarm dimensionality in terms of physical shape. Collision avoidance is usually defined between two members of the swarm and requires active monitoring of agent location and trajectory. Formation control maintains a set swarm formation shape throughout the navigation and path changes of the flight process.

Flocking behavior is thus a combination of dynamic navigation and obstacle avoidance with velocity matching. Such behavior has been observed in nature in swarms of bees, fish, and birds. The global objective is often to move towards a region as a whole. Certain formations may provide more aerodynamic qualities to the swarm reducing wind resistance and providing longer flight durations. Article [96] notes how specific formations of fixed-wing aircraft have a reduced overall drag coefficient. Several formation control methods use a form of hierarchy such as leader-follower, or force methods by using potential fields generated by agents [60]. These are less than optimal solutions that increase complexity due to the addition of collision and obstacle avoidance. Bio-inspired pheromone based funneling of a group of agents by herders as well as behavior-based methods are also prevalent [60].

2.2.7. Formation Control

Multiple formation strategies have been formulated by researchers such as leader-follower and virtual structure [98]. Article [99] mentions the usefulness of high-fidelity biological models in formation control and tracking. The grey wolf tracking strategy uses the four-level division found in wolf packs, alpha, beta gamma, and omega. While alpha agents are decision-making agents in tasks such as directional movement, division of labor, and target pursuit, beta agents are for communication. Delta agents act as sentries for the pack. Extending this concept, ref. [99] develops an adaptive formation tracking protocol in a pursuit scenario inspired by wolf behavior. The agents follow the wolf tracking strategy of tracking leaders and formation before encircling the target. Multiple modifications to this control strategy are examined in [100–104]. A centralized control approach may have higher computational costs with an increase in the number of agents. A decentralized formulation such as by [98] called the Markov decision process describes the creation of formations in short time ranges. The authors of [105] define a waypoint-based formation control protocol called mission based UAV-swarm coordination protocol (MUSCOP). The master UAV follows a set of defined waypoints during mission planning. The follower UAV stays in formation between waypoints along with the master. Once a waypoint is reached the master synchronizes its position with the follower UAV. Relative location is used by the master to define formations. The center UAV is the master and follows mission parameters whereas the surrounding UAV follows offset coordinates. The protocol scales well in with large UAV swarms and has a negligible time delay in terms of the synchronization wait required at each waypoint.

2.3. Search-and-Rescue (SAR)

SAR missions are usually defined as an exploration problem. Exploration approaches can be used in a wide range of applications [21]. Target search applications can include searching for an intended target such as an entity in danger or need of medical attention or surveying the aftermath of an accident. The search function can also be expanded to include other agents in the UAV swarm that might have malfunctioned and crashed. Two major applications of the search function are discussed.

1. Target search and tracking for entities that are not a part of the swarm.
2. Track and search for agents of the swarm to open further conditional processes related to mission progress.

Rescue activity has additional decision parameters. If a crashed agent were located by the swarm, the cost of additional time and fuel that would be required to recover the agent should be incorporated into the model as a function that opposes the primary mission function. Additional conditional statements must be programmed to gauge agent failure in the first place. If the agent has failed due to a locally present disruption, deploying additional agents may result in their loss too.

The search function is an application of UAV where the aerial vehicle hunts for a particular target using vision ability or location information. Searching for a target using an UAV has various challenges. Depending on the hardware used, UAVs may have limited sensing and communication hardware onboard. Capturing raw footage and sending it to the ground station to be processed is computationally costly and may introduce delay. Two types of search algorithms are often used to enable autonomous search: (1) visual search using learning-based detection algorithms [106], and (2) location-based search using active or passive onboard sensor arrays. Search algorithms conventionally divide the area by using a probability index of where the target is most likely to be present [107,108].

Similar to functions discussed in Sections 2.2.1 and 2.3, UAV swarms allocate multiple agents in a cooperative problem formulation to each search a part of the grid. Path planning and formation control modules keep agents from colliding while preventing explored areas from being searched multiple times. Article [21] proposes a nonlinear MPC solution for searching an area. Citing communication and delay concerns for the control, each UAV optimizes self-control instead of processing at a ground station. A PSO algorithm is used to find problem inputs such as airspeed and roll angle. A cost function is also associated with the probability map to determine the effect of the search function on mission progress. Article [109] proposed a vision-based search function for mobile ground targets by an UAV swarm. Equipment challenges exist such as the agents having cameras with a limited field of view. With vision-based functions, targets may also not be recognized accurately if low-resolution equipment captures inputs from higher altitudes. Dealing with mobile ground targets requires additional considerations as compared to a fixed target. While formulating the target search, they use cooperative coverage control which is categorized under area coverage in Section 2.2.1. For the probability map update, each agent is assigned its map for the whole region and takes individual measurements of its assigned area for examination before updating it on the other agents. The time delay that might occur due to this is not addressed in their paper. Cooperative agent vision-based inputs creating dynamic probability maps have a distributed nature, where complete network connectivity is not required. The probability map uses a fusion of the Bayesian function with consensus from multiple agents. A pigeon-inspired optimization proposed by [110] is inspired by birds using geographic magnetic fields and landmarks to reach destinations [111]. Article [112] uses a profit-driven adaptive moving targets search (PAMTS) which uses the familiar decomposition technique for dividing the search area into equal cells for each agent. An observation history map is created for each cell by the agent surveying it. This can then be shared with other agents to create a global knowledge base. The two objectives here have a tradeoff, the explore action wants to reach newer cells whereas the following action needs to track already detected objects. Moving target search problems is computationally intensive, as they are classified as NP-hard. A recent observation table allows agents to change their behavior weights which in turn balances the tradeoff. The reward for each cell creates the decision metric, where greater rewards are given to cells with a higher probability of the target being located there. The neighboring cells thus require priority in being explored. Similar strategies are used for the area coverage models discussed in Section 2.2.1. The adaptive framework in [112] consists of 5 components: module-sensor inputs, information merging, operation adjustment, profit calculation, and path planning. Location-based search functions rely on the last known location of the targets that are to be tracked. However, they alone may be insufficient to pinpoint the target location. Typical application scenarios are to locate fallen swarm agents. Agents failing due to any internal or external disruptions have to be located and rescued. Often finding failed agents may give an insight into the disruptions that caused them to fail in the first place, whether it be due to territorial birds damaging the aircraft, entanglement of flight equipment with foreign materials, or some other cause.

Location data is usually from active location broadcasting equipment that is present onboard the agents. In several cases, the agent may be lost in the vicinity of where it last broadcasted its location from. Swarm agents may collect this information and home in on the location. Based on the accuracy of location equipment, a switch can be made to visually search the area to locate the fallen agent. In case an active location is no longer available due to loss of critical flight power, triangulation methods can be used. Based on swarm formation, the nearest neighbors of the fallen agent can provide a rough area in which the probability of agent crash is highest. Visual search patterns may be used to further refine the search. Agents can be equipped with location tagging solutions that run on independent power such as Apple AirTags [113] or other similar products. These are light and low-cost sensors that can be added to the agents externally. They trigger location alerts using other active equipment present nearby such as cellular towers, mobile phones, and modems to find the agent. UAV target rescue scenarios are usually application-specific and combined with target search scenarios such as in disaster management. Since the goal of this study is to examine approaches for generalized resilient UAV swarms, developments are necessary to facilitate retrieving fallen agents. There is a lack of research on the use of swarm agents to recover fallen agents. Due to the nature of operational space, it is often necessary that other agents with heterogenous capabilities are required. For example, in scenarios in which a swarm contains both UAVs and unmanned water surface vehicles (UWSVs) such as in [114], if the UAV agent fails, the water surface agents take up the responsibility of recovering the fallen agent. On the ground, an unmanned ground vehicle (UGV) can be deployed to try and recover the lost UAV agent. Apart from the physical recovery of agents, there should be re-connectivity protocols in place that attempt to establish connections with lost agents that have been found on the ground.

2.4. Security

Swarm security is divided into two main categories. Physical agent security and protecting the swarm from cyber threats. Both are discussed in the following sections.

2.4.1. Physical Security

The physical security of agents deals with the detection of threats that might physically impede swarm progress. Additionally, defense or escape countermeasures should be designed as part of securing any multi-agent system. The counteraction from swarm agents is a response of the agents once a threat is detected and can include the following.

- Counteraction of UAV swarms against malicious agents trying to take down agents in the swarm
- Counterattack of UAV swarms against malicious agents trying to enter a restricted airspace

In either case, a threat classifier is needed for UAV swarms to detect and recognize potential threats. One defense approach to incoming malicious agents is to engage a swarm of counter-attack drones to intercept intruders. The approach by [115] follows a similar method by deploying a swarm of drones that approaches the intruder UAV. The deployed defense agents form a cluster around the intruder and restrict its movement while attempting to herd it to a non-threatening location. The assumption is that the enemy agent is aware of other agents surrounding it and will take steps to prevent collision with them. However, if an enemy agent is not equipped with such abilities, there is a high possibility that it may collide with one or more of the defender agents and result in damage or loss. UAV agents can also be used to jam network connections of enemy agents and stop their crucial operations. A GPS spoofing attack was proposed in [116] where it attempts to take control of an enemy agent. This is considered accomplished when it can successfully artificially specify the enemy agent's perceived position and velocity. By controlling the agent and providing false data it is possible to disable the enemy. For example, the spoofer used in this study earlier demonstrated similar actions whereby it falsely produced ascending actions on a captured UAV that was hovering. To compensate, the agent started a descent and would have been catastrophic unless precautionary manual control took over [116,117]. However, such solutions which involve the deployment of additional drones for counterattack can be an expensive process considering the physical interactions that might take place among these agents. Replacement and repair of damaged defender aircraft can be costly. Such approaches should be deployed only, when necessary, when the main swarm is deemed incompetent to defend itself. Other solutions exist that can be accommodated onboard existing UAVs without the need for additional agents. Enemy agent ability jamming and evasive maneuvers are often advised in UAV swarm defense [97]. For recognition of enemy agents, detection methods are necessary, especially to differentiate between swarm agents and foreign agents. While this can be done using software in-loop, network recognition, and unique hardware identifiers, vision-based frameworks are also being studied. For example, ref. [118] uses a vision-based object detection method backed by deep learning to detect and track a potential enemy UAV. In addition to detection, a tracking system is implemented to keep the detected agent in a local bounding box and follow it.

2.4.2. Network Security

A large portion of control tasks for UAVs is dependent on a network structure. Any device, node, or link over a network is susceptible to cyber-attacks. With recent developments of malicious agents attempting to dissuade swarms from functioning, damaging the network capabilities of an UAV swarm is the primary method of executing attacks [57]. Moreover, the remote nature of UAVs, combined with limited battery power, fast switching routing, formation topologies, and small onboard computing power has made securing drone networks a challenge. As with any network, FANETs are susceptible to attacks as well, more so because of their high mobility and reduced computational powers. Onboard agents have reduced computational power, and most of the processor load might be dedicated to functions such as flight, navigation, and mission tasks. Any security measures thus need to consume as little energy and computation resources as possible [51].

Although attacks such as eavesdropping can be prevented using encryption for transmission, other attacks may still penetrate UAV networks. The encryption keys must be secure themselves to guarantee performance. Current security and management deployments assume that UAV swarms may accomplish tasks on a single charge. If refueling is needed, the subtraction of old agents and the addition of new swarm agents should be considered. Article [45] creates a swarm broadcast protocol that accounts for rapid changes in swarm numbers, such as the addition of new drones. It also accounts for agents leaving the swarm for activities such as fuel recharge. The addition of new agents may require them to use the encryption keys used to secure network transmissions. However, challenges to this approach include offset delays that may be caused by validating new agents, the transmission of keys, and unstable networks interrupting the key transfer process. A loss of key transmission could cause new agents to be unable to decrypt transmissions [45]. The requirements of the key management scheme proposed are also similar to the IDS requirements discussed in the next section. A lightweight management scheme that consumes low network and CPU resources is desirable. Secure broadcast protocols work whereby every time the swarm changes its agent composition a new secure key is created. This is done to prevent the old key from being used by any attackers. The new agents have separate identifiers that recognize them to be a part of the swarm, this identifier unlocks the broadcast packet that contains the new key. The master maintains a list of all agent identifiers actively connected, and every time an agent sends a leave or join request the identifier table gets updated, and a new broadcast key is sent out. To prevent offset delays during key sharing, agents also receive an updated copy of the identifier table. They can then verify that the nearest neighbor is a verified agent and send the key packet. In this way, the master does not have to send the key to every individual agent. Intrusion detection systems are a reliable and efficient way of securing computer networks. Recent studies demonstrate they can be deployed on UAV networks as well [119].

As with a regular IDS, an UAV-IDS is created to detect any suspicious activity on the network and prevent its execution. Intrusion detection systems can be developed based on behaviors or anomalies. Bio-inspired particle detection IDS can also be used [120]. An updated taxonomy of intrusion detection systems has been conducted by the authors in previous research [121]. The IDS monitors network traffic such as data packets, transmission power, and routes, as well as new incoming requests from nodes.

While traditional attacks for computer networks may not directly apply here, IDS protocols can be easily modified to suit FANET requirements. An integration with Hyperledger Fabric, a distributed ledger platform, is possible to maintain unique identifiers for all acceptable nodes thereby blocking any unknown nodes from gaining access [122]. An IDS functionality can also be influenced by its deployment location. Typically, an IDS located at ground control may have access to higher resources than on-board implementations. The placement can be determined depending on factors such as the level of protection required, the type of IDS being used, as well as resource requirements of the IDS itself. Multi-layered IDS are possible but require additional considerations of network and node capacity during system development. A policy-based IDS can be created that sets distinct patterns of behavior that are allowed whereas all others are flagged and sent to a higher-level operation for additional examination. Several research challenges as mentioned by [106] discuss detection latency, IDS computational costs, and implementation overheads. A solution is required that is multi-positioned (agents and ground control both), multi-layered, and comprehensive in terms of attack detection and mitigation. Add to that the challenges faced by current IDS implementations, such as bottlenecks due to higher bandwidth and a lack of concrete defense policies [123], and attack classifiers contribute further challenges to the problem. Articles [124,125] have a rule-based mechanism for the IDS. Detection rules for some of the most notorious attack types are predefined. The location is at the ground station and categorizes each agent in the swarm by its alleged threat. An intrusion report message is generated by each agent and sent to the ground station, which assists in verifying agent status as well as any foreign agents that might seek to infiltrate the swarm. A verification check is conducted by using anomaly detection backed by learning algorithms trained to identify and label behaviors.

The addition of learning algorithms significantly expands the capability of intrusion detection systems, and blockchain-supported network security has also been recently explored as a way of securing UAV swarms. Blockchain is a peer-to-peer (P2P) distributed ledger that is secure, transparent, and flexible. It can store data in a chain of blocks that are tamper-proof. Smart contracts can be developed on the chain to execute specific predefined actions. Blockchain has widespread use in security, finance, and government applications [126]. Blockchain technology is a viable solution to issues with UAV network security.

Current UAV softwarization techniques are summarized in [122] such as the wide range of attacks on control, application, and data plane. Several proposed blockchain-based architectures are a comprehensive solution to these issues. Figure 7 demonstrates a simplified blockchain based method for securing low level swarms. A blockchain instance runs on a cloud and ground-based station. Every agent in the swarm has a hardcoded unique identifier. A state table with the agents and their identifiers is maintained by the blockchain and synced across the station nodes. Agents validate their existence at every iteration and the blockchain is updated accordingly. A spoofed agent with an identifier that is not present on the blockchain is unable to join the swarm, is denied communication services and is recognized immediately.

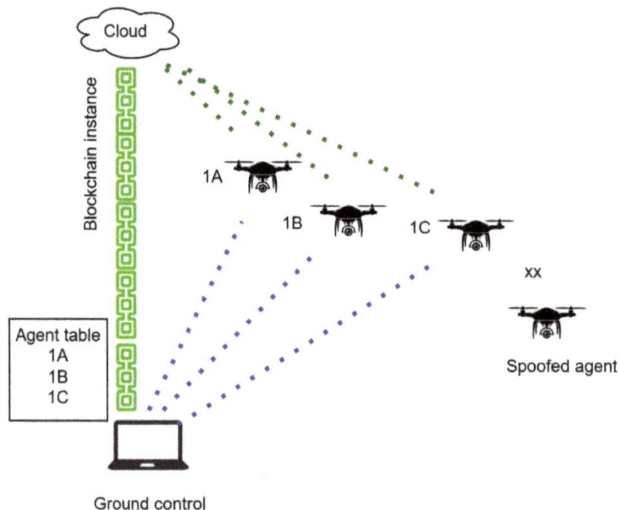

Figure 7. A blockchain based methodology for securing UAV swarms.

Integration benefits include communication data security and transparency in node transactions. However current blockchain technology has scaling and latency issues [127]. A proper framework choice is needed that fits the stochastic requirements of UAV swarms. The first step is the selection of an appropriate consensus algorithm for the blockchain itself. Fast consensus algorithms instead of proof of work are required. Article [53] recommends PBFT (practical byzantine fault tolerance) consensus that has a frequency of 500 Hz. Such higher frequencies can satisfy the intensive demands of routing policies, resource allocation, and mobility management in addition to deploying secure swarms. Article [122] in particular focuses on the security and privacy of communication links for UAVs via blockchain-supported softwarization architecture. By ensuring the authenticity of virtual machines in the virtual infrastructure, it is also capable of protecting against various security threats.

Software-defined networking (SDN) is a technique that offers configurations for network monitoring and performance. By separating control and data planes and providing intelligent control to network hardware, real-time dynamic decision-making is possible. SDN architecture features are described as vendor-neutral and agile [128]. Article [129] examines four areas where blockchain can efficiently secure system-communication, user authentication, device configuration, and legitimacy.

A blockchain-based SDN controller is capable of resisting several attacks that normally plague networked systems. It is possible to integrate fast chains as middleware between the control planes and infrastructure layers. Blockchain ledgers can be used to verify agents in a swarm as well, and a lightweight implementation can hold identifiers for all agents in the swarm. Any additional deployments are checked with values present in the blockchain. It would be difficult for malicious agents to gain access as only agents who are keyed into the blockchain will be allowed to be a part of the swarm. Applications such as those in [130] generate a blockchain receipt for each data record that a drone stores in a traditional cloud server. Labeled as "DroneChain", drone data uploads are captured as a blockchain transaction and hashed. Such developments are easier to deploy as they can be readily integrated with traditionally available services, such as the cloud, merely by adding an additional layer of authentication to data that already had an established pathway. Usage of Drones as a Service (DaaS) is also prevalent. Package delivery services are being explored where package status and tracking data are available on a blockchain [131]. Passive tags on customer endpoints verify packages being dropped in mailboxes. Since these records cannot be tampered with, denying that a package was received is difficult. Package tracking information is available to both the customer and package sender for transparency.

Some interoperability issues with blockchain integrations do exist though. The blockchain needs to work on network hardware that is often vendor specific and may not support it [122]. Data latency issues exist due to a lack of efficient optimization protocols between network routing and coverage protocols interacting with blockchain-provided data sources. Computational complexity may increase due to actions such as validation and block mining. Additional research and testbeds are needed to refine proposed solutions into real-time deployments.

2.5. Resource and Task Handling

Resource allocation and task assignment are terms used by some researchers interchangeably. This is based on the premise that once a task is assigned to an agent or a group of agents, their resources will be locked during the duration of task completion or until a dynamic change is required in the decision-making capacity of the task assignment module. However, not all studies incorporate higher levels into task assignments. The assumption is that each agent in the swarm is a resource that is assigned to complete a particular task. The problem is an interaction between resources and the environment, whereas the allocation scheme is an incentivized function for the agents.

2.5.1. Task Assignment

UAV task assignment is an optimization problem [132]. It is needed to improve computational efficiency and to provide faster solutions. Bio-inspired algorithms such as the wolf pack algorithm (WPA) have been studied in great detail due to their success in reaching optimal solutions at lower time rates [101,103,133]. In a general wolf pack algorithm, each swarm is divided into subgroups that have freedom of evolution and emergent behavior. Sub-swarm interactions are allowed. Common parameters include maximum iterations, population size, and the number of subpopulations. After the initial population solution is created, the fitness function evaluates it before creating the subpopulations. These act independently to create solutions until the termination condition, which is when the best solution is selected from the set of candidate solutions.

Article [133] proposes a modified wolf pack algorithm called the multi-population parallel wolf pack algorithm that performed better than traditional implementations. Their proposed solution allows the creation of virtual sub-populations that are created on the existing layer. This is constructed of the best-derived agents from the overall subpopulation and does not change until the migration stage. Every iteration first optimizes the virtual group and then proceeds to the actual populations. The individuals of this virtual layer are distributed among the actual population and help in accelerated convergence of optimal solutions. Results showed that up to a certain threshold, the simulations with more subpopulations reached convergence faster than those with fewer subpopulations in a general scenario. Improved performance was observed over the genetic algorithm and generic wolf pack.

The wolf pack algorithm formulated in [132] dynamically changes leaders based on certain criteria. This can be scaled to UAV agents to examine the level of computational resources being used by each agent in the swarm at a given moment, and the agent with the most resources available is selected as the leader. In the case that the swarm has a fixed hierarchy, instead of leader shifting, additional tasks can be assigned to the selected agent to create a swarm-wide balance.

Efficient task allocation can allow simple agents to accomplish relatively complex tasks. However, the task allocation problem (TAP) is nonlinear, multi-modal, and highly adversarial. Selecting the right agent to perform a task in the instantaneous decision process is a challenge. Task assignment and resource allocation are interchangeable in terms of usage and considerations for researchers. Task allocation solving algorithms can be centralized or distributed. Bio-inspired algorithms are centrally distributed and include wolf pack, ant colony, and particle swarm optimization. However, they are known to be computationally intensive particularly when the number of agents increases, and they also show poor performance in dynamic environments [134].

Distributed algorithms are top-down and bottom approaches. Top-down approaches decompose problems into smaller optimization problems and solve them using cooperation schemes among multiple contributors. The study uses a fixed response threshold model based on ant colony labor division. It addresses how individual agents determine that a task needs to be performed and how the swarm as a whole exhibits functionality based on different tasks with varying complexities.

Each agent has individual capabilities and a response threshold to tasks. Each task has a stimulus that motivates agents to take the task. Once the task stimuli cross the agent threshold, the agent begins the task. If an agent stops the task, the task priority increases, thus increasing the environmental stimuli.

Article [135] summarize the similarity between ants and a group of UAVs performing tasks. Because of the similarity in their distributed command structure, FRT models are ideal for implementation in UAVs encouraging emergency biological swarm behavior that performs tasks efficiently. Like other bio-inspired algorithms, PSO is also a good candidate to use for task assignments. Article [136] improves on it to overcome the deficits of the original PSO method as it can only be used for continuous space optimization. The improved PSO algorithm divides the swarm into overlapping subgroups. Each particle corresponds to an agent UAV. The second dimension is the number of tasks that are assigned to the swarm. A fitness function is created to evaluate the particles and reach the local solution. These steps are repeated until a termination stage is reached or the max iteration number is reached. However, the number of tasks cannot be assumed as constant. Actual task assignment problems are evolutionary. Factors such as fuel constraints, threats, and mobility have to be considered. These may vary across situations thereby varying the assignment too.

In cases where ground station services are readily available for swarms, the stations can have their strategies for task allocation in addition to decision models on the agents themselves. Both work in coordination to assign and select tasks. Cooperation frameworks combine them to provide efficient task servicing. Implementations such as [137] integrate multi-layered decision models for tasks. These models consider base station capacities to expand drone task maps from onboard to ground control.

Station strategies are examined such as round robin that distributes tasks cyclical to available agents. However, such a simplified strategy may have offsets in agents returning task results as well as bottlenecks in task queuing. Complex decision models consider multiple parameters before an allocation is made. Agent status strategy examines the well-being of an agent in terms of remaining fuel, network strength and proximity to other agents, and ground control before assigning a task.

Article [137] relies on the agent's ability to handle requests. One is distance-based and the other is reward-based. Common factors in both to be considered are agent battery life, the distance between client and drone (in delivery to client specific scenario), and environmental factors such as visibility and pressure. In tasks that deal with delivery to client scenarios such as in [137], the drone can be in either one of the four states: at the station, on route to the client, serving the client, or back at the station.

2.5.2. Resource Allocation

The allocation problem balances shared resources among a group of swarm agents. Each task that a swarm carries out such as path planning, target detection, or surveillance, requires a combination of onboard and ground control resources and computation time. Inefficient task queuing can result in bottlenecks and delays that may damage sensitive operational parameters such as target detection within a particular time frame or instantaneous formation reconfiguration. Multiple functional hardware elements can be combined to make modules that combine capability as well as attempt to solve the resource allocation problem. A novel system proposed in [138] provides a radar module for communication and target detection. It also attempts to solve the resource availability problem in UAVs using a learning-based method to optimize resources. The input parameters are channel, power, and beam resource. A reward scheme is provided to incentivize agents when they select a particular resource, thereby freeing up other resources. Application-specific scenarios such as the one used in [139] examine offloading of computation-intensive tasks to and from edge computing servers. UAVs are used to provide flexible computing services, where such a process can be used to offload intensive UAV tasks such as image processing and target detection in videos to centralized facilities rather than being processed on board.

2.6. Agent Property

The agent property component focuses on individual agents in the swarm. It is possible to increase overall swarm performance by modifying individual agent capabilities [140,141]. The easiest way to do this is to introduce swarm heterogeneity. Heterogeneity is defined as components of a system that are of dissimilar composition or properties. Heterogeneity may be imbibed in a multi-agent system by using a variety of features. The following studies focus on performance effects on a mission by the inclusion of heterogenous agents in a previously homogenous swarm [140–142]. There is a marked increase in performance observed by the introduction of varied agents. This performance might be in terms of time taken to complete tasks or an increase in another measurement metric used to measure swarm resilience. Our previous research provides a classification system for swarms labeled as heterogeneous:

a. By operational space of agents
b. By nature of agents
c. By hardware of agents

In addition to aerial spaces, other varied operational spaces for the unmanned vehicle in a swarm may include ground surfaces, water surfaces, underwater, or even underground. Article [4] deploys a heterogenous team in an enclosed environment. The team comprises robots classified as heterogenous by their operational space—an UGV and an UAV. The authors mention how each vehicle brings different capabilities to the swarm. The UGV is autonomous whereas the UAV is fast and provides greater motion flexibility. The UAV can fly over obstacles that the UGV cannot cross. Challenges to the deployment of such hybrid swarms are also mentioned. Air ground coordination and navigation are the primary challenges. Deployment of such swarms is not limited to enclosed environments only. Indeed, a wider operational space is available by the inclusion of agents on the ground, water surfaces, or even at different altitudes in the air. Current deployments of heterogenous agents have shown a marked increase in resilient behavior [140–142]. Such swarms can be single or multi-space or heterogonous by nature or have hardware capability. In pursuit–evader situations, the pursuit UAV often needs to be agile and lightweight to be capable of faster flight speeds than the target UAV. However, a swarm that features a mix of lightweight pursuit-capable agents along with heavier support agents that carry additional equipment needed to support swarm operations is ideal. A military purpose swarm such as the one in [143] where a mix of differently capable agents is used can be used. Moreover, experiments by [141] observed a marked increase in swarm responsiveness to external stimuli due to the inclusion of fast agents in swarms. Benefits are observed across the whole range of UAV swarm components. A hierarchical structure of mixed agents described by [144] includes high-altitude fixed-wing aircraft providing communication and sensing support to a group of low-altitude quadcopters. They also note a difference in computational resources, energy consumption, and communication systems among the agents. The possibility of evolving capabilities among heterogeneous swarms has also been observed. Implementing a different characteristic trait by modifying operational parameters can change the way homogenous agents behave thereby allowing them to function better. A change in the allowable range that some swarm agents can move from a fixed ground target can allow them to become less cautious of external disruptions, thereby exhibiting different behavior. A coordination scheme by other non-modified agents to balance tasks and resources is observed in such scenarios. The addition of differently capable agents can also provide complementary capabilities [145]. Homogenous agents in a swarm may have a particular weakness that may easily bring the whole swarm down. Such issues can be varied such as bugs in the control system or routing protocols, detection and tracking, or communication range. Such issues can be solved by introducing agents that might fix such gaps in the overall swarm properties. For example, a relay drone that possesses extra hardware to support multiple communication bands can aid in communication gaps and delays. A similar deployment is described in [146], where some robots in the group have higher sensor payload, processing power, and memory capacity. These are labelled as leader robots, whereas child robots have more limited resources. These child robots rely on leader robots for tasks such as localization. Data from the child robots can be used for sensor fusion, global pose, and location estimation, which can then be used to modify the swarm movement. The proposed distributed leader-assisted localization algorithm can provide accurate localizations for child robots even when they are beyond the sensing range of leader agents.

Application-specific improvements such as those in [145] show how a richer dataset is created when a micro aerial vehicle (MAV) collaborates with an UGV in the creation of detailed maps. The MAV recorded the top view dimensions whereas the UGV accounted for the side view spread. Added considerations for implementation need to be accounted for in this type of scenario though. Figure 8 shows how an UAV-UGV coordination can produce higher detailed maps. The UGV maps the ground level dimensions while the UAV maps the top view.

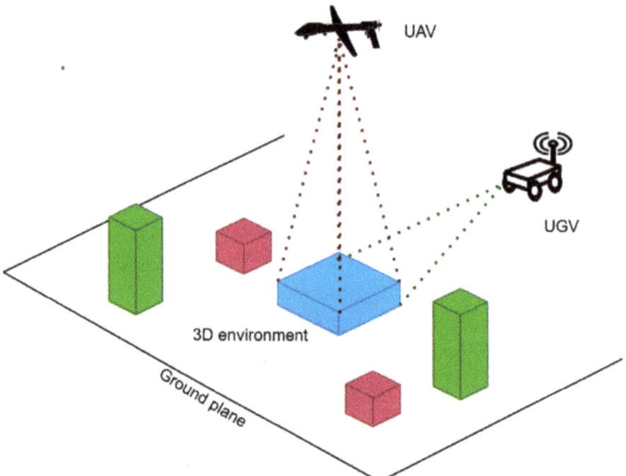

Figure 8. A multi space unmanned vehicle swarm working a mapping mission in tandem.

A study by [147] mentions how most current mission planning and control strategies are for homogenous systems primarily. The added dynamics for different operational spaces that the hetero swarm will be exposed to must be considered. For example, an UAV-UWSV swarm needs to consider air and water environments to accommodate all vehicles. Agent support and coordination protocols need to be updated as well. Marsupial platforms such as [148] use water surface vehicles to recover and recharge UAVs.

Cooperative coevolution algorithms that support the evolution of heterogenous systems suffer from scalability issues. This can be overcome by creating partially heterogeneous systems called hybrid systems where groups of homogenous subswarms are present. Each subswarm has identical controllers. By controlling the agents that need to be evolved, the learning scheme improves scalability [149]. The inclusion of heterogeneous agents is sometimes considered no more than just an upgrade to existing swarms, but it is a necessity to aid in the development and evolution of resilient behavior in multi-agent systems.

Despite the more challenging implementation considerations and a change in development pipelines, the possible benefits are significant.

2.7. Resiliency Evaluation

Cyber-physical systems need testbeds for simulations before real-world deployments. However, due to the extreme software-hardware interdependence, designing them is a challenging task. UAV swarms are among those systems that require thorough testing but are difficult to test in controlled environments. Simulations cannot mimic the varied and extreme variables they might face [150]. Resiliency assessment is a process of observing and evaluating the resilient behavior of UAV swarms to disruptions. An evaluation metric system is needed to accurately recognize performance and its lack thereof. It is important to devise feedback systems and metrics to measure resilience performance in the above modules. Real-time feedback loops may allow decision-making models to follow an iterative process in creating better outputs. Metrics created to evaluate system components such as signal strength, target detection times based on varied inputs, and fuel saved based on optimized path planning can provide insights and recognize tradeoffs.

Evaluation metric design is a less explored branch. It is a strategic process for evaluating multi-pronged decision models by measuring individual decision branches and quantifying their outputs into a single value. Article [151] cites a flaw in the creation of resilience evaluation metrics. Metrics often use the initial swarm performance as a baseline. This may not be accurate, as certain onboard systems may need more time to exhibit full performance capacity. The approach by [151] introduces variables that provide free space for missions. By relaxing the condition that the swarm has to return to performance before it is attacked, the authors argue that a flexible baseline that determines whether a system is performing is needed. The swarm may have to take certain actions during the attack process such as sacrificing a certain agent or reducing the number of agents assigned to cover an area after an attack takes place. Article [151] lays out preliminary dynamic evaluation parameters.

The second problem is that these metrics often use network connectivity as a basis of evaluation. As long as the swarm maintains a connected state in terms of its required signal strength or coverage, the swarm is deemed resilient. However, the presence or absence of network and communication capabilities alone cannot determine the robustness of swarms. Time values for the recovery process should also be measured. For example, in the case of an attack scenario on a swarm described by [151], there are a few issues that the article does not take into account:

- The swarm does not check for agent wellbeing after it determines that the attack has ended.
- In the case that an agent is lost, there are no search and recovery procedures.
- Mission progress may be lost when swarm control completes task re-assignment. In this case, depending upon the scenario tasks such as localization, area decomposition and data collection may need to be restarted after the loss of data is examined.

A basic evaluation metric is the one proposed by [152] which measures performance loss after a disruption with the time it takes for the system to return to normal levels. This creates a baseline through which preliminary system performance can be maintained. Additional application-specific constraints can be established. On the occasion that an agent is lost, the time taken for the mayday signal of the agent to reach central command can be measured along with the time it takes for other agents to locate the lost agent and deploy rescue procedures. A secondary decision process is started where probability models determine if it is safe to try and locate the lost agent or proceed with the task. If the decision to proceed is made, the model then decides if additional agents should be deployed to make up for lost agents or if task reassignment should distribute the workload to other agents. The fundamental metric here is the time value which is measured for every decision to be executed. Multiple time values can be fed into a measurement model to produce one time-based value of system performance.

Article [13] represents resiliency heuristics that are important building blocks of evaluation metrics. Additionally, they note how heuristics complement each other, with some becoming increasingly efficient if others are present. They define these heuristics as a "function of the type of disruption, the type of system being architected, and the type of resilience needed". These heuristics act as a support framework for action statements that can be framed for specific measurements. For UAV swarm systems, they can be collision avoidance, localization and navigation, and fuel discharge rates.

The types defined by them can also help program system responses to a particular disruption. For example, in case an agent fails, additional agents should be on standby to take over the task. Simulation using metrics should be all-encompassing. Researchers develop fast algorithms for changes in swarm formation such as in [60] where computationally efficient algorithms for coverage and flocking are developed. However, their robustness test fails to consider scenarios where additional agents are added to the swarm to make up for lost agents due to disruptions, thereby changing the number of active agents in the swarm. A time-based metric that measures system reaction to the coverage problem is ideal in this case. Accident model metrics such as in [153] can lead to new insights into operational tradeoffs. Section 2.1.1 mentions one such balance between coverage and connectivity. A "slider" metric designed into control algorithms can add autonomy in decisions related to the same processes. Article [13] mentions how such tradeoffs can be recognized. The measurement framework they present is comprehensive and accurately summarizes resiliency metrics into time and cost functions. Additionally, integration into system design must be completed. A time metric that measures the amount of time it takes for an UAV swarm to recover from an attack should have feedback from agents that were affected. If the agent assures its wellbeing, the time measured should be reconfiguration time only. If the agent is unresponsive, the time measured should be the total of reconfiguration plus the time it takes for an additional agent to be deployed from the base. This can be an indication measurement of the adaptability of the system.

Application-specific descriptive metrics are usually designed by researchers to evaluate their respective research. For example, path planning and collision avoidance [18] have metrics such as route length, number of crashes, and frequency of route regeneration. Such metrics scale well in terms of interoperability. However, some specific metrics might not scale due to features being absent in other the research domains being compared. Learning-based systems have different evaluation criteria than game theoretic methods. Similarly, resilient methods for evolutionary algorithms cannot be compared with routing protocols that make static assumptions during the planning stage. If the metric for resilience evolution is based purely on performance over time, let $p(t)$ be the desired performance when no disruptions exist. When a disruption occurs in time t_d, the performance function reduces to $p(t_r)$. Recovery rate is defined as the amount of time taken for the performance function to go back to the original level. Methods to measure the resiliency of interdependent systems based on time and performance scales have been proposed [154]. A comparison metric that compares system data before and after a disruption takes place is also required. A survey conducted by [155] in 2019 shows the lack of research for applying metrics in UAV swarm resiliency. They suggest the development of a metric that reflects differences between performance after a single disruption event compared to the expected performance in the absence of disruptions.

Current metrics are not suitable for decision-making when it comes to completing a mission. The ability to compare current swarm performance with standard performance is missing [155]. This can help decide if the mission should be continued, put on hold, or aborted completely. Article [155] also highlights the necessity of a resilience metric by citing reasons similar to those mentioned in this section. It recognizes shortcomings in current studies and proposes a resilience evaluation method for UAV agents due to external malicious disruptions. Essentially this can be labeled as a network-based approach to measuring UAV swarm security because it does not consider the failure of nodes due to malfunctions other than an attack. However, there are a multitude of reasons that a node can fail other than an attack. There can be failures of agents in areas where they are not under threat from external enemy agents. Such failures include collision with obstacles, other agents and fuel, and communication issues.

The framework described in [156] considers only the number of swarm agents available at a given moment in time due to an attack. Current position, speed, fuel, and availability of secure resources are taken as inputs. The verification engine labels vulnerable agents and herds other agents to safety. Considering a wide range of inputs can substantially improve the accuracy of solvers to label threats as they see them, however, there is an increase in computational time taken to reach this decision. Furthermore, the sacrifice of several agents should be considered. Depending on hardware and location, agents may be difficult to replace and it may be an expensive process without some recovery plans in place.

Similar to the need for dynamic programming in swarm operations, a flexible resilience assessment is beneficial. One such approach is used by [157], where dynamic assessments do not use a fixed set of metrics. Rather, the evaluation parameters change depending on mission progress across the timeline. Article [157] divides the process into four parts, namely, to determine the mission objective, select evaluation attributes, propose an evaluation method, and obtain results. Known parameters of the mission and environment are collected in the first stage. Predefined evaluation attributes for the collected data are selected. These attributes are added to a dynamic resilience method as constraints for the problem. Evaluation results are obtained which can then be used as a frame of reference for simulations and actual flights. The dynamic resilience method dictates that the selected evaluation attributes determine the evaluation methods. Here the goal is to preserve original data when doing so. Different attribute selections will result in a varied method and thereby different results.

Analysis of mobility model performance impacts on mobile ad-hoc networks has been proposed by researchers in as early as 2015 [158]. They mention the need for an accurate evaluation of how mobility models affect routing protocol performance. Correctly deduced, any conclusions drawn on performance based on simulations developed and measured using incorrectly developed metrics will lead to errors. Metrics for the mobility models described are network diameter, the average number of components, average coverage, and average path length. Routing protocol metrics include end-to-end delay and routing overheads. While naming conventions across research might be different, for network and routing, an essential idea of network health needs to be developed. This can be deduced from the number of individual nodes in the swarm. Each node is an UAV agent that is actively participating in swarm activities. Path lengths are measured and compared, and the success rate and time of delivery for a data packet following each path should be noted. For network coverage, signal strengths at proposed network diameters should be measured for expected versus actual values.

3. Open Issues and Future Research Directions

This section addresses current open issues and future research directions related to UAV swarm resiliency. Swarm dynamics are expected to further evolve with advancements in the field of UAVs as well as their associated peripheral equipment. Future developments may resolve some of the issues that are prevalent in current swarm deployments, whereas new issues may emerge.

With the gradual implementation of networks beyond 5G, the inherent ability to communication may upgrade UAV swarm networks in a manner that they are more efficient and secure. 6G networks are envisioned to address issues in node-based networks such as latency, power usage, and transmission quality. Upgrades to infrastructure such as the non-terrestrial components for 6G networks can aid UAV communications as well as make use of UAV swarms to aid in further extending the range and strength of available communication facilities [159]. These upgrades may naturally apply to any devices that use the new protocols as well, with sectors such as UAV swarms benefitting greatly.

Other techniques being explored such as the use of ground-mounted lasers to recharge UAV agent batteries or wireless power transfer methods [36] may solve the low power issue that is present in almost all small- to medium-scale unmanned aerial platforms. Future exploratory studies are needed to know how improvements in such high-priority areas may improve resiliency in swarms. Exploration of UAV swarms for application-specific purposes can also lead to greater insights into UAV swarm resiliency needs and solutions. UAV swarms have huge potential applications in surveillance, disaster management, remote sensing, SAR, agriculture, and ecology monitoring, thus novel applications envisioned for UAV swarms may open up further resiliency requirements for the swarms expected to work in those environments.

Similarities exist between collision and obstacle avoidance in their interchangeability and merging. However, these two issues need to be examined independently by focusing on their definitions. While collision avoidance focuses on maintaining a safe distance between swarm agents, obstacle avoidance deals with mechanisms to avoid in-path static and dynamic obstacles. A similar research gap was noted in swarm coverage in terms of network and area coverages. While this study addresses the coverage of operational space separately into area coverage and network coverage, most other investigations do not. While network coverage addresses the positioning of agents in an operational space such that they maintain a standard connection strength at all times, area coverage focuses on distributing swarm agents to cover the maximum area while making assumptions about communication capacity [48]. The coverage type is often selected depending on the application-specific uses of the swarm, however, resilient measures are needed in both. Certain methods are being developed for UAV swarm network security leading to improved resiliency. SDN and blockchain to secure UAV networks are still in the nascent phase and under research or in the simulation phase. SDN has been widely applied to fixed networks, but it has challenges when being implemented in MANETS and FANETS [53]. Routing protocols continue to improve in dealing with the issue of intermittent connections due to frequent changes in network topology [9]. Dispersed research for resilience engineering for various modules in UAV swarms might be effective for particular scenarios, however, it might be difficult to merge with other modules that are a part of a different study. For example, it is not easy to combine resilient routing protocols with counter-defense mechanisms for physical agent safety against external malicious agents.

Every resiliency integration has potential weaknesses. Blockchain has issues with deployment and latency, whereas multiple decision process models for path planning, navigation, and obstacle avoidance have computational cost and delay problems. There is a development need for comprehensive testbeds and simulators for the testing of multiple resilient framework interactions with each other. Current simulators have hardware or software limitations, lack features, or face interfacing issues [150]. There is also a gap between developed protocols and the availability of synthetic data to test them on. All of these problems need to be addressed before a truly resilient swarm can be created.

4. Discussion and Conclusions

This study presents a systemic breakdown of UAV swarm components that affect overall swarm resilience and discusses current research trends. Although previous studies and experiments have attempted to address major challenges in UAV swarm research, there are still prevailing problems that warrant future research before swarms can be considered truly resilient. A general trend is observed where researchers often focus on one facet of UAV development to create resilient behavior but neglect the rest. Integrating multiple such developments is a novel challenge. Thirty percent of the research examined mentioned that one or more components of swarm resiliency research was beyond the scope of their study. Researchers often fail to consider all disruptions that might affect their swarm implementations, choosing to focus on only some instead. This makes system behavior unpredictable to unanticipated disturbances. Additionally, the lack of standard resiliency evaluation metrics makes it difficult to compare system performance across varied implementations. Multiple studies do not accurately design feedback systems in place. UAV swarm applications are increasing rapidly. Performance efficiency and reduction in mission completion time are two major factors that promote multi-agent systems. Heterogeneous agent swarms acting in multiple operational spaces, (UAV-UGV, UAV-UWSV), etc., further increase the scope of applications. This recognizes the need for a comprehensive resilient swarm solution that is built from the ground up while taking into consideration all of the components discussed above. This study creates an identification and classification taxonomy for components vital to UAV swarm operations. Analysis of the current literature gave insights into the research focuses over the years, as well as current trends in publishing. Research gaps in the current implementations of resiliency in UAV swarms are identified and set the groundwork for future work toward building resilient swarms.

Author Contributions: Conceptualization, A.P. and F.A.M.; investigation, A.P.; original draft preparation, A.P.; review and editing, A.P. and F.A.M. All authors have read and agreed to the published version of the manuscript.

Funding: This research received no external funding.

Data Availability Statement: Data sharing not applicable.

Conflicts of Interest: The authors declare no conflict of interest.

Abbreviations

ABS	Aerial Base Station
BSA	Backtracking Search Algorithm
CNS	Communication, Navigation, Surveillance
DTRP	Dynamic Topology Reconstruction Protocol
DaaS	Drones as a Service
FANET	Flying Ad Hoc Networks
FOA	Fruit fly Optimization Algorithm
GNSS	Global Navigation Satellite System
IDS	Intrusion Detection System
IMU	Inertial Measurement Unit
MANET	Mobile Ad hoc Network
MAS	Multi Agent system
MUSCOP	Mission based UAV Swarm Coordination Protocol
MAV	Micro Aerial Vehicle
MBCAP	Mission Based Collision Avoidance Protocol
PAMTS	Profit Driven Adaptive Moving Target Search
PSO	Particle Swarm Optimization
P2P	Peer to Peer
PBFT	Practical Byzantine Fault Tolerance
QR	Quick Response
RSSI	Received Signal Strength Indication
ROI	Region of Interest
SAI	Surveillance Area Importance
SINR	Signal to Interference plus Noise Ratio
SAR	Search and Rescue
SDN	Software Defined Networking
TAP	Task Allocation Protocol
UAV	Unmanned Aerial Vehicle
UWSV	Unmanned Water Surface Vehicle
UGV	Unmanned Ground Vehicle
VANET	Vehicular Ad Hoc Network
WPA	Wolf Pack Algorithm

References

1. Rubio, F.; Valero, F.; Llopis-Albert, C. A review of mobile robots: Concepts, methods, theoretical framework, and applications. *Int. J. Adv. Robot. Syst.* **2019**, *16*, 172988141983959. [CrossRef]
2. Santa Ana, R. Drones Survey Waning Red Tide. Available online: https://agrilifetoday.tamu.edu/2015/10/22/drones-survey-waning-red-tide-at-south-padre-island/ (accessed on 8 October 2022).
3. Rieucau, G.; Kiszka, J.J.; Castillo, J.C.; Mourier, J.; Boswell, K.M.; Heithaus, M.R. Using unmanned aerial vehicle (UAV) surveys and image analysis in the study of large surface-associated marine species: A case study on reef sharks *Carcharhinus melanopterus* shoaling behaviour. *J. Fish Biol.* **2018**, *93*, 119–127. [CrossRef]
4. Roldán, J.; Garcia-Aunon, P.; Garzón, M.; de León, J.; del Cerro, J.; Barrientos, A. Heterogeneous Multi-Robot System for Mapping Environmental Variables of Greenhouses. *Sensors* **2016**, *16*, 1018. [CrossRef]
5. Tsouros, D.C.; Bibi, S.; Sarigiannidis, P.G. A Review on UAV-Based Applications for Precision Agriculture. *Information* **2019**, *10*, 349. [CrossRef]
6. Anderson, K.; Gaston, K.J. Lightweight unmanned aerial vehicles will revolutionize spatial ecology. *Front. Ecol. Environ.* **2013**, *11*, 138–146. [CrossRef]
7. Vincent, P.; Rubin, I. A Framework and Analysis for Cooperative Search Using UAV Swarms. In Proceedings of the 2004 ACM Symposium on Applied Computing (SAC '04), Nicosia, Cyprus, 14–17 March 2004; Association for Computing Machinery: New York, NY, USA, 2004; pp. 79–86.
8. Aljehani, M.; Inoue, M. Multi-UAV tracking and scanning systems in M2M communication for disaster response. In Proceedings of the 2016 IEEE 5th Global Conference on Consumer Electronics, Kyoto, Japan, 11–14 October 2016; IEEE: New York, NY, USA, 2016; pp. 1–2.
9. Arafat, M.Y.; Moh, S. Routing Protocols for Unmanned Aerial Vehicle Networks: A Survey. *IEEE Access* **2019**, *7*, 99694–99720. [CrossRef]
10. Isufaj, R.; Omeri, M.; Piera, M.A. Multi-UAV Conflict Resolution with Graph Convolutional Reinforcement Learning. *Appl. Sci.* **2022**, *12*, 610. [CrossRef]
11. Woods, D.D. Four concepts for resilience and the implications for the future of resilience engineering. *Spec. Issue Resil. Eng.* **2015**, *141*, 5–9. [CrossRef]
12. Mian, S.; Hill, J.; Mao, Z.-H. Optimal Control Techniques for Heterogeneous UAV Swarms. In Proceedings of the 2020 IEEE/AIAA 39th Digital Avionics Systems Conference (DASC), San Antonio, TX, USA, 11–15 October 2020; IEEE: New York, NY, USA, 2020; pp. 1–10.
13. Madni, A.M.; Jackson, S. Towards a Conceptual Framework for Resilience Engineering. *IEEE Syst. J.* **2009**, *3*, 181–191. [CrossRef]
14. Wears, R.L. Resilience Engineering: Concepts and Precepts. *BMJ Qual. Saf.* **2006**, *15*, 447–448. [CrossRef]
15. Madni, A. Designing for resilience. In *ISTI Lecture Notes on Advanced Topics in Systems Engineering*; ISTI: Tokyo, Japan, 2007; pp. 20–35.
16. Ron, W.; Hollnagel, E.; Woods, D.; Leveson, N. A typology of resilience situations. In *Resilience Engineering: Concepts and Precepts*; CRC Press: Boca Raton, FL, USA, 2006.
17. Wolpert, D.H.; Macready, W.G. No free lunch theorems for optimization. *IEEE Trans. Evol. Comput.* **1997**, *1*, 67–82. [CrossRef]
18. Majd, A.; Ashraf, A.; Troubitsyna, E.; Daneshtalab, M. Using Optimization, Learning, and Drone Reflexes to Maximize Safety of Swarms of Drones. In Proceedings of the 2018 IEEE Congress on Evolutionary Computation (CEC), Rio de Janeiro, Brazil, 8–13 July 2018; IEEE: New York, NY, USA, 2018; pp. 1–8.
19. Macek, K.; Vasquez, D.; Fraichard, T.; Siegwart, R. Safe Vehicle Navigation in Dynamic Urban Scenarios. In Proceedings of the 2008 11th International IEEE Conference on Intelligent Transportation Systems, Beijing, China, 12–15 October 2008; IEEE: New York, NY, USA, 2008; pp. 482–489.
20. Cooley, R.; Wolf, S.; Borowczak, M. Secure and Decentralized Swarm Behavior with Autonomous Agents for Smart Cities. In Proceedings of the 2018 IEEE International Smart Cities Conference (ISC2), Kansas City, MO, USA, 16–19 September 2018; IEEE: New York, NY, USA, 2018; pp. 1–8.
21. Andrade, F.A.A.; Hovenburg, A.; de Lima, L.N.d.; Rodin, C.D.; Johansen, T.A.; Storvold, R.; Correia, C.A.M.; Haddad, D.B. Autonomous Unmanned Aerial Vehicles in Search and Rescue Missions Using Real-Time Cooperative Model Predictive Control. *Sensors* **2019**, *19*, 4067. [CrossRef]
22. Liu, Y.; Qi, N.; Yao, W.; Zhao, J.; Xu, S. Cooperative Path Planning for Aerial Recovery of a UAV Swarm Using Genetic Algorithm and Homotopic Approach. *Appl. Sci.* **2020**, *10*, 4154. [CrossRef]
23. Chen, X.; Tang, J.; Lao, S. Review of Unmanned Aerial Vehicle Swarm Communication Architectures and Routing Protocols. *Appl. Sci.* **2020**, *10*, 3661. [CrossRef]
24. Berger, J.; Lo, N.; Barkaoui, M. Static target search path planning optimization with heterogeneous agents. *Ann. Oper. Res.* **2016**, *244*, 295–312. [CrossRef]
25. Besada, J.A.; Bernardos, A.M.; Bergesio, L.; Vaquero, D.; Campana, I.; Casar, J.R. Drones-as-a-service: A management architecture to provide mission planning, resource brokerage and operation support for fleets of drones. In Proceedings of the 2019 IEEE International Conference on Pervasive Computing and Communications Workshops (PerCom Workshops), Kyoto, Japan, 11–15 March 2019; IEEE: New York, NY, USA, 2019; pp. 931–936.
26. Dimensions. Available online: https://www.dimensions.ai/ (accessed on 19 October 2022).
27. Tahir, A.; Böling, J.; Haghbayan, M.-H.; Toivonen, H.T.; Plosila, J. Swarms of Unmanned Aerial Vehicles—A Survey. *J. Ind. Inf. Integr.* **2019**, *16*, 100106. [CrossRef]

28. Bithas, P.S.; Michailidis, E.T.; Nomikos, N.; Vouyioukas, D.; Kanatas, A.G. A Survey on Machine-Learning Techniques for UAV-Based Communications. *Sensors* **2019**, *19*, 5170. [CrossRef]
29. Schranz, M.; Umlauft, M.; Sende, M.; Elmenreich, W. Swarm Robotic Behaviors and Current Applications. *Front. Robot. AI* **2020**, *7*, 36. [CrossRef]
30. Nawaz, H.; Ali, H.M.; Laghari, A.A. UAV Communication Networks Issues: A Review. *Arch. Comput. Methods Eng.* **2021**, *28*, 1349–1369. [CrossRef]
31. Sun, W.; Tang, M.; Zhang, L.; Huo, Z.; Shu, L. A Survey of Using Swarm Intelligence Algorithms in IoT. *Sensors* **2020**, *20*, 1420. [CrossRef]
32. Zhou, Y.; Rao, B.; Wang, W. UAV Swarm Intelligence: Recent Advances and Future Trends. *IEEE Access* **2020**, *8*, 183856–183878. [CrossRef]
33. Sharma, A.; Shoval, S.; Sharma, A.; Pandey, J.K. Path Planning for Multiple Targets Interception by the Swarm of UAVs based on Swarm Intelligence Algorithms: A Review. *IETE Tech. Rev.* **2022**, *39*, 675–697. [CrossRef]
34. Peng, Q.; Wu, H.; Xue, R. Review of Dynamic Task Allocation Methods for UAV Swarms Oriented to Ground Targets. *Complex Syst. Model. Simul.* **2021**, *1*, 163–175. [CrossRef]
35. Rovira-Sugranes, A.; Razi, A.; Afghah, F.; Chakareski, J. A review of AI-enabled routing protocols for UAV networks: Trends, challenges, and future outlook. *Ad Hoc Netw.* **2022**, *130*, 102790. [CrossRef]
36. Mohsan, S.A.H.; Khan, M.A.; Noor, F.; Ullah, I.; Alsharif, M.H. Towards the Unmanned Aerial Vehicles (UAVs): A Comprehensive Review. *Drones* **2022**, *6*, 147. [CrossRef]
37. Campion, M.; Ranganathan, P.; Faruque, S. UAV swarm communication and control architectures: A review. *J. Unmanned Veh. Syst.* **2019**, *7*, 93–106. [CrossRef]
38. Puente-Castro, A.; Rivero, D.; Pazos, A.; Fernandez-Blanco, E. A review of artificial intelligence applied to path planning in UAV swarms. *Neural Comput. Appl.* **2022**, *34*, 153–170. [CrossRef]
39. Tang, J.; Duan, H.; Lao, S. Swarm intelligence algorithms for multiple unmanned aerial vehicles collaboration: A comprehensive review. *Artif. Intell. Rev.* **2022**. [CrossRef]
40. Suo, W.; Wang, M.; Zhang, D.; Qu, Z.; Yu, L. Formation Control Technology of Fixed-Wing UAV Swarm Based on Distributed Ad Hoc Network. *Appl. Sci.* **2022**, *12*, 535. [CrossRef]
41. Celtek, S.A.; Durdu, A.; Kurnaz, E. Design and Simulation of the Hierarchical Tree Topology Based Wireless Drone Networks. In Proceedings of the 2018 International Conference on Artificial Intelligence and Data Processing (IDAP), Malatya, Turkey, 28–30 September 2018; IEEE: New York, NY, USA, 2018; pp. 1–5.
42. Zhu, Q.; Zhou, R.; Zhang, J. Connectivity Maintenance Based on Multiple Relay UAVs Selection Scheme in Cooperative Surveillance. *Appl. Sci.* **2016**, *7*, 8. [CrossRef]
43. Sahingoz, O.K. Mobile networking with UAVs: Opportunities and challenges. In Proceedings of the 2013 International Conference on Unmanned Aircraft Systems (ICUAS), Atlanta, GA, USA, 28–31 May 2013; IEEE: New York, NY, USA, 2013; pp. 933–941.
44. Zhang, K.; Zhang, W.; Zeng, J.-Z. Preliminary Study of Routing and Date Integrity in Mobile Ad Hoc UAV Network. In Proceedings of the 2008 International Conference on Apperceiving Computing and Intelligence Analysis, Chengdu, China, 13–15 December 2008; IEEE: New York, NY, USA, 2008; pp. 347–350.
45. Secinti, G.; Darian, P.B.; Canberk, B.; Chowdhury, K.R. Resilient end-to-end connectivity for software defined unmanned aerial vehicular networks. In Proceedings of the 2017 IEEE 28th Annual International Symposium on Personal, Indoor, and Mobile Radio Communications (PIMRC), Montreal, QC, Canada, 8–13 October 2017; IEEE: New York, NY, USA, 2017; pp. 1–5.
46. Peng, J.; Gao, H.; Liu, L.; Li, N.; Xu, X. TBM: An Efficient Trajectory-Based Multicast Routing Protocol for Sparse UAV networks. In Proceedings of the 2020 IEEE 22nd International Conference on High Performance Computing and Communications; IEEE 18th International Conference on Smart City; IEEE 6th International Conference on Data Science and Systems (HPCC/SmartCity/DSS), Yanuca Island, Cuvu, Fiji, 14–16 December 2020; IEEE: New York, NY, USA, 2020; pp. 867–872.
47. Rosalie, M.; Brust, M.R.; Danoy, G.; Chaumette, S.; Bouvry, P. Coverage Optimization with Connectivity Preservation for UAV Swarms Applying Chaotic Dynamics. In Proceedings of the 2017 IEEE International Conference on Autonomic Computing (ICAC), Columbus, OH, USA, 17–21 July 2017; IEEE: New York, NY, USA, 2017; pp. 113–118.
48. Yanmaz, E. Connectivity versus area coverage in unmanned aerial vehicle networks. In Proceedings of the 2012 IEEE International Conference on Communications (ICC), Ottawa, ON, Canada, 10–15 June 2012; IEEE: New York, NY, USA, 2012; pp. 719–723.
49. Kaur, M.; Prashar, D.; Rashid, M.; Alshamrani, S.S.; AlGhamdi, A.S. A Novel Approach for Securing Nodes Using Two-Ray Model and Shadow Effects in Flying Ad-Hoc Network. *Electronics* **2021**, *10*, 3164. [CrossRef]
50. Park, S.; Kim, H.T.; Kim, H. Energy-Efficient Topology Control for UAV Networks. *Energies* **2019**, *12*, 4523. [CrossRef]
51. Aznar, F.; Pujol, M.; Rizo, R.; Pujol, F.; Rizo, C. Energy-Efficient Swarm Behavior for Indoor UAV Ad-Hoc Network Deployment. *Symmetry* **2018**, *10*, 632. [CrossRef]
52. Stirling, T.; Floreano, D. Energy-Time Efficiency in Aerial Swarm Deployment. In *Distributed Autonomous Robotic Systems: The 10th International Symposium, Proceedings of the 10th International Symposium on Distributed Autonomous Robotic Systems (DARS 2010), Lausanne, Switzerland, 1–3 November 2010*; Martinoli, A., Mondada, F., Correll, N., Mermoud, G., Egerstedt, M., Hsieh, M.A., Parker, L.E., Støy, K., Eds.; Springer: Berlin/Heidelberg, Germany, 2013; pp. 5–18.
53. Hu, N.; Tian, Z.; Sun, Y.; Yin, L.; Zhao, B.; Du, X.; Guizani, N. Building Agile and Resilient UAV Networks Based on SDN and Blockchain. *IEEE Netw.* **2021**, *35*, 57–63. [CrossRef]
54. Horvath, D.; Gazda, J.; Slapak, E.; Maksymyuk, T. Modeling and Analysis of Self-Organizing UAV-Assisted Mobile Networks with Dynamic On-Demand Deployment. *Entropy* **2019**, *21*, 1077. [CrossRef]

55. Phadke, A.; Medrano, F.A.; Ustymenko, S. A Review of Vehicular Micro-Clouds. In Proceedings of the 2021 International Conference on Computational Science and Computational Intelligence (CSCI), Las Vegas, NV, USA, 15–17 December 2021; IEEE: New York, NY, USA, 2021; pp. 411–417.

56. Hydher, H.; Jayakody, D.N.K.; Hemachandra, K.T.; Samarasinghe, T. Intelligent UAV Deployment for a Disaster-Resilient Wireless Network. *Sensors* **2020**, *20*, 6140. [CrossRef]

57. Pu, C. Jamming-Resilient Multipath Routing Protocol for Flying Ad Hoc Networks. *IEEE Access* **2018**, *6*, 68472–68486. [CrossRef]

58. Park, M.; Lee, S.; Lee, S. Dynamic Topology Reconstruction Protocol for UAV Swarm Networking. *Symmetry* **2020**, *12*, 1111. [CrossRef]

59. Chen, R.; Xu, N.; Li, J. A Self-Organized Reciprocal Decision Approach for Sensing Coverage with Multi-UAV Swarms. *Sensors* **2018**, *18*, 1864. [CrossRef]

60. Elmokadem, T.; Savkin, A.V. Computationally-Efficient Distributed Algorithms of Navigation of Teams of Autonomous UAVs for 3D Coverage and Flocking. *Drones* **2021**, *5*, 124. [CrossRef]

61. Reynolds, C.W. Flocks, Herds and Schools: A Distributed Behavioral Model. *SIGGRAPH Comput. Graph.* **1987**, *21*, 25–34. [CrossRef]

62. Acar, E.U.; Choset, H.; Rizzi, A.A.; Atkar, P.N.; Hull, D. Morse Decompositions for Coverage Tasks. *Int. J. Robot. Res.* **2002**, *21*, 331–344. [CrossRef]

63. Choset, H.; Pignon, P. Coverage Path Planning: The Boustrophedon Cellular Decomposition. In *Field and Service Robotics*; Zelinsky, A., Ed.; Springer: London, UK, 1998; pp. 203–209.

64. Huang, W.H. Optimal line-sweep-based decompositions for coverage algorithms. In Proceedings of the 2001 ICRA, IEEE International Conference on Robotics and Automation (Cat. No.01CH37164), Seoul, Korea, 21–26 May 2001; IEEE: New York, NY, USA, 2001; Volume 21, pp. 27–32.

65. Gonzalez, E.; Alvarez, O.; Diaz, Y.; Parra, C.; Bustacara, C. BSA: A Complete Coverage Algorithm. In Proceedings of the 2005 IEEE International Conference on Robotics and Automation, Barcelona, Spain, 18–22 April 2005; IEEE: New York, NY, USA, 2005; pp. 2040–2044.

66. Sun, Y.; Tan, Q.; Yan, C.; Chang, Y.; Xiang, X.; Zhou, H. Multi-UAV Coverage through Two-Step Auction in Dynamic Environments. *Drones* **2022**, *6*, 153. [CrossRef]

67. Ahmed, N.; Pawase, C.J.; Chang, K. Distributed 3-D Path Planning for Multi-UAVs with Full Area Surveillance Based on Particle Swarm Optimization. *Appl. Sci.* **2021**, *11*, 3417. [CrossRef]

68. Shi, K.; Zhang, X.; Xia, S. Multiple Swarm Fruit Fly Optimization Algorithm Based Path Planning Method for Multi-UAVs. *Appl. Sci.* **2020**, *10*, 2822. [CrossRef]

69. Liu, H.; Ge, J.; Wang, Y.; Li, J.; Ding, K.; Zhang, Z.; Guo, Z.; Li, W.; Lan, J. Multi-UAV Optimal Mission Assignment and Path Planning for Disaster Rescue Using Adaptive Genetic Algorithm and Improved Artificial Bee Colony Method. *Actuators* **2021**, *11*, 4. [CrossRef]

70. Liu, X.; Yan, C.; Zhou, H.; Chang, Y.; Xiang, X.; Tang, D. Towards Flocking Navigation and Obstacle Avoidance for Multi-UAV Systems through Hierarchical Weighting Vicsek Model. *Aerospace* **2021**, *8*, 286. [CrossRef]

71. Yu, T.; Tang, J.; Bai, L.; Lao, S. Collision Avoidance for Cooperative UAVs with Rolling Optimization Algorithm Based on Predictive State Space. *Appl. Sci.* **2017**, *7*, 329. [CrossRef]

72. Fabra, F.; Calafate, C.T.; Cano, J.C.; Manzoni, P. A collision avoidance solution for UAVs following planned missions. In Proceedings of the 2018 IEEE Wireless Communications and Networking Conference Workshops (WCNCW), Barcelona, Spain, 15–18 April 2018; IEEE: New York, NY, USA, 2018; pp. 55–60.

73. Lao, M.; Tang, J. Sense selection strategy of collision avoidance for cooperative UAVs sharing airspace. In Proceedings of the 2016 IEEE Information Technology, Networking, Electronic and Automation Control Conference (ITNEC), Chongqing, China, 20–22 May 2016; IEEE: New York, NY, USA, 2016; pp. 11–15.

74. Chou, F.-Y.; Yang, C.-Y.; Yang, J.-S. Support vector machine based artificial potential field for autonomous guided vehicle. In Proceedings of the SPIE, Fourth International Symposium on Precision Mechanical Measurements, Hefei, Anhui, China, 25–29 August 2008; SPIE-International Society for Optical Engineering: Bellingham, WA, USA, 2008; p. 71304J.

75. Masoud, A.A. Managing the Dynamics of a Harmonic Potential Field-Guided Robot in a Cluttered Environment. *IEEE Trans. Ind. Electron.* **2009**, *56*, 488–496. [CrossRef]

76. Huang, H.; Zhou, H.; Zheng, M.; Xu, C.; Zhang, X.; Xiong, W. Cooperative Collision Avoidance Method for Multi-UAV Based on Kalman Filter and Model Predictive Control. In Proceedings of the 2019 International Conference on Unmanned Systems and Artificial Intelligence (ICUSAI), Xi'an, China, 22–24 November 2019; IEEE: New York, NY, USA, 2019; pp. 1–7.

77. Nordlund, P.-J.; Gustafsson, F. *Probabilistic Conflict Detection for Piecewise Straight Paths*; Linköping University Electronic Press: Linköping, Sweden, 2008.

78. Li, D.; Cui, D. Air traffic control conflict detection algorithm based on Brownian motion. *J. Tsinghua Univ.* **2008**, *48*, 477.

79. Gan, X.; Wu, Y.; Liu, P.; Wang, Q. Dynamic Collision Avoidance Zone Modeling Method Based on UAV Emergency Collision Avoidance Trajectory. In Proceedings of the 2020 IEEE International Conference on Artificial Intelligence and Information Systems (ICAIIS), Dalian, China, 20–22 March 2020; IEEE: New York, NY, USA, 2020; pp. 693–696.

80. De Sá, A.O.; Nedjah, N.; Mourelle, L.d.M. Distributed and resilient localization algorithm for Swarm Robotic Systems. *Appl. Soft Comput.* **2017**, *57*, 738–750. [CrossRef]

81. De Haag, M.U.; Huschbeck, S.; Huff, J. sUAS Swarm Navigation using Inertial, Range Radios and Partial GNSS. In Proceedings of the 2019 IEEE/AIAA 38th Digital Avionics Systems Conference (DASC), San Diego, CA, USA, 8–12 September 2019; IEEE: New York, NY, USA, 2019; pp. 1–10.

82. Nallanthighal, R.S.; Chinta, V. Improved Grid-Scan Localization Algorithm for Wireless Sensor Networks. *J. Eng.* **2014**, *2014*, 628161. [CrossRef]

83. Bulusu, N.; Heidemann, J.; Estrin, D. GPS-less low-cost outdoor localization for very small devices. *IEEE Pers. Commun.* **2000**, *7*, 28–34. [CrossRef]

84. Li, N.; Becerik-Gerber, B.; Krishnamachari, B.; Soibelman, L. A BIM centered indoor localization algorithm to support building fire emergency response operations. *Autom. Constr.* **2014**, *42*, 78–89. [CrossRef]

85. Yun, S.; Lee, J.; Chung, W.; Kim, E.; Kim, S. A soft computing approach to localization in wireless sensor networks. *Expert Syst. Appl.* **2009**, *36*, 7552–7561. [CrossRef]

86. Kim, S.-Y.; Kwon, O.-H. Location estimation based on edge weights in wireless sensor networks. *J. Korean Inst. Commun. Inf. Sci.* **2005**, *30*, 938–948.

87. Son, J.; Kim, S.; Sohn, K. A multi-vision sensor-based fast localization system with image matching for challenging outdoor environments. *Expert Syst. Appl.* **2015**, *42*, 8830–8839. [CrossRef]

88. DJI. DJI Tello Drone Shop. Available online: https://m.dji.com/product/tello-edu?from=shop (accessed on 18 October 2022).

89. Rampinelli, M.; Covre, V.B.; De Queiroz, F.M.; Vassallo, R.F.; Bastos-Filho, T.F.; Mazo, M. An Intelligent Space for Mobile Robot Localization Using a Multi-Camera System. *Sensors* **2014**, *14*, 15039–15064. [CrossRef]

90. Causa, F.; Fasano, G.; Grassi, M. GNSS-aware Path Planning for UAV swarm in complex environments. In Proceedings of the 2019 IEEE 5th International Workshop on Metrology for AeroSpace (MetroAeroSpace), Turin, Italy, 19–21 June 2019; IEEE: New York, NY, USA, 2019; pp. 661–666.

91. Rabaey, C.S.J.; Langendoen, K. Robust positioning algorithms for distributed ad-hoc wireless sensor networks. In Proceedings of the USENIX Technical Annual Conference, Monterey, CA, USA, 10–15 June 2002; pp. 317–327.

92. De Sá, A.O.; Nedjah, N.; de Macedo Mourelle, L. Distributed efficient localization in swarm robotic systems using swarm intelligence algorithms. *Neurocomputing* **2016**, *172*, 322–336. [CrossRef]

93. Rashid, A.T.; Frasca, M.; Ali, A.A.; Rizzo, A.; Fortuna, L. Multi-robot localization and orientation estimation using robotic cluster matching algorithm. *Robot. Auton. Syst.* **2015**, *63*, 108–121. [CrossRef]

94. Choi, S.-C.; Hussen, H.R.; Park, J.-H.; Kim, J. Geolocation-Based Routing Protocol for Flying Ad Hoc Networks (FANETs). In Proceedings of the 2018 Tenth International Conference on Ubiquitous and Future Networks (ICUFN), Prague, Czech Republic, 3–6 July 2018; IEEE: New York, NY, USA, 2018; pp. 50–52.

95. Vicsek, T.; Zafeiris, A. Collective motion. *Collect. Motion* **2012**, *517*, 71–140. [CrossRef]

96. Zhu, C.; Liang, X.; He, L.; Liu, L. Demonstration and verification system for UAV formation control. In Proceedings of the 2017 3rd IEEE International Conference on Control Science and Systems Engineering (ICCSSE), Beijing, China, 17–19 August 2017; IEEE: New York, NY, USA, 2017; pp. 56–60.

97. Brust, M.R.; Danoy, G.; Bouvry, P.; Gashi, D.; Pathak, H.; Goncalves, M.P. Defending Against Intrusion of Malicious UAVs with Networked UAV Defense Swarms. In Proceedings of the 2017 IEEE 42nd Conference on Local Computer Networks: Workshops (LCN Workshops), Singapore, 9 October 2017; IEEE: New York, NY, USA, 2017; pp. 103–111.

98. Azam, M.A.; Mittelmann, H.D.; Ragi, S. UAV Formation Shape Control via Decentralized Markov Decision Processes. *Algorithms* **2021**, *14*, 91. [CrossRef]

99. Xie, Y.; Han, L.; Dong, X.; Li, Q.; Ren, Z. Bio-inspired adaptive formation tracking control for swarm systems with application to UAV swarm systems. *Neurocomputing* **2021**, *453*, 272–285. [CrossRef]

100. Madden, J.D.; Arkin, R.C.; MacNulty, D.R. Multi-robot system based on model of wolf hunting behavior to emulate wolf and elk interactions. In Proceedings of the 2010 IEEE International Conference on Robotics and Biomimetics, Tianjin, China, 14–18 December 2010; IEEE: New York, NY, USA, 2010; pp. 1043–1050.

101. Zhang, S.; Zhou, Y.; Li, Z.; Pan, W. Grey wolf optimizer for unmanned combat aerial vehicle path planning. *Adv. Eng. Softw.* **2016**, *99*, 121–136. [CrossRef]

102. Dewangan, R.K.; Shukla, A.; Godfrey, W.W. Three dimensional path planning using Grey wolf optimizer for UAVs. *Appl. Intell.* **2019**, *49*, 2201–2217. [CrossRef]

103. Duan, H.; Yang, Q.; Deng, Y.; Li, P.; Qiu, H.; Zhang, T.; Zhang, D.; Huo, M.; Shen, Y. Unmanned aerial systems coordinate target allocation based on wolf behaviors. *Sci. China Inf. Sci.* **2018**, *62*, 14201. [CrossRef]

104. Yao, P.; Wang, H.; Ji, H. Multi-UAVs tracking target in urban environment by model predictive control and Improved Grey Wolf Optimizer. *Aerosp. Sci. Technol.* **2016**, *55*, 131–143. [CrossRef]

105. Fabra, F.; Zamora, W.; Reyes, P.; Calafate, C.T.; Cano, J.-C.; Manzoni, P.; Hernandez-Orallo, E. An UAV Swarm Coordination Protocol Supporting Planned Missions. In Proceedings of the 2019 28th International Conference on Computer Communication and Networks (ICCCN), Valencia, Spain, 29 July–1 August 2019; IEEE: New York, NY, USA, 2019; pp. 1–9.

106. Opromolla, R.; Inchingolo, G.; Fasano, G. Airborne Visual Detection and Tracking of Cooperative UAVs Exploiting Deep Learning. *Sensors* **2019**, *19*, 4332. [CrossRef]

107. Bertuccelli, L.F.; How, J.P. Robust UAV search for environments with imprecise probability maps. In Proceedings of the 44th IEEE Conference on Decision and Control, Seville, Spain, 15 December 2005; IEEE: New York, NY, USA, 2005; pp. 5680–5685.

108. Yang, Y.; Minai, A.A.; Polycarpou, M.M. Decentralized cooperative search by networked UAVs in an uncertain environment. In Proceedings of the 2004 American Control Conference, Boston, MA, USA, 30 June–2 July 2004; IEEE: New York, NY, USA, 2004; pp. 5558–5563.

109. Hu, J.; Xie, L.; Xu, J.; Xu, Z. Multi-Agent Cooperative Target Search. *Sensors* **2014**, *14*, 9408–9428. [CrossRef]

110. Duan, H.; Qiao, P. Pigeon-inspired optimization: A new swarm intelligence optimizer for air robot path planning. *Int. J. Intell. Comput. Cybern.* **2014**, *7*, 24–37. [CrossRef]

111. Li, L.; Xu, S.; Nie, H.; Mao, Y.; Yu, S. Collaborative Target Search Algorithm for UAV Based on Chaotic Disturbance Pigeon-Inspired Optimization. *Appl. Sci.* **2021**, *11*, 7358. [CrossRef]

112. Li, X.; Chen, J.; Deng, F.; Li, H. Profit-Driven Adaptive Moving Targets Search with UAV Swarms. *Sensors* **2019**, *19*, 1545. [CrossRef] [PubMed]

113. Apple. Lose Your Knack for Losing Things. Available online: https://www.apple.com/airtag/?afid=p238%7Csk3e2cSut-dc_mtid_1870765e38482_pcrid_569595868773_pgrid_125218674714_pntwk_g_pchan__pexid__&cid=aos-us-kwgo-btb--slid---product- (accessed on 18 October 2022).

114. Mendonca, R.; Marques, M.M.; Marques, F.; Lourenco, A.; Pinto, E.; Santana, P.; Coito, F.; Lobo, V.; Barata, J. A cooperative multi-robot team for the surveillance of shipwreck survivors at sea. In Proceedings of the OCEANS 2016 MTS/IEEE Monterey, Monterey, CA, USA, 19–23 September 2016; IEEE: New York, NY, USA, 2016; pp. 1–6.

115. Brust, M.R.; Danoy, G.; Stolfi, D.H.; Bouvry, P. Swarm-based counter UAV defense system. *Discov. Internet Things* **2021**, *1*, 2. [CrossRef]

116. Kerns, A.J.; Shepard, D.P.; Bhatti, J.A.; Humphreys, T.E. Unmanned Aircraft Capture and Control Via GPS Spoofing: Unmanned Aircraft Capture and Control. *J. Field Robot.* **2014**, *31*, 617–636. [CrossRef]

117. Shepard, D.P.; Bhatti, J.A.; Humphreys, T.E.; Fansler, A.A. Evaluation of smart grid and civilian UAV vulnerability to GPS spoofing attacks. In Proceedings of the 25th International Technical Meeting of The Satellite Division of the Institute of Navigation (ION GNSS 2012), Nashville, TN, USA, 17–21 September 2012; pp. 3591–3605.

118. Akhloufi, M.A.; Arola, S.; Bonnet, A. Drones Chasing Drones: Reinforcement Learning and Deep Search Area Proposal. *Drones* **2019**, *3*, 58. [CrossRef]

119. Choudhary, G.; Sharma, V.; You, I.; Yim, K.; Chen, I.-R.; Cho, J.-H. Intrusion Detection Systems for Networked Unmanned Aerial Vehicles: A Survey. In Proceedings of the 2018 14th International Wireless Communications & Mobile Computing Conference (IWCMC), Limassol, Cyprus, 25–29 June 2018; pp. 560–565.

120. Sharma, M.; Saini, S.; Bahl, S.; Goyal, R.; Deswal, S. Modified Bio-Inspired Algorithms for Intrusion Detection System. In Proceedings of the International Conference on Innovative Computing and Communications, Delhi, India, 20–21 February 2021; Gupta, D., Khanna, A., Bhattacharyya, S., Hassanien, A.E., Anand, S., Jaiswal, A., Eds.; Springer: Singapore, 2021; pp. 185–201.

121. Phadke, A.; Ustymenko, S. Updating the Taxonomy of Intrusion Detection Systems. In Proceedings of the 2021 IEEE 45th Annual Computers, Software, and Applications Conference (COMPSAC), Madrid, Spain, 12–16 July 2021; pp. 1085–1091.

122. Kumari, A.; Gupta, R.; Tanwar, S.; Kumar, N. A taxonomy of blockchain-enabled softwarization for secure UAV network. *Comput. Commun.* **2020**, *161*, 304–323. [CrossRef]

123. Tan, X.; Su, S.; Zuo, Z.; Guo, X.; Sun, X. Intrusion Detection of UAVs Based on the Deep Belief Network Optimized by PSO. *Sensors* **2019**, *19*, 5529. [CrossRef]

124. Li, L.; Zhang, H.; Peng, H.; Yang, Y. Nearest neighbors based density peaks approach to intrusion detection. *Chaos Solitons Fractals* **2018**, *110*, 33–40. [CrossRef]

125. Sedjelmaci, H.; Senouci, S.M.; Ansari, N. A Hierarchical Detection and Response System to Enhance Security Against Lethal Cyber-Attacks in UAV Networks. *IEEE Trans. Syst. Man Cybern. Syst.* **2018**, *48*, 1594–1606. [CrossRef]

126. Phadke, A.; Medrano, F.A.; Ustymenko, S. Applications of Blockchain in E-government. In Proceedings of the 2022 International Symposium on Electrical, Electronics and Information Engineering (ISEEIE), Chiang Mai, Thailand, 25–27 February 2022; pp. 157–164. [CrossRef]

127. Jensen, I.J.; Selvaraj, D.F.; Ranganathan, P. Blockchain Technology for Networked Swarms of Unmanned Aerial Vehicles (UAVs). In Proceedings of the 2019 IEEE 20th International Symposium on "A World of Wireless, Mobile and Multimedia Networks" (WoWMoM), Washington, DC, USA, 10–12 June 2019; pp. 1–7.

128. Bhatia, J.; Dave, R.; Bhayani, H.; Tanwar, S.; Nayyar, A. SDN-based real-time urban traffic analysis in VANET environment. *Comput. Commun.* **2020**, *149*, 162–175. [CrossRef]

129. Singh, M.; Singh, A.; Kim, S. Blockchain: A game changer for securing IoT data. In Proceedings of the 2018 IEEE 4th World Forum on Internet of Things (WF-IoT), Singapore, 5–8 February 2018; pp. 51–55. [CrossRef]

130. Liang, X.; Zhao, J.; Shetty, S.; Li, D. Towards data assurance and resilience in IoT using blockchain. In Proceedings of the MILCOM 2017—2017 IEEE Military Communications Conference (MILCOM), Baltimore, MD, USA, 23–25 October 2017; pp. 261–266. [CrossRef]

131. Alladi, T.; Chamola, V.; Sahu, N.; Guizani, M. Applications of blockchain in unmanned aerial vehicles: A review. *Veh. Commun.* **2020**, *23*, 100249. [CrossRef]

132. Lu, Y.; Ma, Y.; Wang, J.; Han, L. Task Assignment of UAV Swarm Based on Wolf Pack Algorithm. *Appl. Sci.* **2020**, *10*, 8335. [CrossRef]

133. Lu, Y.; Ma, Y.; Wang, J. Multi-Population Parallel Wolf Pack Algorithm for Task Assignment of UAV Swarm. *Appl. Sci.* **2021**, *11*, 11996. [CrossRef]

134. Novoa-Hernández, P.; Corona, C.C.; Pelta, D.A. A software tool for assisting experimentation in dynamic environments. *Appl. Comp. Intell. Soft Comput.* **2015**, *2015*, 302172. [CrossRef]

135. Wu, H.; Li, H.; Xiao, R.; Liu, J. Modeling and simulation of dynamic ant colony's labor division for task allocation of UAV swarm. *Phys. A Stat. Mech. Its Appl.* **2018**, *491*, 127–141. [CrossRef]

136. Jiang, X.; Zhou, Q.; Ye, Y. Method of Task Assignment for UAV Based on Particle Swarm Optimization in logistics. In Proceedings of the 2017 International Conference on Intelligent Systems, Metaheuristics & Swarm Intelligence, Hong Kong, China, 25–27 March; 2017; pp. 113–117.

137. Alwateer, M.; Loke, S.W. A Two-Layered Task Servicing Model for Drone Services: Overview and Preliminary Results. In Proceedings of the 2019 IEEE International Conference on Pervasive Computing and Communications Workshops (PerCom Workshops), Kyoto, Japan, 11–15 March 2019; pp. 387–390.

138. Wang, M.; Chen, P.; Cao, Z.; Chen, Y. Reinforcement Learning-Based UAVs Resource Allocation for Integrated Sensing and Communication (ISAC) System. *Electronics* **2022**, *11*, 441. [CrossRef]

139. Chen, S.; Shi, L.; Ding, X.; Lv, Z.; Li, Z. Energy Efficient Resource Allocation and Trajectory Optimization in UAV-Assisted Mobile Edge Computing System. In Proceedings of the 2021 7th International Conference on Big Data Computing and Communications (BigCom), Deqing, China, 13–15 August 2021; pp. 7–13. [CrossRef]

140. Scheutz, M.; Schermerhorn, P.; Bauer, P. The utility of heterogeneous swarms of simple UAVs with limited sensory capacity in detection and tracking tasks. In Proceedings of the 2005 IEEE Swarm Intelligence Symposium, 2005. SIS 2005, Pasadena, CA, USA, 8–10 June 2005; pp. 257–264. [CrossRef]

141. Kwa, H.L.; Tokić, G.; Bouffanais, R.; Yue, D.K.P. Heterogeneous Swarms for Maritime Dynamic Target Search and Tracking. In Proceedings of the Global Oceans 2020: Singapore—U.S. Gulf Coast, IEEE/MTS OCEANS 2020, Singapore, 5–30 October 2020; pp. 1–8.

142. Gade, S.; Joshi, A. Heterogeneous UAV swarm system for target search in adversarial environment. In Proceedings of the 2013 International Conference on Control Communication and Computing (ICCC), Thiruvananthapuram, India, 13–15 December 2013; pp. 358–363.

143. Ramana Makkapati, V.; Tsiotras, P. Apollonius Allocation Algorithm for Heterogeneous Pursuers to Capture Multiple Evaders. *arXiv* **2020**, arXiv:2006.10253.

144. Xu, C.; Zhang, K.; Jiang, Y.; Niu, S.; Yang, T.; Song, H. Communication Aware UAV Swarm Surveillance Based on Hierarchical Architecture. *Drones* **2021**, *5*, 33. [CrossRef]

145. Dewan, A.; Mahendran, A.; Soni, N.; Krishna, M. Optimization Based coordinated uGV-MAV exploration for 2D augmented mapping. In Proceedings of the 2013 International Conference on Autonomous Agents and Multi-Agent Systems, St. Paul, MN USA, 6–10 May 2013; pp. 1125–1126.

146. Wanasinghe, T.R.; Mann, G.K.I.; Gosine, R.G. Distributed Leader-Assistive Localization Method for a Heterogeneous Multirobotic System. *IEEE Trans. Autom. Sci. Eng.* **2015**, *12*, 795–809. [CrossRef]

147. Wang, J.; Jia, G.; Lin, J.; Hou, Z. Cooperative Mission Planning for Heterogeneous UAVs with the Improved Multi-objective Quantum-behaved Particle Swarm Optimization Algorithm. In Proceedings of the 2019 Chinese Control and Decision Conference (CCDC), Nanchang, China, 3–5 June 2019; pp. 3740–3745.

148. Zhang, H.; He, Y.; Li, D.; Gu, F.; Li, Q.; Zhang, M.; Di, C.; Chu, L.; Chen, B.; Hu, Y. Marine UAV–USV Marsupial Platform: System and Recovery Technic Verification. *Appl. Sci.* **2020**, *10*, 1583. [CrossRef]

149. Gomes, J.; Mariano, P.; Christensen, A.L. Cooperative Coevolution of Partially Heterogeneous Multiagent Systems. 9. Available online: https://dl.acm.org/doi/abs/10.5555/2772879.2772919 (accessed on 8 October 2022).

150. Kumar, P.S.; Emfinger, W.; Karsai, G. A testbed to simulate and analyze resilient cyber-physical systems. In Proceedings of the 2015 International Symposium on Rapid System Prototyping (RSP), Amsterdam, The Netherlands, 8–9 October 2015; pp. 97–103.

151. Sun, Q.; Li, H.; Zhang, Y.; Xie, Y.; Liu, C. A Baseline Assessment Method of UAV Swarm Resilience Based on Complex Networks. In Proceedings of the 2021 IEEE 19th World Symposium on Applied Machine Intelligence and Informatics (SAMI), Herl'any, Slovakia, 21–23 January 2021; pp. 83–86.

152. Tierney, K.; Bruneau, M. Conceptualizing and measuring resilience: A key to disaster loss reduction. *TR News* **2007**, *17*, 14–15.

153. Leveson, N.G. *System Safety Engineering: Back to The Future*; Aeronautics and Astronautics Massachusetts Institute of Technology: Cambridge, MA, USA, 2002; Manuscript in preparation; Available online: http://sunnyday.mit.edu/book2.pdf (accessed on 18 October 2022).

154. Nan, C.; Sansavini, G. A quantitative method for assessing resilience of interdependent infrastructures. *Reliab. Eng. Syst. Saf.* **2017**, *157*, 35–53. [CrossRef]

155. Bai, G.; Li, Y.; Fang, Y.; Zhang, Y.-A.; Tao, J. Network approach for resilience evaluation of a UAV swarm by considering communication limits. *Reliab. Eng. Syst. Saf.* **2020**, *193*, 106602. [CrossRef]

156. Jakaria, A.H.M.; Rahman, M.A. Formal Analysis of k-Resiliency for Collaborative UAVs. In Proceedings of the 2018 IEEE 42nd Annual Computer Software and Applications Conference (COMPSAC), Tokyo, Japan, 23–27 July 2018; pp. 583–592.

157. Li, H.; Sun, Q.; Ren, K.; Xie, Y.; Liu, C.; Zhang, Y. Dynamic Resilience Assessment of UAV Swarm for Battlefield Surveillance Mission. In Proceedings of the 2021 IEEE International Conference on Unmanned Systems (ICUS), Beijing, China, 15–17 October 2021; pp. 472–477. [CrossRef]

158. Medjo Me Biomo, J.-D.; Kunz, T.; St-Hilaire, M.; Zhou, Y. Unmanned Aerial ad Hoc Networks: Simulation-Based Evaluation of Entity Mobility Models' Impact on Routing Performance. *Aerospace* **2015**, *2*, 392–422. [CrossRef]

159. Khan, M.A.; Kumar, N.; Mohsan, S.A.H.; Khan, W.U.; Nasralla, M.M.; Alsharif, M.H.; Zywiolek, J.; Ullah, I. Swarm of UAVs for Network Management in 6G: A Technical Review. *IEEE Trans. Netw. Serv. Manag.* **2022**. [CrossRef]

 **drones**

Article

A Distributed Task Rescheduling Method for UAV Swarms Using Local Task Reordering and Deadlock-Free Task Exchange

Jie Li *, Runfeng Chen and Ting Peng

College of Intelligence Science and Technology, National University of Defense Technology, Changsha 410073, China

* Correspondence: lijie09@nudt.edu.cn

Abstract: Distributed task scheduling is an ongoing concern in the field of multi-vehicles, especially in recent years; UAV swarm performing complex tasks endows it with new characteristics, such as self-organization, scalability, reconfigurability, etc. This requires the swarm to have distributed rescheduling capability to dynamically include as many unassigned tasks or new tasks as possible, while satisfying tight time constraints. As one of the most advanced rescheduling methods, the Performance Impact (PI)-MaxAss algorithm provides an important reference for this paper. However, its task exchange-based strategy faces the deadlock problem, and the task rescheduling method should not be limited to this. To this end, a new distributed rescheduling method is proposed for UAV swarms, which combines the local task reordering strategy and the improved task exchange strategy. On the one hand, based on the analysis of the fact that the scheduler is unreasonable for individuals, this paper proposes a local task reordering strategy denoted as PI-Reorder, which simply adds the reordering strategy to the recursive inclusion phase of the PI-MinAvg algorithm, so that unassigned tasks or new tasks can be included without relying on the task exchange. On the other hand, from the phenomenon that two or more vehicles occasionally get caught in an infinite cycle of exchanging the same tasks, the deadlock problem of PI-MaxAss is analyzed, which is then solved by introducing a deadlock-free task exchange strategy, where some defined counters are used to detect and isolate the deadlocks. Then, a rescue scenario is used to demonstrate the performance of the proposed methods, PI-Hybrid compared with PI-MaxAss. Monte Carlo simulation results show that, compared with PI-MaxAss, this method can not only increase the number of allocations to varying degrees, but also reduce the average waiting time, while ensuring deadlock avoidance. The methods can be used not only for the secondary optimization of the existing task exchange scheduling algorithms to escape local optima, but also for task reconfiguration of swarm tasks after adding or removing tasks.

Keywords: performance impact method; multi-agent scheduling; task rescheduling; distributed method; deadlock problem; search and rescue; UAV swarms

Citation: Li, J.; Chen, R.; Peng, T. A Distributed Task Rescheduling Method for UAV Swarms Using Local Task Reordering and Deadlock-Free Task Exchange. *Drones* **2022**, *6*, 322. https://doi.org/10.3390/drones6110322

Academic Editors: Mou Chen, Xiwang Dong and Fei Gao

Received: 30 September 2022
Accepted: 21 October 2022
Published: 27 October 2022

Publisher's Note: MDPI stays neutral with regard to jurisdictional claims in published maps and institutional affiliations.

1. Introduction

In recent years, UAV swarms have attracted much attention due to their low cost, scalability, wide-area distribution, etc. [1–3]. A team of homogeneous or heterogeneous specialized UAVs can cover more ground and be more resilient to failures than a single all-purpose UAV [4]. They have broad application prospects in post-disaster search and rescue, pollution traceability, pick-up and delivery, logistics, and any scenario that requires many urgent jobs to be completed in a minimum time by multiple UAVs. Different from simple clustering behavior, which can be achieved with a stigmergy-based approach, the complex tasks cannot be accomplished by a swarm without the assignment of tasks and the arrangement of their timing, that is, multi-vehicle scheduling. It is nothing new, but the swarm boom in recent years activates it again [5,6], and endows it with new characteristics, such as self-organization, scalability, reconfigurability, etc.

Multi-vehicle scheduling is a proven NP-hard combinatorial optimization problem; therefore, obtaining the global optimal solution is computationally intensive [7]. Earlier studies used a centralized approach to generate and distribute plans for all vehicles by using a central server capable of gathering system-wide information (situational awareness), either on the ground (the fleet's ground control terminal) or on one of the vehicles selected as the master. In this approach, the overall objective function of the problem concerned can be explicitly minimized or maximized based on the collection of all vehicle information on a central server. Therefore, the main advantage lies in the ability to optimize the overall objective function (and thus obtain the approximate optimal solution) by some heuristic algorithms (such as genetic algorithm, ant colony algorithm, etc.). However, there are several disadvantages [8]: First, the computational requirements are high because the computational operations are placed on a central server for global optimization of the multi-vehicle scheduling problem; second, each vehicle is required to communicate with the central server, and all vehicles need to communicate their situational awareness (SA) to the central server, which will bring a heavy communication burden. Third, centralized approaches are prone to single points of failure.

Given the individual low cost and self-organization of the swarm, the individual cannot afford the computationally expensive centralized scheduling, and the central server faces the communication pressure brought by information collection. To this end, the swarm can only resort to distributed task scheduling [9–11], which is roughly divided into three implementation approaches: redundant central computing, distributed parallel optimization and market-based bidding. Redundant central computing instantiates the centralized scheduler on each vehicle in order to achieve distribution, as well as eliminate the single point of failure [12–14]. These approaches often assume perfect communication links with unlimited bandwidth, since every vehicle must have the same SA. Inconsistencies in SA can lead to allocation conflicts because each vehicle will perform a centralized optimization using a different information set. Distributed parallel optimization divides the global optimization problem into sub-problems, which are solved by parallel computation to obtain higher optimization performance and computational efficiency [14–18]. These methods rely on high-frequency and high-speed data exchange and seem to be more suitable for gigabit connected server farms. Market-based bidding is where vehicles bid on tasks in a greedy strategy, the higher bidder wins the task, and the suboptimal solution can be obtained through multiple rounds of bidding [7,19,20]. In these types of algorithms, vehicles bid on tasks with values based solely on their own SA. Even if there are inconsistencies in their SA, they can naturally converge to a conflict-free solution. In comparison, market-based bidding is more suitable for the low cost, self-organization, and wide-area distribution of swarms.

The scalability of the swarm cannot be ignored; it means that the UAV can flexibly join or withdraw. As the number of tasks and UAVs increases, it is usually too computationally expensive to consider each combination of tasks for each UAV in order to find the optimal solution. The computational limitations are particularly relevant in search and rescue scenarios in which time and resources could be limited. Thus, heuristic methods are employed to speed up the process of task scheduling while maintaining an efficient and scalable algorithm. Consensus-Based Bundle Algorithm (CBBA) [19] and Performance Impact (PI-MinAvg) [7,21] are two of the most representative heuristic methods, and many methods are derived from them. CBBA is a popular solution that utilizes a heuristic greedy strategy locally to select tasks and resolves task conflicts among multiple agents through consensus rules. It has been shown that CBBA produces the same solution as the centralized sequential greedy procedures, and guarantees 50% optimality, which means that CBBA is still trapped into local optimization where each robot is striving for its own optimal scheduling and cannot give up some local optimal tasks for the global optimization. PI-MinAvg is an extension of CBBA, which introduces the 'significance' to prioritize task assignments and updates the significance of the follow-up tasks in the respective task list after any task is included or removed. It was shown empirically to solve time-critical task scheduling problems that CBBA could not, and was shown to find a lower average start time compared with CBBA. Despite the improved performance, PI-MinAvg still fails to solve some problems that are solvable due to converging to locally optimal but globally suboptimal solutions. It can be seen that heuristic methods to solve combinational optimization problems are prone to finding a local optimum and lack the flexibility to escape from it [22]; nonetheless, a second search can perturb the first phase solution out of local optima to reach an enhanced solution closer to a nondominated global optimum. The procedure follows a two-phase task scheduling strategy that starts from a solution generated with an existing scheduling algorithm that minimizes average waiting time; a rescheduling method needs to be used in the second stage for maximizing task allocations.

In addition to the above features, the reconfigurability of swarms is meant to support the dynamic addition or removal of tasks, which also requires the swarm to have the ability to reschedule tasks. For instance, CBBA with Partial Replanning (CBBAPR) [20] enables CBBA to respond quickly to new tasks that appear online. PI-MaxAss [23] can increase the allocation number by exchanging tasks among robots and creating a feasible time slot for unallocated tasks. Moreover, preliminary experiments running PI-MaxAss showed that two or more UAVs occasionally get caught in an infinite cycle of exchanging the same tasks, i.e., a deadlock problem. This phenomenon is intolerable for the swarm and will cause the system to stagnate. Still taking search and rescue as an example, when multiple UAVs conduct distributed scheduling, without the truncation of the algorithm, they will not be able to jump out of the infinite loop of the algorithm, resulting in the failure to obtain and execute the schedules in time, and ultimately leading to the failure of survivors obtaining rescue. The paper [22] proposed a solution to limit the number of times that a robot can remove the same task before it no longer attempts to include it. However, this method relies heavily on the reasonableness of the maximum number of removals, with a lower limit causing premature truncation of the algorithm, and a higher limit causing multiple invalid bids. The above facts indicate that solving the deadlock problem arising from the exchange mechanism and further improving the performance of task rescheduling methods are worthy of research.

It can be seen that task rescheduling can be used not only for the secondary optimization of the algorithm to escape local optima, but also for task reconfiguration after adding or removing tasks. This paper proposes a novel distributed rescheduling method for UAV swarms using local task reordering and deadlock-free task exchange. First, in terms of the search and rescue scenario, a distributed tasks rescheduling problem is formulated as a mathematical optimization problem coupled with some time, space or task constraints. Second, based on suboptimal solutions obtained with a scheduling method such as PI-MinAvg, a novel local task reordering strategy denoted as PI-Reorder is designed, which simply adds the strategy to the recursive inclusion phase of PI-MinAvg, so that unassigned tasks or new tasks can be included as soon as possible without relying on the task exchange strategy (such as PI-MaxAss). Third, from the phenomenon that shifting tasks occasionally falls into an infinite cycle, the deadlock problem of PI-MaxAss is analyzed, which is then solved by introducing a deadlock-free task exchange strategy, where a removed task set, an included task set, a task removed counter, a task included counter and a collection of task included counters are introduced together to detect and isolate deadlocks. Fourth, the local task reordering strategy and the deadlock-free task exchange strategy is integrated into an algorithm called PI-Hybrid. Monte Carlo simulation results show that the proposed local task reordering strategy PI-Reorder and the improved exchanging strategy PI-Hybrid can both increase the number of allocated tasks and reduce the total task waiting time to different degrees in the swarm search and rescue task. Pi-Reorder provides a new boost to the result performance of the PI-MinAvg algorithm. Compared with PI-MaxAss method, PI-Hybrid has a small improvement in the results, and overcomes the intolerable deadlock problem of swarm distributed scheduling.

The rest of this paper is organized as follows. In Section 2, a formal description of a typical task rescheduling problem is given and a classical distributed scheduling algorithm is introduced. Section 3 presents a distributed rescheduling method—Section 3.1 presents a local task reordering strategy, Section 3.2 presents an improved task exchange strategy and Section 3.3 presents an integrated reordering and exchange method. Numerical simulation and analysis are presented in Section 4. Finally, some conclusions are drawn in Section 5, followed by expectations for the future.

2. Preliminaries

2.1. Problem Formulation

Consider a variety of heterogeneous UAVs performing a search and rescue mission; the purpose is to cooperate with each other to rescue a number of survivors discovered, who require food supplies or medical provisions [22,24]. Each survivor must be visited by one UAV in order to be deemed rescued. Every UAV always follows a minimum time path to visit all of the unvisited survivors allocated to it, while not being required to return to its initial location. The scenarios have similarities with the traveling salesman problem (TSP), a well-known NP-hard combinatorial optimization problem. The objectives considered here are comparable to the constraints of two variants of the TSP: the team orienteering problem with time windows (TOPTW) [25] and the K-traveling repairmen problem (K-TRP) [26]. These objectives and constraints are applicable to a variety of scenarios such as those found in target tracking, pick-up and delivery, logistics and any scenario that requires many urgent jobs to be completed in a minimum time by multiple drones. One challenge in using a swarm is to schedule them to perform tasks while optimizing one or more objectives, for example, to minimize the average waiting time before their rescue, and to maximize the number of rescued survivors.

To formulate the problem mathematically, a set of n heterogeneous UAVs is defined by $\mathbf{V} = [v_1, v_2, \ldots, v_n]$, and a set of m tasks waiting to be conducted is defined by $\mathbf{T} = [t_1, t_2, \ldots, t_m]$. Each UAV v_i selects its appropriate tasks from the task set $\mathbf{T}$ and arranges them in sequence to form its scheduler $\mathbf{p}_i$ respecting time constraints. Whether a task is appropriate for a UAV depends on two aspects: one is whether the UAV can undertake this type of task, which is limited by the fact that low cost UAVs usually only have single task capability; another aspect is whether the task is within the limited range of the drone. The optimization objective J of the scheduling problem shown in the following formula is to minimize the total waiting time C as much as possible under the premise of allocating as many tasks N as possible.

$$J = \min C(\max N) \tag{1}$$

where

$$C = \sum_{i=1}^{n} \sum_{k=1}^{|\mathbf{p}_i|} c_{i,k}(\mathbf{p}_i), N = \sum_{i=1}^{n} |\mathbf{p}_i| \tag{2}$$

in which the time cost of a task t_k in $\mathbf{p}_i$, defined as $c_{i,k}(\mathbf{p}_i)$, is the predicted waiting time taken by the drone to arrive at the location of the task t_k. This time includes the duration of earlier tasks in $\mathbf{p}_i$ and travel time to and from those earlier tasks, but does not include the duration of the execution of t_k. The objective of minimizing waiting time measures the cost of a task scheduling as the time it takes to start serving the task from the start of the drone's schedule, i.e., the total time the survivor must wait before being attended to. $|\mathbf{p}_i|$ is the number of assigned tasks for v_i. Maximizing N measures the number of tasks assigned, that is, the number of survivors rescued.

The corresponding constraints are listed below:

$$\begin{cases} |\mathbf{p}_i| \leq N_i, & \forall i \in \{1, \ldots, n\} \\ \mathbf{p}_i \cap \mathbf{p}_j = \varnothing, & \forall i \neq j \in \{1, \ldots, n\} \\ \bigcap_i^z \mathbf{p}_i \subseteq \mathbf{T}, & \forall i, z \in \{1, \ldots, n\} \\ c_{i,k}(\mathbf{p}_i) \leq s_{\mathbf{p}_{i,K}}, & \forall i \in \{1, \ldots, n\}, \forall k \in \{1, \ldots, m\} \\ h_{i,k} \in H, & \forall i \in \{1, \ldots, n\}, \forall k \in \{1, \ldots, m\} \end{cases} \tag{3}$$

where the first constraint represents the capacity N_i at which the UAV v_i can perform tasks. The second constraint ensures that each task is assigned to at most one UAV, or is left unassigned because it is beyond the capabilities of all UAVs. The validity of scheduling is promised by the third constraint where any scheduler's combination is a subset of the task set **T**. The fourth constraint guarantees that the time $c_{i,k}(\mathbf{p}_i)$ UAV v_i to initiate the kth task in its scheduler $\mathbf{p}_i$ would be before their tasks' deadlines $s_{\mathbf{p}_{i,K}}$, after which it is too late for the task to be executed successfully. It is therefore necessary to determine whether a drone can arrive at the location of a task t_k before the latest start time $s_{\mathbf{p}_{i,K}}$. The last constraint in the form of a compatibility matrix H expresses that each UAV can only undertake one corresponding type of task, where $h_{i,k}$ is a Boolean variable.

Assume that a suboptimal solution in which additional tasks cannot be directly included without violating time constraints is obtained with the scheduling method; the UAVs rely on a decentralized rescheduling method to coordinate a rescue plan over multiple iterations to increase the allocation number simply. The main challenge is to reach an optimal allocation where the allocation number is maximized and the waiting time minimized while respecting time constraints.

2.2. Scheduling with PI-MinAvg

Whitbrook et al. [21] and Zhao et al. [7] proposed a distributed scheduling method as an extension of CBBA; it was shown empirically to solve time-critical task allocation problems such as search and rescue. In this method, a quantity called performance impact is introduced to determine the priority of task allocation for the robot, which takes into account not only the cost of task allocation, but also the impact of the task allocation on other allocation costs in the robot scheduler. The method employs a heuristic greedy strategy to enable robots to build a bundle of tasks sequentially and to optimize one or more objectives indirectly with a consensus rule; for instance, the average waiting time can be reduced by shifting tasks between robots. With PI-MinAvg, a robot does not release a task until it is reassigned elsewhere at a lower cost; i.e., once a task is assigned it does not become unassigned. Like CBBA, PI-MinAvg iterates over a task inclusion phase, a communication and conflict resolution phase.

During the task inclusion phase, robots select a task at a time to include in their schedulers until no more tasks can be added. Before including a task, the algorithm computes the inclusion performance impacts (abbreviated as IPI-MinAvg) of all candidate tasks t_q at each position l according to Equation (4), where candidate tasks are those compatible with v_i's capabilities and not already in $\mathbf{p}_i$. Here, the IPI-MinAvg of a task represents the minimum impact of a task addition on the average waiting time of the existing tasks in $\mathbf{p}_i$.

$$w_q^{\oplus}(\mathbf{p}_i, t_q) = \min_{l=1}^{|\mathbf{p}_i|+1} \left\{ \sum_{z=l}^{|\mathbf{p}_i|+1} c_{i,z}(\mathbf{p}_i \oplus_l t_q) - \sum_{z=l}^{|\mathbf{p}_i|} c_{i,z}(\mathbf{p}_i) \right\} \tag{4}$$

where $c_{i,z}(\mathbf{p}_i)$ denotes the start time of the task at position z in v_i's scheduler; $\oplus_l$ indicates that the task t_q is inserted into the l_{th} position of $\mathbf{p}_i$. Since the inserted task t_q only affects the start times of its subsequent tasks in $\mathbf{p}_i$, the formula calculates the cumulative time delay of the l_{th} task and the tasks after it. A list to store the IPI-MinAvg of each task is kept on each robot and is defined as $\gamma_i^{\oplus} = \left[w_1^{\oplus}, \ldots, w_m^{\oplus} \right]$ for robot v_i.

After the IPI-MinAvg of all candidate tasks have been computed, v_i selects for inclusion the task whose IPI-MinAvg can improve upon the task's removal performance impacts (abbreviated as RPI-MinAvg) the most. The maximum difference between the RPI-MinAvg of all tasks and the IPI-MinAvg of all tasks is computed as $\max_{q=1}^{m}\{\gamma_{i,q} - \gamma_{i,q}^{\oplus}\} > 0$. An IPI-MinAvg of t_q in $\mathbf{p}_i$ lower than t_q's RPI-MinAvg in another robot's task list $\mathbf{p}_j$ indicates that the global cost can be reduced if t_q is reallocated to v_i.

$$w_k^{\ominus}(\mathbf{p}_i, t_k) = \sum_{z=b}^{|\mathbf{p}_i|} c_{i,z}(\mathbf{p}_i) - \sum_{z=b+1}^{|\mathbf{p}_i|} c_{i,z}(\mathbf{p}_i \ominus t_k) \tag{5}$$

where b is the position of task t_k in v_i's task list, $c_{i,z}(\mathbf{p}_i)$ denotes the time cost of the task at position z in $\mathbf{p}_i$'s scheduler and $\mathbf{p}_i \ominus t_k$ denotes $\mathbf{p}_i$ with t_k removed. The RPI-MinAvg of a task t_k is referred to formally as $w_k^{\ominus}$ and each robot stores the vector $\gamma_i = \left[w_1^{\ominus}, \ldots, w_m^{\ominus} \right]$.

During the communication and conflict resolution phase, robots share their schedulers with neighboring robots and resolve conflicting allocations. As two or more robots may be assigned the same task, the consensus procedure introduced in [19] is used to resolve these conflicting assignments. A lower RPI-MinAvg indicates a more optimal schedule; therefore robots with a higher RPI-MinAvg for a conflicting assignment release the task. The above two phases are alternately repeated until the task assignment is agreed upon within the swarm.

2.3. Rescheduling with PI-MaxAss

Turner et al. [22] proposed a distributed rescheduling method called PI-MaxAss as an extension of PI-MinAvg. Starting from a suboptimal assignment in which additional tasks cannot be directly included without violating time constraints, PI-MaxAss is able to reschedule tasks to increase the allocation number simply through the cost-independent allocation. The idea is to attribute a high cost to an assigned task when the release of this task can permit an additional task to be inserted within the free time created. A schedule is considered optimal and without cost if the release of any task does not permit another task to be scheduled within the free time created.

PI-MaxAss follows the same algorithm process of PI-MinAvg, iterating over a task inclusion phase, a communication and conflict resolution phase. However, it adopts different definitions of inclusion performance impact and removal performance impact, which are denoted as IPI-MaxAss and RPI-MaxAss, respectively. A task's IPI-MaxAss is set to be without cost if it can be included in a scheduler and satisfy time constraints.

$$w_q^{\oplus}(\mathbf{p}_i, t_q) = 0, \quad \exists l \; \forall t_z \in \left\{ \mathbf{p}_i \oplus_l t_q \right\}, c_{i,z}(\mathbf{p}_i \oplus_l t_q) \leq \min(s_z, f_i), t_q \in \psi_i \tag{6}$$

where the candidate tasks for inclusion in $\mathbf{p}_i$ are formally defined as $\psi_i = \{ t_q | t_q \notin \mathbf{p}_i, w_q^{\ominus} > 0 \}$, which are those compatible with the robot's capabilities and not already in $\mathbf{p}_i$, and with an RPI-MaxAss greater than 0. The RPI-MaxAss of a task is formally defined as:

$$w_q^{\ominus} = \left\{ \begin{array}{ll} U, & if \quad t_q \notin \mathbf{p}_i \\ 0, & if \quad t_q \in \mathbf{p}_i \end{array} \right. \tag{7}$$

where the formula shows that unassigned tasks are initially set to have a fixed highest RPI-MaxAss, a constant defined as U, and the assigned tasks t_k are initially set to have 0.

3. Proposed Rescheduling Method

In general, distributed scheduling is not necessarily reasonable for individuals; it sacrifices local optimization to obtain a conflict-free global suboptimal solution while limiting the number of assigned tasks. Moreover, the exchange strategy is prone to deadlock, which is intolerable for the swarm and must be resolved. Below, we propose solutions to the above two aspects. In this section, we apply two rescheduling strategies to improve these problems: (1) One is the cost-dependent rescheduling strategy, which simply adds a local task reordering strategy in the recursive inclusion phase of PI-MinAvg to include unassigned tasks or new tasks directly; (2) the other is the cost-independent rescheduling strategy, that is, the improved deadlock-free PI-MaxAss method, which relies on the task exchange strategy without cost to create free time slots to contain unassigned tasks or new tasks indirectly. Finally, we incorporate the task reordering mechanism of strategy 1 into the recursive inclusion phase of strategy 2 to form a hybrid task rescheduling method.

3.1. A Novel Local Task Reordering Strategy

For search and rescue missions, PI-MinAvg prioritizes minimizing the average start time of all tasks, and then considers the number of tasks that can be accommodated, which makes it difficult to further increase the total number of allocated tasks and is not the most desirable for rescue. In order to achieve a consistent schedule, there is a mutual compromize among agents, although each adopts a heuristic greedy strategy. This may lead to the local scheduler possibly not being reasonable for individuals; either there is a task to look far away, or there is a necessity to wait. In other words, on the individual level, there is room for improvement in the local scheduler. Moreover, exchange-based methods, such as CBBA and PI-MinAvg, have a common feature that once included tasks that are not removed from the swarm in subsequent bidding rounds; they will only be exchanged within the swarm and obtained by other UAVs at a lower price. Actually, it is possible for the UAVs to form new slots to contain unassigned tasks or new tasks simply by rearranging their respective tasks without violating time constraints. To this end, we propose a novel distributed rescheduling method (denoted as PI-Reorder), which simply adds the local task reordering strategy to the recursive inclusion phase of PI-MinAvg, so that unassigned tasks or new tasks can be included as soon as possible without relying on the task exchange strategy (such as PI-MaxAss). The following is an example where the heuristic greedy strategy leads to non-ideal individual scheduling, and PI-MaxAss is unable to further improve the scheduling results through cost-independent task exchange.

Example 1. *An illustrative example of two mobile UAVs v_1, v_2 and five tasks $t_1, \ldots, t_5$ distributed in a common partitioned workspace is depicted in Figure 1a. Driven by the greedy strategy of PI-MinAvg running on v_1, the UAV will firstly select t_1 due to its lower IPI-MinAvg and then t_3. Similarly, v_1 will choose to undertake t_2 and t_4 subsequently. However, the remaining task t_5 will not be selected by any UAV because neither v_1 nor v_2 can contain task t_5 without abandoning the existing task. Figure 1b depicts the timeline corresponding to the scheduling result in the left figure. Intuitively, it is not possible to create a feasible slot containing t_5 with a task exchange mechanism such as PI-MaxAss. For example: v_1 tries to pass t_1 to v_2 by raising t_1's significance, thereby creating a slot that can contain t_5; however, v_2 is currently unable to take on additional tasks, and t_5 cannot be inserted into v_2's task list. That is, there is no way to obtain a boost in the number of tasks by swapping methods.*

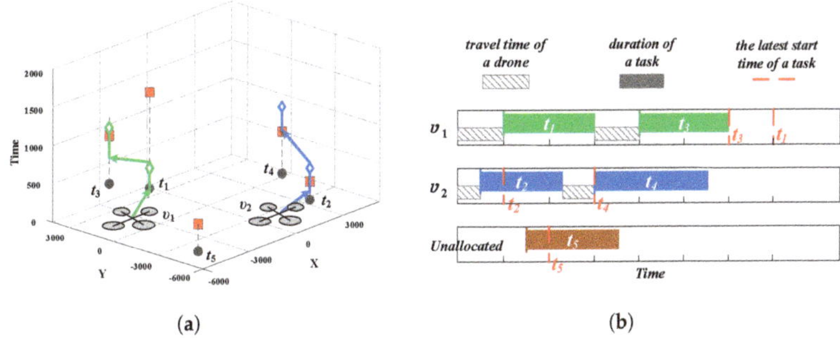

Figure 1. Illustration of task schedules in Example 1, where v_1 executes tasks t_1, t_3 in turn, v_2 executes t_2, t_4, with remaining task t_5 unallocated, and the rescheduling of PI-MaxAss cannot change this situation. (**a**) Schedulers generated by PI-MinAvg, where the dots represent tasks, dashed lines represent available time periods for tasks, red rectangles represent deadlines for task start, solid lines with arrows represent task schedulers and diamond patterns represent the duration of task execution. (**b**) Corresponding task timelines, where the red dotted lines indicate the tasks' latest start time, the diagonal boxes indicate the transition time of the drones, the color-filled box indicates the duration of task execution.

The following describes the basic principles of the task reordering strategy, and the specific procedure is shown in Algorithm 1.

First, the UAV v_i tries to sort its local scheduler in ascending order, according to the deadlines of tasks in the scheduler (Algorithm 1, line 2).

$$\tilde{\mathbf{p}}_i = sort(\mathbf{p}_i), s.t. \ \forall j \in \{1, \ldots, |\mathbf{p}_i|\}, s_{j,K} < s_{j+1,K} \tag{8}$$

Second, the UAV v_i updates the time costs $c_{i,k}(\mathbf{p}_i)$ of tasks in $\tilde{\mathbf{p}}_i$ (Algorithm 1, line 2). By adding t_q to $\mathbf{p}_i$, v_i will perform its remaining task assignments later. The time cost of tasks in the task list earlier than t_q is not affected by increasing t_q.

Next, following PI-MinAvg, mentioned in Section 2.2, the UAV v_i selects a task at a time to include in its updated task lists $\tilde{\mathbf{p}}_i$ until no more tasks can be added (Algorithm 1, lines 4–8).

Once tasks are recursively included, subsequent work follows the same method as PI-MinAvg, which is not described here.

Algorithm 1 Task Inclusion Procedure with Reordering running on v_i.

1: **while** $|\mathbf{p}_i| \leq N_i$ **do**
2:    Reorder $\mathbf{p}_i$ according to tasks' deadlines: $\tilde{\mathbf{p}}_i \leftarrow sort(\mathbf{p}_i)$.
3:    Update the time costs $c_{i,k}(\tilde{\mathbf{p}}_i)$ for the tasks in $\tilde{\mathbf{p}}_i$.
4:    Compute the marginal significance list $\gamma_{i,q}^{\oplus}, \forall q \in \{1, \ldots, m\}$.
5:    **if** $\max_{q=1}^{m} \{\gamma_{i,q} - \gamma_{i,q}^{\oplus}\} > 0$ **then**
6:      Insert the task t_{q^*} into $\tilde{\mathbf{p}}_i$ at position l^*.
7:      Update $\gamma_{i,q^*}, \ \beta_{i,q^*}, \ c_{i,K}(\tilde{\mathbf{p}}_i)$.
8:    **end if**
9:    $\mathbf{p}_i \leftarrow \tilde{\mathbf{p}}_i$.
10: **end while**
11: Update significance list $\gamma_{i,k}, \forall t_k \in \mathbf{p}_i$.

Continuing on Example 1, it is explained how the task reordering strategy achieves the further incorporation of t_5 while keeping the existing tasks $t_1, \ldots, t_4$ unchanged when PI-MinAvg and PI-MaxAss are incapable.

Example 2. *Based on the schedulers $\{t_1, t_3\}$ generated by PI-MinAvg, the UAV v_1 reorders its local scheduler according to tasks' deadlines as shown in Figure 2a and it can create a new feasible slot for t_5. Thus, v_1 is able to add t_5 into its scheduler, increasing the allocation numbers further. The task timelines corresponding to Example 1 (continued) are illustrated in Figure 2b. Intuitively, utilizing the task reordering strategy, v_1 reorders its local scheduler from $\{t_1, t_3\}$ to $\{t_3, t_1\}$, despite increasing time cost. The change can create a feasible slot for unassigned tasks, so that v_1 adds a task into its scheduler as $\{t_5, t_3, t_1\}$.*

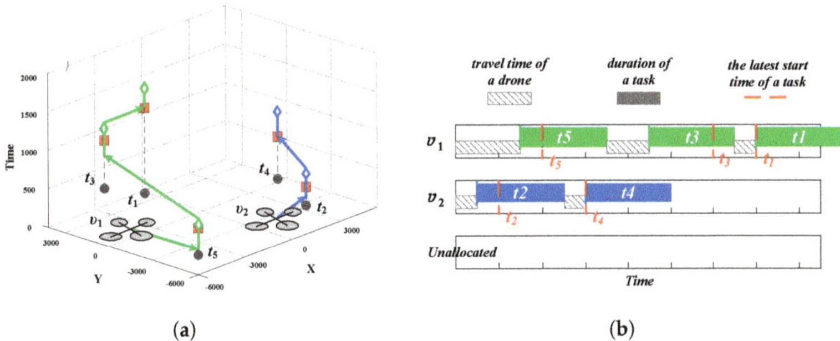

Figure 2. Illustration of task schedules in Example 2, where v_1 executes tasks t_5, t_3, t_1 in turn, v_2 still executes t_2, t_4, with no remaining task. (**a**) Schedulers generated by PI-Reorder, where v_1 creates a new time slot to include t_5 by reordering its existing task list $\{t_1, t_3\}$. (**b**) Corresponding task timelines, where t_5 can be inserted into a new time slot after $\{t_1, t_3\}$ is reordered, although t_1 can be executed with the latest start time allowed.

The above example shows that without a swapping task, it is possible for each UAV to further increase the allocation number by exploiting its local schedule potential.

3.2. A Deadlock-Free Task Exchange Strategy

The effect of simply using the task reordering strategy is limited, and the task exchange strategy is still needed to obtain better schedules. Starting from a suboptimal assignment in which additional tasks cannot be directly included without violating time constraints, PI-MaxAss presented in [22] is able to reschedule tasks to increase the allocation number simply through the cost-independent allocation. The idea is to attribute a high cost (RPI-MaxAss) to an assigned task when the release of this task can permit an additional task to be inserted within the free time created, to further increase the total allocation number. However, PI-MaxAss occasionally falls into an infinite cycle of shifting the same task, i.e., the deadlock problem. In this section, we first analyze the deadlock phenomenon, and then propose a deadlock-free task exchange strategy.

(1) Deadlock Analysis for PI-MaxAss. To explain the deadlock problem for PI-MaxAss intuitively here an example of deadlock is exhibited in Table 1 and a similar situation remains problematic.

Example 3. *A representative example of three mobile UAVs v_1, v_2, v_3 suffering from the deadlock problem when bidding for six tasks $t_1, \ldots, t_6$ with PI-MaxAss is depicted in Table 1, wherein each bidding round task consensus (abbreviated as TC), task removal (abbreviated as TR), task's RPI-MaxAss update (abbreviated as RU) and task inclusion (abbreviated as TI) work in turn, but the stage in which the intermediate results do not change is omitted here for simplification. With PI, v_1 obtains two tasks t_5, t_3, v_2 also obtains two tasks t_2, t_4, v_3 obtains the task t_6 and the remaining task t_1 cannot be executed by any UAV. Next, the rescheduling process of PI-MaxAss is shown in Table 1, where the symbol in the form of $t_k(c), k \in \{1, \ldots, m\}$ indicates that t_k is rescheduled to a UAV with its RPI-MaxAss equal to c. For an unassigned task (abbreviated as Un task), such as t_1, its RPI-MaxAss is set to a high value $U - n \cdot r$, where r is a decay factor; that is, each time a task is rescheduled, its RPI value is subtracted by r. A task is regarded as an unassigned task if its RPI is larger than a preset value such as $U - 3r$; that is, rescheduling is allowed for t_1 at most three times.*

Table 1. Deadlock phenomenon for PI-MaxAss.

Rounds	Phases	v_1	v_2	v_3	Unassigned
0	—[1]	$t_5(0), t_3(0)$	$t_2(0), t_4(0)$	$t_6(0)$	$t_1(U)$
1	TC	$t_1^{\varnothing}(U), t_2^2(0), t_3^1(0), t_4^2(0), t_5^1(0), t_6^3(0)$			
	RU	$t_5(U-r), t_3(0)$	$t_2(0), t_4(0)$	$t_6(U-r)$	$t_1(U)$
2	TC	$t_1^{\varnothing}(U), t_2^2(0), t_3^1(0), t_4^2(0), t_5^1(U-r), t_6^3(U-r)$			
	RU	$t_5(U-r), t_3(0)$	$t_2(U-2r), t_4(U-2r)$	$t_6(U-r)$	$t_1(U)$
3	TC	$t_1^{\varnothing}(U), t_2^2(U-2r), t_3^1(0), t_4^2(U-2r), t_5^1(U-r), t_6^3(U-r)$			
	TI	~[2]	~	$+t_2(U-3r)$	~
	RU	$t_5(U-r), t_3(0)$	$t_2(U-2r), t_4(U-2r)$	$t_6(U-3r), t_2(U-3r)$	$t_1(U)$
4	TC	$t_1^{\varnothing}(U), t_2^3(U-3r), t_3^1(0), t_4^2(U-2r), t_5^1(U-r), t_6^3(U-3r)$			
	TR	~	$-t_2(U-2r)$	~	~
	TI	~	$+t_2(0)$	~	~
	RU	$t_5(U-r), t_3(0)$	$t_2(0), t_4(0)$	$t_6(U-3r), t_2(U-3r)$	$t_1(U)$
5	TC	$t_1^{\varnothing}(U), t_2^2(0), t_3^1(0), t_4^2(0), t_5^1(U-r), t_6^3(U-3r)$			
	TR	~	~	$-t_2(U-3r)$	~
	RU	$t_5(U-r), t_3(0)$	$t_2(0), t_4(0)$	$t_6(U-r)$	$t_1(U)$
6	G2	$t_1^{\varnothing}(U), t_2^2(0), t_3^1(0), t_4^2(0), t_5^1(U-r), t_6^3(U-r)$			

[1] Not applicable. [2] No tasks have been removed or included.

According to the schedules of PI-MinAvg given in the first row of Table 1, the rescheduling process of three UAVs bidding for six tasks is described in detail, which leads to an infinite deadlock loop.

In round 1, since v_1 and v_2 can add the unassigned task t_1 to their local scheduler on the premise that their tasks t_5 and t_6 are removed, respectively, they both actively increase t_1's RPI-MaxAss to $U-r$, so that t_1 may be bid by other UAVs in the next round. The RPI-MaxAss of the remaining tasks in the individual UAV schedulers remain the same because even if they intend to remove any tasks from their schedulers, they cannot add any unassigned tasks.

In round 2, after communicating with each other, v_2 will find that the RPI-MaxAss of tasks t_5 and t_6 belonging to v_1 and v_3, respectively, have increased. At this point, v_2 has the opportunity to bid t_6 by removing t_2 or t_4, while t_5 is not considered because it cannot satisfy v_2's time constraints. Similar to round 1, the RPI-MaxAss of t_2 and t_4 need to be raised to $U-2r$, so that they can be outbid by other UAVs.

In round 3, after communication and consistency, v_3 finds that v_2 increases the RPI of t_2, so it can add t_2 to its scheduler without removing any tasks. In the RPI-MaxAss update stage, v_3 finds that t_4 can be added on the premise of removing t_6, so t_6's RPI-MaxAss is updated to $U-3r$.

In round 4, v_2 has to remove t_2 which was outbid by v_3. After removing t_2, v_2 selects the task with the lowest cost among the current candidate tasks, which happens to be still t_2 here. Then, when v_2 updates the RPI-MaxAss, it can still obtain t_6 by increasing the RPI-MaxAss of t_2 or t_4 as in round 2, but at this time, since t_6' RPI-MaxAss for v_3 has reached the threshold value $U-3r$, for this reason v_2 cannot obtain t_6 with a lower RPI-MaxAss.

In round 5, v_3 has to remove t_2 which was outbid by v_2. After removing t_2, v_3 finds that there is a time margin to include t_1, so it updates t_6's RPI-MaxAss to $U-r$, which goes back to round 2 (abbreviated as G2) and gets stuck in an endless loop.

Preliminary experiments running PI-MaxAss show that two or more UAVs occasionally get caught in an infinite cycle shifting the same tasks. For a swarm system, this phenomenon is intolerable and will cause the system to stagnate. For example, in a search and rescue mission, when multiple UAVs perform distributed task scheduling, in the absence of algorithm truncation, they will not be able to get out of the algorithm's infinite loop, resulting in the inability to obtain schedules in time to carry out actions, and ultimately causing survivors to be unable to obtain rescue. Therefore, it is necessary to find a way to avoid this deadlock problem.

(2) A Deadlock-free Task Exchange Strategy. Ref. [22] proposed a solution to limit the number of times that a UAV can remove the same task from its list before it no longer attempts to include it. A removal task set $Y_i \in \mathbb{Z}_+$ is created to record removed tasks, and a counter vector ω_i is used to store the times each task has been removed from a UAV v_i's task list. The precaution $\omega_{i,k} \leq \sigma$, $\forall t_k \in \mathbf{p}_i$ can prevent those tasks that are being repeatedly swapped from being scheduled optimally, but it ensures that the system can converge. However, this method relies heavily on the reasonableness of the maximum number σ of removals, with a lower limit causing premature truncation of the algorithm, and a higher limit causing multiple invalid bids. Here, based on the exchange mechanism of PI-MaxAss, we introduce a removed task set Y_i, an included task set Γ_i, a task removed counter $\omega_{i,q}$, a task included counter $\vartheta_{i,q}$ and a collection of task included counters Ξ_i to achieve deadlock-free task exchange. The specific process is as follows:

As with PI-MaxAss, unallocated tasks are set initially to have a fixed highest RPI-MaxAss, a constant defined as U, such that if t_q is unassigned then $w_q^\ominus = U$. The RPI-MaxAss of assigned tasks t_k are initially set to 0, such that $w_k^\ominus = 0$.

In the task inclusion phase as shown in Algorithm 2, UAVs select tasks to include in their schedulers until no more tasks can be added. The candidate task t_q is the one compatible with the functionality of v_i and not already in $\mathbf{p}_i$, and with an RPI-MaxAss greater than 0. In addition, t_q either does not belong to the removed task set Y_i, or it belongs to Y_i but $\omega_{i,q} < \sigma$ (Algorithm 2, line 4). This condition indicates that t_q has never been removed in the previous iteration, or the task removed counter $\omega_{i,q}$ is less than the set threshold σ. The candidate tasks for inclusion in $\mathbf{p}_i$ are formally defined as

$$\psi_i = [t_1, \ldots, t_\varsigma], t_q \notin \mathbf{p}_i, w_q^\ominus > 0, \quad if \quad t_q \notin Y_i \; or \; (t_q \in Y_i \; and \; \omega_{i,q} < \sigma) \tag{9}$$

If the condition in Algorithm 2, line 6, returns true for at least one position l, the IPI-MaxAss $w_q^\oplus$ of t_q in $\mathbf{p}_i$ is recorded such that,

$$w_q^\oplus(\mathbf{p}_i, t_q) = 0, \quad \exists l \; \forall t_z \in \{\mathbf{p}_i \oplus_l t_q\}, c_{i,z}(\mathbf{p}_i \oplus_l t_q) \leq \min(s_z, f_i), t_q \in \psi_i \tag{10}$$

Next, v_i selects for inclusion the task whose IPI-MaxAss can improve upon the task's current RPI-MaxAss the most as shown in the following formula. An IPI-MaxAss of t_q in $\mathbf{p}_i$ lower than t_q's RPI-MaxAss in another UAV's task list $\mathbf{p}_j$ indicates that the global cost can be reduced if t_q is reallocated to v_i.

$$\max_{q=1}^{m}\{\gamma_{i,q} - \gamma_{i,q}^\oplus\} > 0 \tag{11}$$

After the most appropriate task t_{q^*} is included (Algorithm 2, line 11), t_{q^*} needs to be added to the included task set Γ_i if $t_{q^*} \notin \Gamma_i$, the corresponding task which included counter ϑ_{i,q^*} is instantiated with its initial value to be 1; otherwise, only the task which included count needs to be incremented by one (Algorithm 2, line 13).

$$\begin{cases} \Gamma_i \leftarrow \Gamma_i \cup \{t_{q^*}\} \; and \; \vartheta_{i,q^*} \leftarrow 1, & if \; t_{q^*} \notin \Gamma_i \\ \vartheta_{i,q^*} \leftarrow \vartheta_{i,q^*} + 1, & if \; t_{q^*} \in \Gamma_i \end{cases} \tag{12}$$

Before the end of the task inclusion procedure, the RPI-MaxAss list, which keeps track of which UAV is assigned to which task, needs to be updated; this differs from PI-MaxAss in that additional conditions need to be met (Algorithm 2, line 15). Refer to Algorithm 3 for the specific process.

Algorithm 2 Task Inclusion Procedure running on v_i.

1: **while** $|\mathbf{p}_i| \leq N_i$ **do**
2:      $w_q^\oplus \leftarrow$ highest permissible cost, $w_q^\oplus \in \gamma_i^\oplus$.
3:      Identify candidate tasks ψ_i satisfying $t_q \notin Y_i$ or $(t_q \in Y_i$ *and* $\omega_{i,q} < \sigma)$ according to (11).
4:      **for** each task t_q in ψ_i **do**
5:          **if** $\mathbf{p}_i \oplus_l t_q$ is feasible **then**
6:             Record the IPI-MaxAss $w_q^\oplus$ of t_q in $\mathbf{p}_i$ according to (10).
7:          **end if**
8:      **end for**
9:      **if** $\max_{q=1}^m \{\gamma_{i,q} - \gamma_{i,q}^\oplus\} > 0$ **then**
10:       Insert task t_{q^*} into $\mathbf{p}_i$ at position l^*.
11:       Update γ_{i,q^*}, β_{i,q^*}, $c_{i,\kappa}(\mathbf{p}_i)$.
12:       $\Gamma_i \leftarrow \Gamma_i \cup \{t_{q^*}\}$ and $\vartheta_{i,q^*} \leftarrow 1$ if $t_{q^*} \notin \Gamma_i$, else $\vartheta_{i,q^*} \leftarrow \vartheta_{i,q^*} + 1$.
13:      **end if**
14: **end while**
15: Update significance $\gamma_{i,k}$ with Algorithm 3 satisfying $\vartheta_{i,k} > \lambda$ or $\omega_{i,k} = 0$, $\forall t_k \in \mathbf{p}_i$.

In this paper, task t_k's RPI-MaxAss $\gamma_{i,k}$ is the maximum significance change brought by the insertion of a candidate task $t_q \in \Omega_{i,k}$ shifted from other UAVs in local scheduler $\mathbf{p}_i$ after t_k is removed, where $\Omega_{i,k}$ is the candidate tasks set consisting of unassigned tasks, and tasks that would be released from other UAVs' schedulers.

$$\Omega_{i,k} = \left\{ t_q \in \bar{\psi}_i | \exists l \; \forall t_z \in \{\mathbf{p}_i^{\ominus k} \oplus_l t_q\}, c_{i,z}(\mathbf{p}_i^{\ominus k} \oplus_l t_q) \leq \min(s_z, f_i) \right\} \tag{13}$$

where $\bar{\psi}_i$ is a list of the candidate tasks defined as follows:

$$\bar{\psi}_i = [t_1, \ldots, t_\zeta], \quad t_q \notin \mathbf{p}_i, 0 < \delta < \gamma_{i,q}, \quad if \; \Xi_{i,k} > \lambda \; or \; \omega_{i,k} = 0 \tag{14}$$

The candidate tasks follow the same constraints as the candidates in (11) with the added constraint that the candidate task's RPI-MaxAss is greater than δ. This constraint is used to limit the number of reassignments permissible to allocate an additional task. In addition, tasks that do not belong to the removed task set Y_i or with a task included counter $\vartheta_{i,k}$ greater than the threshold λ can only be updated with the RPI-MaxAss. That is, if a task t_k has never been removed, it can be freely updated with the RPI-MaxAss; if a task t_k has a removed record and the task included counter $\Xi_{i,k}$ exceeds the threshold λ after being included in Γ_i, then the task can be updated with the RPI-MaxAss.

If a task t_k in $\mathbf{p}_i$ can be replaced by two or more candidate tasks t_q with different RPI-MaxAss, the highest RPI-MaxAss is recorded. The RPI-MaxAss of t_q in $\mathbf{p}_i$ is formally defined as

$$w_k^\ominus(\mathbf{p}_i, t_k) = \max_{q=1}^{|\Omega_{i,k}|} \{w_q^\ominus - r\}, \quad t_q \in \Omega_{i,k}, r \in \mathbb{R}_+ \tag{15}$$

where the decay factor r subtracted is used for limiting the number of reschedules.

Algorithm 3 Significance Update Procedure running on v_i.

1: Maintain tasks in $\mathbf{p}_i$ with the lowest RPI-MaxAss to 0: $\gamma_{i,k} \leftarrow 0$, *if* $t_k \in q_i$.
2: Identify Candidate Tasks $\bar{\psi}_i$ satisfying $\Xi_{i,k} > \lambda$ or $\omega_{i,k} = 0$ according to (14).
3: **for** each task t_k in $\mathbf{p}_i$ **do**
4:     Create a temporary scheduler $\mathbf{p}_i^{\ominus k} = \mathbf{p}_i \ominus t_k$.
5:     Update $c_{i,z}(\mathbf{p}_i^{\ominus k})$ for tasks after t_k.
6:     Identify the list of candidate tasks $\Omega_{i,k}$ according to (13).
7:     Compute t_k's RPI-MaxAss $w_k^{\ominus}(\mathbf{p}_i, t_k)$ according to (15).
8: **end for**

In the communication and conflict resolution phase as shown in Algorithm 4, except for the following differences, this paper follows the same task removal process for PI-MinAvg introduced in Section 2.2. When a task t_{k*} has been removed from the UAV v_i's scheduler $\mathbf{p}_i$ for the first time, it will be added to $\mathbf{Y}_i$, and $\omega_{i,k*}$ is initialized to 1; otherwise, only the removal count needs to be incremented by one.

$$\begin{cases} \mathbf{Y}_i \leftarrow \mathbf{Y}_i \cup \{t_{k*}\} \text{ and } \omega_{i,k*} \leftarrow 1, & \text{if } t_{k*} \notin \mathbf{Y}_i \\ \omega_{i,k*} \leftarrow \omega_{i,k*} + 1, & \text{if } t_{k*} \in \mathbf{Y}_i \end{cases} \tag{16}$$

Once the current round of removal while-cycle ends, the tasks just removed need to be immediately deleted from the inclusion set Γ_i.

$$\Gamma_i \leftarrow \Gamma_i - \{t_k\}, \quad \forall\, t_k \in \Gamma_i \cap \mathbf{Y}_i \tag{17}$$

The corresponding task included counter $\vartheta_{i,k} \in \Xi_i$ also needs to be cleared.

$$\Xi_i \leftarrow \Xi_i - \{\vartheta_{i,k}\}, \quad \forall\, \vartheta_{i,k} \in \Xi_i \text{ and } t_k \notin \Gamma_i \tag{18}$$

For the detailed removal process of PI-MinAvg, please refer to [7].

Algorithm 4 Task Removal Procedure running on v_i.

1: UAV v_i forms its conflict tasks set $\mathbf{d}_i \leftarrow \mathbf{p}_i[\beta_i[\mathbf{p}_i] \neq i]$.
2: **while** $\max_{k=1}^{|\mathbf{d}_i|}\{\gamma_i^{\diamond}[\mathbf{d}_i] - \gamma_i[\mathbf{d}_i]\} > 0$ and $\mathbf{d}_i \neq \varnothing$ **do**
3:     Remove task $t_{k*} \leftarrow \arg\max_{k=1}^{|\mathbf{d}_i|}\{\gamma_i^{\diamond}[\mathbf{d}_i] - \gamma_i[\mathbf{d}_i]\}$ from $\mathbf{p}_i$ and $\mathbf{d}_i$.
4:     Update costs $c_{i,\kappa}(\mathbf{p}_i)$ and significance $\gamma_i^{\diamond}$ for the rest of the tasks in $\mathbf{p}_i$ and $\mathbf{d}_i$.
5:     $\mathbf{Y}_i \leftarrow \mathbf{Y}_i \cup \{t_{k*}\}$ and $\omega_{i,k*} \leftarrow 1$ if $t_{k*} \notin \mathbf{Y}_i$, else $\omega_{i,k*} \leftarrow \omega_{i,k*} + 1$.
6: **end while**
7: Remove task t_k from the included task set Γ_i for any $t_k \in \Gamma_i \cup \mathbf{Y}_i$.
8: Remove the task included counter $\vartheta_{i,k}$ from Ξ_i satisfying $t_k \notin \Gamma_i$.
9: Update the local list $\beta_i(\mathbf{d}_i) \leftarrow i$ belongs to UAV v_i.

3.3. The Integrated Method with Task Reordering and Task Exchange

Here, a decentralized rescheduling method for each UAV running independently is presented, denoted as PI-Hybrid, which combines the local reordering strategy with the modified task exchange strategy to improve optimization and avoid deadlocks. The pseudocode of the entire iterative process is shown in Algorithm 5.

Before entering the loop iteration, the parameters $\mathbf{Y}_i, \omega_{i,k}, \sigma, \Gamma_i, \vartheta_{i,k}, \Xi_i, \lambda$ need to be initialized. After entering the while loop, Algorithm 1 is called to reorder the tasks in the respective scheduler $\mathbf{p}_i$, and the unassigned tasks are added according to the greedy strategy to allocate as many tasks as possible. Next, Algorithm 2 is called to try to exchange the local task to other UAVs to free time slots for unassigned tasks. Subsequently, through mutual communication, the UAV v_i obtains the consistent assignments of its scheduler β_i and significance list γ_i using consensus rules [19]. This information is used to guide the call to Algorithm 4 to remove local conflicting tasks. The above phases repeatedly alternate until a global conflict-free task allocation is agreed upon by all UAVs.

Algorithm 5 Distributed Scheduling Method running on v_i

1: Initialize timer $T \leftarrow 1$ and *converged* $\leftarrow$ *false*.
2: Initialize parameters $\mathbf{Y}_i, \varpi_{i,k}, \sigma, \mathbf{\Gamma}_i, \vartheta_{i,k}, \mathbf{\Xi}_i, \lambda$.
3: **while** *converged* is false **do**
4:     Call Task Reordering Procedure with Algorithm 1.
5:     Call Task Inclusion Procedure with Algorithms 2 and 3, in which.
6:     Communication and Conflict Resolution using consensus rules.
7:     Call Task Removal Procedure with Algorithm 4.
8:     *converged* $\leftarrow$ *Check Convergence*.
9:     $T \leftarrow T + 1$.
10: **end while**

---

4. Numerical Results

This section presents the numerical simulations conducted to test the performance of the proposed PI-Hybrid compared with that of PI-MinAvg and PI-MaxAss. In addition, PI-Reorder is also compared to independently reflect its impact on the overall performance. Among the above methods, PI-MaxAss, PI-Reorder and PI-Hybrid are all initialized with the PI-MinAvg solution. The results of simulations are analyzed from optimization and deadlock-free, respectively.

4.1. Scenario and Simulation Setup

The scenarios adopted in this paper are consistent with those of the literature [7,21]. The setup uses a rescue team equally split into two UAVs rescuing two types of survivors, who are likewise equally split into those requiring food and those requiring medicine. Parameters related to the scenario are summarized in Table 2 and are illustrated as follows. The UAVs are randomly distributed in 10 km $\times$ 10 km $\times$ 0 km ground space, where the UAVs' speeds are assumed to be constant and are set to 30 m·s^{-1} and 50 m·s^{-1}, respectively. The tasks take place in a 3D space spanning 10 km $\times$ 10 km $\times$ 1 km, with coordinates drawn from uniform distributions randomly. The medicine tasks last for a duration of 300 s and the food tasks last 350 s. The deadlines for starting each rescue are uniformly distributed on a timeline between 0 and 2000 s. Given the random initialization of tasks and UAV locations and deadlines, it is sometimes impossible for some tasks to be started by any UAV before their deadline. In these simulations, all task information is available to all UAVs up front.

Table 2. Parameters related to the simulation scenario.

Properties	Medicine	Food
UAV speed	30 m·s^{-1}	50 m·s^{-1}
UAV initial position	Random distribution in 10 km $\times$ 10 km $\times$ 0 km	
Task duration	300 s	500 s
Task deadline	Random distribution between 0 s and 2000 s	
Task location	Random distribution in 10 km $\times$ 10 km $\times$ 1 km	

4.2. Results and Analysis

As elaborated here, the paper uses the Monte Carlo method to implement random simulations, where each method is tested 1000 times for each pair of $p \in \{2, 3, 4, 5\}$ and $n \in \{2, 4, \ldots, 16\}$. p is the ratio of the number of tasks to the number of UAVs, and n is the number of UAVs.

4.2.1. Optimization

First of all, the total allocation number obtained with four methods is compared and analyzed, which can be said to be the most important optimization indicator. For example, in search and rescue missions, saving one more survivor may be more important than other things. Figure 3 depicts the allocation numbers using four methods in the same scenario. When the proportion p is the same, as n increases, the total number of assigned tasks tends to increase. PI-MinAvg as a scheduling method has the lowest allocation number, and other rescheduling methods initialized with the PI-MinAvg solution all produce a boost in the allocation numbers. PI-Reorder can increase the limited allocation number by reordering UAVs' local scheduler. PI-MaxAss as a typical task-swap method is able to obtain an appreciable effect on the allocation number. PI-Hybrid has the best improvement for the allocation numbers. Moreover, it should be explained that the deadline is randomly distributed between 0 and 2000 s, which means that some tasks have a short random deadline and it is impossible to complete them. Therefore, it is impossible to schedule all the tasks in such a situation.

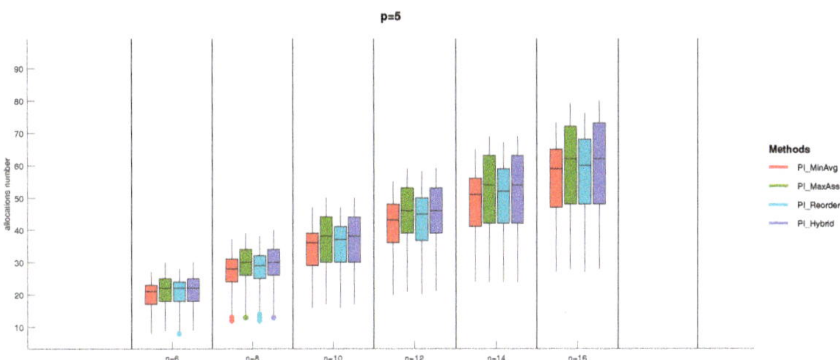

Figure 3. Box plot comparison of total allocation numbers of four methods at $p = 5$ and $n \in \{6, 8, \ldots, 16\}$.

Since Figure 3 can only present the variation trend of the number of scheduled tasks as increases and the approximate differences between methods, Figure 4 shows the box plots of the allocation numbers under the conditions of two groups of $p = 2, n = 6$ and $p = 5, n = 16$, respectively.

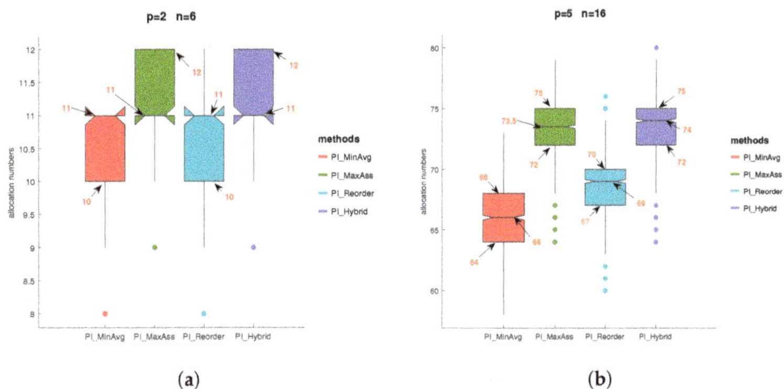

Figure 4. Detailed comparison of the allocation number box plots for four methods at $p = 5$ and $n = 16$. (**a**) $p = 2, n = 6$; (**b**) $p = 5, n = 16$.

It can be seen from Figure 4a that PI-Reorder cannot increase the number of scheduled tasks through local reordering based on the solution of PI-MinAvg, while PI-MaxAss allocates one more task as a whole through task exchange. Similarly, PI-Hybrid's total allocation number is the same as PI-MaxAss. The reason for the above phenomenon is that when the total number of tasks to be assigned is low and the number of UAVs is small, the optimization of the PI-MinAvg solution is close to optimal, and the room for improvement is small. At this time, since the local task switching cannot obtain a better solution, PI-Hybrid actually works only by its task switching, which is equivalent to PI-MaxAss. It can be seen from Figure 4b that as p and n increase, the median task assignment of PI-Reorder increases from 66 to 69 compared to PI-MinAvg, which is due to the fact that more UAVs have more tasks that can be chosen so that the total number of tasks for the problem has more room for improvement. Of course, PI-Reorder that only relies on local tasks reordering cannot obtain the effect of the PI-MaxAss method using task exchange, but PI-Hybrid combining the two mechanisms can obtain a better solution than PI-MaxAss. For example, the median assigned tasks increased from 73.5 to 74.

Since the solution performance of PI-MinAvg and PI-MaxAss is already very good, it is not obvious from the improvement of the allocation numbers obtained in the above experiments. Next, the number of times increasing allocation numbers in 1000 tests with different combinations of p and n is counted as shown in Figure 5. Removing unnecessary elaboration, here mainly PI-Hybrid and PI-MaxAss are compared. It can be found that, when p is larger, more solutions of PI-MaxAss can be improved by PI-Hybrid. This is because the larger p is, the larger the space for reordering tasks will be, and the hybrid mechanism can achieve a better optimization effect that cannot be reached with a solo task swap.

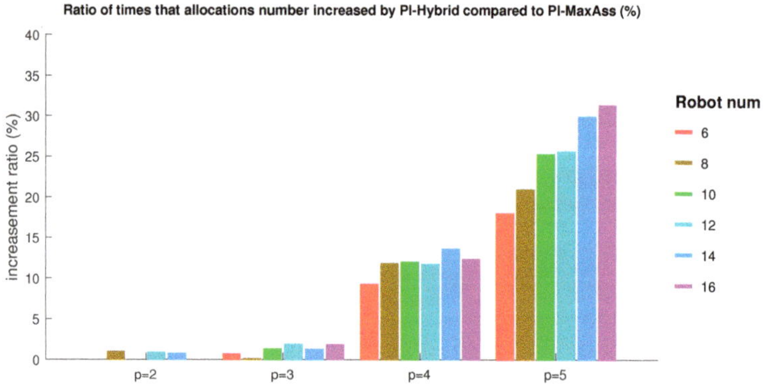

Figure 5. Ratio of the allocation number increased by PI-Hybrid compared to PI-MaxAss.

Figure 6 summarizes the average increase of scheduled tasks in the improved solutions in which the mean and standard deviation are shown. Obviously, PI-Hybrid has a good effect on the increase of the number of allocated tasks compared with PI-MaxAss, where the increase rate has reached about 7% when p and n are equal to 3 and 6, respectively. When the number of UAVs increases, the space for swapping tasks is also enlarged, making the difference between the two methods smaller. That is why PI-Hybrid has a weaker effect compared with PI-MaxAss under the condition of n increasing but p remaining unchanged, which can be seen in Figure 6.

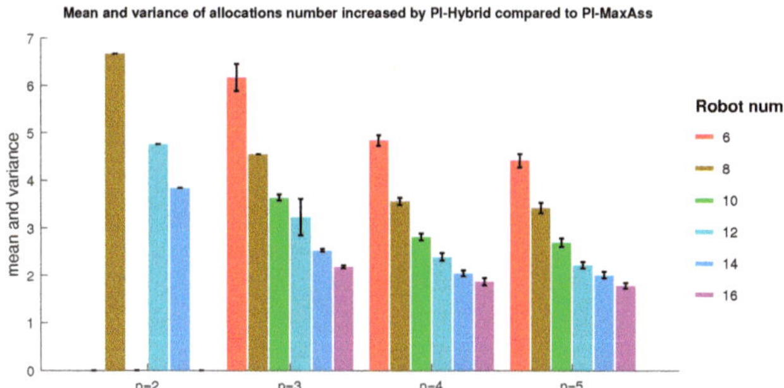

Figure 6. Mean and standard deviation of allocation number increased by PI-Hybrid compared with PI-MaxAss.

Next, let us look at another important indicator. Under the same total number of tasks scheduled, the shorter the average waiting time, the better the method. Figure 7 is a graph about the average waiting time of various methods to complete tasks in the same scenarios. It is observed that PI-MinAvg has the shortest completion time because it also completes the least number of tasks. PI-Reorder has the second short travel time, which is also reasonable for its completed task number exceeds PI-MinAvg but not others. In addition, carefully comparing PI-MaxAss and PI-Hybrid, whether it is the upper quartile, the median or the lower quartile, PI-Hybrid is lower than PI-MaxAss to varying degrees, which fully shows that the proposed method not only increases the total number of the scheduled tasks but also reduces the average waiting time.

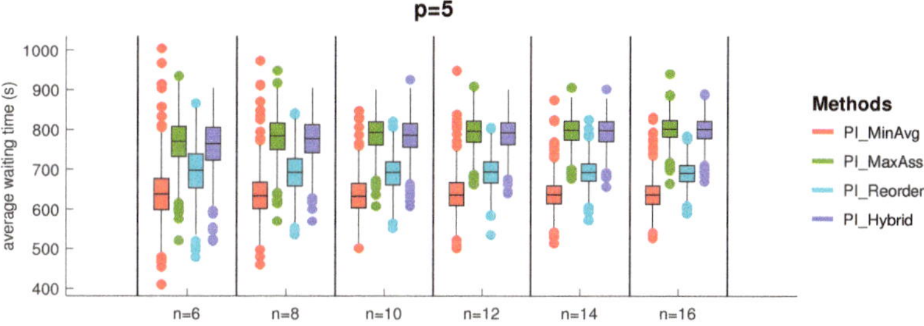

Figure 7. Box plot comparison of average waiting time of four methods at $p = 5$ and $n \in \{6, 8, \ldots, 16\}$.

Figure 8 shows the boxplot statistics of the average waiting time of the four methods under the conditions of $p = 2, n = 6$, and $p = 6, n = 16$, respectively. It shows the boxplot statistics of the average waiting time of the four methods under the conditions of $p = 2, n = 6$, and $p = 6, n = 16$, respectively. As the number of assigned tasks increases, the average waiting time of PI-MinAvg, PI-Reorder and PI-MaxAss also increases. It is worth remembering that from the median, upper quartile and lower quartile of average waiting time, PI-Hybrid is able to obtain less average waiting time while allocating more total tasks. This performance improvement is obtained by mixing two strategies of the task reordering and the task exchange, which is bound to be crucial for the swarm search and rescue.

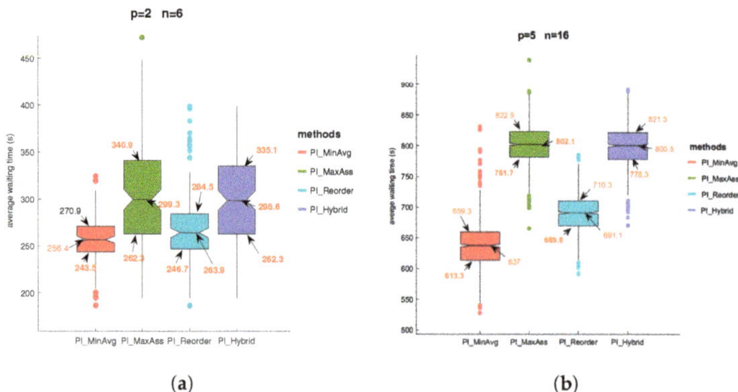

Figure 8. Detailed comparison of the average waiting time box plots for four methods at $p = 5$ and $n = 16$. (**a**) $p = 2, n = 6$; (**b**) $p = 5, n = 16$.

From the above analyses, it can be seen that the combination of the local task reordering and the task exchange can not only improve the allocation number but also have better average waiting time.

4.2.2. Deadlock Avoidance

Another important issue addressed in this paper is to avoid the algorithm occasionally falling into an infinite cycle of shifting the same task. Figure 9 shows the statistical deadlock situation of different methods with various values of p, where the abscissa is the ratio of task and UAV numbers, and the ordinate is the deadlock rate. The deadlock rate of a method with each is its average under various conditions of n from 6 to 16, where each situation is tested 1000 times with random initialization of UAVs and tasks.

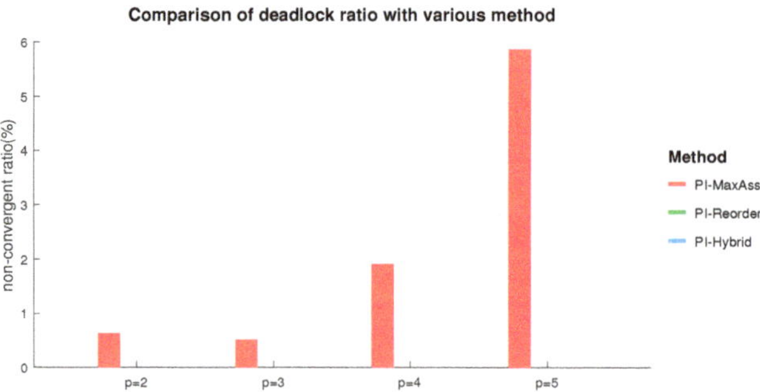

Figure 9. Deadlock ratio of different methods with various values of p.

It can be found that the deadlock situation only occurs in PI-MaxAss, and all the other proposed methods do not exist in any deadlock. For the PI-MaxAss algorithm, when p is equal to 2 or 3, its deadlock rate is very low, having not reached 1%. However, its deadlock rate increases rapidly with the increase of p, which comes up to 6% approximately at the condition of $p = 5$. This is because more participating UAVs and tasks will lead to more conflict between UAVs, while PI-MaxAss does not consider the potential deadlock avoidance. By contrast, PI-Hybrid is completely deadlock-free no matter what the values of p and n are.

5. Conclusions

Aiming at the new characteristics of multi-vehicle scheduling given by UAV swarm, such as self-organization, scalability and reconfiguration, this paper proposes a task rescheduling method that integrates local task reordering and improved task exchange strategy, with the goal of secondary optimization of task scheduling suboptimal solution and task reconfiguration. As a supplement to the task exchange-based strategy, local task reordering is a new cost-related strategy to increase the number of allocation tasks. To improve the task exchange-based strategy, deadlock detection and isolation mechanism are introduced to avoid occasional task exchange falling into the deadlock problem. Based on the swarm rescue mission, a large number of numerical simulation experiments show that the integrated method can not only improve the number of allocation tasks but also have a better average waiting time.

In the future, we will carry out research on scheduling and rescheduling methods under the condition of ad hoc network communication. For large-scale clusters, we will focus on architecture design, interaction mechanisms and algorithms. Field experiments will be conducted in combination with the real UAV swarm system.

Author Contributions: Conceptualization, J.L. and R.C.; methodology, J.L.; software, J.L.; validation, R.C.; investigation, T.P.; writing—original draft preparation, J.L.; writing—review and editing, T.P. All authors have read and agreed to the published version of the manuscript.

Funding: This work was supported by the Science and Technology Innovation 2030-Key Project of 'New Generation Artificial Intelligence' under Grant 2020AAA0108200.

Data Availability Statement: Not applicable.

Conflicts of Interest: The authors declare no conflict of interest.

References

1. Chung, S.J.; Paranjape, A.A.; Dames, P.; Shen, S.; Kumar, V. A Survey on Aerial Swarm Robotics. *IEEE Trans. Robot.* **2018**, *34*, 837–855. [CrossRef]
2. Brambilla, M.; Ferrante, E.; Birattari, M.; Dorigo, M. Swarm Robotics: A review from the swarm engineering perspective. *Swarm Intell.* **2013**, *7*, 1–41. [CrossRef]
3. Hadaegh, F.Y.; Chung, S.-J.; Manohara, H.M. On development of 100-gram-class spacecraft for swarm applications. *IEEE Syst. J.* **2014**, *10*, 673–684. [CrossRef]
4. Cao, Y.; Yu, W.; Ren, W.; Chen, G. An overview of recent progress in the study of distributed multi-agent coordination. *IEEE Trans. Ind. Inform.* **2013**, *9*, 427–438. [CrossRef]
5. Bossens, D.M.; Tarapore, D. Qed: Using quality-environmentdiversity to evolve resilient robot swarms. *IEEE Trans. Evol. Comput.* **2021**, *25*, 346–357. [CrossRef]
6. Atn, G.M.; Duan, M.S.; Voulgaris, P.G. A swarm-based approach to dynamic coverage control of multi-agent systems. *Automatica* **2020**, *112*, 108637.
7. Zhao, W.; Meng, Q.; Chung, P.W. A heuristic distributed task allocation method for multiUAV multitask problems and its application to search and rescue scenario. *IEEE Trans. Cybern.* **2015**, *46*, 902–915. [CrossRef]
8. Zhang, K.; Collins, E.G., Jr.; Shi, D. Centralized and distributed task allocation in multi-robot teams via a stochastic clustering auction. *ACM Trans. Auton. Adapt. Syst.* **2012**, *7*, 21:1–21:22. [CrossRef]
9. Chen, C.-H.; Chou, F.-I.; Chou, J.-H. Optimization of robotic task sequencing problems by crowding evolutionary algorithms. *IEEE Trans. Syst. Man Cybern. Syst.* **2021**, *52*, 6870–6885. [CrossRef]
10. Kim, S.; Moon, I. Traveling salesman problem with a drone station. *IEEE Trans. Syst. Man Cybern. Syst.* **2018**, *49*, 42–52. [CrossRef]
11. Xin, B.; Wang, Y.; Chen, J. An efficient marginal-returnbased constructive heuristic to solve the sensor–weapon–target assignment problem. *IEEE Trans. Syst. Man Cybern. Syst.* **2018**, *49*, 2536–2547. [CrossRef]
12. Chen, W.; Wang, D.; Li, K. Multi-user multi-task computation offloading in green mobile edge cloud computing. *IEEE Trans. Serv. Comput.* **2018**, *12*, 726–738. [CrossRef]
13. Mnif, S.; Elkosantini, S.; Darmoul, S.; Said, L.B. An immune network based distributed architecture to control public bus transportation systems. *Swarm Evol. Comput.* **2019**, *50*, 100478. [CrossRef]
14. Mudrova, L.; Hawes, N. Task scheduling for mobile robots using interval algebra. In Proceedings of the 2015 IEEE International Conference on Robotics and Automation (ICRA), Seattle, WA, USA, 26–30 May 2015; pp. 383–388.
15. Testa, A.; Rucco, A.; Notarstefano, G. Distributed mixedinteger linear programming via cut generation and constraint exchange. *IEEE Trans. Autom. Control* **2019**, *65*, 1456–1467. [CrossRef]

16. Camisa, A.; Notarnicola, I.; Notarstefano, G. Distributed primal decomposition for large-scale milps. *IEEE Trans. Autom. Control* **2021**, *67*, 413–420. [CrossRef]

17. Alighanbari, M.; How, J.P. Decentralized task assignment for unmanned aerial vehicles. In Proceedings of the 44th IEEE Conference on Decision and Control, Seville, Spain, 15 December 2005; pp. 5668–5673.

18. Patel, R.; Rudnick-Cohen, E.; Azarm, S.; Otte, M.; Xu, H.; Herrmann, J.W. Decentralized Task Allocation in Multi-Agent Systems Using a Decentralized Genetic Algorithm. In Proceedings of the 2020 IEEE International Conference on Robotics and Automation (ICRA), Paris, France, 31 May–31 August 2020; pp. 3770–3776. [CrossRef]

19. Choi, H.-L.; Brunet, L.; How, J.P. Consensus-based decentralized auctions for robust task allocation. *IEEE Trans. Robot.* **2009**, *25*, 912–926. [CrossRef]

20. Buckman, N.; Choi, H.-L.; How, J.P. Partial replanning for decentralized dynamic task allocation. In Proceedings of the AIAA Scitech 2019 Forum, San Diego, CA, USA, 7–11 January 2019; p. 915.

21. Whitbrook, A.; Meng, Q.; Chung, P.W. Reliable, distributed scheduling and rescheduling for time-critical, multiUAV systems. *IEEE Trans. Autom. Sci. Eng.* **2017**, *15*, 732–747. [CrossRef]

22. Turner, J.; Meng, Q.; Schaefer, G.; Whitbrook, A.; Soltoggio, A. Distributed task rescheduling with time constraints for the optimization of total task allocations in a multiUAV system. *IEEE Trans. Cybern.* **2017**, *48*, 2583–2597. [CrossRef]

23. Turner, J.; Meng, Q.; Schaefer, G. Increasing allocated tasks with a time minimization algorithm for a search and rescue scenario. In Proceedings of the 2015 IEEE International Conference on Robotics and Automation (ICRA), Seattle, WA, USA, 26–30 May 2015; pp. 3401–3407.

24. Parker, J.; Nunes, E.; Godoy, J.; Gini, M. Exploiting spatial locality and heterogeneity of agents for search and rescue teamwork. *J. Field UAV* **2016**, *33*, 877–900. [CrossRef]

25. Vansteenwegen, P.; Souffriau, W.; Oudheusden, D.V. The orienteering problem: A survey. *Eur. J. Oper. Res* **2021**, *209*, 1–10. [CrossRef]

26. Fakcharoenphol, J.; Harrelson, C.; Rao, S. The K-traveling repairmen problem. *ACM Trans. Algorithms* **2007**, *3*. [CrossRef]

Article

A Group Maintenance Method of Drone Swarm Considering System Mission Reliability

Jinlong Guo [1], Lizhi Wang [2] and Xiaohong Wang [1,*]

[1] Institute of Reliability Engineering, Beihang University, Beijing 100083, China
[2] Unmanned System Institute, Beihang University, Beijing 100083, China
* Correspondence: wxhong@buaa.edu.cn

Abstract: Based on the characteristics of drone swarm such as low cost, strong integrity, and frequent information exchange, as well as the high cost of timely maintenance of traditional units. This paper proposes a swarm maintenance method based on the reliability assessment of multi-layer complex network missions, which combines the multi-layer complex network system evaluation method with the group maintenance method. On the basis of considering the problem of maintenance grouping cost, the failure mechanism of drones in different modes and the impact of drone maintenance on the system are studied. According to the failure model of single node, the complex network method is used to establish the swarm system's topology model and evaluate the system mission reliability. The maintenance grouping strategy is optimized by using the multi-objective planning of cost and system mission reliability. Compared with the existing just in time maintenance methods, this method can greatly reduce the total maintenance cost of the swarm system maintenance under the condition of ensuring the high mission robustness of the swarm. In addition, a universal drone swarm mission scenario is used to illustrate the method, and the results verify the feasibility and effectiveness of the method.

Keywords: swarm maintenance; drone swarm; complex network; system reliability; multi-objective optimization

Citation: Guo, J.; Wang, L.; Wang, X. A Group Maintenance Method of Drone Swarm Considering System Mission Reliability. *Drones* **2022**, *6*, 269. https://doi.org/10.3390/drones6100269

Academic Editor: Carlos Tavares Calafate

Received: 14 August 2022
Accepted: 16 September 2022
Published: 22 September 2022

Publisher's Note: MDPI stays neutral with regard to jurisdictional claims in published maps and institutional affiliations.

1. Introduction

The drone swarm has the characteristics of many nodes and complex interactions. The completion of the internal missions of the system mainly depends on the control structure and information exchange, which is dependent on mission networking. Drone swarm is widely used in various fields to perform public security [1], military [2], and industry due to its high flexibility, strong adaptability, and controllable economic cost. With the improvement of swarm usage requirements, swarm maintenance has become the focus of attention, and many maintenance theories and optimization design methods are applied to this [3,4]. The research on the failure mode of the drone system in the swarm [5] has been relatively mature. However, the concept of low-cost unmanned aerial vehicles has been born in recent years. Under the swarm system, the problem of the drone maintenance is no longer the primary factor limiting the ability of swarm missions. At the same time, performing single-point maintenance on many units in the swarm according to the timely maintenance method of traditional unit equipment will face problems such as large maintenance magnitude and difficulty in taking into account the mission networking connection. How to gradually transition from unit system maintenance to swarm system maintenance has become the focus of researchers [6].

To solve the problem of swarm system maintenance, researchers first introduced group maintenance into swarm systems based on the similarity of multi-level systems. The group maintenance method is of great significance for ensuring system safety and restoring the ability of multi-level systems [7]. The existing research's maintenance models generally conduct many analyses at the unit and system levels [8–10], group the maintenance information at the unit level, and make system-level maintenance decisions. The general research method of unit maintenance [11] uses the unit-level failure rate to propose a grouping strategy and combine the system maintenance cost to model the unit maintenance model. In addition, there is a degradation model [12] used to describe intuitively. the health and working conditions of the unit level, and then combined with the system-level maintenance cost to carry out maintenance planning for the system. However, unlike other unit systems, drone swarm communicate more frequently at the unit level and in the system. The swarm network that relies on information transmission is likely to affect the reliability of missions due to the maintenance of the unit level. The performance parameters are also closer to the topology parameters of complex networks. It is difficult to guarantee the execution of actual swarm missions based only on unit-level degradation parameters and system-level cost functions. Therefore, in this study, we consider introducing complex network-related methods to evaluate system-level reliability, As another optimization indicator for swarm maintenance, the mission capability of the swarm is fully taken into account.

The complex network method constructs a network model of mutual information interaction in the form of nodes and edges. According to this model, the system's topology structure and evolution law is studied from the perspective of the nodes, edges, and their network evolution. Then the critical points of the system can be targeted. The system is analyzed by vulnerability cascading effect, mission reliability, and other vital issues [13]. With the development trend of complex systems and unmanned cooperative systems in recent years, complex network methods are widely used in swarm-like systems such as transportation systems [14] and circuit systems [15]. At present, the complex network methods mainly apply to cascade failure [16], seepage, phase transition [17], propagation dynamics [18], and other mechanisms to evaluate network vulnerability [19], elasticity [20], mission reliability and other characteristics.

To solve the optimization problem under multi-parameter constraints, the currently recognized method in the academic circles is the multi-objective optimization theory [21]. Whether it is based on the traditional linear weighting method or the particle swarm optimization algorithm based on the evolutionary algorithm, it has been widely and standardized in various fields.

Therefore, because of the combination of mission capability and maintenance planning of swarm systems, this paper proposes a maintenance method for drone swarm, which adopts multi-objective optimization theory and a swarm system-level mission reliability evaluation model to solve. This method provides a cognitive basis for the drone swarm research of related projects, and also provides technical support for the follow-up more complex cluster maintenance model research.

The remainder of this paper is as follows: Section 2 describes the mission characteristics of drone swarm and the theoretical basis for evaluating the reliability of swarm systems using complex network methods. Section 3 puts forward the basic assumptions of the application scenarios for the swarm maintenance method proposed in this paper and gives specific practical methods. Section 4 uses the drone swarm network model to verify the feasibility and superiority of the method by comparing it with the traditional method of timely maintenance of single aircraft. Section 5 illustrates the effectiveness of the application of this method with a general case model as an example. In Section 6, we summarize this paper and propose future work.

2. Preliminaries

2.1. Mission Characteristics of Drone Swarm

A drone swarm is a system unit composed of multiple drones with system mission capabilities. However, the drone swarm is not a simple combination of numerous drones. It also includes many structures such as data links between drones, mission network, ground control platform, etc., as shown in Figure 1.

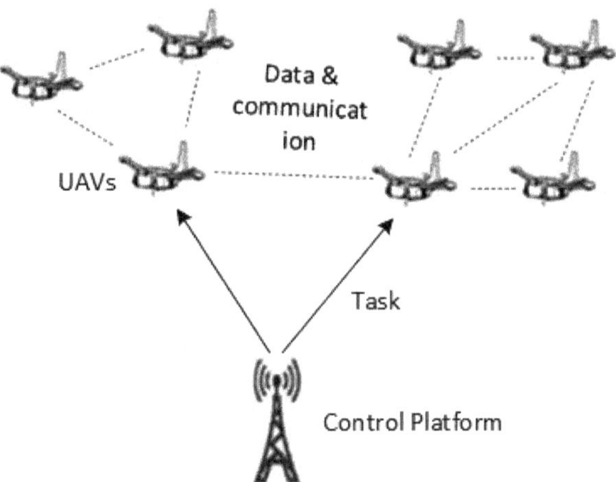

Figure 1. Composition of drone swarm.

The distance between the individual drones is controlled by formation to ensure a specific data transmission and communication capability. The ground control system or the drone of the overall mission independently assigns the swarm mission. The submissions are transmitted through the data link to form the mission execution. Links form an internal mission network, and each drone performs the corresponding sub-missions according to the assigned instructions and comprehensively completes the final mission of the swarm.

2.2. Complex Network Model of Swarm

In the theory of complex network topology, the network is composed of nodes and edges, where edge represents the attribute parameters of edges in the form of weights. When modeling drone swarm systems with complex networks, a drone is usually regarded as a node, and the data link connection between drones is considered an edge. Both nodes and edges are given corresponding attributes according to the actual swarm capability values. Finally, transform the complex drone swarm system into a hierarchical network model composed of interacting nodes for analysis.

Considering there are factors at different levels in the drone's mission execution process, such as communication interference, structural attack, mission system failure, etc., all of which are related to the degradation and unexpected collapse of the corresponding system level of the drone. Therefore, multi-node network is introduced, considering the changes in the capability indicators of different types of drones and thoroughly combining the follow-up corresponding maintenance strategies. According to the characteristics of drone and swarm interaction, the drone node is divided into three nodes of communication, structure, and mission, corresponding to the three-layer system of communication data link, drone carrier, and mission load in the network. Then, the relationship between the layers is described on this basis, and the process of establishing a multi-layer complex network is shown in Figure 2.

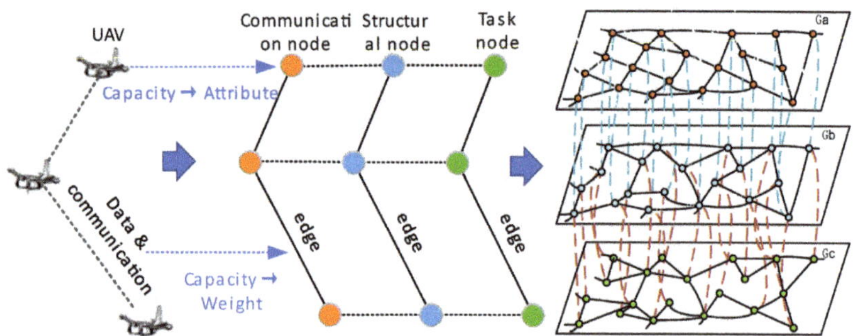

Figure 2. Establishment of multi-layer complex network structure.

2.3. Relationship between Swarm Maintenance and Network

Swarm maintenance is based on stand-alone maintenance. The premise of stand-alone maintenance is that the reliability of a certain level of single-machine structure cannot meet the requirements of subsequent missions. Therefore, it is necessary to recall the drone to be repaired during the mission execution process. At this time, the remaining swarm continue to perform the predetermined mission. That is, the maintenance of a drone is equivalent to the disappearance of the corresponding node in the swarm network. No matter what level (communication, structure, and mission) fails for a drone, the entire drone needs to be recalled for maintenance. The failure of a single-level node is equivalent to the disappearance of the corresponding nodes at the three levels, As shown in Figure 3.

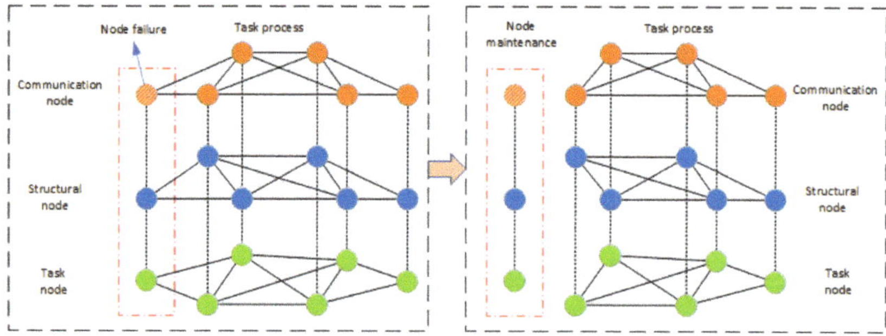

Figure 3. Relationship between swarm maintenance and network.

We pay more attention to the mission completion ability of the drone swarm, so the node failures in the communication layer and the structure layer are mapped to the disappearance of the corresponding nodes in the mission layer in the swarm network. At this time, the reliability assessment for the mission network is to consider all levels.

2.4. Swarm Mission Reliability Evaluation Method

Network reliability refers to the ability of the network to maintain functional and structural integrity when nodes and edges in the network are attacked. In the actual maintenance scenario, the maintenance of a drone can also be regarded as removing the mission node in the corresponding network model, and the corresponding connection edges disappear. Based on this, the network reliability can be used to evaluate the reliability of swarm missions.

The concepts of basic network parameters: node degree, average path length, and swarm coefficient are introduced first. to establish network reliability evaluation indicators.

The node degree k_i is defined as the number of edges that node i is directly connected to other nodes, and the average of the degrees of all nodes in the network is called the average network degree, denoted as $\langle k \rangle$.

The shortest path between two nodes i and j in the network refers to the way with the least number of edges connecting these two nodes, and the distance d_{ij} between node i and j are defined as the edge of the shortest path connecting these two nodes. Calculating the average path length L of the network is defined as the average of the distances between any two nodes.

$$L = \frac{1}{\frac{1}{2}N(N-1)} \sum_{i \geq j} d_{ij} \tag{1}$$

where N is the number of network nodes.

The swarm coefficient C of the network is defined as the average of the swarm coefficients of all nodes in the network.

$$C = \frac{1}{N} \sum_{i=1}^{N} C_i \tag{2}$$

Among them, the swarm coefficient C_i of a node i with a degree k_i in the network is defined as:

$$C_i = \frac{E_i}{\frac{k_i(k_i-1)}{2}} \tag{3}$$

Here E_i and $\frac{k_i(k_i-1)}{2}$ are the actual number of edges and the possible maximum number of edges between node i and its k_i neighbor nodes, respectively.

Based on the above three fundamental indicators of a complex network, reference [22] proposes network reliability indicators:

Assuming that the swarm system consists of N drones, the swarm system is randomly removed by considering the network characteristics, and the reliability index value $s_k(k = 1, 2, 3, \ldots, m)$ of the remaining network is calculated at the same time.

For each index, a set of changes of the index with successive removal of nodes $\Delta s_{ki}(i = 1, 2, 3, \ldots, N)$ can be obtained. Then the variance of the change of the index can be calculated, which is expressed as follows:

$$S^2 = \frac{\sum_{i=1}^{N} \left(\Delta s_{ki} - \overline{\Delta s_{ki}} \right)^2}{N}, \ k = 1, 2, 3, \ldots, m \tag{4}$$

Furthermore, by repeating this process n times, the mean value of the variance of the variation of the indicator for n simulations can be obtained, which can be expressed as follows:

$$\overline{S_k^2} = \frac{1}{n} S_k^2, \ k = 1, 2, 3, \ldots, m \tag{5}$$

By comprehensively comparing the network reliability index and network performance index after sensitivity analysis, the weight of each index in the comprehensive reliability evaluation index can be determined according to the mean value of the variance of the index variation. The relationship between the weight and the mean variance is expressed as follows:

$$w_k = \frac{\overline{S_k^2}}{\sum_{k=1}^{m} \overline{S_k^2}}, \ k = 1, 2, 3, \ldots, m \tag{6}$$

Finally, the index processing method of the literature [23] is introduced, and the comprehensive evaluation index of network reliability can be obtained, which is expressed as follows:

$$R = \sum_{k=1}^{m} w_k \frac{s_k - \min(s_k)}{\max(s_k) - \min(s_k)} \tag{7}$$

Among them, s represents the structural topology parameters and network performance parameter indicators, such as the average degree of the entire network node, the network efficiency of the communication layer, etc. w_k Represents the weight of the indicator, and its value can be determined according to Equation (7); R represents the reliability of the network, and its value range is [0, 1], and the closer the value is to 1, the better the network reliability.

3. Swarm Maintenance Method

Aiming at the problem of how to formulate a swarm maintenance plan considering both the reliability of swarm missions and the maintenance cost, this chapter proposes a swarm maintenance method based on the reliability assessment of multi-layer complex network missions. The framework of the method is shown in Figure 4. This method integrates the primary method of group maintenance and a complex network analysis method. Based on the existing group maintenance, the drone level uses the degradation prediction method to evaluate the reliability of the drone, and the system level introduces the mission network and mission reliability evaluation index, and finally uses the multi-level maintenance method. The objective optimization method optimizes for maintenance cost and mission reliability. The method includes four steps: unit reliability prediction based on prevention, maintenance grouping strategy and cost, mission reliability evaluation, and optimize grouping decision.

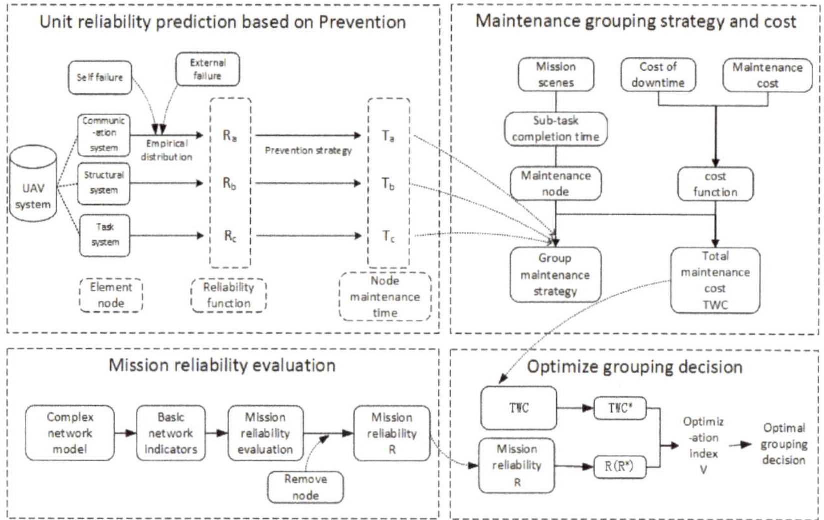

Figure 4. Basic theoretical flow chart.

3.1. Basic Assumptions

This paper makes the following assumptions about the mission scenarios and maintenance behaviors of drone swarm applying this method:

Assumption 1. *The theoretical maintenance time considering the reliability of system components is based on a preventive strategy, allowing a lag in maintenance within a specific time frame.*

Assumption 2. *Failure maintenance includes preventive maintenance of two kinds of failures: accidental failure maintenance and degradation failure maintenance. This method does not consider other unknown failure maintenance caused by complex environmental stress during the mission.*

Assumption 3. *The drone in a maintenance state only affects the reliability of the swarm system in the next mission stage.*

Assumption 4. *After the maintenance is completed, the return swarm link remains unchanged. It is assumed that the state parameters after the maintenance are the same as the initial parameters.*

Based on the above assumptions, this paper introduces the basic theory of the maintenance method decision in Sections 3.2–3.5.

3.2. Unit Reliability Prediction Based on Prevention

According to experience, for determining the drone reliability function distribution of the system, empirical distributions, such as exponential distribution, normal distribution, Weibull distribution, etc., can be used to represent the reliability function distribution under its degradation failure. For the accidental failure in the drone mission, it is clarified that based on the known empirical data, the drone with unexpected failure can also be obtained using the observed distribution.

According to the above distribution theoretical basis, combined with the observation data of the single machine, fit and estimate the reliability prediction, and obtain the single machine life. For example, if the single machine structure life distribution obeys an exponential distribution, according to the reliability function formula of the exponential distribution: $R(t) = \exp(-\lambda t)$. Where λ is the failure rate constant, determined by empirical data.

From the above analysis, we can predict any single node's life considering accidental and self-degradation failure. For drone stand-alone equipment, preventive maintenance is generally used. In the case of prevention, a small estimation of the lifespan can be obtained to obtain the stand-alone maintenance time T based on prevention.

Considering the maintenance time T of a single machine based on prevention, a single machine failure may occur before time T and after time T. If a single machine failure occurs before time T, the single machine will not be able to perform the mission immediately and enter the maintenance state. According to the reliability function, at time T, The probability of failure before and after time T is respectively $R(T)$ and $F(T) = 1 - R(T)$. Next, the total maintenance cost is further calculated.

Because there is a demarcation point at time T, the cost of preventive maintenance is different before and after time T. C_p and C_{dp} are defined as the maintenance cost and downtime cost of maintenance before time T, respectively, and C_f and C_{df} are defined as maintenance costs after time T respectively. Maintenance cost and downtime cost, t is the actual maintenance time. Therefore, the equivalent stand-alone maintenance cost at time t is expected to be

$$TWC(t) = \frac{\left(C_p + C_{dp}\right) R(t) + \left(C_f + C_{df}\right) F(t)}{\frac{t}{T}} \tag{8}$$

It can be seen from the calculation formula of TWC that the equivalent maintenance cost brought by earlier maintenance is relatively large. It is consistent with the actual maintenance situation of the drone swarm; in addition, the formula does not involve accidental failure and self-degradation failure. The difference can be applied to the maintenance cost calculation in both cases.

In Section 2.2, we propose considering the different failure modes of a single drone. The above life prediction and cost calculation methods of a single aircraft are also applied to different structural ways of a single plane. The three nodes of communication, structure, and mission are used for node life prediction and cost calculation. Calculated to get the prevention-based node maintenance time and maintenance cost.

3.3. Maintenance Grouping Strategy and Cost

From Section 3.2, it can be seen that any node failure in a drone system requires drone maintenance. Single-machine maintenance will inevitably lead to increased maintenance costs, downtime, and reduced reliability at the swarm system level. Rest for maintenance results in extremely high maintenance costs and low mission reliability for the swarm.

We propose a swarm maintenance method based on mission completion time. The swarm system decomposes the total mission into sub-missions for different drone groups through mission allocation because of the characteristics of the completion time interval between sub-missions and little influence on mutual system reliability. It is assumed that there is a maintenance opportunity after each submission is completed, and the maintenance time is determined by the mission time. The maintenance opportunity divides the mission time of the entire swarm into multiple time intervals $t_1, t_2, \ldots, t_n$. According to the comparison between the preventive maintenance time T of the stand-alone node in Section 3.2 and the corresponding time t of the maintenance opportunity (that is, the completion time of the submission), select the drone near the maintenance time to enter the maintenance state at the corresponding maintenance time t, as shown in Figure 5. Unit 1 chooses to be repaired at the time t_3. Unit 2 and unit 3 can choose to be repaired at the time t_1 And unit 4 and unit 5 can choose to be repaired at the time t_2.

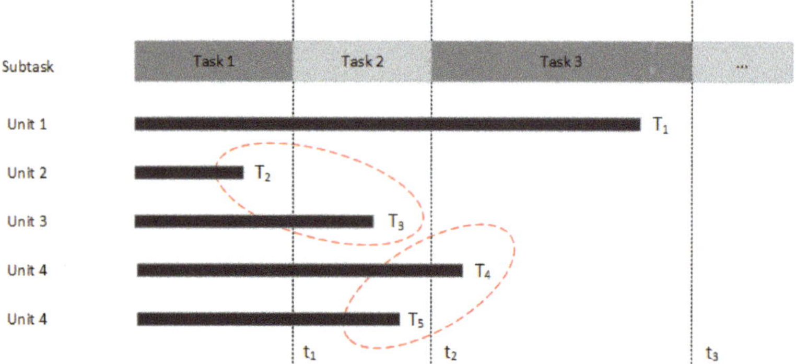

Figure 5. Maintenance grouping strategy.

At this time, according to Formula (8), the total cost of the equivalent maintenance of the actual single machine is

$$TWC_i(t_j) = \frac{\left(C_p + C_{dp}\right)R(t_j) + \left(C_f + C_{df}\right)F(t_j)}{\frac{T_i}{t_j}} \tag{9}$$

The total maintenance cost of all single-node nodes is accumulated to obtain the total system maintenance cost TWC under this grouping strategy.

$$TWC = \sum_{j=1}^{n}\sum_{i=1}^{N} TWC_i(t_j) \tag{10}$$

where N represents the total number of nodes and n represents the total number of maintenance opportunities.

The cost calculation function includes maintenance cost and downtime cost. The cost function is related to the grouping strategy selection, the maintenance opportunity, and the actual maintenance time of a single machine. In terms of maintenance strategy selection, each node can choose to maintain in advance or lag. At this time, different combinations form a variety of maintenance strategies.

3.4. Mission Reliability Evaluation

For different maintenance grouping situations, it is necessary to perform swarm mission reliability evaluation on all maintenance opportunities under the grouping strategy in the entire mission cycle. For example, for the maintenance grouping situation shown in Figure 5, at maintenance time t_1, remove maintenance nodes 2 and 3, evaluate the mission reliability R_{11} of the remaining nodes to form a swarm network, remove maintenance nodes 4 and 5 at maintenance time t_2, and evaluate the remaining nodes. The mission reliability R_{12} of the swarm network is formed, and so on, to obtain the swarm mission reliability $R_{11}, R_{12}, \ldots \ldots, R_{1n}$ under all maintenance opportunities under the first group of maintenance strategies.

Among them, the swarm mission reliability $R_{11}, R_{12}, \ldots \ldots, R_{1n}$ under all maintenance, opportunities may exist if the swarm mission reliability is lower than the mission completion threshold at a particular maintenance time due to the centralized maintenance of multiple nodes. To be eliminated, the n groups of data for grouping decisions without abnormal data can be weighted and averaged, respectively, according to the weight value of maintenance opportunities, and the final swarm mission reliability R_i the group strategy of this group can be obtained:

$$R_i = w_1 R_{i1} + w_2 R_{i2} + \cdots + w_n R_{in} \tag{11}$$

Among them, w_i is the weight value of swarm robustness under the maintenance opportunity i, $w_1 + w_2 + \cdots + w_n = 1$.

The swarm mission reliability index under each grouping strategy is obtained and combined with the total maintenance cost obtained according to the cost function under each maintenance plan in Section 3.3. It provides an index basis for the subsequent comprehensive optimization index and grouping decision.

3.5. Optimize Grouping Decision

In this method, the swarm maintenance grouping decision is a multi-objective optimization problem that finally considers the total maintenance cost and the reliability of the swarm mission. For the optimization problem of the two objectives of this method, we use the linear weighting method to solve the problem and propose the final optimization index V, which can be expressed as:

$$V_i = w_1 \varphi(O_{1i}) + w_2 \varphi(O_{2i}) + \cdots + w_n \varphi(O_{ni}) \tag{12}$$

Among them, w_i is the weight value of the optimization target, $w_1 + \cdots + w_n = 1$.

$\varphi(O_{1i}) \sim \varphi(O_{ni})$ Represents the parameter value of each optimization objective. There are only two parameter values in this method, but the value ranges of different optimization objectives are different, such as the mission reliability index $R \in [0,1]$, But the total cost $TWC \in (0, \infty)$. Therefore, it is necessary to normalize the optimization target parameter values.

$$O_{ni} = \frac{\theta_{ni}}{\max(\theta_{n1}, \theta_{n2}, \ldots, \theta_{nk})} = \frac{\theta_{ni}}{\theta_n^*} \tag{13}$$

Among them, θ_{ni} is the parameter value of the policy i of the n optimization objective:

$$V_i = w_1 \varphi\left(\frac{\theta_{1i}}{\theta_1^*}\right) + w_2 \varphi\left(\frac{\theta_{2i}}{\theta_2^*}\right) + \cdots + w_n \varphi\left(\frac{\theta_{ni}}{\theta_n^*}\right) \tag{14}$$

In the normalization process of the above indicators, we default that the larger the optimization target value, the better. For example, the closer the mission reliability index is to 1, the better. However, not all indicators obey this assumption, such as the total cost TWC in this method. Therefore, we add the φ function, which is defined as follows:

$$\varphi(O_{ni}) = \begin{cases} \dfrac{\theta_{ni}}{\max(\theta_{n1}, \theta_{n2}, \ldots, \theta_{nk})}, & \varphi \propto \theta \\[2mm] -\dfrac{\theta_{ni}}{\max(\theta_{n1}, \theta_{n2}, \ldots, \theta_{nk})}, & \varphi \propto \left(\dfrac{1}{\theta}\right) \end{cases} \tag{15}$$

Based on the above analysis, we establish the comprehensive optimization index of system mission reliability index and total maintenance cost as:

$$V_i = w_1 R_i - w_2 \varphi \left(\frac{TWC_i}{TWC*} \right) \tag{16}$$

where w_1 and w_2 are the index weights of system mission reliability and total maintenance cost respectively and $w_1 + w_2 = 1$. When $w_1 > w_2$, the system mission reliability index is the dominant factor in the final decision. When $w_1 < w_2$, the total maintenance cost is the dominant factor in the final decision. When $w_1 = w_2$, the two have equal status.

4. Simulation Analysis

To verify the effectiveness of this method applied to drone swarm maintenance, taking a typical case as an example, the total maintenance cost and swarm system reliability under the application of this method is compared with the total cost and system reliability under the timely maintenance strategy.

Assuming that multiple drone swarm with heterogeneous drones perform the overall mission, the assumed sub-mission completion times are 500 min, 650 min, 800 min, and 950 min, respectively. Taking a swarm composed of 10 drones as an example, the establishment of the swarm model structure and communication link situation of the swarm model are shown in Figure 6, and each drone reliability parameters are estimated based on experience shown in Table 1.

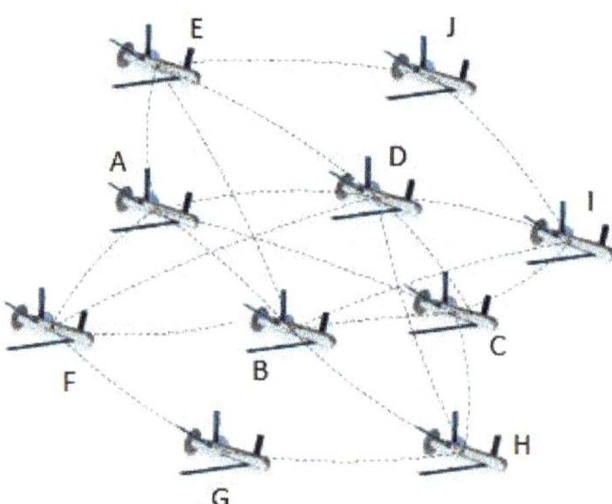

Figure 6. Method to verify the 10-node swarm model.

Table 1. Parameters of each node.

Nodes	Life Distribution Model	Reliability Function	Cost ($)			
			C_p	C_{dp}	C_f	C_{df}
A, B, C, D	Index distribution	$R(t) = e^{-(e^{6.75 \times 10^{-4}t} - 1)}$	2840	150	26	150
E, F	normal distribution	$R(t) = P\left(z > \frac{t-800}{200}\right)$	1200	150	3670	150
I, H	Weibull distribution	$R(t) = e^{-(\frac{t}{700})^3}$	960	150	7630	150
J, G	Inverse Gaussian distribution	$\mu = 900, \lambda = 8 \times 10^5$	1984	150	18	150

The established drone node is the minor node of the system and does not continue to divide the three-layer network. In addition, the life distribution of the node is given randomly, covering different empirical distribution functions as much as possible to verify the universality of the method.

4.1. Maintenance Time and Grouping Strategy

The reliability function under the empirical distribution of each node is given in the Table 1. Under the condition that a large number of nodes obey the same distribution, the maintenance time based on prevention can be generated by a random method. According to the empirical data, we use its reliability for a single node to decrease. The preventive maintenance time of the node is calibrated at the corresponding time of 0.6, and the preventative maintenance time under the four life distributions is 611 min, 748 min, 559 min, and 936 min, respectively. Considering that each drone unit does not start the mission in the best state, there is an actual maintenance time ahead of schedule. Due to the possibility of theoretical maintenance time, the exact maintenance time of a single machine earlier than the academic maintenance time is randomly generated, as shown in Table 2.

Table 2. Maintenance time of nodes based on prevention.

Nodes	A	B	C	D	E	F	G	H	I	J
Actual maintenance time(min)	584	603	536	555	689	720	930	496	523	871

Each maintenance opportunity is allocated according to the completion time of the submissions, and the following maintenance grouping scheme will be obtained when considering the maintenance in advance.

For each grouping strategy in the Table 3, the cost function can obtain the total maintenance cost under the maintenance strategy. The complex network method is used to evaluate the mission reliability of the swarm and adjust the maintenance plan.

Table 3. Swarm maintenance group.

Maintenance Plan	500 min	650 min	800 min	950 min
①	I, H, C	A, B, D	E, F	J, G
②	I, H, D, C	B, A, E	F, J	G
③	I, H, D	A, B, C	E, F	J, G

4.2. Mission Reliability and Cost of a Swarm System

As shown in Section 2.4, the mission reliability assessment of a complex network system requires establishing a complex network model. For the model structure in Figure 6, the established complex network model is shown in Figure 7.

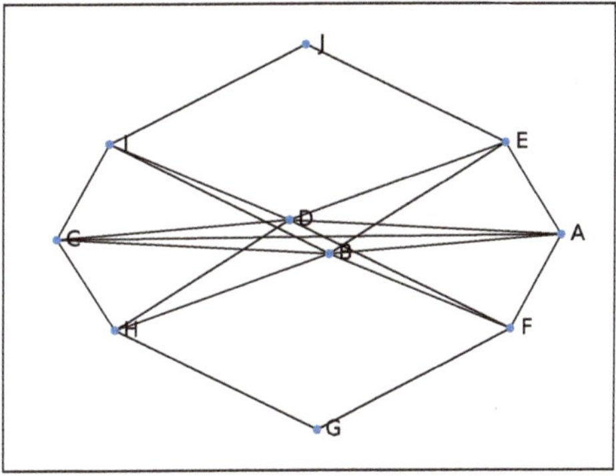

Figure 7. Complex network model.

Based on the complex network evaluation method, the corresponding nodes are removed from the swarm network under the related maintenance opportunity, and the reliability changes of the system missions are evaluated. As shown in the Figure 7, since the initial swarm network is not fully connected, the initial network mission of the evaluation If the reliability is not 1, it is normalized. The mission reliability index values and total maintenance cost results under each maintenance strategy are obtained as shown in the Table 4, and the comparison of each maintenance scheme pair over time is shown in Figure 8.

Table 4. Mission reliability and total maintenance cost under each maintenance strategy.

Maintenance Strategy	Mission Reliability				Total Maintenance Cost ($)
	500 min	650 min	800 min	950 min	
①	0.803	0.735	0.892	0.933	23,620
②	0.718	0.779	0.917	0.970	22,343
③	0.819	0.714	0.906	0.937	22,864

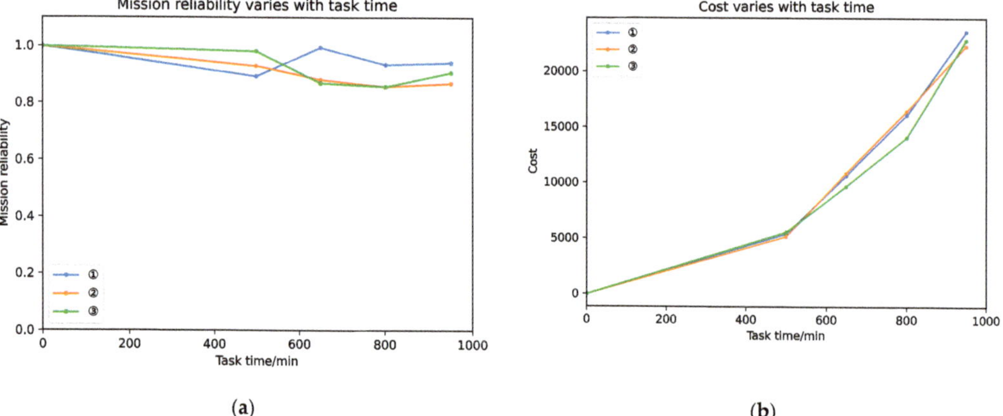

Figure 8. Graph of the change of indicators over time under different maintenance strategies for 10 nodes: (**a**) describes the change of mission reliability with mission time; (**b**) describes the change of the total cost of swarm maintenance with mission time.

4.3. Comparison Summary

In the following, for the maintenance scenario considering timely maintenance, the swarm model in Figure 6 is used to analyze the cost and system mission reliability. Under the condition that convenient maintenance is considered, every drone needs to be shut down immediately for maintenance at the expected maintenance time. Under the condition of viewing the downtime cost, the complex network model is used to remove the nodes corresponding to the four-time points, and the number of nodes with different numbers of nodes is obtained. Figures 9–11 compares the system mission reliability changes and the total maintenance cost with the results of group maintenance.

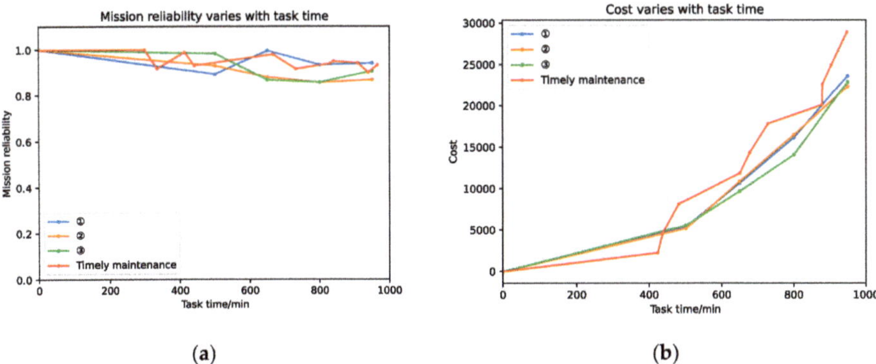

(a)

(b)

Figure 9. Comparison of timely maintenance and group maintenance strategies under 10 nodes: (**a**) describes the change of system mission reliability with mission time; (**b**) describes the change of total maintenance cost with mission time.

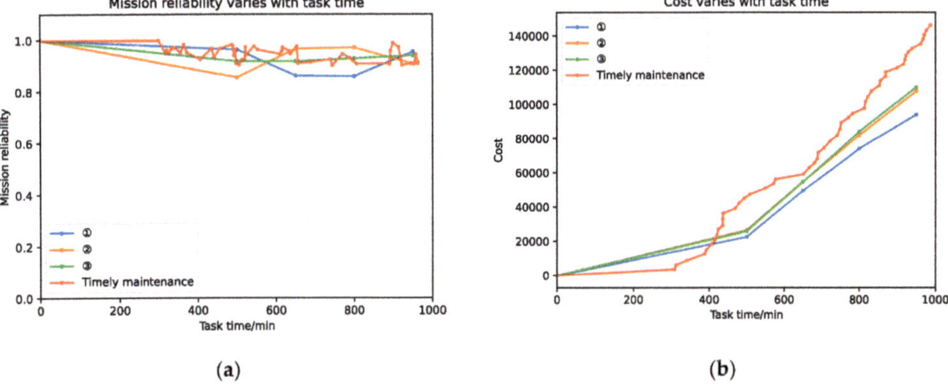

(a)

(b)

Figure 10. Comparison of timely maintenance and group maintenance strategies under 50 nodes: (**a**) describes the change of system mission reliability with mission time; (**b**) describes the change of total maintenance cost with mission time.

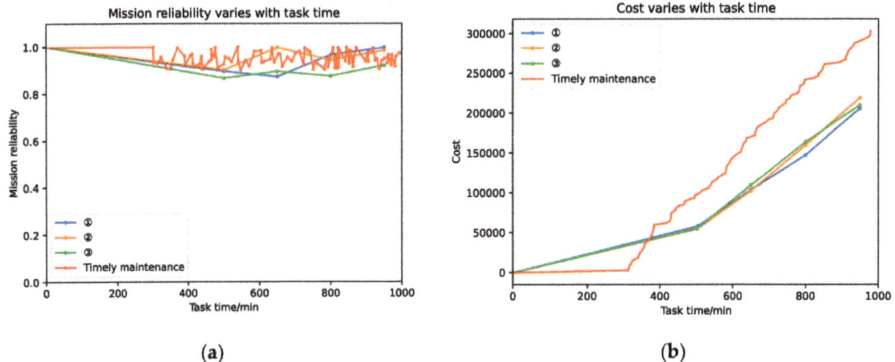

(a)

(b)

Figure 11. Comparison of timely maintenance and group maintenance strategies under 100 nodes: (**a**) describes the change of system mission reliability with mission time; (**b**) describes the change of total maintenance cost with mission time.

According to the comparison of the system mission reliability index and maintenance cost under the timely maintenance strategy and the group maintenance strategy, the following conclusions can be drawn:

(1) In terms of system reliability: (**a**) As the scale of the swarm increases, the impact of node maintenance brought by timely maintenance strategy and group maintenance strategy on the reliability of swarm missions will gradually decrease. This is because there are many nodes, and the maintenance and removal of nodes have relatively little impact on the reliability of the remaining swarm missions; (**b**) The average mission reliability under the timely maintenance strategy is generally higher than that under any group maintenance. Because timely maintenance only repairs one node at a time, it has less impact on the reliability of the remaining swarm missions; (**c**) The reliability of system missions fluctuates under the timely maintenance strategy. Because the system is forced to be repaired during non-mission completion time and many repair opportunities, the system mission reliability fluctuates wildly. Still, the system mission reliability fluctuation will weaken with the expansion of the swarm scale.

(2) In terms of maintenance costs: (**a**) Considering the cost of downtime, the maintenance cost of timely maintenance is far greater than the maintenance cost of group maintenance under any strategy. The cost has a specific impact, that is, the cost of maintenance in advance is reduced, whereas the cost of maintenance increases; (**b**) As the scale of the swarm continues to expand, the cost of timely maintenance is higher than the cost of any grouped maintenance. The scale has grown exponentially.

Based on the above conclusions, the group maintenance method considering mission reliability has great advantages in maintenance cost control, but has no obvious advantages in mission reliability. When considering the method selection in specific scenarios, the timely maintenance strategy should be applied in scenarios with high requirements for cluster task reliability, and the group maintenance method considering mission reliability should be applied in scenarios with high requirements for maintenance cost.

In addition, under the group maintenance strategy, different group strategies can be selected to meet the different requirements for maintenance cost and system mission reliability during mission execution.

The time and space complexity of this method is calculated as follows: Assuming that there are n drones in the whole mission cycle of the swarm that need to be maintained. Then the nodes of the swarm need to be removed in sequence for a single calculation of the robustness parameter under a limited number of maintenance opportunities, and the time complexity is $O(n) * O(n)$, so the total time complexity of the algorithm is $O(1)$. The process needs to establish a one-dimensional data space, and the space complexity is $O(n)$.

5. Case Study

This section will take a widely used drone swarm performing a long-endurance tour mission as an analysis case and formulate an optimal group maintenance strategy to maximize the mission reliability of the swarm system.

In this case, the drone swarm adopts the control structure of the distributed pilot method (wolves). The swarm system is divided into five sub-mission swarm, each with 20 single drones (including one central drone with communication relay and distributed decision center functions and 19 mission execution drones), totaling 100. According to the theory of each multi-layer complex network. There are a total of 300 unit nodes. The distributed link relationship of the communication network between every single node is shown in Figure 12.

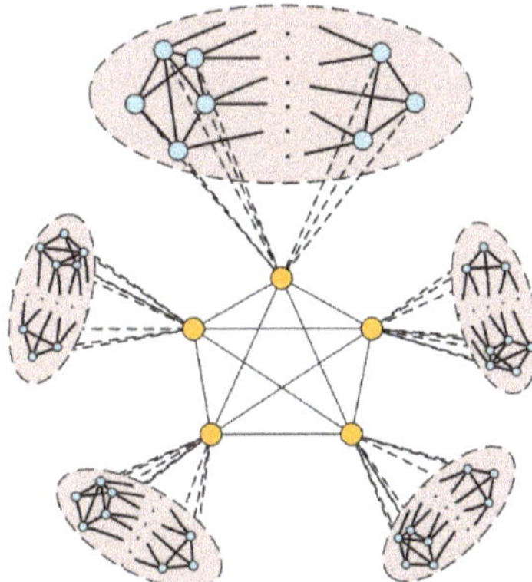

Figure 12. Distributed connection diagram between individual nodes.

Among them, one communication relay drone in each group of drones establishes a communication link connection with all drones in the sub-swarm. The drones in the group establish connections randomly, and five communication relay drones are used in pairs. Establish connections to each other to form the network topology of the entire swarm.

The three-layer complex network model established according to the drone node connection diagram is shown in Figure 13.

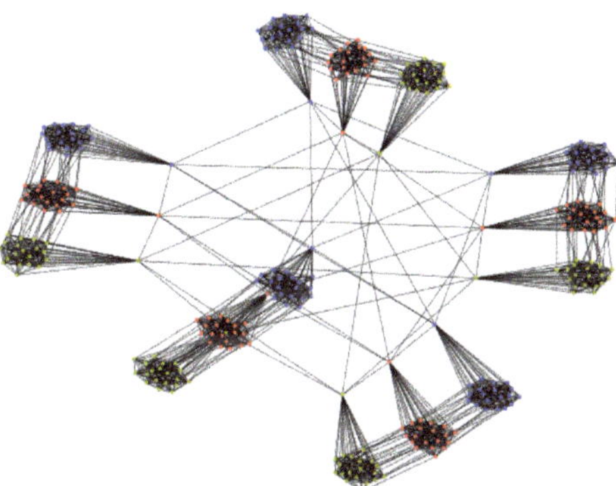

Figure 13. Three-layer complex network model.

This drone swarm performs three missions in its life cycle, with three mission profiles. The missions will be performed in a particular order. The sequence is shown in Figure 14, and missions 1, mission 2, and mission 3 are executed each cycle. Among them, the single execution time of mission 1 is 60 min, the single execution time of mission 2 is 120 min, and the single execution time of mission 3 is 180 min. The number of working cycles is 10, so the working time limit is 3600 min, and the maintenance opportunities of the drone swarm system are 60 min, 180 min, 360 min, . . . , 3600 min, a total of 30 maintenance nodes.

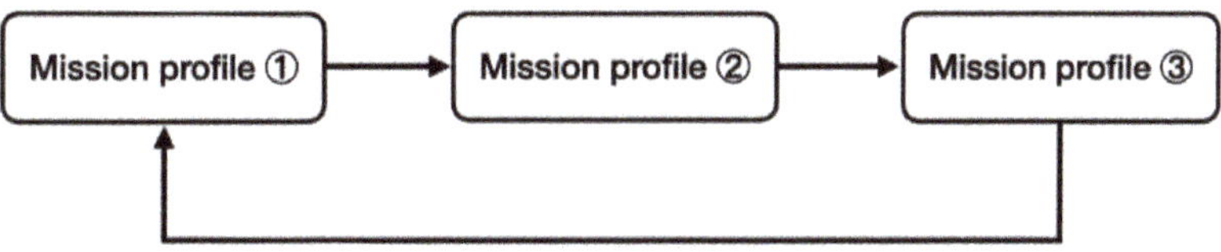

Figure 14. Mission profile sequence.

According to empirical data and observation data, it is assumed that the structural layer nodes, communication layer nodes, and mission layer nodes of the drone swarm obey exponential distribution, Weibull distribution, and inverse Gaussian distribution, respectively. Figure 15 shows the single-node prevention-based maintenance time distribution map generated by random distribution.

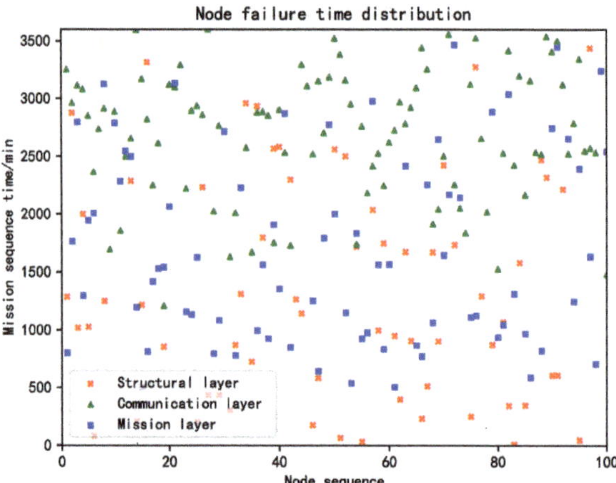

Figure 15. The distribution of maintenance time based on prevention for a single node.

According to the prevention-based maintenance node distribution and maintenance node time of each node, multiple groups of maintenance grouping decisions can be obtained. Taking a group decision as an example, the number of drones for maintenance under each maintenance opportunity is shown in Figure 16.

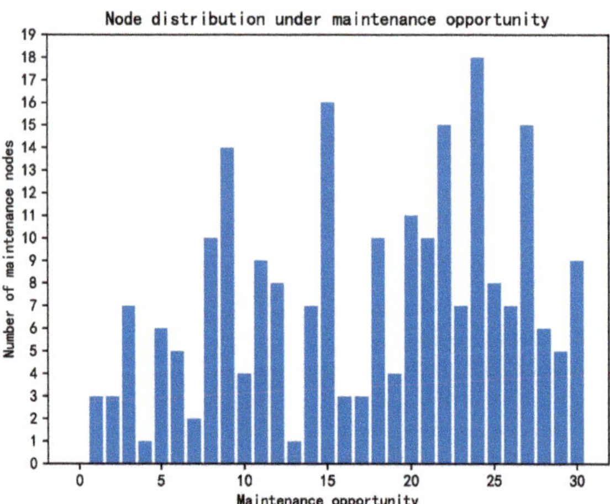

Figure 16. Distribution of maintenance nodes corresponding to maintenance opportunities.

In each maintenance opportunity, the maintenance node is regarded as the node removed. Under the nodes with different maintenance times, the mission reliability value of the swarm system with the maintenance node removed is considered. The mission reliability change diagram of the swarm system under the total mission time is obtained, as shown in Figure 17.

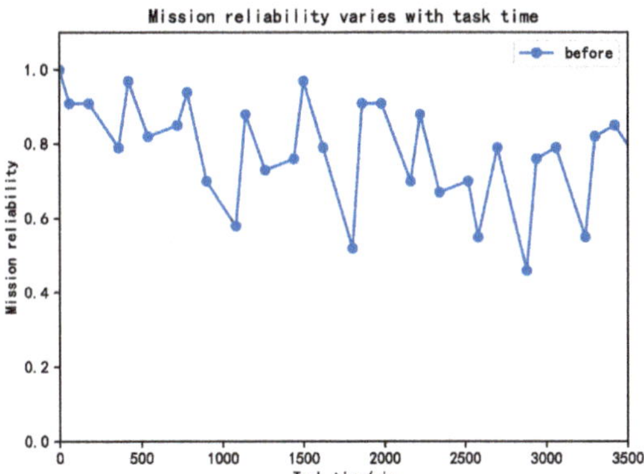

Figure 17. Changes in system mission reliability under this grouping strategy.

According to the mission reliability evaluation results in the above figure, the maintenance plan at 1800 min and 2880 min after the mission start leads to a sudden drop in the mission reliability of the swarm system.

Given this problem, the diversity of group maintenance schemes is used to repair some nodes in advance or lag in forming an optimized maintenance scheme and evaluate the reliability of system missions, as shown in Figure 18.

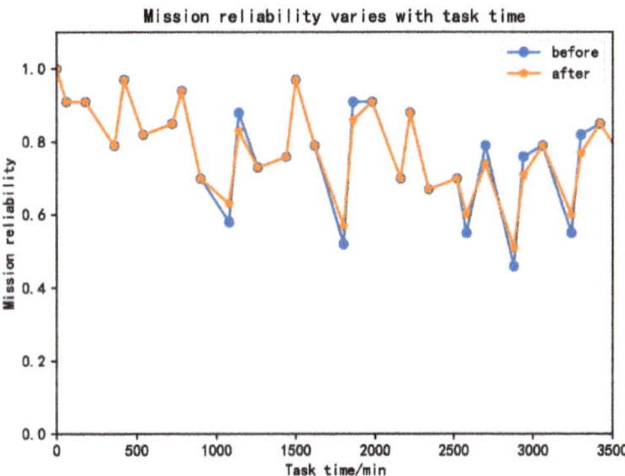

Figure 18. Changes in mission reliability of maintenance plan before and after optimization.

Optimizing the maintenance plan has a significant positive impact on the reliability of the drone swarm system. The maintenance plan can be further optimized for different drone mission requirements, reflecting the critical advantages of swarm group maintenance. Combined with the maintenance cost of the drone swarm, the optimal solution can be formulated and applied to the actual swarm maintenance research.

6. Conclusions

Aiming at the problem that the traditional unit maintenance method cannot meet the maintenance requirements of the complex swarm system, a mission-oriented system maintenance grouping method is proposed in this paper. Based on many maintenance groups proposed based on the failure model of drone functional nodes, complex network modeling and evaluation methods are used as tools. According to the topology information of the swarm network, the maintenance effects of the drone are spread to the mission capability changes of the swarm system network. The cost and swarm system mission reliability is the final objective function, and the grouping strategy is optimized.

The method validation and case analysis results show that the complex network method can effectively evaluate system-level reliability, solve the trade-off between cost and system reliability, and ensure the drone swarm system maintenance during mission execution. Swarm missions can be carried out effectively when the node part of the drone maintenance is in progress. Compared with the traditional maintenance method, the maintenance method can significantly save maintenance costs and reduce the number of downtimes while ensuring that the reliability of system missions is less fluctuated. It has the advantage of reducing maintenance costs. The cost increases exponentially. The method is more effective in larger and more complex swarm scenarios.

There are still many challenges to be solved in the future for the research and improvement of this method. For example, when setting the accidental failure model, the mission's real-time mission environment and accidental conditions can be considered, the failure model information can be updated in real time, and the strategy can be adjusted.

Author Contributions: Conceptualization, J.G. and L.W.; methodology, J.G.; software, J.G.; validation, J.G., L.W. and X.W.; formal analysis, L.W.; investigation, J.G.; resources, X.W.; writing—original draft preparation, J.G.; writing—review and editing, L.W. and X.W. All authors have read and agreed to the published version of the manuscript.

Funding: This research was funded by the Science and Technology Innovation 2030-Key Project of "New Generation Artificial Intelligence" under Grant 2020AAA0108201 and the National Natural Science Foundation of China (62176015).

Data Availability Statement: This paper did not use data from other sources.

Conflicts of Interest: The authors declare no conflict of interest.

References

1. Xiao, W.; Li, M.; Alzahrani, B.; Alotaibi, R.; Barnawi, A.; Ai, Q. A Blockchain-Based Secure Crowd Monitoring System Using UAV Swarm. *IEEE Netw.* **2021**, *35*, 108–115. [CrossRef]
2. He, D.; Yang, G.; Li, H.; Chan, S.; Cheng, Y.; Guizani, N. An Effective Countermeasure Against UAV Swarm Attack. *IEEE Netw.* **2021**, *35*, 380–385. [CrossRef]
3. Xu, J.; Sun, K.; Xu, L. Integrated system health management-oriented maintenance decision-making for multi-state system based on data mining. *Int. J. Syst. Sci.* **2016**, *47*, 3287–3301. [CrossRef]
4. Nzukam, C.; Voisin, A.; Levrat, E.; Sauter, D.; Iung, B. Opportunistic maintenance scheduling with stochastic opportunities duration in a predictive maintenance strategy. *IFAC-PapersOnLine* **2018**, *51*, 453–458. [CrossRef]
5. Gonçalves, P.; Sobral, J.; Ferreira, L. Establishment of an initial maintenance program for UAVs based on reliability principles. *Aircr. Eng. Aerosp. Technol.* **2017**, *89*, 66–75. [CrossRef]
6. Wang, H. A survey of maintenance policies of deteriorating systems. *Eur. J. Oper. Res.* **2002**, *139*, 469–489. [CrossRef]
7. Gandhi, K.; Schmidt, B.; Ng, A.H.C. Towards data mining based decision support in manufacturing maintenance. *Procedia CIRP* **2018**, *72*, 261–265. [CrossRef]
8. Batun, S.; Azizoğlu, M. Single machine scheduling with preventive maintenances. *Int. J. Prod. Res.* **2009**, *47*, 1753–1771. [CrossRef]
9. Koochaki, J.; Bokhorst, J.C.; Wortmann, H.; Klingenberg, W. The influence of condition-based maintenance on workforce planning and maintenance scheduling. *Int. J. Prod. Res.* **2013**, *51*, 2339–2351. [CrossRef]
10. Liu, X.; Wang, W.; Peng, R. An integrated preventive maintenance and production planning model with sequence-dependent setup costs and times. *Qual. Reliab. Eng. Int.* **2017**, *33*, 2451–2461. [CrossRef]
11. Tian, Z.; Liao, H. Condition based maintenance optimization for multi-component systems using proportional hazards model. *Reliab. Eng. Syst. Saf.* **2011**, *96*, 581–589. [CrossRef]
12. Martinod, R.M.; Bistorin, O.; Castañeda, L.F.; Rezg, N. Maintenance policy optimization for multi-component sys-tems considering degradation of components and imperfect maintenance actions. *Comput. Ind. Eng.* **2018**, *124*, 100–112. [CrossRef]
13. Ma, L.; Zhang, M. Research Progress on War Complex System of Systems Modeling Based on Complex Network. *J. Syst. Simul.* **2015**, *27*, 217–225. (In Chinese)
14. Qu, X.; Liu, E.; Wang, R.; Ma, H. Complex Network Analysis of VANET Topology With Realistic Vehicular Traces. *IEEE Trans. Veh. Technol.* **2020**, *69*, 4426–4438. [CrossRef]
15. Gao, Z.; Wang, H.; Dang, W.; Li, Y.; Hong, X.; Liu, M.; Chen, G. Complex Network Analysis of Wire-Mesh Sensor Measurements for Characterizing Vertical Gas–Liquid Two-Phase Flows. *IEEE Trans. Circuits Syst. II Express Briefs* **2020**, *67*, 1134–1138. [CrossRef]
16. Ren, W.; Wu, J.; Zhang, X.; Lai, R.; Chen, L. A Stochastic Model of Cascading Failure Dynamics in Communication Networks. *IEEE Trans. Circuits Syst. II Express Briefs* **2018**, *65*, 632–636. [CrossRef]

17. Zhao, Z. Study on the Network Traffic Abnormal Detection Based on EEMD. In Proceedings of the 2016 2nd International Conference on Advances in Mechanical Engineering and Industrial Informatics (AMEII 2016), Hangzhou, China, 9–10 April 2016.
18. Yang, L.; Draief, M.; Yang, X. Heterogeneous virus propagation in networks: A theoretical study. *Math. Methods Appl. Sci.* **2017**, *40*, 1396–1413. [CrossRef]
19. Duan, J.; Zheng, H. Vulnerability Analysis Method for Complex Networks Based on Node Importance. *Control Eng. China* **2020**, *27*, 692–696. (In Chinese)
20. Cheng, C.; Bai, G.; Zhang, Y.-A.; Tao, J. Resilience evaluation for UAV swarm performing joint reconnaissance mission. *Chaos* **2019**, *29*, 053132. [CrossRef]
21. Peng, C.; Huang, X.; Wu, Y.; Kang, J. Constrained Multi-Objective Optimization for UAV-Enabled Mobile Edge Computing: Offloading Optimization and Path Planning. *IEEE Wirel. Commun. Lett.* **2022**, *11*, 861–865. [CrossRef]
22. Wang, L.; Lu, D.; Zhang, Y.; Wang, X. A Complex Network Theory-Based Modeling Framework for Unmanned Aerial Vehicle Swarm. *Sensors* **2018**, *18*, 3434. [CrossRef] [PubMed]
23. Van Mieghem, P.; Doerr, C. *A Framework for Computing Topological Network Robustness*; Delft University of Technology: Delft, The Netherlands, 2010; pp. 1–11.

Article

Bearing-Based Distributed Formation Control of Unmanned Aerial Vehicle Swarm by Quaternion-Based Attitude Synchronization in Three-Dimensional Space

Muhammad Baber Sial [1,*], Yuwei Zhang [1], Shaoping Wang [1,2,3], Sara Ali [4,5], Xinjiang Wang [1,2,3], Xinyu Yang [6], Zirui Liao [1] and Zunheng Yang [1]

[1] School of Automation Science and Electrical Engineering, Beihang University, Beijing 100191, China
[2] Beijing Advanced Innovation Center for Big Data-Based Precision Medicine, Beihang University, Beijing 100191, China
[3] Ningbo Institute of Technology, Beihang University, Ningbo 315800, China
[4] School of Mechanical and Manufacturing Engineering, National University of Sciences and Technology, Islamabad 44000, Pakistan
[5] Human-Robot Interaction (HRI) Lab, School of Interdisciplinary Engineering & Science (SINES), Islamabad 44000, Pakistan
[6] School of Energy and Power Engineering, Beihang University, Beijing 100191, China
* Correspondence: babersial@buaa.edu.cn

Abstract: Most of the recent research on distributed formation control of unmanned aerial vehicle (UAV) swarms is founded on position, distance, and displacement-based approaches; however, a very promising approach, i.e., bearing-based formation control, is still in its infancy and needs extensive research effort. In formation control problems of UAVs, Euler angles are mostly used for orientation calculation, but Euler angles are susceptible to singularities, limiting their use in practical applications. This paper proposed an effective method for time-varying velocity and orientation leader agents for distributed bearing-based formation control of quadcopter UAVs in three-dimensional space. It combines bearing-based formation control and quaternion-based attitude control using undirected graph topology between agents without the knowledge of global position and orientation. The performance validation of the control scheme was done with numerical simulations, which depicted that UAV formation achieved the desired geometric pattern, translation, scaling, and rotation in 3D space dynamically.

Keywords: formation control; UAV swarm; quadrotor UAVs; VTOL UAVs; attitude synchronization; orientation estimation; bearing-based formation control

Citation: Sial, M.B.; Zhang, Y.; Wang, S.; Ali, S.; Wang, X.; Yang, X.; Liao, Z.; Yang, Z. Bearing-Based Distributed Formation Control of Unmanned Aerial Vehicle Swarm by Quaternion-Based Attitude Synchronization in Three-Dimensional Space. *Drones* **2022**, *6*, 227. https://doi.org/10.3390/drones6090227

Academic Editors: Xiwang Dong, Mou Chen, Xiangke Wang and Fei Gao

Received: 27 July 2022
Accepted: 26 August 2022
Published: 30 August 2022

1. Introduction

Nature has always inspired many great scientific triumphs and countless feats in the advancement of technology. Observation of natural formations of creatures such as a flock of birds, a school of fish, and formations of ants makes us realize that each entity in the formation controls its place in the formation just by aligning itself with respect to each other's bearing angle without knowledge of its global position or orientation. The same is true for aerobatic displays of piloted jet aircrafts in formation flying, where each aircraft is flown at a specific angle with other aircrafts of the formation. Design and control of distributed agent formations have become a keystone to solving multifarious complex applications such as coordination of mobile robots [1], satellite formation flying [2], and search and rescue [3]. A group or formation of UAVs, also referred to as a UAV swarm, has received compelling attention in military and civilian applications [4–6]. The process in which a group of agents obtains and maintains a predetermined geometric shape in space is called formation control [7].

UAV formations are all set to become one of the most essential tools for future military and civilian operations [8]. Formation control strategies, in which spatial constraints are defined among agents, are a powerful instrument in multi-robot systems [9]. Many kinds of consensus algorithms related to formation control for multi-agent systems can be found in the literature, mainly categorized as leader–follower [10], virtual structure [11], and potential field methods [12]. The leader–follower configuration is a widespread technique in formation control literature [13]. This configuration is ideal yet nontrivial in the case of distributed multi-agent systems, where a central controller such as a ground station is not present to centrally control all the agents. The leader follows a specific path or reference trajectory while all the followers are bound to adjust their position with respect to the leader. Because only knowledge about neighboring robots is required to define the formation, the leader–follower architecture best fits distributed schemes by enabling formation control relative to agent poses [14].

The current formation control methods, as per sensed and controlled variables, can be categorized into three groups: (1) position-based, (2) distance-based, and (3) bearing-based [9]. Position-based methods are currently most employed because they utilize the fact that each agent in the formation can obtain its position with respect to the global coordinate frame [15–17]. This means the agents rely on the global positioning system (GPS) or other related sensor information, forcing them to rely on external information to help conform themselves in a formation. However, in many situations, such as urban indoor or subterranean environments, the external signals cannot be obtained and position accuracy is uncertain, making it inadvisable to rely on such information. It is preferred to rely on an agent's onboard sensors rather than external sources for measurements. Trinh et al. [18,19] have verified that distance-based rigid formation control could not achieve global stability; furthermore, flip ambiguities commonly occur in distance-rigid graphs [20]. Additionally, compared to other approaches, the bearing-only approach has some advantageous features, such as relying less on the sensing ability of each robot [21]. The problem of bearing-based formation control of non-holonomic robots was considered by Li et al. [21] in 3D space using the Euler–Lagrange model using Euler angles to express 3D rotations of agents. Initial research on bearing-based formation control [22,23] was restricted to 2D space and primarily intended to control the bearing between agents to achieve the desired formation configuration. As per the proposed bearing rigidity theory, an almost globally stable control law was proposed for the single integrator robot with or without the inertial reference frame [24]. It is also important to mention that most of the reported research [24–26] has modeled their agents as a single or double integrator with randomly controlled velocity and acceleration. Using bearing rigidity-based control architecture also uniquely determines the formation's shape [24]. Additionally, it is paramount that orientation dynamics are independent of the position dynamics but not the other way around [24]. The bearing can also be calculated by employing the agent's onboard cameras [27] or vision sensors and sensor arrays [28,29].

Complete formation maneuver control and time-varying formation control using bearing-only measurements have not been realized yet [21]. Bearing rigidity theory to solve nonlinear robotic systems has generally only used the Euler–Lagrange model [21], where systems subjected to non-holonomic constraints are discussed. Quaternions are preferred over Euler angles because the latter are prone to gimbal lock when two out of three axes align during interpolation. Simple linearization using Euler angles overcompensates the errors in environments susceptible to unknown errors. Furthermore, if the inclination in disturbances is too large, the linear conditions are not met. However, using quaternions instead of Euler angles, even hard inclinations can also be sustained. Robustness against external disturbances is also a key factor for preferences of quaternions over Euler angles. Furthermore, it has also been observed that conversion of quaternions into a matrix is also efficient. Their mathematical simplicity comes from the fact that for modeling rigid body dynamics, no trigonometric functions are required [30]. As per the reviewed literature, directed graph topology takes less of a toll on the overall computation complexity of the formation but is not as robust as undirected graphs. In undirected graphs, bidirectional control of relative bearing measurements makes the formation more robust.

Motivated by the above observations, the significance of this article is such that we proposed a singularity-free novel quaternion-based relative attitude synchronization control scheme to reinforce undirected bearing-based control of a UAV swarm. Each agent's relative attitude and bearing were measured locally with its neighbor; hence, the dependence on the global coordinate frame was eliminated. The proposed approach validated its effectiveness on time-varying velocity and time-varying orientation leader agents in 3D space.

The main theoretical contributions of this article are:

1. A novel cascaded approach for distributed formation control of quadcopter UAVs was presented, consisting of an undirected bearing-based controller and a quaternion-based attitude synchronization controller working together in unison.
2. The distributed attitude synchronization and bearing-based formation control law were designed for 3D formation control as compared to [22,23], which have only designed bearing-based controllers for 2D space. Moreover, the proposed scheme uses quaternion-based attitude control, which is much more robust than research that has used Euler angles such as [21,31].
3. This work investigated and implemented the distributed formation control for time-varying velocity and time-varying orientation leader agents, which has not been accomplished yet in the domain of bearing-based formation control as compared to [21,31,32].
4. We designed our control method based on dynamic models of UAVs and undirected graph topology, a more robust technique as compared to [21,24,31], which only have used directed graph communications and kinematic models. The practical validation of the model was done using numerical simulations in MATLAB.

The remainder of this paper is organized as follows. In Section 2, preliminaries and problem formulation are given. Section 3 presents the system model and proposed control design. Section 4 covers the simulations and analysis, whereas discussion and the conclusions are drawn in Sections 5 and 6, respectively.

2. Preliminaries

In this section, we discuss some necessary background concepts about quaternions, the UAV quadcopter dynamical model, graph theory, and bearing rigidity theories that form a basis for problem formulation and design of our proposed control scheme.

2.1. Quaternions

This section briefly covers the mathematical background of quaternions, which are four-dimensional algebraic constructs that extend the concept of complex numbers. While quaternions are less comprehensible than Euler angles, quaternions lead to more efficient and accurate computation of rotations [33]. A quaternion is expressed formally as $q = q_0 + q_1 i + q_2 j + q_3 k$, where q_0 represents the real part or the scalar part and $q_1 i + q_2 j + q_3 k$ represents the vector part in R^3. Similarly, a pure quaternion's real part is zero. The conjugate of a quaternion is $q^* = q_0 - q_1 i - q_2 j - q_3 k$, whereas its norm is

$$\| q \| = \sqrt{q \otimes q^*} = \sqrt{q_0{}^2 + q_1{}^2 + q_2{}^2 + q_3{}^2} \tag{1}$$

The norm is calculated by taking the Kronecker product of a simple quaternion and its conjugate. Similarly, the quaternion inverse can be given by

$$q^{-1} = \frac{q^*}{\| q \|} \tag{2}$$

The quaternion conjugate can also be expressed as $q^* = q_0 - q$, and similarly, the quaternion inverse can be obtained by $q^{-1} = \frac{q^*}{\|q\|}$ hence $q^{-1} = q^*$. We assumed that only unit quaternions are used for quadrotor attitude representation for this work. Rotation from one coordinate frame A to another coordinate frame B can be expressed by conjugate operation; a quaternion expresses a rotation q_R with an added condition that its norm is equal to 1. Therefore, if q_A is a quaternion expressed in frame A, then the same quaternion can be expressed in frame B as:

$$q_B = q_R q_A q_R^*$$
(3)

We compute the multiplication of quaternions to change a coordinate frame

$$q_R q_A q_R^* = (q_0 + q_1 i + q_2 j + q_3 k)(xi + yj + zk)(q_0 - q_1 i - q_2 j - q_3 k)$$
(4)

The product collected in one quaternion gives us

$$\begin{aligned}
q_R q_A q_R^* = &(x(q_0^2 + q_1^2 - q_2^2 - q_3^2) + 2y(q_1 q_2 - q_0 q_3) + 2z(q_0 q_2 + q_1 q_3))i + \\
&(2x(q_0 q_3 + q_1 q_2) + y(q_0^2 - q_1^2 + q_2^2 - q_3^2) + 2z(q_2 q_3 - q_0 q_1))j + \\
&(2x(q_1 q_3 - q_0 q_2) + 2y(q_0 q_1 + q_2 q_3) + z(q_0^2 - q_1^2 - q_2^2 + q_3^2))k
\end{aligned}$$
(5)

Quaternions in formation control circulated around simply using quaternions for representation of the orientation of each agent with respect to a global frame [34] or one of the other agents serving as an orientation reference [35]. As per the Euler theorem for rigid bodies [32], the rotation of a body around an axis in R^3 can be expressed in quaternions. The attitude of the ith quadrotor defined by the unit quaternion is given as

$$Q = \begin{bmatrix} q \\ \eta \end{bmatrix} = \begin{bmatrix} \| e \| \sin \frac{\theta}{2} \\ \cos \frac{\theta}{2} \end{bmatrix}$$
(6)

where $\| e \|$ represents a unit axis in Euclidean space on which the agent is rotated, and θ is the magnitude of rotation. As the quaternion norm is equal to 1, it is used as the rotation operator. q represents the vector part and gives the magnitude of rotation, and η represents the scalar part and gives the axis of rotation. While some conventions also use the representation where the rotation is expressed later than the axis such as $Q = \begin{bmatrix} \eta & q \end{bmatrix}^T$, it is of less significance and varies from one method to another. The unit quaternion Q can also be transformed into its equivalent rotation matrix by the Rodrigues formula which is

$$R(Q) = (\eta^2 - q^\top q)I_3 + 2qq^\top - 2\eta S(q)$$
$$S(q) = \begin{pmatrix} 0 & -q_3 & q_2 \\ q_3 & 0 & -q_1 \\ -q_2 & q_1 & 0 \end{pmatrix}$$
(7)

where $S(.)$ is the skew-symmetric matrix operator.

Quadcopter UAV Attitude Dynamics

Considering a swarm of n number of UAVs, the dynamics of the ith agent can be given as in [36].

$$\begin{aligned}
\dot{p}_i &= v_i \\
m_i \dot{v}_i &= m_i g \hat{e}_3 - \Gamma_i R(Q_i)^\top \hat{e}_3 \\
\dot{Q}_i &= \tfrac{1}{2} T(Q_i) \omega_i \\
J_i \dot{\omega}_i &= \tau - S(\omega_i) J_i \omega_i
\end{aligned}$$
(8)

For $i \in N := \{1, 2 \ldots \ldots, n\}$, p_i denotes the position, v_i denotes the velocity, m_i denotes the mass of the ith UAV, $Q_i = \begin{pmatrix} q_i^T & \eta_i \end{pmatrix}^T$ is the agent's orientation, and $\omega_i = [\omega^x, \omega^y, \omega^z]^T \in R^3$ is the angular velocity denoted by the skew-symmetric matrix operator from R^3 to a matrix in $R^{3 \times 3}$. The positive scalar Γ denotes the total thrust by all four rotors in the direction $\hat{e} = \begin{pmatrix} 0 & 0 & 1 \end{pmatrix}^{T}$ unit vector in the body coordinate frame, and τ is the control input torque. $T(Q)$ can be given as

$$T(Q) = \begin{pmatrix} \eta I_3 + S(q) \\ -q^{\top} \end{pmatrix} \tag{9}$$

For a UAV leader–follower configuration, the relative attitude between the ith agent (leader) and jth agent (follower) expressed by the unit quaternion $Q_{ij} = \begin{pmatrix} q_{ij}^{T} & \eta_{ij} \end{pmatrix}^{T}$ can be defined as

$$Q_{ij} = Q_j^{-1} \odot Q_i \tag{10}$$

The relative attitude between the two agents can be expressed as

$$\dot{Q}_{ij} = \tfrac{1}{2}T(Q_{ij})\omega_{ij}, \quad T(Q_{ij}) = \begin{pmatrix} \eta_i I_3 + S(q_{ij}) \\ -q_{ij}^{\top} \end{pmatrix} \tag{11}$$

where ω_{ij} is the relative angular velocity of the ith agent's body frame with respect to the jth agent's body frame expressed in the ith agent's body frame given as

$$\omega_{ij} = \omega_i - R\left(Q_{ij}\right)\omega_j \tag{12}$$

where the rotation matrix $R\left(Q_{ij}\right)$ represents the rotation from the jth agent's body frame to the ith agent's body frame such as

$$R\left(Q_{ij}\right) = R(Q_i)R\left(Q_j\right)^{\top} \tag{13}$$

Figure 1 shows the representation of a UAV in the inertial reference frame O_W and the body-fixed frame O_B. Figure 1a shows a dynamic model of a quadrotor UAV in Euler angles and depicts the different actuator level entities affecting flight dynamics. Each of the four propellers rotates with an angular speed ω_i, producing thrust force F_i upwards and with opposite rotor spins. Figure 1b shows the quaternion representation of the same UAV in 3D space.

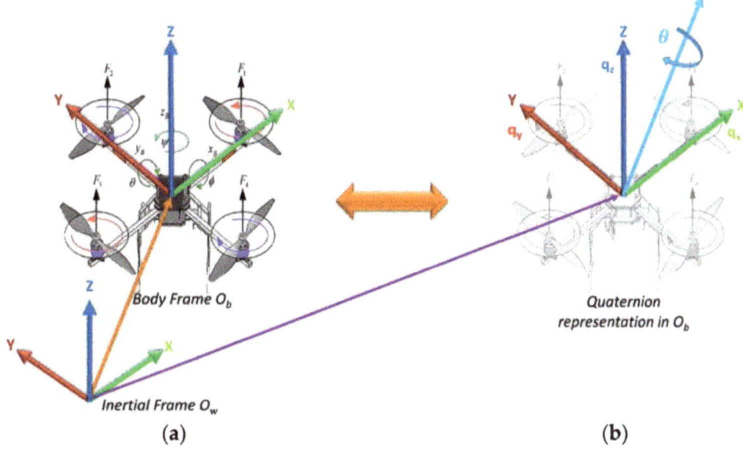

Figure 1. (a) UAV dynamical model expressed in inertial and body reference frames with Euler angles. (b) UAV model expressed with quaternions.

2.2. Graph and Bearing Rigidity Theories

Consider an individual UAV n in which ($n \geq 2$) can be considered as a swarm of UAVs. An undirected graph $G = (V, E)$ characterizes a dynamic undirected interaction network among multiple UAVs in the swarm representing a set of nodes $V = \{1, 2, \ldots, n\}$ representing UAVs, and the interaction among UAVs is represented by a set of edges $E = \{e_{ij}: i = 1, 2, \ldots, n, j \in N_i\}$, with the neighbor set N_i of UAVs. The graph is directed if $(v_i, v_j) \in E$, $(v_j, v_i) \notin E$ and undirected if otherwise.

It is problematic to inspect the distinctiveness of the formation shape determined by distance rigidity because the rank condition of infinitesimal distance rigidity cannot assure the formation shape to be unique [37]. However, in bearing rigidity theory, the rank condition and formation shape distinctiveness are considered adequate. In any coordinate frame, $p_i(t) = [p_i^x, p_i^y, p_i^z]^T \in R^3$ being the position of the ith UAV $C = [p_1^T, p_2^T \ldots \ldots p_n^T]^T \in R^3$ shows the configuration of the formation, and similarly, the desired configuration can be expressed as $C^* = [p_1^{*T}, p_2^{*T} \ldots \ldots p_n^{*T}]^T \in R^3$. A UAV formation, represented by (G, C), is a blend of graph G and a configuration C, where every $v_i \in V$ is related to a position p_i in the configuration [38]. Therefore, for (G, C), define

$$
\begin{aligned}
e_{ij} &\triangleq p_j - p_i \\
g_{ij} &\triangleq \frac{e_{ij}}{\|e_{ij}\|}
\end{aligned}
\tag{14}
$$

where e_{ij} denotes an edge vector and g_{ij} represents a unit vector, which gives the bearing from p_j to p_i. This unit vector representation represents both the azimuth angle and altitude angle in R^3. The objective of a UAV formation is to transform into a desired geometrical shape or final configuration by controlling the bearing constraints of its agents, where bearing constraints can be defined as

$$
\beta_G = \left\{ \frac{g_{ij}^* = \left(p_j^* - p_i^*\right)}{\| p_j^* - p_i^* \|} (v_j, v_i) \in E \right\}
\tag{15}
$$

In a UAV swarm, the desired distance between two agents is given by $d_{21}^* = \| p_1^* - p_2^* \|$, while β_G is a set of bearing constraints, and the target position of the next agent can be defined as $p_2^* = p_1^* - d_{21}^* g_{21}^*$; here, d, p, and g represent distance, position, and bearing of the agents, respectively. Inspired by [37], an orthogonal projection operator $P_{g_{ij}}$ is introduced to geometrically project any vector at the orthogonal compliment of x; moreover, the $\text{Null}(P_{g_{ij}}) = \text{span}\{x\}$ and the eigenvalues of $P_{g_{ij}}$ are $\{0, 1^{(d-1)}\}$, such that for any vector $x > 0$, $x \in R^d (d \geq 2)$, the operator $P_{g_{ij}} : R^d \to R^{d \times d}$ is defined as

$$
\begin{aligned}
P_{g_{ij}} &\triangleq I_d - \left(\frac{x}{\|x\|}\right)\left(\frac{x}{\|x\|}\right)^T \\
P_{g_{ij}} &\triangleq I_3 - g_{ij} g_{ij}^T
\end{aligned}
\tag{16}
$$

The orthogonal projection matrix provides an efficient way to define the parallel vectors in bearing rigidity theory. To define the target formation, we introduce a bearing Laplacian matrix as introduced in [24], which is

$$
[B(G(p^*))]_{ij} = \begin{cases} 0_{d \times d}, & i \neq j, \ (i, j) \notin \varepsilon \\ -P_{g_{ij}^*}, & i \neq j, \ (i, j) \in \varepsilon \\ \sum_{k \in N_i} P_{g_{ik}^*}, & i = j, i \in V \end{cases}
\tag{17}
$$

This bearing Laplacian matrix B describes the inter-agent topology and bearings between agents. Similarly, the bearing Laplacian matrix can be explained as

$$B = \begin{bmatrix} B_{LL} & B_{LF} \\ B_{FL} & B_{FF} \end{bmatrix} \tag{18}$$

where every part can be explained as $B_{LL} \in R^{dn_L \times dn_L}$, $B_{LF} \in R^{dn_L \times dn_F}$, $B_{FL} \in R^{dn_F \times dn_L}$, and $B_{FF} \in R^{dn_F \times dn_F}$ is significant and useful, being symmetric positive semidefinite. We also need to ensure that the framework is unique and rigid; therefore, we employ the infinitesimal bearing rigidity theory introduced by [24]. This theory states that: if a framework (G, C) is infinitesimally rigid, it depicts two vital properties: (1) the positions of the vertices can be distinctively calculated up to a translational and a scaling factor, and (2) the configurations are infinitesimally bearing rigid in a d-dimensional space if and only if the bearing rigidity matrix satisfies: $\text{Null}\left(G_{ij}^B(C)\right) = \text{span}\{1_n \otimes I_d, p\}$ or

$$\text{rank}\left(G_{ij}^B(C)\right) = dn - d - 1 \tag{19}$$

where n number of agents in a swarm are expressed in R^d, comprising dn coordinates, d specifying the centroid, and 1 specifying the scale and being subtracted, and if the resulting value is equal to the rank of $G_{ij}^B(C)$, it means that the formation is infinitesimal bearing rigid. Due to these two properties [24], infinitesimally bearing rigid configurations not only have unique geometric shapes but can also be mathematically inspected. The centroid and scale [38] can be defined as

$$c(p^*(t)) = \frac{1}{n} \sum_{i=1}^{n} p_i^*(t)$$
$$s(p^*(t)) = \sqrt{\frac{1}{n} \sum_{i=1}^{n} p_i^*(t) - c(p^*(t))^2} \tag{20}$$

For bearing rigidity, it is vital to determine if two given bearings are equal; thus, the orthogonal projection operator provides an upfront approach. Figure 2 illustrates the difference between rigid and non-rigid graphs and shows the topology used in this work.

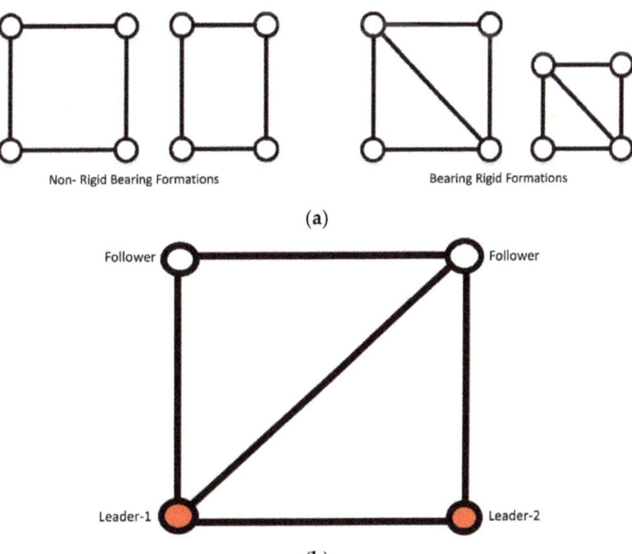

Figure 2. (a) Examples of non-rigid and rigid bearing formations. (b) Proposed Formation Topology.

2.3. Problem Formulation

We designed a cascaded model for formation control of a UAV swarm, considering the dynamic model of quadrotor UAVs. In the upper cascade, we gave the bearing-based law, and in the lower cascade, we designed the attitude synchronization controller that improves the performances of [21,24,31] by the use of quaternions instead of Euler angles. This complete cascaded structure was then used to simulate a UAV swarm's translation, scaling, and rotation in 3D space.

We formulated the problem as follows.

Problem 1. *Consider a UAV swarm with n number of agents in R^3 under assumptions 1–3, where the positions and velocities of leader agents are time-varying. Based upon relative bearing measurements g_{ij} , relative distance measurements such as d_{ij} , and relative velocity measurements $v_{ij}(t)$, design an acceleration input $u_i(t)$ for each agent such that $g_{ij}^i(t) \to g_{ij}^*$ exponentially as $t \to \infty, \forall_i = 1, 2, \ldots, n$.*

Assumption 1. *In this work, we assume that all UAVs are equipped with sensor packages, such as onboard-calibrated vision-based sensors, rate gyroscopes, accelerometers, and magnetometers, for accurate orientation calculation and also with communication modules to communicate with neighboring UAVs.*

Assumption 2. *Only the leader agent has the right to use the inertial reference frame; therefore, we assume other agents do not have this information. Another limitation on the leader agent is that it can only use that data to calculate its attitude in Euclidean space.*

Problem 2. *Consider a UAV swarm with n number of agents in R^3 with $\{p(0)\}_{i \in V}$ as initial positions and $\{Q(0)\}_{i \in V}$ as initial orientations under assumptions 1–2, and design an attitude synchronization law based on control inputs based on relative attitude and angular velocities of agents such that $\{q_{ij}(t)\}_{i \in V} \to 0$, $q_i \to q_j$ and $\omega_i \to \omega_j$ exponentially as $t \to \infty, \forall_i = 1, 2, \ldots, n$.*

3. Proposed Control Scheme

We designed a control scheme utilizing the information from both bearing and attitude controllers to control a swarm of quadcopter UAVs to form a specific formation shape, translate, and scale in 3D Euclidean space.

Figure 3 illustrates the block diagram of the proposed control structure depicting the flow of control inputs and outputs as well as their interaction with both controllers. The overall architecture of the control scheme and the designated operations of all UAVs are depicted in Figure 4.

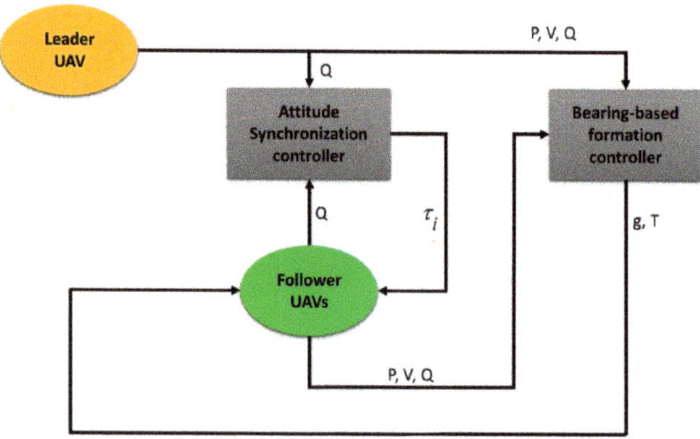

Figure 3. Block Diagram of Proposed Structure.

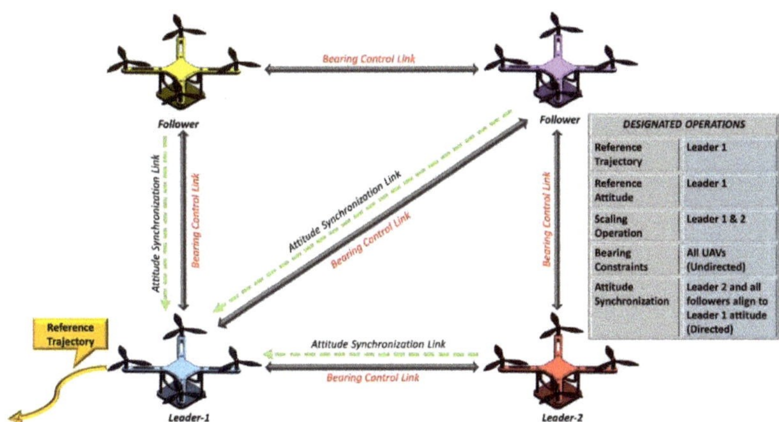

Figure 4. Overall Architecture and Designated Operations.

The interaction between each UAV, the flow of information, and designed operational tasks of all UAVs in a distributed manner are also explained in Figure 4. The formation was designed in such a way that each agent in the formation aligns itself with the spawned body frame of the leader agent. All control actions in 3D space such as formation acquisition, translation, scaling, and rotation were achieved by employing both controllers in unison. Therefore, bearing control and attitude synchronization were achieved in seamless harmony.

3.1. Bearing-Based Controller

To compute relative bearing between quadrotors in Euclidean space, considering assumptions 1–2, the position and velocity errors of agents are given as

$$\delta_p(t) = p_i(t) - p_i^*(t), \delta_v(t) = v_i(t) - v_i^*(t) \tag{21}$$

Given problem 1, the control objective was to design a control law for all agents in formation to make the complete formation do translational, rotational, and scaling maneuvers by enforcing $\delta_p(t) \to 0$ and $\delta_v(t) \to 0$ as $t \to \infty, \forall i = 1, 2, \ldots, n$. It should be noted that only leaders know the desired translational, rotational, and scaling maneuvering information.

To accomplish the control objective, we propose a control structure where the target formation is tracked with time-varying velocity and time-varying orientation leaders. In this sub-section, we only consider the case of time-varying velocity leaders, and in the following sub-section (attitude controller), we consider the time-varying orientation leaders. According to [26], when the leader's velocity $v_l(t)$ is time-varying, formation tracking errors might not converge to zero; therefore, a supplementary acceleration feedback term is required to be added in the controller. The following controller is proposed for the time-varying velocity leader case

$$u_i = -\xi_i^{-1} \sum_{j \in N_i} P_{g_{ij}^*}[k_p(p_i - p_j) + k_v(v_i - v_j) - \dot{v}_j] \tag{22}$$

where $\xi_i = \sum_{i \in N_i} P_{g_{ij}^*}$ and $P_{g_{ij}^*} = I_d - g_{ij}^*\left(g_{ij}^*\right)^T$ was defined earlier as an orthogonal projection matrix, while k_p and k_v are position and velocity positive control gains, respectively, and $\dot{v}_j$ is the acceleration of the neighboring agent. The controller in (22) was inspired by consensus algorithms proposed in [39]. It can be proved that ξ_i is non-singular because the target formation to be tracked is unique.

Lemma 1. *The constant matrix ξ_i is non-singular for all follower agents if the acquired formation is distinct and unique.*

Proof of Lemma 1. Firstly, the matrix ξ_i is singular when the bearings g_{ij}^* are aligned because for any $x \in R^d, x^T \xi_i x = 0 \Leftrightarrow \sum_{j \in N_i} x^T P_{g_{ij}^*} x = 0 \Leftrightarrow P_{g_{ij}^*} x = 0, \forall j \in N_i$. $\text{Null}\left(P_{g_{ij}^*}\right) = \text{span}\left\{g_{ij}^*\right\}$; therefore, $x^T \xi_i x = 0$ when x and g_{ij}^* are aligned. If g_{ij}^* is aligned, the follower position p_i^* cannot be estimated because p_i^* moves on the straight line aligned with g_{ij}^*. Resultantly, it can be established that ξ_i is singular.

The stability analysis of control law (22) is given; hereby, $\square$

Theorem 1. *For the time-varying velocity leader, the position and velocity errors defined in (21) converge exponentially to zero.*

Proof of Theorem 1. Multiply ξ_i on both hand sides of the control law (22). $u_i = \dot{v}_i$; therefore,

$$
\begin{aligned}
\xi_i\left(\dot{v}_i - \dot{v}_j\right) &= \xi_i \xi_i^{-1} \sum_{j \in N_i} P_{g_{ij}^*}\left[-k_p(p_i - p_j) - k_v(v_i - v_j)\right] \\
\sum_{j \in N_i} P_{g_{ij}^*}\left(\dot{v}_i - \dot{v}_j\right) &= \sum_{j \in N_i} P_{g_{ij}^*}\left[-k_p(p_i - p_j) - k_v(v_i - v_j)\right]
\end{aligned}
\tag{23}
$$

In terms of the bearing Laplacian matrix form,

$$
\begin{aligned}
B_{FF}\dot{v}_F + B_{FL}\dot{v}_L &= -k_p(B_{FF}p_F + B_{FL}p_L) - k_v(B_{FF}v_F + B_{FL}v_L) \\
&= -k_p B_{FF}\delta_p - k_v B_{FF}\delta_v
\end{aligned}
\tag{24}
$$

With this, it can be shown that $\dot{v}_F = -k_p\delta_p - k_v\delta_v - B_{FF}^{-1}B_{FL}\dot{v}_L$, and therefore, the error terms are $\dot{\delta}_p = \delta_V$ and $\dot{\delta}_V = \dot{V}_F + B_{FF}^{-1}B_{FL}\dot{v}_L = -k_p\delta_p + k_v\delta_v$, which can be shown in state space form as

$$
\begin{bmatrix} \dot{\delta}_p \\ \dot{\delta}_v \end{bmatrix} = \begin{bmatrix} 0 & I \\ -k_p I & -k_v I \end{bmatrix} \begin{bmatrix} \delta_p \\ \delta_v \end{bmatrix}
\tag{25}
$$

The eigenvalue of this state matrix is $\lambda = \left(-k_v \pm \sqrt{k_v^2 \pm 4k_p}\right)/2$, which proves to be in the left-half plane for any $k_p, k_v > 0$. Therefore, convergence is achieved. $\square$

3.2. Attitude Synchronization Controller

The orientation of follower UAVs is determined with respect to the orientation of the leader agent. As per problem 2, our objective was to guarantee attitude synchronization when $\omega_{ij} \to 0$, $Q_{ij} \to \pm Q_I$ and $R\left(Q_{ij}\right) \to I_3 \,\forall\, i,j \in N$. As $Q_{ij} = \left(q_{ij}^T \,\, \eta_{ij}\right)^T$ is a unit vector representing the attitude from the ith agent (leader) and jth agent (follower), its inverse or conjugate can be written as

$$
Q_{ij}^{-1} = \begin{pmatrix} -q_{ij} \\ \eta \end{pmatrix}
\tag{26}
$$

such that

$$
Q_{ij} \odot Q_{ij}^{-1} = Q_{ij}^{-1} \odot Q_{ij} = Q_I
\tag{27}
$$

where Q_I is the unit quaternion identity and can be expressed as

$$
Q_I = \begin{bmatrix} 0_3 \\ I \end{bmatrix}
\tag{28}
$$

Similarly, it can be seen that $R\left(Q_{ij}^{-1}\right) = R(Q_{ij})^{T}$. It is adequate to say that when $q_{ij} \to 0$, it implies that the attitude synchronization or alignment between agents has taken place. Furthermore, the relative attitude approximation is based on special orthogonal groups SO(3), which guarantees accurate approximation for all UAVs in the formation. For translation of a UAV swarm in 3D space, the UAVs must track a reference trajectory; therefore, it is necessary to define an attitude tracking error. To do this, we define the desired attitude $Q_d = \left(q_d^T \quad \eta_d\right)^T$ with components of the unit quaternion described as

$$\dot{Q}_d = \frac{1}{2}T(Q_d)\omega_d \tag{29}$$

where $T(Q_d)$ is defined similarly to Equation (9). The attitude tracking error $\tilde{Q}_i = \left(\tilde{q}_i^T \quad \tilde{\eta}_i\right)^T$ can be defined as

$$\tilde{Q}_i = Q_d^{-1} \odot Q_i \tag{30}$$

Therefore, the relative attitude tracking can be written similarly to equation (11) as

$$\dot{\tilde{Q}}_i = \tfrac{1}{2}T\left(\tilde{Q}_i\right)\tilde{\omega}_i, \quad T(\tilde{Q}_i) = \begin{pmatrix} \tilde{\eta}_i I_3 + S(\tilde{q}_i) \\ -\tilde{q}_i^T \end{pmatrix} \tag{31}$$

where the angular velocity tracking vector can be defined as

$$\tilde{\omega}_i = \omega_i - R\left(\tilde{Q}_i\right)\omega_d \tag{32}$$

The rotation matrix associated to $\tilde{Q}_i$ is given as

$$R\left(\tilde{Q}_i\right) = R(Q_i)R(Q_d)^{T} \tag{33}$$

For attitude synchronization and alignment of all UAVs in the swarm, the control input of each UAV has to be based upon relative attitudes and relative angular velocities among neighboring agents. Inspired by [40] and with the aim that all UAVs in the swarm align their attitudes and angular velocities in an undirected graph, the following attitude synchronization controller is proposed

$$\tau_i = \omega_i \times J_i \omega_i - J_i \sum_{j=1}^{n} a_{ij}\left[k_q q_{ij} + k_\omega \left(\omega_i - \omega_j\right)\right] \tag{34}$$

where a_{ij} is the value of a weighted adjacency matrix representing information exchange between UAVs, k_q and k_ω are positive scalar gains, and the value of inertia matrices $J \in R^{3 \times 3}$ should be known for all UAVs in the swarm, which means that the controller can be implemented on heterogeneous quadcopter UAV swarms.

Theorem 2. *For a time-varying orientation leader under the action of control law (34), the relative attitude and angular velocity between two neighboring UAVs should reach $q_{ij} \to 0$, $q_i \to q_j$, and $\omega_i \to \omega_j$ asymptotically as $t \to \infty$, $\forall_i = 1, 2, \ldots, n$.*

Proof of Theorem 2. Select a Lyapunov candidate function, such as:

$$V = \frac{1}{2}\sum_{i=1}^{n}\sum_{j=1}^{n} a_{ij}k_{ij} \parallel q_{ij} - q_I \parallel^2 + \frac{1}{2}\sum_{i=1}^{n} \omega_i^T \omega_i \tag{35}$$

Under the dynamics of unit quaternions, the derivative of V becomes

$$\dot{V} = \frac{1}{2} \sum_{i=1}^{n} \sum_{j=1}^{n} a_{ij} k_{ij} (\omega_i - \omega_j)^T q_{ij} + \sum_{i=1}^{n} \omega_i^T (\tau_i - \omega_i \times J_i \omega_i) \tag{36}$$

As $\omega_i^T (\omega_i \times J_i \omega_i) = 0$, and under the fact that in an undirected graph $a_{ij} = a_{ji}$,

$$\begin{aligned}
\dot{V} &= \frac{1}{2} \sum_{i=1}^{n} \sum_{j=1}^{n} a_{ij} k_{ij} (\omega_i - \omega_j)^T q_{ij} \\
&= \frac{1}{2} \sum_{i=1}^{n} \omega_i^T \left(\sum_{j=1}^{n} a_{ij} k_{ij} q_{ij} \right) - \frac{1}{2} \sum_{i=1}^{n} \sum_{j=1}^{n} a_{ij} k_{ij} \omega_j^T q_{ij} \\
&= \frac{1}{2} \sum_{i=1}^{n} \omega_i^T \left(\sum_{j=1}^{n} a_{ij} k_{ij} q_{ij} \right) - \frac{1}{2} \sum_{i=1}^{n} \sum_{j=1}^{n} a_{ji} k_{ji} \omega_j^T q_{ij} \\
&= \frac{1}{2} \sum_{i=1}^{n} \omega_i^T \left(\sum_{j=1}^{n} a_{ij} k_{ij} q_{ij} \right) + \frac{1}{2} \sum_{j=1}^{n} \sum_{i=1}^{n} a_{ji} k_{ji} \omega_j^T q_{ij} \\
&= \frac{1}{2} \sum_{i=1}^{n} \omega_i^T \left(\sum_{j=1}^{n} a_{ij} k_{ij} q_{ij} \right) + \frac{1}{2} \sum_{j=1}^{n} \omega_j^T \left(\sum_{i=1}^{n} a_{ji} k_{ji} q_{ij} \right) \\
&= \sum_{i=1}^{n} \omega_i^T \left(\sum_{j=1}^{n} a_{ij} k_{ij} q_{ij} \right)
\end{aligned} \tag{37}$$

Resultantly, Equation (36) becomes

$$\dot{V} = \sum_{i=1}^{n} \omega_i^T \left(\sum_{j=1}^{n} a_{ij} k_{ij} q_{ij} + \tau_i \right) \tag{38}$$

$\sum_{i=1}^{n} \omega_i^T \sum_{j=1}^{n} a_{ij} k_{ij} (\omega_i - \omega_j) = \frac{1}{2} \sum_{i=1}^{n} \sum_{j=1}^{n} a_{ij} k_{ij} \| \omega_i - \omega_j \|^2$; therefore, the derivative of V becomes negative semidefinite

$$\dot{V} = -\frac{1}{2} \sum_{i=1}^{n} \sum_{j=1}^{n} a_{ij} k_{ij} \| \omega_i - \omega_j \|^2 \leq 0 \tag{39}$$

By LaSalle's invariance principle, it is established that $q_{ij} \to 0$, $q_i \to q_j$, and $\omega_i \to \omega_j$ asymptotically. $\square$

4. Simulation Results

In this section, we share numerical simulation results demonstrating the effectiveness of our proposed model on a swarm of quadrotor UAVs. This swarm of UAVs contained four quadrotors depicting an undirected leader–follower topology. The formation consisted of two leaders and two follower UAVs. To highlight the operations of the formation in a simple way, a square-shaped geometric configuration was selected, and the communication topology is described in Figure 3, which depicts leader agents as $V_L = \{1, 2\}$ and followers as $V_F = \{3, 4\}$. The model information and specifications are given in Table 1.

In this work, we assumed that the formation encounters various kinds of obstacles in its path while translating in an underground environment, e.g., narrow passages, pipes, tunnels, etc., and negotiates those obstacles while keeping the formation intact. Figure 5 depicts the entire time-lapse of the formation translation and different maneuvers. The formation was designed to carry out four distinct actions, and the case-wise details of all actions and maneuvers achieved by the formation are given below.

Table 1. Model Information and specifications.

Parameter	Value
m	0.80
$J(kgm^2)$	[1,0.1,0.1; 0.1,0.1,0.1; 0.1,0.1,0.9]
a_{ij}	[0,1,1,1; 1,0,1,1; 1,1,0,1; 1,1,1,0]
k_q	1
k_ω	10
k_p	0.5
k_v	2

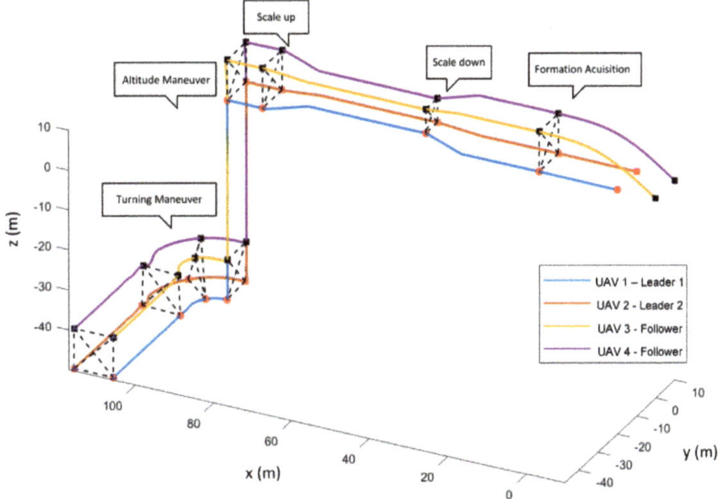

Figure 5. Time-lapse of complete formation operation.

4.1. Case 1—Formation Acquisition

(1) Objective: a swarm of four UAVs at random positions takes off and acquires a specific square shape under the control of proposed laws.

(2) Results: the target formation formed a designated square shape and was attained by implementing pre-defined bearing constraints between the agents as $g_{21}^* = -g_{12}^*$, $g_{31}^* = -g_{13}^*$, $g_{12}^* = \begin{bmatrix} -1 & 0 & 0 \end{bmatrix}^T$, $g_{13}^* = \begin{bmatrix} 0 & 0 & -1 \end{bmatrix}^T$, $g_{14}^* = \begin{bmatrix} \frac{-\sqrt{2}}{2} & 0 & \frac{-\sqrt{2}}{2} \end{bmatrix}^T$, $g_{41}^* = -g_{14}^*$, $g_{23}^* = \begin{bmatrix} \frac{\sqrt{2}}{2} & 0 & \frac{-\sqrt{2}}{2} \end{bmatrix}^T$, $g_{32}^* = -g_{23}^*$, $g_{42}^* = \begin{bmatrix} 0 & 0 & 1 \end{bmatrix}^T$, and $g_{24}^* = -g_{42}^*$. The formation trajectories are given in Figure 5. The formation tracking error $\| \delta_i \|$ is shown in Figure 6 (section highlighted in blue), which asymptotically converged to zero from t = 0 to 20 s.

4.2. Case 2—Formation Scaling

(1) Objective: to verify that formation can scale down (decrease size) and scale up (increase size) while translating in 3D space by still keeping formation-bearing constraints, inter-agent distances, and heading direction intact.

(2) Results: the formation continued translation on the *x*-axis, scaled down at t = 40 s, and scaled up at t = 80 s to negotiate imaginary obstacles. This was achieved by adjusting and altering the distance and velocities of two leaders. Figure 5 depicts both scaling operations, and Figure 6 shows the convergence of formation tracking errors to zero (highlighted with yellow color for scaling down and with green color for scaling up).

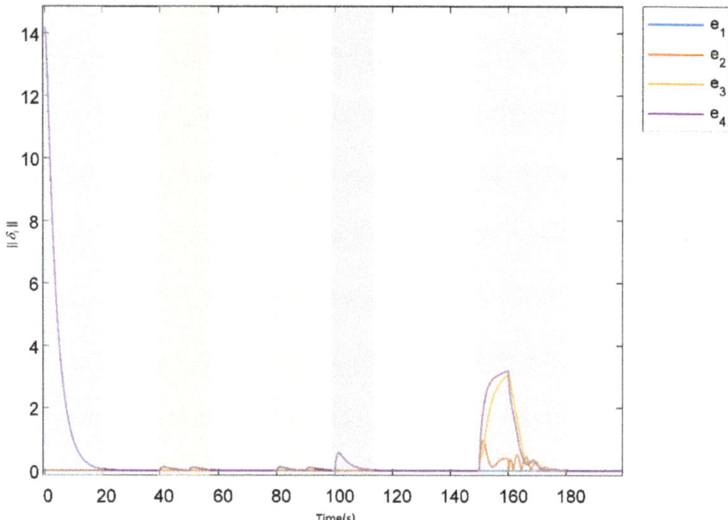

Figure 6. Formation tracking errors.

4.3. Case 3—Altitude Maneuver

(1) Objective: to verify that UAVs in the formation can also make an altitude descent while staying in the desired formation to negotiate an obstacle or follow a specific trajectory involving sudden altitude descent.

(2) Results: after the scaling operation while translating in the x-axis direction, the formation abruptly descended its altitude in the z-axis direction in 3D space at t = 100 to 110 s by altering the velocity of leaders. The trajectory plot of UAVs is given in Figure 5, and the formation tracking error converged to zero asymptotically as shown in Figure 6 (highlighted with grey color).

4.4. Case 4—Formation Translational Rotation

(1) Objective: to verify that formation while translating in 3D space can rotate its heading direction by altering the velocity of agents such that the swarm stays dynamically intact.

(2) Results: in Figure 5 at t = 150 to 180 s, it can be seen that the final formation was rotated from the initial formation heading direction by altering the leader's orientation so that the formation takes a translational rotation. The formation tracking error also converged to zero as shown in Figure 6 (section highlighted in orange color).

Both the bearing-based controller and attitude controller ensured the performance of the formation during the entirety of the operation. The attitude controller aligned the attitude of all follower UAVs as per the attitude of leader UAVs at every stage of formation operation as shown in Figure 7. As can be seen in the figures, the different cases of formation operations are shown at different time intervals such as formation acquisition (t = 0 to 20 s), scale-down (t = 0 to 20 s), scale-up (t = 0 to 20 s), altitude descend (t = 0 to 20 s) and translation maneuver (t = 0 to 20 s).

The linear velocity of all follower UAVs achieved consensus as per the linear velocity of the leader UAV, as shown in Figure 8 for all cases. Figure 8a,b illustrates the linear velocity profile of all agents in the x and y-axis where it can be noticed that Leader-2 (agent 2) had the maximum deviation; this is because as per configuration, agent 2 lay at the farthest end and had to align itself with the rest of the agents. Therefore, the controller action forced agent 2 to rapidly align with the rest of the formation, ensuring the formation configuration is intact. Similarly, in Figure 8c, it can be noticed that the velocity profile of agents 1 and 2 and agents 3 and 4 were identical; this is because of the formation configuration as can be seen in Figure 5. The angular velocity of all followers converged to that of the leader UAV, as can be seen in Figure 9 for all cases, while the formation tracking error remained at zero despite the hard inclination in maneuvers.

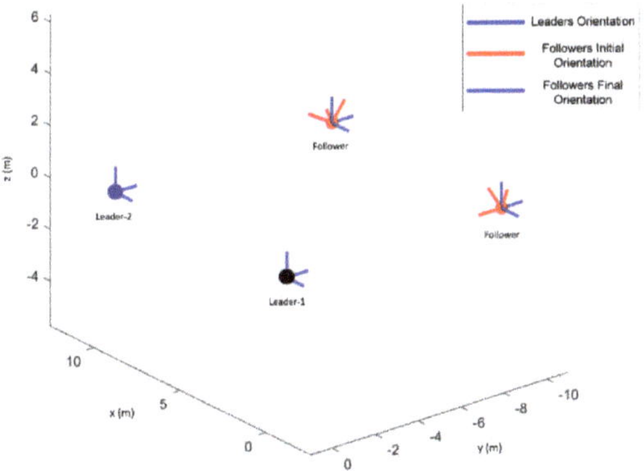

Figure 7. Attitude synchronization.

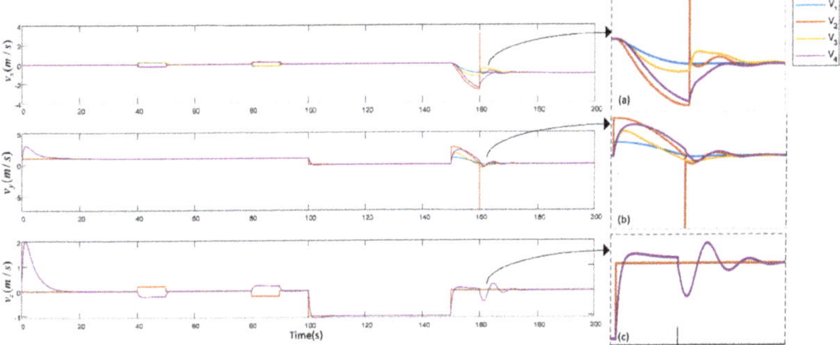

Figure 8. Linear velocity profile. (**a**) Magnified view of velocity peaks in x-axis at t = 160 s. (**b**) Magnified view of velocity peaks in y-axis at t=160 s. (**c**) Magnified view of velocity peaks in z-axis at t = 160 s.

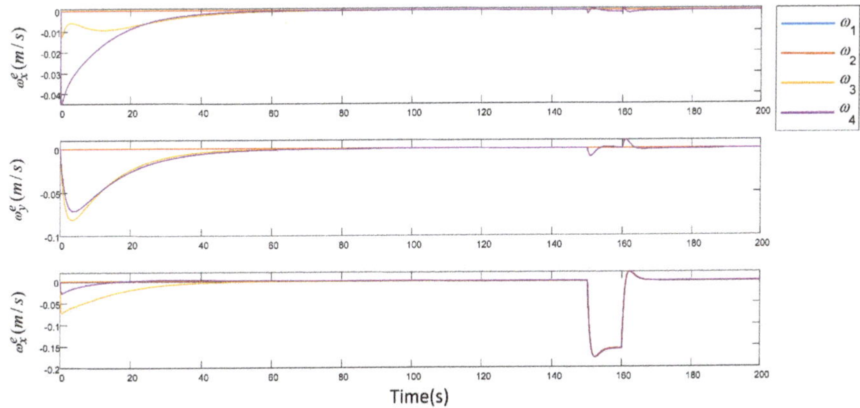

Figure 9. Angular velocity profile.

The orientation of the leader is time-varying because of changing maneuvers; therefore, in Figure 10, it can be seen that all followers aligned their orientation to that of the leader as per changing maneuvers; hence, the attitude error was maintained at zero. The same is represented in Figure 11 in terms of roll, pitch, and yaw angles.

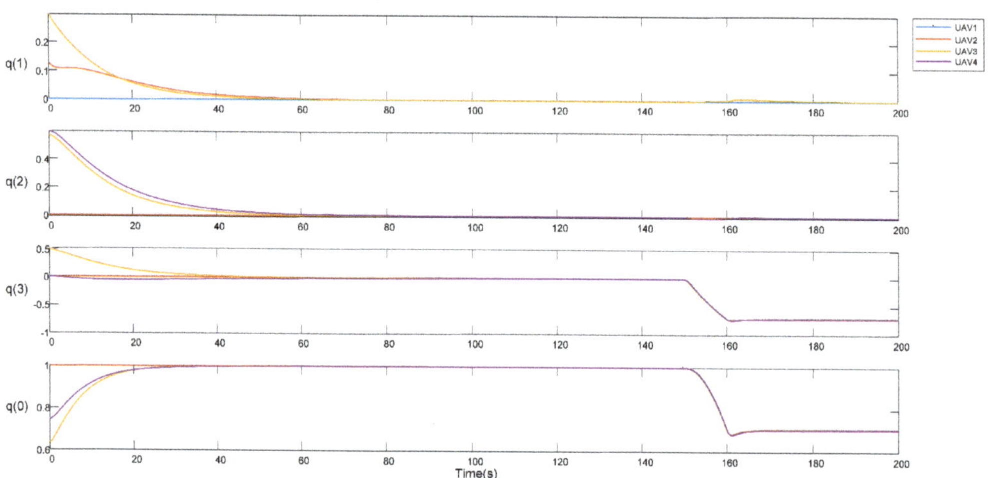

Figure 10. Attitude synchronization in quaternions.

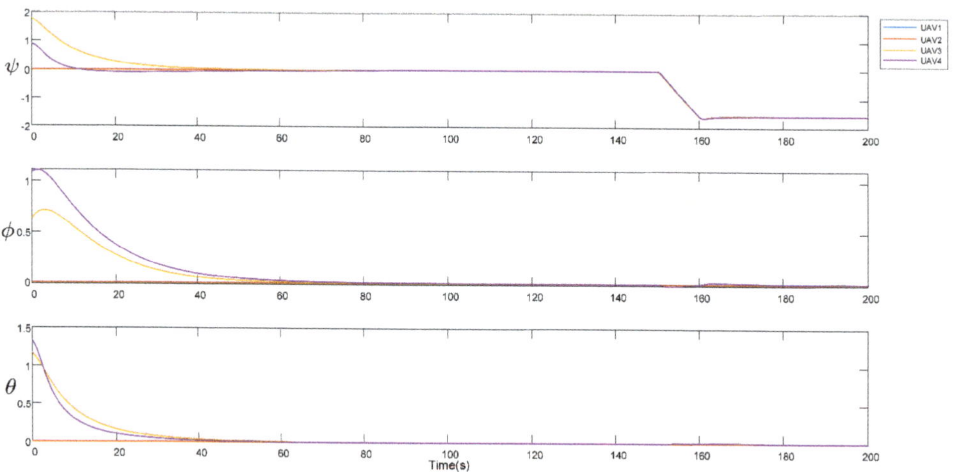

Figure 11. Attitude synchronization as per yaw, pitch, and roll angles.

5. Discussion

From the results of the case studies and simulations above, it can be established that the formation carried out the specified tasks efficiently. The leader aligned its body frame with the inertial reference frame and moved in the 3D Euclidean space without any further constraints while followers followed the leaders as per the designed topology. All UAVs in the swarm were responsible for maintaining bearing vectors; hence, the desired formation and maneuvers were done in an undirected manner. The formation could avoid narrow obstacles and pass through tight corners and obstacles because the scale, orientation, translation, and velocity could be adjusted. Moreover, the attitude synchronization controller was designed in such a way that it can be implemented not only on homogeneous quadrotor formations but can also support heterogeneous quadrotor formations. For bearing-based UAV formation control problems, the orientation parameters are often neglected or calculated using a primary de facto method of Euler angles for attitude representation. The quaternion-based orientation approximation provides robust, unambiguous, and computationally efficient attitude calculation. Attitude calculations by quaternions ensure that agents in the swarm do not suffer from gimbal lock and singularities, improving the control scheme's overall robustness. Calculating attitude and bearing in the local body frames of each agent is advantageous because agents do not have to depend on the global frame, as GPS signals may be faulty in subterranean environments (e.g., indoors, underwater, deep space, etc.). Many previous works such as [31,41] have assumed that the body frame of the UAV should coincide with the center of mass, while we suggested that it should coincide at the geometrical center of the UAV for accurate position and orientation measurement.

6. Conclusions

This work investigated the joint operation of bearing-based and attitude synchronization controllers to control a quadcopter UAV swarm in 3D space by using undirected graph topology. This combination of controllers added to the overall robustness of formation during complicated maneuvers. Since this work focused on the distributed formation control by depicting various motions of the formation in 3D space, its limitation is that obstacle avoidance was not considered during practical implementation. Further performance improvement for more complex maneuvers such as curved trajectories and implementation of this work on experimental platforms are treated as future works.

Author Contributions: Conceptualization, M.B.S. and Y.Z.; methodology, M.B.S. and Y.Z.; software, M.B.S. and Z.Y.; investigation, X.W. and X.Y.; writing—original draft preparation, M.B.S. and Z.L.; writing—review and editing, M.B.S. and S.A.; supervision, S.W. All authors have read and agreed to the published version of the manuscript.

Funding: This study was co-supported by the National Science and Technology Major Project of China (Nos. 2017-V-0010-0060, 2017-V-0011-0062), National Natural Science Foundation of China (Nos. 51675019, 51620105010), and the fellowship of China Postdoctoral Science Foundation (Grant No.2022M710305).

Institutional Review Board Statement: Not applicable.

Informed Consent Statement: Not applicable.

Data Availability Statement: Not applicable.

Acknowledgments: The authors would like to thank all the teachers at the School of Automation Science and Electrical Engineering, Beihang University for their support and advice in the completion of this work.

Conflicts of Interest: The authors declare no conflict of interest.

References

1. Bai, H.; Wen, J.T. Cooperative Load Transport: A Formation-Control Perspective. *IEEE Trans. Robot.* **2010**, *26*, 742–750. [CrossRef]
2. Liu, G.P.; Zhang, S. A Survey on Formation Control of Small Satellites. *Proc. IEEE* **2018**, *106*, 440–457. [CrossRef]
3. Dong, X.; Zhou, Y.; Ren, Z.; Zhong, Y. Time-Varying Formation Tracking for Second-Order Multi-Agent Systems Subjected to Switching Topologies With Application to Quadrotor Formation Flying. *IEEE Trans. Ind. Electron.* **2017**, *64*, 5014–5024. [CrossRef]
4. Zhang, J.; Xing, J. Cooperative Task Assignment of Multi-UAV System. *Chin. J. Aeronaut.* **2020**, *33*, 2825–2827. [CrossRef]
5. Wang, Y.; Zhang, T.; Cai, Z.; Zhao, J.; Wu, K. Multi-UAV Coordination Control by Chaotic Grey Wolf Optimization Based Distributed MPC with Event-Triggered Strategy. *Chin. J. Aeronaut.* **2020**, *33*, 2877–2897. [CrossRef]
6. Zhen, Z.; Zhu, P.; Xue, Y.; Ji, Y. Distributed Intelligent Self-Organized Mission Planning of Multi-UAV for Dynamic Targets Cooperative Search-Attack. *Chin. J. Aeronaut.* **2019**, *32*, 2706–2716. [CrossRef]
7. Zhang, P.; De Queiroz, M.; Cai, X. Three-Dimensional Dynamic Formation Control of Multi-Agent Systems Using Rigid Graphs. *J. Dyn. Syst. Meas. Control Trans. ASME* **2015**, *137*, 111006. [CrossRef]
8. Sial, M.B.; Wang, S.; Wang, X.; Wyrwa, J.; Liao, Z.; Ding, W. Mission Oriented Flocking and Distributed Formation Control of UAVs. In Proceedings of the 16th IEEE Conference on Industrial Electronics and Applications, ICIEA 2021, Chengdu, China, 1–4 August 2021.

9. Oh, K.K.; Park, M.C.; Ahn, H.S. A Survey of Multi-Agent Formation Control. *Automatica* **2015**, *53*, 424–440. [CrossRef]
10. Lin, Z.; Ding, W.; Yan, G.; Yu, C.; Giua, A. Leader-Follower Formation via Complex Laplacian. *Automatica* **2013**, *49*, 1900–1906. [CrossRef]
11. Lewis, M.A.; Tan, K.H. High Precision Formation Control of Mobile Robots Using Virtual Structures. *Auton. Robot.* **1997**, *4*, 387–403. [CrossRef]
12. Hsieh, M.A.; Kumar, V.; Chaimowicz, L. Decentralized Controllers for Shape Generation with Robotic Swarms. *Robotica* **2008**, *26*, 691–701. [CrossRef]
13. Mesbahi, M.; Egerstedt, M. *Graph Theoretic Methods in Multiagent Networks*; Princeton University Press: Princeton, NJ, USA, 2010.
14. Giribet, J.I.; Colombo, L.J.; Moreno, P.; Mas, I.; Dimarogonas, D.V. Dual Quaternion Cluster-Space Formation Control. *IEEE Robot. Autom. Lett.* **2021**, *6*, 6789–6796. [CrossRef]
15. CAI, Z.; WANG, L.; ZHAO, J.; WU, K.; WANG, Y. Virtual Target Guidance-Based Distributed Model Predictive Control for Formation Control of Multiple UAVs. *Chin. J. Aeronaut.* **2020**, *33*, 1037–1056. [CrossRef]
16. Zou, Y.; Zhou, Z.; Dong, X.; Meng, Z. Distributed Formation Control for Multiple Vertical Takeoff and Landing UAVs with Switching Topologies. *IEEE/ASME Trans. Mechatron.* **2018**, *23*, 1750–1761. [CrossRef]
17. CHEN, H.; WANG, X.; SHEN, L.; CONG, Y. Formation Flight of Fixed-Wing UAV Swarms: A Group-Based Hierarchical Approach. *Chin. J. Aeronaut.* **2021**, *34*, 504–515. [CrossRef]
18. Trinh, M.H.; Pham, V.H.; Park, M.C.; Sun, Z.; Anderson, B.D.O.; Ahn, H.S. Comments on "Global Stabilization of Rigid Formations in the Plane [Automatica 49 (2013) 1436–1441]". *Automatica* **2017**, *77*, 393–396. [CrossRef]
19. Mou, S.; Belabbas, M.A.; Morse, A.S.; Sun, Z.; Anderson, B.D.O. Undirected Rigid Formations Are Problematic. *IEEE Trans. Automat. Contr.* **2016**, *61*, 2821–2836. [CrossRef]
20. Cai, X.; Queiroz, M. De Adaptive Rigidity-Based Formation Control for Multirobotic Vehicles with Dynamics. *IEEE Trans. Control Syst. Technol.* **2015**, *23*, 389–396. [CrossRef]
21. Li, X.; Wen, C.; Chen, C. Adaptive Formation Control of Networked Robotic Systems with Bearing-Only Measurements. *IEEE Trans. Cybern.* **2021**, *51*, 199–209. [CrossRef]
22. Zhao, S.; Lin, F.; Peng, K.; Chen, B.M.; Lee, T.H. Distributed Control of Angle-Constrained Cyclic Formations Using Bearing-Only Measurements. *Syst. Control Lett.* **2014**, *63*, 12–24. [CrossRef]
23. Basiri, M.; Bishop, A.N.; Jensfelt, P. Distributed Control of Triangular Formations with Angle-Only Constraints. *Syst. Control Lett.* **2010**, *59*, 147–154. [CrossRef]
24. Zhao, S.; Zelazo, D. Bearing Rigidity and Almost Global Bearing-Only Formation Stabilization. *IEEE Trans. Automat. Contr.* **2016**, *61*, 1255–1268. [CrossRef]
25. Li, Z.; Tnunay, H.; Zhao, S.; Meng, W.; Xie, S.Q.; Ding, Z. Bearing-Only Formation Control With Prespecified Convergence Time. *IEEE Trans. Cybern.* **2020**, *52*, 620–629. [CrossRef]
26. Zhao, S.; Zelazo, D. Translational and Scaling Formation Maneuver Control via a Bearing-Based Approach. *IEEE Trans. Control Netw. Syst.* **2017**, *4*, 429–438. [CrossRef]
27. Tron, R.; Thomas, J.; Loianno, G.; Daniilidis, K.; Kumar, V. A Distributed Optimization Framework for Localization and Formation Control: Applications to Vision-Based Measurements. *IEEE Control Syst.* **2016**, *36*, 22–44. [CrossRef]
28. Gurbuz, A.C.; Cevher, V.; McClellan, J.H. Bearing Estimation via Spatial Sparsity Using Compressive Sensing. *IEEE Trans. Aerosp. Electron. Syst.* **2012**, *48*, 1358–1369. [CrossRef]
29. Chen, Y.M.; Lee, J.H.; Yeh, C.C.; Mar, J. Bearing Estimation Without Calibration for Randomly Perturbed Arrays. *IEEE Trans. Signal Process.* **1991**, *39*, 194–197. [CrossRef]
30. Carino, J.; Abaunza, H.; Castillo, P. Quadrotor Quaternion Control. In Proceedings of the 2015 International Conference on Unmanned Aircraft Systems, ICUAS 2015, Denver, Colorado, USA, 9–12 June 2015.
31. Zhang, Y.; Wang, X.; Wang, S.; Tian, X. Distributed Bearing-Based Formation Control of Unmanned Aerial Vehicle Swarm via Global Orientation Estimation. *Chin. J. Aeronaut.* **2021**, *35*, 44–58. [CrossRef]
32. Goldstein, H.; Poole, P.C.; Safko, J.L. *Pearson Education—Classical Mechanics*; Pearson: London, UK, 2002.
33. Dantam, N.T. Robust and Efficient Forward, Differential, and Inverse Kinematics Using Dual Quaternions. *Int. J. Rob. Res.* **2021**, *40*, 1087–1105. [CrossRef]
34. Qiuling, J.; Guangwen, L.; Jingchao, L. Formation Control and Attitude Cooperative Control of Multiple Rigid Body Systems. In Proceedings of the ISDA 2006: Sixth International Conference on Intelligent Systems Design and Applications, Jinan, China, 16–18 October 2006; Volume 2.
35. Wang, P.K.C.; Hadaegh, F.Y.; Lau, K. Synchronized Formation Rotation and Attitude Control of Multiple Free-Flying Spacecraft. *J. Guid. Control Dyn.* **1999**, *22*, 28–35. [CrossRef]
36. Abdelkader Abdessameud, A.T. *Motion Coordination for VTOL Unmanned Aerial Vehicles: Attitude Synchronisation and Formation Control*; Springer Science & Business Media: Berlin/Heidelberg, Germany, 2013.
37. Zhao, S.; Zelazo, D. Bearing Rigidity Theory and Its Applications for Control and Estimation of Network Systems: Life beyond Distance Rigidity. *IEEE Control Syst.* **2019**, *39*, 66–83. [CrossRef]
38. Trinh, M.H.; Zhao, S.; Sun, Z.; Zelazo, D.; Anderson, B.D.O.; Ahn, H.S. Bearing-Based Formation Control of a Group of Agents with Leader-First Follower Structure. *IEEE Trans. Automat. Contr.* **2019**, *64*, 598–613. [CrossRef]
39. Ren, W. On Consensus Algorithms for Double-Integrator Dynamics. *IEEE Trans. Automat. Contr.* **2008**, *53*, 1503–1509. [CrossRef]

40. Ren, W. Distributed Attitude Alignment in Spacecraft Formation Flying. *Int. J. Adapt. Control Signal Process.* **2007**, *21*, 95–113. [CrossRef]
41. Huang, Y.; Meng, Z. Bearing-Based Distributed Formation Control of Multiple Vertical Take-Off and Landing UAVs. *IEEE Trans. Control Netw. Syst.* **2021**, *8*, 1281–1292. [CrossRef]

Article

Multi-UAV Collaboration to Survey Tibetan Antelopes in Hoh Xil

Rui Huang, Han Zhou, Tong Liu and Hanlin Sheng *

School of Energy and Power Engineering, Nanjing University of Aeronautics and Astronautics, Nanjing 210016, China
* Correspondence: dreamshl@nuaa.edu.cn; Tel.: +86-18963646736

Abstract: Reducing the total mission time is essential in wildlife surveys owing to the dynamic movement of animals throughout their migrating environment and potentially extreme changes in weather. This paper proposed a multi-UAV path planning method for counting various flora and fauna populations, which can fully use the UAVs' limited flight time to cover large areas. Unlike the current complete coverage path planning methods, based on sweep and polygon, our work encoded the path planning problem as the satisfiability modulo theory using a one-hot encoding scheme. Each instance generated a set of feasible paths at each iteration and recovered the set of shortest paths after sufficient time. We also flexibly optimized the paths based on the number of UAVs, endurance and camera parameters. We implemented the planning algorithm with four UAVs to conduct multiple photographic aerial wildlife surveys in areas around Zonag Lake, the birthplace of Tibetan antelope. Over 6 square kilometers was surveyed in about 2 h. In contrast, previous human-piloted single-drone surveys of the same area required over 4 days to complete. A generic few-shot detector that can perform effective counting without training on the target object is utilized in this paper, which can achieve an accuracy of over 97%.

Keywords: complete coverage path planning; few-shot object counting; multi-drone collaboration; Tibetan antelopes; Hoh Xil nature reserve

Citation: Huang, R.; Zhou, H.; Liu, T.; Sheng, H. Multi-UAV Collaboration to Survey Tibetan Antelopes in Hoh Xil. *Drones* **2022**, *6*, 196. https://doi.org/10.3390/drones6080196

Academic Editors: Xiwang Dong, Mou Chen, Xiangke Wang and Fei Gao

Received: 13 July 2022
Accepted: 4 August 2022
Published: 6 August 2022

Publisher's Note: MDPI stays neutral with regard to jurisdictional claims in published maps and institutional affiliations.

1. Introduction

Located between the Tanggula Mountains and the Kunlun Mountains, Hoh Xil is one of the main water sources of the Yangtze River and the Yellow River [1]. It is a significant habitat for wildlife, such as Tibetan antelopes, black-necked cranes, lynxes, and wild yaks. Hoh Xil National Nature Reserve (HXNNR) was established in 1995. It was majorly divided into four national nature reserves, the Qiangtang, the Arjinshan, the Sanjiangyuan, and the Hoh Xil, located in Qinghai province. Furthermore, HXNNR joined the World Heritage List in 2017 for its unique biodiversity and environmental conditions [2], of which Tibetan antelopes constitute a highly representative population.

However, wild Tibetan antelopes have been listed as endangered by the International Union for Conservation of Nature (IUCN) [3]. Facing this heated issue, they are protected by the Wildlife Protection Law of China, and the Convention on International Trade in Endangered Species of Wild Fauna and Flora (CITES). Over the past century, hunting and grazing have been the main threats to the survival rate of the Tibetan antelope population. Worse, Tibetan antelopes' poachers killed large numbers of antelopes for their hides and skins due to the enormous financial profits. From 1900 to 1998, the Tibetan antelope population declined from a million to less than 70,000, with a mortality rate of nearly 85%. In recent two decades, the Chinese government and wildlife conservation organizations have acted to constrain poaching, which has gained progress. Still, the deterioration of the antelopes' natural habitat and environmental degradation continue to threaten their survival [4]. Owing to the shortage of funds and labor, reserve staff can only monitor limited areas but not the entire reserve. However, artificial ground investigation caused potential damage to the protection of fragile regions [5].

Traditional survey methods mainly include field surveys by vehicle or on foot. Nevertheless, Tibetan antelopes, especially females in pregnancy, are sensitive to the appearance of humans and vehicles, so surveys are usually conducted from a distance. This method can be appropriate during the non-migratory season when small groups are evenly distributed across the plateau. Still, it may lead to biased data where terrain features block sight lines, or animals are herded [6]. Moreover, some population surveys were conducted by land-based rovers remotely piloted within survey sites [7], but this method is not suitable for surveys in larger areas or uneven, rocky, and steeply sloping terrain. It is arduous to determine the population size with ground survey methods, as animal movements can cause violent fluctuations in density estimates. In addition, severe weather conditions in unpopulated areas often create much uncertainty at high altitudes, making it difficult for observers to access investigation areas [8]. Moreover, some zoologists tracked individuals through collars with global positioning system devices to identify calving grounds, migration corridors, and suitable habitats [1,9]. However, they cannot accurately and quickly estimate the number of animals or their distribution range, so new observation methods are needed.

In recent years, aerial surveys by UAVs in unpopulated areas, poor-developed transportation, and a fragile environment have become promising options [5,6]. Compared to full-size helicopters, small UAVs produce less noise flying at altitudes that can effectively obtain ground information and significantly reduce the disturbance to wildlife [10]. Fixedwing UAVs have also been applied to wildlife surveys [11]. However, they are not suitable for the large-area survey missions in this survey, considering the take-off and landing environments and portability. Moreover, fixed-wing UAVs may also disturb wildlife as they may be perceived as avian predators [12]. In contrast, rotary-wing UAVs are more controllable and can obtain higher-resolution images, as UAVs can fly at lower speeds and altitudes without disturbing Tibetan antelopes [6]. Besides, using commercial rotary-wing UAVs can reduce the difficulty of flight operations for scientific investigations and allow researchers to focus on data analysis [13]. Moreover, UAV airports have been successfully applied in multiple fields recently [14], making regular unmanned aerial surveys possible. Utilizing multiple UAVs in an autonomous system can reduce manual requirements and conduct surveys with higher frequency and efficiency with the help of the short window period of weather, environments, and animal migration.

The efficiency of UAVs in conducting aerial surveys depends largely on the strategies the area coverage paths are planned. Generally, the complete coverage path-planning (CCPP) problem of a single UAV can be solved by sweep-style patterns, which guide the UAV back and forth over a rectangular space [15]. Other similar spatial sweep-style patterns include spiral patterns [16] and Hilbert curves [17]. Robotics designed to solve cyclic paths may encounter many tracebacks in sweep-style patterns [18]. Meanwhile, some tracebacks also occur in graphic search methods, such as wavefront methods [19,20]. Approaches that focus on minimizing the number of turns are often improved based on the minimum energy method [21,22], which is more suitable for fixed-wing or fast-moving aircraft. In contrast, we pay more attention to rotary-wing UAVs, consuming less energy during low-speed turns. In these path planning methods, the distance between the starting and ending points of the trajectories is often long. At the same time, unnecessary tracebacks may exist if the take-off and landing positions are far from the survey area. Moreover, it is time-consuming and inefficient for multiple sorties by a single UAV to perform coverage survey tasks in large areas. Presently, it is becoming a trend for multiple UAV formations to accomplish tasks jointly [23].

Most multi-UAV CCPP algorithms are extended based on the single-UAV area coverage research [24]. The common multi-UAV CCPP can be divided into two ways according to whether the task assignment is performed or not. In the multi-UAV CCPP without task assignment, each UAV may avoid collision with others by communicating with or regarding them as moving obstacles. After that, they complete area coverage tasks wholly and jointly. For instance, the path planning of each UAV is randomly generated by random coverage methods [25] and with a high traceback rate. There is also an improvement in the spanning tree coverage algorithm that employs multiple UAVs for complete coverage of a task area where there is acquired environmental information [26]. However, this method has poor robustness, and a planned trajectory's quality depends on the robot's initial position. In contrast, the research idea of the task assignment-based multi-UAV CCPP method is to divide a task area according to a certain way and assign the multiply divided sub-regions to each UAV. Then, each UAV plans its sub-region coverage trajectory within the region by a single UAV CCPP algorithm. For instance, task assignments of the sweep-style pattern [18] and polygonal area coverage [27] have been widely studied with high backtracking rates. Moreover, many of these methods are often used to find UAVs that satisfy appointed duration requirements. Moreover, many of these methods are often used to find UAVs that meet some endurance requirements. However, this paper mainly emphasizes generating optimal paths that satisfy a maximum path length constraint decided by UAV's endurance conditions.

Therefore, we formulate the multi-robot CCPP problem according to the satisfiability modulo theory (SMT), which is a generalization of Boolean satisfiability because it allows us to encode all the specific survey-related constraints. Compared to the traditional traveling salesman problem (TSP), our method focuses more on finding feasible paths that satisfy a maximum path length constraint, allowing for vertex revisitation. Because the UAV's battery constraints dictate the maximum path length, we can find solutions tailored to the limitations of the chosen UAV rather than finding a vehicle that would need to meet some endurance requirement. Although, our planning method is inherently a non-deterministic polynomial complete problem. The coverage problem can also be encoded as a MILP [28,29], it takes 4 to 6 days to solve problems for the same area, even using the most advanced solver on a powerful workstation. In contrast, the SMT method can generate available solutions in a few hours on a laptop [30].

The raw image shot by UAVs were assembled to form a large patchwork map. Then, this larger image is adopted for analysis. For example, ecologists leverage it to count flora and fauna for population analysis [31] or wildfire risk assessment in forests and grasslands [32]. However, most current counting methods can only be applied to specific types of objects, such as people [33], plants [34], and animals [31]. Meanwhile, they usually require images with tens of thousands or even millions of annotated object instances for training. Some works tackle the issue of expensive annotation cost by adopting a counting network trained on a source domain to any target domain using labels for only a few informative samples from the target domain [35]. However, even these approaches require many labeled data in the source domain. Moreover, obtaining this type of annotation is costly and arduous by annotating millions of objects on thousands of training images, especially when the photos of endangered animals are scarcer. The few-shot image classification task only requires learning about known or similar categories to classify images without training in a test, which is a promising solution [36,37]. Most existing works for few-shot object counting involve the dot annotation map with a Gaussian window of a fixed size, typically 15×15, generating a smoothed target density map for training the density estimation network. However, there are huge variations in the size of different target objects, and using a fixed-size Gaussian window will lead to significant errors in the density map. Therefore, to solve this problem, this paper uses Gaussian smoothing with adaptive window size to generate target density maps suitable for objects of different sizes. Once UAV aerial survey data have been collected and images have been analyzed to acquire accurate survey data, which can provide information on species distribution and population size, it also enables a deeper understanding of animals' seasonal migrations and habitat changes. That is critical for preparing proper wildlife conservation policies and provides affluent data for ecologists. The focus of this paper is on the development of a multi-UAV system for conducting aerial population counting.

In this paper, we considered a path planning method with a set of aerial robots for population counting, which plans the trajectories of these robots for aerial photogrammetry. UAVs acquire a set of images covering the entire area with overhead cameras. It requires the images to have higher resolution and ensure enough overlap to stitch the images further together. Besides, aerial surveys should be conducted as soon as possible, owing to changing weather, light, and dynamic ground conditions. So it is necessary to require all UAVs to perform their tasks simultaneously. The collected UAV data and analyzed images will promote the sustainable development of population ecology. This paper focuses on developing a multi-UAV system for conducting aerial surveys to provide an autonomous multi-robot solution for UAV surveys and planning suitably for most survey areas.

This research is aimed at: (1) developing and testing a multi-UAVs CCPP algorithm for fast and repeatable unmanned aerial surveys on a large area that optimizes mission time and distance. Meanwhile, it compensates for the short battery life of small UAVs by minimizing tracebacks when investigating large areas. (2) Developing and testing a population counting method applicable to various species that does not require pre-training and large data sets.

The method in this paper is demonstrated in an extensive field survey of Tibetan antelope populations at the HXNNR, China. In this task, the celerity and repeatability of aerial surveys were conducted under extreme conditions on a large area. So, it can demonstrate the effectiveness of our system in facilitating the development of population ecology. In this paper, an investigating system of four UAVs observed a herd containing approximately 20,000 Tibetan antelopes distributed over an area of more than 6 km^2, which reduced the survey time from 4 to 6 days to about 2 h. The method put forward in this paper can complete aerial surveys in shorter distance and less mission time. For instance, this method reduced the total distance and mission time by 2.5% and 26.3%, respectively. Furthermore, the effectiveness increased by 5.3% compared to the sweep-style pattern over an area of the same size for the Tibetan antelope herd survey described in this paper.

2. Materials and Methods

2.1. Survey Area

Situated in southwest Qinghai Province, China (top right of Figure 1), the unpopulated areas of Hoh Xil are 83,000 square kilometers with an average altitude of 4800 m. The Zonag Lake and Sun Lake in northwest HXNNR (lower left of Figure 1) have been identified as the current principal calving grounds for Tibetan antelopes. The birth season reaches its peak from mid-June to July every year, when tens of thousands of female Tibetan antelopes from the TRHR (Three-Rivers Headwater Region), the Qiangtang Plateau of Tibet, and Hoh Xil migrate about 300 km southwest to the Zonag Lake and the Sun Lake for parturition [4]. Therefore, the Zonag Lake is also known as the "great delivery room of antelopes". After calving, the antelope population will return to the northeast and arrive at the major overwintering pastures in mid-August.

Zonag Lake most frequently sees poachers. It is difficult to hunt antelopes because they usually are strong and vigorous. Most poachers decide to besiege Tibetan antelopes during the calving period around the Lake and kill them, which seriously affects the ecological balance of Tibetan antelope reproduction and leads to a sharp decline in population quantity. Tibetan antelopes are almost extinct.

One of the main calving grounds is presented in Figure 2. The water area of the Zonag Lake shrank drastically by 39% after the flood occurred in 2011. The western, southern, and eastern shorelines of the lake significantly altered, and the calving grounds of Tibetan antelopes on the southwestern shore also changed [2]. Recently, the government and organizations have gradually paid attention to conserving Tibetan antelopes. However, specific conservation efforts are often accompanied by the danger of entry into unpopulated areas, which requires high human costs and may cause unavoidable human interventions on the antelopes. In order to reduce costs in antelope population conservation and human interventions, we arranged a team of four rotary-wing UAVs to conduct aerial surveys and population counting in one of the main calving grounds of Tibetan antelopes located on the southwestern shore of the Zonag Lake. Meanwhile, we can learn more about the Tibetan antelope population's reproduction and changes in calving grounds.

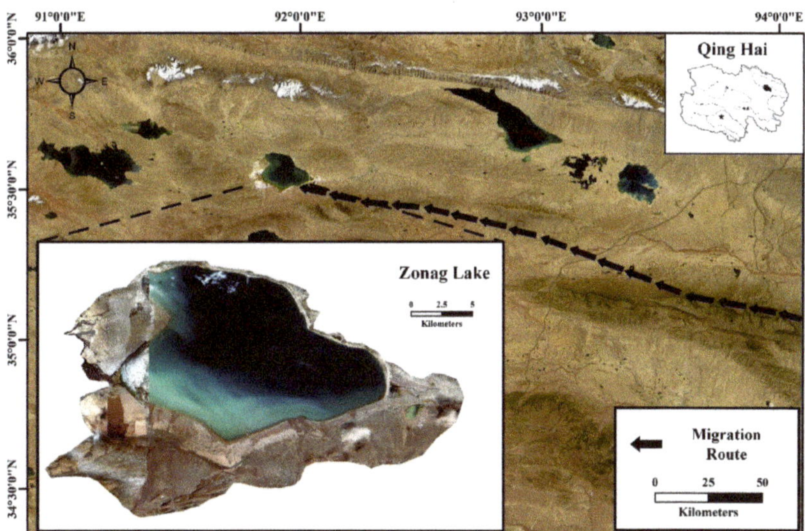

Figure 1. Survey area. Location of Zonag Lake and the migration route of Tibetan Antelopes.

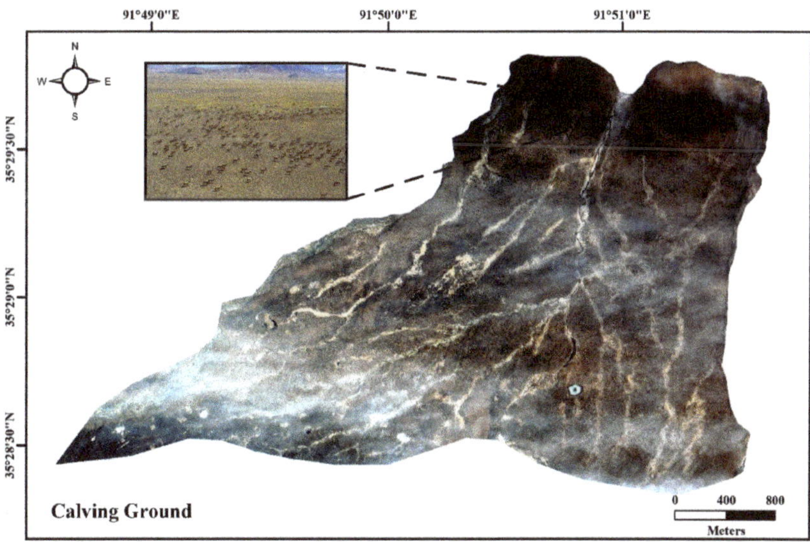

Figure 2. One of the calving grounds of Tibetan Antelopes.

2.2. CCPP Methods

2.2.1. Path-Planning Problem Encoding

The multi-UAV CCPP problem was described as SMT in this paper, which is a popularization of SAT, as it allows us to encode all the specific survey-related constraints. SMT allows predicate logic in addition to the propositional logic used in the SAT solver.

$B_{i,a,t} \in \{0,1\}$ was set to be a Boolean variable where $i \in \{1,\ldots,N_{robots}\}, a \in A = \{1,\ldots,N_{vertex}\}, and\ t \in \{0,\ldots,T_{max}\}$. The UAV i is at the vertex $a \in A$ when the step is t if $B_{i,a,t} = 1$. The parameter T_{max} is the maximum trajectory allowed by any UAV. For instance, a UAV will pass through a maximum of 20 vertices if $T_{max} = 20$. For simplicity, it was assumed in this paper that all UAVs share the same T_{max} and there were enough UAVs to cover the entire map. The Figure $G(A, E)$ was constructed based on the survey areas, where there are various types of logical statements to describe the expected behaviors allowed by UAVs. The solver attempted to assign a value of 0 (False) or 1 (True) to each Boolean variable. By doing so, the values of these statements formed in the predicate logic are true. These logical statements can be considered constraints, like expression approaches to the optimization problem.

To strengthen the physical constraints of coverage, it is required to define that

$$\forall(i,t)\ \exists a\ B_{i,a,t} = 1 \tag{1}$$

works only for an a. This expression confines the UAV i to the point a at any given step t. Then, the UAV is only allowed to move to the vertex y at the step $t + 1$, given that it is in the vertex a at the step t. Therefore, there is a line $(a \sim y)$ between these two vertices.

$$\neg B_{i,a,t}A_{a\sim y}B_{i,a,t+1} = 1 \tag{2}$$

Only the starting/ending vertex is connected to itself, and all other vertices have no self-loops. Next, it is defined that:

$$\forall a\ \exists(i,t)B_{i,a,t} = 1 \tag{3}$$

It enforces constraint requirements of the coverage task. For each vertex a in the figure, there is at least a step t where there is at least a UAV i occupying the space. If there is only a pair of (i,t) in a spot where the constrains must be satisfied, it will be equivalent to traveling salesman problem. The anti-collision constraint between UAVs may be achieved by, at most, one i as required.

$$\forall a, t\ B_{i,a,t} = 1 \tag{4}$$

At the end, the starting and the ending points were coded in this paper by setting a_i^0 as the starting and ending vertices of the UAV i:

$$B_{i,a_i^0,0} = B_{i,a_i^0,T_{max}} = 1 \tag{5}$$

It is expected that there is a closed-loop trajectory to ensure that UAVs do not have to return to the take-off location with repeated paths through the area. A separate boundary condition can be set if a closed-loop trajectory is not required, for the maximum trajectory length is determined by a UAV's battery use. To find a feasible trajectory means that the UAV has enough power to arrive at the end of the trajectory. In an emergency recall, the UAV will have enough power to return, since the straight-line distance between any point of the trajectory and its ending point is at most the length of its remaining part. This feature allows the UAV to safely return to the starting point of the survey path when the vehicle cannot safely complete the survey (e.g., adverse weather conditions or poor visibility). It also encourages the UAV to end the survey at a spot near the starting point of the trajectory.

2.2.2. Model Optimization

1. Variables reduction

The number of Boolean variables and the computing time can be reduced by leveraging the problem structure although these four constraints have fully encoded the multi-UAV coverage problem on SMT. It is assumed that each UAV can only move to an adjacent vertex at each time step, then the binary variables representing unreachable states can be removed from the problem to reduce the problem-solving time. Algorithm 1 describes the method.

2. Sequential SMT

An SMT instance does not directly optimize the objective function but only finds a feasible solution at any time. The standard descending (or ascending) method was adopted in this paper to solve the objective function until the problem is infeasible to transform the feasibility framework into an optimization framework. Algorithm 1 performs this iterative process by repeatedly solving an SMT instance to identify the shortest trajectory. Each step is sequentially decremented by $T_{\max}$ according to a certain search dispatch until the user terminates the process or the problem is infeasible. Regarding problem structure, each subsequent iteration guarantees that the generated trajectory is no longer than that in the previous iteration if a solution is found. For instance, linear search dispatch for $T_{\max}$ is provided to Algorithm 1. However, in effect, any reasonable search dispatch is applicable.

3. Parameter optimization

A parameter optimization algorithm (POA) [38] was adopted for the optimal trajectory. In this paper, the variational design of variables was performed by adding a normally distributed random value to each design variable, based on the evolutionary strategy of Schwefel's [39] work. The goal of the POA is to minimize $f(x)$. We use x_i to denote the candidate solution at the i-th generation. For each generation, we randomly mutate x_i to obtain x_i'. If $f(x_i') < f(x_i)$, then we set $x_{i+1} = x_i'$. If $f(x_i') > f(x_i)$, then we use the following logic to determine x_{i+1}. If we had set $x_{k+1} = x_k'$ at the previous generation k at which $f(x_k') > f(x_k)$, then we set $x_{i+1} = x_i'$ with a 10% probability, and $x_{i+1} = x_i$ with a 90% probability. However, if we had set $x_{k+1} = x_k$ at the previous generation k at which $f(x_k') > f(x_k)$, then we set $x_{i+1} = x_i'$ with a 50% probability, and we set $x_{i+1} = x_i$ with a 50% probability. This POA is greedy as it always accepts a beneficial mutation. However, it also includes some exploration in that it sometimes accepts a detrimental mutation. The probability of accepting a detrimental mutation varies, depending on whether or not the previous detrimental mutation was accepted. The algorithm for this POA is shown in Algorithm 1.

AcceptFlag indicates if the previous detrimental mutation replaced the candidate solution. The function of Distance() refers to time steps required to return to two vertices. In a linear grid, the function of Distance() refers to Manhattan distance required to return to two points.

2.2.3. Evaluation Indexes

The focus of this study is to reduce the total survey mission time and the total course flown by UAVs in order to accomplish a survey quickly while saving energy consumption. Therefore, the total course length was regarded as a constraint, rather than the result of a path planner's calculation. The planning aims to satisfy the trajectory coverage requirement while minimizing the distance of tracebacks. The efficiency of a set of planned trajectories R is defined as:

$$\eta_{\text{path}} = \frac{\sum_i^R S_i - \sum_i^R L_i}{\sum_i^R S_i} \tag{6}$$

S_i is the total course length flown by the UAV i, L_i is the sum of the UAV's i repeated flight paths and the distance from the take-off point to the survey area (also called transit distance). When the efficiency is 1, it means that the UAV took off within the survey area and had no repeated flight path.

Algorithm 1 Model optimization algorithm

Initialize x_1 to a random candidate solution
Initialize AcceptFlag to False
For $i = 1, 2, \cdots$

    Mutate x_i to get x_i'
    SMT $\leftarrow$ *Init* (*Survey data*)
        if Variable reduction (SMT) *is* feasible:
            for each UAV i do:
                for each vertex a do:
                    d $\leftarrow$ Distance $\left(a, a_i^0\right)$
                    for $t \in \{0, ..., d-1\}$ do:
                        False $\leftarrow B_{i,a,t}$
                        False $\leftarrow B_{i,a,T_{max}-t}$
      solution $\leftarrow$ *null*
      while *True* do:
        if Solve $\left(\text{SMT}, T_{max}\right)$ *is* feasible:
            solution $\leftarrow$ Get solution (SMT)
            Calculate $f(x)$
            $T_{max} \leftarrow T_{max} - 1$
        else:
            Return solution, $f(x)$
    If $f(x_i') < f(x_i)$:
      $x_{i+1}' \leftarrow x_i'$
    else:
      If AcceptFlag:
        $\Pr(x_{i+1} \leftarrow x_i') = 0.1$, and $\Pr(x_{i+1} \leftarrow x_i) = 0.9$
      else:
        $\Pr(x_{i+1} \leftarrow x_i') = 0.5$, and $\Pr(x_{i+1} \leftarrow x_i) = 0.5$
      AcceptFlag $\leftarrow (x_{i+1} \leftarrow x_i')$
Next i

2.3. Counting Methods

2.3.1. Adaptation and Matching Counting Network

In Figure 3, the network architecture of the animal counting algorithm adopted in this paper is depicted. Input to the network is the image $X \in \mathbb{R}^{H \times W \times 3}$ and several examples describing bounding boxes of the objects to be counted from the same image. Output from the network is the predicted density map $Z \in \mathbb{R}^{H \times W}$, and the counts of the interesting objects are obtained by summing up all density values. The network consists of two crucial modules: (1) a multi-scale feature extraction module; (2) a density prediction module. These two modules were designed in this paper for them to handle new categories in a test. ImageNet was adopted to extract features for such a network that can handle a wide range of visual categories. The feature extractor consists of a multi-layer convolution operation that maps the input into d-channel features. For the image X, it outputs a downsampled feature map $F(X) \in \mathbb{R}^{d \times h \times w}$. For an exemplar box a, the output feature map is further processed with global average pooling to form a feature vector $F(a) \in \mathbb{R}^d$. The density prediction module was designed to be incognizable to visual categories. The multi-scale feature extraction module consists of the first four blocks from a pre-trained ResNet-50 backbone [40]. Images were represented in this paper by convolutional feature mapping in the third and fourth blocks. In addition, ROI pooling was performed in this paper for the convolutional feature mapping from the third and fourth Resnet-50 blocks to obtain the examples' multi-scale features.

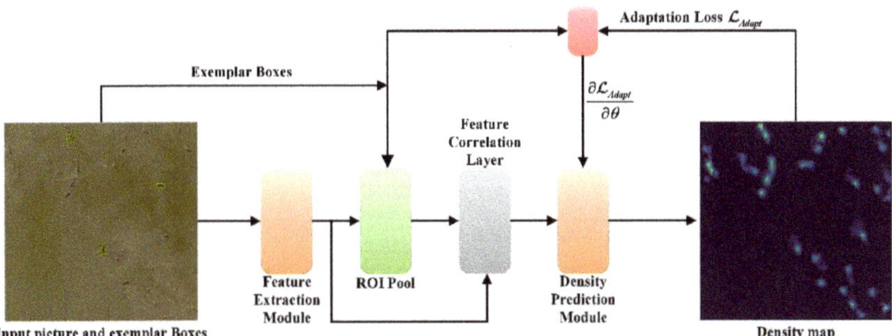

Figure 3. Few-shot adaptation and matching network.

The features obtained from the feature extraction module were not directly adopted in density prediction, in order to make the density prediction module incognizable to visual categories. Instead, the correlation mapping between sample features and image features was input to the density prediction module. The sample features were resized to different scales given the target objects at different scales should be considered. The resized sample features were associated with the image features to obtain multiple correlation maps, each of which corresponds to one scale. Scales of 0.9 and 1.1, as well as the original scales, were adopted in all experiments. The correlation maps were connected in series' and fed to the density prediction module. The density prediction module consists of five convolutional blocks and three upsampling layers placed after the first, second, and third convolutional layers, respectively. The last layer is a 1×1 convolutional layer that predicts the two-dimensional density map. The size of the predicted density map is the same as the size of the input image.

2.3.2. Adaptive Window

The training images of a dataset were leveraged to train the network. Each training image contains multiple target objects. Moreover, most objects were only annotated with dots, while only the example objects were annotated with bounding boxes. However, it is difficult to train the density prediction network based on the training losses defined by dotted annotations. Most existing visual counting efforts, especially crowd counting [41], use a fixed size Gaussian window (usually 15×15) to convolve dotted annotated maps, thus generating smooth target density maps for training the density prediction network. The selected training dataset consists of 147 different categories in which the object size varies considerably [42]. Therefore, Gaussian smoothing with a self-adaptive window size was employed to generate the target density map. Firstly, the object size was estimated with dotted annotations. In a given dotted annotated map, each dot is nearly located at the center of the object. The horizontal distance x_i and vertical distance y_i between each dot i and its nearest neighbor were calculated. With the modes $\hat{x}, \hat{y}$ of X and

$$X, Y (x_i \in X, y_i \in Y, i = 1, 2, 3 \ldots n), \text{ the distance } d = \sqrt{\left(\hat{x}\right)^2 + \left(\hat{y}\right)^2} \text{ can be calculated as}$$

the size of the Gaussian window to generate the target density map. The Gaussian standard deviation was set to be one quarter of the window size. The mean square error between the density prediction map and the ground truth density map was minimized to train the network. To match it, we adopted Adam optimizer with a learning rate of 10^{-5} and a batch size of 1. The height of each image was fixed at 384, and the width was accordingly adjusted to maintain the aspect ratio of the original image.

2.3.3. Loss Function

The above network structure only uses the samples' example bounding boxes to extract their feature information. It does not fully use the position information provided by the exemplar boxes, while annotations by human eyes are crucial in this process. So we define the combination of two losses here to correct the model's errors during the test.

- Count Loss

For each example bounding box, a, the sum of the density values within the box should be at least 1 because the exemplar box may also contain parts of other objects or overlap other objects with the same type to some extent. The following count loss was defined to quantify the number of constraint violations.

$$\mathcal{L}_{count} = \sum_{a \in A} \max(0, 1 - \|Z_a\|_1) \tag{7}$$

In this equation, A is the set of bounding boxes, and $\|Z_a\|_1$ the sum of all values in Z_a.

- Perturbation Loss

The density prediction map Z in essence is a correlation map between a sample and an image. Therefore, the density values near the sample should be similar to Gaussian distribution in an ideal circumstance. $G_{h \times w}$ was set to be a two-dimensional Gaussian window with the size of $h \times w$. The perturbation loss was defined as follows:

$$\mathcal{L}_p = \sum_{a \in A} \|Z_a - G_{h \times w}\|_2^2 \tag{8}$$

- The Combined Adaptation Loss

The self-adaptation loss for testing is a weighted combination of the minimum count loss and the perturbation loss.

$$\mathcal{L}_{Adapt} = \lambda_1 \mathcal{L}_{count} + \lambda_2 \mathcal{L}_p \tag{9}$$

In the equation, λ_1 and λ_2 are coefficients of the count loss and the perturbation loss respectively. In testing, we referred to parameters setting [42], so as to perform 100 gradient descent steps for each test image and optimize the weighted loss of Equation (9). The learning rate adopted in this paper is 10^{-7}. The values of λ_1 and λ_2 are, respectively, 10^{-9} and 10^{-4}. The self-adaptation loss was only employed in testing. The loss function during training is the root mean square error between the density prediction map and the actual density map.

2.3.4. Evaluation indexes

The mean absolute error (MAE) and the root mean square error (RMSE) were adopted in this paper to measure the accuracy of counting methods, which are defined as follows.

$$\text{MAE} = \frac{1}{n} \sum_{i=1}^{n} |c_i - \hat{c}_i| \tag{10}$$

$$\text{RMAE} = \sqrt{\frac{1}{n} \sum_{i=1}^{n} (c_i - \hat{c}_i)^2} \tag{11}$$

In these equations, n refers to the quantity of test images, c_i and $\hat{c}_i$ are the ground truth and the predicted counts, respectively.

3. Results

3.1. Multi-UAVs CCPP

3.1.1. Summary of the Surveys

Considering the altitude of the animal survey area is around 5000 m, we chose a high-altitude low noise propeller. In this way, the UAVs can better adapt to the flight conditions at high altitudes and meet the noise limits of animal surveys. During the execution of the survey, the path spacing and flight altitude of the UAV are set according to the camera parameters and image stitching requirements. During non-mission execution (transit distance transit distance to and from the takeoff point and mission start point), the flight speed of the UAV is 10 m/s, which is determined by the environmental conditions and the specifications of the UAV. We want the speed to be as fast as possible under the premise of ensured safety and low noise. In addition, considering the safety withdrawal requirements and the actual hovering test results in abnormal weather, we set the safe flight time of the UAV at 28 min. Its detailed parameters are shown in Table 1.

Table 1. Technical parameters of the aircraft.

Technical Parameters	Description	Technical Parameters	Description
Safe endurance time	28 min	Survey area flight speed	5 m/s
Maximum flight time	55 min	Transit distance flight speed	10 m/s
Maximum takeoff weight	9 kg	Maximum flight altitude	7000 m
Body weight with battery	6.3 kg	Flight height from ground	200 m
Autopilot	DJI flight control unit	Size (folded, including paddles)	$430 \times 420 \times 430$ mm (L $\times$ W $\times$ H)

During each survey, more than 6000 images were taken by a group of UAVs and stitched together to form a large, stitched image covering the entire survey area, as presented in Figure 4d. In this paper, these stitched images were employed to identify the number of Tibetan antelopes for estimating the population density of these groups. In our work, CCPP methods and population counting were emphasized rather than ecology.

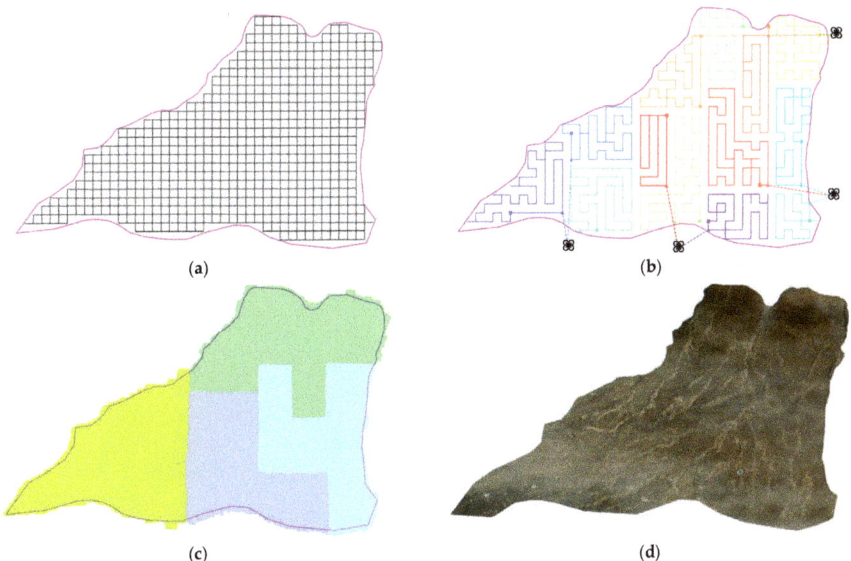

Figure 4. Planning workflow. (**a**) The coverage grides of survey area; (**b**) Planned flight path of UAV. The planned UAV path is a solid line, and the transit route is a dotted line. The drone logo represents the take-off and landing location; (**c**) The resulting photo coverage area allocated to the four UAVs with different starting points in (**b**); (**d**) The last stitched image.

Figure 4 shows the workflow of our system in the survey area (Figure 2). Firstly, we specified the interested area and divided it into grids with the spacing of 100 m between vertices, as shown in Figure 4a. Secondly, the CCPP algorithm is used to generate offline trajectories and optimize the starting and ending positions of the UAVs, as shown in Figure 4b. The four starting points are distributed in the vicinity of the survey area to ensure the optimal total flight distance and mission time for a group of UAVs. Thirdly, UAVs that execute these routes were activated to collect survey images at regular intervals. Figure 4c shows the image coverage area of each UAV. The images are stitched according to the UAVs' time intervals and flight paths. The four different color blocks represent the captured areas by each of the four UAVs. Finally, these images were stitched together to generate a mosaic image, as shown in Figure 4d. The mosaic images will be used for animal counts to estimate population size, as described in Section 3.2.

12~13 flights (the flight number depends on the UAV position) were required to survey the area with four deployed UAVs in advance. It cost 2 days for a single manually piloted UAV to survey. However, in this paper, the survey can be completed in about 2 h in some regions of Zonag Lake by the proposed multi-UAV collaborative survey method. Using the autonomous system reduces the burden on manual requirements and realizes faster surveys with higher frequencies, thus enabling the exploitation of short windows in favor of the survey. That makes it more likely to make arrangements in advance and perform unmanned observation and data collection with the knowledge of animal migration routes.

The paths in Figure 4b were input to universal ground control software (UGCS) [43] for arranging the interface between a route planner and the UAV on-board flight computer. Although a trajectory is generated offline, its self-adaption allows a user to place the starting node at any position on it without recalculating the entire path. The entire course of each path is displayed in Figure 5. The UGCS will send a UAV command after it takes off so that the drone can execute the route and automatically land back at the starting point for recharging, thus preparing for the next flight.

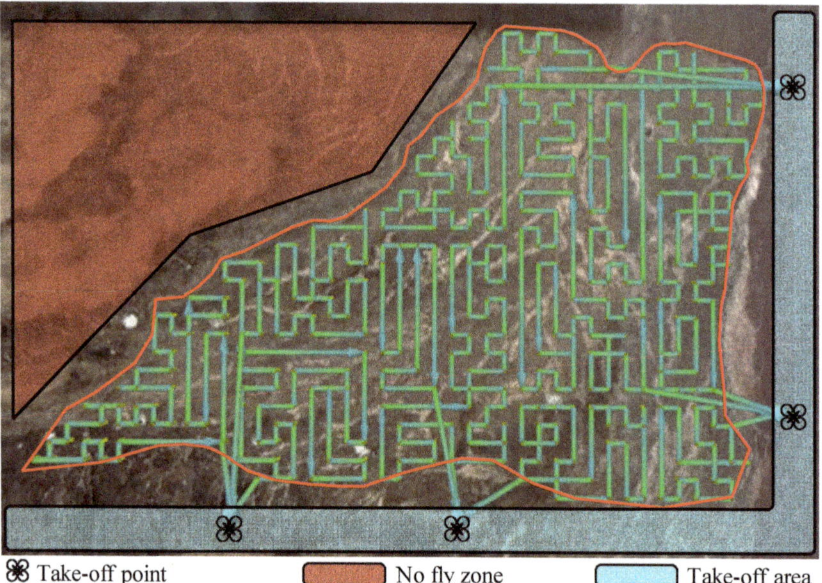

Figure 5. Survey routes. All routes to the survey area are shown here. The blue-green lines are routes, which are planned by the method in this paper, and are executed by four UAVs at different starting points. The blue area is the optional takeoff and landing area. The red no fly zone marks the mountainous area or flight interference area.

3.1.2. Comparisons Ignoring Transit Distance

In this paper, we employed the SMT algorithm employed with the Z3 solver [44], the experimental calculations, which are performed under CPU:12th Gen Intel(R) Core(TM) i9-12900KF 3.20 GHz and the operating system of Windows11. Each computation takes about 15 s, and optimized calculation can get the best result in less than 100 epochs. We developed the algorithms on python and the commercial software ISIGHT, and we eventually ported them to python for ease of deployment and fast computation.

The proposed method in this paper was compared with both the common sweep-style method in the literature [18] and recently cell decomposition polygon coverage path planning method [27]. The sweep-style pattern was planned by using the area coverage function already realized in the UGCS [43]. The polygon coverage method divides the required region into polygons first, then plans coverage routes by using the sweep-style method, and last stitches these paths together with the TSP solver at the end. Moreover, Polygon division allows the solver to locally optimize the sweep direction of each polygon cell, which leads to shorter trajectories in most cases.

The efficiency of the polygon coverage method is higher than that of the simple sweep-style method, but there are still considerable tracebacks in both two methods because the core planning strategy of two methods lies are the sweep-style pattern. We compared routes of our trajectory planning algorithm, the standard sweep-style method, and the polygon division method in Figure 6. Although the sweep-style method has covered enough areas, it is influenced greatly by tracebacks, as presented in Figure 6b.

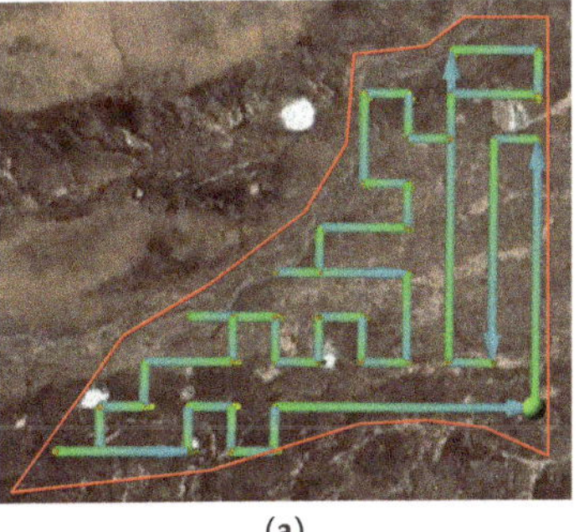 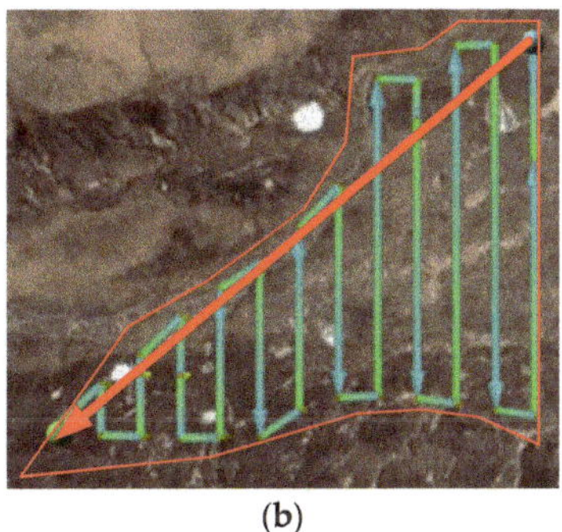 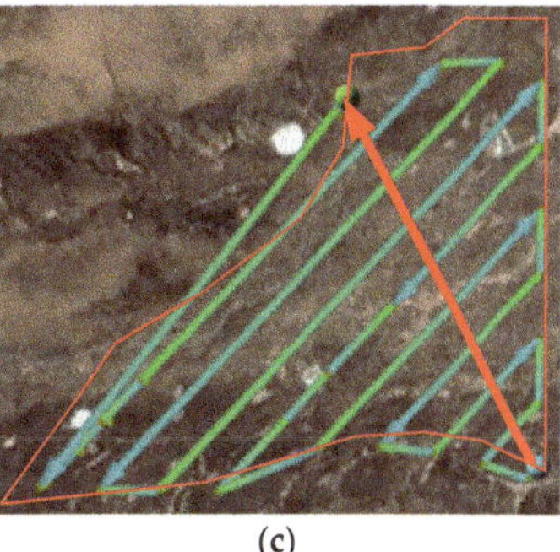

(a)   (b)   (c)

Figure 6. Comparison of route planning methods. (**a**) Ours; (**b**) Sweep-style path; (**c**) Polygon coverage. The red arrow is the flight path of backtracking.

The path efficiency of our method, the sweep-style method, and the polygon method are demonstrated in Table 2. The most effective set of paths is presented in bold. Algorithms in this paper realize the best efficiency in all but one case, with an average efficiency improvement of 8.2% compared to the sweep planner and 8.1% compared to the polygon coverage method. Without considering the transit distance of the planned trajectory, the sweep-style method results in a total course that is about 700 m longer than the trajectory generated by our method. The route generated by the polygon method is 400 m longer than that planned by our method. In all cases, the method proposed in this paper can reduce 366 m and 1.5 min compared to the polygon method, which will be 666 m and 2.2 min, compared to the sweep-style method.

Table 2. Comparison of route efficiency. Bold entries indicate the most efficient set of routes.

Method	Parameters	Zone 1	Zone 2	Zone 3	Zone 4	Zone 5	Zone 6	Average
Ours	Length	5.9 km	4.7 km	4.4 km	3.6 km	2.5 km	1.5 km	-
	N_{vertex}	132	109	102	82	62	56	-
	N_{path}	136	112	104	85	65	61	-
	η_{path}	**0.971**	**0.973**	**0.981**	**0.965**	**0.954**	0.918	**0.960**
Sweep	Length	6.5 km	5.5 km	4.9 km	4.5 km	3.4 km	1.8 km	-
	Backtrack	0.57 km	0.98 km	0.37 km	0.63 km	0.52 km	0.08 km	-
	η_{path}	0.912	0.822	0.924	0.860	0.847	**0.956**	0.887
Polygon	Length	6.8 km	4.7 km	4.5 km	4.3 km	2.9 km	1.6 km	-
	Backtrack	0.80 km	0.51 km	0.38 km	0.61 km	0.33 km	0.17 km	-
	η_{path}	0.882	0.891	0.916	0.858	0.886	0.894	0.888

3.1.3. Comparisons Considering Transit Distance

In an actual flight survey, there is a long distance between the UAV take-off point and the survey area. In the event of regular multi-UAV surveys, the take-off point may be flexibly arranged in advance before the survey window, which can reduce the impact of tracebacks and the consumption of UAV power to a certain extent. The total UAV flight distance was optimized in this paper within a given take-off area to make full use of the advantages of multiple UAVs.

We demonstrated the actual survey results of our method, the sweep-style method, and the polygon method in parts of the Zonag Lake (Figure 2) in Table 3, where the effective set of trajectories are presented in bold. In Table 3, the take-off point is designated at the geometric center of the take-off area, based on empirical judgment. It is evident that the trajectory efficiency and the mission time are improved for different methods by flexibly configuring take-off points (number of starting points, restrictions on the starting area). Differences among the methods are few and finding the optimal set of paths is easy when there are two different take-off points. As the differences among algorithms grow when plans become more flexible, there are more starting points available for choosing. Moreover even the take-off area is no longer restricted to be outside the survey area (which is undesirable in animal surveys because the noise from the UAV may be perceived as a threat). The method proposed in this paper demonstrates the best path's efficiency and the shortest time consumption under all three different plans, with a 22.72% increase in trajectory efficiency and a 34.62% reduction in mission time with unrestricted starting area. The method may reduce at most 54 min of the mission time, when compared to the manual setting method (designate starting points according to experience), which is equivalent to the time about twice as long as the maximum UAV endurance time.

Table 3. Comparison of route parameters (distance, efficiency, and mission time). Set the total flight distance as the optimization parameter. Bold entries indicate the most efficient set of routes. The data are the average of 50 experiments.

Take-Off Plan	Method	Distance	Backtrack	η_{path}	Efficiency Improvement	Mission Time	Time Reduction
Two different take-off points within the area restrictions	Manual setting	94.2 km	20.8 km	0.779	Control group	2.6 h	Control group
	Sweep	94.1 km	20.8 km	0.779	0.00%	2.6 h	0%
	Polygon	**94.0 km**	**20.7 km**	**0.780**	**0.13%**	**2.4 h**	**7.69%**
	Ours	**94.0 km**	**20.7 km**	**0.780**	**0.13%**	**2.4 h**	**7.69%**
Four different take-off points within the area restrictions	Manual setting	94.2 km	20.8 km	0.779	Control group	2.6 h	Control group
	Sweep	91.8 km	18.4 km	0.799	0.00%	1.9 h	26.92%
	Polygon	91.6 km	18.3 km	0.800	0.13%	1.9 h	26.92%
	Ours	**88.7 km**	**15.5 km**	**0.825**	**5.91%**	**1.7 h**	**34.62%**
Four different take-off points without the area restrictions	Manual setting	95.8 km	22.4 km	0.766	Control group	2.6 h	Control group
	Sweep	92.4 km	16.2 km	0.795	3.79%	1.8 h	30.77%
	Polygon	89.0 km	15.6 km	0.825	7.70%	1.8 h	30.77%
	Ours	**78.0 km**	**4.7 km**	**0.940**	**22.72%**	**1.7 h**	**34.62%**

During short-term and quick surveys, the flight time of UAVs is a factor to be fully considered to make full use of the valuable survey window. Shorter aerial survey time can help accomplish more surveys and capture more data throughout the breeding season. Based on the above methods, the total UAV flight distance and total mission time were optimized in this paper within a given planned take-off area.

Table 4 shows the actual test results of our method and other methods in some parts of Zonag Lake. If the configuration of starting points is more flexible (number of starting points, restrictions on the starting area), the metrics of the different methods will improve accordingly. This is similar to the results of the method mentioned above, that optimizes only the total flight distance of a group of aircrafts. The variation between algorithms is increasing as plans become more flexible, starting points become more selective, and even take-off areas are no longer restricted to survey areas. Our method shows the best trajectory efficiency and the shortest task time among all three different schedules, with a 23.63% improvement in path efficiency and a 47.83% reduction in task time as the starting area is unrestricted. Our method's flight distance is unchanged, compared to methods only optimizing the total flight distance while the total mission time decreases from 2.6 h to 1.2 h at most, with an improvement of 19.23%. The method proposed in this paper witnesses an average improvement of 12.82% with three different take-off plans. Meanwhile, it can reduce mission times by up to 66 min, compared to manual setup methods, which are about twice the safe endurance time.

Table 4. Comparison of route parameters (distance, efficiency, and mission time). Set the weighted sum of total flight path and total mission time as optimization parameters. Bold entries indicate the most efficient set of routes. The data are the average of 50 experiments.

Take-Off Plan	Method	Distance	Backtrack	η_{path}	Efficiency Improvement	Mission Time	Time Reduction
Two different take-off points within the area restrictions	Manual setting	94.2 km	21.6 km	0.771	Control group	2.6 h	Control group
	Sweep	95.0 km	21.7 km	0.772	0.13%	2.5 h	3.85%
	Polygon	95.0 km	21.3 km	0.776	0.65%	2.3 h	11.54%
	Ours	**94.4 km**	**20.8 km**	**0.780**	**1.17%**	**2.2 h**	**15.38%**
Four different take-off points within the area restrictions	Manual setting	94.2 km	20.8 km	0.779	Control group	2.6 h	Control group
	Sweep	91.8 km	18.5 km	0.799	0.00%	1.9 h	26.92%
	Polygon	92.7 km	19.0 km	0.795	2.05%	1.7 h	34.62%
	Ours	**89.5 km**	**16.1 km**	**0.820**	**5.26%**	**1.4 h**	**46.15%**
Four different take-off points without the area restrictions	Manual setting	95.8 km	22.4 km	0.766	Control group	2.5 h	Control group
	Sweep	90.8 km	21.2 km	0.866	11.55%	1.5 h	42.31%
	Polygon	80.3 km	7.0 km	0.913	19.19%	1.4 h	46.15%
	Ours	**77.4 km**	**4.1 km**	**0.947**	**23.63%**	**1.2 h**	**53.85%**

3.2. Population Counting

3.2.1. Comparison with Few-Shot Approaches

The performance of counting effects of our method was compared with that of other methods. There are two common baselines, namely (1) always outputting the average object quantity of the training images; (2) always outputting the mid-value counts of the training images. The two popular algorithms are: feature reweighting (FR) few-shot detectors [36] and few-shot object detection (FSOD) [37]. The method adopted in this paper is more advance than others, as presented in Table 5.

Table 5. Comparing our method to two simple baselines (mean, median) and two popular algorithms (FR few-shot detector, FSOD few-shot detector). These are few-shot methods that have been adapted and trained for population counting. Our method has the lowest MAE and RMSE on Tibetan antelope dataset. Bold entries indicate the most efficient method.

Method	MAE	RMSE
Mean	48.7	146.7
Median	47.7	152.4
FR few-shot detector	41.7	141.0
FSOD few-shot detector	35.5	140.7
Ours	**23.2**	**98.4**

3.2.2. Comparison with the Object Detector

One of the other counting methods is to use detectors to count objects after they have been detected, which is only applicable to certain categories of objects with pre-trained detectors. In general, training object detectors requires thousands of examples, so it is not a practical approach for general visual counting. The performance of the network on a validation set with pre-trained object detectors and on categorized subsets of a test set was evaluated in this paper. To be specific, we compared our method with Faster R-CNN [45] and Mask R-CNN [46]. These two pre-trained detectors can be found in the Detectron2 library [47]. The comparison results are shown in Table 6. It can be seen that our method outperforms target detectors that have been pre-trained with thousands of annotated examples.

Table 6. Comparing our method with pre-trained object detectors. Bold entries indicate the most efficient method.

Method	MAE	RMSE
Faster R-CNN	36.5	79.2
Mask R-CNN	35.2	79.8
Ours	**23.9**	**45.4**

3.2.3. Qualitative Analysis

Figures 7–10 shows the visual prediction results of our algorithm. The input was the image to be tested and several bounding boxes of counting objects. In order to analyze the correctness of the density map expediently, we showed the input image and the density map superimposed. Figures 7 and 8 show examples of Tibetan antelope images with different densities from the aerial view of the UAVs, which has proved the algorithm works well in the case of a flat background from an overhead view with more than 97% correct counting rate. Furthermore, the model still has an accuracy rate of more than 97% in the case of complex backgrounds (Figure 9), where both water and land are present in the overhead view and the color of the land is close to that of Tibetan antelope. Figure 10 shows an example of a Tibetan antelope herd image taken from the side view of a UAV flight, which has a correct counting rate of over 96%. The overhead and side view results demonstrate the robustness of the method utilized in this paper for counting Tibetan antelopes photographed from different viewpoints.

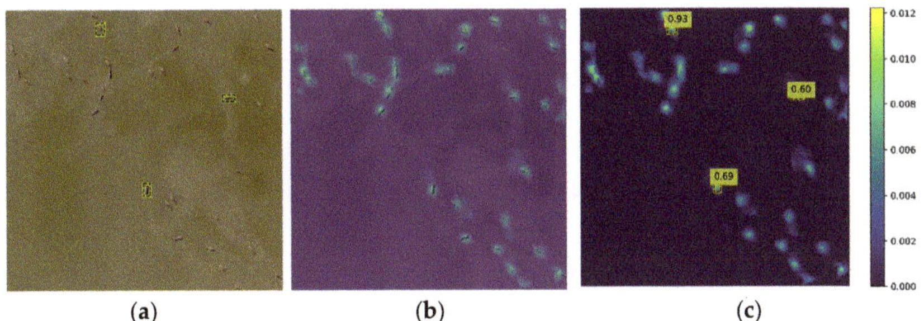

(a)   (b)   (c)

Figure 7. Low-density photos of Tibetan antelope taken by UAVs above. (**a**) Input image and a few exemplar bounding boxes depicting the object to be counted; (**b**) Overlapping of the input image and the predicted density map; (**c**) Predicted density map. The value in the yellow box represents the predicted density value. The ground truth count is 34, the predicted count is 35, and the accuracy is 97.1%.

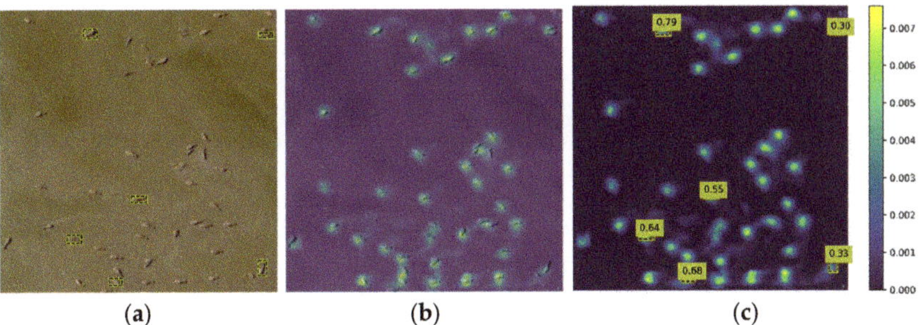

(a)   (b)   (c)

Figure 8. Medium-density photos of Tibetan antelope taken by UAVs above. (**a**) Input image and a few exemplar bounding boxes depicting the object to be counted; (**b**) Overlapping of the input image and the predicted density map; (**c**) Predicted density map. The value in the yellow box represents the predicted density value. The ground truth count is 53, the predicted count is 52, and the accuracy is 98.1%.

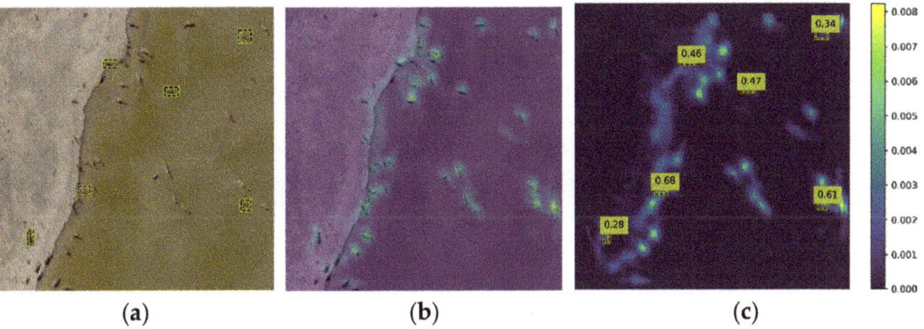

(a)   (b)   (c)

Figure 9. Medium-density photos of Tibetan antelope with complex background taken by UAVs above. (**a**) Input image and a few exemplar bounding boxes depicting the object to be counted; (**b**) Overlapping of the input image and the predicted density map; (**c**) Predicted density map. The value in the yellow box represents the predicted density value. The ground truth count is 45, the predicted count is 46, and the accuracy is 97.8%.

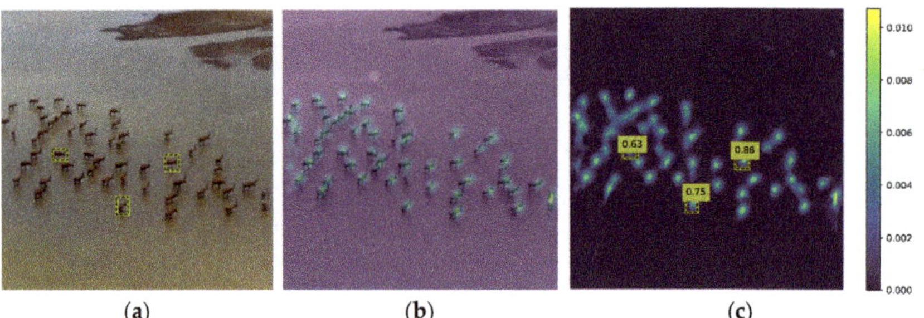

Figure 10. Network prediction results of medium density photos of Tibetan antelope taken by UAVs from the side. (**a**) Input image and a few exemplar bounding boxes depicting the object to be counted; (**b**) Overlapping of the input image and the predicted density map; (**c**) Predicted density map. The value in the yellow box represents the predicted density value. The ground truth count is 55, the predicted count is 53, and the accuracy is 96.4%.

4. Discussion

The study demonstrates that daily monitoring in protected areas is greatly complemented by surveys of wildlife, which are conducted by small, low-cost, multiple rotary-wing UAVs in large areas. However, the influence of UAVs' noise and appearance on wildlife should be considered during surveys. As we only surveyed and analyzed Tibetan antelopes, prior detailed testing on the UAV's influence of noise and movements should be conducted when this method is applied to other animals. Moreover, our method extends the field of UAV application and provides an autonomous multi-robot solution for UAV path planning, which is suitable for any size of the survey area. It can also select the appropriate path according to the existing UAV parameters and allows them to optimize the take-off point of each flight.

The flexible deployment of multiple UAVs can significantly improve the path efficiency for completing area coverage missions and reducing manual survey time and manpower consumption, especially in remote and unpopulated areas. The results of multi-UAV trajectory planning (Tables 2 and 3) have been visualized in Figures 11–13. Regarding the total distance and the mission time, our method has more edges over the current multi-UAV CCPP methods. In one aspect, the shorter total distance means less energy consumption, which indicates a group of recharged UAVs can complete more flights when deployed at a UAV airport. In another aspect, a shorter mission time means that a quicker survey, allowing more surveys to be done within a window period to reduce da-ta errors and learn about timing-related issues. It can provide ecologists with more options for their survey methods. Compared to the method where only the total distance is done, both the total distance and mission time decline when these two indicators are optimized simultaneously. This is because the planner divides and distributes the entire survey route task to each with different take-off points evenly while acquiring the optimal solution for the area to survey on its own.

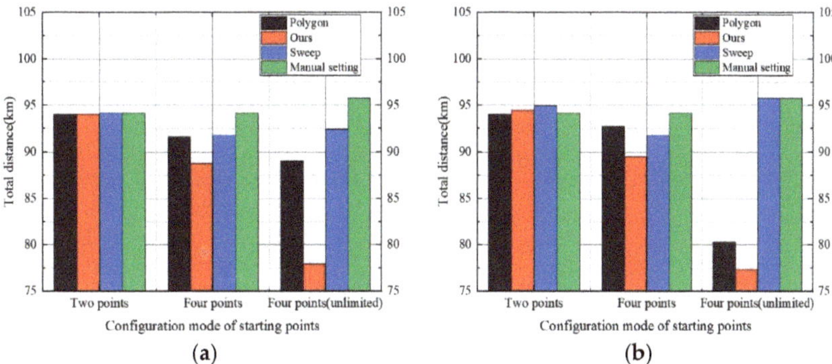

Figure 11. The total flight distance of a group of UAVs. (**a**) Set the total flight distance as the optimization parameter; (**b**) Set the weighted sum of total flight path and total mission time as optimization parameters.

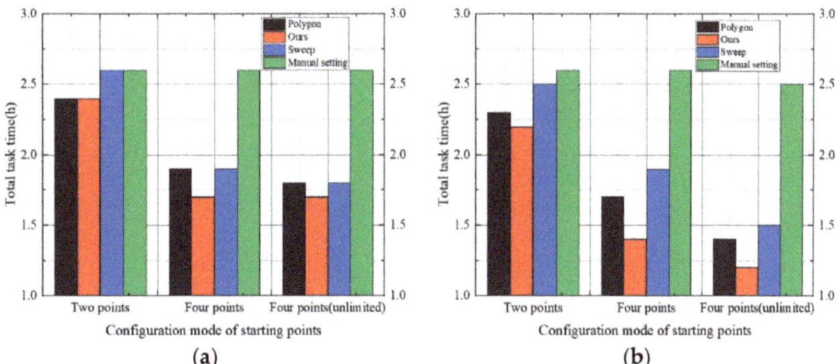

Figure 12. The total mission time of a group of UAVs. (**a**) Set the total flight distance as the optimization parameter; (**b**) Set the weighted sum of total flight path and total mission time as optimization parameters.

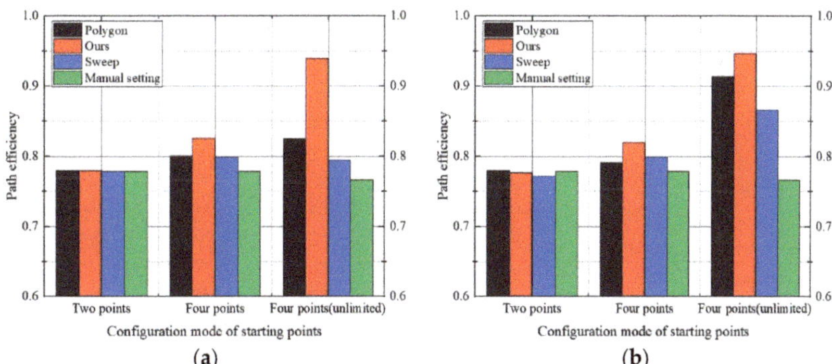

Figure 13. Trajectory efficiency of a group of UAVs. (**a**) Set the total flight distance as the optimization parameter; (**b**) Set the weighted sum of total flight path and total mission time as optimization parameters.

The few-shot object counting method can be used without pre-training the model via images of the target population. This improves the universality of the counting method to a certain extent, and it can be applied to many different fields. When using this method to test the photos of Tibetan antelope taken by UAVs, it has good applicability to images of different densities and perspectives. It can meet the requirements of population counting. A particular animal can be repeatedly photographed when images are stitched together because of the possible animal movements during a multi-UAV survey. Ecologists focusing on data research need to correct the survey results to ensure the accuracy of data results, which has become a necessary research subject [48]. Therefore, it is essential to reduce the mission time of such investigations. The information obtained directly from traditional ground surveys is often one-dimensional. The distribution of all animals may be visually linear (Figure 14) if interpreted from an image perspective when surface observation is conducted on the ground [49]. These factors often lead to significant errors in the results, which is not conducive to animal counting. When compared with traditional field survey methods by vehicle or on foot, UAV surveys have a better field of view (Figures 7–9) and have reduced the impact of animal movements on the data to a large extent, which can improve the survey accuracy.

Figure 14. Tibetan antelope was photographed on the ground. In the sun, four Tibetan antelopes overlap, so there seem to be only three. Therefore, artificial ground measurement is not conducive to accurate counting.

With the current development of commercial multi-rotary-wing UAVs and UAV airports, an untrained layman may acquire the skills to control a rotary-wing UAV quickly. The aerial animal counting system with multi-UAVs, developed by us, aims to enable scientists to focus on experimental design and data analysis rather than flight control skills. The survey deployment can be made before the survey window through a simple setup that determines the surveyed areas, the take-off areas, and the UAVs' information. The current artificial intelligence methods can quickly assist humans in completing many of tedious counting tasks. The approach proposed in this paper can replace humans in performing uninterrupted unmanned count surveys, and we only need to ensure that UAVs will not influence the ecology of the survey areas.

However, more sorties of UAVs are often required for conducting simultaneous surveys of large areas, due to UAV endurance limitations. Besides, excessively high-flight altitudes, which means reducing the impact of UAV noise on animals, may not be conducive to animal counting in feature identification. It does harm to the applicability of the multi-UAV survey method because of camera resolution. UAV aerial surveys will also integrate more features by combining visible light and multispectral and thermal infrared cameras to improve information accuracy further.

5. Conclusions

A general and flexible CCPP method was provided in this paper for a group of UAVs to count animals. This planning algorithm enables scientists to employ multiple drones to conduct regular unmanned surveys in the uninhabited region in fragile environments with poor access. Moreover, multiple UAVs allow faster surveys with higher frequencies, lower costs, and greater resolution than helicopters and better leverage for rare observation opportunities by reducing wildlife disturbance from human activities. We have implemented our algorithms on a set of UAVs and conducted field experiments. The rapid and repeatable aerial surveys of large-scale environments demonstrated the utility of our system in facilitating population ecology. The results unveil that, in this paper, a herd containing approximately 20,000 Tibetan antelopes distributed over an area of more than 6 km^2 was observed by a measuring system of four UAVs, which reduced the survey time from 4 to 6 days to about 2 h. Our method can complete aerial surveys in shorter trajectories and mission time. With this method, the trajectories and time were reduced by 2.5% and 26.3%, respectively, and the effectiveness increased by 5.3% when compared to sweep-style paths over an area of the same size.

Different from the target detector methods commonly used in current counting methods, the current work mainly focuses on one specific category at a time, such as people, cars, and plants. In this paper, we use a general few-shot adaptive matching counting method, which can count effectively without pre-training through the objects to be measured. We tested the photos of Tibetan antelope taken from the UAV, and the accuracy can reach more than 97%. The survey data will facilitate the development of animal protection policies. The results of the later survey can contribute to the development of animal protection policy, which is conducive to protecting and promoting ecological sustainability. However, animals may move in the process of multi-drone surveys, which makes photos of them possible to be duplicated when stitched together. In the future, reasonable corrections based on the survey results can further improve the accuracy of the survey. Other investigations that require fast aerial surveys under unpopulated circumstances can also benefit from our algorithms, such as surveys of other wildlife populations, forests, and shrublands.

Author Contributions: Methodology, H.S., R.H., H.Z. and T.L.; software R.H. and T.L.; Investigation, H.S. and H.Z.; data curation, H.Z. and T.L.; writing—original draft preparation visualization, H.S., R.H., H.Z. and T.L.; funding acquisition, H.S. All authors have read and agreed to the published version of the manuscript.

Funding: This work was supported by Funding of National Key Laboratory of Rotorcraft Aeromechanics (No. 61422202108), National Natural Science Foundation of China (No.51906103, No.52176009).

Data Availability Statement: Not applicable.

Conflicts of Interest: The authors declare no conflict of interest.

References

1. Manayeva, K.; Hoshino, B.; Igota, H.; Nakazawa, T.; Sumiya, G. Seasonal migration and home ranges of Tibetan antelopes (*Pantholops hodgsonii*) based on satellite tracking. *Int. J. Zool. Res.* **2016**, *13*, 26–37. [CrossRef]
2. Lu, S.; Chen, F.; Zhou, J.; Hughes, A.C.; Ma, X.; Gao, W. Cascading implications of a single climate change event for fragile ecosystems on the Qinghai-Tibetan Plateau. *Ecosphere* **2020**, *11*, e03243. [CrossRef]
3. IUCN Website. Available online: www.redlist.org (accessed on 13 July 2022).
4. Lin, X. *Conservation and Monitoring of Tibetan Antelopes in Hoh-Xil Nature Reserve*; BP Conservation Programme BPCP: Beijing, China, 2014.
5. Bollard, B.; Doshi, A.; Gilbert, N.; Poirot, C.; Gillman, L. Drone technology for monitoring protected areas in remote and fragile environments. *Drones* **2022**, *6*, 42. [CrossRef]
6. Hu, J.; Wu, X.; Dai, M. Estimating the population size of migrating Tibetan antelopes *Pantholops hodgsonii* with unmanned aerial vehicles. *Oryx* **2020**, *54*, 101–109. [CrossRef]
7. Le Maho, Y.; Whittington, J.D.; Hanuise, N.; Pereira, L.; Boureau, M.; Brucker, M.; Chatelain, N.; Courtecuisse, J.; Crenner, F.; Friess, B. Rovers minimize human disturbance in research on wild animals. *Nat. Methods* **2014**, *11*, 1242–1244. [CrossRef]

8. Schaller, G.B.; Aili, K.; Xinbin, C.; Yanlin, L. Migratory and calving behavior of Tibetan antelope population. *Acta Theriol. Sin.* **2006**, *26*, 105.

9. Hongping, Z.; Wei, H.; Dong, W.; Jie, J. A wildlife monitoring system based on Tianditu and Beidou: In case of the Tibetan antelope. *Int. Arch. Photogramm. Remote Sens. Spat. Inf. Sci.* **2016**, *41*, 259–262. [CrossRef]

10. Christie, K.S.; Gilbert, S.L.; Brown, C.L.; Hatfield, M.; Hanson, L. Unmanned aircraft systems in wildlife research: Current and future applications of a transformative technology. *Front. Ecol. Environ.* **2016**, *14*, 241–251. [CrossRef]

11. Linchant, J.; Lisein, J.; Semeki, J.; Lejeune, P.; Vermeulen, C. Are unmanned aircraft systems (UAS s) the future of wildlife monitoring? A review of accomplishments and challenges. *Mammal Rev.* **2015**, *45*, 239–252. [CrossRef]

12. McEvoy, J.F.; Hall, G.P.; McDonald, P.G. Evaluation of unmanned aerial vehicle shape, flight path and camera type for waterfowl surveys: Disturbance effects and species recognition. *PeerJ* **2016**, *4*, e1831. [CrossRef]

13. Pirotta, V.; Hocking, D.P.; Iggleden, J.; Harcourt, R. Drone observations of marine life and human–wildlife interactions off Sydney, Australia. *Drones* **2022**, *6*, 75. [CrossRef]

14. Dajiang Airport. Available online: https://www.dji.com/cn/dock (accessed on 13 July 2022).

15. Maza, I.; Ollero, A. Multiple UAV cooperative searching operation using polygon area decomposition and efficient coverage algorithms. In *Distributed Autonomous Robotic Systems 6*; Springer: Tokyo, Japan, 2007; pp. 221–230. [CrossRef]

16. Cabreira, T.M.; Di Franco, C.; Ferreira, P.R.; Buttazzo, G.C. Energy-aware spiral coverage path planning for UAV photogrammetric applications. *IEEE Robot. Autom. Lett.* **2018**, *3*, 3662–3668. [CrossRef]

17. Artemenko, O.; Dominic, O.J.; Andryeyev, O.; Mitschele-Thiel, A. Energy-aware trajectory planning for the localization of mobile devices using an unmanned aerial vehicle. In Proceedings of the 2016 25th international conference on computer communication and networks (ICCCN), Waikoloa, HI, USA, 1–4 August 2016; pp. 1–9. [CrossRef]

18. Avellar, G.S.; Pereira, G.A.; Pimenta, L.C.; Iscold, P. Multi-UAV routing for area coverage and remote sensing with minimum time. *Sensors* **2015**, *15*, 27783–27803. [CrossRef]

19. Nam, L.; Huang, L.; Li, X.J.; Xu, J. An approach for coverage path planning for UAVs. In Proceedings of the 2016 IEEE 14th international workshop on advanced motion control (AMC), Auckland, New Zealand, 22–24 April 2016; pp. 411–416. [CrossRef]

20. Pérez-González, A.; Benítez-Montoya, N.; Jaramillo-Duque, Á.; Cano-Quintero, J.B. Coverage path planning with semantic segmentation for UAV in PV plants. *Appl. Sci.* **2021**, *11*, 12093. [CrossRef]

21. Li, Y.; Chen, H.; Er, M.J.; Wang, X. Coverage path planning for UAVs based on enhanced exact cellular decomposition method. *Mechatronics* **2011**, *21*, 876–885. [CrossRef]

22. Valente, J.; Del Cerro, J.; Barrientos, A.; Sanz, D. Aerial coverage optimization in precision agriculture management: A musical harmony inspired approach. *Comput. Electron. Agric.* **2013**, *99*, 153–159. [CrossRef]

23. Yang, J.; Thomas, A.G.; Singh, S.; Baldi, S.; Wang, X. A semi-physical platform for guidance and formations of fixed-wing unmanned aerial vehicles. *Sensors* **2020**, *20*, 1136. [CrossRef]

24. Shafiq, M.; Ali, Z.A.; Israr, A.; Alkhammash, E.H.; Hadjouni, M.; Jussila, J.J. Convergence analysis of path planning of multi-UAVs using max-min ant colony optimization approach. *Sensors* **2022**, *22*, 5395. [CrossRef] [PubMed]

25. Batalin, M.A.; Sukhatme, G.S. Spreading out: A local approach to multi-robot coverage. In *Distributed Autonomous Robotic Systems 5*; Springer: Tokyo, Japan, 2002; pp. 373–382. [CrossRef]

26. Hazon, N.; Kaminka, G.A. On redundancy, efficiency, and robustness in coverage for multiple robots. *Robot. Auton. Syst.* **2008**, *56*, 1102–1114. [CrossRef]

27. Bähnemann, R.; Lawrance, N.; Chung, J.J.; Pantic, M.; Siegwart, R.; Nieto, J. Revisiting boustrophedon coverage path planning as a generalized traveling salesman problem. In *Field and Service Robotics*; Springer: Singapore, 2021; pp. 277–290. [CrossRef]

28. Yu, J.; LaValle, S.M. Planning optimal paths for multiple robots on graphs. In Proceedings of the 2013 IEEE International Conference on Robotics and Automation, Karlsruhe, Germany, 6–10 May 2013; pp. 3612–3617. [CrossRef]

29. Nedjati, A.; Izbirak, G.; Vizvari, B.; Arkat, J. Complete coverage path planning for a multi-UAV response system in post-earthquake assessment. *Robotics* **2016**, *5*, 26. [CrossRef]

30. Shah, K.; Ballard, G.; Schmidt, A.; Schwager, M. Multidrone aerial surveys of penguin colonies in Antarctica. *Sci. Robot.* **2020**, *5*, eabc3000. [CrossRef]

31. Fettermann, T.; Fiori, L.; Gillman, L.; Stockin, K.A.; Bollard, B. Drone surveys are more accurate than boat-based surveys of bottlenose dolphins (*Tursiops truncatus*). *Drones* **2022**, *6*, 82. [CrossRef]

32. Ouattara, T.A.; Sokeng, V.-C.J.; Zo-Bi, I.C.; Kouamé, K.F.; Grinand, C.; Vaudry, R. Detection of forest tree losses in Côte d'Ivoire using drone aerial images. *Drones* **2022**, *6*, 83. [CrossRef]

33. Fan, J.; Yang, X.; Lu, R.; Xie, X.; Li, W. Design and implementation of intelligent inspection and alarm flight system for epidemic prevention. *Drones* **2021**, *5*, 68. [CrossRef]

34. Rominger, K.R.; Meyer, S.E. Drones, deep learning, and endangered plants: A method for population-level census using image analysis. *Drones* **2021**, *5*, 126. [CrossRef]

35. Ranjan, V.; Wang, B.; Shah, M.; Hoai, M. Uncertainty estimation and sample selection for crowd counting. In Proceedings of the Asian Conference on Computer Vision, Kyoto, Japan, 30 November–4 December 2020. [CrossRef]

36. Kang, B.; Liu, Z.; Wang, X.; Yu, F.; Feng, J.; Darrell, T. Few-shot object detection via feature reweighting. In Proceedings of the IEEE/CVF International Conference on Computer Vision, Seoul, Korea, 27 October–2 November 2019; pp. 8420–8429. [CrossRef]

37. Fan, Q.; Zhuo, W.; Tang, C.-K.; Tai, Y.-W. Few-shot object detection with attention-RPN and multi-relation detector. In Proceedings of the IEEE/CVF Conference on Computer Vision and Pattern Recognition, Seattle, WA, USA, 13–19 June 2020; pp. 4013–4022. [CrossRef]
38. Dan, S. *Evolutionary Optimization Algorithms*; Genetic Programming; John Wiley &Sons, Inc.: Hoboken, NJ, USA, 2013.
39. Strubel, D. Coverage Path Planning Based on Waypoint Optimization, with Evolutionary Algorithms. Ph.D. Thesis, Université Bourgogne Franche-Comté, Besançon, France, 2019.
40. He, K.; Zhang, X.; Ren, S.; Sun, J. Deep residual learning for image recognition. In Proceedings of the IEEE conference on computer vision and pattern recognition, Las Vegas, NV, USA, 27–30 June 2016; pp. 770–778. [CrossRef]
41. Zhang, Y.; Zhou, D.; Chen, S.; Gao, S.; Ma, Y. Single-image crowd counting via multi-column convolutional neural network. In Proceedings of the IEEE conference on computer vision and pattern recognition, Las Vegas, NV, USA, 27–30 June 2016; pp. 589–597. [CrossRef]
42. Ranjan, V.; Sharma, U.; Nguyen, T.; Hoai, M. Learning to count everything. In Proceedings of the IEEE/CVF Conference on Computer Vision and Pattern Recognition, Nashville, TN, USA, 20–25 June 2021; pp. 3394–3403. [CrossRef]
43. SPH Engineering, Universal Ground Control Station (UGCS). Available online: www.ugcs.com (accessed on 13 July 2022).
44. Moura, L.D.; Bjørner, N. Z3: An efficient SMT solver. In Proceedings of the International conference on Tools and Algorithms for the Construction and Analysis of Systems, Budapest, Hungary, 29 March–6 April 2008; pp. 337–340. [CrossRef]
45. Ren, S.; He, K.; Girshick, R.; Sun, J. Faster r-cnn: Towards real-time object detection with region proposal networks. *IEEE Trans. Pattern Anal. Mach. Intell.* **2017**, *39*, 1137–1149. [CrossRef] [PubMed]
46. He, K.; Gkioxari, G.; Dollár, P.; Girshick, R. Mask r-cnn. In Proceedings of the IEEE international conference on computer vision, Venice, Italy, 22–29 October 2017; pp. 2961–2969. [CrossRef]
47. Wu, Y.; Kirillov, A.; Massa, F.; Lo, W.; Girshick, R. Detectron2. 2019. Available online: https://github.com/facebookresearch/detectron2/ (accessed on 13 July 2022).
48. Dunstan, A.; Robertson, K.; Fitzpatrick, R.; Pickford, J.; Meager, J. Use of unmanned aerial vehicles (UAVs) for mark-resight nesting population estimation of adult female green sea turtles at Raine Island. *PLoS ONE* **2020**, *15*, e0228524. [CrossRef] [PubMed]
49. Wen, D.; Su, L.; Hu, Y.; Xiong, Z.; Liu, M.; Long, Y. Surveys of large waterfowl and their habitats using an unmanned aerial vehicle: A case study on the Siberian crane. *Drones* **2021**, *5*, 102. [CrossRef]

Article

A Four-Dimensional Space-Time Automatic Obstacle Avoidance Trajectory Planning Method for Multi-UAV Cooperative Formation Flight

Jie Zhang, Hanlin Sheng *, Qian Chen, Han Zhou, Bingxiong Yin, Jiacheng Li and Mengmeng Li

College of Energy and Power Engineering, Nanjing University of Aeronautics and Astronautics (NUAA), Nanjing 210016, China; dreamjane@nuaa.edu.cn (J.Z.); freeflycq@nuaa.edu.cn (Q.C.); han.zhou@nuaa.edu.cn (H.Z.); mr.yin@nuaa.edu.cn (B.Y.); ileejc@nuaa.edu.cn (J.L.); mengmeng.li@nuaa.edu.cn (M.L.)
* Correspondence: dreamshl@nuaa.edu.cn; Tel.: +86-189-6364-6736

Abstract: Trajectory planning of multiple unmanned aerial vehicles (UAVs) is the basis for them to form the formation flight. By considering trajectory planning of multiple UAVs in formation flight in three-dimensional space, a trajectory planning method in four-dimensional space-time is proposed which, firstly, according to the formation configuration, adopts the Hungarian algorithm to optimize the formation task allocation. Based on that, by considering the flight safety of UAVs in formation, a hierarchical decomposition algorithm in four-dimensional space-time is innovatively put forward with spatial positions and time constraints both considered. It is applied to trajectory planning and automatic obstacle avoidance under the condition of no communication available between UAVs in the formation. The simulation results illustrated that the proposed method is effective in cooperative trajectory planning and automatic obstacle avoidance in advance for multiple UAVs. Meanwhile, it has been tested in a Swarm Unmanned Aerial System project and boasts quite significant value in engineering applications.

Keywords: formation flying; formation transformation; Hungarian algorithm; trajectory planning; task allocation; flight control; automatic obstacle avoidance

Citation: Zhang, J.; Sheng, H.; Chen, Q.; Zhou, H.; Yin, B.; Li, J.; Li, M. A Four-Dimensional Space-Time Automatic Obstacle Avoidance Trajectory Planning Method for Multi-UAV Cooperative Formation Flight. *Drones* **2022**, *6*, 192. https://doi.org/10.3390/drones6080192

Academic Editors: Xiwang Dong, Mou Chen and Xiangke Wang

Received: 5 July 2022
Accepted: 28 July 2022
Published: 31 July 2022

Publisher's Note: MDPI stays neutral with regard to jurisdictional claims in published maps and institutional affiliations.

1. Introduction

In recent years, as the technology of unmanned aerial vehicles (UAVs) advances and artificial intelligence (AI) develops, UAVs have been widely adopted in military fields, such as intelligence reconnaissance and military exercises [1]. Combat by a single UAV has been unable to meet the demand due to the increasing complexity of military combat tasks. Therefore, combat by multi-UAV cooperative formations has gradually become the trend in development [2].

Multi-UAV cooperative trajectory planning [3] is an indispensable part of multi-UAV mission execution, which is essentially an optimization problem. First, the task is assigned, then combined with flight safety UAV performance and other constraints [4], the optimal trajectory [5] is planned for the UAV formation to meet the flight performance and environmental constraints. Compared with single UAVs, trajectory planning for multiple UAVs to form the formation needs to further consider position constraints and cooperative obstacle avoidance among UAVs.

At present, the commonly-used UAV trajectory planning algorithms can be divided into two categories: one is traditional algorithms for classical optimization [6,7], mainly consisting of dynamic planning, rapidly exploring random tree (RRT), Voronoi diagram, A * algorithm [8–13] and Dijkstra algorithm; and the other is modern intelligent optimization algorithms [14], including differential evolution (DE) [15], ant colony optimization (ACO), particle swarm optimization, and whale optimization algorithm (WOA) [16–18]. As plenty of researchers have studied and applied these two types of algorithms, they have suggested a cooperative coevolutionary genetic algorithm in the literature [19], which solves the problem of trajectory planning of multiple UAVs in two-dimensional space. In the literature [20], the artificial bee colony (ABC) algorithm and simulated annealing (SA) algorithm have been combined to solve issues of cooperative trajectory planning in two-dimensional space. A hierarchical potential field algorithm was designed in the literature [21], introducing rotational force for the trajectory planning of multiple UAVs. However, this has the issue that planning becomes slower as hierarchies increase. The Particle Swarm Optimization–Hook Jeeves (PSO-HJ) algorithm was adopted in the literature [22]. As an integrated way of multi-UAV trajectory planning, it improves planning effectiveness to some extent. However, with the increase of the number of UAVs, the calculation efficiency decreases. In the literature [23], UAV formation was configured based on an improved consensus algorithm. It introduced particle swarm optimization (PSO) and a model predictive control (MPC) algorithm to deal with obstacle avoidance, which proved effective by numerical simulation. In the literature [24], a time-varying formation was formed on the basis of a consistency protocol. It then guided formation movement by setting virtual leaders, planned obstacle avoidance by artificial potential field algorithm, and proved the stability of the closed-loop system by using Lyapunov stability theory. In the end, a numerical simulation experiment was performed and verified the effectiveness of obstacle avoidance during formation flight.

It can be seen from the above studies that: (1) some algorithms cover only two-dimensional situations; (2) there is still room for further improvement in trajectory planning and algorithm efficiency; (3) in some literature, only numerical simulations instead of actual experiments are performed to verify algorithm feasibility.

To this end, this paper proposes a trajectory planning approach with automatic obstacle avoidance for multiple UAVs to achieve formation flight, which is a way of prior trajectory planning according to flight tasks. The main innovations of this paper are listed as follows: (1) compared with the literature [19,20], the algorithm in this paper could be applied to trajectory planning in three-dimensional space; (2) this method combined the Hungarian algorithm with the hierarchical decomposition strategy, which is of significantly better efficiency and can achieve planning the optimal trajectory in only a few hundred milliseconds; (3) under the condition of no communication between UAVs in formation, the method proposed in this paper could realize automatic obstacle avoidance and be able to automatically generate an obstacle-free path through collision detection; (4) unlike previous studies [23,24] without actual tests, the method proposed in this paper is verified by actual tests and achieved good experimental results.

Problems related to automatic obstacle avoidance of multi-UAV cooperative formation flying are studied in this paper. Considering the solution of multi-dimensional trajectory cooperative planning and complex constrained optimization problems, the main content of this paper is organized through the following parts: in Part 1, the shortest path and collision judgments during the flight formation are converted into the corresponding mathematical models; in Part 2, the Hungarian algorithm and the four-dimensional spatiotemporal hierarchical decomposition algorithm are studied respectively according to descriptions of the mathematical models; in Part 3, the effectiveness of the method is verified through simulation software and a practical experiment. The results showed that the method is outstandingly adaptable and applicable to multi-UAV cooperative planning.

2. Establishment of the Mathematical Model

Trajectory planning with automatic obstacle avoidance for multi-UAV cooperative flight formation can be described in the form of a mathematical model:

(1) Mission assignment to formations: missions will be assigned to all UAVs according to the initial waypoint information $O(x_1, y_1, z_1, \ldots x_n, y_n, z_n)$ of the given n UAVs and the waypoint information $T(x'_1, y'_1, z'_1, \ldots x'_n, y'_n, z'_n)$ of the target point to plan the shortest trajectory, to minimize the total flight distance on the premise that the mission objective of a formation and flight constraints of the UAVs themselves are satisfied.

The above problem can be translated into the following mathematical description: Take the set of the total distance of trajectories of different waypoints corresponding to different UAVs as $D\{d_1, d_2, \ldots d_n\}$, For any $i \in \{1, 2, \ldots, n\}$, $d_i < d_n$ is always true.

(2) Trajectory planning for formations: when UAVs are performing corresponding tasks in formation, it is necessary to ensure that all the UAVs do not collide with each other when performing the tasks in formation, in addition to meeting the shortest trajectory planning. Therefore, judgments should be made on trajectories assigned by missions to make sure UAVs will not collide or the distance between UAVs is not too small. In this case, the determination of whether UAVs collide is converted into the problem of whether the minimum distance between two routes is greater than or equal to the safe spacing between UAVs. The above problem can be transformed into the following mathematical description:

Take the set of all routes of UAVs in formation as $L\{l_1, l_2, \ldots l_n\}$, for any $i \in \{1, 2, \ldots, n\}$, $j \in \{1, 2, \ldots, n\}$ and $i \neq j$, the minimum distance between any two routes is set as $d_{\min l_i, l_j}$, d_{safe} as the safe distance for UAVs, distance between routes should satisfy the condition of $d_{\min l_i, l_j} \geq d_{safe}$.

3. Method to Plan Multi-UAV Trajectories during Flight Formation

Essentially, the problem with trajectory planning for multi-UAV formation is how to distribute n UAVs to the next target waypoint from one waypoint and, in the meantime, satisfy time constraints and safety constraints of formation flying. To address this problem, a method to plan multi-UAV trajectories is proposed in this paper. Firstly, the Hungarian algorithm is applied to roughly plan the multi-UAV trajectory; secondly, the collision judgment algorithm is utilized to detect the collision situation; thirdly, precise planning combined with the four-dimensional spatiotemporal hierarchical decomposition strategy is contrived to generate the final collision-free flight trajectory. The schematic diagram of the planning method flow is presented in Figure 1.

3.1. Method to Assign Tasks to Multiple UAVs in Formation Based on the Hungarian Algorithm

To enable multiple UAVs to realize the formation flight, trajectory planning according to mission objectives is the first thing to do, which is usually difficult in calculation, time-consuming in planning, and not efficient to meet actual demands. Therefore, it is important to explore a path planning algorithm with simple calculation and high efficiency for multiple UAVs.

Task assignment is of significance for multi-UAV trajectory planning, while for solving assignment problems, the Hungarian algorithm happens to be a very effective method with fast calculation and high efficiency. Therefore, this paper applies the Hungarian algorithm for the task allocation (assignment) of multiple UAVs to achieve the objective of obtaining the shortest total flight distance of all UAVs, as shown in Figure 2.

Trajectory planning for UAVs formation can be carried out based on the theorem of the optimal solution of the Hungarian algorithm. The specific steps are presented as follows:

Step 1: Input the initial position matrix of n UAVs $O(x_1, y_1, z_1, \ldots x_n, y_n, z_n)$ and the target position matrix $T(x'_1, y'_1, z'_1, \ldots x'_n, y'_n, z'_n)$.

Step 2: Calculate the distance between the initial position of n UAVs and n target waypoints to form an $n * n$ distance matrix A. The element A_{ij} represents the distance between the UAVs numbered i and the jth target waypoint.

Step 3: Apply row transformation to the distance matrix A, where the smallest element of each row is subtracted from the elements of that row in matrix A.

Step 4: Apply column transformation to the distance matrix A, where the smallest element of each column is subtracted from the elements of that column in matrix A.

Step 5: The trial assignment:

(1) Find the row and column with the least number of zero elements without lines, that is, traverse all the zero elements without lines to check how many

zeros are in rows and columns where the elements are located, and finally select the zero elements in the row and column with the least number of zero elements;

(2) Find the zero elements without a line in that row and column, which is an independent zero element. Draw a line for the row and column in which those zero elements are located;

(3) Leave elements covered by lines alone for the moment and repeat steps (2) and (3) until there are no lines to draw;

(4) Succeed when finding n independent zero elements according to the number of zero elements found in (3) and execute the next step when the number of independent zero elements is less than n.

Step 6: Draw a line to cover the zeros:

(1) Tick rows without an independent zero element;

(2) Tick the corresponding columns with one or more zero elements in the ticked rows;

(3) Tick the corresponding rows with one or more independent zero elements in all ticked columns;

(4) Repeat (3) and (4) until no more ticks can be made.

Step 7: Update the matrix:

(1) Find the smallest number among those without lines;

(2) Subtract the smallest number from those without lines;

(3) Add the smallest number to the numbers with two lines, ensuring that the position of zero elements remains unchanged.

Step 8: Repeat steps 5–7 until success.

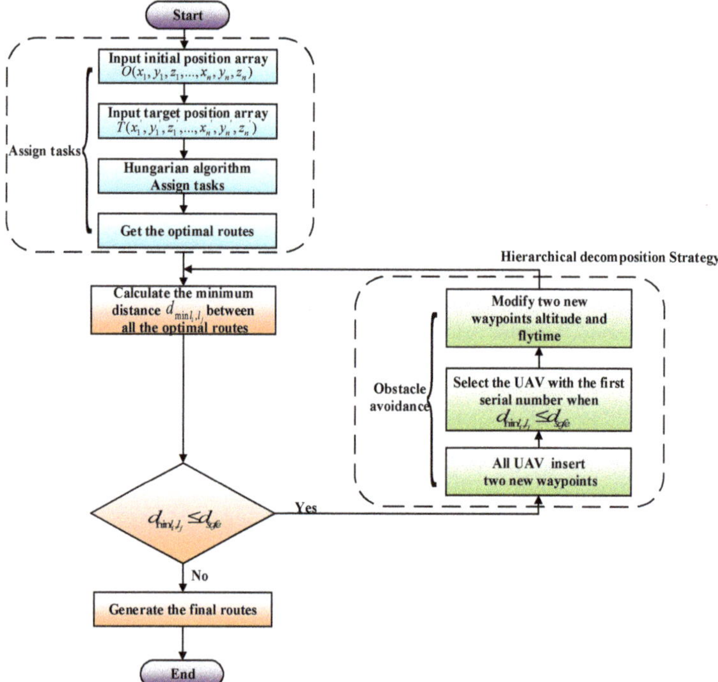

Figure 1. Schematic diagram of the trajectory planning process for multi-UAV coordinated formation flight.

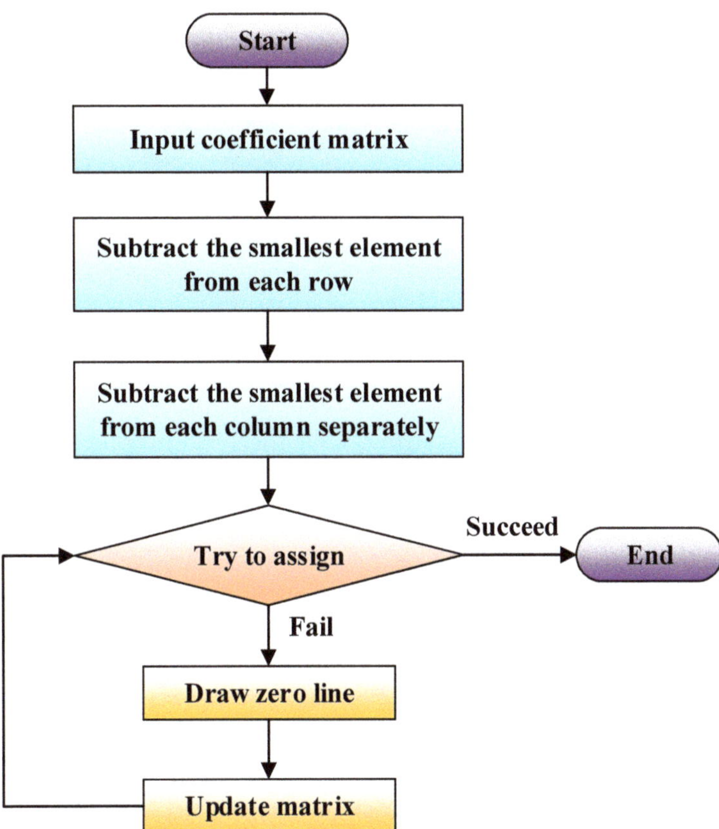

Figure 2. Hungarian algorithm flow block diagram.

The resulting matrix obtained after the above steps is the solution matrix. Replace all the independent zero elements in this matrix with 1 and all the elements except the independent zero elements with 0. Such a 0–1 matrix is the optimal solution to the multi-UAV allocation problem.

Pseudo-code of the task allocation Algorithm 1 is:

Algorithm 1: UAV task assignment algorithm

---

function Edmonds (origin-position, goal -position)

    Input: *origin-position:* $O(x_1, y_1, z_1, ... x_n, y_n, z_n)$,

        Goal-position: $T(x'_1, y'_1, z'_1, ... x'_n, y'_n, z'_n)$,

        origin-position/ goal-position size: n,

        output: *ans X*

  for $i \leftarrow 1$ **to** n **do**

    for $j \leftarrow 1$ **to** n **do**

      Calculate the distance from the origin position to the goal position, and the result is stored in $A[n][n]$

  for ($i \leftarrow 1$ **to** n **do**)

    Subtract the smallest element of each row from $A[n][n]$

  for ($i \leftarrow 1$ **to** n **do**)

    Subtract the smallest element of each column from $A[n][n]$

    r_cov$[n] \leftarrow \{0\}$; c_cov$[n] \leftarrow \{0\}$; $M[n][n] \leftarrow \{0\}$

  for $i \leftarrow 1$ **to** n **do**

    for $j \leftarrow 1$ **to** n **do**

    if ($A[i][j] == 0 \&\& r\_cov[i] == 0 \&\& c\_cov[j] == 0$)

    r_cov[i] $\leftarrow 1$

      c_cov[j] $\leftarrow 1$

      M[i][j] $\leftarrow 1$

      if ($sum(c\_cov) == n$)

  end

    else

      zflag $\leftarrow 1$

    While (zflag $\leftarrow 1$)

    exit_flag $\leftarrow 1$

  While (exit_flag $\leftarrow 1$)

    If ($A[i][j] == 0 \&\& r\_cov[i] == 0 \&\& c\_cov[j] == 0$)

      Draw lines to the row and column where the 0 elements are located;

      Store 0 elements to k

      if k < n then

        Pick the row that has no 0 elements, on this basis, Pick the columns that have 0 elements

        Check the row that contains the independent 0 elements in all checked columns;

      ///Update matrix:

      a $\leftarrow$ find($r_$cov $\leftarrow 0$)

      $b \leftarrow$ find(c_cov $\leftarrow 0$)

      min *val* $\leftarrow \min(\min(A[i][j]))$

      $A(find(r\_cov \leftarrow 1, 0) \leftarrow A(find(r\_cov \leftarrow 1, 0) + \min val$

      $A(: find(c\_cov \leftarrow 0) \leftarrow A(: find(c\_cov \leftarrow 0) - \min val$

  return X

  The resulting matrix obtained after the above steps is the solution matrix. Replace all independent 0 elements in this matrix with 1 and replace all elements except independent 0 elements with 0. Such a 0-1 matrix is a multi-UAV optimal solution to the assignment problem.

  end

3.2. The Multi-UAV Trajectory Planning Method Based on the Four-Dimensional Spatiotemporal Hierarchical Decomposition

According to the task assignment result, a trajectory can be designed for each UAV from the current point to the target point, as shown in Figure 3. A collision situation would occur if the mission is executed through the trajectory. To safely complete the formation flight mission, all UAVs need to take the same time to traverse the trajectory and not collide with each other, which satisfies the time coordination and safety of the formation. Therefore, the planning of the final flight trajectory also needs to meet the time constraints and safety constraints of the UAVs' formation based on the task assignment.

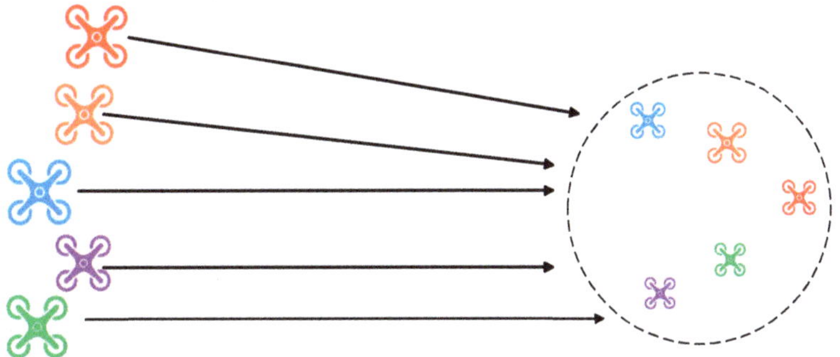

Figure 3. Example diagram of a UAV formation.

(1) Time constraints

Time constraints require all UAVs to arrive at the corresponding target point simultaneously. In the time constraint of this paper, all UAVs are required to reach the navigation point within the time constraint. In a formation mission, it is assumed in this paper that the speed of all UAVs is maintained at a certain level, and the approximate time of arrival at the target point is calculated based on the range of the UAVs to realize the cooperativeness in time for all UAVs during formation flying since each trajectory consists of a series of waypoints which do not contain accurate time information. The approximate time is employed as the benchmark to establish a time-constrained model for UAVs formation.

$$P^i = \left\{ p^i_{\,1}(x_1, y_1, z_1, t_1), p^i_{\,2}(x_2, y_2, z_2, t_2), \ldots, p^i_{\,n}(x_n, y_n, z_n, t_n) \right\} \tag{1}$$

P^i denotes waypoints of UAV i, n the total number of all waypoints of UAV i's trajectory, $p^i_{\,n}(x_n, y_n, z_n, t_n)$ the nth waypoint of the trajectory P^i, (x, y, z) the position coordinates of the trajectory, and t_n the flight time to reach the waypoint.

Based on the above representation of a trajectory, cooperative time constraints on waypoints for a multi-UAV formation can be expressed as:

$$\max(t_n) \tag{2}$$

(2) Safety constraints

Safety constraints require the distance between UAVs to be greater than or equal to the safety distance until they arrive at the target point to avoid collisions. Due to the large granularity of scattered waypoints, not only safety constraints at waypoints but also the safety of the entire trajectory within the same period should be determined. Specifically, $M1N1$ and $C1D1$ are set to represent the trajectories at the first and second waypoints of UAVs i and j, respectively, and the coordinates of point $M1$ are set to be $(x1, y1, z1)$, point $N1(x2, y2, z2)$, point $C1(x3, y3, z3)$, and point $D1(x4, y4, z4)$. Safety constraints between the two UAVs are displayed in the following steps:

Step 1: Set the safety distance between two UAVs to d_{safe} meters according to UAVs' control accuracy and relevant external equipment;

Step 2: Determine whether the shortest distance between the line segments connecting the adjacent waypoints corresponding to UAVs is less than the safe distance d_{safe} in a way shown as follows:

Set the line H to be a point on the line $M1N1$. The coordinates (X, Y, Z) of the point H can be expressed as:

$$\begin{cases} X = x1 + k(x2 - x1) \\ Y = y1 + k(y2 - y1) \\ Z = z1 + k(z2 - z1) \end{cases} \tag{3}$$

When the parameter k satisfies the condition where $0 \leq k \leq 1$, H is a point on the line $M1N1$; when the parameter $k < 0$, H is a point on the extended line of $N1M1$; when the parameter $k > 1$, H is a point on the extended line of $N1M1$.

Set the line Q to be a point on the line $C1D1$ and the coordinates (U, V, W) of the point Q can be expressed as:

$$\begin{cases} U = x3 + k(x4 - x3) \\ V = y3 + k(y4 - y3) \\ W = z3 + k(z4 - z3) \end{cases} \tag{4}$$

When the parameter y satisfies the condition where $0 \leq y \leq 1$, Q is a point on the line $C1D1$; when the parameter $y < 0$, Q is a point on the extended line of $D1C1$; when the parameter $y > 1$, Q is a point on the extended line of $C1D1$.

The distance between the point H and the point Q is:

$$HQ = \sqrt{(X - U)^2 + (Y - V)^2 + (Z - W)^2} \tag{5}$$

The squared distance is:

$$f(k, y) = HQ^2 = (X - U)^2 + (Y - V)^2 + (Z - W)^2 \tag{6}$$

The shortest distance between the line $M1N1$ and the line $C1D1$ is the minimum value of $f(k, y)$.

Deduce the partial derivatives of k, y for $f(k, y)$, respectively, and set the partial derivatives as 0. H falls on line $M1N1$ and Q falls on line $C1D1$, and HQ is the shortest distance when the parameters obtained finally satisfy the condition where $0 \leq k \leq 1$ and $0 \leq y \leq 1$. Otherwise, it will be impossible to find a point H on the line $M1N1$ and a point Q on the line $C1D1$ to minimize the distance $HQ = \sqrt{(X - U)^2 + (Y - V)^2 + (Z - W)^2}$. At this point, deduce the shortest distance from $M1$ to the line $C1D1$, and from $N1$ to the line $C1D1$; the shortest distance from $C1$ to the line $M1N1$, and from $D1$ to the line $M1N1$.

Above is the method to determine whether two trajectories collide or not, and the process of determination is presented in Figure 4.

A four-dimensional spatiotemporal hierarchical decomposition algorithm combined with automatic obstacle avoidance trajectory planning is proposed in this paper to meet the time constraints and safety constraints for UAVs' formation. When the distance between UAVs is too close, the procedure of automatic obstacle avoidance takes effect and plans a route for the UAVs to avoid obstacles. Thenceforth, the UAVs will continue to perform their original tasks. Targeting the specific tasks for formations, collision detection on trajectories of all UAVs is conducted in a combination of time constraints to automatically contrive routes for obstacle avoidance in the future. The specific implementation steps are as follows:

Step 1: Set the safety distance as d_{safe} and the time required for UAVs to fly the safety distance d_{safe} as t_{safe};

Step 2: Step 2: Determine whether the shortest distances between the trajectory connecting the initial and target points of the first UAV and that of the remaining $(n - 1)$ UAVs are greater than or equal to the safe distance s according to the initial point matrix

O and the target point matrix T of a formation. Practice Step 3 if the shortest distance is smaller than the safe distance, and perform Step 4 if it is greater than the safe distance;

Step 3: Select the smaller serial numbers of UAVs, put them into the array E, and conduct Step 4;

Step 4: Determine the distances between the trajectory connecting the initial and target points of the nth UAV and that of the remaining UAVs and repeat Step 3 until all the distances in a formation mission have been determined.

Step 5: Sort the array E in ascending order and copy the array E to the array F;

Step 6: Put all UAVs' serial numbers in the formation mission other than those in the array E into the array G in ascending order;

Step 7: Traverse the array F, copy the position information of all UAVs' initial and target points in the array, and at the same time increase the two points' heights by d_{safe} meters, respectively. Insert the two points with increased heights into the middle of the original points (for example, the initial and target points are at $h1(x1,y1,z1,t1)$ and $h2(x2,y2,z2,t2)$, and there will four waypoints at $h1(x1,y1,z1,t1)$ $h1^+(x1,y1,z1+d_{safe}\cdot count,t1+t_{safe}\cdot count)$, $h2^+(x2,y2,z2+d_{safe}\cdot count,t2+t_{safe}\cdot count)$, $h2(x2,y2,z2,t2)$ after the two new points are inserted. Count denotes the number of times to carry out step 5, which will increase by 1 each time and whose initial value is 0. Practice Step 8;

Step 8: Clear the array E and traverse the array F to redetermine whether the distances between trajectories of all UAVs are greater than the safe distance s. Perform Step 9 after repeating steps 2 to 5;

Step 9: Traverse the array F and modify the altitude information of newly added UAV waypoints in the array F into $h1(x1,y1,z1,t1)$, $h1^+(x1,y1,z1+d_{safe}\cdot count,t1+t_{safe}\cdot count)$, $h2^+(x2,y2,z2+d_{safe}\cdot count,t2+t_{safe}\cdot count)$, $h2(x2,y2,z2,t2)$. Count means the number of times to practice Step 5, which will increase by 1 each time and whose initial value is 0. Repeat steps 8 to 9 until the distances between UAVs in the array F are greater than the safe distance;

Step 10: Traverse the array G and insert two new waypoints corresponding to the nth segment of trajectories of all UAVs in the array G. In this case, the original $h1(x1,y1,z1,t1)$ and $h2(x2,y2,z2,t2)$ become four waypoints of $h1(x1,y1,z1,t1)$, $h1^+(x1,y1,z1,t1+t_{safe}\cdot count)$, $h2^+(x2,y2,z2,t2+t_{safe}\cdot count)$, $h2(x2,y2,z2,t2)$;

Step 11: Clear arrays E, F, and G, and move the positions of the initial and target points to the next section of the trajectory (for instance, set the first and the second waypoints of all UAVs as the initial and target points at the very beginning, and then take the second and the third waypoints as the initial and target points). Repeat steps 1 to 10 until the distances between trajectories of all UAVs have been checked. At this point, the algorithm ends, and multi-UAV trajectory obstacle avoidance is completed.

The schematic diagram of obstacle avoidance based on hierarchical decomposition is shown in Figure 5, where trajectories of UAV1 and UAV2 are $A1B1$ and MN, respectively. There is a crossover between $A1B1$ and MN, so automatic obstacle avoidance is required. According to the algorithm, UAV1 automatically inserts waypoints $A1'$ and $B1'$, and UAV2 automatically inserts waypoints M' and N', that is, the trajectory of UAV1 becomes $A1 - A1' - B1' - B1$; the trajectory of UAV2 trajectory becomes $M - M' - N' - N$. UAV1 has performed a hierarchically decomposed flight and automatically avoided obstacles.

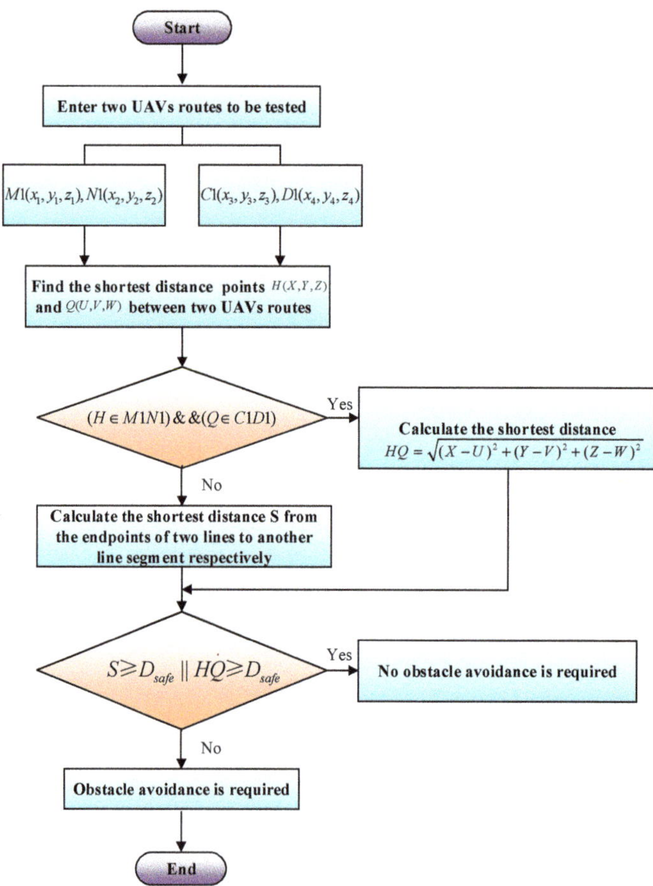

Figure 4. Determine the trajectory collision flowchart.

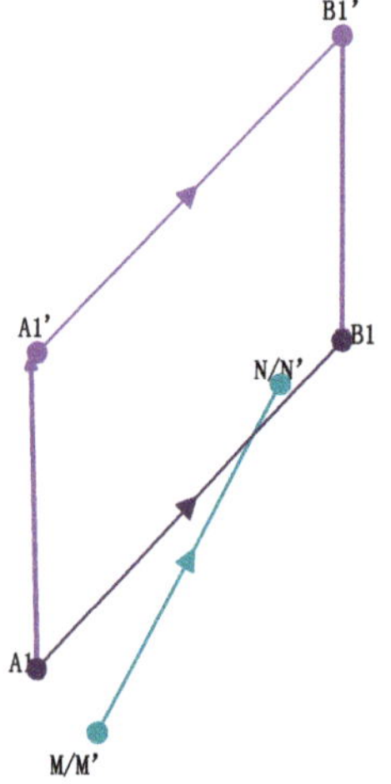

Figure 5. Schematic diagram of hierarchical decomposition obstacle avoidance.

Pseudo-code of the hierarchical decomposition and automatic obstacle avoidance Algorithm 2:

Algorithm 2: High layered realization of UAV obstacle avoidance

---

Input:

  origin-position: $O(x_1, y_1, z_1, \ldots x_n, y_n, z_n)$

  Goal-position: $T(x'_1, y'_1, z'_1, \ldots x'_n, y'_n, z'_n)$,

  UAV-count: n,

  safe distance: d_{safe}

  safe distance flight time: t_{safe},

output: Solution Y

for $i \leftarrow 1$ **to** n **do**

  for $j \leftarrow 1$ **to** $(n\text{-}i)$ **do**

    Calculate the distance between UAV tracks;

  count $= count + 1$;

  if *distance* $< d_{safe}$ **then**

  Select the top UAV serial number and put it in the array E;

  Sort array E *in ascending order, copy array E to array* F;

  Sort the serial numbers of other UAVs in the formation task except in array E *into array* F *in ascending order;*

  for $i \leftarrow 1$ **to** F **do**

    Insert two new positions:

    $h1(x1, y1, z1, t1), h1^+\left(x1, y1, z1 + d_{safe} \cdot count, t1 + t_{safe} \cdot count\right),$

    $h2^+\left(x2, y2, z2 + d_{safe} \cdot count, t2 + t_{safe} \cdot count\right), h2\left(x2, y2, z2, t2\right);$

    empty array E;

  for $i \leftarrow 1$ **to** F **do**

    Calculate the distance between UAV tracks;

    if *distance* $< d_{safe}$ **then**

    modify two new positions:

    $h1^+\left(x1, y1, z1 + d_{safe} \cdot count, t1 + t_{safe} \cdot count\right),$

    $h2^+\left(x2, y2, z2 + d_{safe} \cdot count, t2 + t_{safe} \cdot count\right);$

  for $i \leftarrow 1$ **to** G **do**

    Insert two new positions:

    $h1^+\left(x1, y1, z1, t1 + t_{safe} \cdot count\right),$

    $h2^+\left(x2, y2, z2, t2 + t_{safe} \cdot count\right);$

    empty array $E / F / G$;

  Return *the best solution* Y *in Pop;*

end

---

4. Implementation of Algorithm, Comparison, and Simulation

4.1. Formation Task Allocation Algorithm Comparison

When compared to random matching and auction algorithms is shown in Figures 6–8, to make the advantages of the method proposed in this paper much clearer, for task allocation from "linear" to "circular" formation, we considered only two indicators: total trajectory length and calculation time. In order to verify the effectiveness of the algorithm,

1000 times, 10,000 times, and 100,000 times of calculations were carried out, and the results are shown in Tables 1 and 2 below. By comparing and analyzing the data in the table, we can tell from the results that the proposed method is able to achieve the formation flight planning with the shortest total trajectory length in a comparatively shorter time.

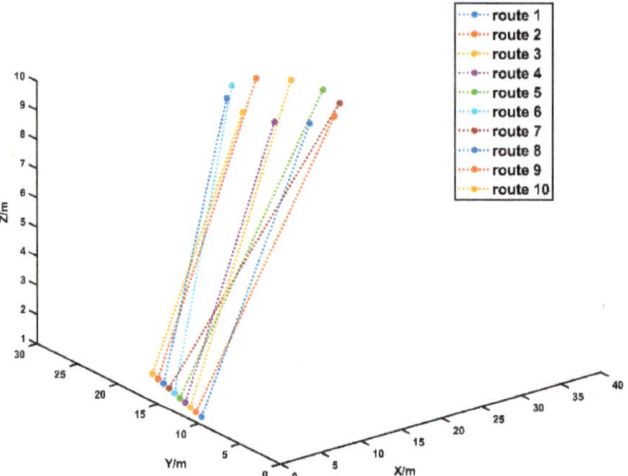

Figure 6. Task assignment based on Hungarian algorithms.

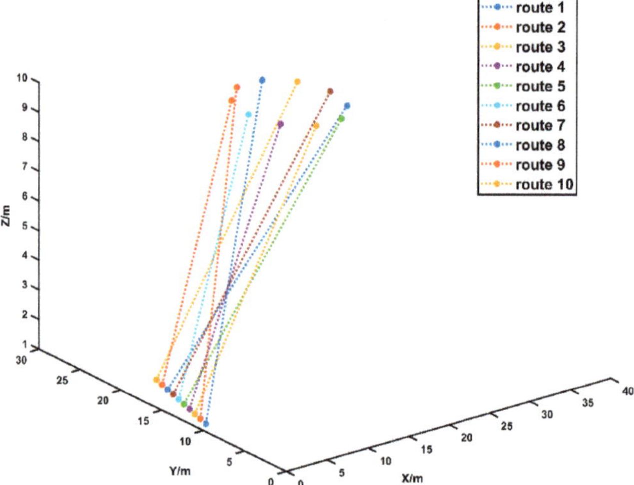

Figure 7. Task assignment based on auction algorithms.

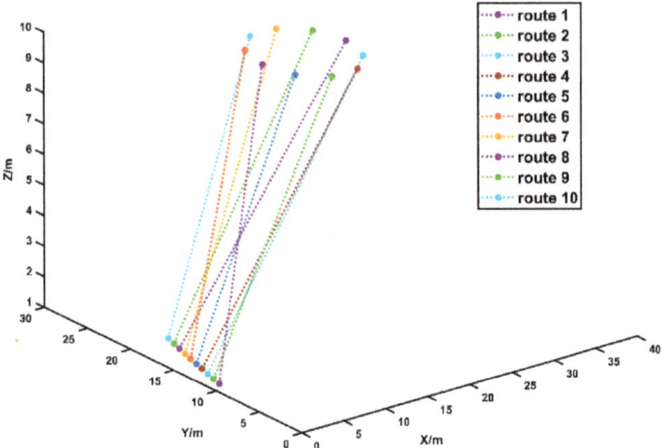

Figure 8. Random task assignment.

Table 1. Average total track length comparison.

Times	1000	10,000	100,000
Hungarian algorithm assignment(m)	163.5636	163.5636	163.5636
auction algorithm(m)	169.2375	168.6958	167.9658
randomly assigned(m)	177.6276	174.3265	174.9328

Table 2. Calculated time comparison.

Times	1000	10,000	100,000
Hungarian algorithm assignment(s)	0.509003	0.5888016	1.3911980
auction algorithm(s)	0.5229998	0.7292022	2.8083982
randomly assigned(s)	0.53226971	0.8156322	2.95687235

4.2. Flight Simulation of Multi-UAV Formation

A simulation test of the multi-UAV formation was completed by writing relevant programs on MATLAB R2019a developed by the Mathwork company of Natick, America in the computers with the main frequency of 3.20 GHz, a memory of 32GB, and the operating system of Windows 10. To verify the effectiveness of algorithms, a formation test was conducted for 10 UAVs, and the initial and target coordinates of the 10 UAVs are shown in Figure 9:

The following Tables 3 and 4 demonstrates the planned trajectory with the shortest distance for 10 UAVs based on the initial and target positions, from "one-line" to "circle" and then to "triangle", through the Hungarian algorithm. Graphical results of the task allocation are displayed below. Row numbers of the matrix represent the UAV numbers, and column numbers are markers of the target points. A 1 element in the matrix represents that the row and the column where the 1 element is located match.

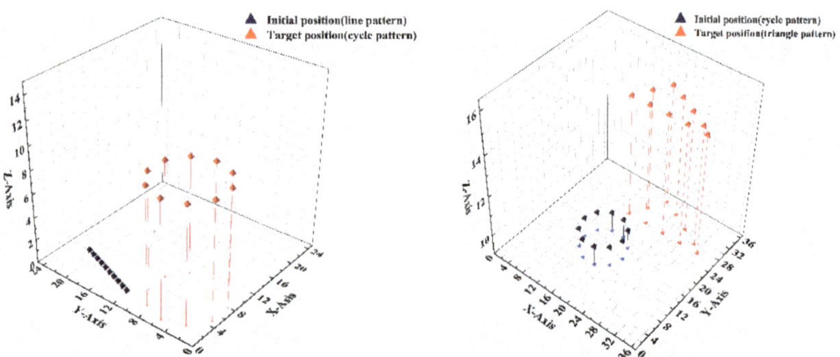

Figure 9. Three initial positions of UAVs formation.

Table 3. The "One-line" to "Circle" task assignment results.

Serial Number	1	2	3	4	5	6	7	8	9	10
1	0	0	0	0	0	0	0	1	0	0
2	0	0	0	0	0	0	0	0	1	0
3	0	0	0	0	0	0	1	0	0	0
4	0	0	0	0	0	0	0	0	0	1
5	0	0	0	0	0	1	0	0	0	0
6	1	0	0	0	0	0	0	0	0	0
7	0	0	0	0	1	0	0	0	0	0
8	0	1	0	0	0	0	0	0	0	0
9	0	0	0	1	0	0	0	0	0	0
10	0	0	1	0	0	0	0	0	0	0

Table 4. The "Circle" to "Triangle" task assignment results.

Serial Number	1	2	3	4	5	6	7	8	9	10
1	0	0	0	0	0	1	0	0	0	0
2	0	0	0	0	0	0	1	0	0	0
3	0	0	0	0	0	0	0	1	1	0
4	0	0	0	0	0	0	0	0	0	1
5	0	0	0	0	0	0	0	0	0	0
6	1	0	0	0	0	0	0	0	0	0
7	0	1	0	0	0	0	0	0	0	0
8	0	0	1	0	0	0	0	0	0	0
9	0	0	0	1	0	0	0	0	0	0
10	0	0	0	0	1	0	0	0	0	0

Based on the results of allocation produced by the Hungarian algorithm, obstacles were avoided through the method of hierarchical decomposition after the target points were determined to ensure that all UAVs did not collide in the air. Results of obstacle avoidance are listed in the following table.

Tables 5 and 6 shows the height hierarchy situation of the UAVs when the formation is transformed from "one-line" to "circle", "circle" to "triangle". Table 5 shows that the UAV

1, 5, the UAV 2, 4, 6, 9, and the UAV 3, 7, 8, 10 are at the same altitude respectively, and the times of UAV rising represents a different height. Table 5 has the same meaning as Table 6.

Table 5. "One-line" to "circle" obstacle avoidance results.

UAV	1	2	3	4	5	6	7	8	9	10
Rising number of UAV	2	1	0	1	2	1	0	0	1	0

Table 6. "Circle" to "triangle" obstacle avoidance results.

UAV	1	2	3	4	5	6	7	8	9	10
Rising number of UAV	2	1	1	1	1	0	0	0	0	0

The above results suggest that the flight altitude of UAVs with different numbers should be planned based on their different results of obstacle avoidance to avoid obstacles in the air.

Trajectory planning is carried out according to the Hungarian algorithm and the four-dimensional spatiotemporal hierarchical decomposition algorithm to realize the formation flying from a "one-line" to the shape "circle", and from the shape of a "circle" to "triangle". The simulation diagram is shown in Figures 10–12, where line segments with the same color indicate flying on the same plane to the target points, while those with different colors imply different flight altitudes.

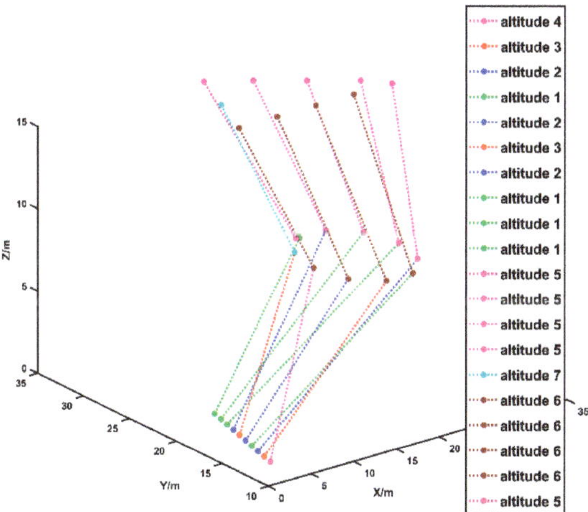

Figure 10. Simulation of the flight of 10 unmanned aerial vehicles in formations of "one-shaped-circle-triangle".

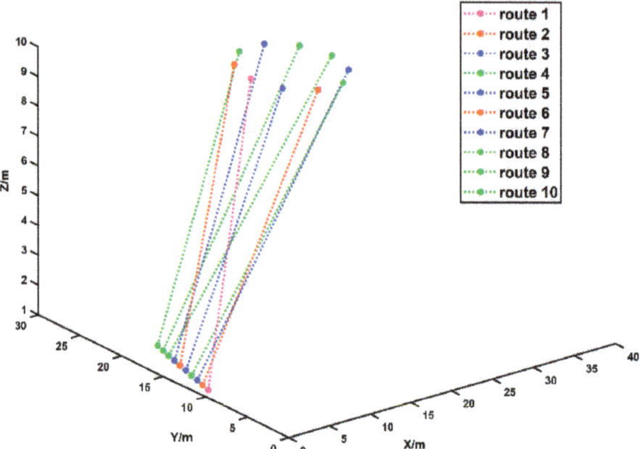

Figure 11. Simulation of 10 UAVs "one-shaped" to "circular" formations.

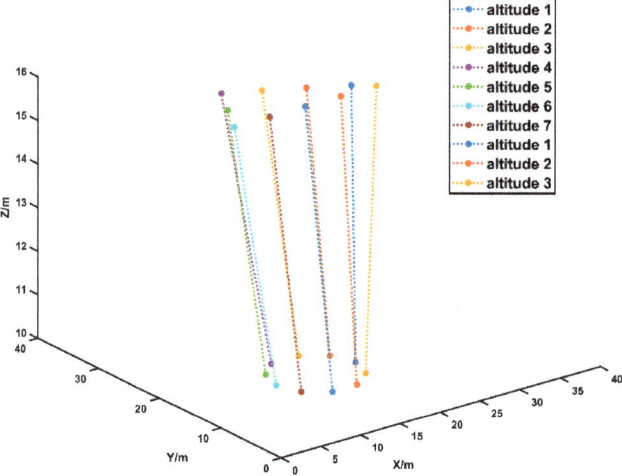

Figure 12. Simulation of 10 UAVs flying in "circular" to "triangle" formations.

4.3. Engineering Application and Flight Test

Quadrotor UAVs were employed to complete a formation flight test in this program to verify the effectiveness and feasibility of algorithms. A quadrotor UAV is demonstrated in the following figure, made from carbon fiber materials with a fixed foldable tripod of an inverted shape, and featuring firmness and stability in Figure 13.

In the flight test, 10 UAVs were selected to complete the UAVs formation, flying in three shapes, namely a "one-line", the shape of a "circle", and a "triangle". Firstly, the initial coordinates of the pattern to be formed and the final coordinates of the UAVs were entered into the simulation software program. The waypoint information of all the UAVs information were output after the program was successfully operated. The trajectory of each UAV was saved in a text file, as shown in Table 7. Then, the trajectory information of all UAVs was input into the ground station software and displayed on a map, as presented in Figure 14. Finally, the information was downloaded to the UAVs through the ground stations, and the UAVs performed the flight plan.

Figure 13. Formation with quadcopter UAVs.

Table 7. Generated information about the location of the drone 1 relative to the track point.

Waypoint Order	Waypoint Type	Relative Latitude (m)	Relative Longitude (m)	Waypoint Altitude (m)	Fly Time (s)
1	1	−11.4839229583740234	4.81873273849487305	20	10
2	2	−11.4839229583740234	4.81873273849487305	30	15
3	2	−23.2751808166503906	24.2204875946044922	30	15
4	2	−23.2751808166503906	24.2204875946044922	20	10
5	2	−23.2751808166503906	24.2204875946044922	30	15
6	2	−16.0270156860351563	37.4085922241210938	30	15
7	2	−16.0270156860351563	37.4085922241210938	20	10

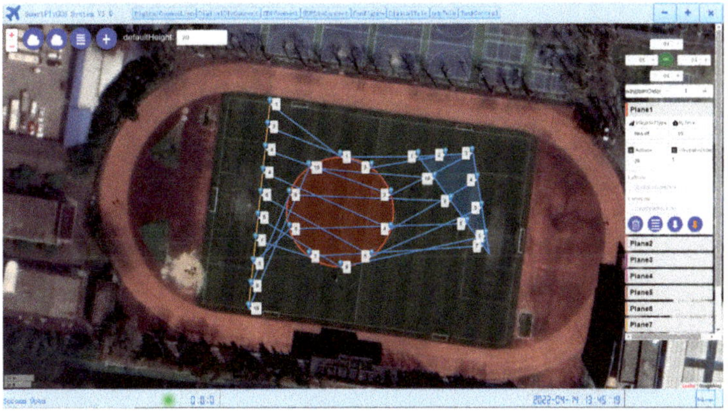

Figure 14. Imported UAV formation map into a ground station.

In Figure 15, each card in the above diagram represents a UAV, and the status related to each UAV (including latitude and longitude information, battery voltage, and flight mode) can be observed clearly.

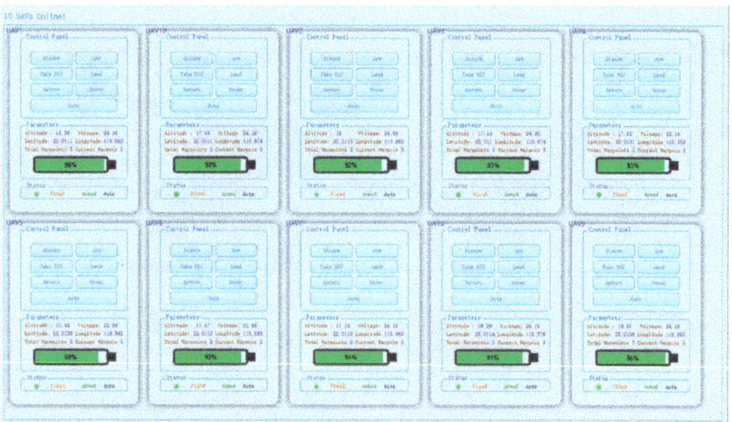

Figure 15. Ground station 10 UAVs' communication.

The actual formation flying of UAVs is displayed in Figure 16. During the flight, each UAV was able to arrive at a place near the designated location, and all of them could form a predetermined pattern. However, the actual positions of the UAVs formation slightly deviated from the target points due to GPS and environmental issues such as wind direction. Nevertheless, the task assigned to each UAV was accurate, and no collision occurred during the flight. Therefore, this experiment proved the feasibility of the algorithms.

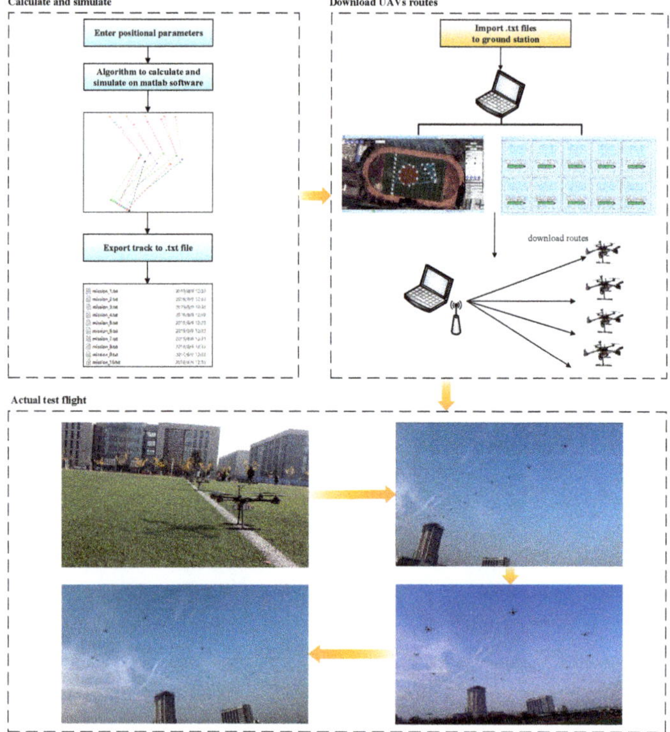

Figure 16. Formation test flight.

5. Conclusions

Targeting three-dimensional space trajectory planning for multi-UAV in cooperative formation, the Hungarian algorithm is adopted in this paper based on formation control under the virtual pilot to minimize costs of the total distance of trajectories during formation transformations, and a four-dimensional spatiotemporal hierarchical decomposition strategy on flight altitude is put forward to avoid collisions during formation flying. This method combines the Hungarian algorithm with the hierarchical decomposition strategy, which is of significantly better efficiency and can plan the optimal trajectory in only a few hundred milliseconds. Compared with the auction algorithm, the total track distance and planning time are reduced by about 3%. Meanwhile, time constraints of the UAVs' formation flying are combined to preplan a route for automatic obstacle avoidance without other communication between UAVs. In the end, verification is conducted through MATLAB simulation and test, and 10 UAVs formation have realized shape transformation from a "one-line" to a "circle", and then to a "triangle". Tested in actual flight, each UAV can reach the corresponding destination quickly and without collision, thus verifying the effectiveness and reliability of this method stated in the paper.

Author Contributions: Conceptualization, J.Z.; Methodology, H.S.; Validation, Q.C.; Formal analysis, B.Y.; Investigation, H.Z.; Data curation, J.L.; Writing—original draft preparation, J.Z.; Writing—review and editing, M.L.; Supervision, H.S. All authors have read and agreed to the published version of the manuscript.

Funding: This work is supported by Funding from the National Key Laboratory of Rotorcraft Aeromechanics (No. 61422202108), and the National Natural Science Foundation of China (No.51906103, No.52176009).

Data Availability Statement: The data presented in this study are available on request from the corresponding author.

Nomenclature

$A = n \cdot n$	distance matrix
$A_{ij} =$	distance between the UAV number and the target waypoint j
$A1' =$	drone waypoint
$A1B1 =$	drone track
$B1' =$	drone waypoint
$count =$	number of waypoint modifications
$C1D1 =$	line segment
$C1, D1, M1, N1 =$	line segment C1D1 and M1N1 endpoints
$D\{d_1, d_2, \ldots d_n\} = n$	UAVs' initial waypoints to target waypoints route distance matrix
$d_{safe} =$	safe distance between UAVs
$d_{\min l_i, l_j} =$	minimum distance between two routes
$E =$	serial number array of UAV whose route distance is less than the safe distance
$F =$	an ascending array of drone serial numbers
$G =$	serial number array of UAV whose route distance is greater than the safe distance
$h1, h2 =$	two waypoints
$h1^+, h2^+ =$	new waypoints
$H, Q =$	line intersection
$i, j =$	serial number of array or matrix
$k, y =$	slope of a line
$L\{l_1, l_2, \ldots l_n\} = n$	UAVs' initial waypoints to target waypoints route matrix
$M1N1 =$	line segment
$MN =$	drone track

$M' =$	drone waypoint
$N' =$	drone waypoint
$n =$	number of UAV
$O(x_1, y_1, z_1, \ldots x_n, y_n, z_n) = n$	UAVs' initial waypoints matrix
$P^i{}_n(x_n, y_n, z_n, t_n) =$	the i-th waypoint information of the UAVs
$T(x'_1, y'_1, z'_1, \ldots x'_n, y'_n, z'_n) = n$	UAVs' target waypoints matrix
$t^i =$	flight time to the i-th waypoint
$t_{safe} =$	time required to fly d_{safe} distance
$(xn, yn, zn), n \in \{1, 2, 3, \ldots\} =$	point coordinates
$(X, Y, Z) =$	intersection coordinates
$(U, V, W) =$	intersection coordinates

References

1. Tong, H.; Andi, T.; Huan, Z.; Dengwu, X.; Lei, X. Multiple UAV cooperative path planning based on LASSA method. *Syst. Eng. Electron.* **2022**, *44*, 233–241.
2. Ni, J.; Tang, G.; Mo, Z.; Cao, W.; Yang, S.X. An improved potential game theory based method for multi-UAV cooperative search. *IEEE Access* **2020**, *8*, 47787–47796. [CrossRef]
3. Wu, W.; Wang, X.; Cui, N. Fast and coupled solution for cooperative mission planning of multiple heterogeneous unmanned aerial vehicles. *Aerosp. Sci. Technol.* **2018**, *79*, 131–144. [CrossRef]
4. Liscouët, J.; Pollet, F.; Jézégou, J.; Budinger, M.; Delbecq, S.; Moschetta, J.M. A methodology to integrate reliability into the conceptual design of safety-critical multirotor unmanned aerial vehicles. *Aerosp. Sci. Technol.* **2022**, *127*, 107681. [CrossRef]
5. Causa, F.; Fasano, G. Multiple UAVs trajectory generation and waypoint assignment in urban environment based on DOP maps. *Aerosp. Sci. Technol.* **2021**, *110*, 106507. [CrossRef]
6. Zhu, Z.; Qian, Y.; Zhang, W. Research on UAV Searching Path Planning Based on Improved Ant Colony Optimization Algorithm. In Proceedings of the 2021 IEEE 3rd International Conference on Civil Aviation Safety and Information Technology (ICCASIT), Changsha, China, 20–22 October 2021; IEEE: Piscataway, NJ, USA, 2021; pp. 1319–1323.
7. Mokrane, A.; Braham, A.C.; Cherki, B. UAV path planning based on dynamic programming algorithm on photogrammetric DEMs. In Proceedings of the 2020 International Conference on Electrical Engineering (ICEE), Istanbul, Turkey, 25–27 September 2020; IEEE: Piscataway, NJ, USA, 2021; pp. 1–5.
8. Wu, X.; Xu, L.; Zhen, R.; Wu, X. Biased sampling potentially guided intelligent bidirectional RRT* algorithm for UAV path planning in 3D environment. *Math. Probl. Eng.* **2019**, *2019*, 5157403. [CrossRef]
9. Huang, J.; Sun, W. A method of feasible trajectory planning for UAV formation based on bi-directional fast search tree. *Optik* **2020**, *221*, 165213. [CrossRef]
10. Junlan, N.; Qingjie, Z.; Yanfen, W. UAV path planning based on weighted-Voronoi diagram. *Flight Dyn.* **2015**, *33*, 339–343.
11. Chen, X.; Chen, X. The UAV dynamic path planning algorithm research based on Voronoi diagram. In Proceedings of the 26th Chinese Control and Decision Conference (2014 CCDC), Changsha, China, 31 May–2 June 2014; IEEE: Piscataway, NJ, USA, 2014; pp. 1069–1071.
12. Mandloi, D.; Arya, R.; Verma, A.K. Unmanned aerial vehicle path planning based on A* algorithm and its variants in 3d environment. *Int. J. Syst. Assur. Eng. Manag.* **2021**, *12*, 990–1000. [CrossRef]
13. Cheng, N.; Liu, Z.; Li, Y. Path Planning Algorithm of Dijkstra-Based Intelligent Aircraft under Multiple Constraints. *Xibei Gongye Daxue Xuebao/J. Northwest. Polytech. Univ.* **2020**, *38*, 1284–1290. [CrossRef]
14. Yuan, G.; Xia, J.; Duan, H. A continuous modeling method via improved pigeon-inspired optimization for wake vortices in UAVs close formation flight. *Aerosp. Sci. Technol.* **2022**, *120*, 107259. [CrossRef]
15. Chai, X.; Zheng, Z.; Xiao, J.; Yan, L.; Qu, B.; Wen, P.; Wang, H.; Zhou, Y.; Sun, H. Multi-Strategy Fusion Differential Evolution Algorithm for UAV Path Planning in Complex Environment. *Aerosp. Sci. Technol.* **2021**, *121*, 107287. [CrossRef]
16. Xianqiang, L.I.; Rong, M.A.; Zhang, S. Improved design of ant colony algorithm and its application in path planning. *Acta Aeronaut. Astronaut. Sin.* **2020**, *41*, 724381.
17. Liu, Y.; Zhang, X.; Guan, X.; Delahaye, D. Adaptive sensitivity decision based path planning algorithm for unmanned aerial vehicle with improved particle swarm optimization. *Aerosp. Sci. Technol.* **2016**, *58*, 92–102. [CrossRef]
18. Jiang, T.; Li, J.; Huang, K. Longitudinal parameter identification of a small unmanned aerial vehicle based on modified particle swarm optimization. *Chin. J. Aeronaut.* **2015**, *28*, 865–873. [CrossRef]
19. Su, Y.; Dai, Y.; Liu, Y. A hybrid hyper-heuristic whale optimization algorithm for reusable launch vehicle reentry trajectory optimization. *Aerosp. Sci. Technol.* **2021**, *119*, 107200. [CrossRef]
20. Miao, H.; Tian, Y.C. Dynamic robot path planning using an enhanced simulated annealing approach. *Appl. Math. Comput.* **2013**, *222*, 420–437. [CrossRef]
21. Ji-yang, D.; Cun-song, W.; Lin-fei, Y.; Bao-jian, Y.; Jun-qiang, X. Hierarchical potential field algorithm of path planning for aircraft. *Control Theory Appl.* **2015**, *32*, 1505–1510.
22. Wenzhao, S.; Naigang, C.; Bei, H.; Xiaogang, W.; Yuliang, B. Multiple UAV cooperative path planning based on PSO-HJ method. *J. Chin. Inert. Technol.* **2020**, *28*, 122–128.

23. Wu, Y.; Gou, J.; Hu, X.; Huang, Y. A new consensus theory-based method for formation control and obstacle avoidance of UAVs. *Aerosp. Sci. Technol.* **2020**, *107*, 106332. [CrossRef]
24. Yu, J.; Dong, X.; Li, Q.; Ren, Z. Practical time-varying output formation tracking for high-order multi-agent systems with collision avoidance, obstacle dodging and connectivity maintenance. *J. Frankl. Inst.* **2019**, *356*, 5898–5926. [CrossRef]

Article

MCO Plan: Efficient Coverage Mission for Multiple Micro Aerial Vehicles Modeled as Agents

Liseth Viviana Campo [1,*], **Agapito Ledezma** [2] **and Juan Carlos Corrales** [1]

[1] Telematics Engineering Group, University of Cauca, Street 5, No. 4-70, Popayan 190003, Colombia; jcorral@unicauca.edu.co
[2] Control Learning and Optimization Group, Universidad Carlos III de Madrid, 28911 Leganés, Madrid, Spain; ledezma@inf.uc3m.es
* Correspondence: liscampo@unicauca.edu.co

Abstract: Micro aerial vehicle (MAV) fleets have gained essential recognition in the decision schemes for precision agriculture, disaster management, and other coverage missions. However, they have some challenges in becoming massively deployed. One of them is resource management in restricted workspaces. This paper proposes a plan to balance resources when considering the practical use of MAVs and workspace in daily chores. The coverage mission plan is based on five stages: world abstraction, area partitioning, role allocation, task generation, and task allocation. The tasks are allocated according to agent roles, Master, Coordinator, or Operator (MCO), which describe their flight autonomy, connectivity, and decision skill. These roles are engaged with the partitioning based on the Voronoi-tessellation but extended to heterogeneous polygons. The advantages of the MCO Plan were evident compared with conventional Boustrophedon decomposition and clustering by K-means. The MCO plan achieved a balanced magnitude and trend of heterogeneity between both methods, involving MAVs with few or intermediate resources. The resulting efficiency was tested in the GAMA platform, with gained energy between 2% and 10% in the mission end. In addition, the MCO plan improved mission times while the connectivity was effectively held, even more, if the Firefly algorithm generated coverage paths.

Keywords: area partitioning; connectivity; coverage mission; firefly algorithm; GAMA platform; heterogeneity; Micro Aerial Vehicles; Voronoi-tessellation

Citation: Campo, L.V.; Ledezma, A.; Corrales, J.C. MCO Plan: Efficient Coverage Mission for Multiple Micro Aerial Vehicles Modeled as Agents. *Drones* **2022**, *6*, 181. https://doi.org/10.3390/drones6070181

Academic Editors: Xiwang Dong, Mou Chen, Xiangke Wang and Fei Gao

Received: 29 June 2022
Accepted: 17 July 2022
Published: 21 July 2022

Publisher's Note: MDPI stays neutral with regard to jurisdictional claims in published maps and institutional affiliations.

1. Introduction

The participation of Unmanned Aerial Vehicles (UAVs) or drones in civil activities has increased due to their versatility in any environment. Most commercial UAVs are classified as Micro Aerial Vehicles (MAVs), restricted to up to 5 kg, a communication range of about 10 km, and a maximum altitude of 250 m [1]; they can be wing-based or multirotor, and their cost is often less than other civil UAVs.

MAVs frequently are deployed to search, surveil, patrol, and perform other activities which generate information to make decisions in agriculture, disaster management, and other processes that require monitoring in space and time. These tasks are framed in the CPP (Coverage Path Planning) problem [2] which usually has three stages; path planning, allocation, and deployment [3]. The key to an efficient coverage plan is to visit all waypoints of a workspace, avoiding obstacles or Zones of Low Interest (ZLIs). However, this type of mission can fail in restricted (and large) workspaces since the MAVs have a limited operational flight, around 15 to 30 min, which decreases by payload, angular acceleration changes, maneuvers against the wind, and continuous deployments.

This limitation has been studied in research projects and some commercial initiatives using energy management through optimal paths and a global positioning [4], with recharge stations in the middle of the course of the MAVs [5] and using a fleet of MAVs [6].

Even though the latter approach has significant challenges, the current proposal is centered on it, since optimizing the mission with a single MAV may not be enough to cover restricted workspaces. The recharge stations require electric infrastructure in the workspace, which is not available for most workspaces such as crops, forests, and other hostile areas.

A fleet of MAVs can be modeled based on the theory of multiple robotic agents that can cooperate to accomplish a coverage mission [7]. The coordination model can be centralized, hierarchical, decentralized, or hybrid. However, the last ones have a significant preference because they are robust to faults and can combine local and high-level control to solve complex tasks [8]. The no-centralized management has two paradigms: the swarm-type and the heterogeneous systems. The coordination of a swarm involves multiple homogeneous agents with behavior inspired by biological societies. For the second paradigm, heterogeneity can be defined in behavior, morphology, performance, size, or cognition. A heterogeneous system exploits the swarming coordination's parallelism, redundancy, and distributed solutions. Moreover, it can include mission specifications because the agents have different skills and payloads [9].

From the previous context, this paper works with a decentralized model to deploy multiple MAV agents for coverage missions in restricted workspaces. However, the multi-drone technology is still not a part of the daily activities of users, possibly because most alternatives work for environments and users in ideal conditions. A typical user does not have the technical knowledge and is unprepared for a dynamic workspace's cost and possible risks. In answer, the first premise of the current solution is to consider simple MAVs, easily acquired and adaptable, for example, Ardupilot, Parrot, and DJI. These commercial MAVs facilitate use and maintenance and mitigate the risk of investment on possible missed missions. From this approach, the question to solve is how to efficiently deploy multiple MAVs for coverage missions conditioned to environments with restricted access.

The present research works on the off-line multi-CPP problem since online planning would require continuous communication and processing resources that simple MAVs and restricted environments cannot support [10]. The solution is a plan for area coverage missions based on the balance of the heterogeneity of MAVs and the connectivity of the fleet (Figure 1). Both are engaged with area partitioning inspired by the Voronoi-tessellation but extended to heterogeneous polygonal sub-areas. The resulting tasks are profiled for agents of type Master, Coordinator, and Operator (MCO). Then, tasks are allocated to the MAVs using an auction mechanism. The main job of an agent is to follow a coverage path, which is calculated with the Firefly algorithm, given as a near-optimal solution to the Traveling Salesman Problem (TSP) for each area partition. The second task is to interact with near neighbors, which is restricted by the role and the communication range.

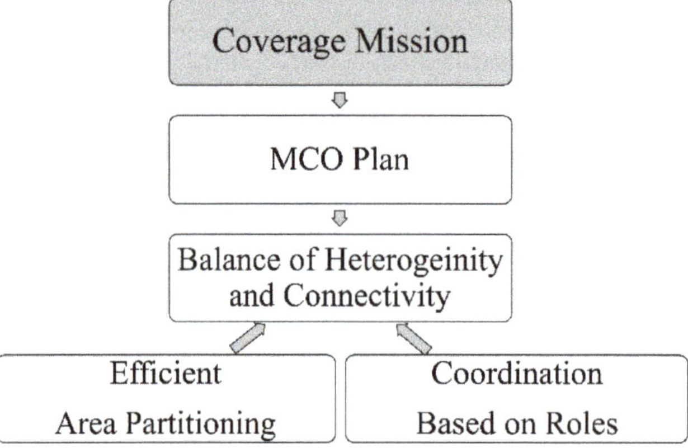

Figure 1. Proposal summary.

The GAMA platform [11] was selected to validate the hypothesis described in Figure 1 due to its extensive features to model physical agents and geographic environments. The results show better resource management of the MCO Plan after comparing it with other alternatives such as Boustrophedon-based and clustering-based planning. The experiments state the performance of the MCO plan in three different workspaces throughout the analysis of scores such as the task standard deviation, remaining energy in the end mission, and the likely number of links. The mean values, 8% for standard deviation, 60% for saved power, and five links per MAV, prove a balanced heterogeneity and connectivity skill.

The remaining paper is organized as follows: Section 2 describes related work on the coverage mission for multiple heterogeneous MAVs. The subsequent problem statement is defined in Section 3. Section 4 states the materials and proposed methods to plan a coverage mission. Section 5 presents and analyzes the simulated results using the GAMA platform. Finally, the conclusion and future research are described in Section 6.

2. Related Work

Some outstanding papers sorted by coverage planning mechanisms for multiple MAVs were selected for this section. The following classification describes a timeline with specified methods of world abstraction, area partitioning, and coordination.

2.1. Polygon-Based Coverage Plans

One of the pioneering works on this topic was found in [12]. The problem they tried to solve was a cooperative search in areas while considering the computational complexity required to implement in near real-time applications. Maza and Ollero defined a polygonal decomposition of the region; every drone was assigned to one of the resulting polygons by a ground coordinator based on deterministic scores. Next, they proposed optimizing the UAV coverage path by minimizing the number of turns during a zigzag pattern, using the optimal sweep direction (flight lines) of fixed-wing MAV for surveillance.

The work in [13] was representative of the 2D coverage solutions. Valente divided the workspace into grid-like cells and planned the sub-areas with the best approximation to the minimum line segment. The off-line planning considered the inherent limitations of a raster and other constraints to be optimized. A heuristic wavefront was extended on a graph generated by the neighborhood adjacency to calculate the coverage paths. Further, the authors proposed task negotiations based on an auction mechanism only at the coverage mission start. Following the line of optimal coverage with MAVs, the authors in [14] decomposed the observed area into that of a regular grid, partitioning the area by equal vertical segments. The area was rotated to find the optimal sweep direction and paths with few flight lines. They compared the spiral, Zamboni, and lawnmower patterns to calculate the coverage paths. The paths were complemented with the Dubbin curves to optimize the agent energy. The work presented did not include a coordination mechanism for the coverage mission; however, [15] proposed a one-to-one coordination algorithm for area partition in patrolling missions that considered limitations of communication. This decentralized strategy was aimed at minimizing the refresh time of the mission; every MAV only exchanged information with nearby neighbors. They proposed a rectangular decomposition and techniques of re-allocation using close links iteratively.

Balampanis et al. in [16] described a novel algorithm for heterogeneous coverage missions in non-convex coastal regions with ZLIs. The Constrained Delaunay Triangulation (CDT) was used to represent the workspace, and the paths were spirals centered in the first cell labeled for an improved wave-front algorithm. The sub-regions were calculated according to the MAV capabilities and a defined start position; then, the paths were adjusted to solve deadlock situations. Ardupilot SITL instances validated the results in the ROS framework. In recent proposals, area partitioning has token relevance for maritime applications [17]. The work uses a polygon decomposition algorithm to carry out complete search coverage. The strategy fixed the start positions on the edge to divide the workspace through line segments that considered the areas and optimized with a 'maximizing-minimum angle' mechanism. The numerical results showed the performance and computational complexity of the proposed algorithm. The coverage path was a parallel sweep search pattern that was improved by decreasing the turns and the traveled distance. A critical approach was found in [18] to surveilling missions. They split the area of interest given the number of MAVs, the requirements for the site to be covered, and the initial position of each MAV. The coverage path was a back-and-forth pattern with a given cross-track separation. The novel partitioning method was based on the directed graph of the triangular sub-regions by a constrained Delaunay triangulation. Each sub-region was adjusted with a pseudo-site to start the mission. The path assignment was centralized and considered few turns. Finally, a generalized proposal of the polygon decomposition for any robotic agent was presented in [19]. The basis of the proposal was to divide a polygon into two parts for given area requirements in terms of the perimeter of the corresponding part. The algorithm continued until it split the workspaces in the given number of vertices in the polygon's border. The authors only worked on the mathematical solution to divide a polygon for any range of applications.

2.2. Clustering-Based Coverage Plans

The coverage mission based on polygons is frequently oriented in the area geometry and ZLIs. Still, clustering is focused on dividing the workspace into homogeneous sub-areas that could be heterogeneous with ZLIs-included. For instance, [20] decomposed the total area using the K-means algorithm. A genetic algorithm calculated the coverage paths for each cluster, and a single MAV was assigned for each group. They proposed two variants of a method for offline and online coverage paths, and the analysis was done for small, medium, and large areas.

Similarly, in [21], an area reconnaissance mission was deployed with offline planning to start the task and a partial online re-planning. The pre-planning was based on the Spanning Tree Coverage algorithm; then, a fuzzy C-means clustering algorithm was executed to calculate the multiple paths and corresponding sub-areas. The online re-planning was designed to distribute tasks in case of MAV failures.

Leng et al. in [22] presented a new proposal to optimize the paths and maximize the ground visibility considering the natural occlusions in forests. The authors changed the traditional grid abstraction by the Voronoi cells to represent the waypoints to be visited. Each MAV was assigned to a cluster of Voronoi cells to carry out surveillance. The coverage path for each MAV was calculated with a custom clustered spiral-alternating algorithm. Following the balance between areas, paths, and number of MAVs, in [23], the coverage mission was defined by solving the Multiple Traveling Salesman Problem (m-TSP). First, collision-free sub-areas were generated by the $\alpha\beta$ swap algorithm. The second part updated the sub-areas to find the partitioning that minimizes the longest MAV path. They allocated paths to MAVs which had nearly equivalent lengths.

The authors in [24] proposed a solution to solve the problem of multi-MAV coverage path planning that divided the region based on Reinforcement Learning with a grid world. The initial cell is random, and each agent has a camera to acquire information. The global control of the fleet is defined by a mechanism of Deep reinforcement learning with Double Q-learning Networks (DDQN). The coverage showed that different start positions are independent of the capacity to cover all spaces. The results proved autonomous collaboration in dynamic environments with energy constraints.

Inspired by the algorithm called Clustering by the Fast Search and Find of Density Peaks (CFSFDP) [25], the regions were classified into clusters and obtained approximate optimal point-to-point paths for UAVs. The simulated UAVs are heterogeneous in their flying speed, energy supply, and scanning width of onboard sensors. The coverage path was calculated under MILP formulation, although the waypoints were efficiently visited with complementary optimization strategies. The Nearest-to-any policy to classify unallocated regions into clusters and an Order optimization strategy to adjust the visiting order of the areas classified into the same group resulted in a reduced task time of MAVs.

Comparable to the previous works considering communication requirements, in [26], a fleet of AI-driven MAVs was modeled to survey urban zones. The proposal mixed Artificial Neural Networks and a modified version of the famous A* pathfinder to solve the coverage path planning. The workspace was clustered by combining K-means and the Voronoi Diagrams for a homogeneous distribution. The results with different complex areas were computed on the Gazebo simulator using the ROS framework.

2.3. Heuristics-Based Coverage Plans

The mentioned works may be classified as medium to low computational requirements for MAVs. In contrast, heuristics-based mechanisms require high computational capacity.

In [27], the overall goal was to quickly build overview mosaics from unknown areas for emergency and disaster response cases. They focused on wireless communication networks for the transmission and control of acquired images. The online path planning was based on the parallel Clarke and Wright savings algorithm and clustering combined with the Christofides algorithm. Specifically, the mission was for area coverage, but they designed an architecture to acquire images with onboard processing, annotating with other sensor data, and transferring by a prioritized scheme. In [28], the coverage planning for multiple UAVs was assumed for a known region and divided into square cells. The spiral algorithm was applied to the search for uncovering cells, and the A* algorithm was used to find the shortest path. The main contribution was a contingency strategy when a UAV failed; it skipped all next path planning for that UAV. The remaining un-surveyed region was automatically assigned to the other UAVs from the base station. In [29], explicit communication was also considered to become a fault-tolerant system. A multi-agent system decentralized for field coverage and weed mapping was introduced with a re-broadcast protocol to account for limited communication ranges. A stochastic exploration strategy based on a reinforced random walk was used, and a mechanism to avoid re-visited areas was defined. Each agent had a local map to store the information acquired from onboard processing or received from other agents through communication. In this work, the results showed that the decentralized, self-organizing nature of the solution led to robustness against faults.

The most recent studies continue to improve the energy consumption of MAVs. In [30], an imagery mission was described using a column generation framework. The authors defined a flight profile to estimate the energy consumption in the mission. The shape was calculated for each coverage path based on an algorithm called RBECOM that calculated an optimal solution by minimizing both the length of a returning path and the number of turns. The sub-regions were calculated by adding a constant to the same model to define a combination of routes that required the least amount of total energy. Another case that considered energy consumption was found in [31] for data collection missions. The paper conserved energy by optimization of the trajectory plan of a cooperative fleet of MAVs. The planning was based on an algorithm called Deep Learning Trained by Genetic Algorithm (DL-GA). The GA received inputs from various scenarios and then the deep neural network was trained while facing familiar scenarios; it could rapidly provide the optimized path which satisfied continuous operations. The solution reached a speed to process a solution better compared with the GA algorithm.

2.4. Conclusion of Coverage Plans

The review of multi-MAV systems for coverage missions denoted that offline planning based on computational geometry had been more often implemented. This result could be because practical work frequently requires campaign workers and infrastructure in the field; therefore, simulated scenarios and/or experiments with few agents could accelerate and cheapen the tests. The scope of the polygonal-based methods was to optimize the division of areas with ZLIs, but it did not consider cooperation. The clustering strategy resulted in an approach to plan optimal coverage with possible collaboration between agents; however, little of the energy consumption heterogeneity was deepened. Finally, the alternative based on heuristics was closer to the goal of autonomous and robust multi-agent systems, but they used high computational resources to compute sub-areas and paths. In addition, continuous communication between all MAVs was delimited. In brief, few works reached the resource balance between task allocation and communication challenges in coverage planning. Further, none of the found results included user restrictions on planning a mission with a MAV fleet.

3. Problem Statement and Principles

The research problem is centered on the coordination of a fleet of MAVs for coverage missions in restricted workspaces. The challenge is to become effective in practical scenarios where it is necessary to think about the investment, the technological usability, and the usefulness of the information to make decisions. The current paper aims to reach a solution considering the user context and the restrictions of workspaces. A user can be a farmer, a security company, a forest manager, a search and rescue squad, an archaeologist, and other persons who make decisions based on remote information. The typical user is not interested in technological knowledge but in making decisions without loss. Therefore, a restriction would be to deploy coverage missions with commercial and adaptable MAVs to facilitate the use, maintenance, and data acquisition, and mitigate the risk in the investment concerning the restricted workspace, including lack of electricity supply, rugged relief, and changing weather.

In order to solve this problem, the research must obtain complete information on using multiple MAVs. Complete information means the acquired data corresponds to the area of interest; it tries the CPP problem in multiple regions with communication-enabled MAVs to support future cooperation. Then, the statement to deploy a coverage mission starts with a group of heterogeneous MAVs and a known polygonal area representing a restricted workspace. The workspace is partitioned into sub-areas with corresponding sub-tasks allocated to the MAVs.

Let $A_1, A_2, \ldots, An$ be a fleet of n MAV agents performing the coverage mission. Let S be the total coverage area, and consider each sub-area as a set of S cells. Then, $S_1, S_2, \ldots, Si$ are sub-areas of S, where Si is assigned to one MAV agent An. In Figure 2, each sub-area Si is decomposed into cells as close as possible to the footprint size of the MAV remote sensor, which is called world abstraction.

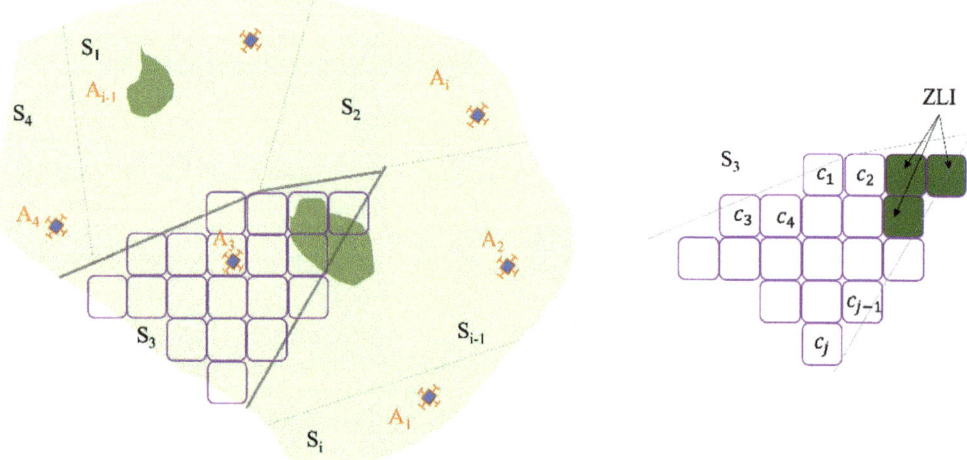

Figure 2. Coverage mission statement.

Each MAV is modeled as an agent with skills to move, perceive its state and the environment, follow a plan, reach a goal, interact, and adapt to a behavior. The MAV agent is modeled with the Belief-Desire-Intention (BDI) control [32], and the interaction is based on FIPA (Foundation for Intelligent Physical Agents) protocols.

According to previous studies, the research starts from the following principles:

P1. The total coverage area S is known, and it should have an extension at least three times greater than the traveled nominal distance of a single MAV agent.

P2. The ZLIs can be identified in S. If a sub-area has ZLIs, the corresponding cells should not be considered on the task. Then, each An should be allocated only with the free cells c_j of each Si (Figure 2).

P3. Each An should have basic communication capabilities to support the interaction. The connectivity skill is required for possible cooperative tasks.

P4. All MAV agents on the mission have the same flight height and speed. In practice, it will not happen, but the current interest is for bi-dimensional coverage and slow cruising speeds.

P5. A Si sub-area corresponding to each MAV agent cannot be disjointed or intersected with another sub-area; this principle is to avoid collisions during the mission.

P6. The number of MAV agents, n, is fixed by the user. The coverage planning should be done with available MAVs without forcing the user to make a more significant investment.

P7. There should be a ground station to monitor agents. Frequently, the station will be close to buildings or an electric supply.

P8. The heterogeneity of MAV agents is given by the adequate flight time, decision capacity, and communication skills.

P9. The MAV agents have the same remote sensor as the payload. The payload is a user's decision to obtain data related to the corresponding business.

4. Coverage Mission Planning

The proposed solution to deploy multiple heterogeneous MAVs is classified as a polygon-based coverage mission plan. The decision is a result of contrasting the works in Section 2. The techniques based on heuristics are discarded because they require high computational capacity and persistent connectivity of the fleet, which could imply an energy excess and additional time to complete the mission. The clustering-based category was rejected because the resulting sub-areas become homogeneous, and the computational complexity could restrict possible practical deployments of coverage missions. However, the strengths of the focuses in the related works are integrated into a novel method of coverage mission planning according to the principles in the problem statement (Section 3).

Consequently, the following content describes the technology to model MAVs as agents and the proposed plan to solve the research problem, called MCO, by the agents' roles (Master, Coordinator, and Operator). The method is stated step-by-step through five components: world abstraction, area partitioning, role allocation, task generation, and task allocation. The MCO Plan is assumed to run at the mission start.

4.1. Agent-Based Simulation

Different alternatives were found to model robotic agents, some more cited such as NetLogo [33] and Repast Simphony [34], and others such as MAS-Planes [35] focused only on UAVs. However, a physical engine and communication skills are required to model MAVs. Both factors are found in the GAMA Platform [11]. The platform has a friendly programming language (GAML), communication protocols for physical agents, complete documentation, and a development community. In addition, the GAMA models are spatially explicit and extended by including GIS (Geographic Information Systems).

Each MAV agent is implemented to move, perceive, follow a plan, interact, and adapt behavior. The control of the skills is designed with BDI architecture as shown in Figure 3. The beliefs are related to predispositions during the execution of a task. They can be updated according to world knowledge and self-knowledge; as an example, the MAV agent can be believed to be at task target (4.At_target). The MAV agent should verify each belief under logical rules from self-knowledge and the allocated task.

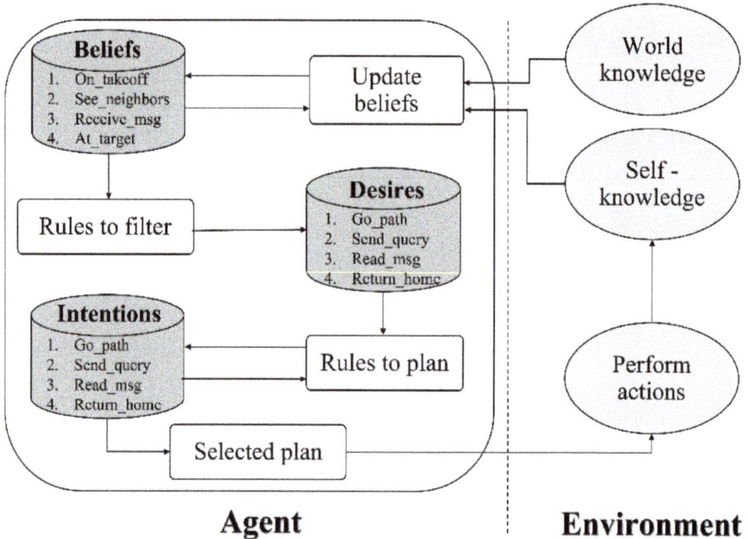

Figure 3. BDI Control for MAV agents.

Consequently, the MAV agent is motivated to reach some single goal; for the previous example, the desire can be a return to home (4.Return_home). However, a MAV agent cannot reach a goal without a plan. The library of plans considers the intentions and desires to define the actions, which is going to affect the environment. From the example, when the desire becomes an intention to 4.Return_home, a cruise flight is a possible plan to execute such action.

Table 1 describes how the beliefs become actions of the MAV agents using steps 1 to 5 after any belief (On_takeoff, See_neighbors, Receive_msg, or At_target) is set. In brief, a belief motivates a desire; then, the desire sets a goal to adopt an intention. The intention is to run a plan that reflects actions.

Table 1. Rules for BDI Control.

	Beliefs	On Takeoff	See Neighbors	Receive Msg	At Target
Step 1	Desire	Go on a path	Send a query	Read message	Return to home
Step 2	Goal	Make a complete path	Update fleet state	Update fleet state	State inactive
Step 3	Intention	Go on a path	Send a query	Read message	Return to home
Step 4	Plan	Follow path steps	Interaction as priority	Interaction as priority	Go to the setpoint
Step 5	Action	Move to next waypoint	Transmit message	Reply message	Fly as cruise mode

Concerning interaction skills, the current proposal only monitors the fleet state and then tests the connectivity between agents. Each MAV agent uses a FIPA Query Interaction protocol with a communicative act called query-if. This type of communication waits by informing the neighboring state [36].

4.2. MCO Plan

MCO Plan is the solution to the problem described in Section 3. The known coverage area (georeferenced raster) with ZLIs, the number of available MAVs, the return of MAVs, and remote sensor parameters are inputs for planning. The outcome of the MCO Plan is allocated roles and coverage tasks to follow. The approach should become scalable and near-optimal, looking for a balance between heterogeneity and connectivity. To reach that goal, five components for coverage mission planning are represented in Figure 4 and are described below.

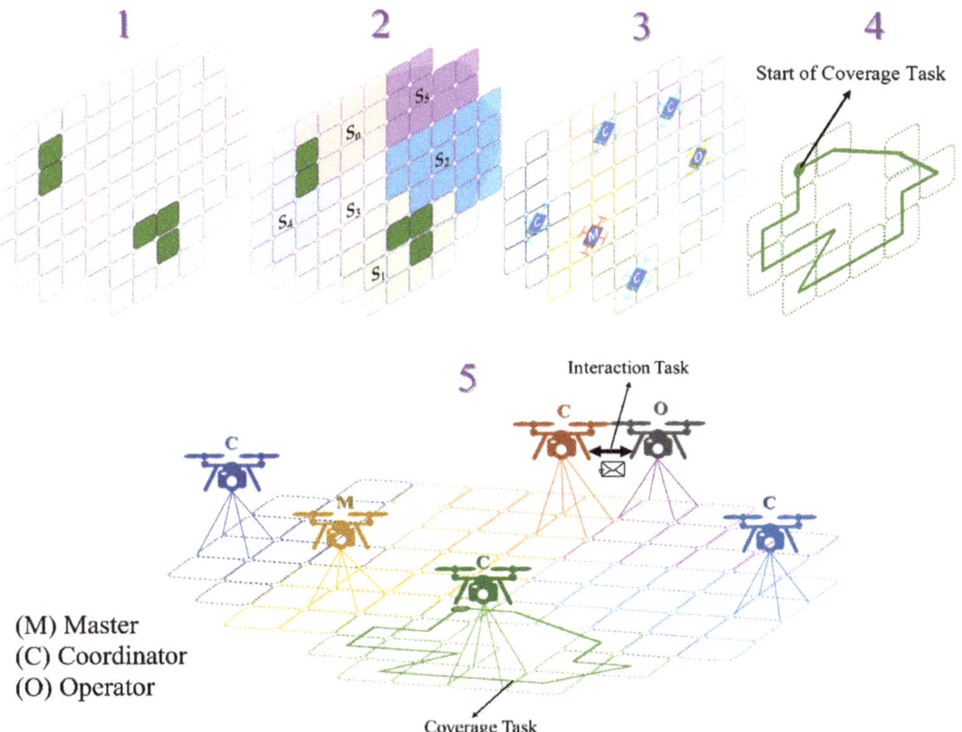

Figure 4. Components of the MCO Plan. (1) world abstraction, (2) area partitioning, (3) role allocation, (4) task generation, and (5) task allocation.

4.2.1. World Abstraction

The first step in coverage mission planning is to define how the MAV agent observes the world. The world abstraction considers a MAV agent acquiring remote information (through cameras, scanners, etc.). This payload type projects a footprint on the observed workspace defining the cell shape to split the area. In the current case, a rectangular shape represents the sensor's approximate range.

Each cell center is the step or waypoint visited by a MAV agent (Figure 5). According to the specification of the coverage mission, the centers should be so close as to overlap more than 60%, according to photogrammetric fundamentals. The cell size is calculated with L_x and L_y from Equation (1), where h is flight height, α is the angle of view from the sensor, and the image size is defined by I_x and I_y. Then, the cell is interpreted into pixels and split by L_x on height and L_y in width, ZLIs included. The final footprint is given by $L_x \cdot (1-p)$ and $L_y \cdot (1-q)$, where p and q are respective percentages (0 to 1) of longitudinal and cross overlap.

$$L_x = 2 \cdot h \cdot tan\left(\frac{\alpha}{2}\right), \; L_y = L_y\left(\frac{I_x}{I_y}\right) \tag{1}$$

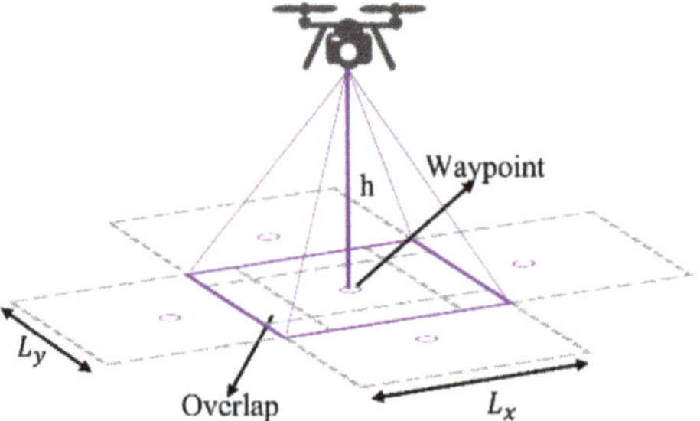

Figure 5. Sensor footprint and overlap.

Some areas from the original workspace could not be included in the rectangular grid, but if this is the case, the cell should be reduced as much as the application allows. The world abstraction is implemented in four steps as seen in Figure 6: a georeferenced raster is loaded, then the user identifies the coverage areas (coverages), and a raster with recognized territories is segmented. Image processing extracts the ZLIs from the raster as in [4]. Finally, the cells and waypoints based on the camera parameters are obtained. A screenshot of the GUI for world abstraction is shown in Figure 6.

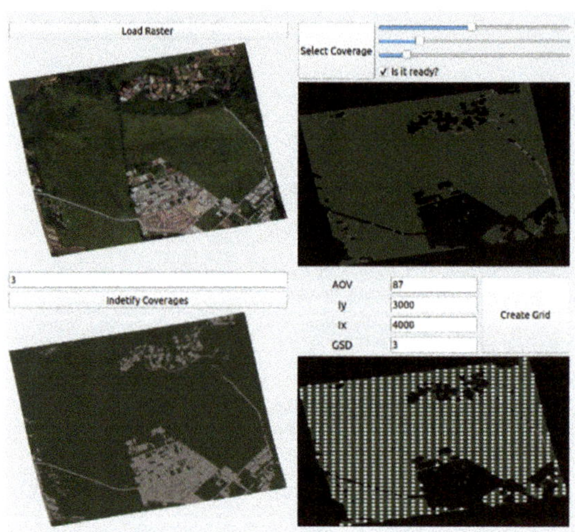

Figure 6. GUI for world abstraction.

4.2.2. Area Partitioning

The focus of offline coverage planning is how to split the total waypoints corresponding with the number of MAVs. The waypoints are the free cells to visit calculated from the world abstraction method.

In Section 2, the planning can be classified as polygonal from related works. The polygonal mechanism uses segments on the area's geometry, which results in heterogeneous sub-areas, while the clustering-based distribution trends to homogeneity. Representants of the polygonal-based partitioning can be the Boustrophedon approach and Voronoi-tessellation, and clustering-based partitioning can be the K-means algorithm. Figure 7 shows partitioning by segments, clustering, and Voronoi-tessellation as an instance for four MAV agents.

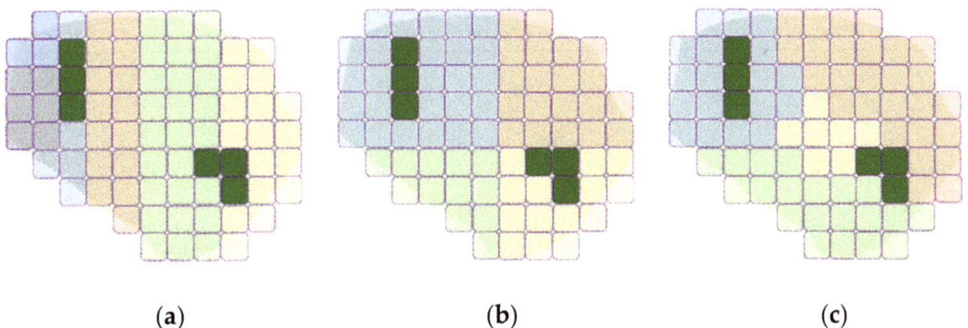

(a)   **(b)**   **(c)**

Figure 7. Area partitioning techniques (**a**) segments, (**b**) clustering, and (**c**) Voronoi-based.

Figure 7c shows Voronoi-tessellation as the initial candidate for this research because the computational complexity can become less than clustering; the areas to connect with neighbors are more than others in some related work. Frequently, the Voronoi-based partitioning results in heterogeneous divisions. However, it is heterogeneous for small numbers; the obtained result can become homogeneous if the number increases. An extension is then proposed in the current research to conserve heterogeneity and solve the Voronoi-tessellation generalization. The phases of the attachment are described below:

Phase I: a takeoff location, called p_0, should be selected to generate the centroids of the Voronoi-tessellation (it could be the center of the coverage area). Other centroids to generate tessellations are centered in p_0 and follow a circumference as in Figure 8. The circumference radius is calculated by dividing the measured width of the workspace between the number of MAV agents, n. The distribution of the initial centroids p_1, p_2, ... , p_{n-1} inside the circumference is random.

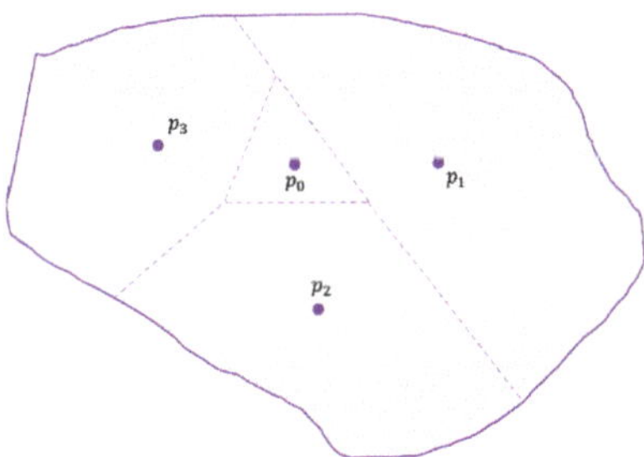

Figure 8. Base partition.

However, after calculating the first Voronoi-tessellation, some partitioning tests for more MAV agents can become a homogeneous distribution. To solve it, a deterministic mechanism is designed on the Base Partition that satisfies the following postulate.

Postulate 1: *a Base Partition can have between two and four divisions. Therefore, a single sub-area can have three connected sub-areas.*

Phase II: If the MAV agents to deploy are more than four ($n > 4$), the centroids of Base Partition are re-distributed based on the centroid p_i of the detected largest sub-area (p_1 is the greatest sub-area in Figure 8). The centroid is moved one-third of the radius closer. Consequently, the selected sub-area becomes larger after rerunning the Voronoi-tessellation.

The next sub-areas result from dividing the largest one using the Phase I method. The new random circumference of centroids is calculated at the center of the largest sub-area. The number of secondary centroids is updated to obtain the required sub-areas (Figure 9). The result of Phase II is named Sub-partition k, which satisfies the following postulate.

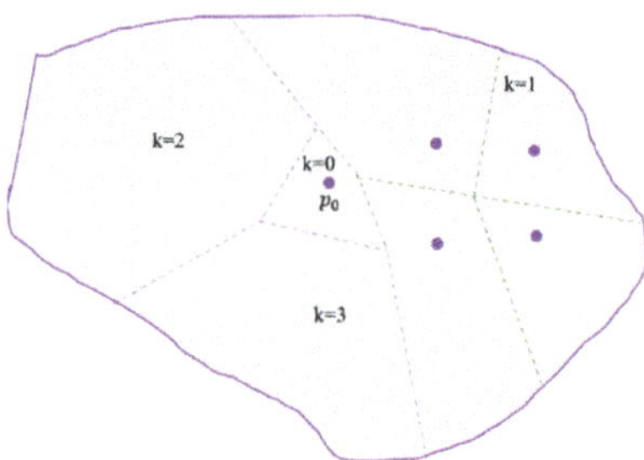

Figure 9. Sub-partition.

Postulate 2: *a Sub-partition can have between two and four sub-areas. Then, if n > 4, the number of divisions for a Sub-partition is (n − 3).*

Phase III: an iterative process is run to determine Sub-partitions following postulate 2. The movement of the secondary centroid of the largest sub-area concerns the closer p centroid, and the cycle continues to complete the area partitioning until there are 16 sub-areas. According to the problem statement for coverage missions, 16 is defined as enough MAV agents. The deployment of 16 simple MAVs can be a great investment for the user, and further, it can become complex to maintain. However, scalability is essential for the research; hence the last phase is proposed below.

Phase IV: if the MAV agents to deploy are more than 16 ($n > 16$), the number of divisions for the complete area, S, is calculated by multiples of 16. It means the number of p_0 to locate is the quotient between n and 16 plus 1 ($n/16 + 1$), called v. For each new p_0, Phases I, II, and III are applied to complete the Partition Levels. Each Partition Level is centered in a new p_0^v (takeoff locations) as shown in Figure 10, satisfying the following postulate.

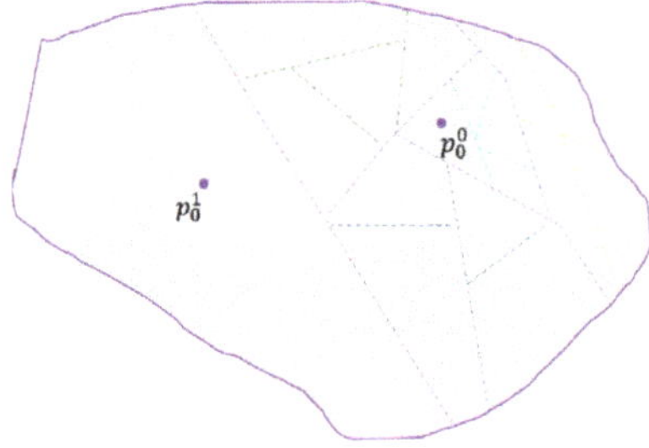

Figure 10. A level of the partition.

Postulate 3: *a Partition Level can have up to 16 sub-areas. Then, a new Partition Level is generated when v ≥ 1.*

4.2.3. Role Allocation

Most work on coverage mission planning describes the method to allocate sub-areas. Nevertheless, the present proposal is projected for MAV agents with some interaction during missions. Consequently, the proposed plan creates roles to define the behavior of the agent model (Section 4.1), creating hierarchical coordination. The model is not designed for continuous connectivity since it would exceed energy consumption. The strategy then becomes managing the fleet's energy during the mission with partial connectivity with few neighbors (at a restricted time). Each MAV agent can only interact when neighbors are in the defined range. The next steps, when MAV agents interact to make decisions to complete a mission, are not tried in the current paper, but they are in development.

The hierarchical model has three roles to manage heterogeneity and connectivity of the fleet, according to the skill scores of flight time, communication, and decision. The Master should have a high score in decision and communication, and an intermediate flight time score. The Coordinators should have a middle score for decision and flight time, and a high score for communication. In case a Master fails, coordinators must support decisions. The Operators should have a low score for decision and communication, and a high score for flight time.

The Master's role is single for one Partition Level. The role is to take principal decisions to complete the mission, and it should be the closest to a ground station (relative to the takeoff location). A Master agent should have a short coverage path to save energy for transmitting and receiving messages from Coordinators. The Coordinator role is for a MAV agent that works as a router of communication between Agents and the Master. A Coordinator can make secondary decisions, and its coverage path should be intermediate. The Coordinators are the four agents closest to the Master. Finally, the Operator role has the hard work of the coverage of the greater peripheral sub-areas. Table 2 summarizes the specifications of roles requested to cover each sub-area and the possible interactions. The roles determine the heterogeneity and connectivity that should be requested from MAV agents for a coverage mission and are the result of the area partitioning method in Section 4.2.2; as an example, Figure 11 shows each role in a partitioned area. To avoid a possible overload of transactions on the network, it is proposed that the number of neighbors for each role is four (as a consequence of the partition phases).

Table 2. Roles for a Partition Level.

Role	Allocated Sub-Area	Interact with
Master	Sub-area closest to the takeoff location	Coordinators and other Masters
Coordinator	Sub-areas closest to the Master	Master, Operators, and other Coordinators
Operator	Sub-areas around the Coordinators	Coordinators and other Operators

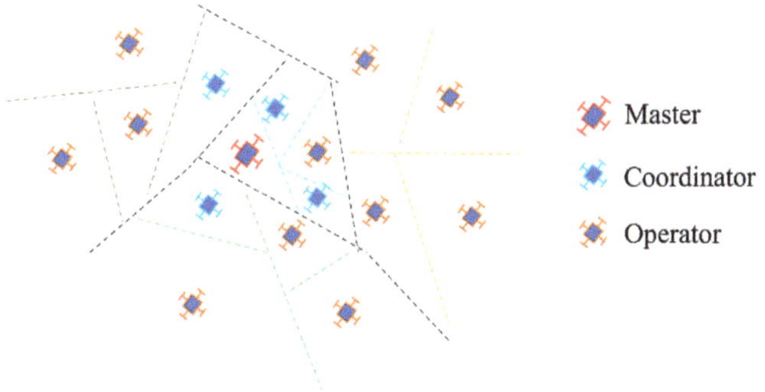

Figure 11. Hierarchical society related to a level partition.

4.2.4. Task Generation

After the requirements for coverage in each sub-area are defined, the next step is to calculate the coverage task. Then, each MAV agent should be allocated with a path, and its completeness should be guaranteed. The path generation component assumes that the MAV agent will acquire data while visiting each waypoint.

According to the review in Section 2, the coverage paths for MAVs are frequently zigzag or lawnmower movements with improvements in the flight line orientation and smooth turns to optimize the task. However, the zigzag movements are indifferent to ZLIs, and they could be inefficient because they do not consider the return home as part of the path. It means more flight time and high redundancy to achieve coverage planning with included ZLIs. To overcome it, some restrictions such as ZLIs, irregular shapes in workspaces, and only a practical runtime to compute missions should be considered. A suitable runtime for the computation of coverage paths is in the range of hours since practical assignments can require continuous deployments.

The literature describes alternatives to compute the coverage paths based on heuristics and metaheuristics that solve the TSP. Some solutions were previously evaluated to decide which accomplishes the previous requirements. The heuristics such as wavefront and spanning tree are suitable for a few waypoints; if the scope increases, the optimization decreases. The found metaheuristics can be classified as trajectory-based or population-based. The first ones are near-optimal, but the runtime was more significant than population-based metaheuristics. Tests of those based on population versus the zigzag movement are in Table 3.

Table 3. Metaheuristics vs. Zigzag movement.

Metaheuristics	Runtime (s)	Visited Waypoints
GA	1998	150
PSO	112188	200
ACO	37	120
BCO	6542	184
CS	0.953	168
FA	0.038	101
Zigzag	0.025	111

The scores used to compare are the runtime of the algorithm and the visited waypoints to reach the start waypoint (target) again. The tests are for an area of 100 free cells homogeneously distributed, an initial population of 500 individuals, and 5000 iterations. Table 3 shows the results of the paths calculated using the general genetic algorithm (GA), Particle Swarm Optimization (PSO), Ant Colony Optimization (ACO), Bee Colony Optimization (BCO), Cuckoo Search (CS), and Firefly Algorithm (FA). The FA [37] was then selected based on the smallest values.

4.2.5. Task Allocation

Task allocation is the last component of the MCO Plan, and looks for each MAV agent to have a role and a coverage path. To develop the method, the current paper considers the taxonomy for task allocation from the reference in [38]. The taxonomy solves multi-robot problems by relating the number of robots with the number and period of the tasks. The current issue is of type single-task (ST); each task can be realized by a single MAV-agent (SR) and an instantaneous allocation is programmed (IA). However, this setup will change because future research wants to support task re-planning based on partial connectivity.

Task allocation adapts the FIPA English Auction Interaction Protocol Specification [39] as in Figure 12. The initiator is a ground station, and the MAV agents are participants. The initiator informs the auction and requires the confirmation of MAV agents to know its bidders. The bid for each MAV agent is calculated with three scores by knowing its resume concerning use history, battery and communication module specifications, type of autopilot, and capacity to process on a small computer board. The initiator takes the information from the role and the flight time to calculate the same three scores. Both calculate three features: flight time, decision capacity, and communication skills, and the initiator gives a value for the first thresholds (subtraction of scores). Table 4 presents the score sources used in Equation (2) for each score.

$$Score = 3 * \frac{X - X_{min}}{X_{max} - X_{max}} \tag{2}$$

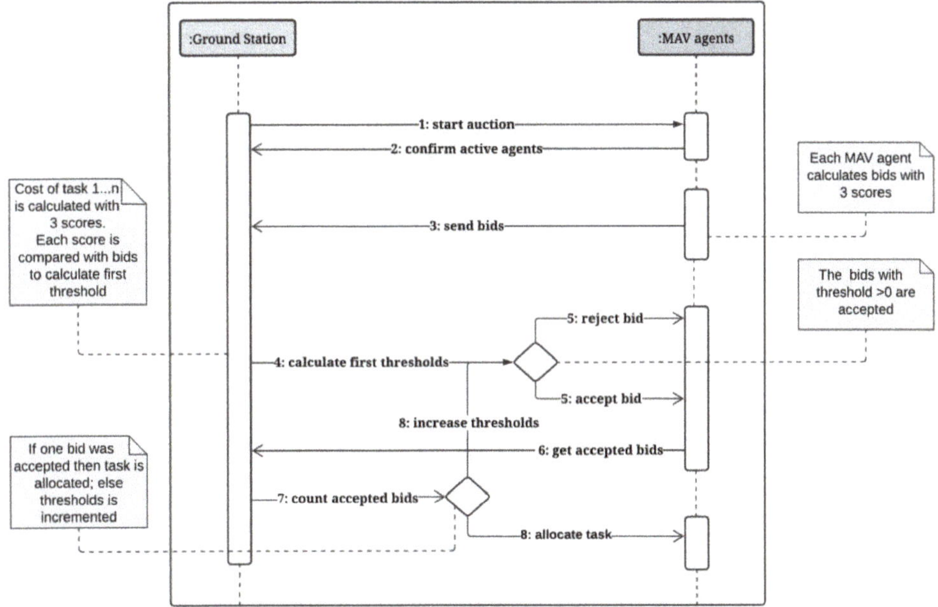

Figure 12. Task allocation method.

Table 4. Task costs and bids.

X Parameter	Ground Station	MAV Agent
Flight time	Estimated coverage time	Estimated time by battery
Communication	Start position for each path with respect to area	Estimated from power of telemetry module
Decision	Start position for each path with respect to area	Autopilot and possible on-board computer

The scores range from zero (0) to three (3), where three means the maximum number of resources. The initiator rejects the participants with Bids more minor than the calculated Task cost and accepts the requests above. The initiator can increase the first thresholds and re-auctions in case of multiple MAV agents with accepted bids. The task is allocated to the agent with the better bid. When one task is assigned, the selected MAV agent saves it in its memory and waits for the call to start the deployment. Then, the process in Figure 12 starts over for another task allocation.

5. Simulations and Discussion

The validation of the proposal was done using three different ZLIs-included workspaces in Cauca, Colombia. They were selected as areas for possible uses with multiple drones (crops and building Zones). The cases of the study were selected to prove the efficiency of three coverage mission plans in areas with different shapes of ZLIs and the number of waypoints in free spaces. Such variations would impact the task heterogeneity and the connectivity skill of the fleet. Figure 13a is the Rejoya Farm with 70 Hectares, has free space in the center and some separate areas, and it is the largest workspace. Naranjos Farm in Figure 13b is 27 Hectares, has a small free area, few ZLIs on edge, and is the smallest area. Finally, the Urban Zone in Figure 13c is approximately 54 Hectares, has the broadest free space, and has one large ZLI on the edge.

(a)

(b)   (c)

Figure 13. Workspaces selected in Cauca, Colombia: (**a**) Rejoya Farm, (**b**) Urban Zone, and (**c**) Naranjos Farm.

Table 5 shows the number of ZLIs and the selected coverage area by image processing as in Section 4.2.1. The number of waypoints was calculated according to camera parameters such as angle of view (87°), resolution (4000 × 3000 pixels), GSD (2 cm), and overlap (75%). The number of minimum and maximum calculated sub-areas was limited from Boustrophedon decomposition because the method does not deliberately allow selecting the number of areas (ZLIs dependence).

Table 5. Selected workspaces to test.

Workspace	ZLI	Coverage	Waypoints	Sub-Areas (Min, Max)
Rejoya Farm	6	Coffee crop	2000	3, 12
Naranjos Farm	3	Mix of crops	315	2, 7
Urban Zone	2	Building lot	870	3, 10

The following results present the performance tests regarding proposed stages of the MCO Plan, specifically, area partitioning, the role and task allocation with a zigzag pattern, and paths generated by the Firefly algorithm. Every approach is analyzed for the resulting coverage missions of the plan based on MCO, Boustrophedon, and K-means. For each plan, the scope to manage heterogeneity and connectivity is tested. Metrics such as standard deviation, heterogeneity trend, active MAVs, task time, battery per agent, and the likely links between neighbors were used for evaluation.

Heterogeneity management is considered the main factor for efficient resource management in working with MAV fleets [40]. The MCO Plan defines the MAVs with three features to manipulate heterogeneity, adequate flight time, decision capacity, and communication skill. Each one represents the primary source of waste energy for a MAV. The proper flight time depends on the battery and the use history; the decision capacity depends on the payload to process data on-board; and finally, the communication skill can change based on the telemetry module and data flow support. Hence, the key is to manage the heterogeneity in the tasks to satisfy the balance of resources for the agents, understanding it as an intermediate effect between the three features in a coverage mission.

The considered heterogeneity in this research is a transparent property for the user since it is natural in practical deployments with MAVs. The users can decide on the investment in at least one equipped MAV and some basic ones, adapt capacities for some basic MAVs, or, as an ideal, have all MAVs equipped with high processing and communication resources. The user should not be limited to homogeneous MAVs for the coverage mission, although they should be adaptable and have a similar remote sensor.

On the other hand, the MCO Plan involves the connectivity from the mission planning. This issue is important since coordination could require interaction mechanisms to solve failures diagnosed during deployment. The estimation of communication resources in the coverage mission efficiently supports detecting any unexpected events and decisions during the mission. The developed MCO Plan involves explicit communication between agents to monitor the allocated tasks and the agent status. The connectivity is partial during a short time while closer neighbors are inside a minimum range. Therefore, the key to achieving possible decisions is that agents continually find neighbors, but interact with limited resource expenditure.

5.1. Area Partitioning

This section shows how the area partitioning proposed in the MCO Plan reaches a balanced heterogeneity of coverage sub-areas. The metrics to compare the MCO plan with the Boustrophedon and the K-means-based plan are the magnitude and trend; both were analyzed in the selected workspaces in Cauca.

Figure 14 shows the resulting sub-areas of each plan, which are the basis for a qualitative analysis of the maximum distribution of waypoints (last column on the right in Table 5), and are grouped by color to differentiate the sub-areas in each coverage. On the left column of the figure, Boustrophedon decomposition projected vertical rectangular shapes in all ranges since ZLI corners are used to segment the geometry of the area. The clustering by K-means on the center column launched pentagonal shapes since the minimal distance between centroids is determined on world abstraction (grid). Non-regular forms are obtained with the MCO Plan on the right column of the figure since it is based on the Voronoi Tessellation that uses the geometric intersection of midwives. In brief, all projections can change geometrically when ZLIs are included in the workspace, resolving in heterogeneous sub-areas.

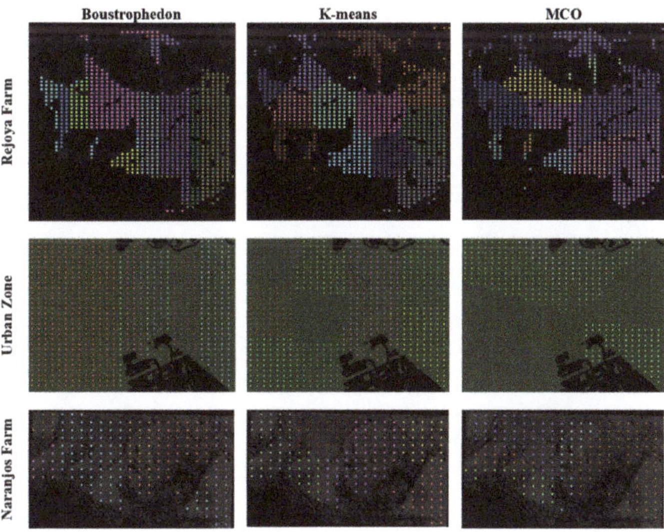

Figure 14. Partitioned workspaces to study.

The next step is the quantitative analysis of the heterogeneity magnitude corresponding to the standard deviation. Then, if the resulting divergence is compared for each workspace as in Figure 15, it is evident that the clustering-based partition (red line) has fewer minor deviations than other methods. Therefore, homogeneous sub-areas increase as the number of MAV agents increases. In contrast, Boustrophedon decomposition shows the highest divergence for the three study cases, as is expected from the literature, achieving a better heterogeneity skill than the MCO plan (yellow line).

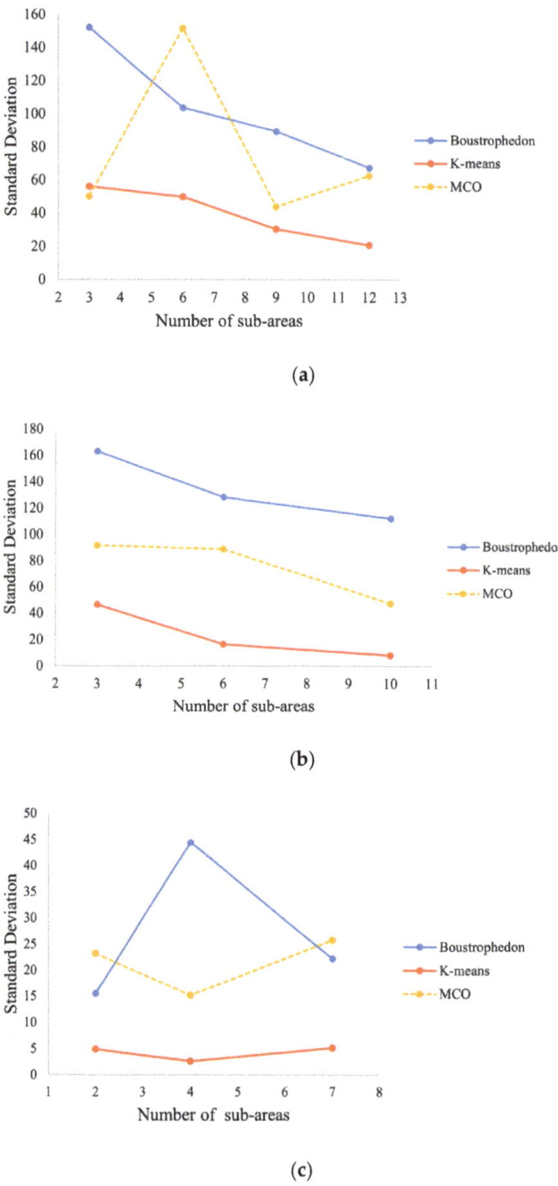

Figure 15. Heterogeneity analysis for each workspace: (**a**) Rejoya Farm, (**b**) Urban Zone, and (**c**) Naranjos Farm.

In the results from Figure 15, the MCO plan states a maximum standard deviation of 7.9% for Naranjos Farm, 9.2% for Urban Zone, and 7.5% for Rejoya Farm. As it is observed, the MCO plan reaches mean values between Boustrophedon decomposition and K-means clustering, although the trend was not coherent between them. The Pearson correlation coefficient was used between the traditional method and the MCO Plan (Tables 6–8) to detail the trend.

Table 6. Correlation between MCO and other methods in Rejoya farm.

Correlation	Number of MAV-Agents			
	3	6	9	12
MCO vs. K-means	0.636	0.362	0.002	0.298
MCO vs. Boustrophedon	−0.472	0.476	0.045	−0.061

Table 7. Correlation between MCO and other methods in Urban Zone.

Correlation	Number of MAV Agents		
	3	6	10
MCO vs. K-means	0.999	0,233	−0.254
MCO vs. Boustrophedon	−0.967	0.104	0.315

Table 8. Correlation between MCO and other methods in Naranjos farm.

Correlation	Number of MAV Agents		
	2	4	7
MCO vs. K-means	1.000	−0.690	−0.708
MCO vs. Boustrophedon	−1.000	−0.242	0.623

In brief, the correlations are higher while the number of MAV agents is slight, such as 2 or 3; it is consistent because, with fewer sub-areas, heterogeneity is not tangible. Meanwhile, the correlations show that the partitioning of the MCO plan is closer to the K-means pattern in most study cases (Figure 15). In addition, in the Naranjos Farm (Table 8), the correlations are higher than in other workspaces, even with the perfect association. The result is given by the expansive workspace that forces the plans' similitude. Another factor could be the non-centered convex ZLIs, which restrict the ability to propagate sub-areas towards the edges as MCO Plan proposes.

According to the observations above, the Boustrophedon decomposition reaches the higher heterogeneity skill in the three evaluated workspaces, with more than 100 waypoints of standard deviation. However, the resulting sub-areas could become small or large, as the ZLIs limited them, which is considered an unbalanced heterogeneity. For instance, the Boustrophedon-based plan for Rejoya Farm (upper left corner of Figure 14) had sub-areas with just three waypoints and others twenty times greater. Consequently, resources are not efficiently managed because, firstly, few MAV agents have greater responsibilities and may make a non-completed mission. On the other hand, the requirement for shorter or longer flight times restricts the user from investing in extra skilled MAVs. Different improvements can be made to achieve a balanced heterogeneity with the Boustrophedon partitioning as the heuristics. However, the effort has a limit due to heterogeneity caused by obstacles, which will consume more computational resources to reach proper partitioning.

The plans' divergences based on K-means show the lowest values between 2 and 50 waypoints (red lines in Figure 15). Despite the low magnitude, the clustering method looks similar to the pattern of Boustrophedon decomposition, which notices a trend to decrease the deviation for both strategies. The Boustrophedon pattern changed notoriously for a smaller number of agents as Figure 15c, which could be a result of partitions dependent on ZLIs. However, after eight divisions, K-means will linearly decrease the heterogeneity, and the Boustrophedon decomposition will stabilize by the fixed ZLIs in the workspace. Hence, it is denoted that the MCO Plan engages the heterogeneity skill as Boustrophedon decomposition without using ZLIs as a reference. It is compensated with the linear decrease without reaching the homogeneity as partitioning based on K-means.

Further, in Figure 14, it is possible to see that the clustering by K-means and Boustrophedon decomposition resulted in disconnected sub-areas with complex ZLIs, such as Rejoya Farm. Independent sub-areas do not support cooperative control. The MCO Plan faces that weakness with sub-areas expanding from a take-off position to the edges. Therefore, the sub-areas would converge, increasing the likelihood of a connection between them.

5.2. Role and Task Allocation with Zigzag Path

The previous description tried to describe heterogeneity according to the workspace and its distribution of waypoints as just "divide and conquer", however, it is necessary to contrast the area partitioning with the MAV agent behavior when a coverage task is allocated. The following results were obtained using the simulation environment (GAMA Platform) and according to the role and task allocation mechanisms. In this experiment, each MAV agent had an allocated role and a coverage path based on back-and-forth movements (a zigzag path) using the method in Section 4.2.5, resulting in a maximum partitioning distribution of each workspace as in Figure 16 (as an example to avoid extra figures).

The objective was to show how the MCO Plan can achieve better energy management during a coverage mission using a balanced heterogeneity (magnitude and trend) and connectivity.

Some metrics were calculated to measure the impact of heterogeneity in each workspace and for the three plans mentioned above. The metrics are the time that each MAV fleet requires to complete the mission (mission time), the number of active MAV agents per time (MAV rate), and the percentage of remaining battery after the task (% remaining battery). Figure 17 states coherence with the partitioning of each plan. Considering the mission times, the allocation based on K-means had the lowest mission times. Still, complex areas like Rejoya Farm expanded the mission times because the sub-areas were broad and homogeneous, as shown in Figure 17a.

Conversely, to have the best mission times, the K-means allocation was based on must-have agents active during longer times, as Figure 17b,d,f show in the red columns. The result is that most MAVs would drain energy faster, risking the completeness of the coverage mission (Tables 9–11). If some event required cooperation, few MAV agents could support it.

The allocation based on Boustrophedon denotes a trend to stabilize mission times in each case of study in Figure 16, due to the resulting tasks around the ZLIs. They varied little and resolved in lighter tasks if the MAV agents increased. As an advantage, the Boustrophedon allocation has the least number of MAVs per time unit, which is a consequence of calculating many small tasks and only three or four long ones (with longer task times). The Boustrophedon allocation transmits a message that most MAVs could have significant excess energy at the mission end. At the same time, few MAV agents would reach their limit of energy before the mission ended.

Concerning the MAV rates in Figure 17b,d,f, the allocation using the MCO plan has a trend of managing active agents during less time than K-means allocation, despite the high correlation demonstrated in the previous section. This triumph is reached because the sub-area sizes increase exponentially for the MCO plan, while in the K-means allocation, the growth is linear (compare the yellow columns in Figure 17b,d,f). The interpretation of the result is that the MCO plan can balance managing missions using MAV agents with minimal to intermediate resources, especially in those with few tasks, as shown in Figure 17f.

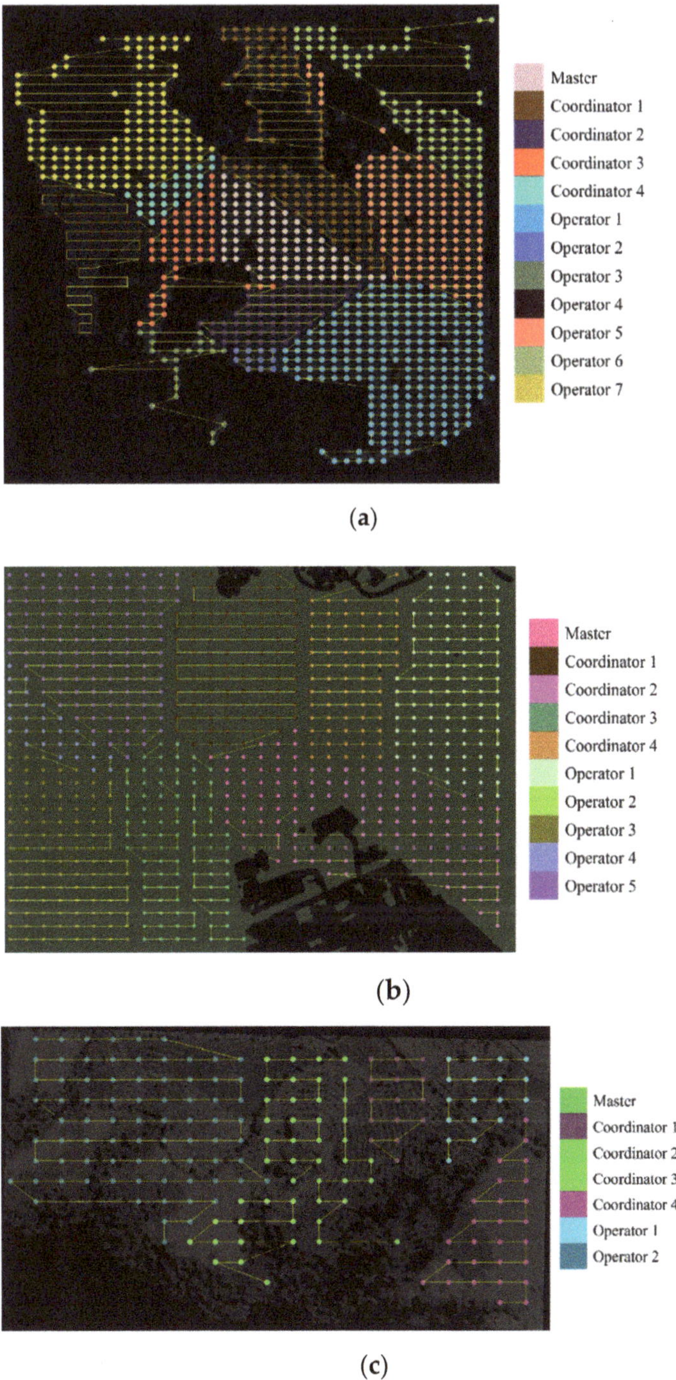

Figure 16. Role and task allocation with zigzag paths for (**a**) Rejoya Farm, (**b**) Urban Zone, and (**c**) Naranjos Farm.

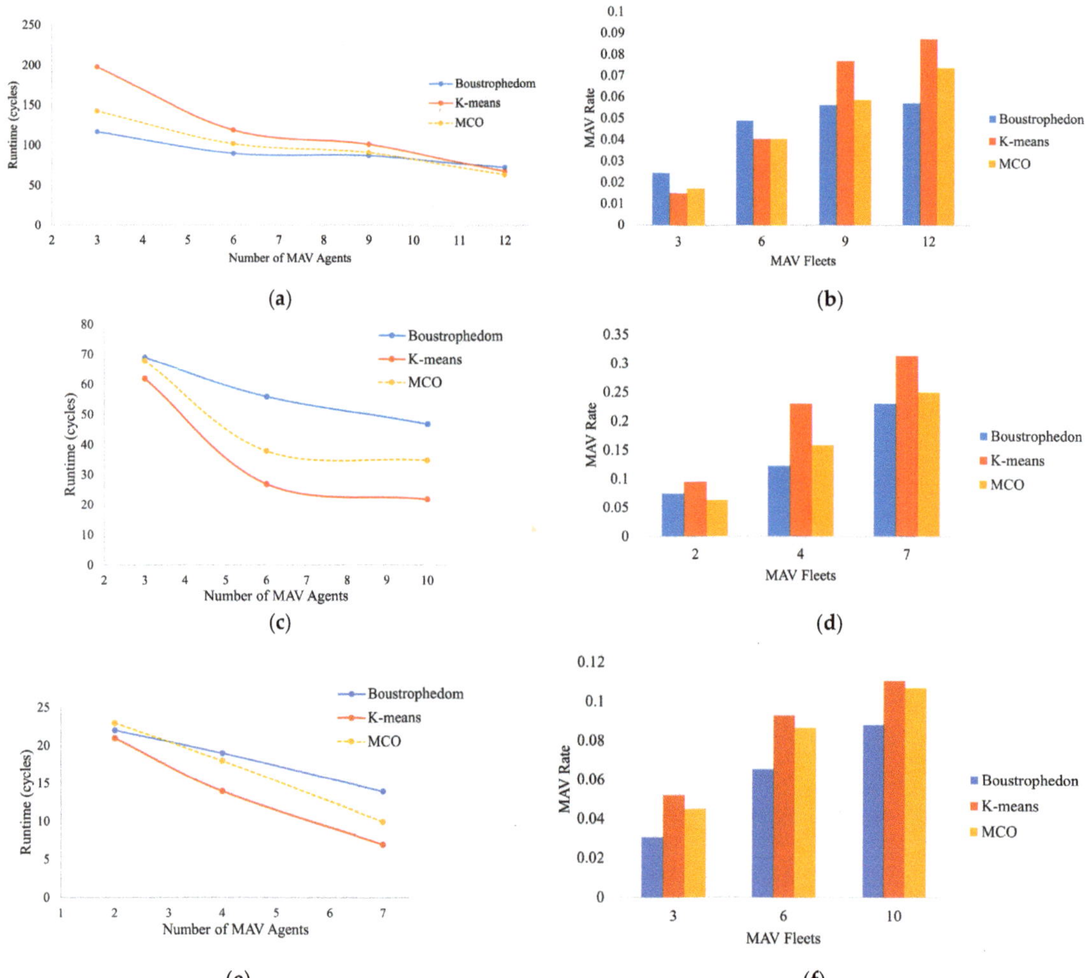

Figure 17. Mission times and active MAV rate in mission: (**a**,**b**) Rejoya Farm, (**c**,**d**) Urban Zone, and (**e**,**f**) Naranjos Farm.

Table 9. Remaining battery in Rejoya Farm.

MAV-Agents	Boustrophedon (%)	K-Means (%)	MCO (%)
1	99.98	93.46	99.47
2	99.98	93.52	97.44
3	99.62	93.2	95.9
4	99.5	93.22	93.88
5	97.2	93.82	93.49
6	96.7	91.3	92.55
7	93.26	90.9	85.83
8	91.49	88	86.83
9	89.20	83.23	84.07
10	88.83	82.31	84.08
11	66.88	79.28	83.09
12	66.61	70.83	82.52

The value of highlighted battery corresponds to Master for MCO Plan.

Table 10. Remaining battery in Urban Zone.

MAV-Agents	Boustrophedon (%)	K-Means (%)	MCO (%)
1	94.01	84.87	94.63
2	92.6	83.81	92.03
3	91.57	84.17	91.94
4	86.59	82.11	89.53
5	86.54	79.95	86.35
6	81.57	79.31	86.03
7	77.93	79.4	85.82
8	71.05	76.63	79.72
9	43.39	73.38	69.65
10	30.24	72.87	63.87

The value of highlighted battery corresponds to Master for MCO Plan.

Table 11. Remaining battery in Naranjos Farm.

MAV-Agents	Boustrophedon (%)	K-Means (%)	MCO (%)
1	89.92	87.91	89.11
2	89.33	87.38	89
3	88.85	87.35	88.25
4	66.33	86.81	87.38
5	66.2	86.17	87.33
6	63.74	86.13	85.51
7	62.96	86.07	82.18

The value of highlighted battery corresponds to Master for MCO Plan.

Tables 9–11 show the remaining battery after the completed mission. This indicator is essential for the current research because it allows MAVs with intermediate skills to cooperate with their energy savings. The averages of saved energy obtained from all agents for the Rejoya farm were 90.77%, 87.76%, and 89.93% with the Boustrophedon, K-means, and MCO allocation, respectively. In the Urban Zone, the mean percentages were 75.55%, 79.65%, and 83.96%, corresponding with the mentioned last order of methods. In the same order, the percentages for Naranjos Farm were 75.33%, 86.83%, and 86.96%. Therefore, the MCO plan reached higher values than other allocations, distributing the resources efficiently regardless of workspace restrictions such as obstacles in the middle of the free space. Moreover, the comparison of the methods shows a battery decrease without exceeding 40% for the MCO plan, while Boustrophedon finished with an upper reduction of 60%. It demonstrates that the heterogeneity pattern identified in the area partitioning section effectively influences the removal of the energy consumption of the fleet of MAV agents.

As a final inference, a result of the MCO Plan is that most MAV agents would have sufficient remaining energy to be used in case of cooperation for non-completed tasks. Although, area partitioning proved above that the chance of faults would be minimal unless environmental causes exist. In contrast, the Boustrophedon allocation obtains a higher case of defects in the deployment of missions by overloading a few agents. Finally, the energy decrease for MAV agents with K-means allocation was up to 22.63%, with the chance of faults lower than the Boustrophedon method and MCO. However, the result would be subject to homogeneous agents with intermediate to high resource levels to complete the mission.

After evaluating the effect of heterogeneity on task allocation, the following analysis shows how the MCO Plan involves connectivity by studying the possible network topology and the number of links when the MAV agents find neighbors in a defined range. For the simulation experiment, the range was calculated with the fourth part of the diagonal segment over space.

Figure 18 shows a likely network topology on the resultant sub-areas for each mission planning and workspace with the maximum MAV agents to deploy. Previously, the

Boustrophedon-based plan resulted in adjacent rectangular shapes (left column), and consequently, the possible topology resembles a Bus. This means that an agent would probably connect with side neighbors. The resultant topology for the K-means planning resembles a mesh (center column), as an evident outcome from the homogeneous waypoint distributions. The topology for the MCO plan has a hybrid focus based on multiple star topologies centered on the centroids of the area partitioning method.

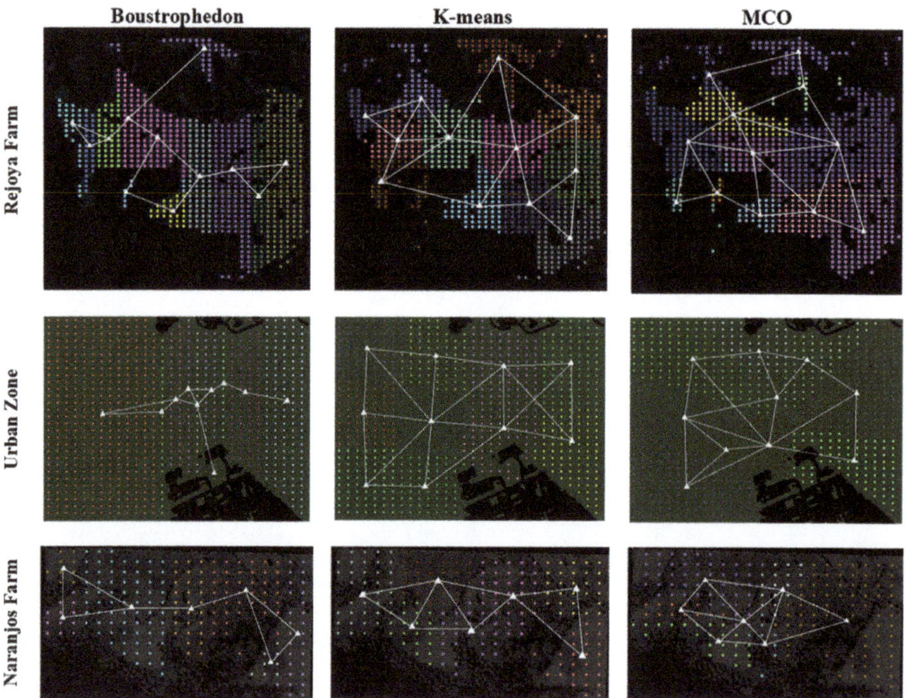

Figure 18. Static network topology for each studied workspace and each planning method.

Figure 18 is only an estimation of the likely links between MAV agents. Still, this trace allows for detecting the skill to reduce faults during the mission and resolve them without affecting the overall goal. The connectivity is low for the allocation based on Boustrophedon, although it can have more connection time when it flies the waypoints at the intersections. The K-means plan is better at managing cooperation, however, dynamic deployments in restricted workspaces can have high data redundancy if a persistent connection is held.

The MCO wants to overcome these issues. The observed topology in the right column of Figure 18 has characteristics that at least one MAV agent can have an overload of links as a star, and the allocated agents on the edge can connect as a tree. In this order, that MAV agent with an excess of communication has taken the role of Master, and the peripheral MAVs have a role as Operators in the hierarchical society proposed in Section 4.2.3. As a complement, Figure 19 shows the network topology at one instant of the simulation on the GAMA Platform for each coverage area with the MCO Plan The numerical results of total links between all fleets are in Tables 12–14, with the Master highlighted in each allocation option to manage high data flow and shorter coverage paths. These tests confirm that the efficient consumption of the fleet can be supported by data routing based on the heterogeneity of the fleet.

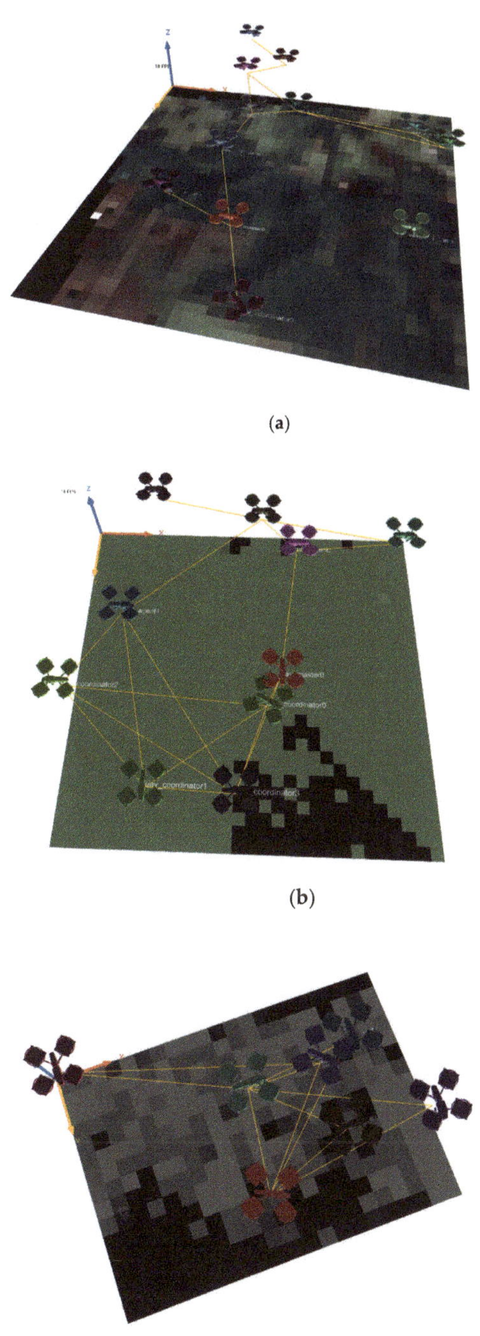

(a)

(b)

(c)

Figure 19. Visualization on GAMA Platform of resulting networks with MCO in (**a**) Rejoya Farm, (**b**) Urban Zone, and (**c**) Naranjos Farm.

Table 12. Total links in Rejoya Farm.

MAV-Agents	Boustrophedon	K-Means	MCO
1	49	337	146
2	102	70	105
3	91	57	141
4	137	107	243
5	66	123	124
6	22	55	55
7	105	224	140
8	62	73	125
9	56	58	57
10	39	58	75
11	56	123	73
12	64	95	279

The number of highlighted links corresponds to Master for MCO Plan.

Table 13. Total links in Urban Zone.

MAV-Agents	Boustrophedon	K-Means	MCO
1	72	49	125
2	98	52	85
3	46	50	97
4	20	54	45
5	30	80	130
6	58	102	60
7	73	69	35
8	58	41	51
9	2	34	47
10	2	57	25

The number of highlighted links corresponds to Master for MCO Plan.

Table 14. Total links in Naranjos Farm.

MAV-Agents	Boustrophedon	K-Means	MCO
1	20	28	75
2	0	23	29
3	33	39	24
4	38	45	52
5	33	39	18
6	15	40	28
7	9	23	17

The number of highlighted links corresponds to Master for MCO Plan.

The inferences above are confirmed by the number of links for each agent and the average connectivity time during the mission. The Boustrophedon-based allocation for Rejoya Farm, Urban Zone, and Naranjos Farm has the lowest link averages with 70.7, 47.9, and 20.3 (blue areas in Figure 20) fronting to 130.2, 70.1, and 30.7 from the MCO plan, respectively (yellow areas in Figure 20). On the other hand, the clustering method holds link averages similar to the MCO Plan but with shorter mission time as it waits. The K-means trend is to have a similar number of links for each MAV agent as Tables 12–14 show. At the same time, the MCO plan proposes a centralized data flow with a few agents, resulting in a substantial difference in interactions between the collaborators. As Tables 9–11 show, those MAVs with the overload have enough energy to resolve it since the MCO Plan manages to have fewer waypoints to visit with roles as Master or Coordinator. These are near the take-off location, facilitating the monitoring with data flows around the ground station and giving more control to the user.

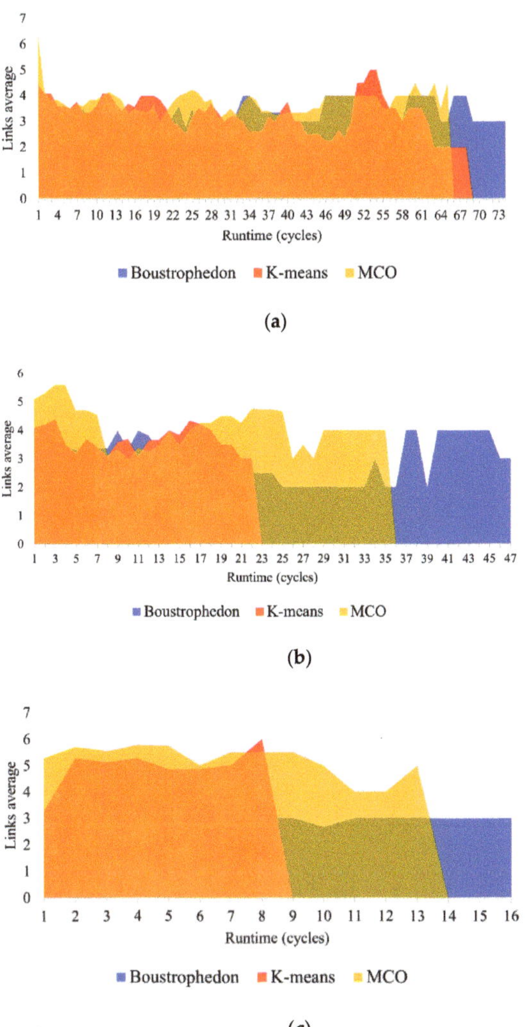

Figure 20. Links average during mission time for (**a**) Rejoya Farm, (**b**) Urban Zone, and (**c**) Naranjos Farm.

In brief, the MCO Plan has the highest link averages in Figure 18 for each workspace, with values between 4 and 5 per time unit because of its connected sub-areas on a hybrid topology. The power of a fleet with coordinated roles can be seen in the Rejoya Farm case in Figure 20a, with ZLIs in the middle. The MCO plan can handle less mission time while the connectivity of the fleet is held. The MCO plan converges to a connectivity skill intermediate with larger free spaces such as the Urban Zone and Naranjos Farm.

5.3. Role and Task Allocation with the Firefly Path

Resource management of the proposed plan in previous sections illustrated the roadmap to efficiently deploying MAVs fleet in outdoor workspaces, considering some user requirements and the fieldwork. However, current research is working further to manage the energy of the coverage path. Based on Section 4.2.4, it is possible to optimize the coverage paths of the mission through the Firefly algorithm [37]. The motivation to advance the study is because the zigzag movement used above and in most reviewed literature and mapping tools frequently includes the ZLIs in the cruise flight, acquiring possible useless or redundant data to make decisions, in addition to increasing the data processing costs.

Figure 21 shows the task allocation with Firefly paths for the experiment with maximum partitioning for each case of study with 12, 10, and 7 tasks, respectively (as an example to avoid extra figures).

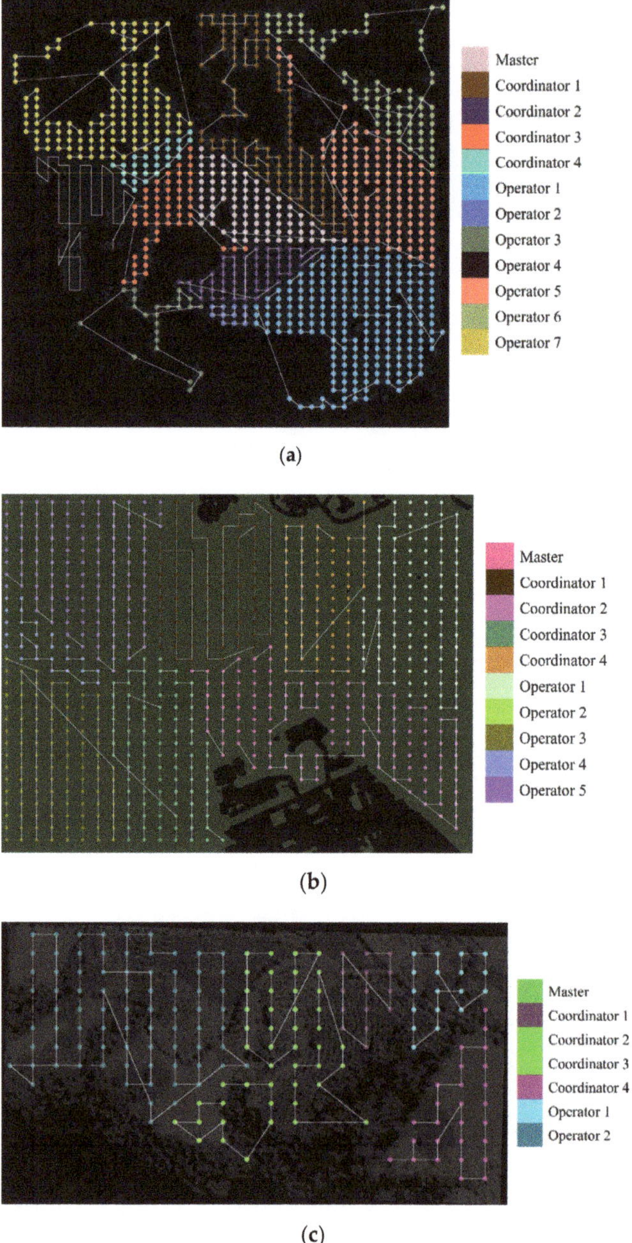

Figure 21. Task allocation with firefly paths for (**a**) Rejoya Farm, (**b**) Urban Zone, and (**c**) Naranjos Farm.

Figure 22 shows that energy consumption can improve with the coverage paths calculated with the Firefly algorithm. In Rejoya Farm, the gained energy reached up to 10% (Figure 22a); in Urban Zone, it gained up to 2.5% (Figure 22c); and in Naranjos Farm, it gained up to 1.9% (Figure 22e). Such performance seen in the bars of the figures is consistent with the heterogeneity of the role allocation shown above. However, just the Firefly algorithm can resolve tasks with small sub-areas more efficiently than the zigzag pattern, gaining some cycles of time to complete the mission. As a complementary analysis, Figure 22b,d,f displays graphs to show the link averages at mission time. The MCO Plan has a higher number of alleged links per time unit, which can be explained because the Firefly paths start to visit the peripheral waypoints of the sub-area and continue until they are as close as they can to the starting location again, therefore increasing the chance of connectivity.

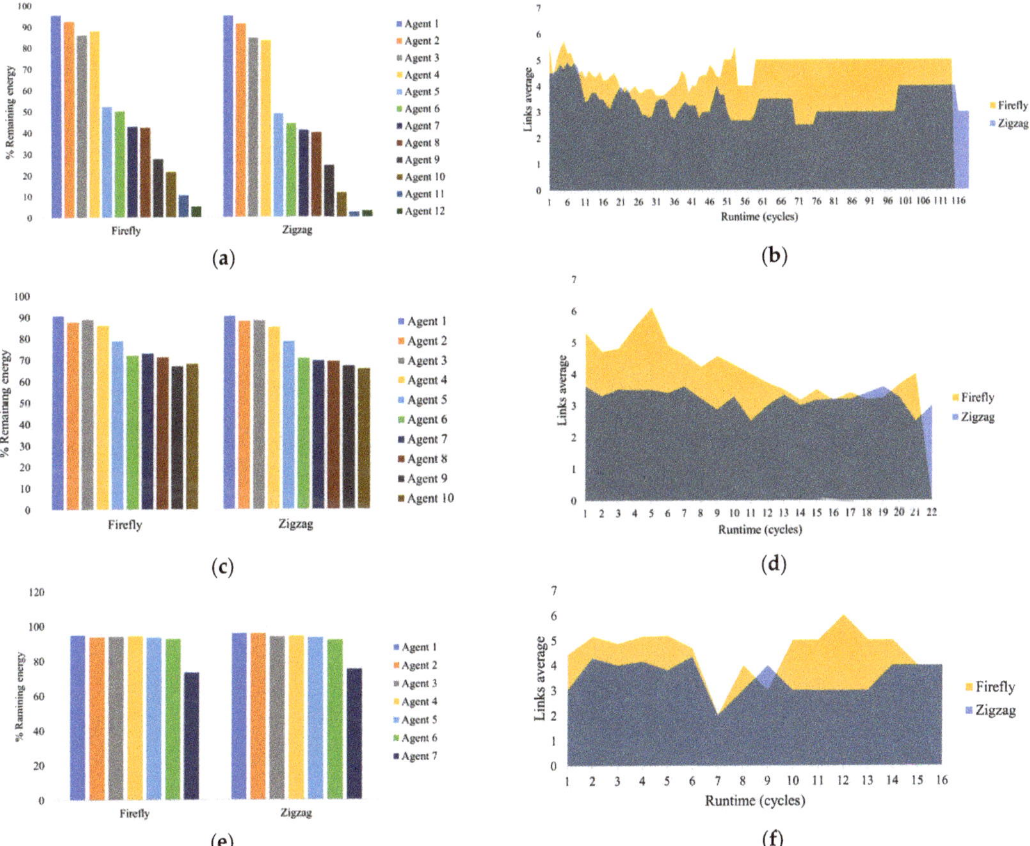

Figure 22. Energy and connectivity management for (**a**,**b**) Rejoya Farm, (**c**,**d**) Urban Zone, and (**e**,**f**) Naranjos Farm.

The gained mission time is also argued in Tables 15–17 which show the active MAV agents during the mission. The path based on the Firefly algorithm completed the missions saving 2%, 9%, and 25% of released MAV agents in Naranjos Farm, Urban Zone, and Rejoya Farm, respectively. This observation is highlighted in Tables 15–17 to contrast that runtime changes the action of the MCO Plan.

Table 15. Number of active agents during Rejoya Farm mission.

Method	Runtime											
	5	15	25	35	45	55	65	75	85	95	105	115
Firefly	11	8	8	6	4	2	1	1	1	1	1	0
Zigzag	11	8	8	7	4	3	2	1	1	1	1	1

The highlighted numbers denote the main changes for active agents between Firefly and Zigzag pattern.

Table 16. Number of active agents during Urban Zone mission.

Method	Runtime						
	3	6	9	12	15	18	21
Firefly	10	9	7	6	5	3	0
Zigzag	10	9	7	6	5	5	2

The highlighted numbers denote the main changes for active agents between Firefly and Zigzag pattern.

Table 17. Number of active agents during Naranjos Farm mission.

Method	Runtime				
	3	6	9	12	15
Firefly	6	1	1	1	1
Zigzag	5	2	1	1	1

The highlighted numbers denote the main changes for active agents between Firefly and Zigzag pattern.

6. Conclusions and Future Work

The MCO Plan is focused on resolving an efficient deployment of the MAV fleet, considering user expectations and restrictions of the workspaces. An efficient deployment means that the coverage mission should be completed with balanced resources. A novel area partitioning method was designed to include heterogeneous MAVs, data flow close to the take-off location, and partial communication between neighbors to satisfy these purposes.

The proposed plan created a hierarchical society with roles defined by adequate flight time, communication skills, and decision capability. Master, Coordinator, and Operator roles were allocated together with the coverage path through an auction mechanism. The integrated strategy was tested in three different coverage areas to show the scope of practical deployments.

The plan's advantages were evident compared with traditional coverage plans such as Boustrophedon decomposition and clustering by K-means. MCO achieved a magnitude and trend of heterogeneity balanced between both methods, directly related to the intermediate mission times reached during the deployment tests in the GAMA Platform. Further, the plan managed fleet energy by decreasing the rate of active MAV agents during missions and increasing the chance to connect with neighbors. Such likelihood was even higher because the resultant coverage paths calculated with the Firefly algorithm followed movement patterns that started on the peripherical waypoints until the start position was found again. Therefore, the connectivity skill was incremented, beyond the clues, indicating reduced energy consumption compared to the zigzag movements.

Consequently, the presented coverage mission planning can improve resource management even more if wind, MAV turns, and energy consumption from communications are considered. Future research should incorporate new alternatives to generate coverage paths beyond managing fault resolution. These facts contribute to the future design of fault-tolerant cooperative MAVs for large and restricted workspaces.

Author Contributions: Conceptualization, L.V.C. and J.C.C.; methodology, L.V.C.; validation, L.V.C., J.C.C. and A.L.; formal analysis, L.V.C.; investigation, L.V.C. All authors have read and agreed to the published version of the manuscript.

Funding: This research was funded by the Ministry of Science Technology and Innovation of Colombia (MinCiencias). Grant number 647.

Institutional Review Board Statement: Not applicable.

Informed Consent Statement: Not applicable.

Data Availability Statement: Not applicable.

Acknowledgments: The authors are grateful to the Telematics Engineering Group (GIT) of the University of Cauca, the Ministry of Science Technology and Innovation of Colombia (Minciencias) for the Ph.D. support granted to Liseth Viviana Campo, as well as Project "Incremento de la oferta de prototipos tecnológicos en estado pre-comercial derivados de resultados de I + D para el fortalecimiento del sector agropecuario en el departamento del Cauca" funding by SGR (BPIN 2020000100098).

Conflicts of Interest: The authors declare no conflict of interest.

References

1. Dalamagkidis, K. Definitions and Terminology. In *Handbook of Unmanned Aerial Vehicles*; Springer: Berlin, Germany, 2015; pp. 43–55.
2. Bähnemann, R.; Lawrance, N.; Chung, J.J.; Pantic, M.; Siegwart, R.; Nieto, J. Revisiting Boustrophedon Coverage Path Planning as a Generalized Traveling Salesman Problem. In *Field and Service Robotics*; Ishigami, G., Yoshida, K., Eds.; Springer: Singapore, 2021; pp. 277–290.
3. Boccardo, P.; Chiabrando, F.; Dutto, F.; Tonolo, F.G.; Lingua, A. UAV Deployment Exercise for Mapping Purposes: Evaluation of Emergency Response Applications. *Sensors* **2015**, *15*, 15717–15737. [CrossRef] [PubMed]
4. Campo, L.; Ledezma Espino, A.; Corrales, J. Optimization of Coverage Mission for Lightweight Unmanned Aerial Vehicles Applied in Crop Data Acquisition. *Expert Syst. Appl.* **2020**, *149*, 113227. [CrossRef]
5. Hoseini, S.A.; Hassan, J.; Bokani, A.; Kanhere, S.S. Trajectory Optimization of Flying Energy Sources Using Q-Learning to Recharge Hotspot UAVs. In Proceedings of the IEEE INFOCOM 2020—IEEE Conference on Computer Communications Workshops (INFOCOM WKSHPS), Toronto, ON, Canada, 6–9 July 2020; pp. 683–688.
6. Skorobogatov, G.; Barrado, C.; Salamí, E. Multiple UAV Systems: A Survey. *Unmanned Syst.* **2020**, *8*, 149–169. [CrossRef]
7. Oliveira, E.; Fischer, K.; Stepankova, O. Multi-Agent Systems: Which Research for Which Applications. *Robot. Auton. Syst.* **1999**, *27*, 91–106. [CrossRef]
8. Verma, J.K.; Ranga, V. Multi-Robot Coordination Analysis, Taxonomy, Challenges and Future Scope. *J. Intell. Robot. Syst.* **2021**, *102*, 10. [CrossRef] [PubMed]
9. Sanjay Sarma, O.V.; Parasuraman, R.; Pidaparti, R. Impact of Heterogeneity in Multi-Robot Systems on Collective Behaviors Studied Using a Search and Rescue Problem. In Proceedings of the 2020 IEEE International Symposium on Safety, Security, and Rescue Robotics (SSRR), Abu Dhabi, United Arab Emirates, 4–6 November 2020; pp. 290–297.
10. Fevgas, G.; Lagkas, T.; Argyriou, V.; Sarigiannidis, P. Coverage Path Planning Methods Focusing on Energy Efficient and Cooperative Strategies for Unmanned Aerial Vehicles. *Sensors* **2022**, *22*, 1235. [CrossRef] [PubMed]
11. Drogoul, A.; Amouroux, E.; Caillou, P.; Gaudou, B.; Grignard, A.; Marilleau, N.; Taillandier, P.; Vavasseur, M.; Vo, D.-A.; Zucker, J.-D. GAMA: A Spatially Explicit, Multi-Level, Agent-Based Modeling and Simulation Platform. In Proceedings of the Advances on Practical Applications of Agents and Multi-Agent Systems, Salamanca, Spain, 22–24 May 2013; Demazeau, Y., Ishida, T., Corchado, J.M., Bajo, J., Eds.; Springer: Berlin/Heidelberg, Germany, 2013; pp. 271–274.
12. Maza, I.; Ollero, A. Multiple UAV Cooperative Searching Operation Using Polygon Area Decomposition and Efficient Coverage Algorithms. In *Distributed Autonomous Robotic Systems 6*; Alami, R., Chatila, R., Asama, H., Eds.; Springer: Tokyo, Japan, 2007; pp. 221–230.
13. Valente, J.; Barrientos, A.; Cerro, J.; Rossi, C.; Colorado, J.; Sanz, D.; Garzon, M. *Multi-Robot Visual Coverage Path Planning: Geometrical Metamorphosis of the Workspace through Raster Graphics Based Approaches*; Springer: Berlin/Heidelberg, Germany, 2011; p. 73. ISBN 978-3-642-21930-6.
14. Araujo, J.; Sujit, P.; Sousa, J.B. Multiple UAV Area Decomposition and Coverage. In Proceedings of the 2013 IEEE Symposium on Computational Intelligence for Security and Defense Applications (CISDA), Singapore, 16–19 April 2013; pp. 30–37.
15. Acevedo, J.J.; Arrue, B.C.; Diaz-Bañez, J.M.; Ventura, I.; Maza, I.; Ollero, A. One-to-One Coordination Algorithm for Decentralized Area Partition in Surveillance Missions with a Team of Aerial Robots. *J. Intell. Robot. Syst.* **2014**, *74*, 269–285. [CrossRef]
16. Balampanis, F.; Maza, I.; Ollero, A. Area Partition for Coastal Regions with Multiple UAS. *J. Intell. Robot. Syst.* **2017**, *88*, 751–766. [CrossRef]
17. Xing, S.; Wang, R.; Huang, G. Area Decomposition Algorithm for Large Region Maritime Search. *IEEE Access* **2020**, *8*, 205788–205797. [CrossRef]

18. Skorobogatov, G.; Barrado, C.; Salamí, E.; Pastor, E. Flight Planning in Multi-Unmanned Aerial Vehicle Systems: Nonconvex Polygon Area Decomposition and Trajectory Assignment. *Int. J. Adv. Robot. Syst.* **2021**, *18*, 1729881421989551. [CrossRef]
19. Skorobogatov, G.; Barrado, C.; Salamí, E. Multi-Robot Workspace Division Based on Compact Polygon Decomposition. *IEEE Access* **2021**, *9*, 165795–165805. [CrossRef]
20. Yanmaz, E.; Kuschnig, R.; Quaritsch, M.; Bettstetter, C.; Rinner, B. On Path Planning Strategies for Networked Unmanned Aerial Vehicles. In Proceedings of the 2011 IEEE Conference on Computer Communications Workshops (INFOCOM WKSHPS), Shanghai, China, 10–15 April 2011; pp. 212–216.
21. Long, G.Q.; Zhu, X.P. Cooperative Area Coverage Reconnaissance Method for Multi-UAV System. *Adv. Mater. Res.* **2012**, *383–390*, 4141–4146. [CrossRef]
22. Leng, G.; Qian, Z.; Govindaraju, V. Multi-UAV Surveillance over Forested Regions. *Photogramm. Eng. Remote Sens.* **2014**, *80*, 1129–1137. [CrossRef]
23. Ann, S.; Kim, Y.; Ahn, J. Area Allocation Algorithm for Multiple UAVs Area Coverage Based on Clustering and Graph Method. *IFAC-PapersOnLine* **2015**, *48*, 204–209. [CrossRef]
24. Perez-imaz, H.I.A.; Rezeck, P.A.F.; Macharet, D.G.; Campos, M.F.M. Multi-Robot 3D Coverage Path Planning for First Responders Teams. In Proceedings of the 2016 IEEE International Conference on Automation Science and Engineering (CASE), Fort Worth, TX, USA, 21–25 August 2016; pp. 1374–1379.
25. Chen, J.; Du, C.; Zhang, Y.; Han, P.; Wei, W. A Clustering-Based Coverage Path Planning Method for Autonomous Heterogeneous UAVs. *IEEE Trans. Intell. Transp. Syst.* **2021**, 1–11. [CrossRef]
26. Sanna, G.; Godio, S.; Guglieri, G. Neural Network Based Algorithm for Multi-UAV Coverage Path Planning. In Proceedings of the 2021 International Conference on Unmanned Aircraft Systems (ICUAS), Athens, Greece, 15–18 June 2021; pp. 1210–1217.
27. Mersheeva, V.; Friedrich, G. Routing for Continuous Monitoring by Multiple Micro AVs in Disaster Scenarios. In *ECAI 2012*; IOS Press: Amsterdam, The Netherlands, 2012; pp. 588–593. [CrossRef]
28. Gupta, S.K.; Dutta, P.; Rastogi, N.; Chaturvedi, S. A Control Algorithm for Co-Operatively Aerial Survey by Using Multiple UAVs. In Proceedings of the 2017 Recent Developments in Control, Automation Power Engineering (RDCAPE), Noida, India, 26–27 October 2017; pp. 280–285.
29. Albani, D.; Manoni, T.; Nardi, D.; Trianni, V. Dynamic UAV Swarm Deployment for Non-Uniform Coverage. In Proceedings of the 17th International Conference on Autonomous Agents and Multiagent Systems, Stockholm, Sweden, 10–15 July 2018; p. 9.
30. Choi, Y.; Chen, M.; Choi, Y.; Briceno, S.; Mavris, D. Multi-UAV Trajectory Optimization Utilizing a NURBS-Based Terrain Model for an Aerial Imaging Mission. *J. Intell. Robot. Syst.* **2020**, *97*, 141–154. [CrossRef]
31. Pan, Y.; Yang, Y.; Li, W. A Deep Learning Trained by Genetic Algorithm to Improve the Efficiency of Path Planning for Data Collection With Multi-UAV. *IEEE Access* **2021**, *9*, 7994–8005. [CrossRef]
32. Dominguez, M.H.; Hernández-Vega, J.-I.; Palomares-Gorham, D.-G.; Hernández-Santos, C.; Cuevas, J.S. A BDI Agent System for the Collaboration of the Unmanned Aerial Vehicle. *Res. Comput. Sci.* **2016**, *121*, 113–124. [CrossRef]
33. Tisue, S.; Wilensky, U. Netlogo: A Simple Environment for Modeling Complexity. In Proceedings of the International Conference on Complex Systems, Boston, MA, USA, 16–21 May 2004; Volume 21, pp. 16–21.
34. North, M.; Howe, T.; Collier, N.; Vos, J. Repast Simphony Runtime System. In Proceedings of the Agent 2005 Conference on Generative Social Processes, Models, and Mechanisms, Chicago, IL, USA, 13–15 October 2005; Volume 1.
35. Pujol-Gonzalez, M.; Cerquides, J.; Meseguer, P. MAS-Planes: A Multi-Agent Simulation Environment to Investigate Decentralized Coordination for Teams of UAVs. In Proceedings of the 13th International Conference on Autonomous Agents and Multiagent Systems, AAMAS 2014, Paris, France, 5–9 May 2014; Volume 2, pp. 1695–1696.
36. Juneja, D. A Review of FIPA Standardized Agent Communication Language and Interaction Protocols. Available online: https://www.semanticscholar.org/paper/A-Review-of-FIPA-Standardized-Agent-Communication-Juneja/c0a8120520bda9b4470dfcea01dedfa862f826bf (accessed on 4 May 2022).
37. Yang, X.-S.; He, X. Firefly Algorithm: Recent Advances and Applications. *Int. J. Swarm Intell.* **2013**, *1*, 36–50. [CrossRef]
38. Gerkey, B.P.; Matarić, M.J. A Formal Analysis and Taxonomy of Task Allocation in Multi-Robot Systems. *Int. J. Robot. Res.* **2004**, *23*, 939–954. [CrossRef]
39. Alarabiat, Z.; Alarabeyyat, A.; AlHeyasat, O. Development of FIPA English Auction Interaction Protocol for Multi-Agent Systems. *Contemp. Eng. Sci.* **2014**, *7*, 1905–1917. [CrossRef]
40. Chen, J.; Ling, F.; Zhang, Y.; You, T.; Liu, Y.; Du, X. Coverage Path Planning of Heterogeneous Unmanned Aerial Vehicles Based on Ant Colony System. *Swarm Evol. Comput.* **2022**, *69*, 101005. [CrossRef]

Article

Aircraft Carrier Pose Tracking Based on Adaptive Region in Visual Landing

Jiexin Zhou [†]**, Qiufu Wang** [†]**, Zhuo Zhang and Xiaoliang Sun** *

College of Aerospace Science and Engineering, National University of Defense Technology, Changsha 410073, China; zhoujiexin11@nudt.edu.cn (J.Z.); wangqiufu@nudt.edu.cn (Q.W.); zhangzhuo14@nudt.edu.cn (Z.Z.)
* Correspondence: alexander_sxl@nudt.edu.cn
† These authors contributed equally to this work.

Abstract: Due to its structural simplicity and its strong anti-electromagnetic ability, landing guidance based on airborne monocular vision has gained more and more attention. Monocular 6D pose tracking of the aircraft carrier is one of the key technologies in visual landing guidance. However, owing to the large range span in the process of carrier landing, the scale of the carrier target in the image variates greatly. There is still a lack of robust monocular pose tracking methods suitable for this scenario. To tackle this problem, a new aircraft carrier pose tracking algorithm based on scale-adaptive local region is proposed in this paper. Firstly, the projected contour of the carrier target is uniformly sampled to establish local circular regions. Then, the local area radius is adjusted according to the pixel scale of the projected contour to build the optimal segmentation energy function. Finally, the 6D pose tracking of the carrier target is realized by iterative optimization. Experimental results on both synthetic and real image sequences show that the proposed method achieves robust and efficient 6D pose tracking of the carrier target under the condition of large distance span, which meets the application requirements of carrier landing guidance.

Keywords: landing guidance; airborne monocular vision; large distance span; scale adaptive; pose tracking

Citation: Zhou, J.; Wang, Q.; Zhang, Z.; Sun, X. Aircraft Carrier Pose Tracking Based on Adaptive Region in Visual Landing. *Drones* **2022**, *6*, 182. https://doi.org/10.3390/drones6070182

Academic Editor: Abdessattar Abdelkefi

Received: 10 June 2022
Accepted: 17 July 2022
Published: 21 July 2022

Publisher's Note: MDPI stays neutral with regard to jurisdictional claims in published maps and institutional affiliations.

1. Introduction

As a necessary means to guarantee the onboard performance of carrier-borne aircrafts, automatic landing technology plays a crucial role in promoting the development of the carrier-borne aircraft [1]. Traditional landing guidance systems mainly rely on inertial navigation, radar navigation, satellite navigation, and photoelectric navigation. These methods require the support of carrier-borne equipment and aircraft–carrier communication link, which are easy to be interfered with, leading to guidance delay or even failure in complex electromagnetic environment.

Thanks to the rapid development of machine vision in recent years, visual landing has gradually emerged and has become one of the important means of automatic carrier landing [2]. According to different installation positions of the camera, visual guidance methods can be divided into carrier-borne visual guidance method and airborne visual guidance method [3]. The airborne visual landing guidance method mainly relies on the image acquisition and processing equipment carried on the carrier-borne aircraft to complete the calculation of the relative pose parameters between the carrier deck and the aircraft. Then, the guidance information is transmitted to the aircraft control system, so as to realize the autonomous navigation of the aircraft. Independent from the support of the radar, photoelectric, and other external equipment, the landing guidance method based on airborne vision has the advantages of simple structure, strong autonomy, high precision, and electromagnetic interference resistance.

Due to the large distance span of landing process, airborne monocular vision is generally adopted in airborne visual landing guidance schemes. Many researchers have carried out relevant work, which can be summarized into two categories: the cooperative mode and the non-cooperative mode. In the cooperative mode, cooperative signs are arranged on the carrier deck, such as corner reflex mirror cross array [4], infrared cooperative sign lamp array [5,6], infrared cooperative sign circle array [7], T-shaped infrared thermal radiation sign [8], and other asymmetric cooperative signs [9,10]. By extracting point and line features of cooperative signs, the relative pose between the aircraft and the carrier are calculated according to the geometric characteristics and imaging relationship of the cooperative sign so as to guide the aircraft. The guidance methods based on the cooperative sign require the installation of additional signs such as light array on the carrier deck, which will introduce modification to the original design of the carrier deck with poor concealment. When the fixed-wing carrier-borne aircraft is landing, the angle between the optical axis of the airborne camera and the deck surface is small. As the light array arranged in the carrier deck area is narrow, the distribution of cooperative sign lights arranged on the deck is extremely compact in the image, which makes it difficult to solve the pose. Therefore, this paper focuses on the non-cooperative guidance mode based on airborne monocular vision.

The existing research work using non-cooperative mode mainly relies on the carrier's own characteristics to solve the pose, such as carrier deck runway line features [11–13], structural point/line features in the carrier's interior [14,15], etc. However, the carrier-borne aircraft landing process can be started from as far as more than 5 kilometers. For precise guidance, the runway line or internal structure features in images need to be clearly visible, making these methods [11–15] unable to adapt to the large distance span. In robotic operation, which is a similar application scenario involving large-span visual guidance, a monocular pose tracking method based on the target's own 3D model information is often used to guide the manipulator to grasp objects [16]. The non-cooperative carrier-guiding process based on airborne monocular vision mainly includes the carrier target detection, the carrier pose detection, and the carrier pose tracking. This paper focuses on the pose tracking part, aiming to propose a monocular pose tracking method merely based on the 3D model of the carrier to track carrier target pose parameters with high precision and stability.

According to the different image information used, the existing 3D-model-based monocular pose tracking methods can be classified into four categories [17]: direct method, feature-based method, edge-based method, and region-based method. In the direct method [18], the photometric consistency of continuous frames is the basis, and the pose change of the target is estimated through direct image alignment. Therefore, this method is extremely sensitive to dynamic light, noise, etc. For feature-based pose tracking methods, sufficient and clear textures are needed to provide robust feature points or feature lines [19]. However, when the carrier-borne aircraft is far away from the carrier, the carrier target in the airborne camera image is too small for this method to be applied. The edge-based method [20,21] generally samples a group of control points along the projected edge of a 3D model. Then, a 1D search is conducted on each sampling point along the normal direction to determine the corresponding relationship. The pose is tracked by minimizing the distance between the sampling edge point and its corresponding point. The region-based method [22–24] combines pose tracking with image segmentation. The pose parameters are optimized iteratively to maximize the segmentation energy. The region-based method performs well in dealing with clutter background, dynamic illumination, motion blur, and defocus. In the process of automatic landing guidance of carrier-borne aircraft, the aircraft gradually approaches the carrier target from a far end. Affected by the change of visual range, the carrier target in airborne imaging varies greatly in the whole process, as shown in Figure 1. Both the edge-based GOS algorithm [21] and the traditional region-based RBOT algorithm [24] fail to track the carrier in the scenario of large visual range changes in the landing application.

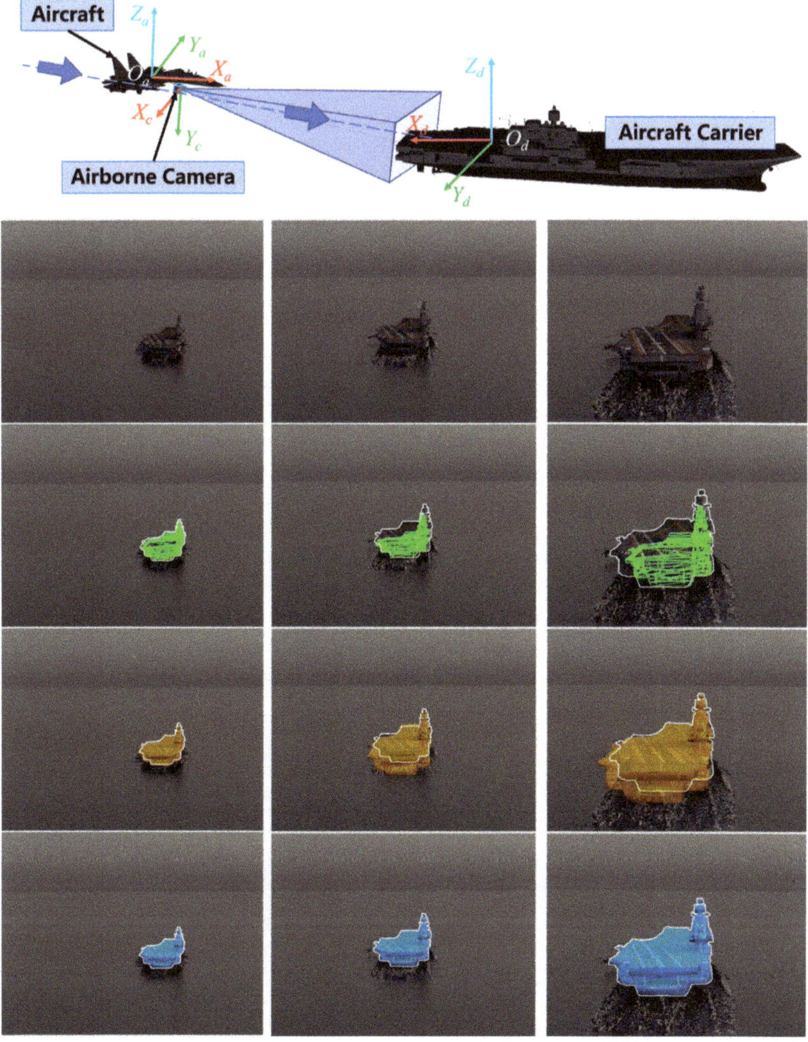

Figure 1. Pose tracking of aircraft carrier target during airborne monocular vision-guided landing. Top: Schematic diagram of carrier landing based on airborne monocular vision guidance. First row: Aircraft carrier targets in images captured by airborne camera. Second row: Pose tracking results of the GOS method [21]. Third row: Pose tracking results of the RBOT method [24]. Forth row: Pose tracking results of the proposed method.

Aiming at the problems illustrated above, this paper proposes a pose tracking algorithm based on scale-adaptive local region to stably track the pose of carrier target with large scale changes in the process of landing with large range span. The proposed method adopts the non-cooperative mode only using the 3D model of the carrier target, relying on the information of the local regions on the peripheral contour of the carrier, which is most stable in the landing process. The proposed method gives full consideration to the scale change of the carrier target, adaptively updating local region model parameters to realize robust pose tracking of the carrier target under the interference of wave background, sea reflection, and imaging blur.

The contributions of this paper are as follows: (1) A new monocular pose tracking method based on scale-adaptive local region is proposed, which achieves robust pose tracking for the carrier target with large scale variation in landing application scenarios, and achieves better performance than existing algorithms. (2) An innovative updating mechanism of local region model parameters, considering the target scale change in the image, is established to better maintain the continuity of color histogram distribution of the sampled local region near the target contour in continuous frames.

2. Scale-Adaptive Local Region-Based Monocular Pose Tracking Method

Focusing on the application of visual guidance for carrier-borne aircraft landing, this paper proposes a scale-adaptive local region-based pose tracking method to solve the problem of robust 6D pose tracking of the carrier target in the large range span scenario. As shown in Figure 2, on the premise that initial pose is given, firstly, the 2D contour of the carrier is rendered based on its 3D model with the initial pose. Then, a number of circular local regions are established from the uniformly sampled 2D image points on the projected contour. Meanwhile, the local region radius parameters are adjusted according to the pixel scale of the projected contour. Thus, a scale-adaptive local region pose tracking model constrained by pose parameters is constructed. Finally, the pose parameters are solved by iterative optimization so as to optimize the segmentation of the carrier target by maximizing the energy function and the 6D pose tracking of the carrier target in the consecutive frames with the cycle repeating.

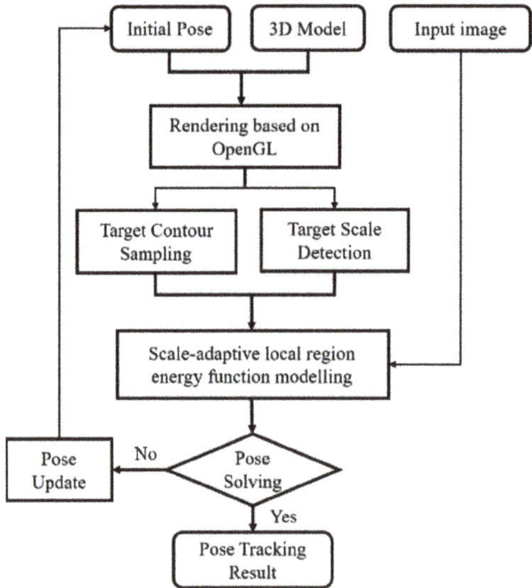

Figure 2. Flow chart of monocular pose tracking based on scale-adaptive local region.

In this section, the 6D pose tracking problem of the carrier is described in detail first. Then, the proposed pose tracking model based on scale-adaptive local region is introduced. Finally, the pose optimization solution process is deduced for the energy function.

2.1. Monocular Pose Tracking Problem

In the monocular pose tracking problem, the initial pose of the carrier is considered to be given. With the 3D CAD model of the carrier target, the problem is to continuously estimate and update the 6D pose of the target in the subsequent video frames. Monocular pose tracking is shown in Figure 3. The carrier's 3D CAD model used in this paper is a dense surface model (triangular mesh) composed of 3D vertices, which can be expressed as $X_m = \left(x_m, \ y_m, \ z_m \right)^{\mathrm{T}} \in \mathbb{R}^3, m = 1, \ldots, n$.

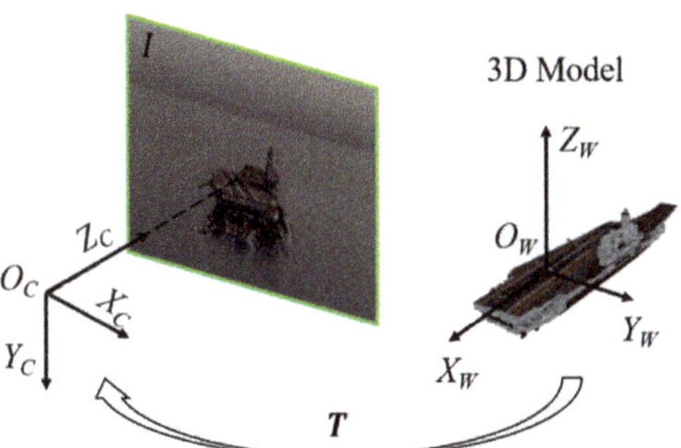

Figure 3. Coordinate system transformation in monocular pose tracking problem.

In Figure 3, I stands for the input image. T is the rigid body transformation matrix from model coordinate $O_W - X_W Y_W Z_W$ to the camera coordinate $O_C - X_C Y_C Z_C$, which is defined by rotation matrix $R \in \mathbb{SO}(3)$ and translation vector $t \in \mathbb{R}^3$.

$$T = \begin{bmatrix} R & t \\ 0 & 1 \end{bmatrix} \in \mathbb{SE}(3) \tag{1}$$

In the continuous image sequence, the rigid body transformation matrix changes constantly with the movement of the model relative to the camera, and pose tracking is actually an iterative update of the transformation matrix. If the pose of the k-th frame is $T_k = \Delta T T_{k-1}$, $\Delta T = T_k T_{k-1}^{-1}$ is the pose update from the previous frame to the k-th frame. During pose tracking, the pose of each frame can be obtained merely by calculating the pose update between frames. In order to facilitate the nonlinear optimization of the transformation matrix solution in continuous frames, the transformation matrix is represented by 6D pose vector p, represented by Lie algebra [25].

$$p = \begin{bmatrix} w \\ v \end{bmatrix} = (\omega_1, \omega_2, \omega_3, v_1, v_2, v_3)^{\mathrm{T}} \in \mathbb{R}^6 \tag{2}$$

The pose update of Lie algebra pose vector $\hat{p}$ is mapped to the transformation matrix ΔT as follows:

$$\Delta T = \exp(\hat{p}) \in \mathbb{SE}(3), \, \hat{p} = \begin{bmatrix} \hat{w} & v \\ 0 & 0 \end{bmatrix} \in se(3), \hat{w} \in so(3), v \in \mathbb{R}^3 \tag{3}$$

where $\hat{w} = \begin{bmatrix} 0 & -w_3 & w_2 \\ w_3 & 0 & -w_1 \\ -w_2 & w_1 & 0 \end{bmatrix}$ is the corresponding antisymmetric matrix of w.

Assume that the camera has been pre-calibrated and its intrinsic parameters are fixed. The intrinsic parameter matrix K is shown below:

$$K = \begin{bmatrix} f_x & 0 & C_x \\ 0 & f_y & C_y \\ 0 & 0 & 1 \end{bmatrix} \in \mathbb{R}^{3 \times 3} \tag{4}$$

After nonlinear distortion correction, the input image is considered undistorted. Perspective projection from 3D vertices to 2D image points can be described as

$$x = \pi \left(K \left(T\tilde{X} \right)_{3 \times 1} \right) \tag{5}$$

where x is the 2D projected point of the 3D model vertex X. $\tilde{X} = (x, y, z, 1)^{\mathrm{T}}$ is the homogeneous extension of $X = (x, y, z)^{\mathrm{T}} = (\tilde{X})_{3\times 1}$, and $\pi(X) = (x/z, y/z)^{T}$. Based on the initial pose, the projection mask I_s is generated by rendering the 3D model. Then, the whole image can be divided into foreground region and background region by extracting the contour on the projection mask. Constrained by the 3D model of the carrier target, when the 6D pose parameters are correct, the projected contour C can perfectly segment the carrier target in the image.

2.2. Pose Tracking Model based on Scale-Adaptive Local Region

In the case of a given initial pose, region-based monocular pose tracking updates the transformation matrix T_k by continuously solving the pose change ΔT in successive frames. Existing region-based methods adopt a fixed local region model parameter, which is difficult to adapt to the application concerned in this paper. In view of this, a new pose tracking model based on scale-adaptive local region is proposed. The specific structure of the proposed model is shown in Figure 4. For the local region-based method, the target in the image is represented by the level set embedding function (or signed distance function), as shown in Equation (6). The signed distance function $\Phi(x)$ computes the (signed) Euclidean distance between a pixel x in an image and its nearest contour point. Then, the projected contour C is defined as a set of zero elevation points $C = \{x | \Phi(x) = 0\}$.

$$\Phi(x) = \begin{cases} -d(x, C), \forall x \in \Omega_f \\ d(x, C), \forall x \in \Omega_b \end{cases} \tag{6}$$

where, $d(x, C) = \min_{c \in C} \|c - x\|_2$.

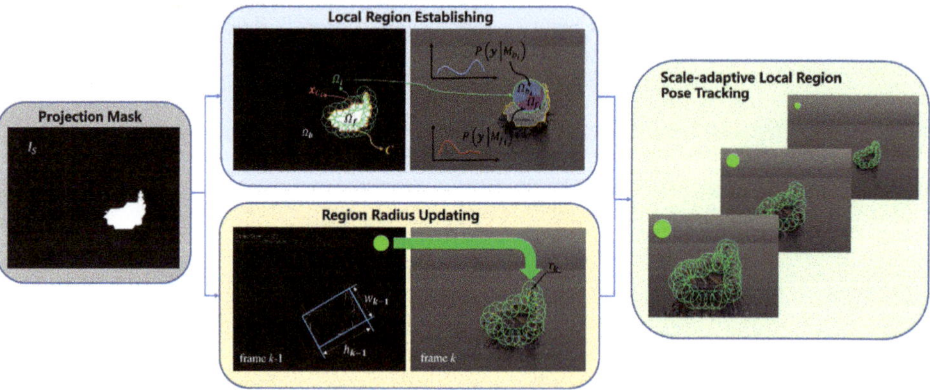

Figure 4. Pose tracking model based on scale-adaptive local region.

We define the region of $\Phi(x) < 0$ in the image as the foreground region Ω_f, while the region of $\Phi(x) > 0$ is the background region Ω_b, and the foreground and the background regions are divided by the projected contour C, as shown in the left graph of "Local Region Establishing" in Figure 4. In order to ensure the correspondence of local regions in the image sequence, we screen the region centers x_{C_i} in the 3D vertices X_m of the target model near the projected contour C. The regional centers are selected to be as evenly distributed as possible on the contour, so as to determine the local circular regions Ω_i. Local circular regions Ω_i are then divided into the foreground part Ω_{f_i} and the background part $\Omega_{b_i} = \Omega_i \backslash \Omega_{f_i}$. For each local region Ω_i, the foreground and background statistical characteristics in the region are usually represented by foreground color appearance model $P\left(y \mid M_{f_i}\right)$ and background color appearance model $P(y \mid M_{b_i})$, where $y = I(x)$ is the RGB value of a pixel x in the image. The color appearance models describe the probability that the pixel color y meets the color distribution of the foreground local region and the background local region, respectively, which are represented by the region RGB color histogram, as shown in the right graph of "Local Region Establishing" in Figure 4.

For each local region, counting the foreground/background color appearance model of the region, the posterior probability of local area pixels can be obtained, which is calculated as follows:

$$P_{f_i}(x) = P\left(M_{f_i} \mid y\right) = \frac{P\left(y \mid M_{f_i}\right)}{\eta_{f_i} P\left(y \mid M_{f_i}\right) + \eta_{b_i} P(y \mid M_{b_i})} \tag{7}$$

$$P_{b_i}(x) = P(M_{b_i} \mid y) = \frac{P(y \mid M_{b_i})}{\eta_{f_i} P\left(y \mid M_{f_i}\right) + \eta_{b_i} P(y \mid M_{b_i})} \tag{8}$$

where

$$\eta_{f_i} = \sum_{x \in \Omega_i} H_e(\Phi(x)), \quad \eta_{b_i} = \sum_{x \in \Omega_i} (1 - H_e(\Phi(x))) \tag{9}$$

$$H_e(\Phi(x)) = \frac{1}{\pi}\left(-atan(s \cdot \Phi(x)) + \frac{\pi}{2}\right) \tag{10}$$

$P_{f_i}(x)$ and $P_{b_i}(x)$ are the foreground pixelwise posterior probability and the background pixelwise posterior probability of the pixel x in the i-th local region Ω_i, respectively. H_e is a smoothed Heaviside step function. s is the slope index; here, $s = 1.2$.

In the iteration of pose parameters in the consecutive frame sequence, the color histogram of the foreground and background corresponding to the local region of the same center x_{C_i} possesses continuity in the adjacent frames. Based on this, we refer to the recursive updating strategy of region appearance model in [13,15] in the practical implementation. After the $(k-1)$-th frame is successfully tracked, the color appearance models $P^K\left(y \mid M_{f_i}\right)$ and $P^K(y \mid M_{b_i})$ of the local background in the k-th frame are established as partial inheritance of the color appearance model of the corresponding region in the previous frame, and the update strategy is as follows:

$$P^k\left(y \mid M_{f_i}\right) - \left(1 - \alpha_f\right)P^{k-1}\left(y \mid M_{f_i}\right) + \alpha_f P^k\left(y \mid M_{f_i}\right) \tag{11}$$

$$P^k(y \mid M_{b_i}) = (1 - \alpha_b)P^{k-1}(y \mid M_{b_i}) + \alpha_b P^k(y \mid M_{b_i}) \tag{12}$$

α_f and α_b are the learning rate of the foreground region and background region. To effectively cope with imaging jitter and avoid the jump of pose parameters to improve the tracking result, we set $\alpha_f = 0.1$, $\alpha_b = 0.2$.

In the fusion of statistical models of all local regions, pixels near the target contour will belong to several local regions at the same time due to the overlap between different local regions. Similar to [15], we choose to calculate the average posterior probability of pixels in all local regions to which they belong. In the k-th frame, for each pixel x in the image, the mean posterior probability of foreground and mean posterior probability of background are calculated as follows:

$$\overline{P}_f^k(x) = \frac{1}{\sum_{i=1}^n B_i^k(x, x_{C_i})} \sum_{i=1}^n P_{f_i}^k(x)B_i^k(x, x_{C_i}) \tag{13}$$

$$\overline{P}_b^k(x) = \frac{1}{\sum_{i=1}^n B_i^k(x, x_{C_i})} \sum_{i=1}^n P_{b_i}^k(x)B_i^k(x, x_{C_i}) \tag{14}$$

where

$$B_i^k(x, x_{C_i}) = \begin{cases} 1, & |x - x_{C_i}| < r^k \\ 0, & else \end{cases} \tag{15}$$

r^k is the radius of the local region Ω_i^k in the k-th frame. x_{C_i} is the corresponding region center.

In order to adapt to the change of carrier target scale caused by the variation of visual distance between carrier target and airborne camera, the size of the local region is set to increase with the increase of target scale to maintain the continuity of foreground background sampling in the image sequence within the regions. Too-small region radius will not be able to cope with large target pose changes, while too-large area radius will lead to poor pose convergence accuracy. Therefore, the adjustment of the region radius needs to be within a certain range. Considering the complexity of the 3D structure of the carrier target, the target scale is defined as the pixel length of the shorter side of the minimum enclosing rectangle of the target's projected contour. Assuming that the carrier target is successfully tracked, the target scales in adjacent frames can be regarded as approximately unvaried. First, the target scales of the carrier target in the previous frame are detected, as shown in the "Region Radius Updating" part in Figure 4. Then, according to the detected target scale, the region parameters are updated in the pose optimization of the current frame.

Sigmoid function is utilized to design the updating mechanism of radius parameters, which can not only limit the variation range of region radius, but also ensure the approximate linear relationship between local region radius and the change of target scale within a certain range. The specific form is as follows:

$$r^k = \frac{c}{1 + e^{-a[\min(w^{k-1},h^{k-1})-b]}} + d \tag{16}$$

w^{k-1} and h^{k-1} are the width and height of the minimum enclosing rectangle of the projected contour of the target 3D model in the $(k\text{-}1)$-th frame. a is the sensitivity coefficient. b is the scale offset. The higher the sensitivity coefficient is, the more drastic the radius changes with the target scale. The sensitivity coefficient is normally set between 0.02 to 0.2. Here, $a = 0.04, b = 150$. c is the length of radius variation interval, and d is the minimum radius parameter. Here, according to the specific application scenario, the radius change interval is set at $[10, 70]$.

Then, the local region-based energy function for pose tracking can be expressed as

$$E(p) = -\sum_{x \in \Omega} \log\left[H_e(\Phi(x(p)))\overline{P}_f(x) + (1 - H_e(\Phi(x(p))))\overline{P}_b(x)\right] \tag{17}$$

This energy function takes the target pose parameters p as independent variables and quantitatively describes the target segmentation performance under the constraints of the 3D model, that is, the degree of coincidence between the rendered 2D template and the target region in the image. When the energy is maximum, the segmentation result of the target region is best.

2.3. Pose Optimization

According to the construction of energy function given in Section 2.2, in order to solve the optimal pose, this paper adopts the Gauss–Newton pose optimization method to solve this complex nonlinear optimization problem by referring to [24]. First, Equation (17) is reconstructed into a nonlinear iterative reweighted least squares problem of the following form:

$$E(p) = \frac{1}{2}\sum_{x \in \Omega} \psi(x)F^2(x,p) \tag{18}$$

where

$$F(x,p) = -\log\left[H_e(\Phi(x(p)))\overline{P}_f(x) + (1 - H_e(\Phi(x(p))))\overline{P}_b(x)\right], \; \psi(x) = \frac{1}{F(x,p)} \tag{19}$$

In order to optimize the energy function, we first consider the weight coefficient $\psi(x)$ as the fixed weight and use the Gauss–Newton method to solve the pose parameters p, then use the optimized pose parameters p to calculate the weight coefficient $\psi(x)$ to update the weight value and enter the next iterative solution.

Under the assumption of fixed weights, the derivative of the energy function with respect to the pose p is obtained, and the gradient function is as follows:

$$
\begin{aligned}
\frac{\delta E(p)}{\delta p} &= \frac{1}{2} \sum_{x \in \Omega} \psi(x) \frac{\delta F^2(x,p)}{\delta p} \\
&= \sum_{x \in \Omega} \psi(x) F(x,p) \frac{\delta F(x,p)}{\delta p}
\end{aligned}
\tag{20}
$$

Since $\psi(x)F(x,p) = 1$, the Jacobian is computed as follows:

$$
\begin{aligned}
J(x) &= \frac{\partial F(x,p)}{\partial p} \\
&= -\frac{\overline{p}_f - \overline{p}_b}{H_e(\Phi(x))\overline{p}_f + (1 - H_e(\Phi(x)))\overline{p}_b} \times \frac{\partial H_e(\Phi(x))}{\partial p}
\end{aligned}
\tag{21}
$$

$$
\frac{\partial H_e(\Phi(x))}{\partial p} = \frac{\partial H_e}{\partial \Phi} \frac{\partial \Phi}{\partial x} \frac{\partial x}{\partial p}
\tag{22}
$$

in which, $\frac{\partial H_e}{\partial \Phi} = \delta_e(\Phi)$ is the smoothed Dirac function. According to Equation (10),

$$
\delta_e(\Phi(x)) = \frac{\partial H_e(\Phi(x))}{\partial \Phi} = -\frac{s}{\pi(1 + s^2\Phi^2(x))}
\tag{23}
$$

For the derivative of the level set embedding function with respect to pixel position $\frac{\partial \Phi}{\partial x}$, the central difference method is used:

$$
\frac{\partial \Phi}{\partial x} = \left[\frac{\partial \Phi}{\partial x}, \frac{\partial \Phi}{\partial y}\right] = \left[\begin{array}{c} \frac{\Phi(x+1,y) - \Phi(x-1,y)}{2} \\ \frac{\Phi(x,y+1) - \Phi(x,y-1)}{2} \end{array}\right]^{\mathrm{T}}
\tag{24}
$$

Since $x(p) = \pi\left(K\left(\exp(\hat{p})T\widetilde{X}\right)_{3\times1}\right)$, $\frac{\partial x}{\partial p}$ can be deduced. Suppose that after tiny motion, we perform piecewise linearization of the matrix exponential in each iteration, and obtain $\exp(\hat{p}) \approx I_{4\times4} + \hat{p}$:

$$
\begin{aligned}
\frac{\partial x}{\partial p} &= \frac{\partial \pi}{\partial K(\exp(\hat{p})T\widetilde{X})_{3\times1}} \frac{\partial K(\exp(\hat{p})T\widetilde{X})_{3\times1}}{\partial p} \\
&= \left[\begin{array}{ccc} \frac{f_x}{z'} & 0 & -\frac{x'f_x}{(z')^2} \\ 0 & \frac{f_y}{z'} & -\frac{y'f_y}{(z')^2} \end{array}\right] \cdot \left[\begin{array}{cccccc} 0 & z' & -y' & 1 & 0 & 0 \\ -z' & 0 & x' & 0 & 1 & 0 \\ y' & -x' & 0 & 0 & 0 & 1 \end{array}\right]
\end{aligned}
\tag{25}
$$

where $X' = (x', y', z')^{\mathrm{T}} = (T\widetilde{X})_{3\times1}$.

Through the derivation of the partial derivatives of the items in the Jacobian matrix above, the Hessian matrix and the pose change can be obtained as follows,

$$
H(x) = \psi(x)J(x)^{\mathrm{T}}J(x)
\tag{26}
$$

$$
\Delta p = -\left(\sum_{x \in \Omega} H(x)\right)^{-1} \sum_{x \in \Omega} J(x)^{\mathrm{T}}
\tag{27}
$$

By iteratively solving the pose update Δp for each input frame of the image sequence, the pose tracking of the target is realized.

3. Experimental Evaluation

3.1. Experiment Setting

To validate the pose tracking performance of the proposed method in this paper on the carrier target, this section simulates the imaging of the carrier target by the airborne camera in the process of carrier landing. During landing, the flight attitude of the carrier-borne aircraft needs to be controlled with high precision. In order to ensure a smooth landing and a successful rope hanging, the aircraft flies at a fixed angle of descent, which is around 3–5°, to maintain the angle of attack. Based on this, this section uses a synthetic simulation image sequence generated by a simulation rendering software 3Ds max and a real image sequence captured through a scale physical simulation system to verify the effectiveness of the proposed algorithm in this paper.

We compared the proposed algorithm with the existing representative region-based pose tracking method RBOT algorithm [24] and the edge-based pose tracking method GOS algorithm [21] on the two image sequences, respectively. For the comparison algorithms, we adopted the default parameter settings suggested in their papers. All experiments were carried out on a laptop equipped with an AMD Ryzen 7 4800H processor (8-core) @ 2.9 GHz, NVIDIA GeForce GTX1650 GPU and 16 GB RAM. The GPU was only used for the rendering of the 3D model. The rest of the algorithm runs on the CPU. The proposed algorithm is implemented in C++. For the synthetic image sequence with a resolution of 800×600 and the real image sequence with a resolution of 800×800, the proposed algorithm can reach the processing speed of 30 fps, which can basically meet the requirements of high efficiency of real-time pose tracking for the carrier target in the process of carrier-borne aircraft landing.

3.2. Experiment on Synthetic Image Sequence

This section uses simulation rendering software to generate a synthetic image sequence to simulate the imaging of the carrier target by airborne camera during the landing process. To quantitatively measure the pose tracking accuracy, the metrics and corresponding results and analysis are given in this section.

3.2.1. Synthetic Sequence Settings

The simulation software used in this section is Autodesk 3Ds Max, which is often used in 3D modeling, animation, and scene rendering. In the simulation rendering of carrier-borne aircraft landing, the carrier's 3D model is the "Varyag" model downloaded on the internet. Table 1 shows the 3D model size parameters and 3Ds Max simulation parameters.

Table 1. Parameter settings in 3Ds Max simulation.

3D Model Size	Focal Length	Field Angle	Frame Size
94.36 m × 66.01 m × 304.51 m	12 mm	24° × 18°	800 × 600

Through the settings of the virtual camera parameter and its trajectory, as well as the rendering of the of dynamic sea surface scene including dynamic light, sea waves, and sea horizon simulation, we generated an image sequence consisting of 350 frames. The synthetic sequence simulates aircraft approaching the ship deck from about 1120 m. With the shortening of the distance, the ship target in the image becomes larger and larger, with the details of the ship target clearer and clearer. The camera intrinsic parameters K, initial pose T_0, and the ship 3D model were used as input to initialize the pose tracking algorithm. The proposed algorithm optimizes the pose variation of the ship target between successive frames by establishing local region models on the projected contour of the 3D model to maximize the energy function for the optimal segmentation, so that the pose parameters of the ship target can be continuously solved in the image sequence to achieve pose tracking. Examples of the synthetic images (the 2nd, 147th, 208th, 290th, and 335th frames) are shown in the first row of the upper half in Figure 5.

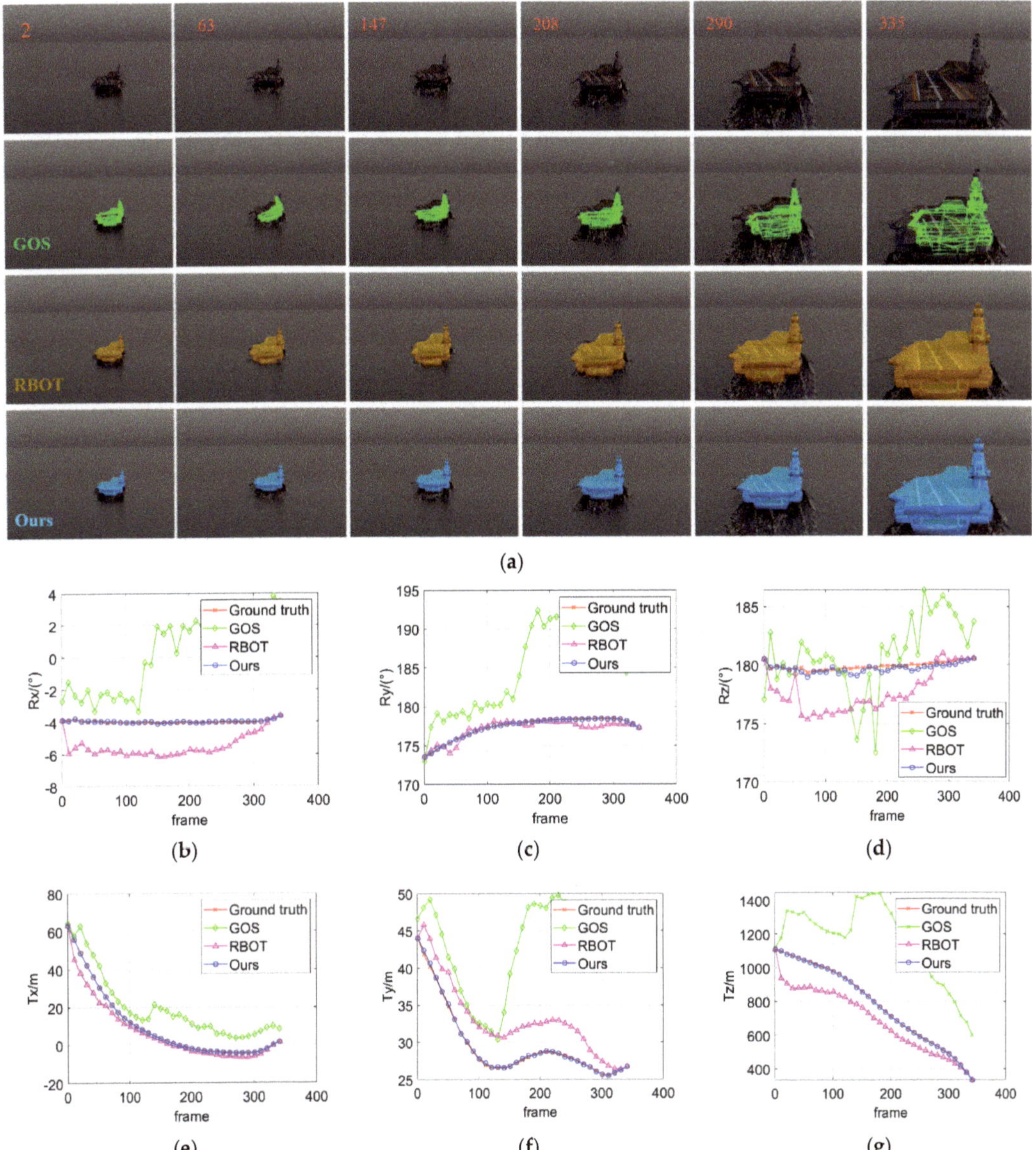

Figure 5. Pose tracking results of carrier target in synthetic image sequence. (**a**) The reprojected results (the first row shows the examples of the synthetic images (the 2nd, 63rd, 147th, 208th, 290th, and 335th frame). The second to the fourth rows show the pose tracking results of GOS method, RBOT method, and the proposed method). (**b**–**g**): Curves of three Euler angles and the translation parameters.

3.2.2. Experimental Results and Analysis

For visualization, the reprojected mask and the pose curves of the pose tracking results are presented in Figure 5. The second to fourth rows are the reprojected results of GOS algorithm [21], RBOT algorithm [24], and the proposed algorithm for the ship target pose tracking. The bottom part of Figure 5 shows the pose curves of the three algorithms. As can be seen from Figure 5, in the synthetic image sequence, the proposed method achieved whole-process stable pose tracking of the ship target. Owing to the ship target's complex interior structure, and its low color distinctiveness under the background of dynamic sea surface, the GOS algorithm [21] failed to track the pose of the ship from the fifth frame when the ship target is small. With the pose error accumulated in the process of tracking, although the target increased gradually, the pose tracking result kept becoming worse. The GOS algorithm [21] searches the corresponding points between frames on the normal line segment for discrete points at the edge of the ship target. This method is based on edge features, and strongly relies on the gradient information on the normal line segment to achieve stable pose tracking for untextured or weakly textured targets. For the ship target with complex internal textures, the 1D search on the normal lines in the GOS algorithm [21] is easily disturbed by the maximum value of internal gradient response, resulting in mismatching and errors in pose solution. The RBOT algorithm [24] also tended to fail tracking in the first five frames of the image sequence. However, according to the reprojected results, the problem centered on the segmentation of the stern part of the ship target, while the segmentation of the bow and tower part behaved well. Firstly, the interference of the aft waves of the ship weakens the foreground–background distinctiveness in the stern regions. Secondly, although the RBOT algorithm [24] applies the region features around the target contour, it adopts the local regions with fixed radius, which fails to adapt to the scale variation of the ship target in the landing application.

Aiming at the specific scenario for carrier landing, we considered the scale variation of the ship target caused by drastic changes in visual range. The proposed pose tracking method adaptively adjusts the local region radius parameter according to the scale of the ship target in the image, ensuring the continuity of the color histograms of corresponding local regions between adjacent frames. In this way, the energy function optimization in image sequence achieves better target segmentation effect by using the continuity information between frames more effectively. Figure 6 shows the process of local region radius adjustment with ship target scale. When the ship target is small, the smaller region radius segments the ship more accurately in terms of the wave background interference. When the scale of the ship target gradually largens with the decrease of the visual distance, the proposed algorithm increases the region radius to better maintain the color distribution characteristics of the front background in the local regions along the contour. As shown in Figure 5, the blue reprojected mask corresponds to the ship target in the whole process. Additionally, pose curves coincide with the ground truth curves. It is indicated that the proposed algorithm realizes robust pose tracking of the ship target, adapting to the target scale changes, the sea background, and the light interference.

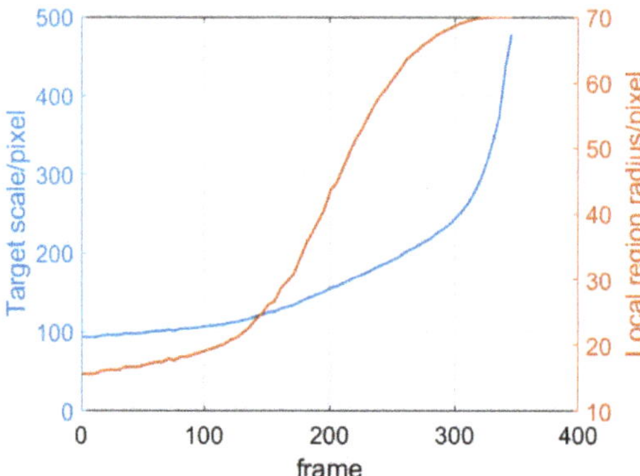

Figure 6. Target scale and local region radius change curve in the synthetic image sequence.

To quantify the accuracy of pose tracking results, the pose tracking results of the algorithm were compared with the pose truth value to calculate the angular error e_R and relative position error $e_{t\_relative}$, as shown below:

$$e_R(t_k) = \cos^{-1}\left(\frac{trace(\boldsymbol{R}(t_k)^{\mathrm{T}}\boldsymbol{R}_{GT}(t_k) - 1}{2}\right) \tag{28}$$

$$e_{t\_relative}(t_k) = \frac{\|\boldsymbol{t}(t_k) - \boldsymbol{t}_{GT}(t_k)\|_2}{\|\boldsymbol{t}_{GT}(t_k)\|_2} \tag{29}$$

$\boldsymbol{R}(t_k)$ and $\boldsymbol{t}(t_k)$ are the rotation matrix and translation vector of the pose tracking result in t_k moment. $\boldsymbol{R}_{GT}(t_k)$ and $\boldsymbol{t}_{GT}(t_k)$ are the ground truth value in t_k moment.

Since the landing process is a large span approaching process between carrier-borne aircraft and carrier target, the absolute displacement error is greatly affected by the distance, so the relative error is chosen to measure the accuracy. The error of pose tracking results is shown in Figure 7. In the synthetic image sequence test, the angle error of pose tracking of the proposed algorithm is no more than 1° in the whole process, and the relative displacement error does not exceed 1%. Synthetic image sequence experiments verify the effectiveness of the proposed algorithm in pose tracking of the carrier target with high precision during the landing process.

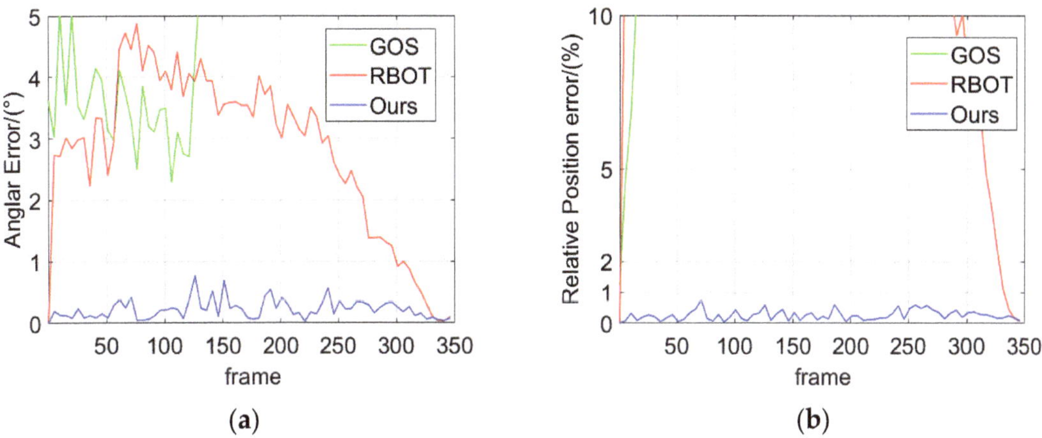

Figure 7. Error of the pose tracking results. (**a**) Angular error. (**b**) Relative position error.

3.3. Experiment on Real Image Sequence

3.3.1. Scale Physical Simulation Platform

In the synthetic image sequence, although the dynamic sea surface scene is simulated, there is still a certain gap between the virtual synthetic images and the practical images taken by the industrial camera, which includes noise interference, defocus blur, and motion blur. In order to further test the performance of the algorithm in the actual image sequence, according to the fixed slide angle setting in real carrier landing, we built the scale simulation sliding platform of the carrier landing, as shown in Figure 8.

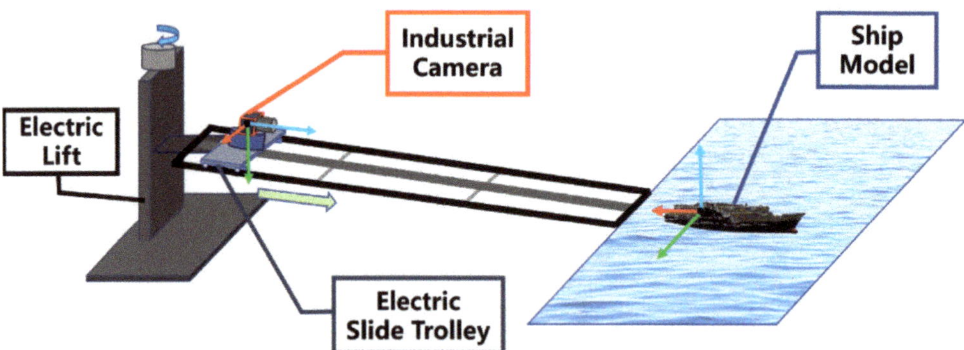

Figure 8. Scale simulation experiment platform of carrier landing.

The industrial color camera is fixed on the electric slide trolley. One end of the track is close to the seascape sand table with the carrier scale model. The other end is supported on the electric lifting platform. The electric lift platform can adjust the angle of descent of the glide track by changing the height of one end of the track. In this experiment, the track length is about 4 m, and the gliding angle of the track is set as 3°. The ratio of actual carrier size to carrier model is 700:1. The scale simulation parameter settings are shown in Table 2.

Table 2. Experimental settings of scale simulation.

Carrier Model Scale	Track Length	Focal Length	Frame Rate	Frame Size
700:1	4 m	12 mm	30 fps	800×800

The CAD model of the carrier model was scanned and reconstructed by high-precision 3D scanning equipment, and the intrinsic parameters of the industrial camera were calibrated in advance using Zhang's calibration method [26]. A total of 1190 frames of images were captured by the industrial camera during the sliding process of the electric trolley with a constant speed. In order to obtain the initial pose parameters of the carrier target, a number of 2D feature points are selected in the image, and then the corresponding 3D coordinates of the feature points are obtained by using the total station and converting into the carrier model coordinate system. With the 2D–3D points correspondence, the initial pose is solved by the PnP algorithm [27]. Similar to Section 2.2, initial pose and camera intrinsic parameters were used as the input to initialize the algorithm. Since the GOS algorithm [21] performs poorly, we only compare RBOT algorithm with our algorithm in the real image sequence test in this section. The reprojected results and pose curves are shown in Figure 9.

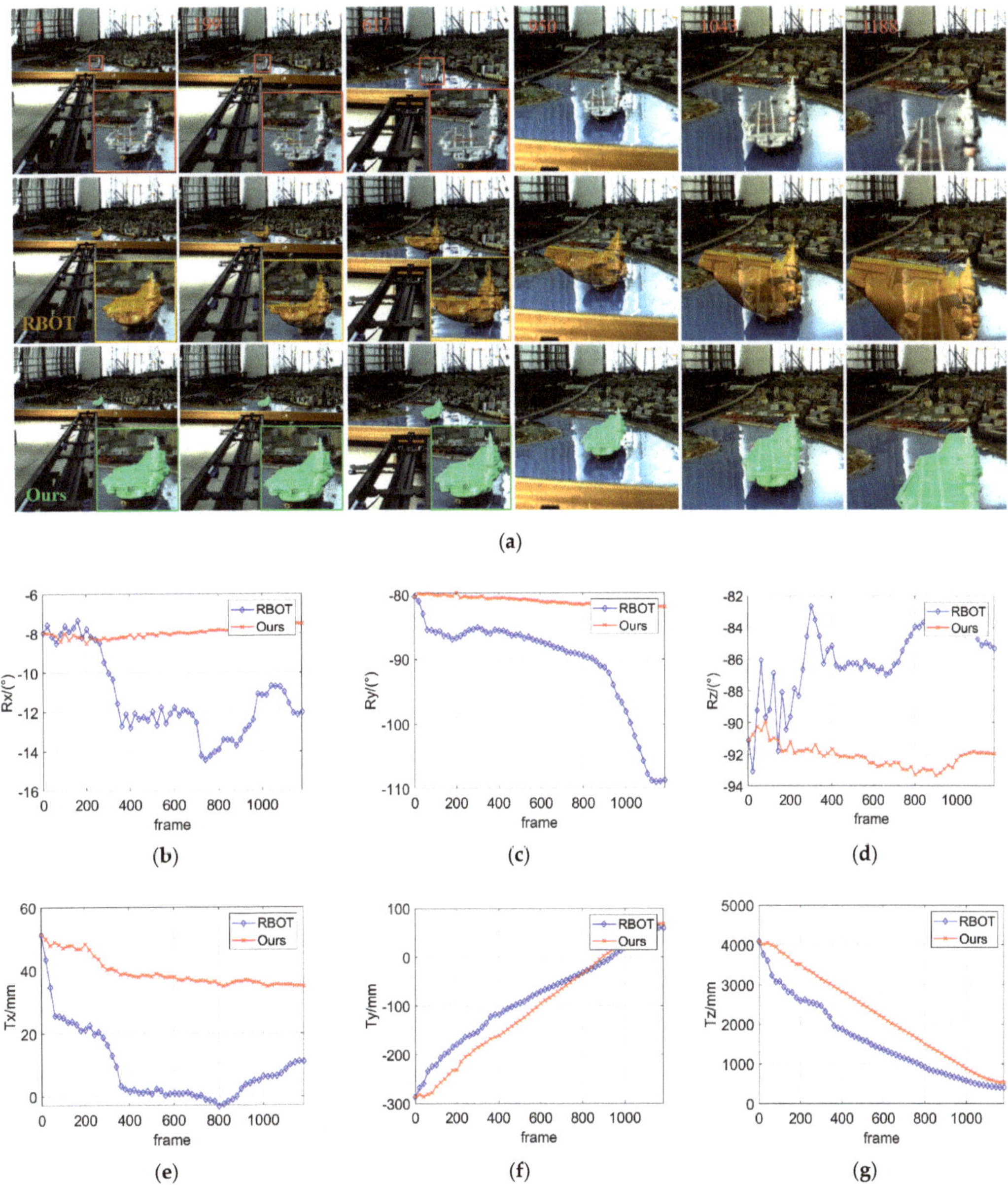

(a)

(b)

(c)

(d)

(e)

(f)

(g)

Figure 9. Pose tracking results of carrier target in real image sequence. (**a**) The reprojected results (the first row shows the examples of the synthetic images (the 4th, 199th, 617th, 950th, 1043rd, and 1188th frame). The second and the third rows show the pose tracking results of the RBOT method, and the proposed method, respectively). (**b–g**): Curves of three Euler angles and the translation parameters.

3.3.2. Experimental Results and Analysis

As shown in Figure 9, in the real image sequence of scale physical simulation, the actual imaging is degraded by defocus blur, motion blur, and environmental interference, such as "sea surface" illumination reflection and cluttered background. According to the pose tracking results, the reprojected mask of the proposed algorithm can accurately match the carrier target in the whole process of image sequence. However, for the RBOT algorithm [24], the pose tracking results showed deviation after initialization immediately. As the carrier target gradually become larger, it failed to timely correct the accumulated tracking error, and the pose tracking deviation also further expanded. Figure 10 shows the process of local area radius changing with the carrier target scale. In the real image sequence, due to the large range span, the scale of the carrier target varies from 50 pixels to 600 pixels. The scale variation of the carrier target is larger than that in the synthetic image sequence. In terms of the large scale variation, the proposed algorithm updates the local area radius in real time according to the target scale, and still realizes the whole-process stable 6D pose tracking of the carrier target. The scale physical simulation experiment again verifies the effectiveness of the proposed algorithm in achieving robust pose tracking for the carrier target in carrier landing application.

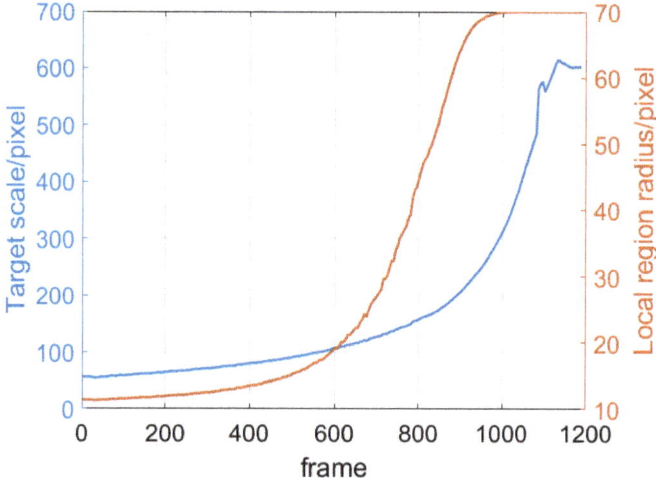

Figure 10. Target scale and local region radius change curve in the real image sequence.

4. Conclusions

Aiming at the carrier landing scenario with monocular airborne visual guidance, this paper proposes a carrier target pose tracking method based on scale-adaptive local region. By combining the carrier target scale variation characteristics caused by visual distance change in the landing process, a pose tracking model based on local regions is set up with the local region radius updating adaptively. Then, using the Gauss–Newton optimization algorithm, the 6D pose tracking of the carrier target is achieved. Experiments on both synthetic image sequence and real image sequence verify that the proposed method can achieve accurate and robust pose tracking for the carrier target in the process of carrier aircraft approaching and landing from far to near.

Author Contributions: Conceptualization, J.Z. and X.S.; methodology, J.Z. and Q.W.; software, J.Z.; validation, J.Z., Q.W. and Z.Z.; formal analysis, J.Z., Q.W.; investigation, X.S.; resources, J.Z.; data curation, J.Z. and Z.Z.; writing—original draft preparation, J.Z.; writing—review and editing, Q.W., and X.S.; visualization, J.Z., Q.W. and Z.Z.; supervision, X.S.; project administration, X.S.; funding acquisition, J.Z. and X.S. All authors have read and agreed to the published version of the manuscript.

Funding: This research was funded by Postgraduate Scientific Research Innovation Project of Hunan Province (CX20200024, CX20200025), and National Natural Science Foundation of China (62003357).

Data Availability Statement: Not applicable.

Conflicts of Interest: The authors declare no conflict of interest.

References

1. Wu, H.; Luo, F.; Shi, X.; Liu, B.; Lian, X. Analysis on the Status Quo and Development Trend of Automatic Carrier Landing Technology. *Aircr. Des.* **2020**, 1–5. (In Chinese) [CrossRef]
2. Zhen, Z.; Wang, X.; Jiang, J.; Yang, Y. Research progress in guidance and control of automatic carrier landing of carrier-based aircraft. *Acta Aeronaut. Astronaut. Sin.* **2017**, *38*, 22. (In Chinese)
3. Wei, Z. Overview of Visual Measurement Technology for Landing Position and Attitude of Carrier-Based Aircraft. *Meas. Control. Technol.* **2020**, *39*, 2–6. (In Chinese)
4. Xu, G.; Cheng, Y.; Shen, C. Key Technology of Unmanned Aerial Vehicle's Navigation and Automatic Landing in All Weather Based on the Cooperative Object and Infrared Computer Vision. *Acta Aeronaut. Astronaut. Sinica* **2008**, *2*, 437–442. (In Chinese)
5. Wang, X.-H.; Xu, G.-L.; Tian, Y.-P.; Wang, B.; Wang, J.-D. UAV's Automatic Landing in All Weather Based on the Cooperative Object and Computer Vision. In Proceedings of the 2012 Second International Conference on Instrumentation, Measurement, Computer, Communication and Control, Harbin, China, 8–10 December 2012.
6. Gui, Y.; Guo, P.; Zhang, H.; Lei, Z.; Zhou, X.; Du, J.; Yu, Q. Airborne Vision-Based Navigation Method for UAV Accuracy Landing Using Infrared Lamps. *J. Intell. Robot. Syst.* **2013**, *72*, 197–218. [CrossRef]
7. Wang, G.; Li, H.; Ding, W.; Li, H. Technology of UAV Vision-guided Landing on a Moving Ship. *J. China Acad. Electron. Inf. Technol.* **2012**, *7*, 274–278. (In Chinese)
8. Chen, L. Research on the Autonomous Landing Flight Control Technology for UAV Based on Visual Servoing. Master's Thesis, Nanjing University of Aeronautics and Astronautics, Nanjing, China, 2009. (In Chinese).
9. Hao, S.; Chen, Y.M.; Ma, X.; Wang, T.; Zhao, J.T. Robust corner precise detection algorithm for visual landing navigation of UAV. *J. Syst. Eng. Electron.* **2013**, *35*, 1262–1267. (In Chinese)
10. Wei, X.H.; Tang, C.Y.; Wang, B.; Xu, G.L. Three-dimensional cooperative target structure design and location algorithm for vision landing. *Syst. Eng.-Theory Pract.* **2019**, *39*, 9. (In Chinese)
11. Zhuang, L.; Han, Y.; Fan, Y.; Cao, Y.; Wang, B.; Qin, Z. Method of pose estimation for UAV landing. *Chin. Opt. Lett.* **2012**, *10*, S20401–S320404. [CrossRef]
12. Anitha, G.; Kumar, R.N.G. Vision Based Autonomous Landing of an Unmanned Aerial Vehicle. *Procedia Eng.* **2012**, *38*, 2250–2256. [CrossRef]
13. Zhou, L.M.; Zhong, Q.; Zhang, Y.Q.; Lei, Z.H.; Zhang, X.H. Vision-based landing method using structured line features of runway surface for fixed-wing unmanned aerial vehicles. *J. Natl. Univ. Def. Technol.* **2016**, *38*, 9. (In Chinese)
14. Coutard, L.; Chaumette, F.; Pflimlin, J.M. Automatic landing on aircraft carrier by visual servoing. In Proceedings of the 2011 IEEE/RSJ International Conference on Intelligent Robots and Systems, San Francisco, CA, USA, 25–30 September 2011.
15. Bi, D.; Huang, H.; Fan, J.; Chen, G.; Zhang, H. Non-cooperative Structural Feature Matching Algorithm in Visual Landing. *J. Nanjing Univ. Aeronaut. Astronaut.* **2021**, *53*, 7. (In Chinese)
16. Zhou, J.; Wang, Z.; Bao, Y.; Wang, Q.; Sun, X.; Yu, Q. Robust monocular 3D object pose tracking for large visual range variation in robotic manipulation via scale-adaptive region-based method. *Int. J. Adv. Robot. Syst.* **2022**, *19*, 7778–7785. [CrossRef]
17. Lepetlt, V.; Fua, P. *Monocular Model-Based 3D Tracking of Rigid Objects*; Now Publishers Inc.: Delft, The Netherlands, 2005.
18. Seo, B.K.; Wuest, H. A Direct Method for Robust Model-Based 3D Object Tracking from a Monocular RGB Image. In Proceedings of the European Conference on Computer Vision 2016 Workshops (ECCVW), Amsterdam, The Netherlands, 8–10 and 15–16 October 2016.
19. Rosten, E.; Drummond, T. Fusing points and lines for high performance tracking. In Proceedings of the Tenth IEEE International Conference on Computer Vision, Beijing, China, 17–21 October 2005.
20. Seo, B.K.; Park, H.; Park, J.-I.; Hinterstoisser, S.; Ilic, S. Optimal local searching for fast and robust textureless 3D object tracking in highly cluttered backgrounds. *IEEE Trans. Vis. Comput. Graph.* **2013**, *20*, 99–110.
21. Wang, G.; Wang, B.; Zhong, F.; Qin, X.; Chen, B. Global optimal searching for textureless 3D object tracking. *Vis. Comput.* **2015**, *31*, 979–988. [CrossRef]
22. Prisacariu, V.A.; Reid, I.D. PWP3D: Real-time segmentation and tracking of 3D objects. *Int. J. Comput. Vis.* **2012**, *98*, 335–354. [CrossRef]
23. Hexner, J.; Hagege, R.R. 2D-3D pose estimation of heterogeneous objects using a region-based approach. *Int. J. Comput. Vis.* **2016**, *118*, 95–112. [CrossRef]
24. Tjaden, H.; Schwanecke, U.; Schömer, E.; Cremers, D. A region-based gauss-newton approach to real-time monocular multiple object tracking. *IEEE Trans. Pattern Anal. Mach. Intell.* **2019**, *41*, 1797–1812. [CrossRef] [PubMed]
25. Vidal, R.; Ma, Y. A unified algebraic approach to 2-D and 3-D motion segmentation and estimation. *J. Math. Imaging Vis.* **2006**, *25*, 403–421. [CrossRef]
26. Zhang, Z. A Flexible New Technique for Camera Calibration. *IEEE Trans. Pattern Anal. Mach. Intell.* **2000**, *22*, 1330–1334. [CrossRef]
27. Li, S.; Chi, X.; Ming, X. A Robust O(n) Solution to the Perspective-n-Point Problem. *IEEE Trans. Pattern Anal. Mach. Intell.* **2012**, *34*, 1444–1450. [CrossRef] [PubMed]

Article

VSAI: A Multi-View Dataset for Vehicle Detection in Complex Scenarios Using Aerial Images

Jinghao Wang, Xichao Teng *, Zhang Li, Qifeng Yu, Yijie Bian and Jiaqi Wei

College of Aerospace Science and Engineering, National University of Defense Technology, Changsha 410073, China; 13147837880@163.com (J.W.); zhangli_nudt@163.com (Z.L.); yuqifeng_nudt@126.com (Q.Y.); charlesbyj@126.com (Y.B.); weijiaqi18@163.com (J.W.)
* Correspondence: tengari@buaa.edu.cn

Abstract: Arbitrary-oriented vehicle detection via aerial imagery is essential in remote sensing and computer vision, with various applications in traffic management, disaster monitoring, smart cities, etc. In the last decade, we have seen notable progress in object detection in natural imagery; however, such development has been sluggish for airborne imagery, not only due to large-scale variations and various spins/appearances of instances but also due to the scarcity of the high-quality aerial datasets, which could reflect the complexities and challenges of real-world scenarios. To address this and to improve object detection research in remote sensing, we collected high-resolution images using different drone platforms spanning a large geographic area and introduced a multi-view dataset for vehicle detection in complex scenarios using aerial images (VSAI), featuring arbitrary-oriented views in aerial imagery, consisting of different types of complex real-world scenes. The imagery in our dataset was captured with a wide variety of camera angles, flight heights, times, weather conditions, and illuminations. VSAI contained 49,712 vehicle instances annotated with oriented bounding boxes and arbitrary quadrilateral bounding boxes (47,519 small vehicles and 2193 large vehicles); we also annotated the occlusion rate of the objects to further increase the generalization abilities of object detection networks. We conducted experiments to verify several state-of-the-art algorithms in vehicle detection on VSAI to form a baseline. As per our results, the VSAI dataset largely shows the complexity of the real world and poses significant challenges to existing object detection algorithms. The dataset is publicly available.

Keywords: dataset; vehicle detection; UAV; complex scenes

Citation: Wang, J.; Teng, X.; Li, Z.; Yu, Q.; Bian, Y.; Wei, J. VSAI: A Multi-View Dataset for Vehicle Detection in Complex Scenarios Using Aerial Images. *Drones* **2022**, *6*, 161. https://doi.org/10.3390/drones6070161

Academic Editor: Diego González-Aguilera

Received: 29 May 2022
Accepted: 25 June 2022
Published: 27 June 2022

Publisher's Note: MDPI stays neutral with regard to jurisdictional claims in published maps and institutional affiliations.

1. Introduction

Objection detection, as one core task in computer vision, refers to localized objects of interest; predicting their categories is becoming increasingly popular among researchers because of the extensive range of applications, e.g., smart cities, traffic management, face recognition, etc. The contributions of many high-quality datasets (such as PASCAL VOC [1], ImageNet [2], and MS COCO [3]) are immeasurable as part of the extensive elements and efforts leading to the rapid development of object detection technology.

In addition to the above-mentioned conventional datasets, the datasets collected by camera-equipped drones (or UAVs) for object detection have been widely applied in a great deal of fields, including agricultural, disaster monitoring, traffic management, military reconnaissance, etc. In comparison to natural datasets, where objects are almost directed upward because of gravity, object instances in aerial images under oblique view generally exist with arbitrary directions relying on the view of the flight platform and scale transformation due to oblique aerial photography, as illustrated in Figure 1.

Numerous research studies significantly contributed to object detection in remote sensing images [4–12], taking advantage of the latest advances in computer vision. Most algorithms [6,8,9,12] experimented by converting object detection in natural scenes to the aerial image fields. It is not surprising that object detection in ordinary images is not applicable to aerial images, as there are many differences (target sizes, degraded images, arbitrary orientations, unbalanced object intensity, etc.) between the two. Overall, it is more challenging for object detection in aerial images.

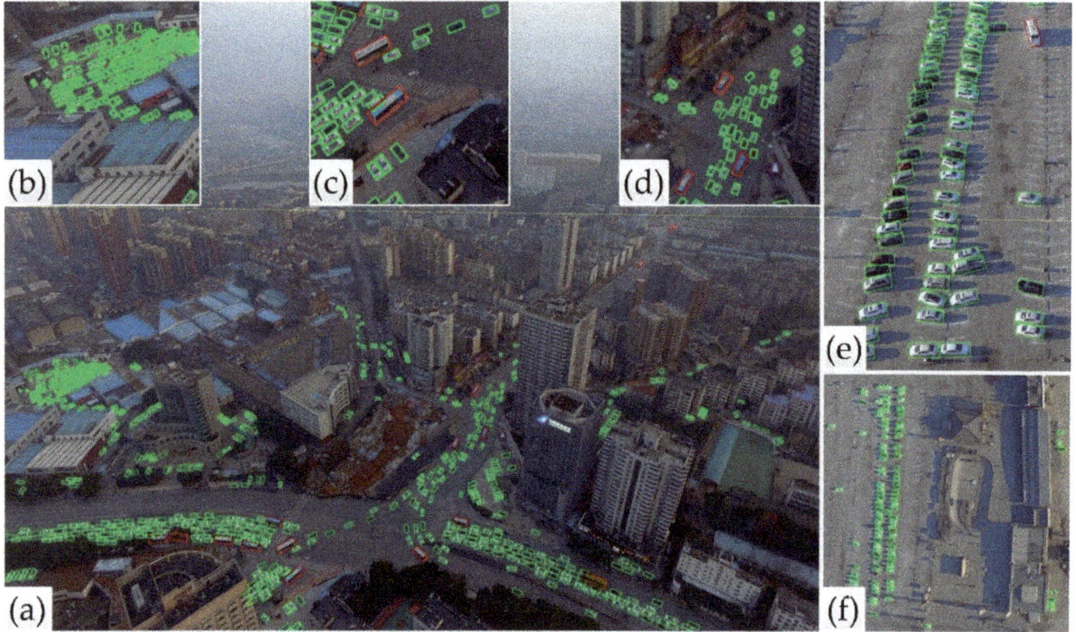

Figure 1. Examples of labeled images taken from VSAI (green box: small vehicles, red box: large vehicles). (**a**) Typical image under slope view in VSAI including numerous instances; examples exhibited in (**b–d**) are cut out from the original image (**a**). (**b**) Represents dense and tiny instances; (**c**) diagram of various instance orientations; (**c,d**) exhibition of the scale change caused by oblique aerial photography; (**e,f**) illustrate the distinctions of the same scene from different perspectives.

Figure 1 illustrates that object detection in aerial images is facing many challenges (such as image degradation, uneven object intensity, complex background, various scales, and various directions) distinguished from conventional object detection tasks:

- Large size variations of instances: this almost depends on the different spatial resolutions of the cameras, which are related to the camera pitch angles and flight heights of UAVs.
- Degraded images: The load carried by a small UAV platform is subject to severe limitations, with respect to the size and battery. Complex external weather variations (e.g., fog, rain, cloud, snow, light, etc.) and rapid UAV flights have led to vague UAV imagery, namely image degradation [13].
- Plenty of small instances: Ground objects with areas smaller than 32×32 pixels (MS COC dataset's definition of small objects) account for the majority of all objects in UAV images, as illustrated in Figure 1. Owing to the less diverse features of small targets, they may yield more errors and miss detection objects.
- Unbalanced object density: Uneven densities of captured objects are extremely prevalent in UAV images. In the same image, some objects may be densely arranged, while others may have sparse and uneven distribution, which are prone to repeated detection and missed detection, respectively.

- Arbitrary orientations: objects in aerial images usually appear in any direction, as shown in Figure 1.

In addition to these challenges, the research on object detection in UAV images is also plagued by the dataset bias problem [14]. The generalization ability (across datasets) is often low due to some preset specified conditions, which cannot fully reflect the task's complexity, e.g., fixed flight altitude [15,16], fixed camera pitch angle [15,17,18], narrow shooting area [16,19], clear background [20,21], etc. To improve the generalization ability of an object detection network, a dataset that adapts to the demands of practical applications needs to be created.

Moreover, compared to the object detection from a nadir image, the ability to identify objects with multi-view (off-nadir) imagery enables drones to be more responsive to many applications, such as disaster monitoring, emergency rescue, and environmental reconnaissance. To further unleash the potential of a drone's multi-view observations, this paper introduces a multi-view dataset for vehicle detection in complex scenarios using aerial images (VSAI), to highlight the object detection research based on drones. We collected 444 aerial images using different drone platforms from multi-view imaging. The resolutions of the pictures included 4000 × 3000, 5472 × 3648, and 4056 × 3040. These VSAI images were annotated by specialists in aerial imagery interpretation, including two categories (small vehicle and large vehicle). The fully labeled VSAI dataset consists of 49,712 instances, 48,925 of which are annotated by an oriented bounding box. The rest are marked with arbitrary quadrilateral bounding boxes for instances at image boundaries, rather than horizontal bounding boxes generally utilized as object labels in natural scenes. The major contributions of this paper are as follows:

- To our knowledge, VSAI is the first vehicle detection dataset annotated with varying camera pitch angles and flight heights (namely multi-view) rather than almost-fixed heights and camera angles of other datasets for object detection. It can be useful for evaluating object detection models in aerial images under complicated conditions closer to real situations.
- Our dataset's images c massive complex scenes (in exception for multi-view information) from many Chinese cities, such as backlights, the seaside, brides, dams, fog, ice and snow, deserts, tollbooths, suburbs, night, forest, Gobi, harbors, overhead bridges, crossroads, and mountainous regions, as shown in Figure 1.

This paper also evaluated state-of-the-art object detection algorithms on VSAI, which can be treated as the baseline for future algorithm development. We accomplished a cross-dataset generalization with the DOTA [22] dataset to evaluate the generalization capability of the VSAI dataset.

2. Related Work

In recent years, computer vision technology based on drones has gained much attention in many fields. As drones are excellent for acquiring high-quality aerial images and collecting vast amounts of imagery data, different datasets have been created for learning tasks, such as object detection, tracking, and scene understanding. Among these tasks, object detection is considered a fundamental problem; datasets for object detection are very important subsets of drone-based datasets. However, many drone-based datasets mainly use nadir imagery (i.e., images taken by a camera pointing to the ground vertically) for object detection and other computer vision tasks, without considering multi-view observations; the objects in scenes with high complexities are also insufficient, as they do not fully reflect complex real-world scenes.

In this section, we firstly review the relevant drone-based benchmarks and then vehicle target benchmarks collected by drones in object detection fields, similar to VSAI.

2.1. Drone-Based Datasets

To date, there are few drone-based datasets in the object detection field. Barekatain [23] proposed the Okutama-Action dataset for human action detection with the drone platform.

It consists of 43 min of completely annotated video sequences, including 77,365 representative frames with 12 action types. The benchmarking IR dataset for surveillance with aerial intelligence (BIRDSAI) [24] is an object detection and tracking dataset captured with a TIR camera equipped on a fixed-wing UAV in many African protected areas. It consists of humans and animals (with resolutions of 640 × 480 pixels). The UAVDT dataset [25] is a large-scale vehicle detection and tracking dataset, which consists of 100 video sequences and 80,000 representative frames, overlapping various weather conditions, flying heights, and multiple common scenarios, including intersections, squares, toll stations, arterial roads, highways, and T-junctions. The VisDrone2018 [26] dataset is a large-scale visual object detection and tracking dataset, which includes 263 video sequences with 179,264 representative frames and 10,209 static images captured by multiple camera devices, using various drones, in over 14 Chinese cities. VisDrone2018 covered some common object types, such as cars, bicycles, pedestrians, and tricycles. VisDrone2019 [18,27], when compared to VisDrone2018, increased 25 long-term tracking video sequences with 82,644 frames in total, 12 of them were taken during the day and the rest at night.

2.2. Vehicle Object Datasets

Hsieh et al. [15] proposed a dataset (CARPK) for car counting, which contained 1448 images shot in parking lot scenes with aerial views (with 89,777 annotated instances). Multi-scale object detection in a high-resolution UAV images dataset (MOHR) [17] is a large-scale benchmark object detection dataset gathered by three cameras with resolutions of 5482 × 3078, 7360 × 4912, and 8688 × 5792, respectively. MOHR incorporated 90,014 object instances with five types, including cars, trucks, buildings, flood damages, and collapses. The UAV-based vehicle segmentation dataset (UVSD) [28] is a large-scale benchmark object detection–counting–segmentation dataset, which owns various annotation formats containing OBB, HBB, and pixel-level semantics. The drone vehicle dataset [18] is a large-scale object detection and counting dataset with both optics and thermal infrared (RGBT) images shot by UAVs. The multi-purpose aerial dataset (AU-AIR) [29] is a large-scale object detection dataset from multimodal sensors (including time, location, IMU, velocity, altitude, and visual) captured by UAVs, which are composed of eight categories—person, car, bus, van, truck, bike, motorbike, and trailer—under different lighting and weather conditions. The largest existing available aerial image dataset for object detection is DOTA [22], composed of 2806 images with 15 categories and about 188,282 bounding boxes annotated with Google Earth and satellite images. The EAGLE [30] dataset is composed of 8820 aerial images (936 × 936 pixels) gained by several flight campaigns from 2006 to 2019 at different times of the day and year with various weather and lighting conditions. It has 215,986 vehicle instances (including large vehicles and small vehicles). To our knowledge, it is the largest aerial dataset for vehicle detection.

2.3. Oriented Object Detection

Significant advances have been made in the last decade in detecting objects in aerial images, which are often allocated with large changes and random directions. However, most current methods are based on heuristically-defined anchors with various scales, angles, and aspect ratios, and typically undergo severe misalignments between anchor boxes (ABs) and axis-aligned convolution features, leading to the usual inconsistency between the category score and localization correctness.

To solve this issue, a single-shot alignment network (S^2A-Net) [31] is proposed, which contains two units: a feature alignment module (FAM) for generating high-quality anchors and adaptively aligning the convolutional features, and an oriented detection module (ODM), with the goal of generating orientation-sensitive and orientation-invariant features to reduce the discrepancy between the localization and accuracy classification score.

To address the misalignment, the feature refinement module of the R3Det re-encodes the location parameters of the existing refined bounding box to the corresponding feature points through pixel-wise feature interpolation to accomplish feature reestablishment and alignment.

Meanwhile, another solution named the ROI transformer is put forward to address the above-mentioned problems. The key point of the ROI transformer is to exert spatial transformations on regions of interest (ROIs) and to learn the conversion parameters under the supervision of oriented bounding box (OBB) ground truth labels. To our knowledge, there is no specific algorithm for object detection under multiple perspectives of aerial images. So, we chose and altered the ROI transformer [32] as our baseline due to its higher localization accuracy for oriented object detection. Its specific principle will be introduced in Section 5.

Instead of directly regressing the four vertices, gliding vertices [33] regress four length ratios, describing the relative gliding offset on each resultant side, which can simplify the offset learning and avert ambiguity of sequential annotation points for oriented objects.

In general, there are abundant research studies [34–36] on down-view oriented object detection, but multi-view object detection is still in its infancy, which is also one of the areas we focus on in our follow-up research.

3. Overview of VSAI

In this section, we mainly explain the collection details of the entire VSAI dataset, the basis for category selection (small vehicle or large vehicle), and the annotation methods of the VSAI dataset.

3.1. Image Collection

Our dataset consists of 444 static images (specifically for vehicle detection tasks). Images in our dataset were collected from DJI Mavic Air, DJI Mavic 2 pro, Phantom 3 Pro, Phantom 4, and a 4 RTK drone platform with a high-resolution camera; partial critical technical parameters (including image sensor size, camera field angle, and imagery resolution) of these drones are exhibited in Table 1.

Table 1. Some technical parameters of UAVs used in the VSAI dataset.

Version	CMOS	Field Angle	Resolution
Mavic air	1/2.3 inch	85°	4056 × 3040
Mavic 2 pro	1 inch	77°	5472 × 3648
Phantom 3 Pro	1/2.3 inch	94°	4000 × 3000
Phantom 4	1/2.3 inch	94°	4000 × 3000
Phantom 4 RTK	1 inch	84°	5472 × 3648

To increase the divergence of data and overlay a wider geographical area, the VSAI dataset gathered images taken in most Chinese cities (including Shenyang, Weihai, Yantai, Weifang, Jinan, Lianyungang, Shanghai, Fuzhou, Xiamen, Zhengzhou, Luoyang, Yichang, Changsha, Guangzhou, Yinchuan, Guyuan, Xian, Delingha, Bayingolin from east to west, from north to south, etc.), as illustrated in Figure 2.

For shooting months shown in Figure 3a, this dataset covers the whole year. We captured the images with all-weather conditions, even the rare ice and snow scenarios as exhibited in Section 4.2. As for the shooting time displayed in Figure 3b, our dataset also basically covers the time range from 7 to 24 o'clock, except for 9 to 10 o'clock and 21 to 23 o'clock. Therefore, the VSAI dataset owns different images of light conditions as illustrated in Section 4.2, such as backlight, daylight, and night.

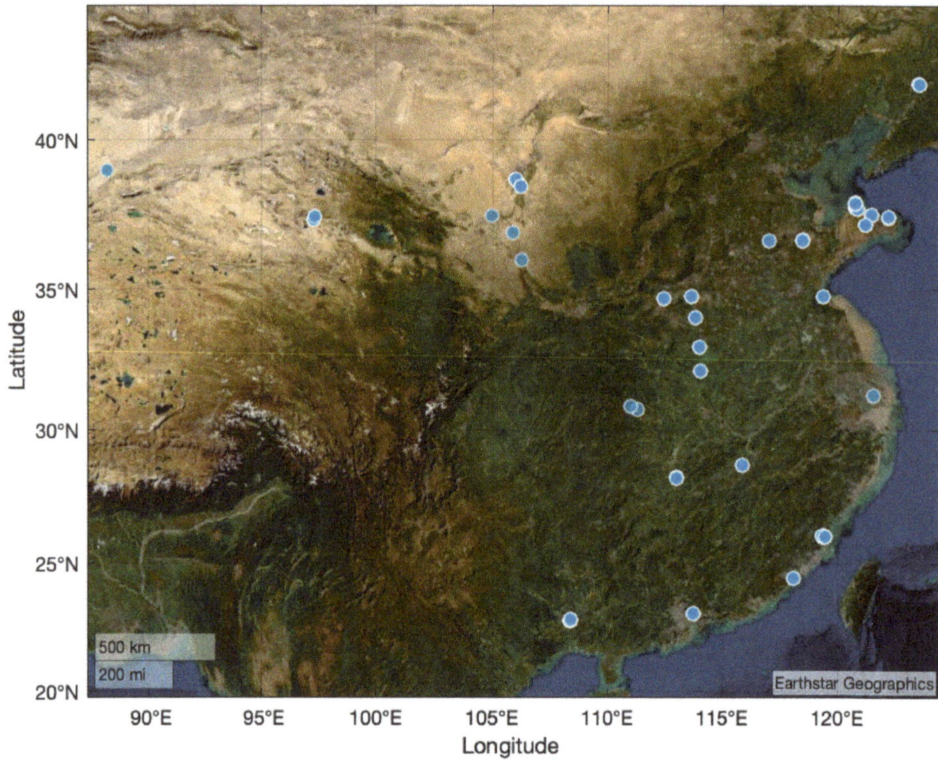

Figure 2. Distribution of image acquisition locations over China.

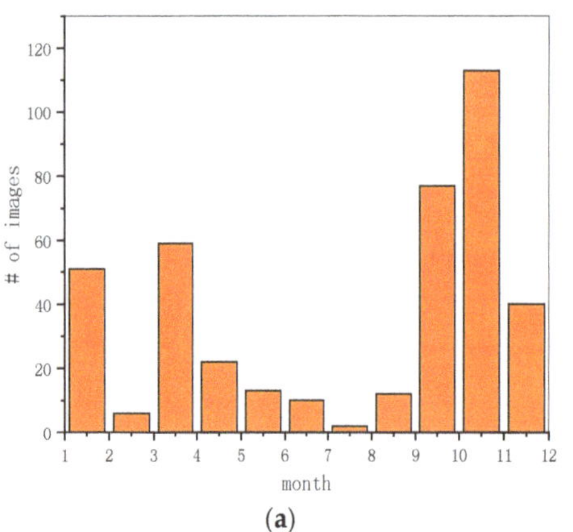

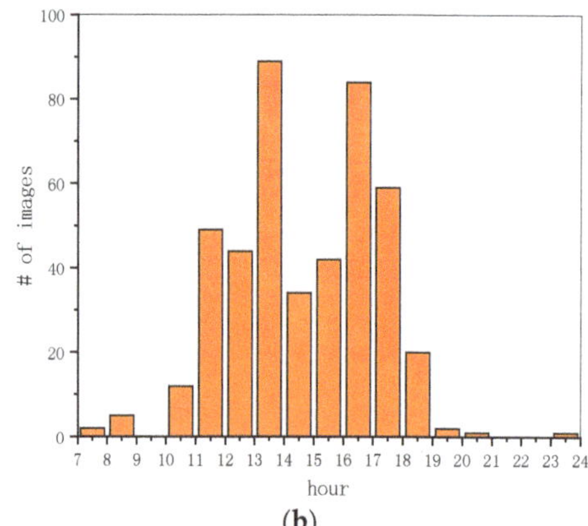

Figure 3. Image statics information. (**a**) The statistical histogram of shooting month; (**b**) the statistical histogram of shooting time from 7 to 24 o'clock.

3.2. Category Selection

Since vehicles photographed at high altitudes are difficult to classify, the VSAI dataset focuses on the vehicle category, which constitutes two categories, as shown in Figure 4, small vehicles (SVs include cars, minibuses, pickups, small trucks, taxis, and police cars) and large vehicles (LVs, such as buses and large trucks), similar to DOTA and EAGLE. The VSAI dataset contains 47,519 small vehicles and 2193 large vehicles, which confirms the uneven distribution of vehicles in the real world.

Figure 4. Samples of annotated images in VSAI (**left** to **right**, **top** to **bottom**). Large trucks belong to LV, a large truck (LV), a bus marked with an arbitrary quadrilateral bounding box (LV), a car labeled using an arbitrary quadrilateral bounding box (SV), cars densely arranged and mutually blocked (SV), cars partially occluded by vegetation (SV), a taxi (SV), small trucks (SV), pickup (SV), car (SV), SUV (SV), police car (SV), box truck (SV), minibus (SV).

3.3. Annotation Method

This paper considers several methods of annotating. In computer vision, many visual concepts (including objects, region descriptions, relationships, etc.) are labeled with bounding boxes (BB) [37]. A popular presentation of bounding boxes is (x_c, y_c, w, h), where (x_c, y_c) is the central location and (w, h) are the width and height of the bounding box, respectively.

However, the BB method cannot precisely annotate and outline the crowded objects with many orientations in aerial images because of the large overlap between bounding boxes. To settle this, we required searching for an annotation method adapted to oriented objects.

A choice for labeling oriented objects is the oriented bounding box, which is adopted in some text detection benchmarks [38], namely (x_c, y_c, w, h, θ), where θ refers to the angle from the horizontal direction of the normal bounding box. In fact, the VSAI dataset uses the θ-based oriented bounding boxes (OBB) to annotate objects in the aerial images due to excellent adaptability to rotating targets.

Another alternative is arbitrary quadrilateral bounding boxes (QBB), which can be defined as $\{(x_i, y_i), i = 1, 2, 3, 4\}$, where (x_i, y_i) refers to the positions of the bounding box apexes in the image. The vertices are arranged in clockwise order, choosing the left front vertices of vehicles as starting points, namely (x_1, y_1). This way is widely adopted in oriented text detection benchmarks [39]. In comparison with θ-based-oriented bounding boxes, arbitrary quadrilateral bounding boxes could compactly enclose oriented objects with large deformations among different parts; the latter will also consume more time in labeling due to a higher amount of parameters. Therefore, we only adopted QBB for the instances at the image edges as illustrated in Figure 4, and chose the time-efficient way (OBB) for the rest.

4. Properties of VSAI

This section depicts the major characteristics of the proposed dataset VSAI, which consists of multi-view UAV images, object visibility information, and more instances in each image. These properties (in comparison to other datasets) are sequentially described.

4.1. Multi-View

The original sizes of the images in VSAI were 4000 × 3000, 4056 × 3040, and 5472 × 3648 pixels, which are particularly huge in comparison to regular natural datasets (e.g., PASCAL-VOC and MSCOCO are no more than 1 × 1 k). To approach the real application scenario, the images in VSAI were shot at various camera pitch angles and flight altitudes in the range of 0° to −90° (0° indicates that the camera points in the forward direction of the UAV; −90° refers to the bird's-eye view) and from 54.5 to 499.4 m, respectively. As far as we know, extant drone datasets for object detection are rarely dedicated to collecting and labeling pictures from multiple views, namely distinct camera pitch angles and flight heights. This paper draws comparisons among MOHR [17], VisDrone2019 [27], Drone Vehicle [18], Okutama-Action [23], and VSAI to show the differences (Table 2). Note that, compared with our dataset's multi-view drone images, for facilitating data acquisition, the current UAV dataset is mostly fixed with several heights and camera pitch angles.

Table 2. Comparison of camera pitch angles and flight heights among VSAI and other object detection datasets based on UAV.

Dataset	Camera Pitch Angles	Flight Heights
MOHR [17]	−90°	About 200, 300, 400 m
VisDrone2019 [27]	Unannotated	Unannotated
Drone Vehicle [18]	−90°	Unannotated
Okutama-Action [23]	−45°, −90°	10–45 m
EAGLE [30]	−90°	Between 300 and 3000 m
VSAI	From 0° to −90°	55–500 m

We graphed the distribution histogram of the camera pitch angles and flight heights of our dataset. As shown in Figure 5, due to careful selection, the distributions of the camera pitch angles were relatively uniform. However, because of the law restricting flight above 120 m, in most Chinese cities, the flight altitudes were generally concentrated between 100 and 200 m in VSAI. In contrast, images taken from 200 to 500 m were mainly centered in the suburbs, accounting for a relatively low proportion of VSAI. Moreover, due to the scale and shape changes of objects attributed to multi-view UAV images as shown in Figure 1, object detection tasks are closer to reality, but they simultaneously become extremely difficult.

In the VSAI dataset, the instances with line of sight (LOS) angles of (−30°, −25°) were the largest, as illustrated in Figure 6. Overall, the LOS angle distribution of the number of instances was not balanced, mainly concentrating on small observation angles in the range of (−45°, −15°).

The main reason for this distribution is that the camera pitch angle decreases; that is, as the camera's line of sight gradually approaches the horizontal plane, the larger the ground scene range corresponding to the image area of the same size, the farther the observation distance, and the more object instances can be included, resulting in more objects corresponding to the smaller oblique line of sight angles when there is no significant difference in the number of images at different pitch angles (Figure 7). This shows that the object instances are unevenly distributed with the observation line of sight angle under the multi-view observation condition. At the same time, this observation method will lead to large object scale variations and image blurring, which increases the difficulty of object recognition.

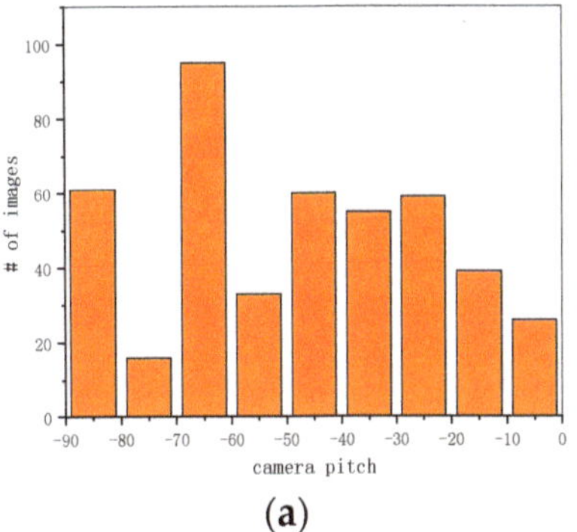

(a)

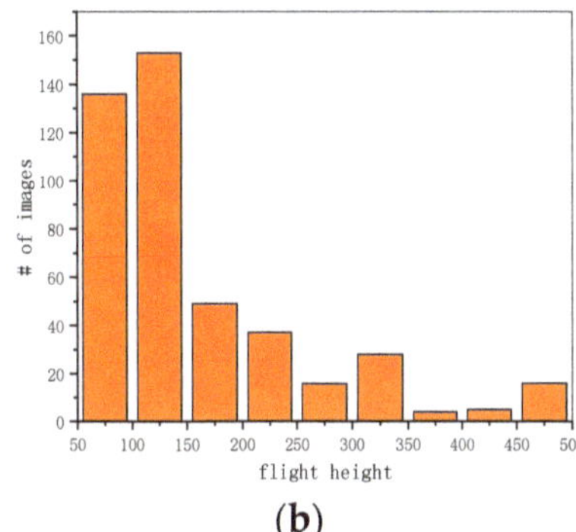

(b)

Figure 5. Image view statics information in VSAI: (**a**) distribution histogram of camera pitch angles; (**b**) distribution histogram of flight heights.

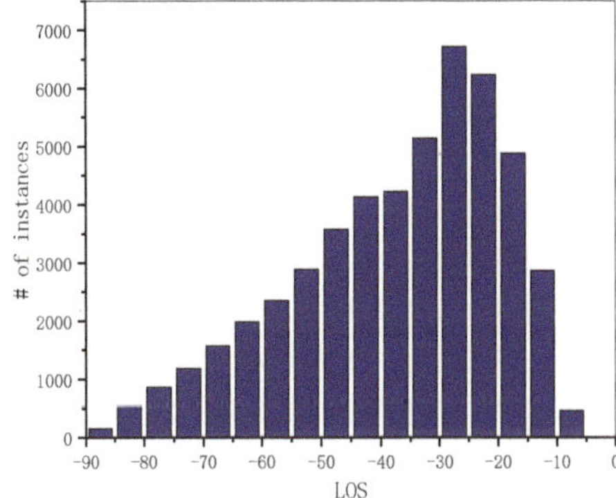

Figure 6. Statics histogram of the instances' line of sight angles (LOS) in VSAI.

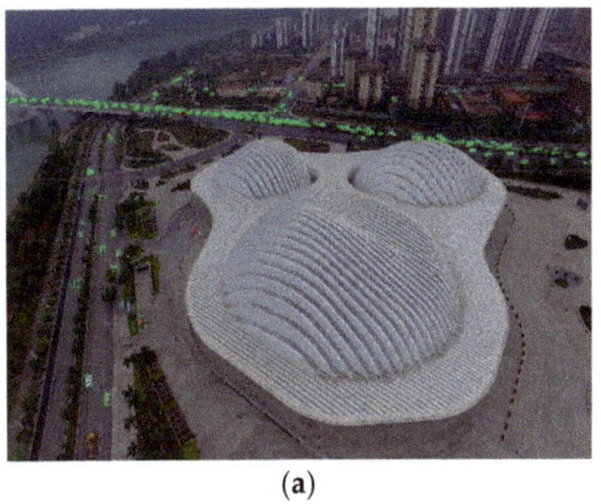

(a)

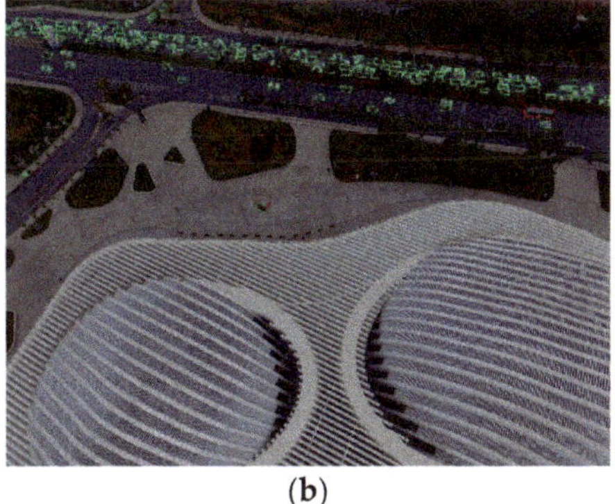

(b)

Figure 7. Examples of multi-view aerial images under the same scenario in the VSAI dataset: (**a**) the view of a higher altitude and smaller observation angle; (**b**) the view of a lower altitude and larger observation angle.

4.2. Complex Scenarios

Apart from the more extensive regional distribution, as shown in Figure 8, VSAI also covers six complicated scenes throughout China, including the desert, city, mountain, suburb, riverside, and seaside, as illustrated in Figure 8. The six scenarios also contain many subsets, such as cities, including the overhead bridge, crossroad, stadium, riverside embracing dam, bridge, etc. Observing Figure 8, VSAI essentially covers the vast majority of real-world complex scenarios, rather than the single urban scenario of other datasets. Meanwhile, aerial images of the VSAI dataset in multiple views are totally different from traditional down-view airborne imageries, because the former have more small targets, instances of occlusion, and larger-scale transformations of targets (as exhibited in Figure 8), which are closer to the complexities of the real-world.

Figure 8. Examples of multi-view annotated images from VSAI with complex scenes and distinct terrains (**left** to **right**, **top** to **bottom**): seaside (120 m, −8.4°); bridge (208.6 m, −31.3°); desert (106.9 m, −41.9°); suburb (114.8 m, −49.9°); Forest (291.5 m, −57.2°); harbor (104 m, −37.1°); overhead bridges (112.2 m, −46.7°); crossroads (203 m, −69.6°); dam (118.8 m, −6.7°); tollbooth (202.2 m, −89.9°); Gobi (356.6 m, −54.6°); mountainous region (409.2 m, −35.4°). The images in the first three lines have resolutions of 4000 × 3000 pixels; the resolution of the last line is 5472 × 3648 pixels.

The statistical histogram of the VSAI scene distribution is shown in Figure 9. It is obvious that the urban scenario accounts for half of the VSAI dataset. The other five scenarios make up the other half. The histogram of the VSAI scene distribution exhibits the complexities of the VSAI dataset.

Except for the weakness of a single scene, most existing datasets also ignored the influence of the natural environment and variations in illumination. However, the VSAI dataset considered complicated scenarios with diverse lighting conditions (such as daylight, backlight, and night) and interference from harsh natural environments (fog, snow cover, and sea ice), some examples of labeled images are shown in Figure 10.

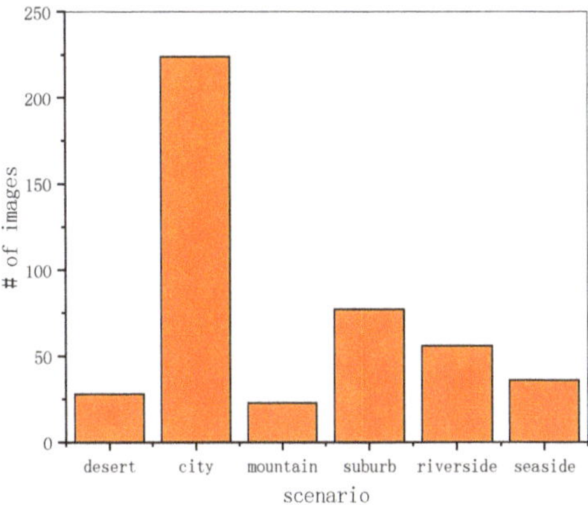

Figure 9. Distribution histogram of six complex scenes, including the desert, city, mountain, suburb, riverside, and seaside.

Figure 10. Examples of multi-view annotated images from VSAI in complex scenes (**left** to **right**, **top** to **bottom**): daylight; backlight that can never appear in a down-view aerial image; night; fog; snow cover; sea ice.

4.3. Vehicle Statistics

We collected statistical information about the vehicles, including the vehicle's orientation angles, instance length, and vehicle aspect ratio, as illustrated in Figure 11. Because of careful selection, we gained relatively uniform distributions of rotation angles, as shown in Figure 11a. Noting Figure 11b, the lengths of the vehicles were concentrated in the range of 0 to 75 pixels, signifying that there were numerous small instances in the VSAI dataset. At the same time, there was a considerable scale change in VSAI, as shown in Figure 11b. In addition, distinct perspectives also resulted in a wider range of the vehicle aspect ratio rather than the aspect ratio of 2 or so in traditional down-view aerial images.

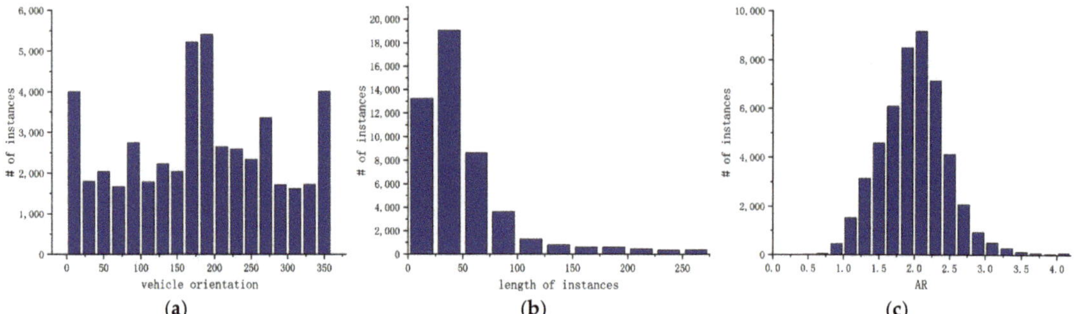

Figure 11. Vehicle statistics information: (**a**) the distribution of the vehicle's orientation angles; (**b**) statics histogram of the instances' lengths; (**c**) the distribution of the vehicle's aspect ratio (AR).

4.4. Object Occlusion Ratio

Additionally, VSAI provides useful annotations with respect to the occlusion ratio (the distribution of the occlusion ratio is shown in Figure 12). In this case, we used the proportion of vehicles being blocked to represent the occlusion ratio and define four levels of occlusions: no occlusion (occlusion ratio 0%), small occlusion (occlusion ratio < 30%), moderate occlusion (occlusion ratio 30~70%), and large occlusion (occlusion ratio > 70%), mainly for better reflecting the instance density of the instance location. The examples of different occlusion ratios are exhibited in the second line of Figure 13.

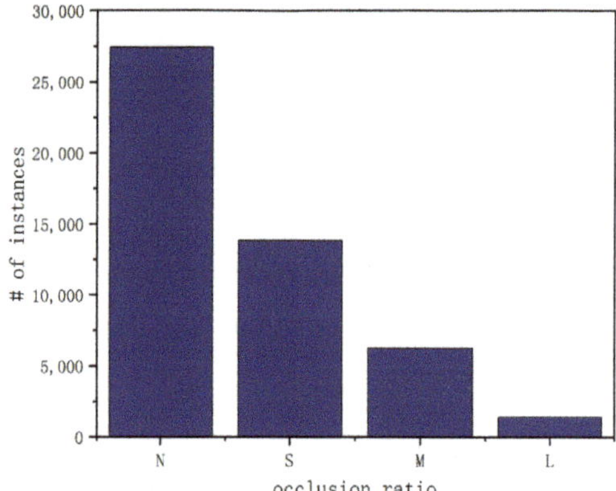

Figure 12. Statics histogram of the instances' occlusion ratios in VSAI.

Reasons for occlusion in multi-view and down-view aerial images are completely disparate. There are a couple of block types in multi-view aerial images that will never exist in down-view airborne imageries, such as occluded by a building, being blocked by other vehicles, or being sheltered by shafts, such as flags (first line of Figure 13). Due to more types of occlusions, there are more hardships for multi-view aerial images to detect objects accurately in comparison with the down-view ones, which means the former is closer to real-world complicacies.

Figure 13. Examples of vehicle occlusion. The first line demonstrates different block reasons of instances, from **left** to **right**, occluded by a building, blocked by other vehicles, sheltered by shafts, such as flags, and occluded by vegetation, respectively. The second line illustrates different occlusion ratios, from **left** to **right**, no occlusion, small occlusion, moderate occlusion, and large occlusion.

4.5. Average Instances

It is common for UAV images to include plenty of instances (but seldom for general images). However, for aerial datasets, UAVDT images [25] only have 10.52 instances on average. DOTA has 67.10. Our dataset VSAI is much larger in instances per image, which can be up to 111.96. Figure 14 shows the histogram of the number of instances per image in our VSAI dataset. Although the numbers of images and instances of VSAI are less than most other datasets, the average number of instances in each image is much greater than most other datasets, as illustrated in Table 3, except for DLR-3K-Vehicle [40], which only has 20 images.

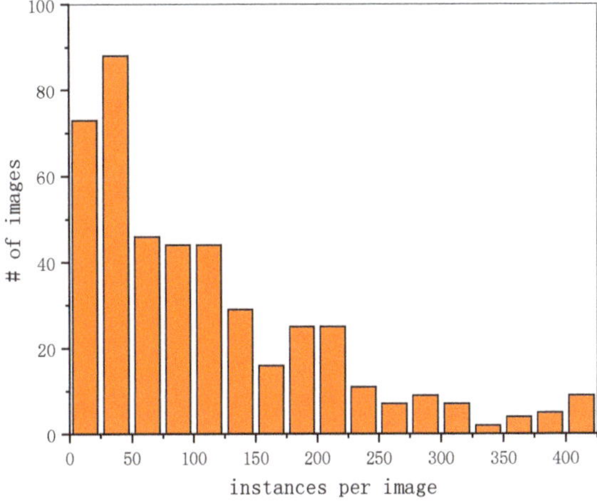

Figure 14. Histogram of the number of annotated instances per image in VSAI.

Table 3. Comparison of statistics between VSAI and other object detection benchmarks.

Dataset	Vehicle Instances per Image	No. of Images	No. of Instances	Instances per Image	Image Width (Pixels)
UAVDT [25]	841,500	80,000	841,500	10.52	1080
DOTA [22]	43,462	2806	188,282	67.10	300–4000
EAGLE [30]	215,986	8280	215,986	26.09	936
DLR-3K-Vehicle [40]	14,232	20	14,232	711.6	5616
VSAI	49,712	444	49,712	111.96	4000, 4056, 5472

5. Method

We benchmarked the current object detection methods based on OBB with VSAI in the evaluation section (below). Moreover, we selected and altered the ROI transformer [32] as our baseline because of its higher localization accuracy for oriented object detection.

When detecting dense objects in aerial images, algorithms based on horizontal proposals for natural object detection always lead to mismatches between regions of interest (ROIs) and objects. The ROI transformer is proposed for addressing this; it contains two parts, RROI learner and RROI warping. In this section, we briefly introduce two parts of the ROI transformer and the ResNeSt backbone, the alternative to ResNet in this paper.

5.1. RROI Learner

The purpose of the RROI learner is to learn to rotate ROIs (RROIs) from the feature map of horizontal ROIs (HROIs). We have HROIs in the form of (x, y, w, h) for predicted 2D coordinates and the width and height of a HROI; the corresponding feature maps are defined as $\{F_i\}$. Because ideally a single HROI is the circumscribed rectangle of the RROI, the ROI learner attempts to infer the geometric parameters of RROIs from F_i by fully connected layers with dimensions of 5, regressing the offsets of rotated ground truths (RGTs) relative to HROI; the regression targets are shown as

$$
\begin{aligned}
t_x^{gt} &= \tfrac{1}{w^r}\left(\left(x^{gt} - x^r\right)\cos\theta^r + \left(y^{gt} - y^r\right)\sin\theta^r\right), \\
t_y^{gt} &= \tfrac{1}{h^r}\left(\left(y^{gt} - y^r\right)\cos\theta^r - \left(x^{gt} - x^r\right)\sin\theta^r\right), \\
t_w^{gt} &= \log\tfrac{w^{gt}}{w^r}, t_h^{gt} = \log\tfrac{h^{gt}}{h^r}, \\
t_\theta^{gt} &= \tfrac{1}{2\pi}\left(\left(\theta^{gt} - \theta^r\right)\bmod 2\pi\right)
\end{aligned}
\tag{1}
$$

where $(x^r, y^r, w^r, h^r, \theta^r)$ represents the location, width, length, and rotation of a RROI and $(x^{gt}, y^{gt}, w^{gt}, h^{gt}, \theta^{gt})$ stands for the ground truth parameters of an OBB. For deriving Equation (1), the ROI learner utilizes the local coordinate systems bound to RROIs instead of the global coordinate system bound to the image.

The output vector $(t_x, t_y, t_w, t_h, t_\theta)$ of the fully connected layer is represented as follows

$$
t = \mathcal{C}(\mathcal{F}; \Theta)
\tag{2}
$$

where $\mathcal{C}$ is the fully connected layer, $\mathcal{F}$ is the feature map for every HROI, and Θ represents the weight parameters of $\mathcal{C}$.

Once an input HROI matches with a ground truth of OBB, t^{gt} is set by the description in Equation (1). The smooth L1 loss function [41] is used for the regression loss. The predicted t is decoded from offsets to the parameters of RROI. In other words, the RROI learner learns the parameters of RROI from the HROI feature map $\mathcal{F}$.

5.2. RROI Warping

Based on the RROI parameters learned by the RROI learner, RROI warping extracts the rotation-invariant deep features for oriented object detection. The module of the rotated position sensitive (RPS) ROI align [32] is proposed as the specific RROI warping, which divides the RROI into $K \times K$ bins and exports a feature map $\mathcal{Y}$ of the shape (K, K, C); for the bin's index $(i, j)(0 \leq i, j < K)$ of the output channel $c(0 \leq c < C)$, we have

$$\mathcal{Y}_c(i, j) = \sum_{(x,y) \in bin(i,j)} \frac{D_{i,j,c}(\mathcal{T}_\theta(x, y))}{n} \tag{3}$$

where $D_{i,j,c}$ is the feature map from the $K \times K \times C$ feature maps. The $n \times n$ denotes the number of sampling locations in the bin. The $bin_{(i,j)}$ represents the coordinates set $\left\{ i\frac{w_r}{k} + (s_x + 0.5)\frac{w_r}{k \times n}; s_x = 0, 1, \ldots n - 1 \right\} \times \left\{ j\frac{h_r}{k} + (s_y + 0.5)\frac{h_r}{k \times n}; s_y = 0, 1, \ldots n - 1 \right\}$. Moreover, each $(x, y) \in bin(i, j)$ is transformed to (x', y') by $\mathcal{T}_\theta$, where

$$\begin{pmatrix} x' \\ y' \end{pmatrix} = \begin{pmatrix} \cos\theta & -\sin\theta \\ \sin\theta & \cos\theta \end{pmatrix} \begin{pmatrix} x - w_r/2 \\ y - h_r/2 \end{pmatrix} + \begin{pmatrix} x_r \\ y_r \end{pmatrix} \tag{4}$$

Equation (3) is realized by bilinear interpolation.

The combination of the RROI learner and RROI warping replaces the normal ROI warping, which provides better initialization of RROIs. In turn, it achieves better results in rotating object detection.

5.3. Architecture of ROI Transformer

The main architecture of the ROI transformer is composed of three parts—the backbone, neck, and head networks. We chose the ResNeSt50 backbone to replace ResNet50 in our baseline for extracting features, and the batch size was set to 2. FPN was selected as the neck network to integrate the feature output of the backbone efficiently, whose input channels were set as $Cin = [256, 512, 1024, 2048]$, output channels $Cout = 256$. The head network includes the RPN head and the ROI transformer head. We used five scales $\{32^2, 64^2, 128^2, 256^2, 512^2\}$ and three aspect ratios $\{1/2, 1, 2\}$, yielding $k = 20$ anchors for the RPN head network initialization. The ROI head adopted the ROI transformer described in the previous subsection; we used the Smooth L1 loss [41] function for the bounding box regression loss and the cross-entropy loss function for the category loss. The IOU threshold was set as 0.5. We trained the model in 40 epochs for VSAI. The SGD optimizer was adopted with an initial learning rate of 0.0025, the momentum of 0.9, and weight decay of 0.0001. We used the learning rate warm-up for 500 iterations.

5.4. ResNeSt

ResNeSt accomplished an architectural alteration of ResNet, merging feature map split attention within the separate network blocks. Specifically, each block partitioned the feature map into multiple groups (along the channel dimension) and finer-grained subgroups or splits with each group's feature representation determined via a weighted combination of the split representations (with weights determined in accordance with global context information). The resulting unit is defined as a split attention block. By stacking some split attention blocks, we gained ResNeSt (S means "Split").

Based on the ResNeXt blocks [42] that divide the feature into K groups (namely "cardinality" hyperparameter K), ResNeSt (Figure 15) introduces a new "Radix" hyperparameter R, which means a split number within a "cardinality" group. Therefore, the total number of feature-map groups is $G = KR$. The group alteration is a 1×1 convolution layer followed by a 3×3 convolution layer. The attention function is composed of a global pooling layer and two fully connected layers followed by SoftMax in the "cardinal" dimension.

ResNeSt combines the advantage of the "cardinality" group in ResNeXt and the "selective kernel" in SKNets [43], and achieves a state-of-the-art performance compared to all existing ResNet variants, as well as brilliant speed–accuracy trade-offs. Therefore, we chose the ResNeSt backbone to replace ResNet in this paper.

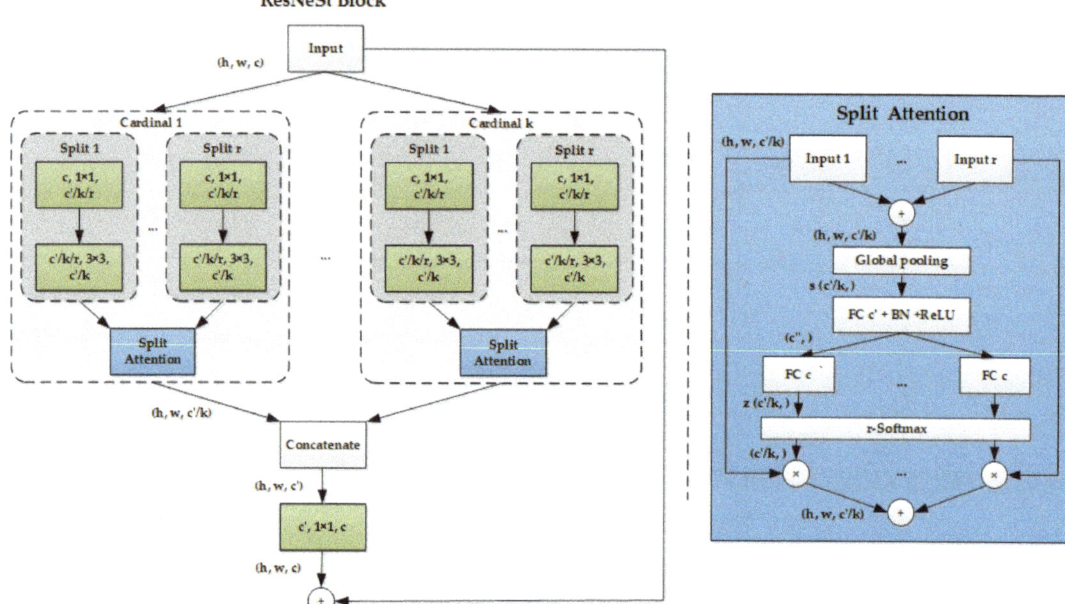

Figure 15. Schematic diagram of the ResNeSt Block and split attention. The left shows the ResNeSt block in a cardinality-major view. The green convolution layer is shown as (no. of in channels, filter size, no. of out channels). The right illustrates the split attention block. s $(c'/k,)$ and z $(c'',)$ stand for channel-wise statistics, as $s \in \mathbb{R}^{c'/k}$ generated by global average pooling and the compact feature descriptor $z \in \mathbb{R}^{c''}$ created by the fully connected layer. FC, BN, and r-SoftMax mean the fully connected layer, batch normalization, and SoftMax in the cardinal dimension. $+$ and $\times$ represent the element-wise summation and element-wise product.

6. Evaluations

6.1. Dataset Split and Experimental Setup

To ensure that the distribution of vehicles in the training, validation, and test sets was approximately balanced, we randomly assigned images with 1/2, 1/6, and 1/3 instances to the training, validation, and test sets, respectively. To facilitate training, the initial images were cropped into the patches with two methods. For the single-scale segmentation method, the size of the patch was 1024 × 1024 pixels, with 200-pixel gaps in the sliding window, leading to 5240, 1520, and 2315 patches of the training, validation, and test sets, respectively, set according to the input size of the DOTA dataset [22]. For the multi-scale segmentation method, the original images were cropped into 682 × 682, 1024 × 1024 and 2048 × 2048 pixels with 500-pixel gaps, resulting in 33,872, 9841, and 14,941 patches of the training, validation, and test sets, respectively. Moreover, this paper evaluated all of the models on NVIDIA GeForce GTX 2080 Ti with PyTorch version 1.6.0.

6.2. Experimental Baseline

We benchmarked the current object detection methods based on OBB with VSAI. In this research, based on the features of VSAI (such as numerous small instances, huge scale changes, and occlusion), we carefully selected rotated Faster R-CNN [44], oriented R-CNN [34], rotated RetinaNet [45], and gliding vertex [33] as our benchmark testing methods due to their wonderful performances on object detection with arbitrary orientations. We chose and altered the ROI transformer [32] as our baseline. We followed the same implementations of these models released by their original developers. Except for the ROI transformer, we modified the backbone network from ResNet50 [46] to ResNeSt50 [47] to obtain better feature extraction results. All codes of the baseline selected in this paper are based on MMRotate [48], available at https://github.com/open-mmlab (accessed on 27 April 2022).

6.3. Experimental Analysis

By exploring the results illustrated in Figure 16, one could see that the OBB detection is still challenging in relation to tiny instances, densely arranged areas, and occlusions in aerial images. In Figure 16, we provide a comparison of small and large vehicle detection with different ROI transformer methods (distinct backbones and split ways). As shown in Figure 16, the unbalanced dataset (the number of small vehicles being much higher than the large vehicles) led to less accuracy of the algorithms in large-vehicle detection compared to small-vehicle detection. Observing the first column in Figure 16, we notice that the models with ResNeSt50, random rotation, and multi-scale split more accurately framed the large vehicle, because the former owns the more powerful feature extraction capability in contrast with ResNet50 and the latter possesses more large-vehicle instances. Whether it is single-scale or multi-scale or ResNet50 or ResNeSt50 models—for large-size vehicles, as shown in the middle column in Figure 16, it precisely detects (even in the shadows and occlusions). As demonstrated in the last column, under reverse light conditions, although the multi-scale split and ResNeSt50 are better than the single-scale split and the ResNet50, the three models completely miss many minor targets. The results are not satisfying, implying the high hardship of this task.

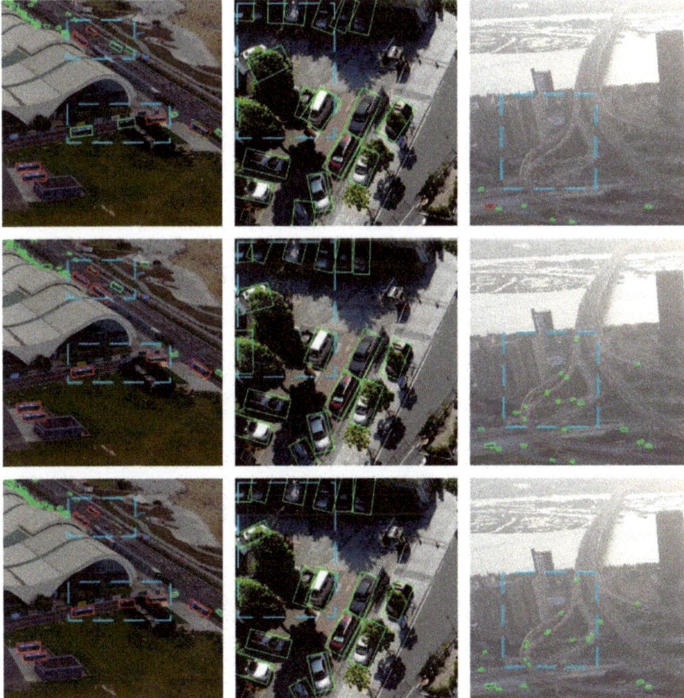

Figure 16. Test prediction samples of the ROI transformer trained on the VSAI dataset. The first row is the result of the model with the ResNet50 backbone and single-scale split, the middle row is the model with the ResNeSt50 backbone and single-scale split, and the third row is the result of the model with ResNeSt50 backbone, random rotation, and multi-scale split. The blue dotted boxes indicate significant differences between the pictures in the rows.

In Table 4, we show the quantitative results of the experiments. Analyzing the results exhibited in Table 4, performances in categories of small vehicles and large vehicles are far from satisfactory, attributed to the former's small size and the scarcity of the latter in aerial images. Overall, a two-stage network is generally better than a one-stage network, except for S^2A-Net and SASM. The former relies on FAM and ODM units to achieve better positioning accuracy by reducing the misalignment between anchor boxes (ABs) and axis-aligned convolution features. The latter obtains better sampling selection results through SA-S and SA-S strategies. Therefore, these two single-stage networks achieve similar results to two-stage networks. It is worth noting that simply replacing the ResNet50 backbone of the ROI transformer with ResNeSt50 improved the mAP by 4.1% and 0.7% for single-scale and multi-scale splits, respectively, which proves the effectiveness of the split-attention module in the ResNeSt50 backbone. An unbalanced dataset contributed to the lower accuracy of all the models in large-vehicle detection in comparison with small-vehicle detection. To summarize, although our baseline of the ROI transformer (adopting ResNeSt50, multi-scale split, and random rotation) achieved the best performances (64.9% mAP, 79.4% average precision (AP) for small vehicles, and 50.4% AP for large vehicles), among the state-of-the-art algorithms used in this paper, object detection in multi-view aerial images was far from satisfactory. Object detection in aerial images under various perspectives needs to be further developed.

Table 4. Benchmark of the state-of-the-art on the rotated bounding box (RBB) detection task trained and tested on VSAI; mAP means mean average precision, higher is better. SS and MS mean single-scale and multi-scale split. RR indicates random rotation. R50 stands for ResNet50, S50 represents ResNeSt50.

Method	Backbone	Split and Rotation	Type	AP [%]		
				SV	LV	Mean
Rotated RetinaNet [45]	R50	SS	One-Stage	67.1	32.6	49.9
R3Det [49]	R50	SS	One-Stage	69.6	38.5	54.0
Gliding Vertex [33]	R50	SS	Two-Stage	70.3	42.5	56.4
Rotated Faster R-CNN [44]	R50	SS	Two-Stage	70.7	44.0	57.3
S^2A-Net [31]	R50	SS	One-Stage	73.6	41.9	57.7
Oriented R-CNN [34]	R50	SS	Two-Stage	76.9	43.1	60.0
SASM [36]	R50	SS	One-Stage	76.7	45.2	60.9
CFA [35]	R50	SS	Two-Stage	77.6	45.0	61.3
ROI Transformer [32]	R50	SS	Two-Stage	77.4	38.4	57.9
	S50	SS	Two-Stage	77.7	46.2	62.0
	R50	MS	Two-Stage	78.9	48.2	63.6
	S50	MS	Two-Stage	78.8	49.8	64.3
	R50	MS, RR	Two-Stage	79.0	49.2	64.1
	S50	MS, RR	Two-Stage	79.4	50.4	64.9

In Figure 17, we provide several examples of fault detection and leak detection with our baseline. As shown in Figure 17, it is still pretty hard for the state-of-the-art methods to gain great detection results due to the complex scenes in the VSAI dataset. The model misidentified the neon lights of buildings and blue blocks of roofs and arches, as large vehicles. Motion blur vehicles at night, buses in close rows, and oblique photography of tiny vehicles in the distance were not successfully detected.

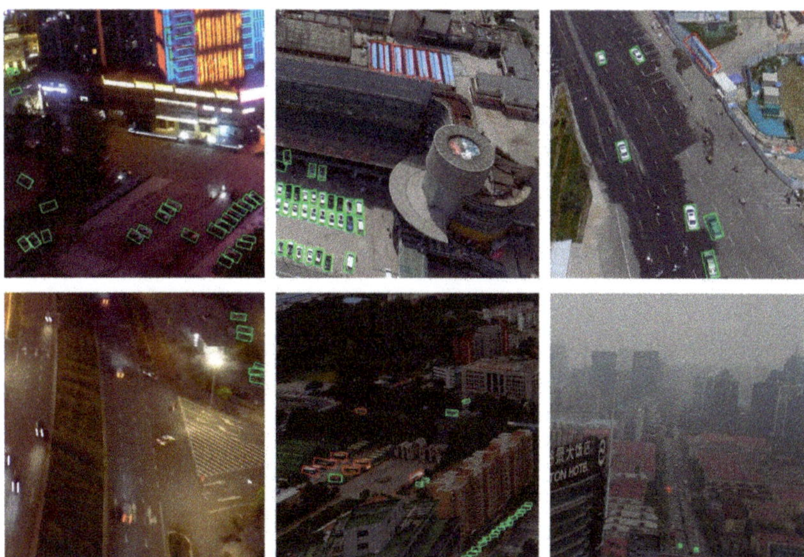

Figure 17. Test prediction samples of our baseline trained on the VSAI dataset. The first row is the result of the fault detection and the second row is the result of the leak detection.

6.4. Cross-Dataset Validation

We completed a cross-dataset generalization experiment to validate the generalization ability of the VSAI dataset. We chose DOTA [22] for comparison and its test set for testing. We selected the ROI transformer models with the baseline of the VSAI dataset for generalization experiments with OBB ground truth. Table 5 shows that a model trained on VSAI generalizes well to DOTA, scoring 10% mAP over a model trained on DOTA and tested on VSAI, which indicates that VSAI contains a wider range of features in comparison to DOTA. At the same time, it reveals that VSAI is particularly more complex and challenging than the current available down-view datasets, which makes it suitable for real-world complicated vehicle detection scenarios.

Table 5. Comparison of results on VSAI and DOTA using the baseline of the VSAI dataset. The comparison is on account of mAP. SL and LV stand for small vehicle and large vehicle, respectively.

Training Set	Test Set	SV	LV	mAP
DOTA	VSAI	17.0	4.5	10.8
VSAI	DOTA	35.5	6.1	20.8

7. Conclusions

We presented VSAI, a UAV dataset for targeting vehicle detection in aerial photography, whose number of instances per image is multiple times higher than existing datasets. Unlike common object detection datasets, we provided every annotated image with camera pitch angles and flight height of drones. We built a dataset highly relevant to real-world scenarios, which included multiple scenarios in aerial images, such as time, weather, illuminative situation, camera view, landform, and season. Our benchmarks illustrated that VSAI is a challenging dataset for the current state-of-the-art orientated detection models; our baseline achieved 64.9% mAP, which is 79.4% the average precision (AP) and 50.4% the AP for small and large vehicles. The cross-dataset validation showed that models trained with pure down-view images could not adapt to multi-angle datasets. On the contrary, VSAI could cover the features of straight-down datasets, such as DOTA. We believe that

VSAI contributes to remote sensing target detection (closer to reality). It also introduces novel challenges to the vehicle detection domain.

Author Contributions: J.W. (Jinghao Wang) and X.T. are the co-first authors of the article. Conceptualization, J.W. (Jinghao Wang) and X.T.; methodology, J.W. (Jiaqi Wei); software, J.W. (Jinghao Wang); validation, Z.L. and Q.Y.; formal analysis, X.T.; investigation, Z.L.; resources, J.W. (Jiaqi Wei); data curation, Y.B.; writing—original draft preparation, J.W. (Jinghao Wang); writing—review and editing, J.W. (Jinghao Wang); visualization, X.T.; supervision, Y.B.; project administration, J.W. (Jiaqi Wei); funding acquisition, Z.L. All authors have read and agreed to the published version of the manuscript.

Funding: This research was funded by the National Natural Science Funding of China (no. 61801491).

Data Availability Statement: The download link for the VSAI dataset is publicly available at www.kaggle.com/dronevision/VSAIv1 (accessed on 31 May 2022).

Conflicts of Interest: The funders had no role in the design of the study; in the collection, analyses, or interpretation of data; in the writing of the manuscript, or in the decision to publish the results.

References

1. Everingham, M.; Van Gool, L.; Williams, C.K.; Winn, J.; Zisserman, A. The pascal visual object classes (voc) challenge. *Int. J. Comput. Vis.* **2010**, *88*, 303–338. [CrossRef]
2. Russakovsky, O.; Deng, J.; Su, H.; Krause, J.; Satheesh, S.; Ma, S.; Huang, Z.; Karpathy, A.; Khosla, A.; Bernstein, M. Imagenet large scale visual recognition challenge. *Int. J. Comput. Vis.* **2015**, *115*, 211–252. [CrossRef]
3. Lin, T.-Y.; Maire, M.; Belongie, S.; Hays, J.; Perona, P.; Ramanan, D.; Dollár, P.; Zitnick, C.L. Microsoft coco: Common objects in context. In Proceedings of the European Conference on Computer Vision, Zurich, Switzerland, 6–12 September 2014; pp. 740–755.
4. Lin, Y.; He, H.; Yin, Z.; Chen, F. Rotation-invariant object detection in remote sensing images based on radial-gradient angle. *IEEE Geosci. Remote Sens. Lett.* **2014**, *12*, 746–750.
5. Liu, Z.; Wang, H.; Weng, L.; Yang, Y. Ship rotated bounding box space for ship extraction from high-resolution optical satellite images with complex backgrounds. *IEEE Geosci. Remote Sens. Lett.* **2016**, *13*, 1074–1078. [CrossRef]
6. Cheng, G.; Zhou, P.; Han, J. Rifd-cnn: Rotation-invariant and fisher discriminative convolutional neural networks for object detection. In Proceedings of the IEEE Conference on Computer Vision and Pattern Recognition, Las Vegas, NV, USA, 27–30 June 2016; pp. 2884–2893.
7. Moranduzzo, T.; Melgani, F. Detecting cars in UAV images with a catalog-based approach. *IEEE Trans. Geosci. Remote Sens.* **2014**, *52*, 6356–6367. [CrossRef]
8. Zhang, F.; Du, B.; Zhang, L.; Xu, M. Weakly supervised learning based on coupled convolutional neural networks for aircraft detection. *IEEE Trans. Geosci. Remote Sens.* **2016**, *54*, 5553–5563. [CrossRef]
9. Wang, G.; Wang, X.; Fan, B.; Pan, C. Feature extraction by rotation-invariant matrix representation for object detection in aerial image. *IEEE Geosci. Remote Sens. Lett.* **2017**, *14*, 851–855. [CrossRef]
10. Wan, L.; Zheng, L.; Huo, H.; Fang, T. Affine invariant description and large-margin dimensionality reduction for target detection in optical remote sensing images. *IEEE Geosci. Remote Sens. Lett.* **2017**, *14*, 1116–1120. [CrossRef]
11. Ok, A.O.; Senaras, C.; Yuksel, B. Automated detection of arbitrarily shaped buildings in complex environments from monocular VHR optical satellite imagery. *IEEE Trans. Geosci. Remote Sens.* **2012**, *51*, 1701–1717. [CrossRef]
12. Long, Y.; Gong, Y.; Xiao, Z.; Liu, Q. Accurate object localization in remote sensing images based on convolutional neural networks. *IEEE Trans. Geosci. Remote Sens.* **2017**, *55*, 2486–2498. [CrossRef]
13. Hong, D.; Yokoya, N.; Chanussot, J.; Zhu, X.X. An augmented linear mixing model to address spectral variability for hyperspectral unmixing. *IEEE Trans. Image Process.* **2018**, *28*, 1923–1938. [CrossRef] [PubMed]
14. Torralba, A.; Efros, A.A. Unbiased look at dataset bias. In Proceedings of the CVPR 2011, Colorado Springs, CO, USA, 20–25 June 2011; pp. 1521–1528.
15. Hsieh, M.-R.; Lin, Y.-L.; Hsu, W.H. Drone-based object counting by spatially regularized regional proposal network. In Proceedings of the IEEE International Conference on Computer Vision, Venice, Italy, 27–29 October 2017; pp. 4145–4153.
16. Li, S.; Yeung, D.-Y. Visual object tracking for unmanned aerial vehicles: A benchmark and new motion models. In Proceedings of the Thirty-First AAAI Conference on Artificial Intelligence, San Francisco, CA, USA, 4–9 February 2017.
17. Zhang, H.; Sun, M.; Li, Q.; Liu, L.; Liu, M.; Ji, Y. An empirical study of multi-scale object detection in high resolution UAV images. *Neurocomputing* **2021**, *421*, 173–182. [CrossRef]
18. Zhu, P.; Sun, Y.; Wen, L.; Feng, Y.; Hu, Q. Drone based rgbt vehicle detection and counting: A challenge. *arXiv* **2020**, arXiv:2003.02437.
19. Robicquet, A.; Sadeghian, A.; Alahi, A.; Savarese, S. Learning social etiquette: Human trajectory prediction in crowded scenes. In Proceedings of the European Conference on Computer Vision (ECCV), Glasgow, UK, 23–28 August 2020.

20. Zhu, H.; Chen, X.; Dai, W.; Fu, K.; Ye, Q.; Jiao, J. Orientation robust object detection in aerial images using deep convolutional neural network. In Proceedings of the 2015 IEEE International Conference on Image Processing (ICIP), Quebec City, QC, Canada, 27–30 September 2015; pp. 3735–3739.

21. Cheng, G.; Han, J.; Zhou, P.; Xu, D. Learning rotation-invariant and fisher discriminative convolutional neural networks for object detection. *IEEE Trans. Image Process.* **2018**, *28*, 265–278. [CrossRef]

22. Xia, G.-S.; Bai, X.; Ding, J.; Zhu, Z.; Belongie, S.; Luo, J.; Datcu, M.; Pelillo, M.; Zhang, L. DOTA: A large-scale dataset for object detection in aerial images. In Proceedings of the IEEE Conference on Computer Vision and Pattern Recognition, Salt Lake City, UT, USA, 18–23 June 2018; pp. 3974–3983.

23. Barekatain, M.; Martí, M.; Shih, H.F.; Murray, S.; Prendinger, H. Okutama-Action: An Aerial View Video Dataset for Concurrent Human Action Detection. In Proceedings of the 2017 IEEE Conference on Computer Vision and Pattern Recognition Workshops (CVPRW), Honolulu, HI, USA, 21–26 July 2017.

24. Bondi, E.; Jain, R.; Aggrawal, P.; Anand, S.; Hannaford, R.; Kapoor, A.; Piavis, J.; Shah, S.; Joppa, L.; Dilkina, B. BIRDSAI: A dataset for detection and tracking in aerial thermal infrared videos. In Proceedings of the IEEE/CVF Winter Conference on Applications of Computer Vision, Snowmass, CO, USA, 1–5 March 2020; pp. 1747–1756.

25. Du, D.; Qi, Y.; Yu, H.; Yang, Y.; Duan, K.; Li, G.; Zhang, W.; Huang, Q.; Tian, Q. The unmanned aerial vehicle benchmark: Object detection and tracking. In Proceedings of the European Conference on Computer Vision (ECCV), Munich, Germany, 8–14 September 2018; pp. 370–386.

26. Zhu, P.; Wen, L.; Bian, X.; Ling, H.; Hu, Q. Vision meets drones: A challenge. *arXiv* **2018**, arXiv:1804.07437.

27. Zhu, P.; Wen, L.; Du, D.; Bian, X.; Hu, Q.; Ling, H. Vision meets drones: Past, present and future. *arXiv* **2020**, arXiv:2001.06303.

28. Zhang, W.; Liu, C.; Chang, F.; Song, Y. Multi-scale and occlusion aware network for vehicle detection and segmentation on uav aerial images. *Remote Sens.* **2020**, *12*, 1760. [CrossRef]

29. Bozcan, I.; Kayacan, E. Au-air: A multi-modal unmanned aerial vehicle dataset for low altitude traffic surveillance. In Proceedings of the 2020 IEEE International Conference on Robotics and Automation (ICRA), Paris, France, 31 May–31 August 2020; pp. 8504–8510.

30. Azimi, S.M.; Bahmanyar, R.; Henry, C.; Kurz, F. Eagle: Large-scale vehicle detection dataset in real-world scenarios using aerial imagery. In Proceedings of the 2020 25th International Conference on Pattern Recognition (ICPR), Milan, Italy, 10–15 January 2021; pp. 6920–6927.

31. Han, J.; Ding, J.; Li, J.; Xia, G.-S. Align deep features for oriented object detection. *IEEE Trans. Geosci. Remote Sens.* **2021**, *60*, 1–11. [CrossRef]

32. Ding, J.; Xue, N.; Long, Y.; Xia, G.-S.; Lu, Q. Learning roi transformer for oriented object detection in aerial images. In Proceedings of the IEEE/CVF Conference on Computer Vision and Pattern Recognition, Long Beach, CA, USA, 15–20 June 2019; pp. 2849–2858.

33. Xu, Y.; Fu, M.; Wang, Q.; Wang, Y.; Chen, K.; Xia, G.-S.; Bai, X. Gliding vertex on the horizontal bounding box for multi-oriented object detection. *IEEE Trans. Pattern Anal. Mach. Intell.* **2020**, *43*, 1452–1459. [CrossRef]

34. Xie, X.; Cheng, G.; Wang, J.; Yao, X.; Han, J. Oriented r-cnn for object detection. In Proceedings of the IEEE/CVF International Conference on Computer Vision, Virtual, 11–17 October 2021; pp. 3520–3529.

35. Guo, Z.; Liu, C.; Zhang, X.; Jiao, J.; Ji, X.; Ye, Q. Beyond bounding-box: Convex-hull feature adaptation for oriented and densely packed object detection. In Proceedings of the IEEE/CVF Conference on Computer Vision and Pattern Recognition, Nashville, TN, USA, 20–25 June 2021; pp. 8792–8801.

36. Hou, L.; Lu, K.; Xue, J.; Li, Y. Shape-Adaptive Selection and Measurement for Oriented Object Detection. In Proceedings of the AAAI Conference on Artificial Intelligence, Virtual, 22 February–1 March 2022.

37. Haag, M.; Nagel, H.-H. Combination of edge element and optical flow estimates for 3D-model-based vehicle tracking in traffic image sequences. *Int. J. Comput. Vis.* **1999**, *35*, 295–319. [CrossRef]

38. Yao, C.; Bai, X.; Liu, W.; Ma, Y.; Tu, Z. Detecting texts of arbitrary orientations in natural images. In Proceedings of the 2012 IEEE Conference on Computer Vision and Pattern Recognition, Providence, RI, USA, 16–21 June 2012; pp. 1083–1090.

39. Karatzas, D.; Gomez-Bigorda, L.; Nicolaou, A.; Ghosh, S.; Bagdanov, A.; Iwamura, M.; Matas, J.; Neumann, L.; Chandrasekhar, V.R.; Lu, S. ICDAR 2015 competition on robust reading. In Proceedings of the 2015 13th International Conference on Document Analysis and Recognition (ICDAR), Tunis, Tunisia, 23–26 August 2015; pp. 1156–1160.

40. Liu, K.; Mattyus, G. Fast multiclass vehicle detection on aerial images. *IEEE Geosci. Remote Sens. Lett.* **2015**, *12*, 1938–1942.

41. Girshick, R.; Donahue, J.; Darrell, T.; Malik, J. Rich Feature Hierarchies for Accurate Object Detection and Semantic Segmentation. In Proceedings of the 2014 IEEE Conference on Computer Vision and Pattern Recognition, Columbus, OH, USA, 23–28 June 2013.

42. Xie, S.; Girshick, R.; Dollár, P.; Tu, Z.; He, K. Aggregated residual transformations for deep neural networks. In Proceedings of the Proceedings of the IEEE Conference on Computer Vision and Pattern Recognition, Honolulu, HI, USA, 21–26 July 2017; pp. 1492–1500.

43. Li, X.; Wang, W.; Hu, X.; Yang, J. Selective kernel networks. In Proceedings of the IEEE/CVF Conference on Computer Vision and Pattern Recognition, Long Beach, CA, USA, 15–20 June 2019; pp. 510–519.

44. Ren, S.; He, K.; Girshick, R.; Sun, J. Faster R-CNN: Towards Real-Time Object Detection with Region Proposal Networks. *IEEE Trans. Pattern Anal. Mach. Intell.* **2017**, *39*, 1137–1149. [CrossRef] [PubMed]

45. Lin, T.-Y.; Goyal, P.; Girshick, R.; He, K.; Dollár, P. Focal loss for dense object detection. In Proceedings of the IEEE International Conference on Computer Vision, Venice, Italy, 22–29 October 2017; pp. 2980–2988.

46. He, K.; Zhang, X.; Ren, S.; Sun, J. Deep residual learning for image recognition. In Proceedings of the IEEE Conference on Computer Vision and Pattern Recognition, Las Vegas, NV, USA, 27–30 June 2016; pp. 770–778.
47. Zhang, H.; Wu, C.; Zhang, Z.; Zhu, Y.; Lin, H.; Zhang, Z.; Sun, Y.; He, T.; Mueller, J.; Manmatha, R. Resnest: Split-attention networks. *arXiv* **2020**, arXiv:2004.08955.
48. Zhou, Y.; Xue, Y.; Zhang, G.; Wang, J.; Liu, Y.; Hou, L.; Jiang, X.; Liu, X.; Yan, J.; Lyu, C.; et al. MMRotate: A Rotated Object Detection Benchmark using PyTorch. *arXiv* **2022**, arXiv:2204.13317.
49. Yang, X.; Yan, J.; Feng, Z.; He, T. R3det: Refined single-stage detector with feature refinement for rotating object. In Proceedings of the AAAI Conference on Artificial Intelligence, Virtual, 2–9 February 2021; pp. 3163–3171.

 *drones*

Article

Distance-Based Formation Control for Fixed-Wing UAVs with Input Constraints: A Low Gain Method

Jiarun Yan, Yangguang Yu and Xiangke Wang *

College of Intelligence Science and Technology, National University of Defense Technology,
Changsha 410073, China; jryan@nudt.edu.cn (J.Y.); yuyangguang11@nudt.edu.cn (Y.Y.)
* Correspondence: xkwang@nudt.edu.cn

Abstract: Due to the nonlinear and asymmetric input constraints of the fixed-wing UAVs, it is a challenging task to design controllers for the fixed-wing UAV formation control. Distance-based formation control does not require global positions as well as the alignment of coordinates, which brings in great convenience for designing a distributed control law. Motivated by the facts mentioned above, in this paper, the problem of distance-based formation of fixed-wing UAVs with input constraints is studied. A low-gain formation controller, which is a generalized gradient controller of the potential function, is proposed. The desired formation can be achieved by the designed controller under the input constraints of the fixed-wing UAVs with proven stability. Finally, the effectiveness of the proposed method is verified by the numerical simulation and the semi-physical simulation.

Keywords: multi-UAV formation; velocity constraints; fixed-wing UAV

1. Introduction

Compared with the single unmanned aerial vehicle (UAV), multiple unmanned aerial vehicle (multi-UAV) formations have several advantages, including improved execution efficiency and capability, better fault tolerance and robustness and etc. [1–3]. In reality, the multi-UAV formations have been frequently used in light shows, disaster relief, and communication maintenance [4,5]. Thus, the study of multi-UAV formation has arisen much attention in recent years.

To achieve the multi-UAV formation, a variety of control methods have been proposed. The survey [6] classified the formation control from perception capabilities into position-based [7,8], displacement-based [9,10], and distance-based [11,12]. Among them, the distance-based formation control requires less individual perception capability. More concretely, it can help design formation control laws in agents' local coordinate frames, which neither requires global position measurements nor the alignment of agents' local coordinate frames [11]. In practical applications the global coordinates are sometimes not available (GPS-denied) and the alignment of coordinates is difficult for the multi-UAV system. Due to the facts mentioned above, the distance-based formation control problem has become a research hotpot recently. Reference [12] derived a gradient controller from the potential function based on an undirected infinitesimal rigidity graph. Then the work [12] proved that the infinitesimal rigidity is a sufficient condition for the local asymptotic stability of the equilibrium manifold. Based on the work of reference [12], a new design strategy for formation control was proposed in reference [13], which can achieve the local asymptotic stability for general infinitesimal rigid formations and the global asymptotic stability for triangular infinitesimal formations. Reference [14] investigated the local asymptotic stability of n-dimensional undirected formations with single and double integrator models, and revealed that a rigid formation is locally asymptotically stable even though the formation is not infinitesimally rigid. Reference [15] integrated the different formation control laws proposed by the previous works into a unified convergence analysis framework, and considered the case of minimally rigid target formation as well

Citation: Yan, J.; Yu, Y.; Wang, X. Distance-Based Formation Control for Fixed-Wing UAVs with Input Constraints: A Low Gain Method. *Drones* **2022**, *6*, 159. https://doi.org/10.3390/drones6070159

Academic Editor: Abdessattar Abdelkefi

Received: 17 May 2022
Accepted: 23 June 2022
Published: 27 June 2022

Publisher's Note: MDPI stays neutral with regard to jurisdictional claims in published maps and institutional affiliations.

as non-minimally rigid target formation. Then the authors of [15] proved the exponential stability of the formation system under a generalized controller. Besides the work mentioned above, different cases were studied for specific considerations, such as control with disturbances [11,16–18], optimal formation control [19,20], and the formation control combined with flocking [21–23].

Although a variety of distance-based formation control methods have been proposed in many studies including the works mentioned above, most of them model the dynamics of the agents in the system as a single integrator or double integrator. As a consequence, when applying to the UAV system, the control method proposed in these works is unsuitable because the UAV cannot move in any direction and the velocity in the head direction must be greater than zero. More specifically, the dynamics of the UAVs are under-actuated and input-constrained. In the current study of the formation control for the fixed-wing UAVs, the kinematics of the fixed-wing UAV are modeled as a unicycle model, which is a nonholonomic system. Thus, the study of formation control with nonholonomic constraints and input saturation is of full meaning in practice. Existing nonholonomic constraint studies can be found in references [23–26], whereas input saturation studies can be found in references [27–31]. It is worth mentioning that the robust backstepping approach or the sliding mode approach is a powerful approach for controlling the nonholonomic system with input constraints [32–34]. However, there may be some problems such as introducing more complex structures, relying on more system information, etc. Most of them are based on the leader-follower structure, which is a simple and clear control architecture but highly dependent on the motion of the leader agent. Reference [31] solved the distance-based formation control problem under the nonholonomic constraint and the velocity saturation constraints by employing the time-varying projection matrix and time-varying scalar. However, the approach in reference [31] requires a minimum linear velocity to be less than zero, which is unsuitable for the fixed-wing UAVs. Therefore, the problem of the distance-based formation control for the fixed-wing UAVs is still an open problem.

The low gain design technique has been proved to be an effective idea in coping with input-constrained problems of linear systems [35–38]. Although distance-based formation control is considered a complex nonlinear problem, the idea of low gain techniques can still bring new perspectives or new thinking. Meanwhile, for the multi-agent formation control problem, it is usually a popular approach to design a controller based on the constructed potential function [13–15].

Different from the previous works on the formation control problem, in this paper, the dynamics of the UAV is modeled as a unicycle model with linear and angular velocity constraints while the coordinates of the UAVs are not required to be aligned. Due to the dynamic property of the fixed-wing UAV [28,30], the angular velocity is saturated while its linear velocity is bounded within a positive interval. Taking both the linear and the angular velocity constraints into consideration, the distance-based formation problem for fixed-wing UAVs becomes more challenging. A potential-function-based controller is then designed by utilizing the low gain design technique. Stability analysis is also provided. Finally, the effectiveness of the proposed method is verified by using both the numerical and the semi-physical simulations.

In summary, the main contributions of this article are as follows.

(1) We present a novel problem formulation for distance-based formation control of fixed-wing UAVs. For fixed-wing UAVs with minimum forward velocity, we modify the problem description of the general unicycle model, i.e., the formation is required to keep moving at a uniform velocity simultaneously.

(2) We design a low-gain formation controller, which can keep the input of the system from saturation. The proposed controller is a general gradient controller with a low gain coefficient, which is designed based on the distance-based potential function. Furthermore, we give the complete stability analysis to prove that the desired distance-based formation can be achieved while the input constraints of each UAV are satisfied.

(3)   We simulate our proposed controller, including numerical simulation and semi-physical simulation, and verify that the proposed method can effectively solve the distance-based formation control problem under the input constraints of fixed-wing UAVs.

The rest of the paper is organized as follows. In Section 2, the problem of distance-based formation control of fixed-wing UAVs is formulated. Section 3 proposes the control law with input constraints and gives the stability analysis. The simulation results are presented in Section 4, followed by a conclusion of the paper in Section 5.

2. Problem Formulation

2.1. UAV Modeling

Consider a formation of N fixed-wing UAVs. For $i = 1, \ldots, N$, the kinematic model of UAV i is described by

$$
\begin{aligned}
\dot{x}_i &= v_i \cos \theta_i, \\
\dot{y}_i &= v_i \sin \theta_i, \\
\dot{\theta}_i &= w_i,
\end{aligned}
\tag{1}
$$

where $[x_i, y_i]^T \in \mathbb{R}^2$ and $\theta_i \in (-\pi, \pi]$ are the position and orientation of the i-th UAV in the inertial Cartesian frame, respectively. In this paper, the linear velocity $v_i \in \mathbb{R}$ and the angular velocity $w_i \in \mathbb{R}$ are the control inputs of system (1).

Remark 1. *It is worth noting that the models of UAVs are described in 2D instead of 3D. It is based on the fact that when the fixed-wing UAVs are performing formations, the UAVs usually fly at constant altitudes [16,17,20,23]. For example, in practical implementations, the UAVs are usually controlled to fly at different altitudes to avoid collisions. In this sense, by setting a given altitude, each UAV performs a fixed altitude flight.*

Suppose that the UAV i is subject to the following velocity constraints:

$$
\begin{aligned}
0 &< v_{i,\min} \le v_i \le v_{i,\max}, \\
-w^l_{i,\max} &\le w_i \le w^r_{i,\max},
\end{aligned}
\tag{2}
$$

where $v_{i,\min}$ and $v_{i,\max}$ are the minimum and maximum forward linear velocities of the i-th UAV, respectively, and $w^l_{i,\max}$ and $w^r_{i,\max}$ are the maximum left-turn and right-turn angular velocities, respectively.

Remark 2. *Although there are some existing works that address the distance-based formation control problem of the unicycle model, they do not consider the velocity constraints of fixed-wing UAVs. That is, the velocity constraints (2) are not present in the general unicycle model [23–26]. To tackle this challenge, a novel controller is designed to implement distance-based formation control of the fixed-wing UAVs in this paper.*

Remark 3. *It is worth noting that the velocity constraints can be different for each UAV, which relaxes the requirement to use the same type of the UAV in the formation [30].*

2.2. Desired Formation

In this paper, the undirected graph $\mathcal{G} \triangleq (\mathcal{V}, \mathcal{E})$ is used to represent the interaction of UAVs, where $\mathcal{V} = \{1, 2, \ldots, N\}$ is the set of N vertices and $\mathcal{E} \subset \mathcal{V} \times \mathcal{V}$ is the set of m edges. Each vertex represents a UAV and the neighbor set of vertex i is defined as $\mathcal{N}_i(\mathcal{E}) = \{j \in \mathcal{V} | (i, j) \in \mathcal{E}\}$. The edge $(i, j) \in \mathcal{E}$ means that the UAV i, j can sense the relative position with respect to each other. Then, let $p_i \triangleq [x_i, y_i]^T \in \mathbb{R}^2$, and denote $p_{ij} \triangleq p_i - p_j$ as the relative position between the UAV i and j. The distance and the desired distance between the UAV i and j are denoted by $d_{ij} \triangleq \|p_{ij}\|$ and d^*_{ij}, respectively.

The distance-based formation usually defines the desired formation based on the distances among the UAVs. When the distance between the UAVs reaches the desired distance, the formation goal will be considered to be achieved. However, fixed-wing UAVs cannot stay still after reaching the desired distance and usually have to keep flying at a uniform velocity. Therefore, different from the distance-based formation control in reference [31], the desired formation requires not only that the desired distance between the UAVs be maintained, but also that the UAVs keep moving at a preset uniform velocity.

Therefore, the desired formation control objective can be described as follows:

$$\begin{aligned}
\|p_i(t) - p_j(t)\| \to d_{ij}^* \quad &\text{as } t \to \infty, \quad \forall (i,j) \in \mathcal{E}, \\
\dot{p}_i(t) - \vec{v}_0 \to 0 \quad &\text{as } t \to \infty, \quad i = 1, \ldots, N,
\end{aligned} \tag{3}$$

where $\vec{v}_0 \in \mathbb{R}^2$ is a constant vector.

Figure 1 illustrates the process of achieving the desired formation consisting of three fixed-wing UAVs. It can be observed that the three UAVs maintain the desired distance from their neighbors while moving at the same velocity $\vec{v}_0$.

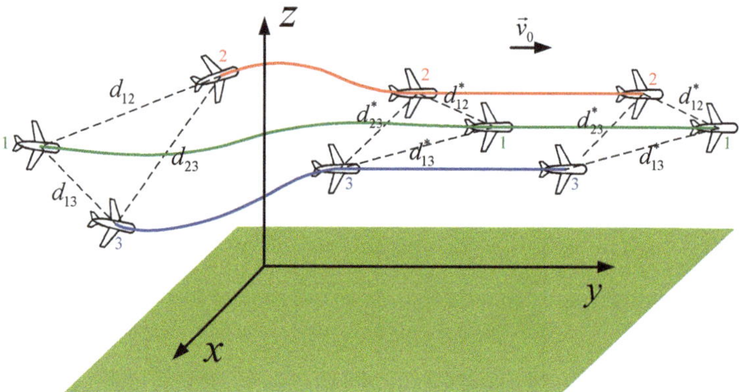

Figure 1. The desired formation of three UAVs.

2.3. Problem Statement

Two assumptions are posed before the problem statement.

Assumption 1. *The velocity constraints of all UAVs have a common range and the uniform velocity $\vec{v}_0$ lies within this velocity range, i.e., $\exists v_{\min}, v_{\max}, w_{\max}^r, w_{\max}^l \in \mathbb{R}^+$, for $\forall i \in \{0, 1, \ldots, N\}$ it holds that*

$$\begin{aligned}
0 < v_{i,\min} \le v_{\min} < \|\vec{v}_0\| < v_{\max} \le v_{i,\max}, \\
-w_{i,\max}^l \le -w_{\max}^l < 0 < w_{\max}^r \le w_{i,\max}^r.
\end{aligned} \tag{4}$$

Assumption 2. *The distances between all UAVs are bounded, and they are all less than a known constant d_M, i.e.,*

$$d_{ij} \le d_M, \quad \forall (i,j) \in \mathcal{E}. \tag{5}$$

Thus, the formation problem is described as follows.

Problem 1. *Under Assumptions 1 and 2, the control inputs v_i and w_i are designed so that each UAV reaches the desired distance from its neighbors while the entire formation maneuvers at a consistent velocity, i.e., Equation (3) holds while the velocity constraint of Equation (2) is satisfied.*

Remark 4. *It is obvious that Assumption 1 is a prerequisite for a formation mission to be achievable. Only if assumption 1 is satisfied, it is possible for all UAVs to be in formation at a uniform velocity.*

Remark 5. *Assumption 2 is reasonable since the communication range of UAVs in reality is usually limited, and once the distance between UAVs is farther than their communication range, their interaction topology will be broken and the formation will not be implemented.*

3. Controller Design

In this section, the concept of distance-based potential function is first proposed, then the designed potential function is used to design the low-gain-based controller so that the velocity constraints can be satisfied. Finally, the stability analysis of the proposed controller is presented.

3.1. Distance-Based Potential Function

Let $e_{ij} = d_{ij} - d_{ij}^*$. Then, the vector $e = [\ldots, e_{ij}, \ldots] \in \mathbb{R}^m$ consists of all e_{ij} where $(i, j) \in \mathcal{E}$.

Definition 1. *For each UAV i, define a distance-based potential function $F_i : \mathbb{R}^{2|\mathcal{N}_i|+1} \to [0, \infty)$ as follows:*

$$F_i(p_i, \ldots, p_j, \ldots) \triangleq \gamma \sum_{j \in \mathcal{N}_i} G(\|p_i - p_j\|), \tag{6}$$

where $\gamma > 0$ and the function G is to be determined such that F_i satisfies the following assumption:

Assumption 3. *The funtion F_i satisfies the following conditions:*

- *$F_i \geq 0$ always holds, where $F_i = 0$ if and only if $\|p_i - p_j\| = d_{ij}^*$ for all $j \in \mathcal{N}_i$;*
- *For the function $g(x) \triangleq \frac{\dot{G}(x)}{x}$, if $\|x\| \leq x_M$, it holds that $\|g(x)\| \leq g_M$, where $\dot{G}(x)$ denotes differentiation of the function G;*
- *Denote*

$$f_i \triangleq -\nabla_{p_i} F_i = -\gamma \sum_{j \in \mathcal{N}_i} g(\|p_{ij}\|) p_{ij}, \tag{7}$$

and there exists r_0 such that $f_i = 0 \Leftrightarrow F_i = 0$ in $\{p : F_i(p) \leq r_0\}$.

In this paper, the function $G(\cdot)$ is designed as

$$G(\|p_{ij}\|) = \frac{1}{2}\left[\left(\|p_{ij}\|^2 + \frac{d_{ij}^{*4}}{\|p_{ij}\|^2}\right) - d_{ij}^{*2}\right]. \tag{8}$$

Correspondingly, the function $g(\cdot)$ is

$$g(\|p_{ij}\|) = 1 - \frac{d_{ij}^{*4}}{\|p_{ij}\|^4}. \tag{9}$$

Remark 6. *The second term of Assumption 3 implicitly implies that the function G is differentiable. Further, the properties of the function F_i are related to the ones of the function G, which means that the function G needs to be suitably selected. In fact, the function G can take many forms which were summarized in reference [15]. Furthermore, similar to Assumption 1 of reference [31], Assumption 3 is satisfied by most of the cooperative control laws including the distance-based formation control law. In addition, r_0 indicates the size of the attraction domain. In other words, it determines whether the system is globally or locally stable.*

Remark 7. *The distance-based potential function is a cornerstone of the controller proposed in this paper. On the one hand, the distance-based potential function can be used as the Lyapunov function candidate for proving the stability of the closed-loop system, as the Lyapunov functions are usually difficult to find for nonlinear systems. On the other hand, for the multi-agent formation control problem, distance-based potential function is more visual and intuitive, which makes it easier to* *understand the action of the controller. In fact, it is a popular approach to design a controller based on the constructed potential function [13–15]. In addition, it has been pointed out by reference [15] that the attractive property of ensuring collision avoidance for the formation system can be obtained by choosing a suitable potential function.*

3.2. Low-Gain-Based Controller

On the basis of the distance-based potential function F_i, the controller for UAV i without velocity constraints is designed as

$$\begin{bmatrix} v_i \\ w_i \end{bmatrix} = \begin{bmatrix} \cos\theta_i & \sin\theta_i \\ -\sin\theta_i & \cos\theta_i \end{bmatrix}(f_i(\gamma) + \vec{v}_0),\tag{10}$$

where $f_i(\gamma)$ is defined in Equation (7) and indicates that f_i is related to the low gain coefficient γ.

For the convenience of notation, let

$$h_i = \begin{bmatrix} \cos\theta_i & \sin\theta_i \end{bmatrix}^T, h_i^{\perp} = \begin{bmatrix} -\sin\theta_i & \cos\theta_i \end{bmatrix}^T, h_0 = \begin{bmatrix} \cos\theta_0 & \sin\theta_0 \end{bmatrix}^T.\tag{11}$$

Thus, Equation (10) can be rewritten as

$$\begin{aligned} v_i &= h_i^T(f_i(\gamma) + \vec{v}_0), \\ w_i &= {h_i^{\perp}}^T(f_i(\gamma) + \vec{v}_0). \end{aligned}\tag{12}$$

Remark 8. *h_i and $h_i^{\perp}$ are widely used in reference [31,39,40], where h_i is the unit vector in the heading direction of the UAV i and $h_i^{\perp}$ is the unit vector pointing to the vertical heading direction to the right. A graphical explanation for h_i and $h_i^{\perp}$ is shown in Figure 2.*

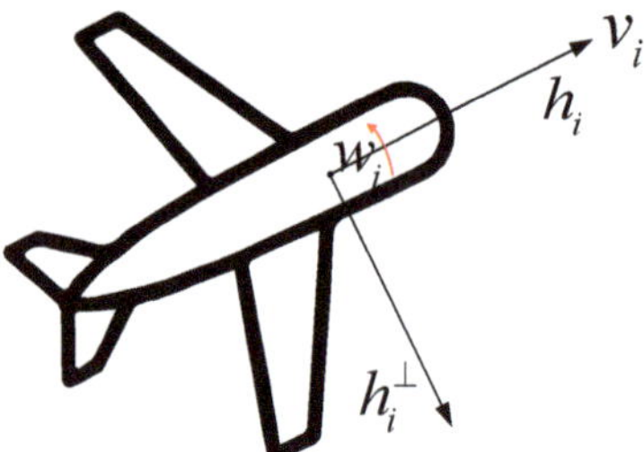

Figure 2. The illustrations of h_i and $h_i^{\perp}$.

Then, based on Equation (12), the low-gain-based controller with velocity constraints is designed as

$$\begin{aligned} v_i &= sat_{v_i}\left(h_i^T(f_i(\gamma) + \vec{v}_0)\right), \\ w_i &= sat_{w_i}\left(sat_{w_{1i}}((h_i^{\perp})^T f_i(\gamma)) + (h_i^{\perp})^T \vec{v}_0\right), \end{aligned}\tag{13}$$

where

$$sat_{v_i}(x) = \begin{cases} v_{i,\min}, & x \in (-\infty, v_{i,\min}) \\ x, & x \in [v_{i,\min}, v_{i,\max}], \\ v_{i,\max}, & x \in (v_{i,\max}, +\infty) \end{cases}$$

$$sat_{w_i}(x) = \begin{cases} -w_{i,\max}^l, & x \in (-\infty, -w_{i,\max}^l) \\ x, & x \in [-w_{i,\max}^l, w_{i,\max}^r], \\ w_{i,\max}^r, & x \in (w_{i,\max}^r, +\infty) \end{cases}\tag{14}$$

$$sat_{w_{1i}}(x) = \begin{cases} -w_{i,1\,\max}^l, & x \in (-\infty, -w_{i,1\,\max}^l) \\ x, & x \in [-w_{i,1\,\max}^l, w_{i,1\,\max}^r], \\ w_{i,1\,\max}^r, & x \in (w_{i,1\,\max}^r, +\infty) \end{cases}$$

and the function $f_i(\gamma)$ is defined in Equation (7) while h_i and $h_i^\perp$ are defined in Equation (11). The variables $w_{i,1\,\max}^l$ and $w_{i,1\,\max}^r$ are the controller parameters satisfying

$$- w_{i,\max}^l \leq -w_{i,1\,\max}^l < 0 < w_{i,1\,\max}^r \leq w_{i,\max}^r. \tag{15}$$

Remark 9. *On the basis of Equation (12), Equation (13) is designed to incorporate the saturation function to accommodate the input constraints. Furthermore, in the later analysis, Equation (13) will degenerate to Equation (12) when the gain γ is small enough, which is why the controller is called "low-gain-based".*

3.3. Stability Analysis

In this part, the stability of the system with the designed controller will be analyzed.

Theorem 1. *Consider a formation of N fixed-wing UAVs with unicycle dynamics (1) and input constraints (2). Suppose that Assumption 3 holds. Then, Problem 1 can be tackled with the contol inputs given by Equation (13). That is, for all init angle $\theta_{i0} \in (-\pi + \theta_0 + \arcsin\frac{w_{i,1\,\max}^r}{\|\vec{v}_0\|}, \pi + \theta_0 - \arcsin\frac{w_{i,1\,\max}^l}{\|\vec{v}_0\|})$, there exists a constant $\gamma^* > 0$ such that, for each given $\gamma \in (0, \gamma^*]$, the following equalities hold*

$$\begin{aligned}\|p_i(t) - p_j(t)\| &\to d_{ij}^* \quad \text{as } t \to \infty, \quad \forall (i,j) \in \mathcal{E}, \\ \dot{p}_i(t) - \vec{v}_0 &\to 0 \quad \text{as } t \to \infty, \quad i = 1,\ldots,N,\end{aligned} \tag{16}$$

where both the linear and the angular velocity constraints are satisfied.

Proof. The proof can be divided into three parts.

Firstly, it will be proved that the angle of each UAV under the angular velocity controller in Equation (13) will converge to a certain region. Consider the derivative of the heading angle of the UAV i

$$\begin{aligned}\dot{\theta}_i = w_i &= sat_{w_i}\left(sat_{w_{1i}}((h_i^\perp)^T f_i) + (h_i^\perp)^T \vec{v}_0\right) \\ &= sat_{w_i}\left(sat_{w_{1i}}((h_i^\perp)^T f_i) - \|\vec{v}_0\|sin(\theta_i - \theta_0)\right)\end{aligned}\overset{.}{} \tag{17}$$

Let $w_{i\delta} = sat_{w_{1i}}((h_i^\perp)^T f_i) \in [-w_{i,1\,\max}^l, w_{i,1\,\max}^r]$, then

$$\dot{\theta}_i = w_i = sat_{w_i}(-\|\vec{v}_0\|sin(\theta_i - \theta_0) + w_{i\delta}). \tag{18}$$

The right-hand side of Equation (18) is regarded as a function of angle θ_i and its graphical explanation is shown in Figure 3.

Without loss of generality, consider $\theta_i \in [-\pi + \theta_0, \pi + \theta_0]$. Then it holds that

$$\dot{\theta}_i = \begin{cases} > 0, & \theta_i \in [-\pi + \theta_0 + \arcsin\frac{w_{i,1\,\max}^r}{\|\vec{v}_0\|}, \theta_0 - \arcsin\frac{w_{i,1\,\max}^l}{\|\vec{v}_0\|}) \\ --, & \theta_i \in [\theta_0 - \arcsin\frac{w_{i,1\,\max}^l}{\|\vec{v}_0\|}, \theta_0 + \arcsin\frac{w_{i,1\,\max}^r}{\|\vec{v}_0\|}] \\ < 0, & \theta_i \in (\theta_0 + \arcsin\frac{w_{i,1\,\max}^r}{\|\vec{v}_0\|}, \pi + \theta_0 - \arcsin\frac{w_{i,1\,\max}^l}{\|\vec{v}_0\|}] \end{cases}, \tag{19}$$

which means that $[\theta_0 - \arcsin\frac{w_{i,1\,\max}^l}{\|\vec{v}_0\|}, \theta_0 + \arcsin\frac{w_{i,1\,\max}^r}{\|\vec{v}_0\|}]$ is an atracting set for $(-\pi + \theta_0 + \arcsin\frac{w_{i,1\,\max}^r}{\|\vec{v}_0\|}, \pi + \theta_0 - \arcsin\frac{w_{i,1\,\max}^l}{\|\vec{v}_0\|})$, i.e., $\theta_i \to [\theta_0 - \arcsin\frac{w_{i,1\,\max}^l}{\|\vec{v}_0\|}, \theta_0 + \arcsin\frac{w_{i,1\,\max}^r}{\|\vec{v}_0\|}]$ as $t \to \infty$ while the symbol "$--$" denotes that the positive or negative sign is uncertain.

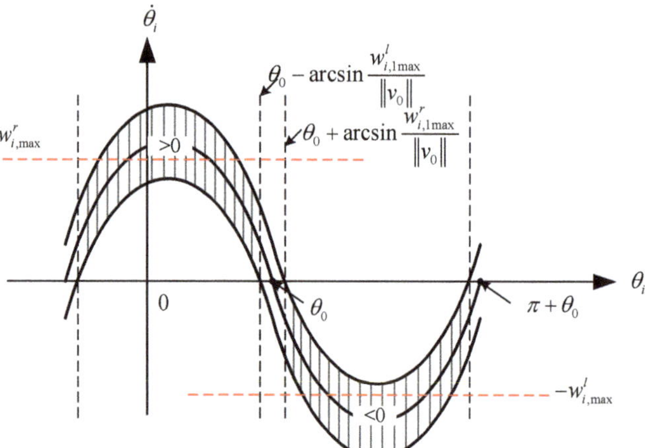

Figure 3. The explanation of Equation (18).

Then, the existence of γ which makes the velocity of the UAVs unsaturated is discussed, when the angle θ_i converges to a certain region.

It can be observed that the linear and angular velocity controllers are the projections of vector $\vec{v}_0$ and vector $f_i(\gamma)$ in the direction of h_i and $h_i^{\perp}$, respectively, with the saturation function added. More intuitively, a scheme of the controller is drawn as Figure 4 for the i-th UAV in the local coordinate frame.

As shown in Figure 4, the two red vertical dashed lines represent the linear velocity constraints in the forward direction, while the two red horizontal dashed lines and the two yellow horizontal dashed lines represent the maximum angular velocity constraints for $\{w_{i,\max}^l, w_{i,\max}^r\}$ and $\{w_{i,1\max}^l, w_{i,1\max}^r\}$, respectively. The blue arrow shows the vector $\vec{v}_0$, and the dark red arrows in four directions show the "shortest" vector $f_i(\gamma)$ that reaches the saturation condition. That is, if the "shortest" vector exists in all four directions, there exists γ that makes all the saturation functions in Equation (13) not work due to the fact that $\lim_{\gamma \to 0} f_i(\gamma) = 0$.

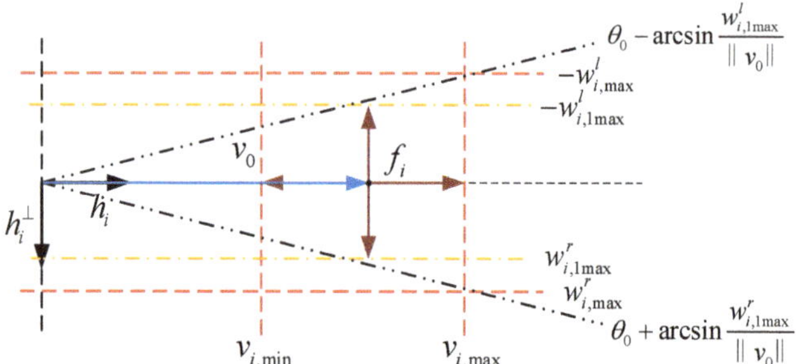

Figure 4. An intuitive presentation about the proposed controller with input constraints.

Let the "shortest" vectors in the four directions be f_i^{up}, f_i^{down}, f_i^{left}, and f_i^{right}, respectively. Clearly as shown in Figure 4, f_i^{right} reaches a minimum when the vector $\vec{v}_0$ and the vector h_i are in the same direction, i.e.,

$$\left\| f_i^{right} \right\|_{\min} = v_{i,\max} - \|\vec{v}_0\|. \tag{20}$$

Obviously, as the angle between the vector $\vec{v}_0$ and the vector h_i changes, the "shortest" vector f_i in each of the four directions changes. Consider the two extreme cases (i.e., the angle reaches its maximum) as shown in Figure 5a,b below:

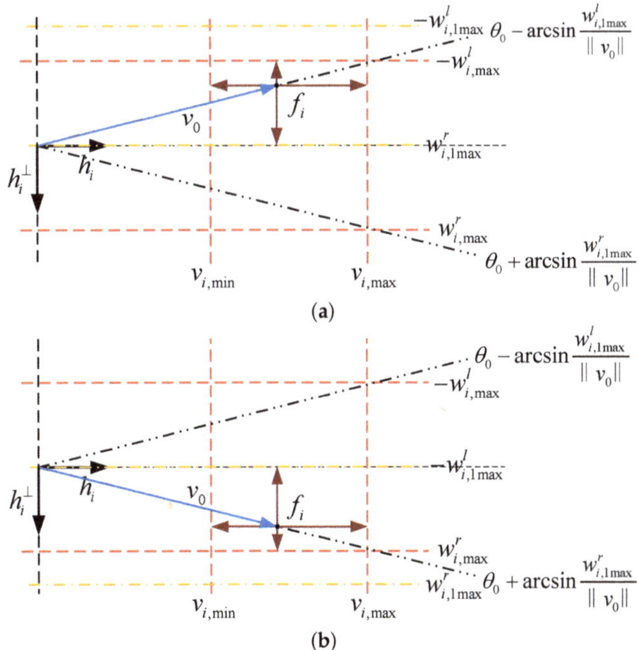

(a)

(b)

Figure 5. Two extreme cases where the angle reaches its maximum. (**a**) An extreme case where θ_i is equal to $\theta_0 - \arcsin\frac{w^l_{i,1\max}}{\|\vec{v}_0\|}$. (**b**) Another extreme case where θ_i is equal to $\theta_0 + \arcsin\frac{w^r_{i,1\max}}{\|\vec{v}_0\|}$.

From Figure 5a, the length of f_i^{up} and f_i^{left} in this case reaches the minimum, respectively. Further, the minimum can be obtained by the following equations, respectively:

$$
\begin{aligned}
\left\|f_i^{up}\right\|_{\min} &= w^l_{i,\max} - w^l_{i,1\max}, \\
\left\|f_i^{left}\right\|_{1\min} &= \|\vec{v}_0\|\cos\left(\arcsin\frac{w^l_{i,1\max}}{\|\vec{v}_0\|}\right) - v_{i,\min}.
\end{aligned}
\tag{21}
$$

Furthermore, from Figure 5b, the length of f_i^{down} and f_i^{left} in this case reaches the minimum, respectively. Further, the minimum can be obtained by the following equations, respectively:

$$
\begin{aligned}
\left\|f_i^{down}\right\|_{\min} &= w^r_{i,\max} - w^r_{i,1\max}, \\
\left\|f_i^{left}\right\|_{2\min} &= v_{i,\max} - \|\vec{v}_0\|\cos\left(\arcsin\frac{w^r_{i,1\max}}{\|\vec{v}_0\|}\right).
\end{aligned}
\tag{22}
$$

Now, let $M_i = \min\{\left\|f_i^{up}\right\|_{\min}, \left\|f_i^{down}\right\|_{\min}, \left\|f_i^{left}\right\|_{\min}, \left\|f_i^{right}\right\|_{\min}\}$, where $\left\|f_i^{left}\right\|_{\min} = \min\{\left\|f_i^{left}\right\|_{1\min}, \left\|f_i^{left}\right\|_{2\min}\}$. Then $M = \min\{\cdots, M_i, \cdots\}$. Obviously $M > 0$. Furthermore, for f_i, since the distance between the UAVs is bounded combined with Assumption 3, it follows that $\sum_{j\in\mathcal{N}_i} g(\|p_{ij}\|)$ is bounded, so there exists a sufficiently small γ such that $f_i(\gamma) < M$ always holds.

Since $f_i(\gamma) < M$ always holds, the following inequality will hold for each UAV i

$$
\begin{aligned}
v_{i,\min} &< h_i^T(f_i + \vec{v}_0) < v_{i,\max}, \\
-w_{i,1\,\max}^l &< (h_i^\perp)^T f_i < w_{i,1\,\max}^r, \\
-w_{i,\max}^l &< (h_i^\perp)^T f_i + (h_i^\perp)^T \vec{v}_0 < w_{i,\max}^r,
\end{aligned}
\tag{23}
$$

where Equation (13) degenerates into Equation (12) which does not involve the saturation function. This means that all velocity constraints are satisfied.

Finally, consider the Lyapunov function candidate

$$
V = \frac{1}{2} \sum_{i \in \mathcal{V}} F_i + \sum_{i \in \mathcal{V}} \|\vec{v}_0\| (1 - \cos(\theta_i - \theta_0)).
\tag{24}
$$

Taking the differential of Equation (24) yields

$$
\begin{aligned}
\dot{V} &= \frac{1}{2} \sum_{i \in \mathcal{V}} \left[\gamma \sum_{j \in \mathcal{N}_i} g(\|p_{ij}\|) p_{ij}^T (\dot{p}_i - \dot{p}_j) \right] + \sum_{i \in \mathcal{V}} \|\vec{v}_0\| \sin(\theta_i - \theta_0) w_i \\
&= \sum_{i \in \mathcal{V}} \left[\gamma \sum_{j \in \mathcal{N}_i} g(\|p_{ij}\|) p_{ij}^T \dot{p}_i \right] - \sum_{i \in \mathcal{V}} \|\vec{v}_0\| {h_i^\perp}^T h_0 w_i \\
&= \sum_{i \in \mathcal{V}} \left[\gamma \sum_{j \in \mathcal{N}_i} g(\|p_{ij}\|) p_{ij}^T \dot{p}_i - \|\vec{v}_0\| {h_i^\perp}^T h_0 w_i \right] \\
&= \sum_{i \in \mathcal{V}} \left[-f_i^T \dot{p}_i - \|\vec{v}_0\| {h_i^\perp}^T h_0 w_i \right] \\
&= \sum_{i \in \mathcal{V}} \left[-f_i^T h_i h_i^T (f_i + \vec{v}_0) - \|\vec{v}_0\| {h_i^\perp}^T h_0 {h_i^\perp}^T (f_i + \vec{v}_0) \right] \\
&= \sum_{i \in \mathcal{V}} \left[-f_i^T h_i h_i^T f_i - f_i^T h_i h_i^T \vec{v}_0 - {h_i^\perp}^T \vec{v}_0 {h_i^\perp}^T f_i - {h_i^\perp}^T \vec{v}_0 {h_i^\perp}^T \vec{v}_0 \right] \\
&= \sum_{i \in \mathcal{V}} \left[-f_i^T h_i h_i^T f_i - f_i^T h_i h_i^T \vec{v}_0 - f_i^T h_i^\perp {h_i^\perp}^T \vec{v}_0 - {h_i^\perp}^T \vec{v}_0 {h_i^\perp}^T \vec{v}_0 \right] \\
&= \sum_{i \in \mathcal{V}} \left[-(h_i^T f_i)^2 - ({h_i^\perp}^T \vec{v}_0)^2 - f_i^T \vec{v}_0 \right] \\
&= \sum_{i \in \mathcal{V}} \left[-(h_i^T f_i)^2 - ({h_i^\perp}^T \vec{v}_0)^2 + \gamma \sum_{j \in \mathcal{N}_i} g(\|p_{ij}\|) p_{ij}^T \vec{v}_0 + \right] \\
&= \sum_{i \in \mathcal{V}} \left[-(h_i^T f_i)^2 - ({h_i^\perp}^T \vec{v}_0)^2 \right] \leq 0,
\end{aligned}
\tag{25}
$$

which implies that the system is stable. The equality $\dot{V} = 0$ yields that $h_i^T f_i = 0$ and ${h_i^\perp}^T \vec{v}_0 = 0$, which further implies that $\theta_i = \theta_0$ or $\theta_i = \theta_0 + \pi$. However, $\theta_i = \theta_0 + \pi$ is impossible because the first part proves that the angle θ_i will converge to a certain region near θ_0. On the other hand, the former can be considered in the following two cases:

1. $f_i = 0$;
2. $h_i^T f_i = 0$ but $f_i \neq 0$.

For the case 1, the desired formation is obviously achieved. The case 2 is discussed below. Consider the dynamics of UAV i's position:

$$\dot{p}_i = h_i v_i$$

$$= h_i sat\left(h_i^T f_i(\gamma) + h_i^T \vec{v}_0\right)$$

$$= h_i h_i^T f_i(\gamma) + h_i h_i^T \vec{v}_0 \qquad , \tag{26}$$

$$= h_i h_i^T f_i(\gamma) + \vec{v}_0 - h_i^\perp h_i^{\perp T} \vec{v}_0$$

$$= \vec{v}_0$$

which means that $\theta_i = \theta_0$, i.e., the vector h_i is invariant. Whereas considering the dynamics of the UAV i's angle, it holds that

$$\dot{\theta}_i = sat_{w_i}\left(sat_{w_{1i}}((h_i^\perp)^T f_i(\gamma)) + (h_i^\perp)^T \vec{v}_0\right)$$

$$= (h_i^\perp)^T f_i(\gamma) + (h_i^\perp)^T \vec{v}_0 \qquad , \tag{27}$$

$$= (h_i^\perp)^T f_i(\gamma)$$

$$\neq 0$$

which means that θ_i is always changing, so the contradiction arises and the case 2 is not valid.

With the discussion mentioned above, the system will converge to the desired formation, i.e., Equation (3) holds and the proof is complete. $\square$

Remark 10. *The effect of low gain γ is to disable all saturation in the controller given by Equation (13). In practice, it works well to protect the fixed-wing UAV from being saturated all the time. Note that all saturation functions will not work after the angle θ_i converges to a certain region.*

Remark 11. *Although Theorem 1 requires the initial angle of all UAVs to be within a certain region, in practice the angle is usually unstable outside of that range, and it has a tendency to enter a certain region. Therefore practically the initial angle can be arbitrary, as can be verified in the experimental results in the following part.*

4. Simulations

In this section, the effectiveness of the proposed low-gain-based controller is verified in numerical and semi-physical simulations, while the corresponding simulation results are analyzed as well.

4.1. Simulation Setup

In the simulation, a formation of five fixed-wing UAVs is considered. Furthermore, the desired formation shape is a regular pentagon whose underlying graph as shown in Figure 6a where $d_{12}^* = d_{23}^* = d_{34}^* = d_{45}^* = d_{15}^* = 2r_d sin(\pi/5)$ and $d_{25}^* = d_{35}^* = 2r_d sin(2\pi/5)$. Additionally, the velocity constraints for each UAV i are given as follows:

$$\begin{aligned} v_{i,max} &= 16 \text{ (m/s)}, \\ v_{i,min} &= 10 \text{ (m/s)}, \\ w_{i,max}^l &= \pi/3 \text{ (rad/s)}, \\ w_{i,max}^r &= \pi/3 \text{ (rad/s)}. \end{aligned} \tag{28}$$

Meanwhile, the direction of the uniform velocity $\vec{v}_0$ is shown by the arrows in Figure 6b with constant magnitude 13 m/s. In Figure 6b, the total time is divided equally into five periods, (t_0, t_1), (t_1, t_2), (t_2, t_3), (t_3, t_4) and (t_4, t_5), and the desired uniform velocity $\vec{v}_0$ during these five periods are $\vec{v}_0^{(0)}, \vec{v}_0^{(1)}, \vec{v}_0^{(2)}, \vec{v}_0^{(3)}$, and $\vec{v}_0^{(4)}$, respectively.

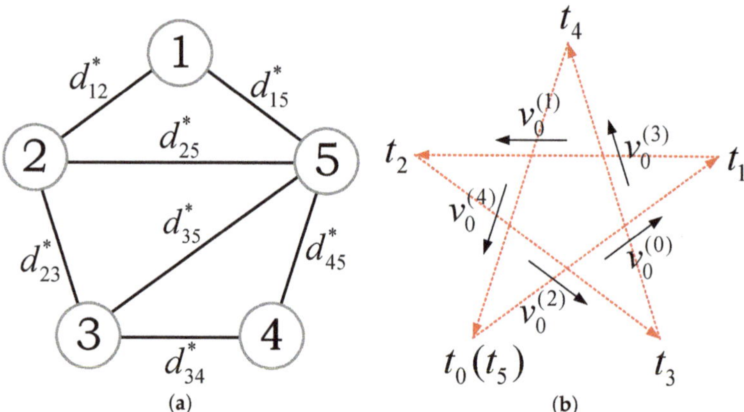

Figure 6. The illustration of some simulation settings. (**a**) The underlying graph of five fixed-wing UAVs. (**b**) The setting of $\vec{v}_0$.

Next, for each UAV i, choose the following potential function:

$$F_i = \gamma \sum_{j \in \mathcal{N}_i} \frac{1}{2} \left[\left(\|p_{ij}\|^2 + \frac{d_{ij}^{*4}}{\|p_{ij}\|^2} \right) - d_{ij}^{*2} \right]. \tag{29}$$

Then

$$f_i = -\nabla_{p_i} F_i = -\gamma \sum_{j \in \mathcal{N}_i} \left(1 - \frac{d_{ij}^{*4}}{\|p_{ij}\|^4} \right) p_{ij}. \tag{30}$$

Let $w_{i,1\max}^l = w_{i,1\max}^r = \pi/4$ for each UAV i. It is then easy to obtain $\left\| f_i^{up} \right\|_{\min} = 0.2618$, $\left\| f_i^{down} \right\|_{\min} = 0.2618$, $\left\| f_i^{left} \right\|_{\min} = 1.9807$, $\left\| f_i^{right} \right\|_{\min} = 4$. Furthermore, note that

$$\sum_{j \in \mathcal{N}_i} \left(1 - \frac{d_{ij}^{*4}}{\|p_{ij}\|^4} \right) \le 4 \left(1 - \frac{d_{ij}^{*4}}{d_M^4} \right) < 4. \tag{31}$$

Assuming that $d_M = 10$, then γ is set as 0.0064.

4.2. Numerical Simulation

In the numerical simulation, the initial positions of the five UAVs are $[20, 8]^T$, $[8, 8]^T$, $[4, 0]^T$, $[14, -10]^T$, $[24, 0]^T\,(m)$, and heading angles are 0 (rad), respectively. The parameter r_d is set to be 10.

To better illustrate the impact of the input constraints on the controller design, we simulate the control algorithm proposed in the reference [23], where the controller parameters are given in the simulation part of reference [23]. It should be noted that reference [23] does not perform stability analysis of the control algorithm in the case of the presence of input constraints. Figure 7c,d illustrate the control inputs v_i and w_i of the algorithm proposed in reference [23] without the velocity constraints. Although the angular velocities can satisfy the constraints after saturating all velocities, it can be observed that the linear velocities still exceed the input constraints defined in Equation (2).

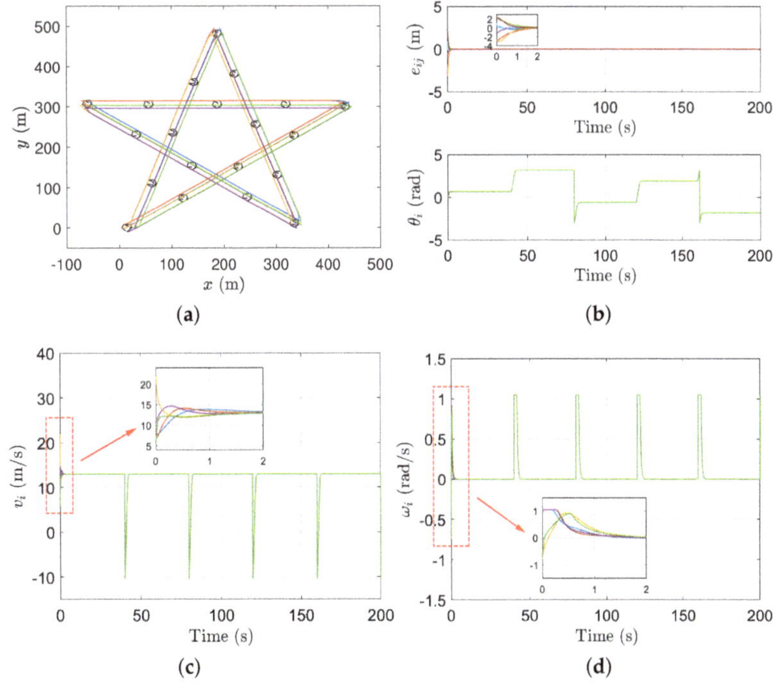

Figure 7. Numerical simulation results of the method proposed in [23] without input constraints. (**a**) The illustration of trajectory of the five UAVs controlled by the method proposed in [23] without input constraints. (**b**) The illustration of distance errors and θ_i controlled by the method proposed in [23] without input constraints. (**c**) The linear velocity inputs v_i of the five UAVs controlled by the method proposed in [23] without input constraints. (**d**) The angular velocity inputs w_i of the five UAVs controlled by the method proposed in [23] without input constraints.

Then, to verify the effectiveness of the control algorithm proposed in this paper, we simulate the method proposed in [23] and our algorithm in the same situation where the constraints (2) are enforced on the UAVs. Figure 8 illustrates the numerical simulation results of the two methods. It can be seen that although the algorithm proposed in reference [23] performs well when there exists no input constraints as shown in Figure 7, the control algorithm fails when the constraints (2) are enforced on the UAVs, as shown in Figure 8. Instead, under the control of our method, the input constraints of each UAV are satisfied while the desired distance-based formation is achieved.

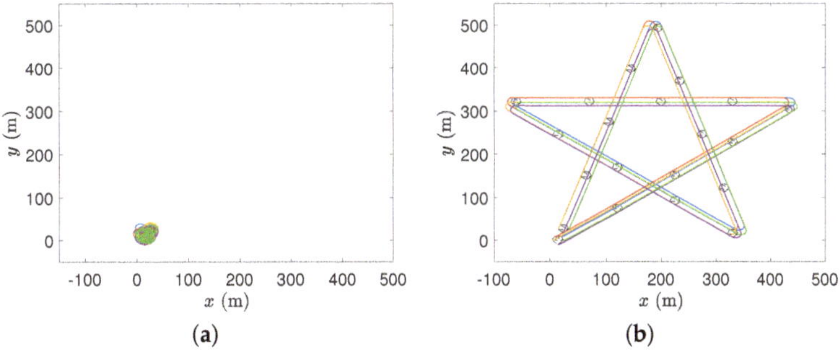

Figure 8. *Cont.*

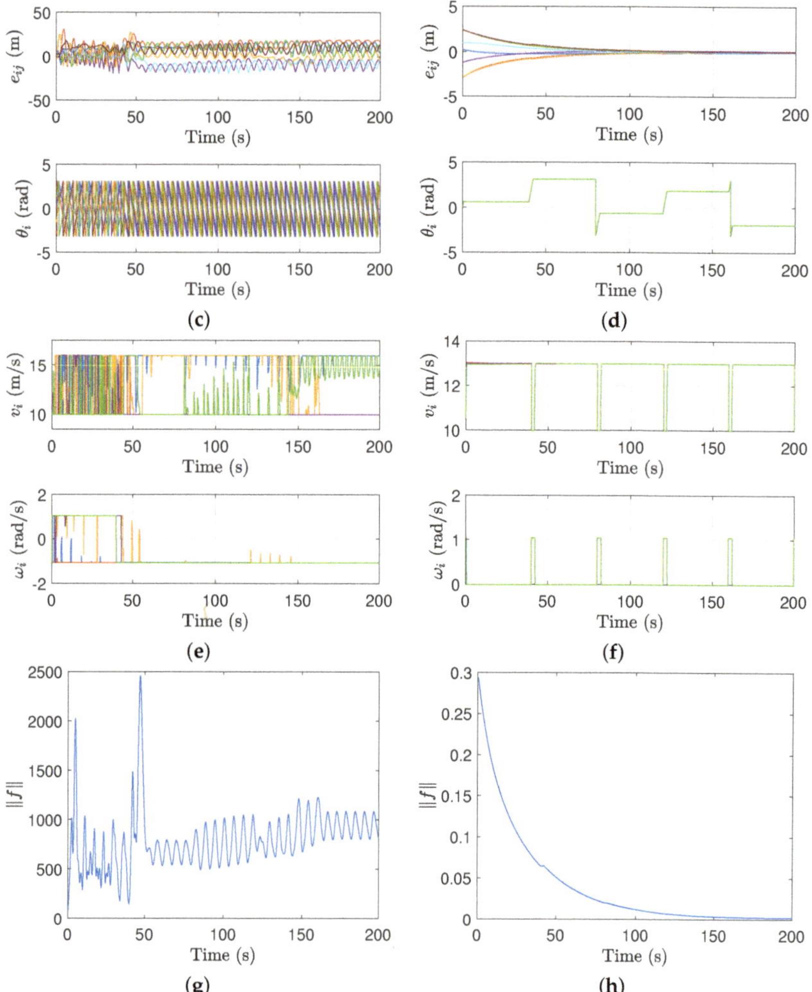

Figure 8. The method in reference [23] vs. our method. (**a**) The illustration of trajectory of the five UAVs controlled by the method proposed in [23]. (**b**) The illustration of trajectory of the five UAVs controlled by our method. (**c**) The illustration of distance errors and θ_i controlled by the method proposed in [23]. (**d**) The illustration of distance errors and θ_i controlled by our method. (**e**) The illustration of control input v_i and w_i controlled by the method proposed in [23]. (**f**) The illustration of control input v_i and w_i controlled by our method. (**g**) The illustration of the norm of f controlled by the method proposed in [23]. (**h**) The illustration of the norm of f controlled by our method.

Figure 9 illustrates that the value of the Lyapunov function converges to zero. It can also be seen that there is a sharp peak in its Lyapunov function when the uniform velocity $\vec{v}_0$ changes. The reason for this phonomenon is that it is the uniform linear velocity rather than the uniform angular velocity that is considered in this paper.

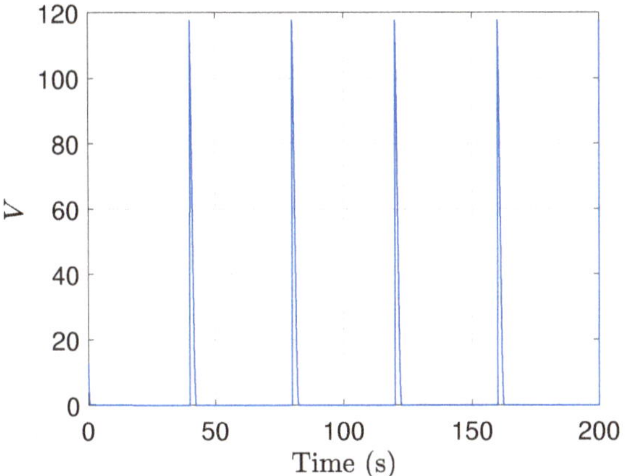

Figure 9. The illustration of Lyapunov function controlled by our method.

To further illustrate that our algorithm can tackle the case of non-identical input constraints, we modify the velocity constraints of each UAV as

$$
\begin{aligned}
9 \le v_1 \le 15, &\quad |w_1| \le 0.95, \\
9.5 \le v_2 \le 15.5, &\quad |w_2| \le 1, \\
10 \le v_3 \le 16, &\quad |w_3| \le 1.05, \\
10.5 \le v_4 \le 16.5, &\quad |w_4| \le 1.1, \\
11 \le v_5 \le 17, &\quad |w_5| \le 1.15,
\end{aligned}
\tag{32}
$$

where the units of velocity and angular velocity are m/s and rad/s, respectively.

Figures 10 and 11 show the simulation results for the case of non-identical input constraints. It can be seen that the input constraints are still satisfied although the convergence time becomes longer. This is the result of a trade-off in the control algorithm.

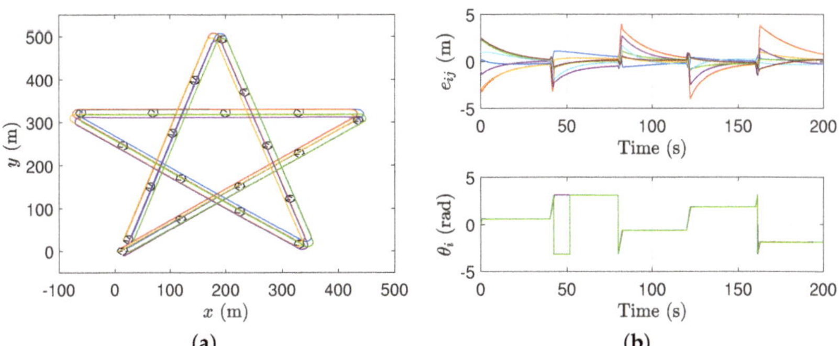

(a)   (b)

Figure 10. Numerical simulation results for the case of non-identical input constraints. (**a**) The illustration of trajectory of the five UAVs controlled by our method with non-identical input constraints. (**b**) The illustration of distance errors and θ_i controlled by our method with non-identical input constraints.

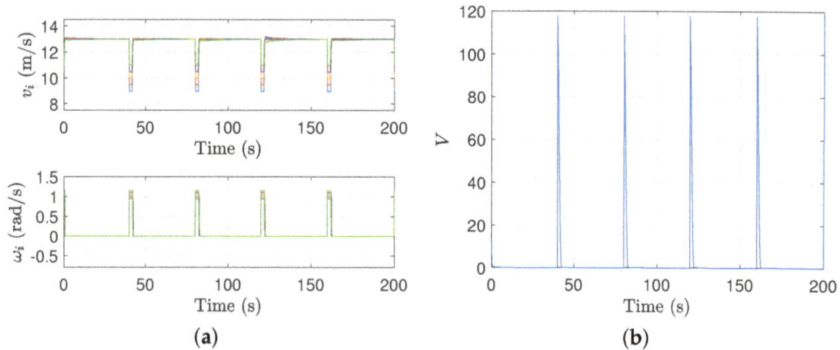

Figure 11. Numerical simulation results for the case of non-identical input constraints. (**a**) The illustration of control input v_i and w_i controlled by our method with non-identical input constraints. (**b**) The illustration of Lyapunov function controlled by our method with non-identical input constraints.

4.3. Semi-Physical Simulation

To prove that our method can be apllied to the physical UAV system, the proposed formation controller is further validated in a semi-physical simulation system.

4.3.1. Semi-Physical Simulation System

The semi-physical simulation system consists of four main parts: onboard computer, autopilot, ground station, and switch. The relationships among them are shown in Figure 12. The functional details of each component are introduced in references [41,42].

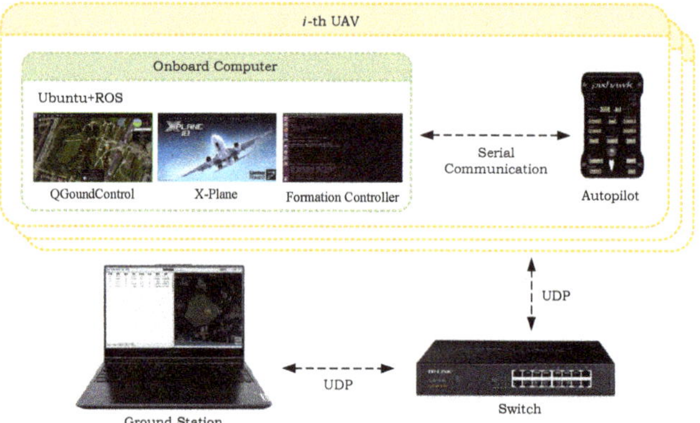

Figure 12. The components of semi-physical simulation system.

In this simulation system, the software X-plane is used to simulate the dynamics of UAVs as well as the flight environment, which is a professional flight simulation software with powerful features, providing high precision dynamics models of UAVs and realistic 3D simulation scenarios. Meanwhile, the autopilot is used for the hardware-in-the-loop (HIL) experiment, which will further narrow the gap between the simulation and the physical reality.

In this paper, we select the HiLStar17 as the model for the semi-physical simulation as shown in Figure 13.

Figure 13. The UAV model used in this paper.

4.3.2. Semi-Physical Simulation Results

In the semi-physical simulation, the parameter r_d will be adjusted to 100 to accommodate the realistic formation formation flight. Then, the semi-physical simulation results are shown as follows.

Figure 14 shows the initial position of the UAV displayed in the ground station and the evolution of the trajectory, and a more detailed trajectory is shown in Figure 15a.

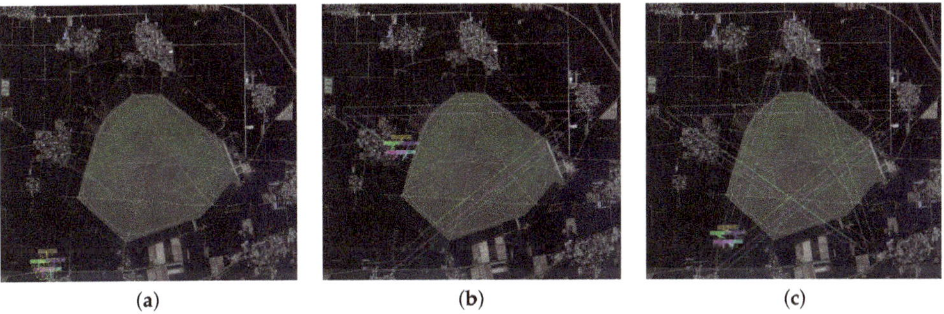

Figure 14. The evolution of the trajectory. (**a**) Initial positions. (**b**) Positions at 583 s. (**c**) Positions at 1248 s.

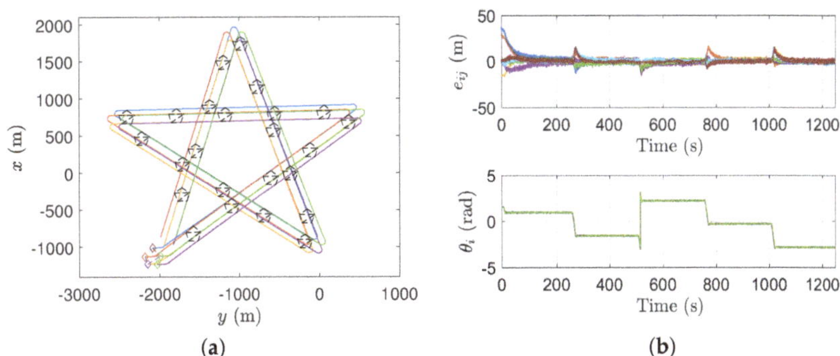

Figure 15. *Cont.*

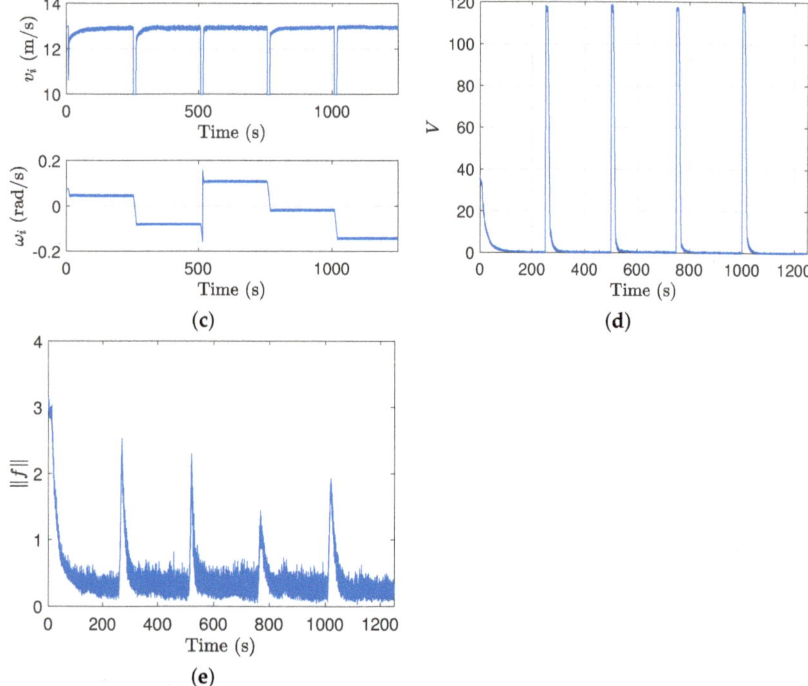

Figure 15. Semi-physical simulation results. (**a**) The illustration of trajectory of the five UAVs in the semi-physical simulation. (**b**) The illustration of distance errors and θ_i in the semi-physical simulation. (**c**) The illustration of control input v_i and w_i in the semi-physical simulation. (**d**) The illustration of Lyapunov function in the semi-physical simulation. (**e**) The illustration of the norm of f in the semi-physical simulation.

Figure 15b shows the distance error and the angle of the UAVs, which converge to the desired values. Figure 15c illustrates the variation of the control input of the UAV labeled by the number 1 during the simulation. Figure 15d then indicates that the value of Lyapunov function converges to zero. Finally Figure 15e indicates that the norm of f is gradually decreasing, which means that the control input energy of the formation is reduced.

5. Conclusions

With the idea of the low gain technique, this paper proposes a low-gain formation controller to solve the formation control problem of distance-based fixed-wing UAVs subject to the input constraints. The proposed controller is designed based on the potential function and can achieve the formation of fixed-wing UAVs while satisfying the velocity constraints. The numerical simulations and the semi-physical simulations are carried out to vefify the effectiveness of the proposed algorithm.

In the future, the formation with the uniform angular velocity will be further considered, and the obstacle avoidance algorithm will also be incorporated to consider the formation and obstacle avoidance problem as a whole.

Author Contributions: Conceptualization, J.Y.; methodology, J.Y.; software, J.Y.; formal analysis, J.Y.; data curation, J.Y.; writing—original draft preparation, J.Y.; writing—review and editing, Y.Y. and X.W.; project administration, X.W.; funding acquisition, X.W. All authors have read and agreed to the published version of the manuscript.

Funding: This research was funded by Natural Science Foundation of Hunan Province under Grant 2021JJ10053 and National Natural Science Foundation of China under Grant 61973309.

Institutional Review Board Statement: Not applicable.

Informed Consent Statement: Not applicable.

Data Availability Statement: Not applicable.

References

1. Chung, S.J.; Paranjape, A.A.; Dames, P.; Shen, S.; Kumar, V. A survey on aerial swarm robotics. *IEEE Trans. Robot.* **2018**, *34*, 837–855. [CrossRef]
2. Chen, S.; Wu, F.; Shen, L.; Chen, J.; Ramchurn, S.D. Decentralized patrolling under constraints in dynamic environments. *IEEE Trans. Cybern.* **2015**, *46*, 3364–3376. [CrossRef] [PubMed]
3. Fathian, K.; Safaoui, S.; Summers, T.H.; Gans, N.R. Robust distributed planar formation control for higher order holonomic and nonholonomic agents. *IEEE Trans. Robot.* **2020**, *37*, 185–205. [CrossRef]
4. Scherer, J.; Yahyanejad, S.; Hayat, S.; Yanmaz, E.; Andre, T.; Khan, A.; Vukadinovic, V.; Bettstetter, C.; Hellwagner, H.; Rinner, B. An autonomous multi-UAV system for search and rescue. In Proceedings of the First Workshop on Micro Aerial Vehicle Networks, Systems, and Applications for Civilian Use, Florence, Italy, 18 May 2015; pp. 33–38.
5. Liu, Z.; Wang, X.; Shen, L.; Zhao, S.; Cong, Y.; Li, J.; Yin, D.; Jia, S.; Xiang, X. Mission-oriented miniature fixed-wing UAV swarms: A multilayered and distributed architecture. *IEEE Trans. Syst. Man Cybern. Syst.* **2022**, *52*, 1588–1602. [CrossRef]
6. Oh, K.K.; Park, M.C.; Ahn, H.S. A survey of multi-agent formation control. *Automatica* **2015**, *53*, 424–440. [CrossRef]
7. Nuno, E.; Loria, A.; Hernández, T.; Maghenem, M.; Panteley, E. Distributed consensus-formation of force-controlled nonholonomic robots with time-varying delays. *Automatica* **2020**, *120*, 109114. [CrossRef]
8. Maghenem, M.; Bautista, A.; Nuño, E.; Loría, A.; Panteley, E. Consensus of multi-agent systems with nonholonomic restrictions via Lyapunov's direct method. *IEEE Control Syst. Lett.* **2018**, *3*, 344–349. [CrossRef]
9. Babazadeh, R.; Selmic, R. Anoptimal displacement-based leader-follower formation control for multi-agent systems with energy consumption constraints. In Proceedings of the 26th Mediterranean Conference on Control and Automation, Zadar, Croatia, 19–22 June 2018; pp. 179–184.
10. de Marina, H.G. Maneuvering and robustness issues in undirected displacement-consensus-based formation control. *IEEE Trans. Autom. Control* **2020**, *66*, 3370–3377. [CrossRef]
11. Mehdifar, F.; Bechlioulis, C.P.; Hashemzadeh, F.; Baradarannia, M. Prescribed performance distance-based formation control of multi-agent systems. *Automatica* **2020**, *119*, 109086. [CrossRef]
12. Krick, L.; Broucke, M.E.; Francis, B.A. Stabilisation of infinitesimally rigid formations of multi-robot networks. *Int. J. Control* **2009**, *82*, 423–439. [CrossRef]
13. Oh, K.K.; Ahn, H.S. Formation control of mobile agents based on inter-agent distance dynamics. *Automatica* **2011**, *47*, 2306–2312. [CrossRef]
14. Oh, K.K.; Ahn, H.S. Distance-based undirected formations of single-integrator and double-integrator modeled agents in n-dimensional space. *Int. J. Robust. Nonlinear Control* **2014**, *24*, 1809–1820. [CrossRef]
15. Sun, Z.; Mou, S.; Anderson, B.D.; Cao, M. Exponential stability for formation control systems with generalized controllers: A unified approach. *Syst. Control Lett.* **2016**, *93*, 50–57. [CrossRef]
16. Van Vu, D.; Trinh, M.H.; Nguyen, P.D.; Ahn, H.S. Distance-based formation control with bounded disturbances. *IEEE Control Syst. Lett.* **2020**, *5*, 451–456. [CrossRef]
17. Mou, S.; Belabbas, M.A.; Morse, A.S.; Sun, Z.; Anderson, B.D. Undirected rigid formations are problematic. *IEEE Trans. Autom. Control* **2015**, *61*, 2821–2836. [CrossRef]
18. Bae, Y.B.; Lim, Y.H.; Ahn, H.S. Distributed robust adaptive gradient controller in distance-based formation control with exogenous disturbance. *IEEE Trans. Autom. Control* **2020**, *66*, 2868–2874. [CrossRef]
19. Babazadeh, R.; Selmic, R. Optimal distance-based formation producing control of multi-agent systems with energy constraints and collision avoidance. In Proceedings of the 59th IEEE Conference on Decision and Control, Nice, France, 11–13 December 2019; pp. 3847–3853.
20. Wang, Y.; Cheng, L.; Hou, Z.G.; Yu, J.; Tan, M. Optimal formation of multirobot systems based on a recurrent neural network. *IEEE Trans. Neural Netw. Learn. Syst.* **2015**, *27*, 322–333. [CrossRef]
21. Sun, Z.; Mou, S.; Deghat, M.; Anderson, B.D.; Morse, A.S. Finite time distance-based rigid formation stabilization and flocking. *IFAC Proc. Vol.* **2014**, *47*, 9183–9189. [CrossRef]
22. Deghat, M.; Anderson, B.D.; Lin, Z. Combined flocking and distance-based shape control of multi-agent formations. *IEEE Trans. Autom. Control* **2015**, *61*, 1824–1837. [CrossRef]
23. Khaledyan, M.; Liu, T.; Fernandez-Kim, V.; de Queiroz, M. Flocking and target interception control for formations of nonholonomic kinematic agents. *IEEE Trans. Control Syst. Technol.* **2019**, *28*, 1603–1610. [CrossRef]

24. Chen, J.; Sun, D.; Yang, J.; Chen, H. Leader-follower formation control of multiple non-holonomic mobile robots incorporating a receding-horizon scheme. *Int. J. Robot. Res.* **2010**, *29*, 727–747. [CrossRef]
25. Dong, W.; Farrell, J.A. Decentralized cooperative control of multiple nonholonomic dynamic systems with uncertainty. *Automatica* **2009**, *45*, 706–710. [CrossRef]
26. Gazi, V.; Fidan, B.; Ordonez, R.; İlter Köksal, M. A target tracking approach for nonholonomic agents based on artificial potentials and sliding mode control. *J. Dyn. Syst. Meas. Control* **2012**, *134*, 061004. [CrossRef]
27. Consolini, L.; Morbidi, F.; Prattichizzo, D.; Tosques, M. Leader–follower formation control of nonholonomic mobile robots with input constraints. *Automatica* **2008**, *44*, 1343–1349. [CrossRef]
28. Yu, X.; Liu, L. Distributed formation control of nonholonomic vehicles subject to velocity constraints. *IEEE Trans. Ind. Electron.* **2015**, *63*, 1289–1298. [CrossRef]
29. Meng, Z.; Zhao, Z.; Lin, Z. On global leader-following consensus of identical linear dynamic systems subject to actuator saturation. *Syst. Control Lett.* **2013**, *62*, 132–142. [CrossRef]
30. Wang, X.; Yu, Y.; Li, Z. Distributed sliding mode control for leader-follower formation flight of fixed-wing unmanned aerial vehicles subject to velocity constraints. *Int. J. Robust. Nonlinear Control* **2021**, *31*, 2110–2125. [CrossRef]
31. Zhao, S.; Dimarogonas, D.V.; Sun, Z.; Bauso, D. A general approach to coordination control of mobile agents with motion constraints. *IEEE Trans. Autom. Control* **2017**, *63*, 1509–1516. [CrossRef]
32. Zheng, Z.; Yi, H. Backstepping control design for UAV formation with input saturation constraint and model uncertainty. In Proceedings of the 36th Chinese Control Conference, Dalian, China, 26–28 July 2017; pp. 6056–6060.
33. Wei, J.; Li, H.; Guo, M.; Li, J.; Huang, H. Backstepping control based on constrained command filter for hypersonic flight vehicles with AOA and actuator constraints. *Int. J. Aerosp. Eng.* **2021**, *2021*, 8620873. [CrossRef]
34. Zhao, M.; Peng, Y.; Wang, Y.; Zhang, D.; Luo, J.; Pu, H. Concise leader-follower formation control of underactuated unmanned surface vehicle with output error constraints. *Meas. Control* **2022**, *44*, 1081–1094. [CrossRef]
35. Su, H.; Chen, M.Z.; Lam, J.; Lin, Z. Semi-global leader-following consensus of linear multi-agent systems with input saturation via low gain feedback. *IEEE Trans. Circuits Syst. I Regul. Pap.* **2013**, *60*, 1881–1889. [CrossRef]
36. Chang, J.L. Robust low gain output feedback sliding mode control design against actuator saturation. *IMA J. Math. Control Inf.* **2019**, *36*, 1237–1253.
37. Xu, J.; Lin, Z. Low gain feedback for fractional-order linear systems and semi-global stabilization in the presence of actuator saturation. *Nonlinear Dyn.* **2022**, *107*, 3485–3504. [CrossRef]
38. Zhao, G.; Wang, Z.; Fu, X. Fully distributed dynamic event-triggered semiglobal consensus of multi-agent uncertain systems with input saturation via low-gain feedback. *Int. J. Control Autom. Syst.* **2021**, *19*, 1451–1460. [CrossRef]
39. Li, H.; Chen, H.; Yang, S.; Wang, X. Standard formation generation and keeping of unmanned aerial vehicles through a potential functional approach. In Proceedings of the 39th Chinese Control Conference, Shenyang, China, 27–29 July 2020; pp. 4771–4776.
40. Li, H.; Chen, H.; Wang, X. Affine formation tracking control of unmanned aerial vehicles. *Front. Inf. Technol. Electron. Eng.* **2022**, 1–11. [CrossRef]
41. Yan, C.; Xiang, X.; Wang, C.; Lan, Z. Flocking and collision avoidance for a dynamic squad of fixed-wing UAVs using deep reinforcement learning. In Proceedings of the 2021 IEEE/RSJ International Conference on Intelligent Robots and Systems, Prague, Czech Republic, 27 September 2021; pp. 4738–4744.
42. Chen, H.; Wang, X.; Shen, L.; Yu, Y. Coordinated path following control of fixed-wing unmanned aerial vehicles in wind. *ISA Trans.* **2022**, *122*, 260–270. [CrossRef]

 *drones*

Article

Multi-UAV Coverage through Two-Step Auction in Dynamic Environments

Yihao Sun [†], Qin Tan [†], Chao Yan, Yuan Chang, Xiaojia Xiang and Han Zhou *

College of Intelligence Science and Technology, National University of Defense Technology,
Changsha 410073, China; sunyihao@nudt.edu.cn (Y.S.); tanqin20@nudt.edu.cn (Q.T.);
yanchao17@nudt.edu.cn (C.Y.); changyuan10@nudt.edu.cn (Y.C.); xiangxiaojia@nudt.edu.cn (X.X.)
* Correspondence: zhouhan@nudt.edu.cn
† These authors contributed equally to this work.

Abstract: The cooperation of multiple unmanned aerial vehicles (Multi-UAV) can effectively solve the area coverage problem. However, developing an online multi-UAV coverage approach remains a challenge due to energy constraints and environmental dynamics. In this paper, we design a comprehensive framework for area coverage with multiple energy-limited UAVs in dynamic environments, which we call MCTA (Multi-UAV Coverage through Two-step Auction). Specifically, the online *two-step auction* mechanism is proposed to select the optimal action. Then, an obstacle avoidance mechanism is designed by defining several heuristic rules. After that, considering energy constraints, we develop the *reverse auction* mechanism to balance workload between multiple UAVs. Comprehensive experiments demonstrate that MCTA can achieve a high coverage rate while ensuring a low repeated coverage rate and average step deviation in most circumstances.

Keywords: multi-UAV; two-step auction; area coverage; obstacle avoidance; energy constraint

Citation: Sun, Y.; Tan, Q.; Yan, C.; Chang, Y.; Xiang, X.; Zhou, H. Multi-UAV Coverage through Two-Step Auction in Dynamic Environments. *Drones* **2022**, *6*, 153. https://doi.org/10.3390/drones6060153

Academic Editor: Abdessattar Abdelkefi

Received: 30 May 2022
Accepted: 17 June 2022
Published: 20 June 2022

Publisher's Note: MDPI stays neutral with regard to jurisdictional claims in published maps and institutional affiliations.

1. Introduction

Unmanned aerial vehicles (UAV) have the characteristics of low operating cost, good maneuverability, and no risk of casualties [1,2]. Accordingly, it has been applied to many fields, such as surveying and mapping [3], surveillance [4], bathymetry [5–9], and disaster rescue [10,11]. As a typical application, area coverage with UAVs has attracted great attention in robotics. Its main objective is to move single UAV or a group of UAVs to cover a given area [12]. Compared with single UAV, area coverage with multiple UAVs has significant advantages. First, due to the workload distribution and collaboration, multi-UAV can efficiently complete the task. Second, it enhances robustness against component failure due to redundancy. Therefore, multiple cooperative UAVs are expected to play an important role in the area coverage field.

Although considerable efforts have been devoted to addressing the area coverage problem for multi-UAV, there are still some challenges that need to be resolved. For instance, most of the existing studies assume that different obstacles are equal and insurmountable [13]. However, the real environment may contain obstacles with different threat levels. In addition, existing studies do not consider energy constraints, and their effectiveness is only verified in simple scenarios. Furthermore, some studies ignore the interaction among multiple UAVs, which may lead to conflicts between UAVs [14].

In light of this, we further investigate multi-UAV for cooperative area coverage. Compared with the previous work, we take the energy constraint and workload balance of multi-UAV into consideration and solve a more challenging coverage problem.

1.1. Related Work

In the field of robotics, much work has been done to solve the area coverage problem. Khamis et al. [15] proposed an auction algorithm for multi-robot systems. Xin et al. [13]

proposed an auction-based spanning-tree coverage algorithm, which ensures the connectivity of each spanning tree and tries to balance the workload between robots. However, the above algorithms are difficult to achieve online planning of area coverage actions due to the lack of global information.

To address the limitation above, Viet et al. [14] proposed an online method based on boustrophedon motions. After reaching the critical point, the intelligent backtracking mechanism based on the suggested A* search was applied to reach the starting point with the shortest collision-free path. Khan et al. [16] used a reputation-based auction mechanism to model the interaction between the UAV operators serving in close-by areas, achieving a strategic balance of control. It should be noted that the environment they considered was too simplistic, making their algorithm unable to adapt to the naturally dynamic environment.

In addition, the setting of dynamic obstacles is also crucial to adapt to the real environment in the field of area coverage. Gabriely and Rimon [17] proposed a spanning tree coverage algorithm based on a grid network. Sonti et al. [18] extended the set of path planning in an environment with static obstacles. Gorbenko et al. [19] studied an effective algorithm for multi-robot forest coverage underweighted terrain, which is suitable for solving synthetic terrain with real-world weight and special hard terrain. Roi Yehoshua et al. [20] divided the obstacles' threat to UAVs into five levels. They proposed a spanning tree-based coverage algorithm to meet the complexity of the real environment and the decision makers requirements for different targets. Nevertheless, most of these studies ignore the influence of energy constraints.

At the same time, the introduction of energy constraints is suitable for practical application scenarios [21–25]. Strimel et al. [26] introduced a new full-coverage planning algorithm that considers the robot's fixed fuel or energy capacity. Dutta et al. [27] proposed a constant-factor approximation algorithm for real-time coverage path planning with energy constraints. They studied the coverage planning problem of mobile robots with a limited energy budget, aimed to minimize the total travel distance covered by the environment and the number of visits to the charging station. However, covering as many low-risk areas as possible under energy constraints is still a challenge.

1.2. Contributions

In this paper, we presents a novel online planning solution of area coverage for multi-UAV in dynamic environments. The main contributions of this paper are as follows:

- We design the comprehensive MCTA framework for area coverage with multiple energy-limited UAVs in dynamic environments.
- We propose the two-step auction mechanism to select the optimal next action and avoid dynamic obstacles.
- We develop a reverse auction mechanism to avoid conflicts and balance workloads between UAVs.

The remainder of this paper is organized as follows: Section 2 formulates the area converage problem. Section 3 presents our MCTA framework. In Section 4, a series of experiments are conducted to evaluate the performance of MCTA. Finally, Section 5 concludes the major findings and outlines the potential direction for future work.

2. Problem Formulation

As illustrated in the left of Figure 1, we employ v energy-limited UAVs to simultaneously cover a given square area with potential threats. The square area is rasterized into m^2 basic square *units* with side length D (Figure 2). Each unit has a threat level $\eta \in [0, 1]$. The unit with $\eta = 0$ means a safe unit, and the UAV can pass freely. The unit with $0 < \eta < 1$ means there is a potential threat to UAVs, but the UAV can still pass. The unit with $\eta = 1$ means an extremely dangerous obstacle which the UAV cannot pass. We assume that each UAV can only acquire the threat level of the units within its *sensing scope*, and it can move in four base directions, i.e., up, down, left, and right. As shown in Figure 2, we define a *module* as a square composed of four units and treat the center point of the module as an equivalent replacement. Then, we define the modules that the UAV can reach in one time step as the *auction scope*.

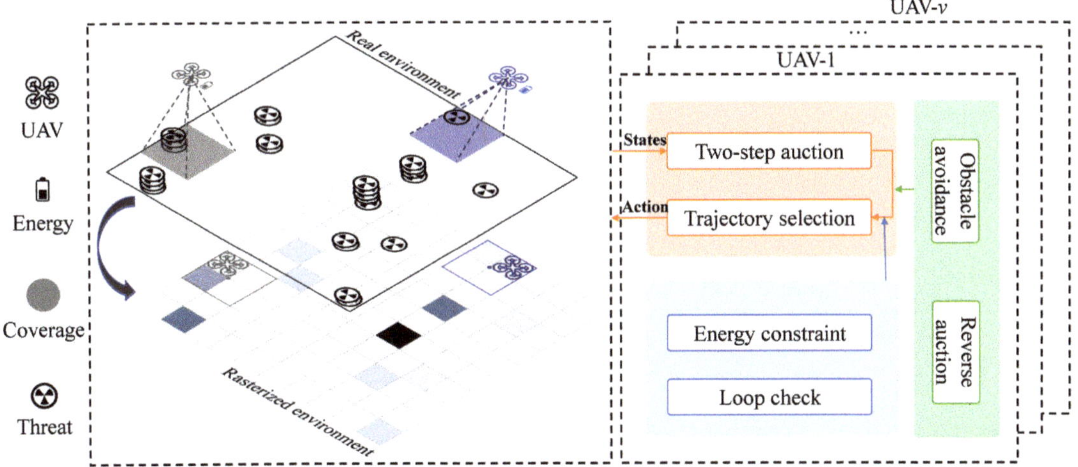

Figure 1. The proposed MCTA framework. The scenario on the left depicts the multi-UAV coverage process. The blocks on the right list key modules of MCTA.

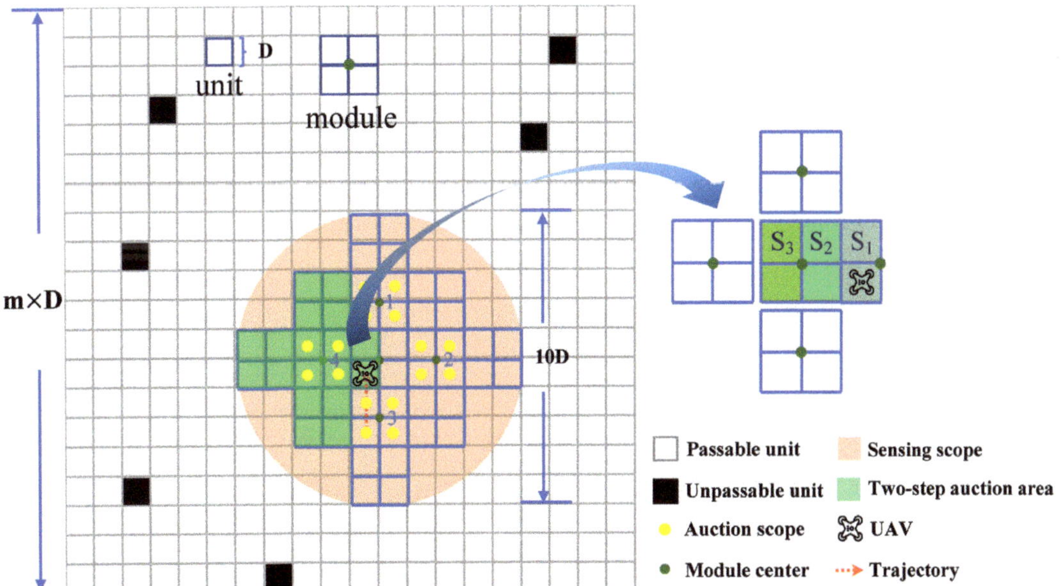

Figure 2. A square area is rasterized into m^2 basic square units with side length D. A module is composed of four basic units.

Let a set of continuous units $\{f_{i,k,1}, f_{i,k,2}, \ldots\}$ denote the trajectory $F_{i,k}$ of UAV$_i$ ($i = 1, 2, \ldots, v$) in the kth step, where $f_{i,k,p}$ is the pth in the kth time step of UAV$_i$'s position on the sequence. Denote the flight mileage (i.e., the number of accumulated passed units) of UAV_i in time step k as $l_{i,k}$. The total flight mileage L_i of UAV_i at the end of the coverage task can be defined as:

$$L_i = \sum_k l_{i,k}. \tag{1}$$

In the coverage process, the remaining energy $B_{i,k}$ of UAV_i in the kth time step is related to its current flight mileage $\sum_{k=0}^{k} l_{i,k}$. The initial energy of each UAV is set as B, i.e., $B_{i,0} = B$. Introducing different initial energy to the problem is sure to be more challenging, but in this paper, we only consider the case where each individual is equivalent to better analyze the completion of the coverage task. The UAV has two modes: work mode and sleep mode. In the process of coverage, the UAV is in work mode. In sleep mode, UAV will not perform the coverage task.

Our goal is to develop an online planning strategy of area coverage that enables multiple energy-limited UAVs to cover as many passable units as possible while avoiding collisions with other UAVs and obstacles. In other words, we expect to complete this coverage task with a higher coverage rate C_r, lower repetitive coverage rate R_r, and lower average flight deviation AD. The optimization objective is defined as:

$$\min \quad \left[-C_r, R_r, AD\right], \tag{2a}$$

$$\text{s.t.} \quad \forall i \in \{1, 2, \ldots, v\}, L_i \leq B, \tag{2b}$$

$$\forall i, j \in \{1, 2, \ldots, v\}, i \neq j, f_{i,k,p} \cap f_{j,k,p} = \varnothing, \tag{2c}$$

where constraint (2b) means the UAV cannot continue searching after its energy is exhausted. Constraint (2c) means no collision between UAVs.

3. MCTA Framework

In this section, we introduce the multi-UAV coverage through two-step auction framework. We first elaborate the two-step auction algorithm, which gives the bidding result of four adjacent modules to the UAV. Then, we focus on obstacle avoidance and multi-UAV conflict resolution. The UAV may take those strategies when the auction process ends. After that, we introduce the energy constraint model and loop-check mechanism. Finally, by assembling those modules above, we develop our framework. The details of each process are described in the following subsections.

3.1. Two-Step Auction

In the online environment, the UAV only knows the local information within its exploration scope. To achieve the optimal coverage result with the limited information, we design the two-step auction algorithm (Algorithm 1).

To distinguish the effect of obstacles in different positions, we divide the auction scope of the UAV into three areas and set different weights W_d ($d = 1, 2, 3$) for areas S_1, S_2, and S_3. As shown in Figure 2, area S_1 is a set of units in the current module that are close to the adjacent module. Area S_2 is a set of basic units in adjacent modules that are closest to the module center where the UAV is located. Similarly, area S_3 is a set of basic units in adjacent modules that are far away from the module center where the UAV is located. Considering that the obstacles which are extremely close or far away from the UAV have relatively little influence on covering the neighbor module, we set $W_2 > W_1 > W_3$.

In order to reduce the risk during the coverage process, the UAV will evaluate the threat level ζ of the module to determine the future flight trajectory [20]. ζ is given as follows:

$$\zeta = \sum (\eta W_d), \tag{3}$$

where η is the threat level of the unit. The UAV will give priority to covering modules with lower ζ and then cover modules with higher ζ to reduce risks.

To make the UAV "see" farther, we design the two-step auction mechanism. In this mechanism, module centers in the auction scope bid to the UAV according to the distribution of obstacles in the corresponding two-step auction area. The auction result of four modules can be written as set $C = \{c_1, c_2, c_3, c_4\}$. Here, we define the bid value c_i of module m_i as:

$$c_i = \frac{1}{\zeta_i + \zeta_m}. \tag{4}$$

The bid value c_i consists of two parts. First, we calculate ζ_i of module m_i according to Equation (3). Then, we assume that the UAV is in module m_i, then calculate the ζ of module m_1, m_2, and m_4 corresponding to module m_i and record the maximum value as ζ_m. To make the UAV preferentially cover modules with lower threat levels, the bidding value of modules should be negatively related to their threat levels. When the bidding price of adjacent module centers is the same, the energy loss will be considered: the greater the turning angle, the greater the energy loss [28]. To reduce the negative effect of turning, we set the module priority level $m_1 > m_2 > m_4 > m_3$.

After obtaining the auction result from all adjacent modules, the UAV will determine the *winning module*, that is, the module corresponding to m_1' (Algorithm 1, line 7). Then, the UAV plans to reach the winning module. In other words, the two-step auction algorithm informs the desired position for the UAV.

Algorithm 1 Two-step Auction.

Input: UAV position p, orientation o
Output: Four models sorted by bidding price c_i and orientation o
1: **for** $i \leftarrow 1$ **to** 4 **do**
2:     $c_i \leftarrow \zeta_i$;
3:     Assume that the UAV is in module m_i;
4:     Based on module m_i, calculate $\zeta_m = \max(\zeta_1, \zeta_2, \zeta_4)$;
5:     $c_i \leftarrow \frac{1}{c_i + \zeta_m}$;
6: **end for**
7: $\{m_1', m_2', m_3', m_4'\} \leftarrow sort(\{c_1, c_2, c_3, c_4\}, o)$;
8: **return** $\{m_1', m_2', m_3', m_4'\}$

3.2. Obstacle Avoidance and Multi-UAV Conflict Resolution

After determining the winning module that it plans to arrive at, the UAV may not actually reach the module due to obstacles or multi-UAV conflicts. Even if a module is accessible, the UAV also has to choose a suitable way to reach it. Therefore, the avoidance of obstacles and the resolution of multi-UAV conflicts should be taken into consideration.

First, we introduce the obstacle avoidance strategy. Assume that the UAV is not able to overcome an obstacle by going over it. As shown in Figure 3, we give the specific path selection method based on obstacles in different locations. When the UAV cannot enter the next module, it does not fly to the corresponding module center. At this time, the UAV will re-select the second-best module in the current auction bidding process. If the UAV can access none of the four module centers involved in the bidding, it enters sleep mode and stops covering.

Then, we develop a conflict resolution strategy to solve this situation in that two or more UAVs bid a certain module at the same time. The key insight behind this strategy is the conflicting module center selects the UAV reversely, which can be regarded as a reverse auction. When the conflict occurs, the module will give priority to choosing the UAV that has encountered "unfair" treatment: the one that has the least flight mileage L. After the UAV is selected, other UAVs will pause for one time step and wait for the winning UAV to pass. Finally, all UAVs continue to execute the two-step auction process, select the module plan to reach, and continue the covering process.

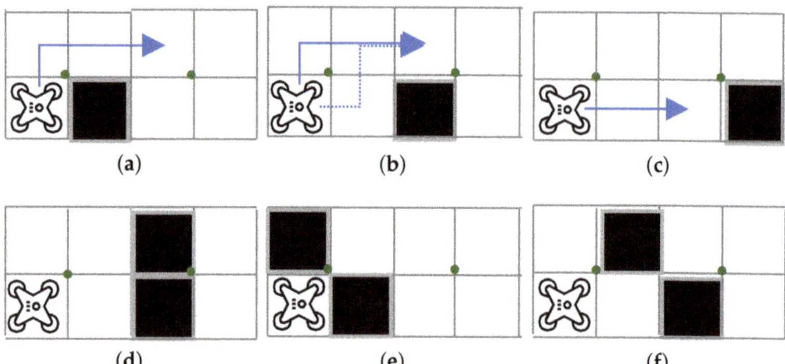

Figure 3. Six obstacle distributions and the UAV's corresponding path selection method. (**a**) The obstacle is located in area S_1. The UAV will turn into the next module. (**b**) The obstacle is located in area S_2. The UAV turns into the next module. (**c**) The obstacle is located in area S_3. The UAV goes straight into the next module. (**d**) Double obstacles located in area S_2. The UAV cannot enter the next module. (**e**) The obstacle is located in the neighbor unit of the unit where the UAV of the current module is located. The UAV cannot fly to the next module. (**f**) The obstacle is located in the neighbor unit in the direction of the winning module and the opposite unit in the current module. The UAV cannot fly to the next module.

3.3. Energy Constraint and Loop-Check

In addition to a higher coverage rate, the framework also has requirements for reducing the repetitive coverage rate and the average flight deviation. If the UAV cannot enter the sleep mode at an appropriate time, the convergence of the framework cannot be guaranteed, and it will also cause a higher repetitive coverage rate and average flight deviation. Based on this, we proposed the energy constraint model and the loop-check mechanism.

In the energy constraint model, the energy of the UAV_i decays as the flight mileage L_i increases. The remaining energy $B_{i,k}$ of UAV_i in the kth time step is defined as:

$$B_{i,k} = B - \sum_{k=0}^{k} l_{i,k}, \tag{5}$$

where B is the initial energy of the UAV, that is, the UAV is allowed to travel at most B units.

Moreover, if the UAV's surrounding obstacles are distributed similarly during the flight process, it may fly back and forth. We define the trajectory generated by the back and forth flying as a loop. This loop increases the repeated coverage rate and leads to energy waste. To avoid the two negative effects, the UAV should do a loop-check. If the UAV enters the loop, it will enter into sleep mode.

In summary, for UAV_i, if its remaining energy $B_{i,k} = 0$, or it is judged to enter the loop, or there are no passable modules, it will enter into sleep mode. That is, the UAV no longer participates in the auction process and stops covering. This theoretically guarantees that our framework will converge.

3.4. MCTA Framework

Combining those mechanisms above, each UAV cooperates to complete the coverage task in a complex dynamic environment (Algorithm 2). During the multi-UAV coverage process, if all UAVs enter into sleep mode, the whole coverage will end. It is worth mentioning that our framework is also suitable for the single-UAV or fixed obstacle situation, which is a degraded version of solving multi-UAV and dynamic environment problems. In the following section, we will conduct related experiments to evaluate the performance of our framework.

Algorithm 2 MCTA.

Input: $p_i[k]$, $orientation_i[k]$, $plan\_poses$, $i = 1, 2, \ldots, v$

1: $\{m_1', m_2', m_3', m_4'\} \leftarrow$ *Two-step Auction* $(p_i[k], o_i[k])$;
2: $plan\_flag_i \leftarrow 0$;
3: **for** $j \leftarrow 1$ **to** 4 **do**
4:     **if** module m corresponding to m_j' is reachable **then**
5:         $orientation_i[k+1] \leftarrow$ direction of module m;
6:         $plan\_flag_i \leftarrow 1$;
7:         **break**
8:     **end if**
9: **end for**
10: **if** $plan\_flag_i = 1$ **then**
11:     Judge if there is a multi-UAV conflict;
12:     **if** no conflict occurs or win the conflict **then**
13:         Check the remaining energy $B_{i,k}$ and the loop;
14:         **if** $B_{i,k} > 0$ and not in the loop **then**
15:             Choose a suitable way to reach module m;
16:             $p_i[k+1] \leftarrow$ the position of module m;
17:         **else**
18:             $mode_i \leftarrow sleep$;
19:         **end if**
20:     **else**
21:         Stay in place for next step;
22:     **end if**
23: **else**
24:     $mode_i \leftarrow sleep$;
25: **end if**

---

4. Experiments and Analysis

This section demonstrates the performance of MCTA by conducting a series of experiments. We first show the effectiveness of MCTA in a typical environment with a single UAV and four UAVs. Then, the adaptability and the scalability are verified through the quantitative experiments with different UAV numbers and energy constraints.

4.1. Qualitative Evaluation

We first evaluate the feasibility of MCTA for only one UAV. In the experiment, we set 10% obstacles in the given environment ($N = 10\% \times M$) and allow obstacles to move randomly. We assume that obstacles of different η have different mobility capabilities: the higher the η, the smaller the obstacle's moving range. Specifically, the position of the obstacle is (x, y) before moving and changes to $(x + m, y + m)$ after moving, where m is a random integer generated from the closed interval $[-a, a]$. The specific mapping relationship between η and a is given in Table 1.

Table 1. The mapping of η and a.

η	0.2	0.4	0.6	0.8	1
a	4	3	2	1	0 (obstacles cannot move)

Figure 4 shows the coverage process with single UAV under a 20×20 dynamic environment. Different colors represent different η of obstacles: the darker the obstacle, the higher the threat level. Figure 4a illustrates the initial situation: obstacles with different η

are distributed in different positions and the UAV is ready to perform its coverage mission. Figure 4b shows the coverage performance when the time step = 20. At this time, there are about 1/5 green units (units covered for the first time) and no brown units (units covered twice or more), which means that the UAV is actively exploring the new area. The coverage result is given in Figure 4c. Obviously, only few units are not covered.

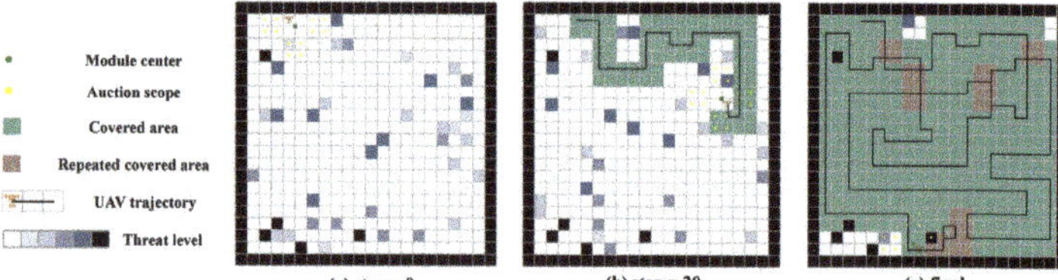

Figure 4. The evolution of coverage with a single UAV. The area covered for the first time is filled in green, and the area covered twice or more is filled in brown. After the UAV performs related flight actions, it will generate its own flight trajectory (the black line).

Then, we test the performance of MCTA with four UAVs in the same environment. In Figure 5a, the position of UAV_2 is $(1, 2)$, and the position of UAV_3 is $(1, 1)$; therefore, they may encounter a conflict in the next time step due to their adjacent positions. However, from the trajectories of two UAVs in Figure 5b, the conflict did not happen, which indicates our reverse auction mechanism works. In addition, compared with Figure 4b, Figure 5b has significantly more green units. It shows the advantages of multi-UAV coverage in terms of efficiency. In the final time step (Figure 5c), we extract and compare the length of each UAV_i's trajectory and find that they are almost the same (Figure 5d). This means the balanced workload between UAVs. Although the coverage rate of four UAVs is almost same with that of the single UAV, the whole coverage process of four UAVs ends faster. As a result, the coverage with multiple UAVs is more efficient.

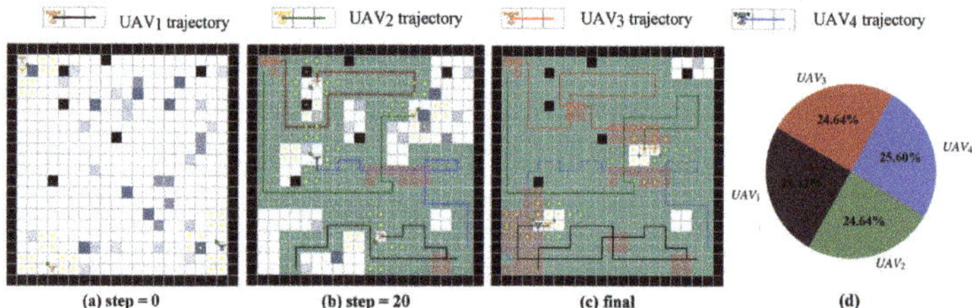

Figure 5. The evolution of coverage with four UAVs. Compared with single UAV, in a similar initial environment (**a**), multiple UAVs can achieve better coverage performance before the framework converges (**b**). In the final time step (**c**), different trajectories are shown in different colors. The pie chart (**d**) illustrates the proportion of each UAV's trajectory length.

In summary, the MCTA framework can obtain a high coverage rate regardless of single-UAV or multi-UAV situations. Moreover, it realizes the avoidance of conflicts and balances the workload between multiple UAVs. This indicates the adaptability and effectiveness of our framework in dynamic and complex environments.

4.2. Quantitative Evaluation

In this subsection, we introduce the following metrics to further evaluate the quantitative performance of MCTA:

• **Coverage rate:** The coverage rate C_r is defined as the ratio between the area of covered modules and the area of the entire environment:

$$C_r = \frac{|F_1 \cup F_2 \cup \cdots \cup F_v|}{M} \times 100\%, \tag{6}$$

where F_i is the set of $F_{i,k}$, $|\cdot|$ represents the number of elements in a set, and $F_1 \cup F_2 \cup \ldots \cup F_v$ represents the union of all UAV flight trajectories.

• **Repeated coverage rate:** The repeated coverage rate R_r is defined as:

$$R_r = \frac{\sum_{i,k}\left(l_{i,k} - |F_{i,k}|\right)}{|F_1 \cup F_2 \cup \cdots \cup F_v|} \times 100\%, \tag{7}$$

where $\sum_{i,k}\left(l_{i,k} - |F_{i,k}|\right)$ represents the difference between the flight mileage of all UAVs and the number of units passed by the flight trajectory, that is, the units that are repeatedly covered are accumulated by the frequency of coverage.

• **Average flight deviation:** To investigate the degree of deviation of the flight path of each UAV from its average path, the average flight deviation AD is defined as follows:

$$AD = \frac{1}{v} \sum_i abs(L_i - \bar{L}), \tag{8}$$

where $\bar{L}$ is the average flight mileage. AD measures whether the workload of each UAV is balanced.

After that, we enrich the environments to challenge the adaptability of our framework quantitatively. As shown in Figure 6, *env-free* is the simplest environment. It is also the environment of the previous subsection; *env-strip* illustrates the staggered distribution of obstacles and its passable routes are less than *env-free*; In *env-convex*, we set a regular border, which brings a challenge to achieve complete coverage; on the basis of *env-convex*, we set up a non-convex *env-ring*. Its passable area is similar to the loop; in *env-honeycomb* and *env-maze*, the border becomes more complicated and both of them increase the risk of UAVs falling into loops.

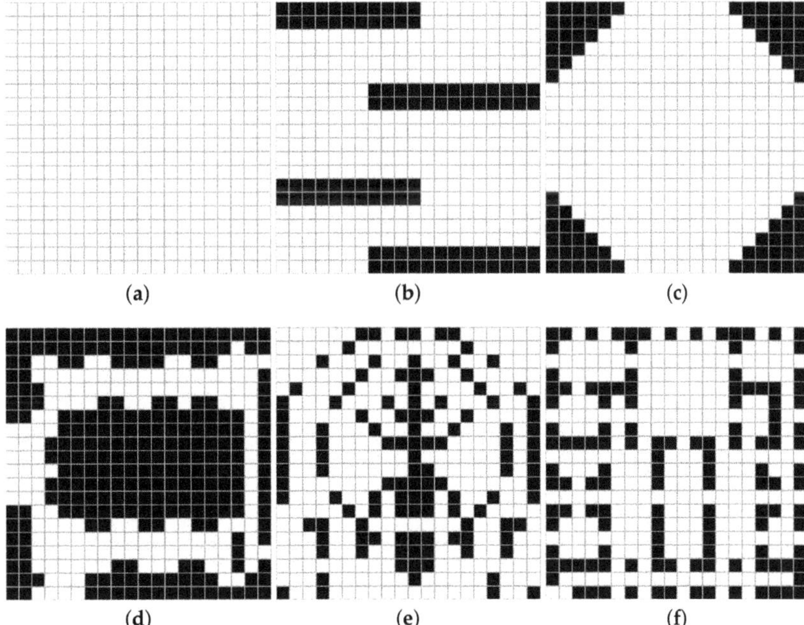

Figure 6. Configuration spaces of the (**a**) free, (**b**) strip, (**c**) convex, (**d**) ring, (**e**) honeycomb, and (**f**) maze.

Table 2 provides part of quantitative experimental results. In simple environments (i.e., *env-free*, *env-strip*, and *env-convex*), increasing the number v of UAVs will increase R_r but makes little contributions to C_r. In this case, if the convergence time is not considered, single UAV performs better than multiple UAVs. However, in complex environments (i.e., *env-ring*, *env-honeycomb*, and *env-maze*), increasing the number of UAVs can increase C_r obviously. At this time, multiple UAVs have more advantages. Then, we compared the C_r, R_r, and AD in static and dynamic environments separately and found that they were almost the same. This shows that the MCTA framework can overcome the negative effects of uncertain factors caused by the dynamic environment. Moreover, most of C_r can reach more than 90%, R_r are generally less than 50%, and AD are about 3. Therefore, our framework can make the UAV complete coverage tasks efficiently in complex environments.

Table 2. C_r, R_r, and AD of MCTA under different configurations.

Environment	Size	v	Static			Dynamic		
			C_r	R_r	AD	C_r	R_r	AD
free	20×20	1	93.54%	15.66%	0	93.06%	14.87%	0
		4	92.18%	23.37%	1.34	91.98%	23.32%	1.44
		8	88.14%	25.11%	1.41	88.52%	24.66%	1.44
	40×40	1	91.06%	19.30%	0	91.74%	19.75%	0
		4	91.09%	23.39%	2.26	91.08%	23.92%	2.41
convex	20×20	1	96.34%	30.32%	0	96.52%	31.40%	0
		4	95.59%	38.91%	1.58	96.28%	40.17%	1.6
		12	92.42%	49.06%	1.4	92.70%	48.36%	1.36
	40×40	1	94.58%	33.51%	0	94.73%	32.47%	0
		8	95.39%	43.23%	3.9	95.40%	43.24%	3.86
		12	95.08%	45.48%	2.74	95.20%	45.78%	2.74
maze	20×20	8	81.89%	33.32%	3.42	80.53%	35.26%	3.4
		12	82.36%	38.42%	2.64	82.67%	41.01%	2.43
	40×40	1	79.47%	29.80%	0	79.75%	29.97%	0
		12	87.07%	45.11%	9.13	87.67%	45.29%	8.93
ring	20×20	1	83.02%	22.78%	0	83.44%	24.26%	0
		12	86.82%	46.64%	1.35	86.43%	47.05%	1.43
	40×40	1	81.27%	33.76%	0	80.81%	32.32%	0
		12	89.09%	49.53%	2.95	88.42%	50.27%	2.97
honeycomb	20×20	1	75.74%	36.19%	0	75.75%	35.62%	0
		8	80.75%	37.57%	3.27	80.45%	37.95%	3.33
strip	20×20	1	92.07%	28.82%	0	93.15%	29.16%	0
		4	93.31%	39.58%	3.26	93.70%	39.52%	2.96
		8	93.30%	45.57%	1.82	92.89%	44.94%	1.85
	40×40	1	87.42%	28.94%	0	87.86%	30.27%	0
		12	92.13%	46.28%	4.83	92.61%	45.93%	4.88

Of course, in individual environments, MCTA performs relatively poorly, such as 20×20 *env-honeycomb* and 40×40 *env-maze*. Regardless of the number v, C_r is typically less than 90%, and Rr is also relatively high. This is probably caused by the complexity and disconnectivity of the environment, where UAVs are prone to fall into the local optimal solution.

Figure 7 shows the final C_r, R_r, and AD of different v under different 20×20 environments. In Figure 7a, when v changes from 4 to 8, C_r increases very obviously. However, when v changes from 8 to 12, C_r does not increase significantly, and sometimes even decreases. At the same time, we also noticed that in the three corresponding environments in Table 2, AD usually increases significantly when the number v increases sharply. We attribute these two phenomena to frequent conflicts caused by excessive number of UAVs, which leads to the reduction of coverage efficiency. As a result, for the environment of a certain size, an appropriate increase in the number of UAVs can effectively improve the coverage rate, but an excessive number of UAVs may have a negative effect, because excess means diseconomy and poor coverage performance. Figure 7b,c show the R_r and AD of four UAVs under different initial energy in 20×20 dynamic *env-free*. From the trend of the two histograms, it is clear that the larger initial energy of the UAV, the larger R_r and AD. This is exactly what we are expected on energy saving: to complete the coverage task, the UAV does not need too much initial energy.

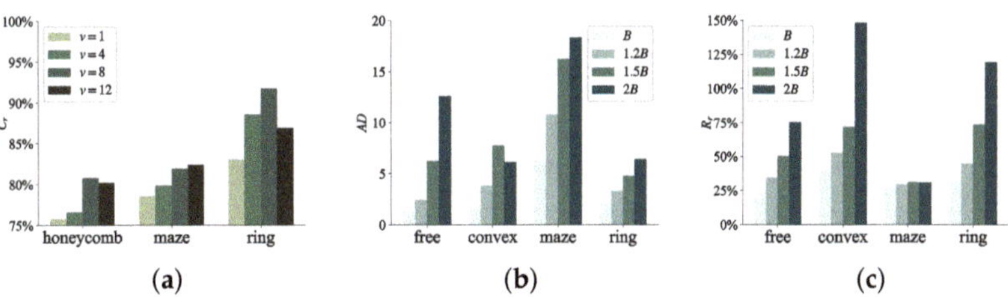

(a)   (b)   (c)

Figure 7. The performance of the MCTA framework under different configurations. (**a**) The coverage rate C_r under three complex environments (*env-honeycomb*, *env-maze*, and *env-ring*) in different numbers of UAVs (1, 4, 8, and 12). (**b,c**) The number v of UAVs is fixed to 4 and the average flight deviation AD and repeated coverage rate R_r are shown with different initial energy (B, $1.2B$, $1.5B$, and $2B$) in two simple environments (*env-free* and *env-convex*) and two complex environments (*env-maze* and *env-ring*).

Considering all factors, to achieve a better search coverage result, for *env-freedom*, *env-convex*, and 20×20 *env-strip*, the best number of UAVs is 1. For 20×20 *env-maze* and *env-honeycomb*, the best number of UAVs is 8. Then, for 40×40 *env-strip*, *env-maze*, and *env-ring*, the best number of UAVs is 12. In general, on the premise of achieving a high coverage rate, MCTA minimizes the repeated coverage rate and average step deviation as much as possible.

5. Conclusions

In this paper, we design an efficient framework for multiple UAVs in dynamic environments. The framework can be executed online and realizes the optimal solution within the exploration scope. A series of experiments have been conducted to demonstrate its superior coverage rate, scalability to the number of UAVs, and adaptivity to the dynamic environments. Note that the physical properties of the UAVs are not specified. The proposed multi-UAV collaborative coverage framework is generally applicable for systems with similar configurations.

In future work, we will take the non-linear change of the remaining battery into account, e.g., a UAV spends more energy to take off than other stages (such as hovering or moving sideways). Furthermore, another promising direction is to explore the upper bound number of UAVs. In addition, we would like to evaluate our framework through field experiments.

Author Contributions: Conceptualization, Y.S. and Q.T.; methodology, Q.T.; software, Y.S.; validation, Y.S. and Q.T.; formal analysis, Y.S., Q.T., C.Y. and Y.C.; investigation, Q.T.; writing—original draft preparation, Y.S. and Q.T.; writing—review and editing, C.Y. and Y.C.; supervision, X.X. and H.Z. All authors have read and agreed to the published version of the manuscript.

Funding: This research received no external funding.

Institutional Review Board Statement: Not applicable.

Informed Consent Statement: Not applicable.

Data Availability Statement: Not applicable.

Acknowledgments: The authors would like to thank the Swarm Observation and Regulation Laboratory, National University of Defense Technology, Hunan Province, China, for the resources provided by them.

Conflicts of Interest: The authors declare no conflict of interest.

References

1. Liu, Z.; Wang, X.; Shen, L.; Zhao, S.; Cong, Y.; Li, J.; Yin, D.; Jia, S.; Xiang, X. Mission-oriented miniature fixed-wing UAV swarms: A multilayered and distributed architecture. *IEEE Trans. Syst. Man Cybern. Syst.* **2020**, *52*, 1588–1602. [CrossRef]
2. Yan, C.; Xiang, X.; Wang, C. Towards real-time path planning through deep reinforcement learning for a UAV in dynamic environments. *J. Intell. Robot. Syst.* **2020**, *98*, 297–309. [CrossRef]
3. Ball, Z.; Odonkor, P.; Chowdhury, S. A swarm-intelligence approach to oil spill mapping using unmanned aerial vehicles. In Proceedings of the AIAA Information Systems-AIAA Infotech@ Aerospace, Grapevine, TX, USA, 9–13 January 2017; p. 1157.
4. Pannozzi, P.; Valavanis, K.P.; Rutherford, M.J.; Guglieri, G.; Scanavino, M.; Quagliotti, F. Urban monitoring of smart communities using UAS. In Proceedings of the 2019 International Conference on Unmanned Aircraft Systems (ICUAS), Atlanta, GA, USA, 11–14 June 2019; pp. 866–873.
5. Alevizos, E.; Oikonomou, D.; Argyriou, A.V.; Alexakis, D.D. Fusion of Drone-Based RGB and Multi-Spectral Imagery for Shallow Water Bathymetry Inversion. *Remote Sens.* **2022**, *14*, 1127. [CrossRef]
6. Bandini, F.; Lopez-Tamayo, A.; Merediz-Alonso, G.; Olesen, D.; Jakobsen, J.; Wang, S.; Garcia, M.; Bauer-Gottwein, P. Unmanned aerial vehicle observations of water surface elevation and bathymetry in the cenotes and lagoons of the Yucatan Peninsula, Mexico. *Hydrogeol. J.* **2018**, *26*, 2213–2228. [CrossRef]
7. Specht, M.; Stateczny, A.; Specht, C.; Widźgowski, S.; Lewicka, O.; Wiśniewska, M. Concept of an Innovative Autonomous Unmanned System for Bathymetric Monitoring of Shallow Waterbodies (INNOBAT System). *Energies* **2021**, *14*, 5370. [CrossRef]
8. Stateczny, A.; Specht, C.; Specht, M.; Brčić, D.; Jugović, A.; Widźgowski, S.; Wiśniewska, M.; Lewicka, O. Study on the Positioning Accuracy of GNSS/INS Systems Supported by DGPS and RTK Receivers for Hydrographic Surveys. *Energies* **2021**, *14*, 7413. [CrossRef]
9. Wang, D.; Xing, S.; He, Y.; Yu, J.; Xu, Q.; Li, P. Evaluation of a New Lightweight UAV-Borne Topo-Bathymetric LiDAR for Shallow Water Bathymetry and Object Detection. *Sensors* **2022**, *22*, 1379. [CrossRef] [PubMed]
10. Lin, L.; Goodrich, M.A. UAV intelligent path planning for wilderness search and rescue. In Proceedings of the 2009 IEEE/RSJ International Conference on Intelligent Robots and Systems, St. Louis, MO, USA, 10–15 October 2009; pp. 709–714.
11. Kohlbrecher, S.; Kunz, F.; Koert, D.; Rose, C.; Manns, P.; Daun, K.; Schubert, J.; Stumpf, A.; von Stryk, O. Towards highly reliable autonomy for urban search and rescue robots. In *Proceedings of the Robot Soccer World Cup*; Springer: Berlin/Heidelberg, Germany, 2014; pp. 118–129.
12. Galceran, E.; Carreras, M. A survey on coverage path planning for robotics. *Robot. Auton. Syst.* **2013**, *61*, 1258–1276. [CrossRef]
13. Gao, G.Q.; Xin, B. A-STC: Auction-based spanning tree coverage algorithm formotion planning of cooperative robots. *Front. Inf. Technol. Electron. Eng.* **2019**, *20*, 18–31. [CrossRef]
14. Viet, H.H.; Dang, V.H.; Laskar, M.N.U.; Chung, T. BA*: An online complete coverage algorithm for cleaning robots. *Appl. Intell.* **2013**, *39*, 217–235. [CrossRef]
15. Khamis, A.; Hussein, A.; Elmogy, A. Multi-robot task allocation: A review of the state-of-the-art. *Coop. Robot. Sens. Netw.* **2015**, *2015*, 31–51.
16. Khan, A.S.; Chen, G.; Rahulamathavan, Y.; Zheng, G.; Assadhan, B.; Lambotharan, S. Trusted UAV Network Coverage Using Blockchain, Machine Learning, and Auction Mechanisms. *IEEE Access* **2020**, *8*, 118219–118234. [CrossRef]
17. Gabriely, Y.; Rimon, E. Spiral-STC: An on-line coverage algorithm of grid environments by a mobile robot. In Proceedings of the 2002 IEEE International Conference on Robotics and Automation (Cat. No. 02CH37292), Washington, DC, USA, 11–15 May 2002; Volume 1, pp. 954–960.
18. Sonti, S.; Virani, N.; Jha, D.K.; Mukherjee, K.; Ray, A. Language measure-theoretic path planning in the presence of dynamic obstacles. In Proceedings of the 2013 American Control Conference, Washington, DC, USA, 17–19 June 2013; pp. 5110–5115.
19. Gorbenko, A.; Popov, V. The multi-robot forest coverage for weighted terrain1. *J. Ambient. Intell. Smart Environ.* **2015**, *7*, 835–847. [CrossRef]
20. Yehoshua, R.; Agmon, N.; Kaminka, G.A. Robotic adversarial coverage of known environments. *Int. J. Robot. Res.* **2016**, *35*, 1419–1444. [CrossRef]
21. Shnaps, I.; Rimon, E. Online coverage of planar environments by a battery powered autonomous mobile robot. *IEEE Trans. Autom. Sci. Eng.* **2016**, *13*, 425–436. [CrossRef]
22. Wei, M.; Isler, V. Coverage path planning under the energy constraint. In Proceedings of the 2018 IEEE International Conference on Robotics and Automation (ICRA), Brisbane, Australia, 21–25 May 2018; pp. 368–373.
23. Wu, C.; Dai, C.; Gong, X.; Liu, Y.J.; Wang, J.; Gu, X.D.; Wang, C.C. Energy-efficient coverage path planning for general terrain surfaces. *IEEE Robot. Autom. Lett.* **2019**, *4*, 2584–2591. [CrossRef]

24. Modares, J.; Ghanei, F.; Mastronarde, N.; Dantu, K. Ub-anc planner: Energy efficient coverage path planning with multiple drones. In Proceedings of the 2017 IEEE International Conference on Robotics and Automation (ICRA), Singapore, 29 May–3 June 2017; pp. 6182–6189.
25. Jensen-Nau, K.R.; Hermans, T.; Leang, K.K. Near-Optimal Area-Coverage Path Planning of Energy-Constrained Aerial Robots with Application in Autonomous Environmental Monitoring. *IEEE Trans. Autom. Sci. Eng.* **2020**, *18*, 1453–1468. [CrossRef]
26. Strimel, G.P.; Veloso, M.M. Coverage planning with finite resources. In Proceedings of the 2014 IEEE/RSJ International Conference on Intelligent Robots and Systems, Chicago, IL, USA, 14–18 September 2014; pp. 2950–2956.
27. Dutta, A.; Sharma, G. A Constant-Factor Approximation Algorithm for Online Coverage Path Planning with Energy Constraint. *arXiv* **2019**, arXiv:1906.11750.
28. Torres, M.; Pelta, D.A.; Verdegay, J.L.; Torres, J.C. Coverage path planning with unmanned aerial vehicles for 3D terrain reconstruction. *Expert Syst. Appl.* **2016**, *55*, 441–451. [CrossRef]

Article

Bioinspired Environment Exploration Algorithm in Swarm Based on Lévy Flight and Improved Artificial Potential Field

Chen Wang [1], Dongliang Wang [1], Minqiang Gu [1], Huaxing Huang [1], Zhaojun Wang [1], Yutong Yuan [2], Xiaomin Zhu [2], Wu Wei [3] and Zhun Fan [1,*]

[1] College of Engineering, Shantou University, Shantou 515063, China; 20cwang2@stu.edu.cn (C.W.); dlwang@stu.edu.cn (D.W.); mqgu@stu.edu.cn (M.G.); 20hxhuang@stu.edu.cn (H.H.); 17zjwang@stu.edu.cn (Z.W.)

[2] College of Systems Engineer, National University of Defense Technology, Changsha 410073, China; 17ytyuan@alumni.stu.edu.cn (Y.Y.); xmzhu@nudt.edu.cn (X.Z.)

[3] School of Automation Science and Engineering, South China University of Technology, Guangzhou 510006, China; weiwu@scut.edu.cn

* Correspondence: zfan@stu.edu.cn

Abstract: Inspired by the behaviour of animal populations in nature, we propose a novel exploration algorithm based on Lévy flight (LF) and artificial potential field (APF). The agent is extended to the swarm level using the APF method through the LF search environment. Virtual leaders generate moving steps to explore the environment through the LF mechanism. To achieve collision-free movement in an unknown constrained environment, a swarm-following mechanism is established, which requires the agents to follow the virtual leader to carry out the LF. The proposed method, combining the advantages of LF and APF which achieve the effect of flocking in an exploration environment, does not rely on complex sensors for environment labelling, memorising, or huge computing power. Agents simply perform elegant and efficient search behaviours as natural creatures adapt to the environment and change formations. The method is especially suitable for the camouflaged flocking exploration environment of bionic robots such as flapping drones. Simulation experiments and real-world experiments on E-puck2 robots were conducted to evaluate the effectiveness of the proposed LF-APF algorithm.

Keywords: bionic algorithm; swarm robotic; environment exploration; distributed control; Lévy flight

Citation: Wang, C.; Wang, D.; Gu, M.; Huang, H.; Wang, Z.; Yuan, Y.; Zhu, X.; Wei, W.; Fan, Z. Bioinspired Environment Exploration Algorithm in Swarm Based on Lévy Flight and Improved Artificial Potential Field. *Drones* 2022, 6, 122. https://doi.org/10.3390/drones6050122

Academic Editors: Xiwang Dong, Mou Chen, Xiangke Wang and Fei Gao

Received: 28 March 2022
Accepted: 5 May 2022
Published: 9 May 2022

Publisher's Note: MDPI stays neutral with regard to jurisdictional claims in published maps and institutional affiliations.

1. Introduction

The exploration problem is an important research area of robotics, which can be applied to various tasks such as military reconnaissance [1], search and rescue [2], foraging [3], and drug delivery [4]. In recent years, exploration using robots in complex environments has attracted widespread attention. Currently, environmental exploration methods often rely on recording the explored area or the marking of the environment using sensors. Among them, some use the odometer method [5,6] (recording the area that has been walked) and pheromone method [7,8] (marking the environment). However, the path information recorded by the agents is often subject to significant errors, and sometimes the agents have difficulties in labelling the environment. Therefore, the elimination of the agent's reliance on complex sensors when searching in an unknown environment, like natural creatures, is crucial.

The ability of a single agent in cognition and action may be inherently limited, and cooperation in a swarm can alleviate the impact of this limitation [9,10]. This kind of problem-solving ability is abundant in nature, for example, swarms of ants search for the shortest path [11] and honeybees choose the best food resources by dancing [12]. Therefore, it is specifically necessary to achieve effective coordination in a swarm robotic system [13,14]. Multi-agent swarm search can search targets in the environment more effectively than is possible with single-agent exploration.

Many studies related to bionic UAVs have been reported in recent years [15–17]. Ramezani et al. used a series of virtual constraints to control an articulated, deformable wing to achieve autonomous flight of a bat robot [16]. EPFL [15] optimized the aerodynamic designs of winged drones for specific flight regimes. Large lifting surfaces provided manoeuvrability and agility like the northern goshawk. Roderick developed a biomimetic robot that can dynamically perch on complex surfaces and grasp irregular objects [17]. These latest studies show that research on bionic robots is crucial in the field of robotics, together with suitable applications. If we assume that a group of biomimetic robots are sent to perform a task of exploring the environment, then all of them as a group should look like birds, including their appearance and movement, but if they fail to behave like a natural cluster, the value of the stealth of the biomimetic robots is largely lost. Therefore, we propose a new swarm intelligence task, so that agents can achieve efficient environmental exploration of an area as much as possible like natural creatures without relying on complex sensors or huge computing power.

To overcome the difficulties mentioned above, random movement, which is a common search pattern for natural creatures, is introduced in this study. In some applications, the agent needs to search for dynamic targets in a complex environment through a random step generation mechanism [18]. Randomness plays a significant role in both swarm intelligence motion control and swarm intelligence optimisation algorithm [19]. In the swarm intelligence environmental exploration task in this study, due to the fact that the target is moving, if the explored path is not repeated properly, that is, if the agent does not go where it has gone, the dynamic target only needs to hide in the area where the agents passed through to avoid being detected. Therefore, the agents must traverse the area with some positions revisited occasionally. However, if a location is repeated many times, the detection efficiency decreases. Therefore, when exploring the environment with dynamic targets, agents must adopt a suitable random walk strategy to traverse an area appropriately.

Natural creatures exhibit two well-known random movement mechanisms: Lévy flight (LF) [18] and Brownian motion (BM) [20]. In animal foraging, when the prey in the environment is abundant, BM is sufficiently efficient [21]. Fredy et al. proposed BM as an exploration strategy for autonomous swarm robots [20]. This solution, to a certain extent, solves the problem of swarm robots realising environmental exploration tasks through bionic motion. As a more sophisticated alternative to BM, LF as a typical random walking strategy has been introduced in many studies [22–25], especially in the literature of agents' environmental exploration [26,27]. Vincenzo Fioriti et al. proved the LF's superiority to the random walk with simulations and applied the LF mechanism in the fish mass model's centre speed according to the Kuramoto equation [26]. Pang et al. pointed out that the mean and variance of steps generated by LF are important parameters effecting the searching efficiency and need to be optimised [25]. Even though these studies have achieved improved results, they are all about single-agent exploration. In biomimetic research of multi-agent exploration, Sutantyo et al. first presented the integration of LF and an artificial potential field method to achieve an efficient search algorithm for multiple agents applications [28]. However, in this study, the agent works in its own way and does not search for targets together with other agents. In some specific task scenarios, it may not be conducive to performing subsequent collaborative tasks such as entrapping after a single agent has discovered the target. It is often too late to call on other agents to collaborate when they are scattered too far away. In some situations, agents need to flock to be prepared to perform following swarm tasks [28]. Therefore, we need to discover how to make agents form flocks.

The artificial potential field (APF) method is widely used to realise the formation control of swarm robots while achieving collision avoidance, such as UAVs [29], wheeled mobile robots [30], and underwater robots [31]. Gabor Vásárhelyi et al. proposed an extensible motion control framework based on an improved APF method that takes into account the motion constraints in swarms, which achieved a good flocking effect in real-world experiments, demonstrating behaviour similar to that of natural creatures [32]. Motivated by [18,32], a method combining LF and an improved APF is proposed in this work for swarm robots to form flocks and explore unknown environments with constraints effectively and efficiently.

In this study, we propose a method that combines the LF mechanism and the improved APF method to make swarms of agents flock and explore the environment in a manner similar to natural organisms. In the swarm, there is an invisible virtual leader in the arena moving with the LF algorithm. The agents follow the virtual leader in groups to find targets in an unknown environment. The swarm system randomly allocates each agent a pre-specified priority. The leader—the agent with the highest priority in the swarm—calculates the position of the virtual leader and broadcasts the information to other agents in the swarm via WiFi. When the leader is destroyed, the agent with the highest priority in the swarm becomes the leader and continues to broadcast the virtual leader's location. An improved APF method is then applied to enable the agents in the swarm to follow the virtual leader in a flocking and to explore the environment. The agents flock to explore the environment without relying on marking the environment and recording the itinerary and achieve efficient exploration of the environment, only relying on a simple random walk mechanism, just like natural creatures. This has the potential to facilitate the stealthy mission of bionic drones. Specifically, this paper contributes the following:

(1) The proposed LF-APF algorithm applies the LF search mechanism at the swarm level. Combining the advantages of LF and APF can enable agents to efficiently explore the environment through simple and natural random walking like natural creatures.
(2) The improved APF method makes agents follow the virtual leader, maintain a certain distance from each other, and move in an orderly manner in the specified task area, autonomously changing their formations to traverse complex obstacles without colliding with them.
(3) Experimental validations on E-puck2 robots are conducted. In particular, the performance of the agent's swarm movement and the fulfilment of environmental exploration tasks are evaluated in comparative studies.

The remainder of this paper is organised as follows. In Section 2, several problems for environmental exploration tasks are defined. In Section 3, we introduce the LF algorithm as the roaming strategy. In Section 4, we describe the flocking speed controller based on an improved APF method. We conduct some simulation experiments and analyse the experimental indicators in Section 5. In Section 6, we report on the real-world experiments based on E-puck2 robots and the completion time of the experiments. Finally, Section 7 concludes the paper.

2. Problem Definition

The central research question in environmental exploration is how to effectively traverse an unknown area. The task of exploring the environment often requires the explorer to have superior target search capabilities and environmental coverage capabilities. In the process of executing the task, the agent needs to consider avoiding collision with other individuals in the swarm and avoid collision with obstacles or boundaries. At the same time, the swarm robot needs to follow the virtual leader. The virtual leader walks randomly with a bionic roaming strategy, and agents follow the virtual leader to achieve the effect of environmental exploration.

Definition 1 (Repulsion). *The distance between agents is maintained within a certain range, and it can be adaptively and dynamically adjusted as the environment changes. When the distance is less than r_{arep}, a repulsion speed is generated. Similarly, when the distance between the agent and target is less than r_{at}, a repulsion speed of the target is generated. When they are far apart from each other, there is no mutual repulsive speed effect.*

Definition 2 (Avoid obstacles and walls). *Agents need to perform tasks within the specified task area. Therefore, agents cannot go out of a specific area in the process of performing tasks, which is equivalent to some virtual walls. In addition, agents need to avoid obstacles. The agent decelerates smoothly when encountering obstacles or walls to avoid colliding with them. The agent needs to slow down smoothly instead of stopping abruptly near obstacles or walls. Specifically, the closer the agent is to the them, the faster it decelerates.*

Definition 3 (Follow the virtual leader). *All agents in the swarm follow the movement of the virtual leader. The virtual leader does not actually exist in the arena, and its position is calculated by the leader. When the agent moves to the position of the virtual leader, it needs to decelerate smoothly as the distance decreases, similar to avoiding obstacles. The closer the agent to its expected stopping point, the faster its speed should decay. When the distance is very close, its speed even needs to decay at the rate of change of the exponential function.*

Definition 4 (Roaming strategy). *The virtual leader traverses the environment with a bionic walking strategy, and the agents in the swarm follow the virtual leader. This traversal strategy should have the following functionalities, i.e., the agents find all targets in the least possible time $t \in R$. In addition, agents travel the arena with the largest possible coverage ratio $r \in (0, 1]$.*

3. Roaming Strategy: Lévy Flight

LF is named after the French mathematician Paul Lévy. It refers to the random walk with a heavy-tailed distribution in the probability distribution of the step length, which means that there is a relatively high probability of large strides in the process of random walking. Natural creatures with LF mechanism tend to traverse a small place by generating many small steps and then move to another area through a large step to continue traversing to obtain higher search efficiency. The Lévy probability distribution is stable with infinite second-order moments and has the following form [18]:

$$P_{\alpha,\gamma}(l) = \frac{1}{\pi} \int_{-\infty}^{\infty} e^{-\gamma q^{\alpha}} \cos(ql) dq \tag{1}$$

The distribution is symmetric with respect to $l = 0$. The parameter α determines the shape of the distribution. The shorter the parameter α ($0 < \alpha < 2$ in Lévy distribution), the bigger the tail region. When the parameter $\alpha = 2$, the distribution changes from a Lévy distribution to a Gaussian distribution. In this study, the parameter $\alpha = 1.5$, and γ is the scaling factor. Equation (1) can be approximated by the following expression [18]:

$$P_{\alpha}(l) \approx l^{-\alpha} \tag{2}$$

Many scholars have proposed an implementation method for generating random numbers subject to Levy distribution, including a method proposed by Mantegna in 1994 [33]. This study adopted the method proposed by Mantegna to calculate the LF step size:

$$z = \frac{u}{|v|^{1/\beta}} \tag{3}$$

where $\beta \in [0.3, 1.99]$; u and v are two normal stochastic variables with standard deviations σ_u and σ_v, respectively:

$$u \sim N\left(0, \sigma_u^2\right) \quad v \sim N\left(0, \sigma_v^2\right)$$

$$\sigma_u = \left\{ \frac{\Gamma(1+\beta)\sin\left(\frac{\pi\beta}{2}\right)}{\beta\Gamma\left(\frac{1+\beta}{2}\right) \cdot 2^{\frac{\beta-1}{2}}} \right\}^{\frac{1}{\beta}} \quad \sigma_v = 1 \tag{4}$$

where $\Gamma(x)$ is the gamma function:

$$\Gamma(x) = \int_0^\infty e^{-t} t^{x-1} dt \tag{5}$$

However, in practical applications, the control scale factor of the LF should be adjusted as the environment changes [33]. We can multiply the stochastic process by an appropriate multiplicative factor $\gamma^{1/\alpha}$. After the linear transformation, the result can be expressed as follows:

$$Z_s = \gamma^{1/\alpha} z \tag{6}$$

From the perspective of the Lévy probability distribution, the LF algorithm produces a large number of small step lengths and a few large step lengths. The agent traverses a local area by generating multiple small steps; a few large steps may cause the agent to jump out of the local area. Based on such a step size generation mechanism, organisms in nature can efficiently traverse the unknown environment without relying on complex sensors.

4. Flocking Based on Improved Artificial Potential Field Method

4.1. Method of Following the Virtual Leader

The leader agent in the swarm continuously calculates the position of the virtual leader and broadcasts its position to other agents in the swarm. When the agents identify a known target point, they need to move to that point in the most reasonable way possible. Gabor proposed a smooth speed decay mechanism through an ideal braking curve $D(.)$ to make their expressed motion resemble natural graceful movements, with constant acceleration at high speeds and exponential approach in time at low speeds [32].

$$D(r,a,p) = \begin{cases} 0 & \text{if } r \leq 0 \\ rp & \text{if } 0 < rp < a/p \\ \sqrt{2ar - a^2/p^2} & \text{otherwise} \end{cases} \tag{7}$$

The parameter r represents the distance between an agent and the expected stopping point, p gain determines the crossover point between the two phases of deceleration, and a is the preferred acceleration of the agent. We introduce $D(.)$ function for the agents' tracking of the virtual leader. Here, v_{li} decreases smoothly as r_{li} decreases. It is easy to understand that as you get closer to your target, you may slow down and stop gradually. When you are far from the target, you need to speed up your pace and catch up. The agent can smoothly approach its target position by the following equation:

$$v_{li} = C^f \cdot D\left(r_{li}, a^f, p^f\right) \cdot \vec{r_{li}} \tag{8}$$

where $r_{li} = |r_l - r_i|$ is the distance between the agent and virtual leader, and a^f is the maximal allowed acceleration in the optimal braking curve used for following the virtual leader. p^f represents the gain of the optimal braking curve. If this value is too large, the braking curve exhibits a constant acceleration characteristic. When this value is small, the final part of braking (at low speeds) with decreasing acceleration is elongated and accompanied by a smooth stop. C^f can linearly adjust the magnitude of the speed item of the agent following the virtual leader. Higher values assume that agents can follow the virtual leader more closely. $\vec{r_{li}} = \frac{r_l - r_i}{|r_l - r_i|}$ represents the agent's moving direction toward the virtual leader.

4.2. Repulsion

Agents must consider executing a task without collision when they move in swarms, like flocks of birds in the sky, which flock but rarely collide. These agents need to consider the collision avoidance between each other in the process of exploring the environment. Agents do not need to worry about individuals who are relatively far away from them in the swarm. On the contrary, every agent should try to avoid all other agents within a certain distance at the same time, and the closer a neighbouring agent is, the stronger the repulsive effect should be from its neighbour. Agents can avoid collision with each other through the following equation:

$$
v_{ij}^{rep} = \begin{cases} p_a^{rep} \cdot (r_{arep} - r_{ij}) \cdot \overrightarrow{r_{ij}} & \text{if } (r_{ij} < r_{arep}) \\ 0 & \text{otherwise} \end{cases}
\tag{9}
$$

where $r_{ij} = | r_i - r_j |$ represents the distance between agent i and agent j, and r_{arep} represents the distance threshold for the speed influence at which agents start to interact and generate repulsion. $\overrightarrow{r_{ij}}$ represents the direction of the speed from agent j to agent i. As a linear gain, p_a^{rep} linearly adjusts the size of repulsion speed term. As the agent may have multiple neighbours, it is necessary to consider the repulsive effects that may be caused by all other agents in the swarm.

$$
v_{it}^{rep} = \begin{cases} p_t^{rep} \cdot (r_{at} - r_{it}) \cdot \overrightarrow{r_{it}} & \text{if } (r_{it} < r_{at}) \\ 0 & \text{otherwise} \end{cases}
\tag{10}
$$

Similar repulsion occurs between the agent and the target. p_t^{rep} linearly adjusts the size of repulsion speed term. When the distance between the two (r_{it}) is less than the desired separation distance (r_{at}), the agent generates a repulsion speed away from the target. The direction of the speed is $\overrightarrow{r_{it}}$, which is from the target to agent i. It is worth noting that this repulsion is one-way, that is, the target will not move away from the agent due to the proximity of the agent. Superimposing the repulsion speeds leads to the speed item of the agent due to repulsion.

$$
v_i^{rep} = \sum_{j \neq i} v_{ij}^{rep} + \sum_{target} v_{it}^{rep}
\tag{11}
$$

4.3. Avoid Obstacles and Avoid Moving out of Boundaries

In some practical tasks, we assume that the agents will explore a certain area, that is, we have defined the boundaries for the agents. They only need to explore such a specific area, and it is not necessary to explore other places. To prevent the agent from moving out of bounds, r_{wall} is the safe distance between the agent and the field boundary. In other words, when the distance between the agent and the boundary is less than r_{wall}, the agent should produce a speed away from the boundary. We place the agents into a square-shaped arena with soft repulsive virtual walls and define virtual agents near the arena walls [32]. Virtual agents are located at the closest point of the given edge of an arbitrarily shaped convex wall polygon relative to agent i. When the distance between the agent and wall is less than r_{wall}, this speed term will take effect:

$$
v_{id}^{wall} = \begin{cases} C^s \cdot (v_{is} - D(r_{is} - r_{wall}, a^s, p^s)) \cdot \overrightarrow{v_{is}} & \text{if } (r_{is} < r_{wall}) \\ 0 & \text{otherwise} \end{cases}
\tag{12}
$$

Convex obstacles inside the arena can be avoided using the same concept. When the agents are far away from the obstacle, they can ignore the influence of the obstacle on their current movement. Here, we assume that when the distance between the agent and obstacle is less than a certain value r_{obs}, the agent will generate the speed away from the obstacle.

$$
v_{id}^{obs} = \begin{cases} C^s \cdot (v_{is} - D(r_{is} - r_{obs}, a^s, p^s)) \cdot \overrightarrow{v_{is}} & \text{if } (r_{is} < r_{obs}) \\ 0 & \text{otherwise} \end{cases}
\tag{13}
$$

In the above two equations, r_s represents the position of the shill agent, located at the closest point of the given edge of an arbitrarily shaped convex wall polygon relative to agent i. $r_{is} = r_i - r_s$ represents the distance between agent i and its closest shill agent. v_s is the speed of the shill agent, pointing perpendicularly to the wall polygon edge inward of the arena ($v_{is} = |v_i - v_s|$). $\overrightarrow{v_{is}}$ is the unit vector difference between the speed of the agent and the shill agent which represents the obstacle avoidance direction of the agent after encountering obstacles. a^s and p^s are the same as a^f and p^f, respectively, but for staying away from the obstacles and walls. C^s adjusts the gain of the two speed terms.

4.4. Final Equation of Desired Speed

When the agents perform environmental exploration tasks in the unknown environment, they may encounter many complex scenarios, so the above speed influencing factors need to be considered at the same time. Agents should have all the velocities mentioned above to produce the desired motion effects. In this way, we take the vectorial sum of all the interaction terms.

$$v_i^{\text{desire}} = v_i^{rep} + v_{li} + v_{is}^{wall} + v_{is}^{obs} \tag{14}$$

Agents should meet motion constraints, that is, their speed can not be unlimited, which does not meet the needs of practical applications. When the speed generated by the above speed controller is too large, the agents should adopt a maximum speed to meet the safety requirements, but the speed direction should not be changed. In this way, after getting the speed generated by our method, we set a cut-off to cope with motion restraint. If the speed v_i^{desire} is over the limit, the direction of the desired speed is maintained but its magnitude is reduced [32]:

$$\tilde{v}_i^{\text{desire}} = \frac{\tilde{v}_i^{\text{desire}}}{|\tilde{v}_i^{\text{desire}}|} \cdot \min\left\{ |\tilde{v}_i^{\text{desire}}|, v_{\text{limit}} \right\} \tag{15}$$

5. Simulation Experiments and Analysis

5.1. Simulation Experiments

In this section, the performance of the proposed LF-APF method is evaluated using simulation based on MATLAB. In the simulation experiments, the agent can obtain the location information of other agents in the swarm through communication and can detect obstacles and calculate the distance from them. In addition, the agent can get the boundary position of the arena. We set the size of the arena as 250 m × 250 m. To reduce the impact of hardware computing power, we assume that the time for the agent to take a step is one second (the true time is related to the computing ability of the computer). Depending on the step size of the arena, it is necessary to adjust the size of the agent's movement. Similar to the albatross and bees in nature, although they both use the LF algorithm to search for food, the step size corresponding to the Lévy distribution should be scaled according to their different athletic abilities. To ensure fairness of comparison, we first optimise the BM step to make the agents perform as well as possible in the 250 m × 250 m arena. Then, we adjusted the parameters γ in LF to let the median step lengths generated by the two algorithms match as closely as possible in the case of having the same size of the arena (250 m × 250 m).

As shown in Figure 1, there are four small isolating islands at sea level in the blue sea. Eight agents flocked to search for two moving targets in the complex obstacle environment. The scenario where the agents are distributed at the initial moment is shown in Figure 1a. Figure 1b shows agents in swarm having found one of the targets. Figure 1c,d shows that the agents adaptively change their formation according to the environment and pass through obstacles without any collision.

There are experimental videos for readers to watch in Appendix A. We can see that the LF-APF method has the following performance on environmental exploration tasks:

(1) The agents do not collide with each other, keep a proper distance from each other, flexibly change their formation, and shuttle in the task area, similar to a natural population.

(2)   The agents can flexibly avoid isolating islands in the ocean. On some special occasions, the agents swim past obstacles in groups or pass through a limited space in a line.

(3)   When the agents move near obstacles, their speed decreases smoothly, which complies more with their dynamic constraints.

(4)   The agents can follow the virtual leader to achieve efficient traversal of the task area.

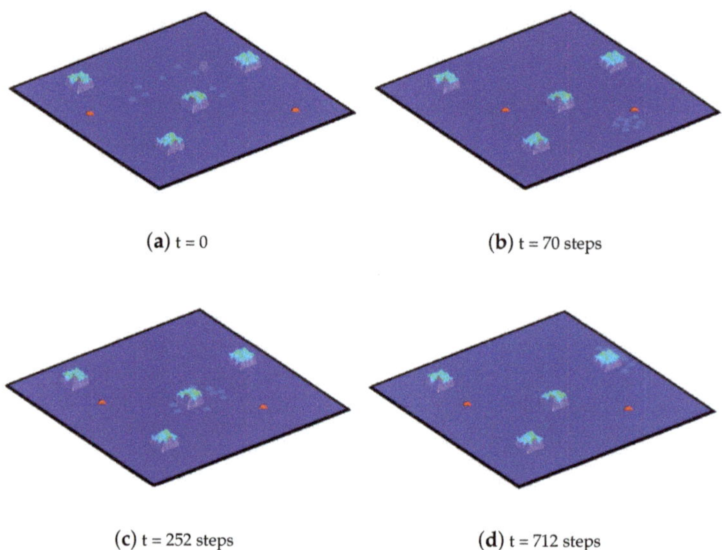

(**a**) t = 0   (**b**) t = 70 steps

(**c**) t = 252 steps   (**d**) t = 712 steps

Figure 1. Eight agents are flocking to explore the environment with our method. There are complex unknown obstacles and two dynamic targets in the sea. Agents are blue, and targets are red. The arena is 250 m on either side. The speed of the virtual leader and targets are 1 m/s and 0.3 m/s, respectively. When the target is found (the distance between agent and target < 8 m), the target becomes stationary.

5.2. Indicator Statistics

In the experimental display discussed in the previous section, we can see from the figure (please also refer to the video in the Appendix A for details) that the method proposed in this study can enable a robotic swarm to achieve a good performance of environmental exploration. Since swarm robots employing BM as the exploration strategy to explore the environment scheme proposed in the previous study [20] also achieved a good environmental exploration effect, we let the swarm agents perform LF and BM for environmental exploration, respectively, and compared the results of both methods. In an unknown environment, since the target is constantly moving, it may move to any reachable place in the environment. Therefore, the evaluation index measures not only the capability of the agents to find all the targets as soon as possible [34] but also their ability to traverse the environment as much as possible. In addition, indicators of the quality of swarm movement to describe different aspects of the agents' motion are needed. The task evaluation indicators used in this study are as follows:

(1)   Time for the swarm to find target;

(2)   The coverage area of the swarm in a period of time;

(3)   The change of agents' area coverage ratio over time;

(4)   The correlation of agents' speed, the average and minimum inter-agent distances while agents are flocking.

We conducted an indicator analysis of the time to find all targets in the arena shown in Figure 2. We counted the time of identifying the target when the virtual leader runs at different speeds in 10 independent runs each. Figure 2 shows that when the agents follow the virtual leader, with the speed of the virtual leader at 3 m/s, the agents find the targets faster. If the speed of the virtual leader is too slow, it may lead to more time for the agents to search for the targets, but if the speed is too fast, it may cause the agents to track not close enough and also lead to more time for the agents to find the target.

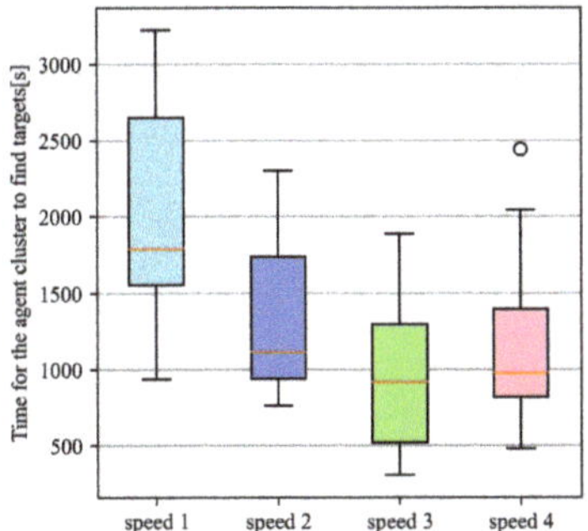

Figure 2. Time for the swarm to find target with LF-APF method for 10 experiments. The figure shows the virtual leader moving at different speeds (speed 1 = 1 m/s, speed 2 = 2 m/s, speed 3 = 3 m/s, speed 4 = 4 m/s) in an area of 250 m × 250 m.

To prove the advantages of LF-APF in exploring unknown regions, we compile statistics on the regions the agents walked. Let the virtual leader move at a speed of 2 m/s under two algorithms (LF and Brownian motion), and the agents follow the virtual leader to search for targets in the arena. We respectively show the coverage area in the arena with obstacles in Figure 3 and without obstacles in Figure 4. In the arena with obstacles, the obstacle areas (marked by yellow boxes) cannot be covered by the agents. Figure 5 shows how the area coverage ratio of agents varies with the time when the virtual leader moves at different speeds with two different algorithms. From Figures 3–5, it can be found that, in general, the LF has better area coverage ability than Brownian motion. At the same time, we find that when the speed of the virtual leader is too fast, the agents do not follow closely (the agents have a speed limit). If the speed is too slow, the time for the agents to traverse the environment increases. In addition, we found that when the virtual leader moves with LF at a speed of 2 m/s, the LF-APF method obtains the best area coverage ability. The above indicators proved the superiority of the LF-APF method to perform tasks in exploring unknown regions.

To make the method deploy successfully in practical applications, it is important to evaluate the effect of flocking. Considering that the speed of the agents and obstacles in the arena will affect the effect of flocking, we utilised some evaluation indicators such as the correlation of speed between agents ϕ^{corr} and the average and minimum of inter-agent distances ($\overline{r}_{ij}^{\min}$ and $\min(r_{ij})$) [32]. N represents the number of agents in the swarm; J_i represents the set of individuals in the swarm except for agent i. The calculation formula of ϕ^{corr} is as follows. We evaluated the flocking effect of LF-APF at different speeds and obstacles.

$$\phi^{\mathrm{corr}} = \frac{1}{T}\frac{1}{N}\int_0^T \sum_{i=1}^N \frac{1}{N_i-1}\sum_{j\in J_i}\frac{v_i \cdot v_j}{|v_i|\cdot|v_j|}dt \qquad (16)$$

In Figure 6, the quality of eight agents flocking to follow the virtual leader with different speeds or obstacles are digitised. In Figure 6a,b, the results are shown when there are no obstacles in the arena. The distance between the agents is kept constant with only very minor fluctuations, and the agents' speed directions are highly correlated most of the time. At some moments, the speed correlation drops due to the virtual leader's sudden turn. When obstacles appear in the arena, the quality of the swarm motion of the agents is affected to a certain extent, as shown in Figure 6c,d. The index $\overline{r}_{ij}^{\min}$ surges in cases when some agents are blocked by an obstacle accidentally in the arena, as shown in Figure 6c. In general, the agents achieve a relatively flocking effect safely.

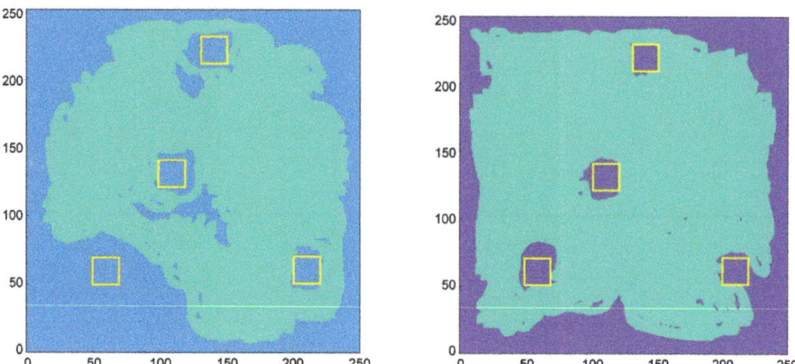

Figure 3. Display of the covered area with eight agents when the virtual leader performs Brownian motion (**left**) and Lévy flight (**right**) at the speed of 2 m/s after 10,000 s. There are some randomly distributed obstacles (yellow line) in the environment. Agents need to avoid obstacles automatically when covering the area.

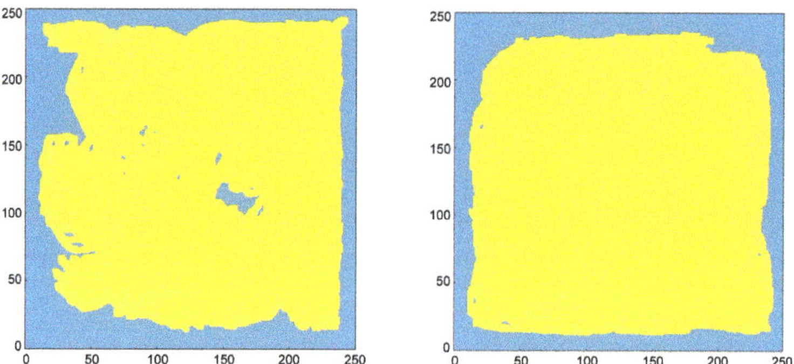

Figure 4. Display of the covered area with eight agents when the virtual leader performs Brownian motion (**left**) and Lévy flight (**right**) at the speed of 2 m/s after 10,000 s. There are no obstacles in the environment.

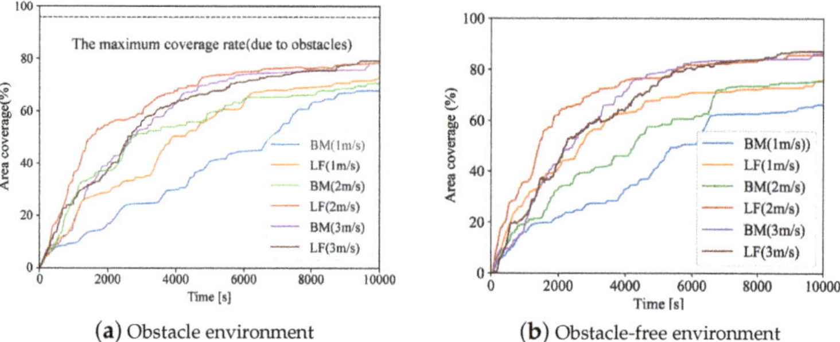

(**a**) Obstacle environment

(**b**) Obstacle-free environment

Figure 5. For the two different algorithms of Lévy flight and Brownian motion, the area coverage ratio of agents varies with the time.

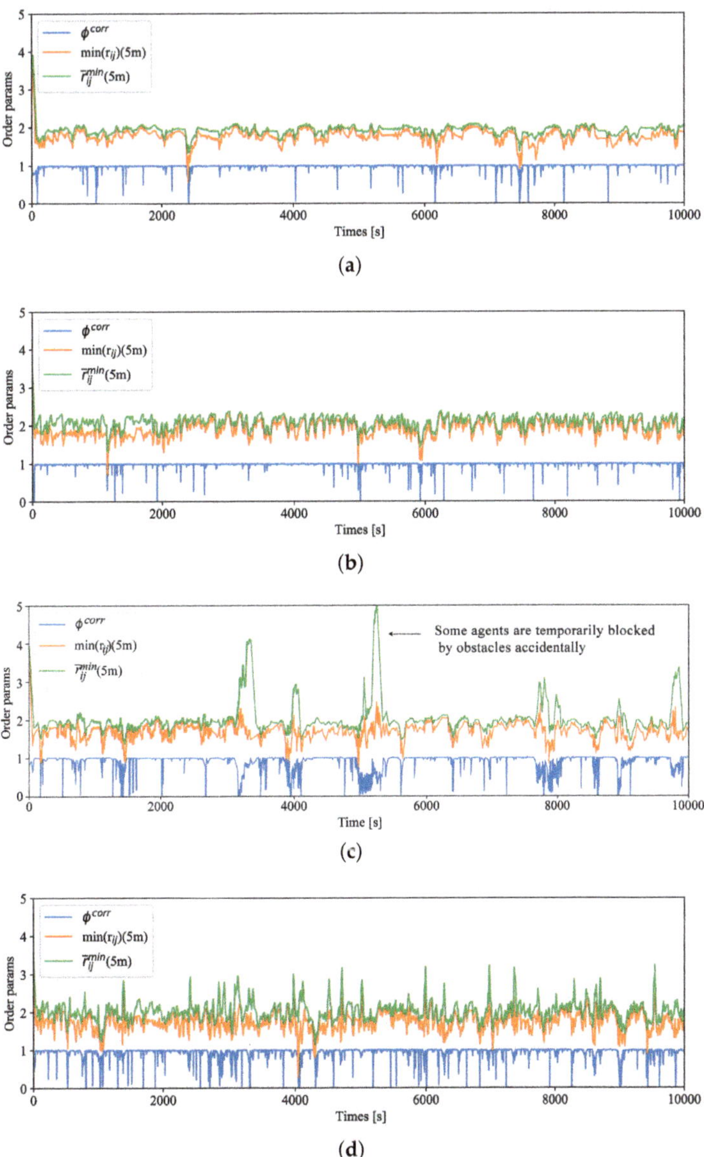

Figure 6. Order parameters as a function of time during experiments with eight agents in 10,000 s. ϕ^{corr} represents the correlation of agent's speed in swarm. $\overline{r_{ij}}^{\mathrm{min}}$ is the average distance of the closest neighbours, whereas $\min{(r_{ij})}$ is the minimum distance of the closest neighbours. The size of the arena is 250 m $\times$ 250 m. (**a**) Agents flocking to explore environment in an obstacle-free arena (1 m/s). (**b**) Agents flocking to explore environment in an obstacle-free arena (2 m/s). (**c**) Agents flocking to explore environment with obstacles in the arena (1 m/s). (**d**) Agents flocking to explore environment with obstacles in the arena (2 m/s).

6. Real-World Experiments

To evaluate the effectiveness of our method in real-world applications, we performed experiments with E-puck2 robots. We added an expansion board with Raspberry Pi to the E-puck2 robots to increase their computing power. The E-puck2 robots communicate with each other via WiFi. In the arena with random obstacles and targets, eight E-puck2 robots searched for two targets with the LF-APF method. The E-puck2 robots obtained global information from the motion capture device above the arena, including the position information of the robots and obstacles of the arena. The E-puck2 robot and the initial scene of the task are shown in Figures 7 and 8, respectively.

We counted the time (Table 1) taken by the E-puck2 robot to search for the targets. From the table, we can see that the E-puck2 robots can always identify all targets and complete the task with the LF-APF method. We selected one representative experiment, as shown in Figure 9. When the E-puck2 robots encounter obstacles, they adjusted formations to bypass the obstacles without any collision. In addition, they can gather in the obstacle-free area and automatically change the formation when it is necessary to disperse. The E-puck2 robots kept a certain distance between each other during the entire flocking without colliding while forming a tight whole and moved orderly in the task area without running out of the boundary of the arena. From the above results and analysis, it can be suggested that the E-puck2 robots can adaptively deal with the environment to perform environmental exploration tasks in the arena using the LF-APF method.

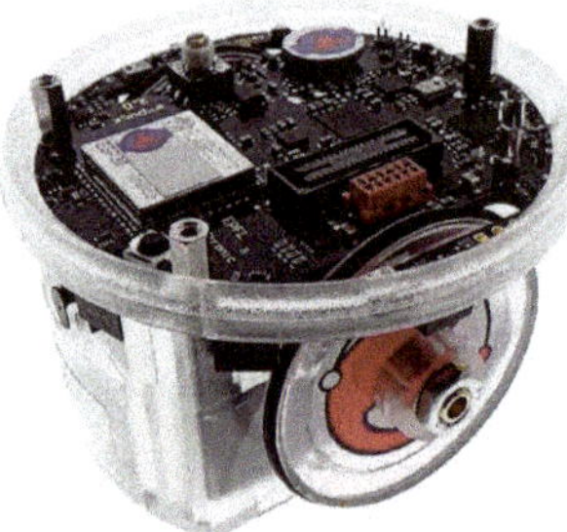

Figure 7. E-puck2 robot platform. E-puck2 is the latest mini mobile robot developed by GCtronic and EPFL, which is an evolution of the successful E-puck robot used in many research and educational institutes.

Table 1. Statistics of the time taken by the agents to find the two targets, respectively, in 6 real-world experimental runs.

Experiments	1st	2nd	3rd	4th	5th	6th
The time of finding one target	174 s	62 s	235 s	419 s	311 s	283 s
The time of finding all targets	432 s	211 s	619 s	847 s	346 s	438 s

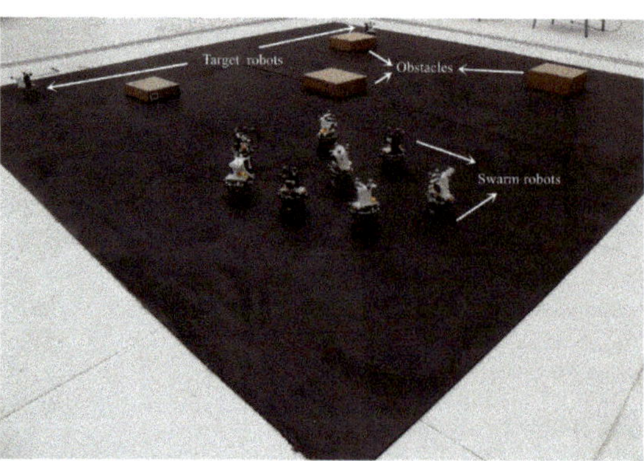

Figure 8. Initial scene of E-puck2 robots searching for the target in the arena.

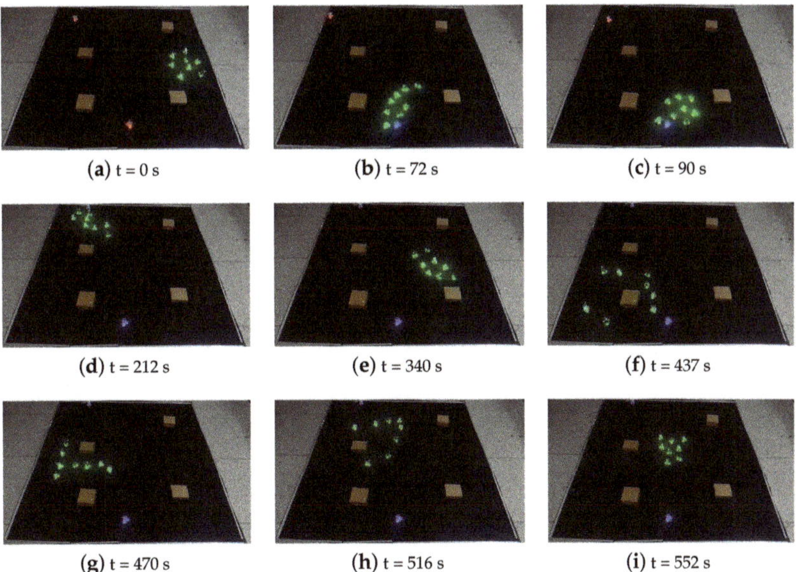

(**a**) t = 0 s　　　　(**b**) t = 72 s　　　　(**c**) t = 90 s

(**d**) t = 212 s　　　　(**e**) t = 340 s　　　　(**f**) t = 437 s

(**g**) t = 470 s　　　　(**h**) t = 516 s　　　　(**i**) t = 552 s

Figure 9. E-puck2 robots flocking to explore the environment. There are complex unknown obstacles and two dynamic targets in the arena (3 m × 3 m). The agents are green, and the targets are red. When the E-puck2 robots find the target (the distance between the agent and the target < 10 cm), the target becomes stationary and its colour turns blue.

7. Conclusions

In this study, we proposed the LF-APF method combining APF and LF mechanisms to achieve environmental exploration in swarms. The proposed method makes agents flock to explore unknown environments, relying on little environmental information. Agents in the swarm determine the position of the virtual leader, who performs LF, and form a flocking to follow the virtual leader to traverse the area searching for the target. In the process, the agents can adjust formations to adapt to the environment and incur no collisions between the agents or between the agents and obstacles. The resulting movements of the swarm robots are similar to those of the natural population, which suggests that the proposed method can be well applied to the exploration task of the bionic robot environment. Several simulations and real-world experiments have validated that the method can achieve effective and efficient environmental exploration.

Author Contributions: Conceptualization, Z.F. and C.W.; methodology, C.W.; software, C.W.; validation, Z.F., C.W., M.G. and H.H.; formal analysis, Z.F., M.G. and C.W.; investigation, Z.F.; resources, Z.F.; data curation, C.W.; writing—original draft preparation, C.W. and D.W.; writing—review and editing, Z.W. and Y.Y.; visualization, C.W. and H.H.; supervision, X.Z., W.W. and Z.F.; project administration, Z.F.; funding acquisition, Z.F. All authors have read and agreed to the published version of the manuscript.

Funding: This research is funded by the National Natural Science Foundation of China (62176147), the Science and Technology Planning Project of Guangdong Province of China, the State Key Lab of Digital Manufacturing Equipment & Technology (DMETKF2019020), and by the National Defense Technology Innovation Special Zone Project (193-A14-226-01-01).

Acknowledgments: The authors would like to thank the Key Laboratory of Digital Signal and Image Processing of Guangdong Province, Shantou University, Guangdong Province, China, for the resources provided by them.

Conflicts of Interest: The authors declare no conflict of interest.

Abbreviations

The following abbreviations are used in this manuscript:

LF	Lévy flight
BM	Brownian motion
APF	Artificial potential field

Appendix A

The video of experiments: Search for two targets with eight agents.

Simulation experiments: https://www.bilibili.com/video/BV1Sr4y1S7Xi?spm_id_from=333.999.0.0.

Real-world experiments: https://www.bilibili.com/video/BV1RT4y1h7vU?spm_id_from=333.999.0.0.

References

1. Chen, J.Y. UAV-guided navigation for ground robot tele-operation in a military reconnaissance environment. *Ergonomics* **2010**, *53*, 940–950. [CrossRef] [PubMed]
2. Waharte, S.; Trigoni, N.; Julier, S. Coordinated search with a swarm of UAVs. In Proceedings of the 2009 6th IEEE Annual Communications Society Conference on Sensor, Mesh and Ad Hoc Communications and Networks Workshops, Rome, Italy, 22–26 July 2009; pp. 1–3.
3. Pitonakova, L.; Crowder, R.; Bullock, S. Information flow principles for plasticity in foraging robot swarms. *Swarm Intell.* **2016**, *10*, 33–63. [CrossRef]
4. Kei Cheang, U.; Lee, K.; Julius, A.A.; Kim, M.J. Multiple-robot drug delivery strategy through coordinated teams of microswimmers. *Appl. Phys. Lett.* **2014**, *105*, 083705. [CrossRef]
5. Wei, C.; Xu, J.; Wang, C.; Wiggers, P.; Hindriks, K. An approach to navigation for the humanoid robot nao in domestic environments. *Appl. Phys. Lett.* **2013**, 298–310.
6. McGuire, K.N.; De Wagter, C.; Tuyls, K.; Kappen, H.J.; de Croon, G.C. Minimal navigation solution for a swarm of tiny flying robots to explore an unknown environment. *Sci. Robot.* **2019**, *4*, eaaw9710. [CrossRef]
7. Herianto; Sakakibara, T.; Koiwa, T.; Kurabayashi, D. Realization of a pheromone potential field for autonomous navigation by radio frequency identification. *Adv. Robot.* **2008**, *22*, 1461–1478. [CrossRef]
8. Tang, Q.; Xu, Z.; Yu, F.; Zhang, Z.; Zhang, J. Dynamic target searching and tracking with swarm robots based on stigmergy mechanism. *Robot. Auton. Syst.* **2019**, *120*, 103251. [CrossRef]
9. Wei, C.; Hindriks, K.V.; Jonker, C.M. Dynamic task allocation for multi-robot search and retrieval tasks. *Appl. Intell.* **2016**, *45*, 383–401. [CrossRef]
10. Alitappeh, R.J.; Jeddisaravi, K. Multi-robot exploration in task allocation problem. *Appl. Intell.* **2022**, *52*, 2189–2211. [CrossRef]
11. Reid, C.R.; Sumpter, D.J.; Beekman, M. Optimisation in a natural system: Argentine ants solve the Towers of Hanoi. *Appl. Intell.* **2011**, *214*, 50–58. [CrossRef]
12. Menzel, R.; Fuchs, J.; Kirbach, A.; Lehmann, K.; Greggers, U. Navigation and Communication in Honey Bees. In *Honeybee Neurobiology and Behavior: A Tribute to Randolf Menzel*; Springer: Dordrecht, The Netherlands, 2012.
13. Wei, C.; Hindriks, K.V.; Jonker, C.M. Altruistic coordination for multi-robot cooperative pathfinding. *Appl. Intell.* **2016**, *44*, 269–281. [CrossRef]
14. Wu, M.; Zhu, X.; Ma, L.; Wang, J.; Bao, W.; Li, W.; Fan, Z. Torch: Strategy evolution in swarm robots using heterogeneous–homogeneous coevolution method. *J. Ind. Inf. Integr.* **2022**, *25*, 100239. [CrossRef]
15. Ajanic, E.; Feroskhan, M.; Mintchev, S.; Noca, F.; Floreano, D. Bioinspired wing and tail morphing extends drone flight capabilities. *Sci. Robot.* **2020**, *5*, eabc2897. [CrossRef] [PubMed]
16. Ramezani, A.; Chung, S.J.; Hutchinson, S. A biomimetic robotic platform to study flight specializations of bats. *Sci. Robot.* **2017**, *2*, eaal2505. [CrossRef] [PubMed]
17. Roderick, W.R.T.; Cutkosky, M.R.; Lentink, D. Bird-inspired dynamic grasping and perching in arboreal environments. *Sci. Robot.* **2021**, *6*, eabj7562. [CrossRef]
18. Viswanathan, G.M.; Afanasyev, V.; Buldyrev, S.V.; Havlin, S.; Da Luz, M.G.E.; Raposo, E.P.; Stanley, H.E. Lévy flights in random searches. *Phys. A Stat. Mech. Appl.* **2000**, *282*, 1–12. [CrossRef]
19. Wei, J.; Chen, Y.; Yu, Y.; Chen, Y. Optimal randomness in swarm-based search. *Mathematics* **2019**, *7*, 828. [CrossRef]
20. Martinez, F.; Jacinto, E.; Acero, D. Brownian motion as exploration strategy for autonomous swarm robots. *J. Anim. Ecol.* **2012**, 2375–2380.
21. Sims, D.W.; Humphries, N.E.; Bradford, R.W.; Bruce, B.D. Lévy flight and Brownian search patterns of a free-ranging predator reflect different prey field characteristics. *J. Anim. Ecol.* **2012**, *81*, 432–442. [CrossRef]
22. Sims, D.W.; Southall, E.J.; Humphries, N.E.; Hays, G.C.; Bradshaw, C.J.; Pitchford, J.W.; James, A.; Ahmed, M.Z.; Brierley, A.S.; Hindell, M.A.; et al. Scaling laws of marine predator search behaviour. *Nature* **2008**, *451*, 1098–1102. [CrossRef]

23. Reynolds, A.M.; Swain, J.L.; Smith, A.D.; Martin, A.P.; Osborne, J.L. Honeybees use a Lévy flight search strategy and odour-mediated anemotaxis to relocate food sources. *Behav. Ecol. Sociobiol.* **2009**, *64*, 115–123. [CrossRef]
24. Zhou, J.; Yao, X. Multi-objective hybrid artificial bee colony algorithm enhanced with Lévy flight and self-adaption for cloud manufacturing service composition. *Appl. Intell.* **2017**, *47*, 721–742. [CrossRef]
25. Pang, B.; Song, Y.; Zhang, C.; Yang, R. Effect of random walk methods on searching efficiency in swarm robots for area exploration. *Appl. Intell.* **2021**, *51*, 5189–5199. [CrossRef]
26. Fioriti, V.; Fratichini, F.; Chiesa, S.; Moriconi, C. Levy foraging in a dynamic environment–extending the levy search. *Int. J. Adv. Robot. Syst.* **2015**, *12*, 98. [CrossRef]
27. Jeddisaravi, K.; Alitappeh, R.J.; Pimenta, L.C.A.; Guimarães, F.G. Multi-objective approach for robot motion planning in search tasks. *Appl. Intell.* **2016**, *45*, 305–321. [CrossRef]
28. Blum, C.; Merkle, D. Swarm Intelligence: Introduction and Applications. In *Applied Intelligence*; Springer Science & Business Media: New York, NY, USA, 2008; pp. 19–20.
29. Chen, Y.B.; Luo, G.C.; Mei, Y.S.; Yu, J.Q.; Su, X.L. UAV path planning using artificial potential field method updated by optimal control theory. *Int. J. Syst. Sci.* **2016**, *47*, 1407–1420. [CrossRef]
30. Sudhakara, P.; Ganapathy, V.; Priyadharshini, B.; Sundaran, K. Obstacle avoidance and navigation planning of a wheeled mobile robot using amended artificial potential field method. *Procedia Comput. Sci.* **2018**, *133*, 998–1004. [CrossRef]
31. Singh, Y.; Sharma, S.; Sutton, R.; Hatton, D.C. Towards use of Dijkstra Algorithm for Optimal Navigation of an Unmanned Surface Vehicle in a Real-Time Marine Environment with results from Artificial Potential Field. *TransNav Int. J. Mar. Navig. Saf. Sea Transp.* **2018**, *12*, 125–131. [CrossRef]
32. Vásárhelyi, G.; Virágh, C.; Somorjai, G.; Nepusz, T.; Eiben, A.E.; Vicsek, T. Optimized flocking of autonomous drones in confined environments. *Sci. Robot.* **2018**, *3*, eaat3536. [CrossRef]
33. Mantegna, R.N. Fast, accurate algorithm for numerical simulation of Levy stable stochastic processes. *Phys. Rev. E* **2000**, *282*, 1–12. [CrossRef]
34. Pal, A.; Tiwari, R.; Shukla, A. Communication constraints multi-agent territory exploration task. *Appl. Intell.* **2013**, *38*, 357–383. [CrossRef]

MDPI

St. Alban-Anlage 66

4052 Basel

Switzerland

www.mdpi.com

Drones Editorial Office

E-mail: drones@mdpi.com

www.mdpi.com/journal/drones